World Geomorphological Landscapes

Series editor

Piotr Migon, Wrocław, Poland

For further volumes:
http://www.springer.com/series/10852

Paolo Billi
Editor

Landscapes and Landforms of Ethiopia

Springer

Editor
Paolo Billi
Physics and Earth Sciences
University of Ferrara
Ferrara
Italy

ISSN 2213-2090 ISSN 2213-2104 (electronic)
World Geomorphological Landscapes
ISBN 978-94-017-8025-4 ISBN 978-94-017-8026-1 (eBook)
DOI 10.1007/978-94-017-8026-1

Library of Congress Control Number: 2015931921

Springer Dordrecht Heidelberg New York London

Printed on acid-free paper

Springer Science+Business Media B.V. Dordrecht is part of Springer Science+Business Media
(www.springer.com)

Series Editor Preface

Landforms and landscapes vary enormously across the Earth, from high mountains to endless plains. At a smaller scale, nature often surprises us by creating shapes which look improbable. Many physical landscapes are so immensely beautiful that they received the highest possible recognition—they hold the status of World Heritage Sites. Apart from often being immensely scenic, landscapes tell stories which not uncommonly can be traced back in time for tens of millions of years and include unique geological events such as meteorite impacts. In addition, many landscapes owe their appearance and harmony not solely to the natural forces. For centuries, and even millennia, they have been shaped by humans who have modified hill-slopes, river courses and coastlines, and erected structures which often blend with the natural landforms to form inseparable entities.

These landscapes are studied by geomorphology—'the science of scenery'—a part of Earth Sciences that focuses on landforms, their assemblages, surface and subsurface processes that moulded them in the past and that change them today. To show the importance of geomorphology in understanding the landscape, and to present the beauty and diversity of the geomorphological sceneries across the world, we have launched a book series *World Geomorphological Landscapes*. It aims to be a scientific library of monographs that present and explain physical landscapes, focusing on both representative and uniquely spectacular examples. Each book will contain details on geomorphology of a particular country or a geographically coherent region. This volume presents the impressive geomorphic legacy of Ethiopia which hosts many landscapes and landforms of global significance. Examples include the otherworldly Afar Depression with its sun-baked volcanoes, high-elevation basalt plateaus, the East African Rift valley with its splendid lakes, to name just a few. Ethiopia is also a country where geomorphology and people have remained in particularly close association since time immemorial. These relationships can be examined by referring to the past—as demonstrated by the chapter on geoarcheology of Aksum, but perhaps more importantly, with the reference to the present-day environmental problems arising from land use, soil erosion, water resources depletion and settlement growth. A number of chapters in this book remind us that geomorphological landscapes are not only beautiful; they are also very fragile if used improperly.

The World Geomorphological Landscapes series is produced under the scientific patronage of the International Association of Geomorphologists (IAG)—a society that brings together geomorphologists from all around the world. The IAG was established in 1989 and is an independent scientific association affiliated with the International Geographical Union (IGU) and the International Union of Geological Sciences (IUGS). Among its main aims are to promote geomorphology and to foster dissemination of geomorphological knowledge. I believe that this lavishly illustrated series, which keeps to the scientific rigour, is the most appropriate means to fulfil these aims and to serve the geoscientific community. To this end, my great thanks go to Prof. Paolo Billi, a person long involved in geomorphological research in Ethiopia, for agreeing to coordinate this timely volume in the series. I am also very grateful to all individual authors who accepted invitations to contribute and, often, delivered stories which contained original, not yet published research.

In contrast to many other countries, Ethiopia is far less known regarding its geomorphology at the local scale and many of its regions are still *terra incognita*. Therefore, it was not feasible to strictly follow the format of previous volumes in the series and offer a wide range of site-specific stories. However, I am sure the readers will value more general presentations of the geomorphic environment of Ethiopia which are not only beautifully illustrated, but also provide an updated, unique source of reference.

For me, to write the preface to the Ethiopia volume is of particular pleasure. In 2008 I was fortunate to join the IAG-organized field trip to the Ethiopian Highlands, expertly run by Paolo Billi, Franco Dramis and Giandomenico Fubelli (all involved in this volume), and became fascinated with the geomorphology of Ethiopia, which lasts until nowadays. As a little evidence of this fascination serves my own modest contribution to this volume, regarding the geomorphic scenery of Aksum. After touring the country for a week I thought I knew it reasonably well. This book in its final shape has told me how much is left to be seen.

Piotr Migoń

Contents

Part I

General Introduction to the Geomorphology of Ethiopia

Geomorphological Landscapes of Ethiopia

1

Paolo Billi

Abstract

This introductory chapter reports about the main geographic and geomorphologic features of Ethiopia. The prevailing soil types are described and soil erosion data are reported and analyzed. A brief description of the natural vegetation is provided as well. The country's main geomorphological landscapes regions are identified as follows: (1) the northern highlands, including (i) the volcanic plug belt of Adwa, (ii) the central highlands and (iii) the southwestern highlands; (2) the Rift Valley, which consists of three main portions, namely the northern, central, and southern trunks and the Afar and Danakil depressions; (3) the southern plateau, which consists of a northern and southern sector and includes also the Ogaden tableland gently descending to Somalia and the Indian Ocean. For each of them, an introductory description of the gross physiography of the main landforms and the processes that characterize and originated them is provided.

Keywords

Ethiopia • Geomorphology • Landforms • Landscapes

1.1 Introduction

Ethiopia is characterized by a wide variety of landscapes and landforms. They were generated by a complex of tectonic, erosive, and depositional processes acting on rocks with different characteristics. The geological evolution of Ethiopia, with alternating phases of orogenesis, peneplanation, crustal updoming, faulting, emplacement of huge amounts of lava, and deep fluvial dissection has imprinted the geomorphological landscapes of the country with specific characteristics, in places unique on Earth. This introductory section is a short summary of the main geographical and landscape regions, with an outline of the most distinctive landforms and associated processes that characterize them.

P. Billi (✉)
Physics and Earth Sciences,
University of Ferrara,
Ferrara, Italy
e-mail: bli@unife.it

1.2 Geography

Ethiopia has a surface area of about 1,127,000 km^2, that is almost twice the size of France. It takes up a large portion of the inner Horn of Africa since it has no border on the Red Sea and the Indian Ocean. The Ethiopian landmass consists of a large, high elevated plateau bisected by the Rift Valley into the northwestern and the southeastern highlands, each with associated lowlands The contrast in relief is remarkable as land elevation ranges between −155 m of Asal Lake in the Afar depression (the lowest point in Africa) to the peak of Mt. Ras Dejen at 4,620 m a.s.l. in the Simen Mountains (Fig. 1.1). The plateau stands between 1,500 and 3,000 m a.s.l. and it is strewn with a number of volcanoes making up high mountain ranges, the highest of which are the Simen in the north and the Bale mountains in the south (Fig. 1.1). The northwestern highlands are considerably more extensive and rugged and are divided into northern, central (centered on the Blue Nile River catchment downstream of Tana lake), and southern sections. The southwestern portion of the plateau (known also as Somali Plateau) is also rugged, but its elevation is

P. Billi (ed.), *Landscapes and Landforms of Ethiopia*, World Geomorphological Landscapes,
DOI 10.1007/978-94-017-8026-1_1, © Springer Science+Business Media Dordrecht 2015

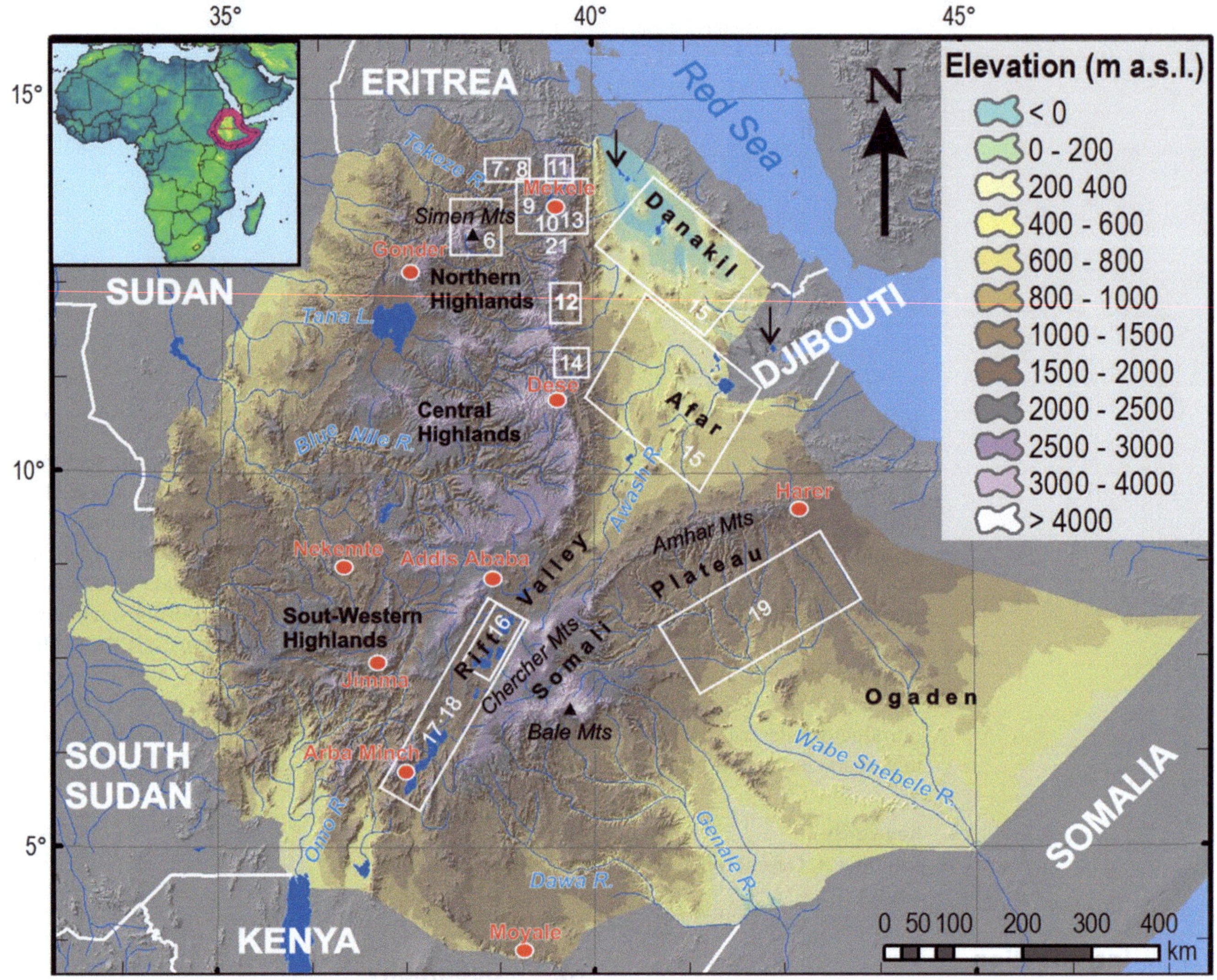

Fig. 1.1 Main geomorphological regions of Ethiopia. *Black triangles* stand for mountain peaks and *downward arrows* indicate lowest points below sea level. The study areas are marked by *white rectangles* and the numbers refer to the specific chapter of this volume

slightly lower (highest peak Mt. Tullu Dimtu, 4,383 m a.s.l.) than in the northern highlands and it can be subdivided into a northern section with the Ahmar and Garamullata mountains, a central section with the Chercher mountains, the southern part with the Bale and Haranna-Mandebo mountains and the Ogaden. Both the northwestern and the southeastern highlands are dissected by deep river valleys, in places as much as 1,500 m deep (e.g., the Blue Nile gorge—Ayalew and Yamagishi 2004), and slope gently toward the Sudan lowlands and the Ogaden and Somali lowlands, respectively, reaching elevations as low as 500 m a.s.l.

The Ethiopian segment of the Great Rift Valley is more than 900 km long between the borders with Kenya and Djibouti and its width varies between 50 and 100 km on average. This extensive fault system is bounded to the north by the Afar triangle and to the south it proceeds beyond the Kenya border and Turkana Lake. The highest points of the Rift floor are between Awasa and Shala lakes (around 1,770 m a.s.l.) and around the Dubeta col (1,670 m s.s.l.) (between Ziway Lake and Koka Reservoir), which forms the divide between the Awash River basin and the endhoreic drainage systems of the Main Ethiopian Rift. The Rift bottom slopes gently to its lowest point at Asal Lake at −155 m b.s.l. in Djibouti and to the 361 m a.s.l. at Turkana Lake in the south. The floor of the Rift Valley is not uniformly flat as scattered volcanoes or volcanic systems, rising for more than 1,000 m (e.g., Zikwala, 2,989, Boset, 2,447, Fantalé, 2,007 m a.s.l.) occur. Seven major lakes, all closed systems, of tectonic or volcano-tectonic origin (Ziway, Langano, Abijata, Shala, Awasa, Abaya—the largest one with a surface area of 1,162 km^2—and Chamo) are located in the main trunk of the Ethiopian Rift. Other lakes are found in the northernmost portion (Abe Lake, 350 km^2, fed by the Awash River) and in the Danakil depression (Afrera Lake, 70 km^2) (Wood and Talling 1988). Large lakes are also present in the northern and central highlands. Tana Lake is by far the largest in Ethiopia with its 3,600 km^2, whereas the other highland lakes such as Hayk (23 km^2) (Demlie et al. 2007) and Ashenge (20 km^2) are much smaller.

The most notable river system is the Blue Nile and its tributaries, the largest of which is the Tekeze/Atbara, joining the Blue Nile in the Sudan territory. Because of the general westward slope of the western highlands, many large rivers are tributaries of the Blue Nile system, which drains an extensive area of their northern and central part. The Blue Nile, the Tekeze, and the Baro account for about half of the country's water outflow. Other important rivers originate from the Somali Plateau (Genale/Juba and Wabe Shebele rivers) and outflow into the Indian Ocean. Several drainage basins of smaller rivers are closed systems and the largest among these rivers are the Awash and Omo.

1.2.1 Soils

Ethiopia presents a large variety of soil types. Berhanu et al. (2013) classified the land mass of Ethiopia into 60 soil types with an area coverage ranging from 1.4 to 208,882 km^2. Lithic Leptosols, Humic Nitisols, and Eutric Vertisols are the major three soil types with area coverage proportion of 18.5, 11.9, and 10.2 %, respectively. These prominent percentages are comparable with those reported by the FAO Soil Database (FAO 1998), though only 45 different soil types were identified by this study.

On the base of the harmonized soil database constructed by FAO (2009), Berhanu produced a textural classification of Ethiopian soils and their relative frequencies of area coverage, which reveals that loam and sandy loam soils are the most common soil textures.

Following the UNDP/FAO (1984) report on the geomorphology and soils of Ethiopia, a short description of the main soil characteristics of different physiographic regions can be outlined.

In western Ethiopia (Gamo Gofa, Ilubabor, Welega, and part of Gojam and Gonder), soils develop mainly on felsic and metamorphic Precambrian rocks and flood basalts. Since in this part of Ethiopia the highlands slope down to the lowlands, soils develop also on alluvial and colluvial deposits. In the highlands, the high rainfall (over 2,000 mm yr^{-1}) is the most important factor in producing very similar soils irrespective of the parent rocks. In western Illubabor, fluvial soils are found on the alluvial plain of the Baro River and its tributaries.

The northern highlands are underlain by Precambrian metamorphic rocks, Upper Paleozoic and Mesozoic sandstones, limestones, and the trap series volcanics. This area shows marked contrasts in topography, rainfall, and land use, hence soils are highly variable and many soil types are present (Nitosols, Vertisols, Andosols, Lithosols, etc.).

The northeastern escarpment of Ethiopia has resulted from strong tectonic activity that produced rugged morphology and significantly influenced soil characteristics. They are, in fact, affected by severe (natural and man induced) erosion processes and show extreme stoniness.

In central Ethiopia highlands, soils developed predominantly on trap basalts and subordinately, on pyroclastic rocks. Rainfall does not vary much spatially (see Chap. 3, this volume) and soils characteristics depend mainly on topography. In deeply incised river valleys, soils are thin and have a high stoniness degree due to very high erosion rates.

In the eastern highlands (Chercher and Ahmar mountains), parent materials consist of Mesozoic sandstones and limestones, flood basalts and, subordinately, of Precambrian basement rocks. In this area, these rocks are very mixed and, due to strong structural influence and intensive cultivation, soils are shallow and similar to those of the northeastern escarpment. To the west, where the structural influence is less marked and rainfall increases, soils show more vertic characteristics.

The northern portion of the Rift Valley includes semidesert areas with the exception of the Awash River valley fill. Parent materials of this area include Tertiary and Quaternary volcanic, alluvial, and colluvial deposits. The very recent age of these parent rocks and the dry weather condition (annual precipitation range between 300 and 500 mm) lead to soils that developed mainly on alluvial and colluvial materials. Eutric and Calcaric Fluvisols are common in the alluvial plain of the Awash River.

By contrast, the geology of the southern portion of the Rift Valley is more complex. Bedrock consists mainly of Tertiary pyroclastics and Quaternary basalts but the Rift floor includes also extensive lacustrine, colluvial, and fan deposits. Vertic and Mollic Andosols prevail in areas underlain by volcanic ash and pumices, whereas Lithosols are more common where Quaternary basalts outcrop.

In the Ogaden, sedimentary rocks prevail, namely limestones, sandstones, and evaporates. In this area, precipitation is scarce (commonly less than 400 mm yr^{-1}), evapotranspiration is high and uniform and topography is comparatively flat, hence it is the parent material to play the most relevant role in soil differentiation and typically Gypsic and Calcic Xerosols and Yermosols occur. In the eastern Ogaden, underlain by sandstones, Cambic Arenosols prevail.

Finally, south of the Bale Mountains, flat landforms dominate and parent rocks consist mainly of deeply weathered rhyolites on which deep Eutric Cambisols occur, whereas steeper slopes are underlain by Precambrian gneisses and granites, and here Lithic phases of the Eutric Cambisols prevail.

1.2.2 Soil Erosion

Soil erosion in Ethiopia varies widely due to the very different physiography, land use, and soil characteristics. Several studies have investigated the rates and causes of soil loss in Ethiopia, but most of quantitative data are derived from plot studies (e.g., Hurni 1985; Soil Conservation Research Project—SCRP—Grunder 1988; Herweg and Ludi 1999) or studies were focused on restricted areas (e.g., Tegene 2000; Nyssen et al. 2004). Bojo and Cassells (1994), however, argued that soil erosion data obtained from plot studies must be corrected including also the delivery ratio factor in order to assess realistic erosion rates, whereas large areas or regional studies are mainly based on soil loss models rather than field measurements.

Soil loss measured in a few sites of the SCRP (Grunder 1988) is reported in Table 1.1. These data show a large range of values given the different physiography, precipitation, soil characteristics, and land use of the regions considered. The plot studies by the SCRP (Grunder 1988) showed also that the traditional conservation practices are the least effective in combating soil erosion. Table 1.2, on the other hand, reports soil erosion data extrapolated for the whole country by different authors and the data range is again large. This is probably due to different reference soil loss values used by the authors.

Berhanu et al. (2013) calculated the erodibility factor (*K*) of different soil types. Unexpectedly, they found relatively low values as *K* ranges between 0.00 and 0.18. Similar results were obtained also by Nyssen et al. (2007) in Tigray and by Shiferaw (2012) in the Borena area of South Welo highlands. These conclusions indicate that in Ethiopia soil erodibility is not an explanation for the magnitude of actual land degradation as the combination of steep slopes, vegetation removal, and erosive rain seems to play a major role. Rainfall intensity of 100 mm in 24 h is calculated by Billi et al. (2015) to have an average return time of 25 year (range 6.3–68.6 years). Land degradation is a common feature of the Ethiopian landscape with severely eroded areas along the rift margins (Fig. 1.2) and in the northern highlands. Other common erosion processes are gullying (Fig. 1.3) and landsliding (Fig. 1.4).

Table 1.1 Soil loss in various parts of Ethiopia (Berehe 1996)

Region	Soil loss (t $ha^{-1}yr^{-1}$)
Gojam	40.2–199.2
South Welo	36.5–53.8
North Shewa	152.4–214.8
Illubabor	18.0–135.3
Harerge	25.5–27.8
Sidamo	41.2–49.5

Table 1.2 Soil loss in Ethiopia

Author/s	Soil loss (t $ha^{-1}yr^{-1}$)
Hurni (1988)	42
Wright and Adamseged (1986)	100
Sutcliffe (1993, cited in Kappel 1996)	40
Bojo and Cassells (1994, cited in Kappel 1996)	20
Stocking (1996)	165
Tamene and Vlek (2008)	14
Average	63

Soil erosion rates obtained from direct measurement of soil dislodgements from slopes is, however, about two orders of magnitude higher than the sediment yield data obtained from river sediment flux measurements. A few authors, in fact, report highly variable sediment yield values ranging from 4 to 3,784 t $km^{-2}yr^{-1}$ (several sources, see Chap. 6 this volume). Nyssen et al. (2004) developed a simple power equation from literature data on 20 river sites in drainage area outside the less-erodible areas in southern Ethiopia as

$$Y_s = 2{,}595\ A^{-0.29} \tag{1.1}$$

in which 56 % of the variability of sediment yield (Y_s in t $km^{-2}yr^{-1}$) is explained by catchment area (A in km^2). This result is partially confirmed by field measurements of Billi (2004) on the Meki River, upstream of the homonymous town (only a few kilometers upstream of the river outlet into Ziway Lake), which gave a sediment yield of about 60 t $km^{-2}yr^{-1}$. By contrast, field measurements by Haregeweyn et al. (2008) on very small rivers in northern Ethiopia, ranging between 0.72 and 24 km^2 in catchment area, resulted in a much higher average sediment yield of 947 t $km^{-2}yr^{-1}$ (range 446–1,817 t $km^{-2}yr^{-1}$). Also these authors provided

Fig. 1.2 Severe land degradation on the main Ethiopian Rift escarpment near Alaba Kulito along the Soddo-Shashemene road. The *top* soil has been completely removed by erosion leaving a badland-type morphology developed into in the soft bedrock consisting of poorly consolidated volcanic ashes and pumices

Fig. 1.3 A deep gully incised into the colluvial deposits mantling the base of the Gade Motta caldera rim (7°57′26″ N–38°39′20″E). This box or U-shaped gully is very deep, with steep flanks the hight of which appreciably decreases downstream as far as the gully splay on the rift *bottom*

Fig. 1.4 Block sliding apart due to lateral spreading from the Amba Aradam Mountain near Antalo (13°17′50″N–39°25′07″ E). Notice the varicolored sub-horizontal, Jurassic Agula shales unconformably overlain by the Cretaceous Amba Aradam sandstones

an interpolation but, conversely to Nyssen et al. (2004), it is expressed by a less significant (R^2 = 0.36) linear equation

$$Y_s = 0.005\,\mathrm{A} + 6 \quad (1.2)$$

which predicts sediment yield to increase with increasing catchment area. This result confirms that in small catchments slope erosion processes are not stationary because of local factors, but are very effective in supplying sediment directly into stream channels. Notwithstanding such wide variations of data from different authors, sediment yield seems to be rather high, especially in the northern highlands (Fig. 1.5) and in the Rift margins as witnessed by a few examples of reservoirs completely filled with sediment (Fig. 1.6).

1.2.3 Natural Vegetation and Land Use

According to Ibrahim (1978), Ethiopia lies within the Sudano-Zambezian phytogeographic region of Africa which comprises one of the largest formations of the continent and can be characterized as tropical, with a long dry season of 4–9 months, annual rainfall between 200 and 1,500 mm, and vegetation typical of Sudanian and Sahelian zones which includes steppe, savanna, and dry (Fig. 1.7) to subhumid woodland and forest. The large differences in physiography, soil types and climatic conditions result in a variety of habitats suitable for evolution of several plant species. Vegetation of Ethiopia, therefore, is very heterogeneous and characterized by considerable endemism which is strongest in the high mountains, in southeastern Ogaden (Soromessa et al. 2004), Borana and Bale lowlands (Woldu 1999 in FAO 2014). In Ethiopia, there are about 6,000 species of higher plants, of which about 10 % are endemic (Kelbessa et al. 1992).

According to White (1965), the Sahel region, which can be characterized as wooded steppe with *Acacia* and *Commiphora* spp. (Fig. 1.8), represents a floristically impoverished western extension of the rich Afro-Oriental domain. *Acacia* species are very common in the Ethiopian Rift Valley and other low rainfall areas but are also found in the southern highlands where they form the basis of a traditional agricultural practice (Poschen 1986). In the arid zone plains, bushed grassland prevails (Fig. 1.9), except some patches of woodland.

The highlands of Ethiopia have climates and vegetations that vary noticeably in relation to altitude, forming part of the Afro-montane region. Ibrahim (1978) includes Ethiopia in the Afro-Oriental domain, which covers also the lowlands of Tanzania, Kenya, and Somalia, and sets an altitudinal limit of 1,100 m for this domain though it extends considerably higher in the Ethiopian Rift Valley.

Fig. 1.5 A mushroom stone witnessing severe soil erosion north of Aksum (14°09′53″N–38°41′58″E)

Fig. 1.6 The Aba Samule Reservoir on the Akaki River, south of Addis Ababa (8°47′15″N–38°42′18″E). The dam was constructed in 1939 but a few decades later it was completely filled with sediment

The forest cover of Ethiopia has been declining rapidly. Most of the remaining forests are confined to the south and southwestern parts of the country (Tilahun et al. 2011).

In these areas, forests are threatened by human activities. Historical documents show that Ethiopia experienced substantial deforestation (Pankhurst 1995; McCann 1997; Dessie and Kleman 2007), soil, and land degradation (Nyssen et al. 2004; Nyssen et al. 2014) over the years. The need for fuelwood, arable land (Fig. 1.10) and grazing areas (Fig. 1.11) are indicated as the main causes for forest degradation (Tilahun et al. 2011). According to Woldu (1999 in FAO 2014), about 34 % of Ethiopia and 57 % of the land above 1,500 m was once covered by dense forests and a further 20 % by wooded savannah. Massive deforestation has reduced these figures to 3.6 % of the total area (Tefera 2011) (Fig. 1.12) and to 9 % of the land above 1,500 m. Widespread deforestation started, particularly in the highlands, at the end of the nineteenth century with the expansion of agriculture.

Fig. 1.7 Example of a natural savannah and dry woodland forest in the southern portion of the Rift Valley, southeast of Konso town

Fig. 1.8 A natural Acacia forest within the Abijata-Shala Park (7°71′34″N–38°39′46″E)

The glorious forests of the past are witnessed by isolated, huge sycamore trees that are present in the Rift and in the highlands below 2,000 m a.s.l. (Fig. 1.13)

Nowadays, the most common trees of Ethiopia are the *Eucalyptus* spp. (Fig. 1.14). This genus was introduced from Australia to East Africa in the late nineteenth and early twentieth century and at that time the largest plantations were in Ethiopia and Rwanda. In Ethiopia, the Eucalyptus genus was introduced in 1894/1895. The purpose was to supply fuelwood and construction timber to the new and

Fig. 1.9 Example of a bushed grassland in "Nech Sar" (*white grass*) Park near Arba Minch (5°57′24″N–37°39′31″E)

Fig. 1.10 Cereals cultivations on the Somali Plateau along the Shashemene-Kofele road. This flat area is underlain by trap basalts and stretches eastward from the Rift margin

growing capital city, Addis Ababa. In the 1970s, the plantation area in the country was about 90,000 ha. Recent estimates indicate the extension of Eucalyptus forests for about 0.5×10^6 ha though huge numbers of trees exist in other land use types, such as homesteads, farm boundaries, and beside roads (Dessie and Erkossa 2011).

In the main Rift Valley, agricultural activity is rapidly expanding and progressive settlement has replaced grazing lands with small (Fig. 1.15) to medium farms, some of which are mechanized (FAO 2014).

1.3 Main Landscape Regions

The geomorphology of Ethiopia is largely controlled by its geological structure, but weathering, erosion, and deposition processes contributed as well in shaping the country into its present association of landforms and landscapes. The crustal evolution, characterized by a marked swelling within one of the most active extensional areas of the planet, resulted in three main morphostructural units, which include also

Fig. 1.11 Pasture on the western portion of the central highlands near Nekemte. Major landforms of this are small volcanic plugs and dome-shaped subvolcanic intrusions

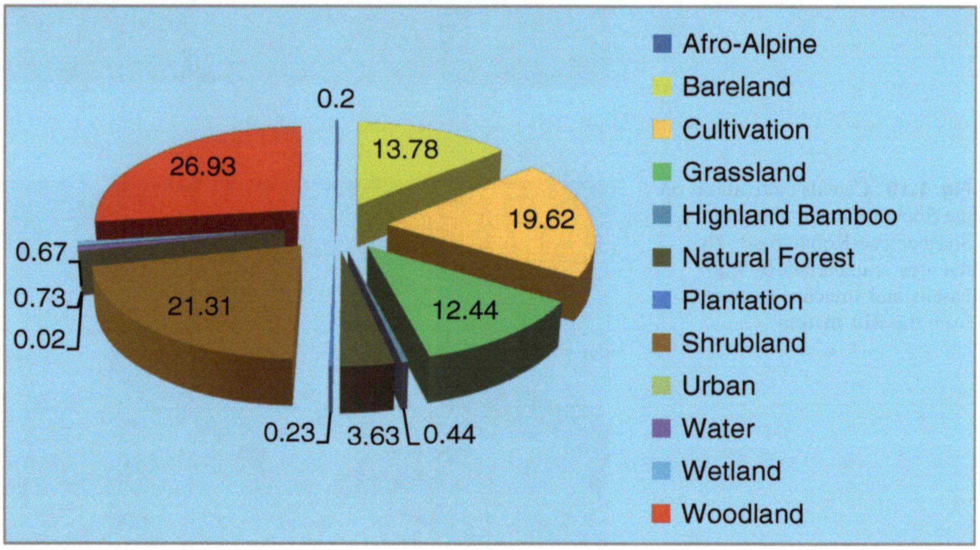

Fig. 1.12 Land cover distribution by percentage of the Ethiopian territory (Woldu 1999)

subregions with specific geomorphic features. These are given as follows: (1) the western plateau that can be divided into (i) the northern highlands, including the volcanic plug belt of Adwa, (ii) the central highlands, and (iii) the south-western highlands; (2) the Rift Valley which consists of three main portions, namely the northern, central, and southern trunks and the Afar and Danakil depressions; (3) the southern plateau which consists of a northern and southern sectors and includes also the Ogaden tableland gently descending to Somalia and the Indian Ocean.

The plateaus originated by the domal uplift of the Arabian-Ethiopian region (Merla et al. 1979) and accumulation of flood basalts, the thickness of which is commonly around 1,000 m but may reach 2,000 m in some regions (Kieffer et al. 2004) or even 3,000 m in the Chercher mountains (Juch 1975) (Fig. 1.1). The most elevated parts of the plateaus stand at 3,000 m a.s.l., but in places they are surmounted by high volcanic mountains, a few of which reach elevations above 4,000 m. Both main plateaus descend radially and gradually to elevations of about 500–600 m around the border with Sudan and 200–400 m close to the border with Somalia. The main topographic gradient is steeper (about 0.004–0.006) near the rift margins and decreases to 0.001–0.002 in the distal parts where granite inselbergs are a common and typical geomorphological feature (Figs. 1.16 and 1.17). By contrast, the transition from the plateau margins to the rift is decidedly abrupt and marked by stepped topography and a number of elongated grabens (Fig. 1.18), with asymmetrical and unpaired horsts and the bottom locally punctuated by recent trachyte plugs (Fig. 1.19). In places, spectacular fault-generated escarpments occur (Fig. 1.20), with long and high exposed fault planes, at the base of which alluvial fans are commonly found (Fig. 1.21).

Fig. 1.13 A huge sycamore tree

Fig. 1.14 Eucalyptus forest on the Somali Plateau along the Shashemene-Kofele road

Fig. 1.15 A small farm on the shore of Abaya Lake, along the Soddo-Arba Minch road (6°30′00″N–37°46′33″E)

Fig. 1.16 Granite inselbergs, with evidence of exfoliation processes and products, in the eastern side of Kassala

1.3.1 Northern Highlands

The geology of the northern highlands is rather complex and includes a variety of formations ranging in age from Precambrian to Quaternary. Such a geodiversity provides the northern highlands with, likely, the most varied association of landforms and landscapes. In the northernmost sector, low grade Proterozoic metamorphic rocks, Lower Paleozoic granites and, subordinately Mesozoic sedimentary sequences occur, whereas the southern sector is mainly underlain by volcanites of the trap series with the typical tabular morphology of the flood basalts (Fig. 1.22). Continuing volcanic activity resulted in the building up of high mountain areas which include a few of the highest peaks of

Fig. 1.17 Granite inselbergs along the road from Mega to Moyale. Notice typical landforms such as tors originated by crystalline rocks weathering

Fig. 1.18 Structural basins within the western plateau escarpment south of Karakore, along the Debre Birhan-Dese road (10°24′35″N–39°56′05″E). This small graben is asymmetric with the western flank (on the *right* in the photo) higher. The graben floor is also slightly tilted and dipped southward (away from the reader)

Ethiopia. The Simen Mountains massif dominates the western plateau and it is the fourth highest range of Africa after Kilimanjaro, Mount Kenya and Rwenzori, with Ras Dejen (4,533 m a.s.l.) being the highest mountain peak of Ethiopia and among the first ten highest peaks of the African continent. In the northern highlands, other notable high mountain peaks are the Abune Yosef (4,284 m a.s.l.), in the Tekeze headwaters, and a number of high mountains, some ranging in elevation between 3,000 and 4,000 m, the most renown of which, for historical reasons, is Amba Alaji (3,438 m a.s.l.) (Fig. 1.23). *Amba* is a local name that is commonly used to indicate flat-topped mountains (see Chap. 9 this volume). Given the tabular structure with an almost negligible slope of the area and the occurrence of hard Mesozoic sedimentary formations such as the Adigrat Sandstone, consisting of 91 % of quartz components (Getaneh 2002), resting on more erodible glacigenic and metamorphic rocks or, like in the case of the Amba Aradam, on softer and easily erodible Agula Shales (Fig. 1.4), the Amba mountain geomorphology is rather common in Tigray (Fig. 1.24).

Drainage of the northern highlands is predominantly to the west-northwest and includes the largest and longest rivers of northern Ethiopia. By contrast, much smaller and commonly ephemeral streams flow to the east into closed basins or disappear in the Danakil lowlands. All the western

Fig. 1.19 Recent (Quaternary?) intrusions of trachytic rocks protruding from the *bottom* of the Kobo-Alamata structural basin (12°07′39″N–39°41′15″E). View is to the north

Fig. 1.20 Impressive fault escarpment in the Termaber basalts with alluvial fans at the base near Wichale, between Dese and Weldiya

Fig. 1.21 Alluvial fan in the footslope of the eastern margin of the Kobo-Alamata structural basin (12°22′25″N–39°41′31″E)

Fig. 1.22 The typical tabular structure of the northern highlands with a deep canyon incised by a tributary of the Mereb River along the Adigrat-Adwa road. The *right-hand side* of the valley is capped by the Paleozoic Enticho sandstones (Abbate, personal communication), whereas in the opposite side the sandstones are overlain by the Tertiary trap basalts

Fig. 1.23 The Amba Alaji peak (3,438 m a.s.l.) seen from the Korem-Mekele road. The pyramidal morphology of the mountain *top* is carved into Miocene rhyolitic ignimbrites

rivers are tributaries of the two main river systems of this area, the Tekeze and the Mereb. Both these rivers flow on the bottom of deeply incised canyons (Fig. 1.22) and mark the border between Eritrea and Ethiopia for a few hundreds of kilometers. At the large scale, in fact, the geomorphology of the northern highlands is mainly the result of erosion, rather than deposition processes, as would be expected given the remarkable uplift of the area that has proceeded until recent times (Merla et al. 1979; Corti and Manetti 2012). However, in a few places underlain by limestone rocks

Fig. 1.24 Flat-topped (*amba*) mountains along the road from Mekele to Adigrat in Tigray. This tabular structure is formed by sub-horizontal, Triassic Adigrat sandstones which in places include dark brown ferruginous/lateritic beds, very resistant to erosion (Enkurie 2010)

Fig. 1.25 The travertine dam of Romanat, a few kilometers NW of Mekele (13°34′24″N–39°25′04″E)

(e.g., Antalo formation), such as in the Mekele outlier, thick deposits of travertine, forming spectacular natural dams, are present (Fig. 1.25).

Among the northern highlands landscapes, the one around Adwa is unique and remarkable. It is characterized by a large number of spires that rise from the eroded surface of the basement to make a sort of stony tree forest (Fig. 1.26). The Axum–Adwa complex is part of a larger magmatic SW–NE belt extending for about 150 km from west of Axum to as far as Senafe in Eritrea. This magmatic complex is set along an uplifted crustal sector overlying the Proterozoic crystalline basement or, in many places, its Mesozoic sedimentary cover (Zanettin et al. 2006; Natali et al. 2013). The magmatic products consist mainly of an alternation of trachyte and

Fig. 1.26 The volcanic plugs of Adwa seen from Axum

Fig. 1.27 Volcanic plugs near Enticho

syenite plugs and domes reaching their maximum density in the area around Adwa. The world renown obelisks of Aksum were carved from the subvolcanic phonolite-syenite domes that dominate the landscape around the town (see Chap. 7 this volume).

The plugs and domes rise from a few tens of meters to about 300 m above the sedimentary substrate or alkaline basalts (Hagos et al. 2010) (Fig. 1.27). A few of the most peaked and isolated pinnacles host monasteries and churches (Fig. 1.28).

Fig. 1.28 The church of St. Pantaleon on *top* of a pointed trachyte plug near Aksum

Fig. 1.29 The central highlands north of Addis Ababa are deeply incised by the Sodoblé River, a tributary of the Blue Nile near Chancho (9°26′11″N–38°38′47″ E). Notice the alternation of softer and harder lithotypes within the trap series on *top*, the sub-vertical fault scarp parallel to the river and cutting the lower portions of smaller transverse divides and the Mesozoic sedimentary formations outcropping in the valley *bottom*

1.3.2 Central Highlands

The central highlands are likely the most monotonous of the Ethiopian highlands since they are almost entirely underlain by basalts of the Tertiary trap series and other, more recent volcanic rocks. Only to the west, Proterozoic metamorphic rocks crop out, whereas the Mesozoic sedimentary sequence is exposed in the Blue Nile valley. The general morphology is therefore that of a tabular structure gently sloping to the west and deeply incised by the Blue Nile and its tributaries

Fig. 1.30 The flat *top* surface of the central highlands north of Addis Ababa with shield volcanoes merged to form a string-shaped ridge in the background

Fig. 1.31 A sharp, sub-vertical fault face, with a hanging river valley, along the road between Karakore and Kombolcha

(Fig. 1.29). The top surface is typically flat and strewn with individual volcanic edifices or groups of a few contiguous apparatuses, commonly arranged in strings along fault lines (Fig. 1.30). Some of these volcanoes may be very large and stand higher on the plateau surface, rising 1,000–1,500 m above it. This is the case of Mt. Guna (4,231 m a.s.l.) near Debre Tabor and Mt. Choke (4,154 m a.s.l.) south of Tana Lake, separated by the Blue Nile which flows to the south and then turns to the west in a large bend right around Mt. Choke. Other high mountains include the Abuye Meda (4,000 m a.s.l.), south of Kombolcha, Amba Farit (3,975 m a.s.l.), west of Dese, and Tulu Welel (3,200 m a.s.l.) in the westernmost portion of the plateau, close to the lowlands across the border with Sudan.

Fig. 1.32 The famous "Afar Window" along the road from Debre Birhan to Debre Sina (9°50′13″N–39°44′30″E). A small fault, transverse to the main rifting system, makes a narrow gap within the plateau edge and from an elevation of about 3,100 m a.s.l. it is possible to see the Afar lowland in the far distance

Fig. 1.33 The rugged landscape of the southwestern highlands. The Omo River valley west of Welkite (8°15′40″N–37°36′09″E)

Apart from volcanoes and their products, the northern highlands gross geomorphology is mainly controlled by faulting and horst and graben structures. Long, high and commonly sub-vertical fault faces are common as well, especially close to the eastern margin (Fig. 1.31). This latter is rather sharp and the relief contrast with the Danakil and Afar lowlands is remarkable (Fig. 1.32). Moving westward, the plateau elevation decreases but the general landscape does not change appreciably, at least as far as Nekemte (Fig. 1.11).

1.3.3 Southwestern Highlands

The southwestern highlands are less even and more rugged compared to the northern and central highlands. In particular in the median part, the landscape assumes a more typical mountainous configuration consisting of deep valleys and mountain groups (Fig. 1.33). The north- and southeastern sides coincide with the margins of the main and the southern Ethiopian Rift Valley, respectively, where the highest mountain peaks are found: Mt Gurage (3,721 m a.s.l.) in the

Fig. 1.34 The main Ethiopian Rift valley floor near Ziway seen from the rim of the Gademotta caldera. The Rift *bottom* is underlain by Quaternary lacustrine, fluvio-lacustrine, volcano-lacustrine, and volcanic ashes

Fig. 1.35 The stepped morphology produced by normal faults on the Rift margin east of Ziway Lake (8°03′20″N–39°07′14″E)

northernmost part, Mt. Guge (4,200 m a.s.l.) west of Arba Minch and Mt. Malgudo (3,390 m a.s.l.) southeast of Jimma. To the west, the southwestern highlands maintain their general mountainous landscape, but the valleys become broader and the mountain and valley bottom elevation decreases gradually as far as the town of Gambela, east of which there is a quite abrupt transition, along a almost north–south alignment, from the 2,000 m a.s.l. of the area close to Fit Makonnen to the tropical lowlands of the western Ilubabor (400–500 m a.s.l. on average), where wetlands are also present (Woldu and Yeshitela 2003).

1.3.4 The Rift Valley

The Great Rift Valley of Africa is one of the largest tectonic structures on Earth and one of the most attractive regions of the whole continent for its geomorphological relevance and

Fig. 1.36 An example of Gilbert-type delta in the recent shore deposits of Shala Lake (7°28′41″N–38°38′14″E)

Fig. 1.37 The Awash River flowing entrenched into the Rift valley floor near Ombole (8°23′22″N–38°45′53″E), with the Zikwala Volcano in the background. The river entrenching is caused by base level drop associated with main Ethiopian Rift lakes shrinkage and Afar tectonics

unique natural environments. The Ethiopian Rift consists of three main sections: (i) the southern portion, from the border with Kenya to the hydrological divide between Abaya and Awasa Lake; (ii) the main portion, commonly indicated also as the Great Lakes Region of Ethiopia, stretching as far as the Awash River; (iii) the northern portion, from the Awash to the apex of the Afar triangle and its northeastern branch, the Danakil depression.

The width of the Rift bottom is relatively constant around 50 km (Fig. 1.34), with the narrowest section being less than 20 km at Arba Minch. By contrast, the top width between the plateaus edges is more variable and depends mainly on

Fig. 1.38 A very small cinder cone along the Nazreth-Metehara road (8°53′57″N–39°48′37″E)

Fig. 1.39 The Fantalè shield volcano seen from the Awash National Park (8°54′17″N–40°02′36″E)

the number and size of the down-faulted blocks making up the stepped morphology of the Rift margins (Fig. 1.35). It is narrowest at Arba Minch and widest at Addis Ababa where it can be as much as 120 km. The Rift margin scarps are mainly cut into the rocks of the trap series in the southern portion and the Danakil, whereas in the Lakes Region and the northern portion (namely around Dire Dawa), more recent volcanic and Mesozoic sedimentary rocks, respectively, prevail.

The Rift floor is generally flat as it is commonly underlain by pyroclastic material and lacustrine and fluvio-lacustrine deposits (Fig. 1.36). Its main geomorphic features include narrow lacustrine terraces (visible especially in the Ziway–Shala lake area), river deltas, and shallow canyons (Fig. 1.37). River entrenchment is mainly due to the base level drop associated with the recent lakes shrinkage (see Chaps. 16 and 17 this volume) and the Afar lowland formation. However, the most distinguished geomorphic features of the Rift bottom are volcanoes and lakes. Volcanoes punctuate the Rift floor as individual or merged edifices ranging in size from small cinder cones (Fig. 1.38) to large strato-volcanoes (Fig. 1.37) or shield volcanoes (Fig. 1.39).

Fig. 1.40 A nice example of maar west of Butajira (8°02′52″N–38°20′58″E)

Calderas are also very common and the largest of them may have a diameter spanning a few tens of kilometers. Calderas host the majority of the main Ethiopian Rift Valley great lakes (namely Langano, Abijata, Shala, and Awasa), whereas a few much smaller lakes are formed within maars (Fig. 1. 40). Most of the larger lakes hold saline water (Abaya, A-bijata, Awasa, Beseka, Chamo, Langno, and Shala) with total dissolved solids concentrations ranging from 771 ppm of Abaya Lake to 56,300 ppm of Beseka Lake (Wood and Talling 1998; Chap. 18, this volume). The Rift lakes make up the base level for all the rivers draining into the Rift. With the exception of the lakes fed by the Awash River, such as Abe Lake, all other lakes are fed by small rivers (Chap. 16, this volume). In the recent decades, the Rift lakes have shown contrasting hydrological trends with some lakes shrinking (e.g., Abijata) and other—Abaya, Awasa and Beseka—expanding (Chap. 16, this volume). Given the remarkable rate of change in water level observed (e.g., Abijata −6 m, Abaya +2 m and Awasa +3 m in 20 years), not paired by an equivalent change in rainfall in the vicinity of these lakes, it is very difficult to ascribe the lake surface variations to climatic factors only. For Awasa Lake, Gebreegziabher (2004) attributes the lake level rise to the combined effects of land use and climate change, whereas Ayenew and Gebreegziabher (Chap. 16, this volume) conclude that water balances of the main Rift lakes are predominantly controlled by groundwater. A special case is offered by fast and large expansion of Beseka Lake that occurred in the last decades. The maps of Fig. 1.41 show clearly an impressive lake surface increase between 1979 and 1995. The water level is still rising leading to the complete drowning of both the road and the railway to Djibouti. Human impact is likely the main controlling factor of such a remarkable change in water level that started after the construction of a large irrigation scheme, fed by the Awash River water, for a newly set sugar plantation (Fig. 1.41).

The Rift becomes progressively more arid as we move to the Afar triangle and its northern branch, the Danakil depression. These areas are described in detail in Chap. 15 of this volume. Here, it is worth recalling briefly that the Danakil depression is a strip of coastal land 60–80 km wide separated from the Red Sea by coastal hills draining into the inland saline lakes. The Danakil depression is the hottest inhabited and one of the least elevated places on Earth and location of unique landscapes, associated with different magmatic activities, and evaporite deposits as much as 2 km thick forming a vast flat area from which salt domes, a few meters high, protrude (Fig. 1.42). Numerous volcanoes and volcanic landscapes occur in the Danakil, including the most active and renown of them, the Erta Ale (see Chap. 15, this volume).

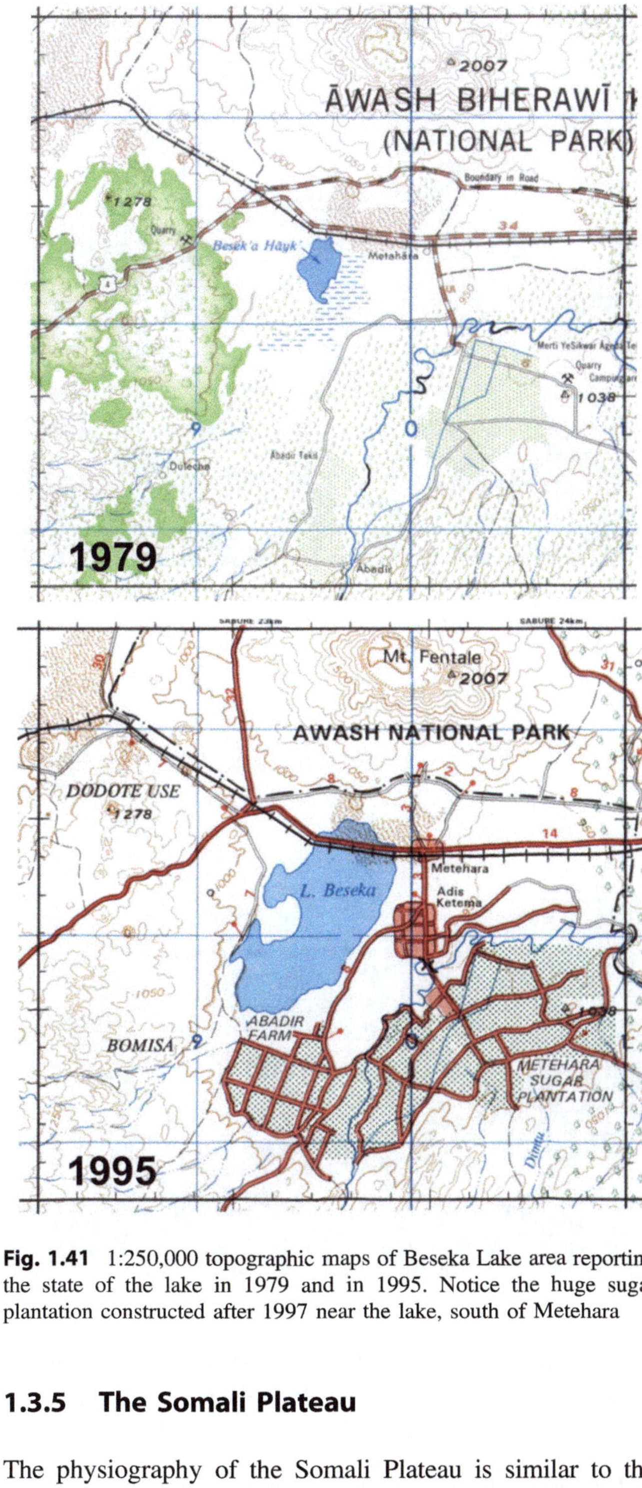

Fig. 1.41 1:250,000 topographic maps of Beseka Lake area reporting the state of the lake in 1979 and in 1995. Notice the huge sugar plantation constructed after 1997 near the lake, south of Metehara

1.3.5 The Somali Plateau

The physiography of the Somali Plateau is similar to the northwestern highlands, but its elevation is slightly lower. The highest mountain ranges, Bale, Chercher, and Ahmar (Fig. 1.1), are aligned with the western margin and close to its stepped escarpment. From here, the plateau gently descends to S-SE and sinks into the Indian Ocean.

In the Bale mountains, we find the highest peaks of the Somali Plateau and among the highest of Ethiopia, Mt. Tullu Dimtu (4,383 m a.s.l.) and Mt. Batu (4,307 m a.s.l.), but other high peaks include Mt. Kaka (4,180 m a.s.l.) and Mt. Bada (4,130 m a.s.l.) in the Chercher range. Moving to the north, along the plateau margin, the mountain height tends to decrease with the highest peak of the Ahmara Mountains, Gara Muleta, being only 3,381 m a.s.l. (Fig. 1.43).

The Somali Plateau is rather flat (Fig. 1.10) and its tabular morphology is interrupted only by deep canyons of the main rivers, i.e., the Wabe Shebele and the Genale-Dawa (Fig. 1.44), and high massifs. Among the latter, the Bale Mountains are decidedly the most prominent and consist of few main volcanoes, that released large amounts of Oligo-Miocene basalts and Quaternary rhyolite ignimbrites and basalt lavas, and fluvio-lacustrine intercalations (Merla et al. 1979). The summit of the Quaternary volcanites is punctuated by volcanic plugs (Fig. 1.45) and glacier erosional and depositional features are present (see Chap. 6, this volume). The latter are particularly evident in the Sanetti plain which is a large plateau at an average elevation of 4,100 m a.s.l., punctuated by small lakes, that has been shaped by an ice cap cover of about 180 km^2 during the last glacial maximum (Osmaston et al. 2005) (Fig. 1.46). Here, erratic boulders, roche mountonnées, and moraines are present, whereas cirques and moraine lakes (tarns) are common in the glaciated valley that radially spread out from the northern side of the plateau (Osmaston et al. 2005).

Evidence of glaciated landforms can be found also on the peaks of Mts. Bada and Kaka in the southern portion of the Chercher mountains (Osmaston et al. 2005), but is missing in the northern, less elevated part in which the most prominent landforms consist of small intermontane, structural basins, especially along the plateau margin (Fig. 1.47).

Most of the central and southern Somali Plateau is known as Ogaden—a large region, which extends from the Chercher and Ahmar mountains (Fig. 1.43) on the northern Somali Plateau margin to the border with Somalia and beyond, to the Indian Ocean. The Ogaden is a relatively flat and geomorphologically rather monotonous tabular land, gently dipping to E-SE. Its main morphological features are associated with large rivers crossing it, namely the Wabe Shebele River and its left side tributaries, the Ramis, Erer, and Fafen rivers (see Chap. 19, this volume). In the most elevated northern portion of the plateau, these rivers have incised deep valleys (Fig. 1.44) which become wider and shallow as the rivers approach the Somali border.

Though most of the Ogaden is underlain by Mesozoic sedimentary rocks, in its northern portion, especially in the Jijjiga area, the flat landscape is interrupted by small

Fig. 1.42 Salt hillocks rising from the Dallol salt plain (14°14′16″N–40°17′51″E)

Fig. 1.43 The Gara Muleta range seen from Kurfa Chele (Curfacelli) village, west of Harar (9°14′11″N–41°49′22″E)

volcanic plugs consisting of Miocene to Pleistocene basalts and rhyolites, whereas the town of Jijjiga stands on about 100 m thick Quaternary fluvio-lacustrine deposits accumulated in a large basin. Small, flat-topped residual hills made of limestone resting on the main gypsum formation rise for a few tens of meters above the surrounding tableland east of the confluence between the Dacata River into the Wabe Shebele (Merla et al. 1979). Elongated basalt hills (see Chap. 19, this volume) runs southward from Jijiga and parallel to the course of the Wabe Shebele, east of Imeey (Merla et al. 1979). Other small, isolated basalt hills occur also in between Warder and the Somali border.

Acknowledgments I am grateful to all the very many farmers, peasants and car drivers that throughout 25 years of field research all across Ethiopia, freely and collaboratively provided substantial help on many occasions. Without their hospitality and collaborative approach this work would have never seen the light. Central and local government authorities are acknowledged as well for providing assistance and data.

Fig. 1.44 The Wabe Shebele canyon near Gasera village, northeast of Robe Bale town (7°23′05″N–40°09′03″E)

Fig. 1.45 A number of volcanic plugs towering on *top* of the Bale Mountains, along the Shashamane-Robe road

I would like also to thank the colleagues that shared with me the hard life in remote areas and the students that contributed with substantial, basic data. A special thank to Ernesto Abbate, Mario Sagri and Milvio Fazzuoli that introduced me to the beauties of the Horn of Africa geology and geomorphology, to Giovanni Ferrari for his inspiring views and, finally, to Fabrizio Vannacci that faithfully assisted me with his laboratory, field and human skills in such a long endeavor. Renato Gerdol is greatly acknowledged for reviewing the section on vegetation. Many thanks also to Carlo Bisci for providing the base map of Fig. 1.1 and to Lorenzo Orioli for the information and photos of the Sanetti Plain.

Fig. 1.46 The Sanetti plain within the Bale Mountains National Park, at an average elevation of 4,100 m a.s.l. (photograph by L. Orioli)

Fig. 1.47 Unnamed mountains around an intermontane structural basin along the Asbe Teferi—Hirna road in the northern part of the Chercher mountains

References

Ayalew L, Yamagishi H (2004) Slope failures in the Blue Nile basin, as seen from landscape evolution perspective. Geomorphology 57:95–116

Berhanu B, Melesse AM, Seleshi Y (2013) GIS-based hydrological zones and soil geo-database of Ethiopia. Catena 104:21–31

Berehe WA (1996) Twenty years of soil conservation in Ethiopia. A personal overview. Regional Soil Conservation Unit/SIDA, Nairobi

Billi P (2004) Sediment yield of a closed river system: the Meki River, western margin of the Ethiopian Rift Valley. In: Proceedings of the IAHS international symposium on sediment transfer through the fluvial system, Moscow, 2–6 Aug 2004, pp 94–100

Billi P, Alemu YT, Ciampalini R (2015) Increased frequency of flash floods in Dire Dawa, Ethiopia: change in rainfall intensity or human impact?. Nat Hazards 75. doi:10.1007/s11069-014-1554-0

Bojo J, Cassells D (1994) Land degradation and rehabilitation in Ethiopia: a reassessment. The World Bank, Environmentally Sustainable Development Division, AFTES Working Paper No. 17, Washington

Corti G, Manetti P (2012) Geologia e paesaggi della rift valley in Ethiopia. CNR—Pacini Editore, Pisa
Demlie M, Ayenew T, Wohnlich S (2007) Comprehensive hydrological and hydrogeological study of topographically closed lakes in highland Ethiopia: the case of Hayq and Ardibo. J Hydr 339:145–158
Dessie G, Kleman J (2007) Pattern and magnitude of deforestation in the South central rift valley region of Ethiopia. Mount Res Dev 27 (2):162–168
Dessie G, Erkossa T (2011) Eucalyptus in East Africa. Socio-economic and environmental issues. Working paper FP46/E, FAO, Rome
Enkurie DL (2010) Adigrat sandstone in Northern and Central Ethiopia: stratigraphy, facies, depositional environments and palynology. Ph.D. thesis, Technischen Universität Berlin
FAO (2014) Country profiles. http://www.fao.org/countryprofiles/en/
FAO/IGADD/Italian Cooperation (1998) Soil and terrain database for northeastern Africa and crop production zones. Land and water digital media series # 2. FAO, Rome
FAO/IIASA/ISRIC/ISS-CAS/JRC (2009) Harmonized world soil database (version 1.1), FAO, Rome, Italy and IIASA, Laxenburg, Austria
Gebreegziabher Y (2004) Assessment of the water balance of Lake Awassa catchment, Ethiopia. M.Sc. thesis, International Institute for Geo-Information Science and Earth Observation, Enschede
Getaneh W (2002) Geochemistry provenance and depositional tectonic setting of the Adigrat Sandstone northern Ethiopia. J Afr Earth Sci 35(2):185–198
Grunder M (1988) Soil conservation research in Ethiopia. Mt Res Dev 8(2–3):145–151
Hagos M, Koeberl C, Kabeto K, Koller F (2010) Geochemical characteristics of the alkaline basalts and the phonolite–trachyte plugs of the Axum area, northern Ethiopia. Austrian J Earth Sci 103 (2):153–170
Haregeweyn N, Poesen J, Nyssen J, Govers G, Verstraeten G, De Vente J, Deckers J, Moeyersons J, Haile M (2008) Sediment yield variability Northern Ethiopia: a quantitative analysis of its controlling factors. Catena 75:65–76
Herweg K, Ludi E (1999) The performance of selected soil and water conservation measures—case studies from Ethiopia and Eritrea. Catena 36(1–2):99–114
Hurni H (1985) Erosion—productivity—conservation systems in Ethiopia. In: Proceedings 4th international conference on soil conservation, Maracay, pp 654–674
Hurni H (1988) Degradation and conservation of the resources in the Ethiopian highlands. Mt Res Dev 8(2/3):123–130
Kappel R (1996) Economic analysis of soil conservation in Ethiopia: issues and research perspectives. University of Berne, Berne, Switzerland, in association with the Ministry of Agriculture, Addis Ababa
Kelbessa E, Demissew S, Woldu Z, Edwards S (1992) Some threatened Endemic plants of Ethiopia. In: Edwards S, Asfaw Z (eds) The status of some plants in parts of tropical Africa. Botany 2000: East and Central Africa NAPRECA monograph series no. 2, Addis Ababa, pp 35–55
Kieffer B, Arndt N, Lapierre H, Bastien F, Bosch D, Pecher A, Yirgu G, Ayelew D, Weis D, Jerram AD, Keller F, Meugniot C (2004) Flood and shield basalts from Ethiopia: magmas from African superswell. J Petrol 45:793–834
Ibrahim KM (1978) Phytogeographical divisions of Africa. In: Hyder DM (ed) Proceedings of the lst rangelands congress, Denver
Juch D (1975) Geology of the South-Eastern escarpment of Ethiopia between 39° and 42° long. East. In: Pilger A, Rosler A (eds) Afar depression of Ethiopia. Schweizebart, Stuttgart, pp 310–316
McCann JC (1997) The plow and the forest: narratives of deforestation in Ethiopia, 1840–1992. Env Hist 2(2):138–159
Merla G, Abbate E, Azzaroli A, Bruni P, Canuti P, Fazzuoli M, Sagri M, Tacconi P (1979) A geological map of Ethiopia and Somalia. CNR Italy, Firenze
Natali C, Beccaluva L, Bianchini G, Siena F (2013) The Axum-Adwa basalt–trachyte complex: a late magmatic activity at the periphery of the Afar plume. Cont Min Petrol 166(2):351–370
Nyssen J, Poesen J, Moeyersons J, Deckers J, Haile M, Lang A (2004) Human impact on the environment in the Ethiopian and Eritrean highlands—a state of the art. Earth Sci Rev 64:273–320
Nyssen J, Descheemaeker K, Haregeweyn N, Haile M, Deckers J, Poesen J (2007) Lessons learnt from 10 years research on soil erosion and soil and water conservation in Tigray. Tigray livelihood papers no. 7, Mekelle Zala-Daget Project, Mekelle University, K.U. Leuven, Relief Society of Tigray, Africa Museum and Tigray Bureau of Agriculture and Rural Development
Nyssen J, Frankl A, Haile M, Hurni H, Descheemaeker K, Crummey D, Ritler A, Portner B, Nievergelt B, Moeyersons J, Munro N, Deckers J, Billi P, Poesen J (2014) Environmental conditions and human drivers for changes to north Ethiopian mountain landscapes over 145 years. Sci Total Env 486–486:164–179
Osmaston HA, Mitchell WA, Osmaston NJA (2005) Quaternary glaciations of the Bale Mountains, Ethiopia. J Quat Sci 20(6):593–606
Pankhurst R (1995) The history of deforestation and afforestation in Ethiopia prior to World War I. Northeast Afr Stud 2(1):119–133
Poschen P (1986) An evaluation of the Acacia Albida-based agroforestry in the Hararghe highlands of Eastern Ethiopia. Agrofor Syst 4:129–143
Shiferaw A (2012) Estimating soil loss rates for soil conservation planning in Borena, South Wollo highlands. Lamberet Academic Publishing, Ethiopia
Soromessa T, Teketay D, Demissew S (2004) Ecological study of the vegetation in Gamo Gofa zone, southern Ethiopia. Trop Ecol 45:209–221
Stocking MA (1996) Soil erosion. In: Adams WM, Goudie AS, Orme AR (eds) The physical geography of Africa. Oxford University Press, Oxford, pp 326–341
Sutcliffe JP (1993) Economic assessment of land degradation in the Ethiopian highlands: a case study. National Conservation Strategy Secretariat, Ministry of Planning and Economic Development, Addis Ababa, Ethiopia
Tamene L, Vlek PL (2008) Soil erosion studies in northern Ethiopia. In: Brahimo AK, Vlek PL (eds) Land use and soil resources. Springer, The Netherlands, pp 73–100
Tefera MM (2011) Land-use/land-cover dynamics in Nonno district, central Ethiopia. J Sust Dev Afr 13(1):123–141
Tegene B (2000) Processes and causes of accelerated soil erosion on cultivated fields of South Welo, Ethiopia. Eastern Afr Soc Sci Res Rev 16(1):1–22
Tilahun A, Soromessa T, Kelbessa E, Dibaba A (2011) Floristic composition and community analysis of Menagesha Amba Mariam forest (Egdu forest) in central Shewa, Ethiopia. Ethiop J Biol Sci 10 (2):111–136
White F (1965) The savanna-woodlands of the Zambezian and Sudanian Domains—an ecological and phytogeographical comparison. Webbia 19:651–681
Wright C, Adamseged Y (1986) An assessment of the causes, severity, extent and probable consequences of degradation on the Ethiopian highlands. FAO/EHRS (Ethiopian highlands reclamation study) working paper 3. FAO/EHRS, Addis Ababa

Woldu Z (1999) Forests in the vegetation types of Ethiopia and their status in the geographical context. In: Edwards S, Demissie A, Bekele T, Haase G (eds) Forest genetic resources conservation: principles, strategies and actions. Institute of Biodiversity Conservation and GTZ, Addis Ababa, pp 1–38

Woldu Z, Yeshitela K (2003) Wetland plants in Ethiopia with examples from Illubabor, south-western Ethiopia. In: Abebe Y, Geheb K (eds) Wetlands of Ethiopia. IUCN Blue Series, Nairobi, pp 49–58

Wood RB, Talling JF (1988) Chemical and algal relationships in a salinity series of Ethiopian inland waters. Hydrobiologia 158:29–67

Zanettin B, Bellieni G, Justin Visentin E (2006) Stratigraphy and evolution of the trachy-rhyolitic volcanism of the Senafe area (Eastern Eritrean Plateau). J Afr Earth Sci 45:478–488

Geology of Ethiopia: A Review and Geomorphological Perspectives

2

Ernesto Abbate, Piero Bruni, and Mario Sagri

Abstract

The Ethiopian region records about one billion years of geological history. The first event was the closure of the Mozambique ocean between West and East Gondwana with the development of the Ethiopian basement ranging in age from 880 to 550 Ma. This folded and tilted Proterozoic basement underwent intense erosion, which lasted one hundred million years, and destroyed any relief of the Precambrian orogen. Ordovician to Silurian fluviatile sediments and Late Carboniferous to Early Permian glacial deposits were laid down above an Early Paleozoic planation surface. The beginning of the breakup of Gondwana gave rise to the Jurassic flooding of the Horn of Africa with a marine transgression from the Paleotethys and the Indian/Madagascar nascent ocean. After this Jurassic transgression and deposition of Cretaceous continental deposits, the Ethiopian region was an exposed land for a period of about seventy million years during which a new important peneplanation surface developed. Concomitant with the first phase of the rifting of the Afro/Arabian plate, a prolific outpouring of the trap flood basalts took place predominantly during the Oligocene over a peneplained land surface of modest elevation. In the northern Ethiopian plateau, huge Miocene shield volcanoes were superimposed on the flood basalts. Following the end of the Oligocene, the volcanism shifted toward the Afar depression, which was experiencing a progressive stretching, and successively moved between the southern Ethiopian plateau and the Somali plateau in correspondence with the formation of the Main Ethiopian Rift (MER). The detachment of the Danakil block and Arabian subcontinent from the Nubian plate resulted in steep marginal escarpments marked by flexure and elongated sedimentary basins. Additional basins developed in the Afar depression and MER in connection with new phases of stretching. Many of these basins have yielded human remains crucial for reconstructing the first stages of human evolution. A full triple junction was achieved in the Early Pliocene when the MER penetrated into the Afar region, where the Gulf of Aden and the Red Sea rifts were already moving toward a connection via the volcanic ranges of northern Afar. The present-day morphology of Ethiopia is linked to the formation of the Afar depression, MER, and Ethiopian plateaus. These events are linked to the impingement of one or more mantle plumes under the Afro-Arabian plate. The elevated topography of the Ethiopian plateaus is the result of profuse volcanic accumulation and successive uplift. This new highland structure brought about a reorganization of the East Africa river network and a drastic change in the atmospheric circulation.

E. Abbate (✉) · P. Bruni · M. Sagri
Dipartimento di Scienze della Terra, Università di Firenze, Via La Pira 4, 50121 Florence, Italy
e-mail: abbate@unifi.it

P. Billi (ed.), *Landscapes and Landforms of Ethiopia*, World Geomorphological Landscapes,
DOI 10.1007/978-94-017-8026-1_2, © Springer Science+Business Media Dordrecht 2015

Keywords
Paleogeography of Ethiopia • Ethiopian volcanites • Planation surfaces • Ethiopian highlands • Main Ethiopian Rift • Afar depression

2.1 Introduction

The spectacular landscape of the Ethiopian region (Fig. 2.1) with the typical flat-topped mountains (Fig. 2.2) (ambas) and deep-incised valleys has fascinated the European travelers since the sixteenth century when Alvarez (1540) visited the fabulous land of Priest John. This fantastic scenario is the result of geodynamic and geomorphic processes which have shaped this territory since the Oligocene. These processes, which came relatively late in the ca. one-billion-year geological history of East Africa, were triggered by the impingement of a mantle plume or plumes under the Afro-Arabian continental crust. The plume action gave rise to extrusion of huge amounts of magma, uplift, and

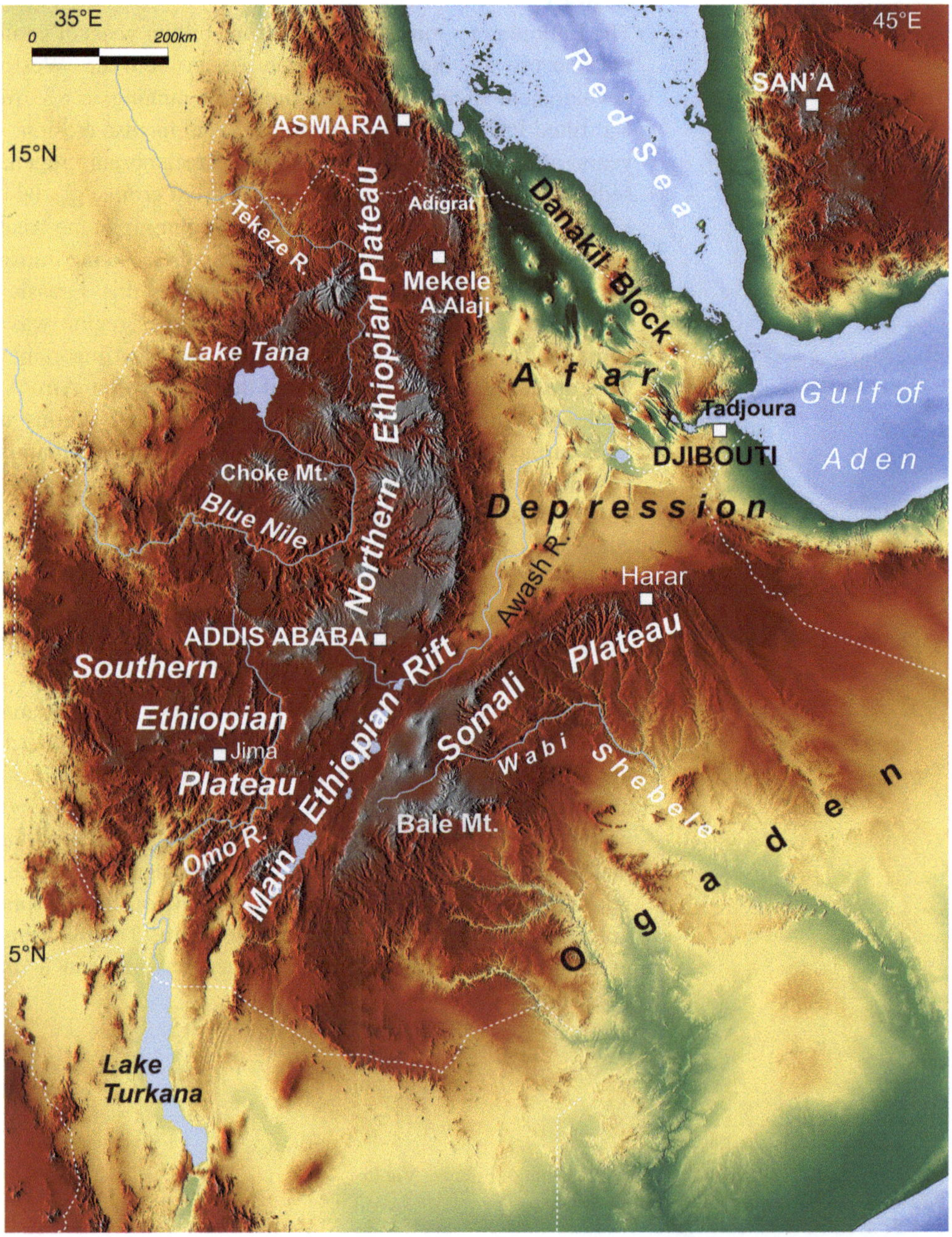

Fig. 2.1 Digital elevation map of Ethiopia (SRTM data) with the main physiographic elements

Fig. 2.2 Flat-topped mountains (*amba*) and deeply incised valleys in the Tigray region (*northern Ethiopia*) exposing Enticho sandstones capped by trap basalts

fragmentation of the continental crust and contributed to the birth of the Red Sea, Gulf of Aden, East Africa Rift valley, and the adjoining Afar depression.

In his evocative book, Mohr (2009) reports the main stages in the geological exploration of East Africa in the nineteenth and early twentieth centuries by outstanding scholars, such as Rüppell (1834), Johnston (1844), Munzinger (1864), Blanford (1870), Baldacci (1891), Dainelli and Marinelli (1912). After these pioneers, Dainelli (1943), in his great synthesis on eastern Africa geology, relied on his own investigations as well as contributions by Stefanini (1933) and Merla and Minucci (1938). Dainelli's three volumes close a research cycle based on a naturalistic approach to the regional geology. After the World War II, researches resumed with a scrupulous re-examination of the available geological data in the light of new investigations by Mohr (1962) and geomorphological contributions by Abul-Haggag (1961) and Merla (1963, and following years) (Fig. 2.3).

In the wake of the plate tectonics theory, eastern Africa has become an invaluable laboratory for understanding passive margin processes, continental fragmentation, and the first stages of oceanization. The Afar region, as a candidate for the formation of a new oceanic crust, has been studied with particular attention using updated analytical methods and facilitated field exploration (e.g., Barberi et al. 1970, 1972; Barberi and Varet 1977; Makris and Rhim 1991). The highland volcanites were investigated from a petrographical, geochemical, and geochronological point of view by Zanettin and Justin-Visentin (1974), Davidson and Rex (1980), Mohr and Zanettin (1988), Woldegabriel et al. (1990), Ebinger et al. (1993), Hofmann et al. (1997), Rochette et al. (1998), Pik et al. (1998), Kieffer et al. (2004), and Beccaluva et al. (2009). The newly collected data made possible the compilation of several geological maps at two million scale very useful for regional syntheses (Mohr 1963; Kazmin 1973; Merla et al. 1973; Tefera et al. 1996).

Information on the sedimentary successions of the Afar and Danakil Alps regions was produced by Vinassa de Regny (1931), Bannert et al. (1970), Tiercelin et al. (1980), Kalb et al. (1982), and Bosworth et al. (2005).

The geology of the Main Ethiopia Rift has been treated, among others, by Mohr (1962), Di Paola (1972), Woldegabriel et al. (1990), Ebinger et al. (1993), Chorowicz et al. (1994), Boccaletti et al. (1998), Le Turdu et al. (1999), Maguire et al. (2006), and Peccerillo et al. (2007). Many of these contributions have been reviewed and summarized by Corti (2009). Specific attention to the sedimentary filling of the Ethiopia Rift has been given by Street (1979), Le Turdu et al. (1999), and Benvenuti et al. (2002).

The possible economic importance of the Proterozoic terranes prompted regional reconnaissance and dedicated surveys (Dainelli 1943; Mohr 1962; Beyth 1972; Kazmin et al. 1978; Kröner 1985; Stern 1994; and Beyth et al. 1997). The Paleozoic to Eocene sedimentary cover of the Proterozoic terranes summarized in the general syntheses by Dainelli (1943) and Mohr (1962) has been thoroughly investigated by Dow et al. (1971), Bosellini et al. (1997, 2001), Hunegnaw et al. (1998), Kumpulainen et al. (2006), and Bussert and Schrank (2007).

The diverse morphology of the Ethiopian region with extended plains dotted by intermittent ponds and lakes, structural corridors along the rift valley surrounded by elevated plateaus, and intense climate changes since the Pliocene has favored human speciation and dispersal toward

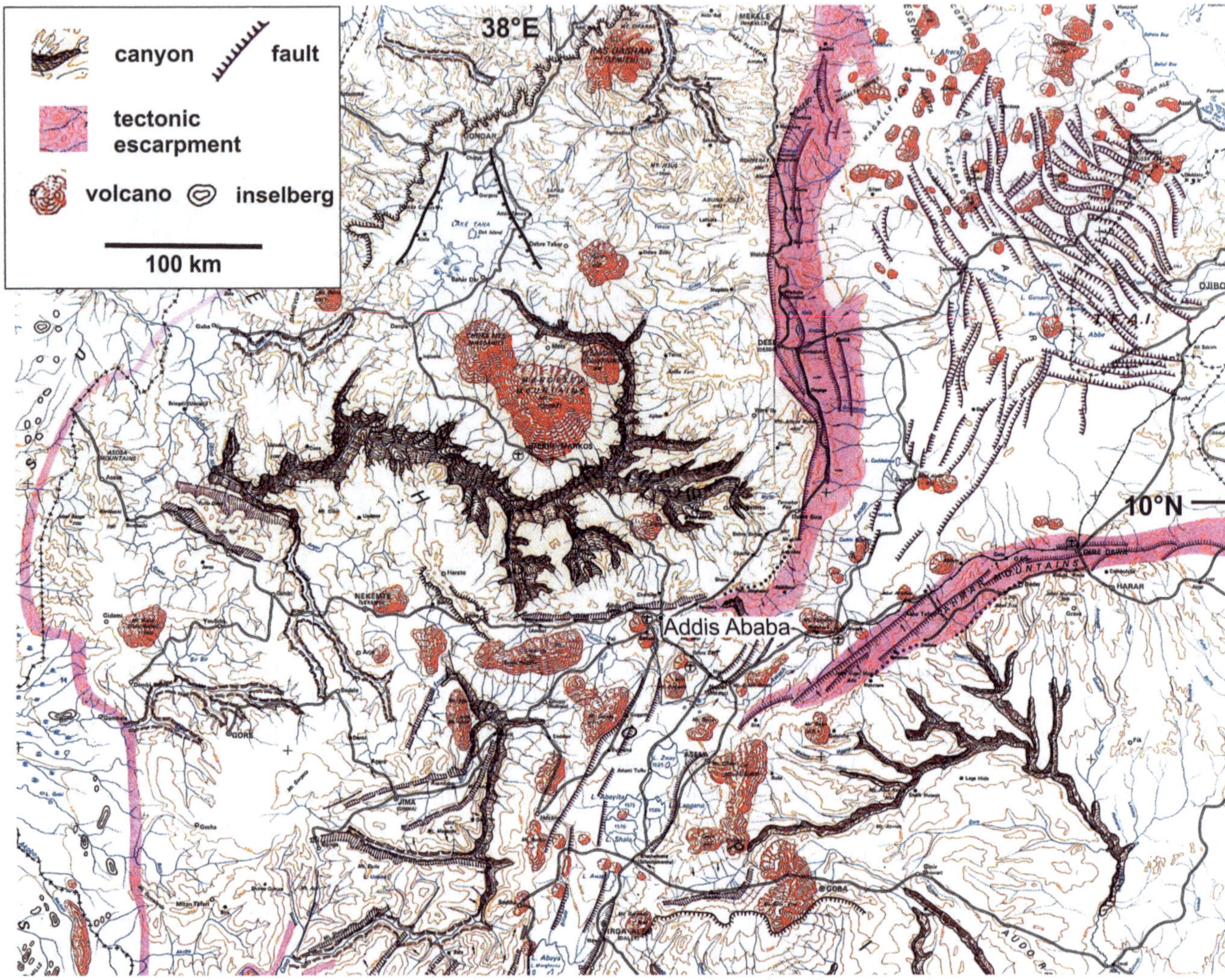

Fig. 2.3 Morphological features of the Blue Nile area, southern Afar depression, Main Ethiopian Rift, and Somali plateau. Excerpt from the "Major landform map of Ethiopia" in Merla et al. (1979), based on a morphological map in Merla (1963)

more suitable territories. The rift valley and Afar sediments contain abundant paleoanthropological evidences and lithic industries connected to these exceptional conditions during the first stages of human evolution. After a few investigations carried out before 1960, a competitive and publicized search for hominid fossils has been carried out in the Omo region since the end of the 1960s (e.g., Chavaillon 1971; Arambourg 1972; Leakey 1974; Walker and Leakey 1978; Johanson and Taieb 1976; Johanson et al. 1976; Lewin 1983), and in the Awash region (e.g., Taieb 1974; Larson 1977; Clark 1985; Kalb 1993; Tiercelin 1986; White and Johanson 1982; Walter 1994). Recent and outstanding results include the discovery of traces of possible human precursors (*Ardipithecus kadabba*, 5.5/5.8 Ma, Haile-Selassie et al. 2009) and ancestors (*A. ramidus*, 4.4 Ma, White et al. 2009) followed by various *Australopithecus* and *Homo* genera (e.g., Haile-Selassie et al. 2007; Johanson and Taieb 1976; Asfaw et al. 2002).

In the following paragraphs, we will summarize and comment on the main sedimentary, volcanic, metamorphic, and structural events recorded in the Ethiopian region. They are related to the continental fragmentation of Gondwana and Afro-Arabian plates and connected basin development, Paleozoic glaciations, and mantle plume activity. The present-day morphology results from their interaction (see chapter 2.7).

2.2 The Closure of the Mozambique Ocean and the Development of the Ethiopian Basement

A Neoproterozoic crystalline basement ranging from 880 to 550 Ma constitutes the crustal backbone of the Ethiopian region with wide exposures in the southern and western Ethiopia and, to a lesser extent, in the northernmost Ethiopia (Fig. 2.4).

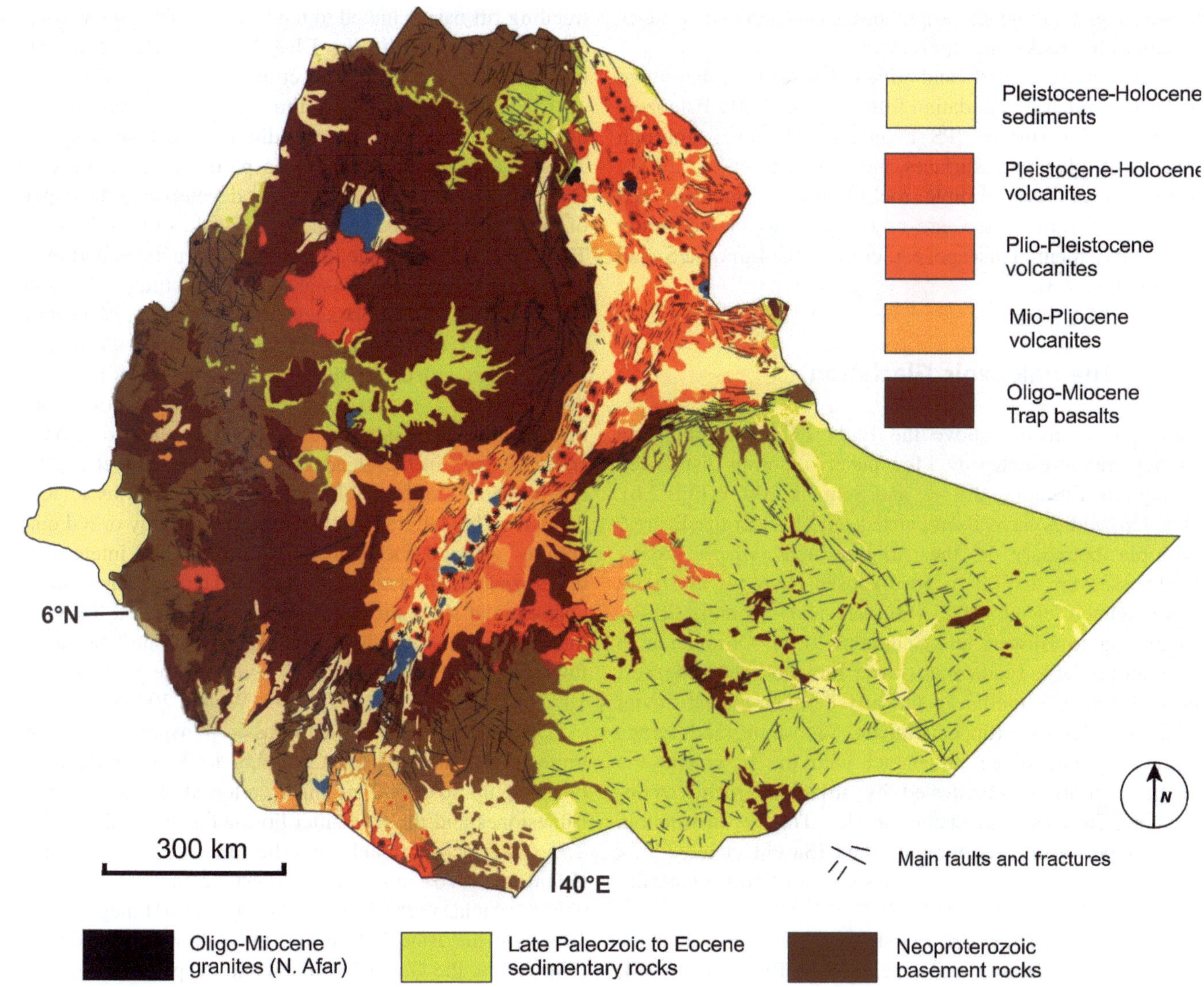

Fig. 2.4 Simplified geological map of Ethiopia modified from Tefera et al. (1996)

The Proterozoic terranes in Ethiopia are related to as the East African Orogen (Stern 1994), a N–S elongated mega-collisional structure stretching from Israel to Madagascar and produced between West and East Gondwana by the closure of the Mozambique ocean (Fig. 2.5). The N–S alignments of the East African Orogen lithic components, sometimes marked by belts of ophiolites, stand out as pronounced present-day morphological expressions. In addition, they influence the trends of the succeeding fragile structures.

In the north, the East African Orogen constitutes the Arabian–Nubian Shield, and in the south the Mozambique Belt. In northern Ethiopia, the Nubian portion of the Shield is prevalent, with dominantly low-grade volcano-sedimentary rocks overlain by metasediments (stromatolitic carbonates and diamictites) associated with "Snowball Earth" (Beyth et al. 2003). In southern Ethiopia, the Mozambique

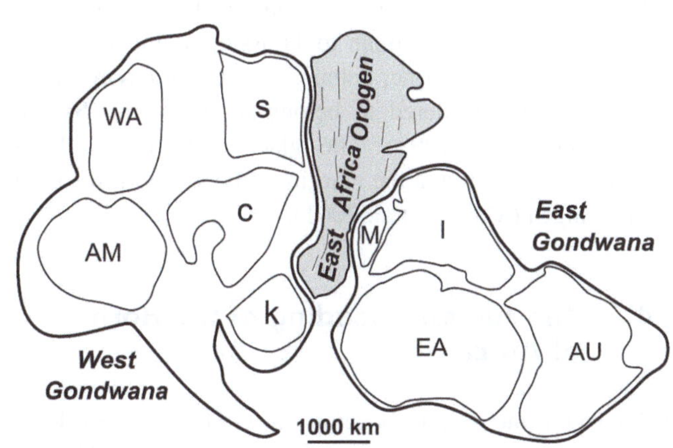

Fig. 2.5 The East Africa Orogen squeezed between West and East Gondwana. *WA* West African craton; *AM* Amazonian craton; *S* Sahara craton; *C* Congo craton; *K* Kalahari craton; *M* Madagascar; *I* India shield; *EA* East Antarctic shield; *AU* Australia craton. Modified from Meert and Lieberman (2008)

Belt exposes abundant amphibolites and granulite facies metamorphic rocks and gneiss terranes.

The tightly folded and tilted Proterozoic basement underwent intense denudation following the Early Paleozoic (first planation surface "PS 1" in Fig. 2.6; for a thorough review of planation surfaces, see Coltorti et al. 2007). Remnants of less erodible rocks, such as granites and gneisses of various age, are now inselbergs in the gently rugged basement landscape, e.g. close to the Ethiopia/Sudan border (Fig. 2.3).

2.3 The Paleozoic Glaciation

The first sediments above the Early Paleozoic planation surface are represented by a few patchy outcrops distributed throughout Ethiopia (Bussert and Schrank 2007) (Fig. 2.6). Detailed studies have been carried out close to the Eritrea/Ethiopia boundary in the Adigrat areas where greater thicknesses up to 500 m are exposed (Dow et al. 1971; Kumpulainen et al. 2006; Bussert and Schrank 2007). According to Bussert and Schrank (2007), Ordovician to Silurian fluviatile sandstones (lower Enticho Sandstone) rest beneath Late Carboniferous to Early Permian glacial fluvio/lacustrine deposits (upper Enticho Sandstone, and Edaga Arbi Glacials). When in contact with the basement, the glacial activity is manifested by striations, roches moutonnées, grooves, and chatter marks. This reconstruction, also supported by petrographic data (Sacchi et al. 2007), could supplant a former hypothesis of an Ordovician/Silurian glaciation (Dow et al. 1971).

Late Paleozoic fluviatile sandstones are also reported in the Blue Nile gorge (Jepson and Athearn 1964; Russo et al. 1994) and near Harar. Many deep oil boreholes in the Ogaden basin have intersected Late Paleozoic sediments of possible glacial to fluvial environment, unconformably resting on the basement and capped by Permian to Triassic continental clastic sediments from lacustrine, deltaic, and fluvial environment. The latter are considered to belong to the Karoo System (Hunegnaw et al. 1998). The glacial and fluvio-glacial deposits were located at the margin of the southern ice sheet of the Pangea (Martini et al. 2001).

2.4 The Jurassic Flooding of the Horn of Africa

Along with the initial breakup of Gondwana, an Early Jurassic regional marine transgression invaded the Horn of Africa from the northeast (Paleotethys) and east (India/Madagascar nascent ocean). The sea covered the NS-trending block-faulted structures connected to the Karoo rift system in southern Ethiopia (Ogaden region), and the NW-trending rift basins linked to the Central Africa shear zone in central Ethiopia (e.g., the Blue Nile rift) (Bosellini 1989; Hunegnaw et al. 1998; Gani et al. 2009) (Fig. 2.7).

The Jurassic marine sedimentation was preceded by deposition of the continental Adigrat Sandstones (Fig. 2.8). These were deposited above the partially peneplained Triassic surface (PS 2 in Fig. 2.6) that developed at the expense of the Permo-Triassic sediments as well as of the basement. The latter was intersected by positive structures and deeply subsiding intracratonic basins which constituted the source areas and accumulation sags for the Adigrat Sandstones, respectively. The Adigrat Sandstones are widespread in Ethiopia and with correlative units in the whole of East Africa and Arabia. In Ethiopia, they commonly rest unconformably on the Paleozoic and basement rocks, but in the some basinal contexts (e.g., in the Ogaden basin, Hunegnaw et al. 1998) are probably conformable with the Permo-Triassic Karoo deposits. The Adigrat Sandstones are light gray or red quartz arenites with interbeds of conglomerates and intensely pedogenized red mudstones (Fig. 2.8). Their thickness is variable even at short distances and reaches 700 m. They were mainly deposited in fluvial or piedmont zones, but also in fluviolacustrine and deltaic environments (e.g., Beauchamp 1977; Bosellini et al. 1997, 2001; Wolela 2008).

The typical facies of the Jurassic transgression is represented by shallow-water Callovian to Kimmeridgian carbonates, up to 1,000 m thick, referred to as the Antalo Limestones and the Hamanlei Formation (Figs. 2.7, 2.8 and 2.9). They conformably overlie the Adigrat Sandstones (Blanford 1870), and their earliest occurrences (Pliensbachian/Aalenian) are found in the Ogaden (Hunegnaw et al. 1998). In the Blue Nile basin, the Adigrat Sandstones are followed by the Gohatsion marls and evaporites of Aalenian to Callovian age (Russo et al. 1994). In the Dire Dawa-Harar and in the Tigray area, a younger widespread marine flooding is recorded since the Bathonian and in the Callovian/Oxfordian, respectively. Within the carbonate succession, characterized by several depositional sequences, the maximum flooding surface is marked by organic-rich marls and shales and is time-transgressive from Oxfordian (Ogaden area) to Tithonian (Tigray) (e.g., Brassier and Guleta 1993; Hunegnaw et al. 1998; Bosellini et al. 1997, 2001). In northern Ethiopia, the carbonate outcrops extend to the Mekele area and the northwestern border of the Afar depression (Figs. 2.8 and 2.9). The westernmost outcrops occur in central Ethiopia in the Blue Nile valley south of Choke Mt., and it is likely that the marine transgression did not cross the 36°E meridian and the 16°N parallel.

The carbonate deposits from Tigray to the Dire Dawa/Harar area are truncated by an erosional surface (PS 3 in Fig. 2.6) due to an Early Cretaceous tectonic event marked by faulting and tilting (Bosellini et al. 2001). The abruptly overlying Amba Aradam Sandstones are fluviatile and

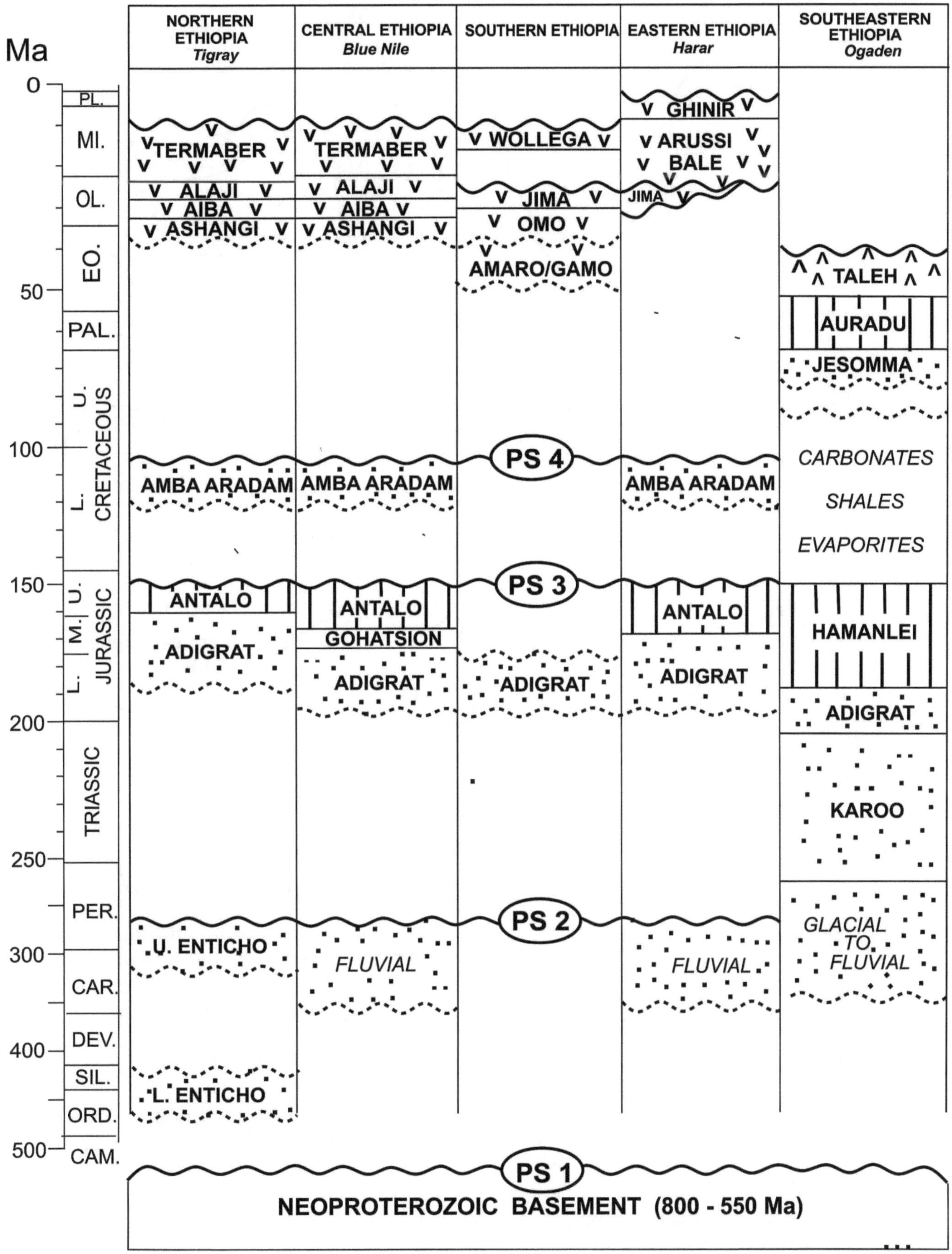

Fig. 2.6 Schematic stratigraphic chart of Ethiopia with the major planation surfaces [PS acronyms as in Coltorti et al. (2007)]

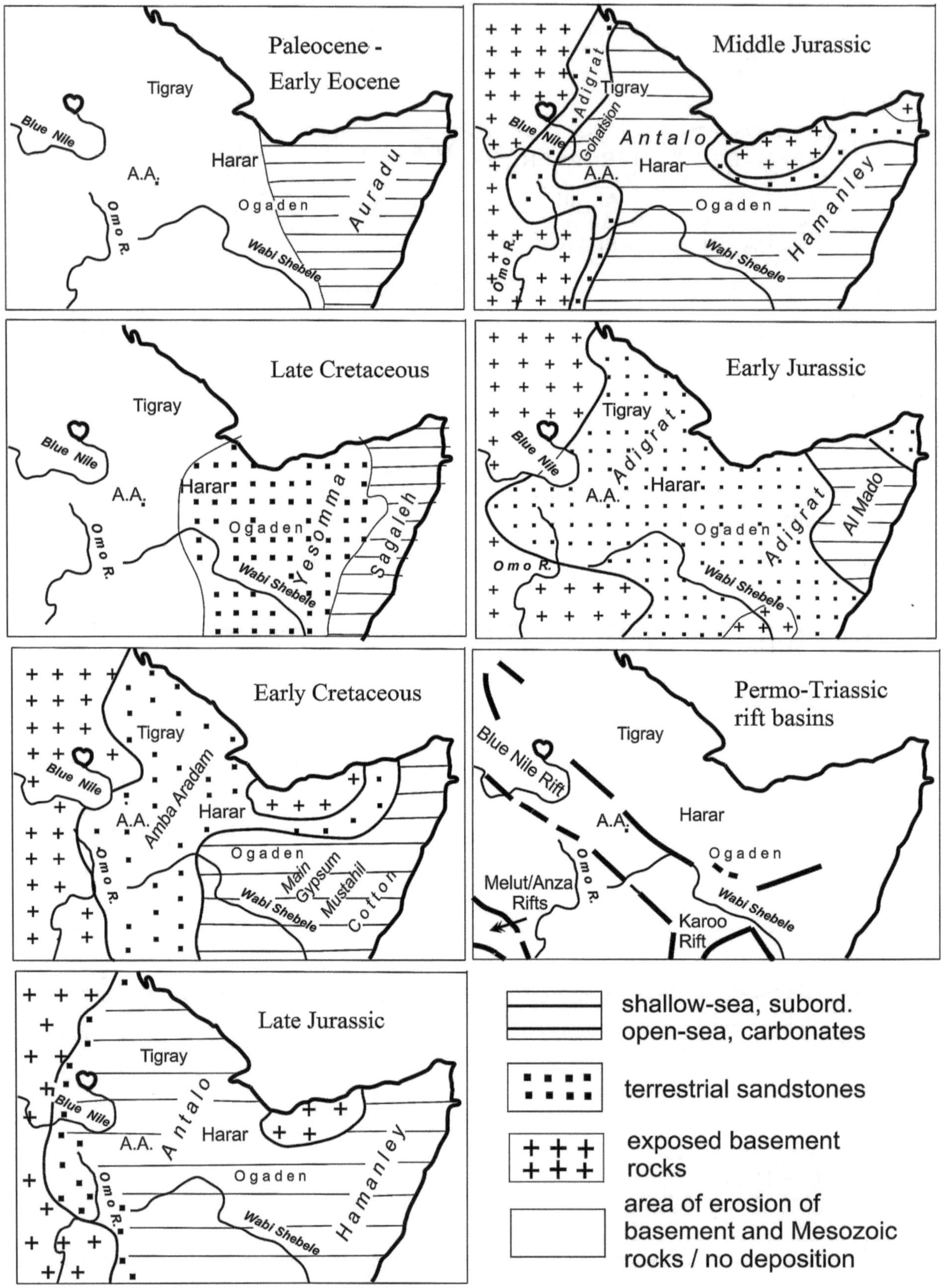

Fig. 2.7 Paleogeographic sketches for the Horn of Africa from Permian to Early Eocene

Fig. 2.8 The reddish Adigrat Sandstones conformably covered by the *gray* Antalo Limestone along the escarpment east of Mekele

Fig. 2.9 The Antalo Limestones at the foot of the escarpment east of Mekele: *horizontal* beds in the foreground and an *anticline* in the background

commonly associated with lenses of quartz conglomerates and red shales. They often exhibit laterites at their base. Their maximum thickness is 200 m, and their age determined on the base of *Orbitolina* findings in the Harar region is Aptian to Albian (Gortani 1973; Bosellini et al. 1999).

By contrast, in the Ogaden basin, marine deposition continues during the Cretaceous until the Turonian or possibly the Campanian (Fig. 2.7). It consists of ca. 1,200-m-thick alternation of transitional, shallow-water marine and open-sea deposits including shales, carbonates, sandstones, and evaporites (e.g., Main Gypsum, Mustahil, Cotton) (Barnes 1976; BEICIP 1985; Hunegnaw et al. 1998). This succession records many sea-level oscillations and represents the final stages of the Mesozoic flooding in the Ogaden basin. An erosional surface truncates the Mesozoic carbonate/evaporite sequence from south to north down to the

Middle Jurassic Hamanlei Formation (Merla et al. 1973). The overlying sediments above this erosive surface are the Jesomma Sandstones, a few-hundred-meter-thick quartzose fluvial deposit with minor conglomerates and siltstones. The age of this unit is poorly constrained and can be assigned to the upper portion of the Late Cretaceous.

A shallow-water marine deposition (Auradu Limestones, ca. 400 m thick) resumes at the beginning of the Paleocene and was connected to a new transgression recorded only in the easternmost portion of the Ogaden basin (Fig. 2.7). It lasts until the Early Eocene, and its deposits were covered by the Early to Middle Eocene Taleh Evaporites which reach 250 m in thickness.

2.5 The Cenozoic Volcanic History: Floods and Volcanoes

Concomitant with the first phases of rifting in the Afro/Arabian plate, a period of a prolific volcanic activity took place predominantly during the Oligocene in the Horn of Africa and southern Arabia which were at that time connected. These volcanic rocks, mainly represented by basalts and traditionally referred to as the trap succession, have estimated to have covered an area in Ethiopia not less than 750,000 km^2 before erosion, with a total volume of ca. 350,000 km^3 (Mohr 1983) (Fig. 2.10). Their great areal extension and volume are due to the exceptional supply of mantle material connected with hot plumes (Schilling 1973; White and McKenzie 1989).

According to Hofmann et al. (1997), the basalt activity was concentrated in the very short time of one million years around 30 Ma, and this rapid outpouring of a huge volume of volcanic rocks has been regarded as a possible cause of climatic deterioration and ensuing mass extinction on a global scale (Courtillot et al. 1988; White and McKenzie 1989; Rochette et al. 1998).

The volcanic succession rests above a peneplained surface (Blanford 1869; the pre-trappean peneplanation of Mohr 1962; PS 4 in Fig. 2.6), marked by laterites, particularly well developed in Eritrea and Tigray (Dainelli and Marinelli 1912; Merla and Minucci 1938) and in southwestern Ethiopia (Davidson and Rex 1980). The laterites are also present in similar contexts in Yemen (Baker et al. 1996) and western Arabia (Overstreet et al. 1997). This extensive pedogenesis is indicative of a long period of morphological stability of the peneplained surface marked by low elevation, little or no vertical deformations, and sediment starvation. The laterization was active until at least 40 Ma (Ar/Ar age, Andrews Deller 2006) and can be related to the Early Eocene climatic optimum.

The volcanic rocks that cover most of Ethiopia have been subdivided into five major provinces on the basis of their lithological development, type of activity, frequency of volcanic centers, and age of effusion (Abbate and Sagri 1980): (1) volcanites of the northern plateau; (2) volcanites of the southern plateau; (3) volcanites of the Somali plateau; (4) Afar volcanites; and (5) Main Ethiopian Rift (MER) volcanites.

The first three groups (Merla et al. 1979) comprise the major part of the Ethiopian volcanites (Fig. 2.10). They have collectively been referred to as "Traps" (Blanford 1870; Kazmin 1973), a general term from an old Swedish word meaning stairs (Fig. 2.11). Fine-grained stratoid fissural Paleogene basalts represent the greater portion of these volcanites. The Afar and the MER volcanites, filling two megastructures related to Neogene continental fragmentation, have a more limited extension (Fig. 2.10) and were referred to as Aden Series (Blanford 1870; Mohr 1962).

2.5.1 The Volcanites of the Northern Ethiopian Plateau

After Blanford's (1870) attempt to subdivide the northern plateau volcanites into a lower Ashangi Group unconformably overlain by a Magdala Group, a more detailed lithostratigraphic approach was proposed in the 1970s by Zanettin and Justin Visentin (1973) and Gregnanin and Piccirillo (1974), with the distinction of the Ashangi and Aiba basalts, Alaji Rhyolites, and Termaber Basalts. These units were incorporated in the Merla et al. (1979) and Tefera et al. (1996) maps. Some authors have pointed out that the lateral heterogeneity of these volcanic rocks, the vertical recurrence of similar lithology, and the different morphological response of basalts with the same petrological or chemical characteristics (e.g., Kieffer et al. 2004) prevent the adoption of the criteria commonly used for sedimentary bodies. However, in the expectancy of further detailed field and petrographic studies, we prefer to maintain the distinctions proposed by the previous authors and also followed by Rochette et al. (1998) in their magnetostratigraphy analyses of the northern Ethiopian traps.

The Ashangi Basalts are composed of transitional to tholeitic olivine basalts, often highly zeolitized, alternating with subordinate tuffs (Mohr and Zanettin 1988). The flows are barely evident owing to their small thickness, reduced horizontal extension, and deep weathering. Their thickness is from 200 to 1,000 m. The Aiba Basalts consist of well-developed columnar massive transitional flood-basalt flows, locally with intervening agglomerate beds. The flows are 15–50 m thick reaching in some cases almost 100 m and

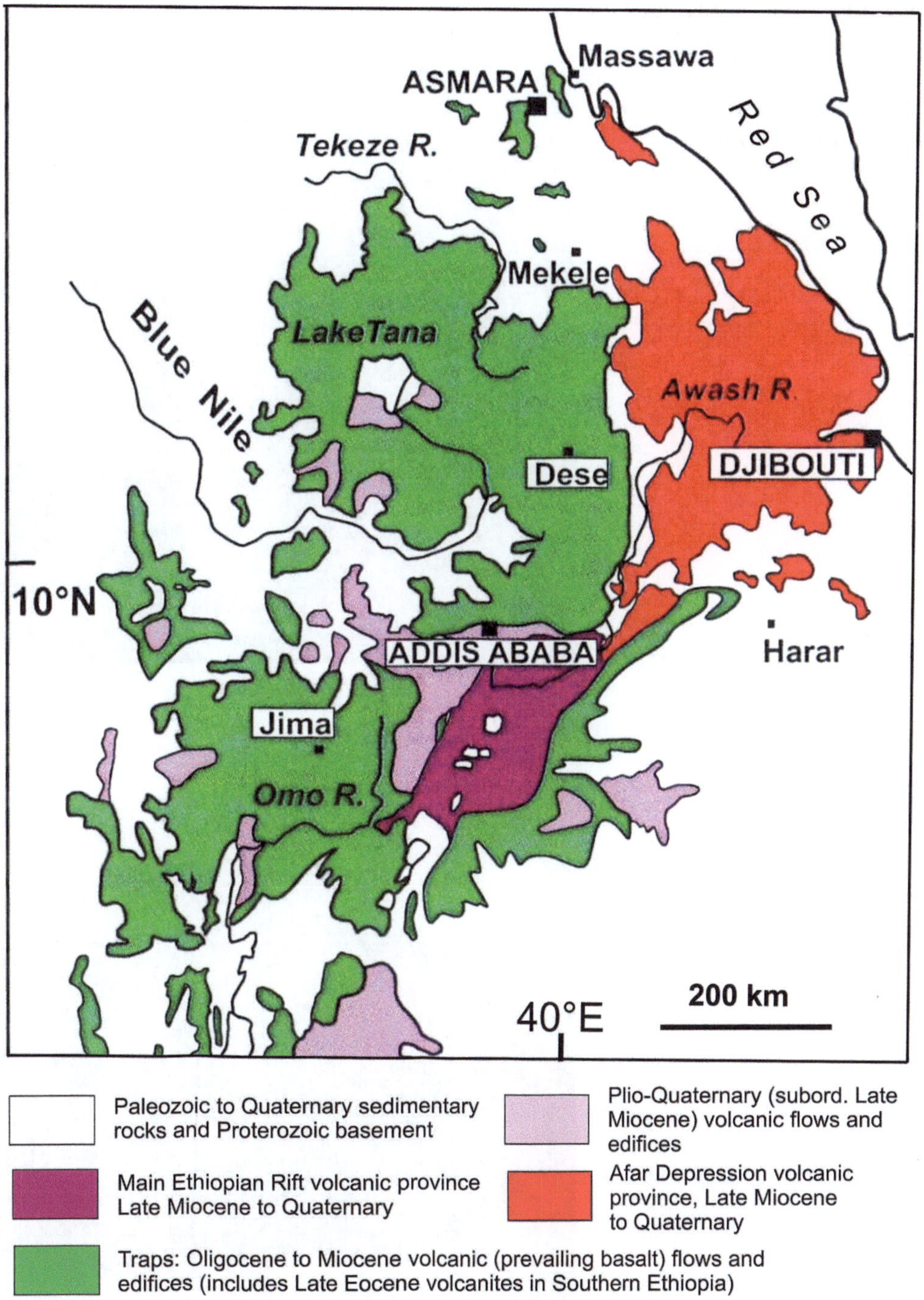

Fig. 2.10 The main volcanic provinces of Ethiopia. After Abbate et al. (2014)

can be traced over long distance. Their total thickness reaches 1,000 m. The above-described fissural basalt units constitute the bulk of the volcanic pile of the northern Ethiopia plateau, locally attaining 2,000 m in thickness. This profuse outpouring took place between 31 and 29 Ma (Hofmann et al. 1997; Pik et al. 1998; Ukstins et al. 2002; Coulié et al. 2003). The Amba Alaji Rhyolites, whose outcrops are limited to the northern portion of the northern Ethiopian plateau, are an alternation of alkaline to peralkaline rhyolites and transitional basalts. The acidic terms are mainly whitish ignimbritic tuffaceous layers that can be followed for a great distance. They form typical landscape elements, such as steep walls and pyramids, well represented in the Amba Alaji peak (Fig. 2.12). Their maximum thickness is 500 m with a decrease to nil about 100 km west of the Afar margin. Their age ranges from Late Oligocene in the northern outcrops to Early Miocene in the south (Zanettin et al. 1974).

Fig. 2.11 Felsic and subordinate basic volcanites of the Semien shield volcano overlie a thick succession of flood basalts. Semien National Park. Photograph courtesy of Frances Williams

Fig. 2.12 The outstanding pyramidal morphology of the Amba Alaji Rhyolites at the Amba Alaji peak

A peculiar feature of the northern Ethiopian plateau is the frequent occurrence of shield volcanoes, about 30 major centers according to Mohr and Wood (1976), some of which reach as much as 100 km in diameter and 1,000–2,000 m in elevation above the plateau (e.g., Semien, Guna, Choke, and Guguftu) (Figs. 2.13 and 2.14). In the regional reviews and maps, they are grouped under the name of Termaber Basalts. They are made of lenticular, often zeolitized, alkali basalts with a large amount of tuffs, scoriaceous lava flows, peralkaline rhyolites, and typical red paleosoils. Dike swarms and acidic extrusions are present. The thickness of the Termaber Basalts reaches 1,000 m close to the volcanic centers.

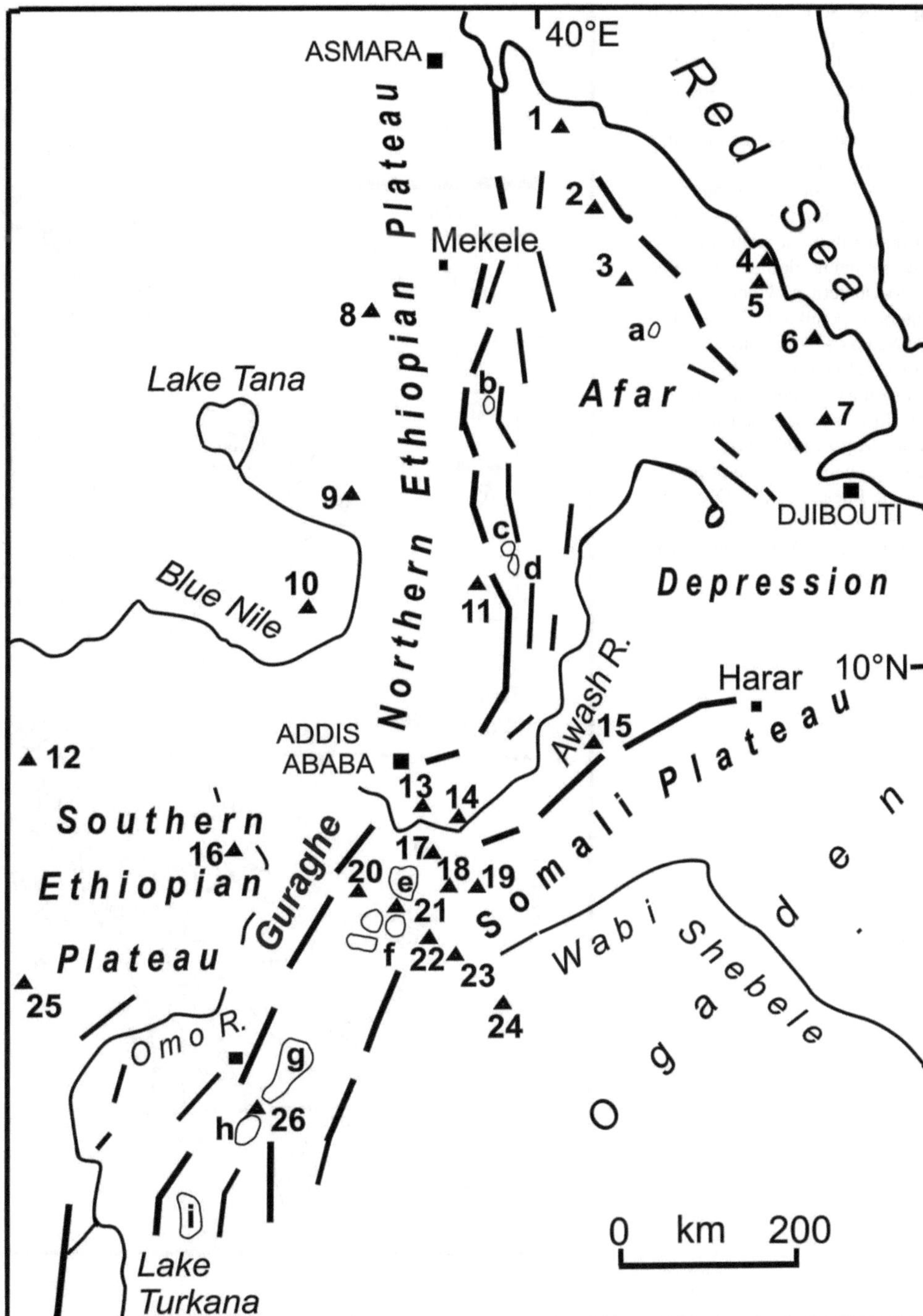

Fig. 2.13 Location map of the volcanic edifices and lakes cited in text. Volcanic edifices: *1* Alid; *2* Dallol; *3* Erta Ale; *4* Dubbi; *5* Sork Ale; *6* Ado Ale; *7* Mussa Ale; *8* Semien; *9* Guna; *10* Choke; *11* Guguftu; *12* Tullu Wellel; *13* Yerer; *14* Boseti Guda; *15* Fantale; *16* Egan; *17* Bora Bericho; *18* Chillalo; *19* Badda; *20* Gademotta; *21* Alutu; *22* Chike; *23* Kecha; *24* Bale; *25* Mizan Tefari; *26* Tosa Sucha. Lakes: *a* Afrera; *b* Ashangi; *c* Haik; *d* Ardibbo; *e* Ziway; *f* Langano, Abiyata, Shala; *g* Abaya; *h* Chamo; *i* Chew Bahir

The ages of the Termaber shield volcanoes are Early and Middle Miocene, ranging from 23 to 11 Ma (e.g., Kieffer et al. 2004), with the exception of the Semien with the base of 30 Ma and the top of 19 Ma.

In addition to the trap basalts and shield volcanoes, some subvolcanic intrusions are an outstanding feature in the Axum–Adwa area producing a landscape of cliffs, pinnacles, and domiform hills with an east–west alignment (Le Bas and Mohr 1968; Abbate and Sagri 1980). They are basalt–trachyte (syenite) plugs and domes (Prior 1900; Merla and Minucci 1938; Mohr 1962; Hagos et al. 2010) and have been dated as Mio-Pliocene by Beyth (1972), and more recently, Natali et al. (2013) determined their Ar/Ar age of 19–15 Ma. The rocks of this volcanic complex have been quarried for the famous ancient Axumite obelisks.

On the western portion of the northern Ethiopian plateau, the Tana rift basin is located in a structural complex area at the junction of three grabens (Chorowicz et al. 1999). Quaternary

Fig. 2.14 Structural sketch map of the Ethiopian and Somali plateaus, Afar depression, and Main Ethiopian Rift (Northern, Central, and Southern MER) with the main transversal tectonic line and major volcanic edifices. Legend: *1* Miocene volcanic centers of Termaber unit in the northern Ethiopian plateau; *2* Pliocene acidic domes and plugs; *3* Plio-Quaternary volcanic centers; *4* peralkaline plugs; *5* Red Sea axis; *6* Flexures; *7* Horizontal lava flows and sedimentary sequences; *8* tilted lava flows and sedimentary sequences; *9* block faulting; *10* major graben; *11* transversal tectonic lines. A–B, C–D cross sections shown in Fig. 2.23. Modified from Abbate and Sagri (1980)

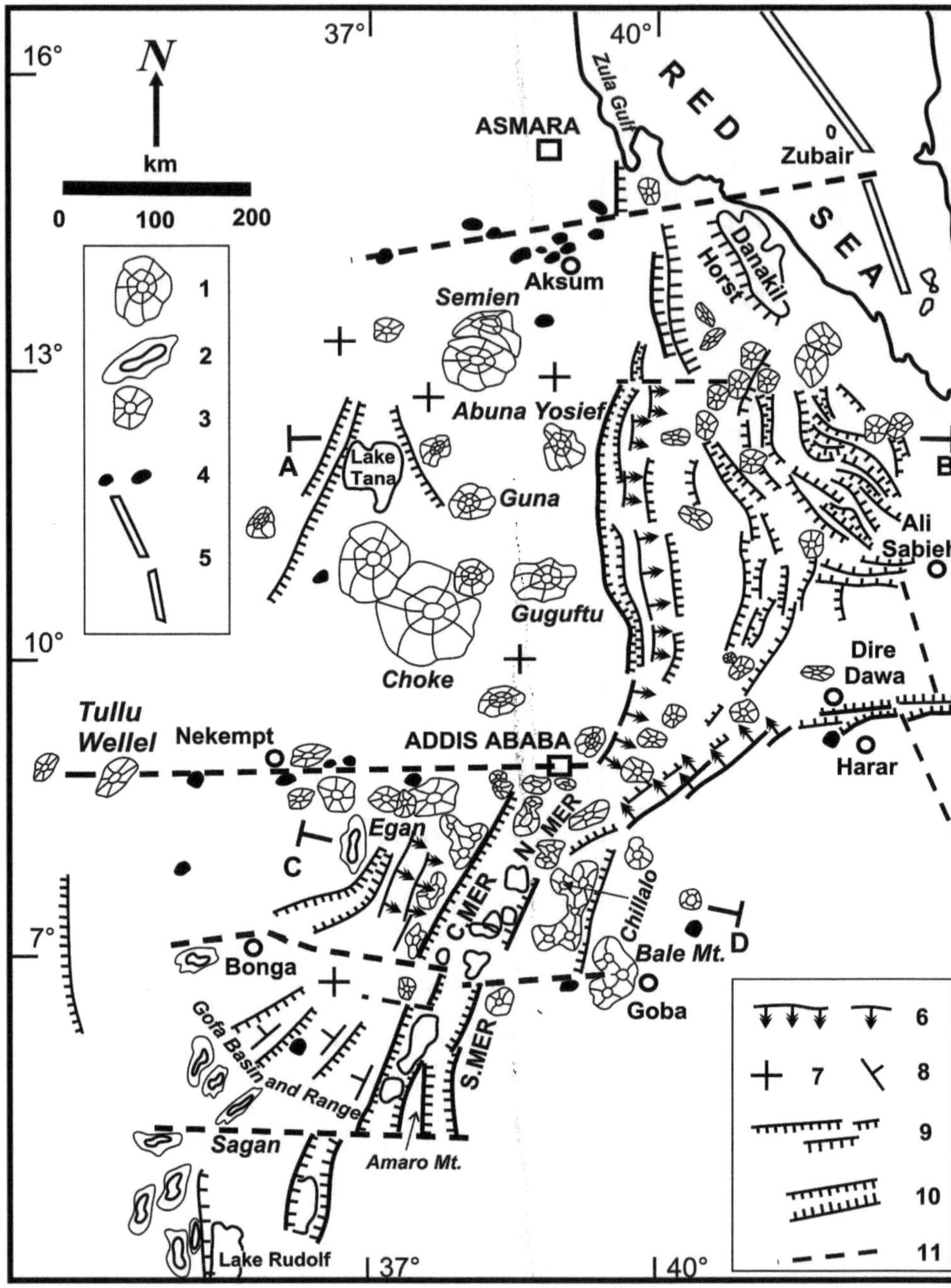

olivine–basalts and subordinate phonolites cover much of the southern portion of the Tana rift (Jepson and Athearn 1964) with well-preserved lavas, plugs, and spatter cones.

The transition to the southwestern plateau volcanites is marked by a 700-km-long and 80-km-wide east–west-trending volcano-tectonic alignment (Fig. 2.14) (Addis Ababa-Nekempt line in Abbate and Sagri 1980; Yerer-Tullu Wellel alignment, Abebe et al. 1998) consisting of a line of Late Miocene to recent volcanic centers gradually shifting in age from Late Miocene in the west to Quaternary in the east (Abebe et al. 1998). These edifices have basic lava flows at their base with acid domes, plugs, and pyroclastic deposits in their evolved end members.

2.5.2 The Volcanites of the Southern Ethiopian Plateau

For the southern plateau volcanites, there are reports of limited volumes of basalts (Amaro and Gamo basalts) with ages between 45 and 35 Ma (Davidson and Rex 1980; Ebinger et al. 1993; George et al. 1998). They predate the

Fig. 2.15 The Jima Volcanites at the top of the Guraghe escarpment (Fig. 2.13). Gently inclined rhyolite ignimbrites incised by a channel in turn filled by horizontally laminated ignimbrite flows. New road on the western flank of the MER between Butajira and Welkite

northern Ethiopian traps by as much as 15 Ma (Rogers 2006). The southern plateau traps are less thicker than those of the northern plateau and are characterized by much thicker and more widespread siliceous rocks. For this volcanic sequence, resting directly on the crystalline basement or, more rarely, on the Eocene basalts and Mesozoic sandstones, we follow the distinctions proposed in Merla et al. (1979). The succession begins with a hundred meters of mildly alkaline basalts (Omo Basalts), capped by a thick unit, up to 1,000 m, of rhyolites, acidic tuffs, and subordinate basalts (Jima Volcanites). The Omo Basalts are commonly fine grained with columnar flows up to 10 m thick alternating with minor tuffs and red paleosols. There are some sparse dates ranging around 30 Ma (Merla et al. 1979; Davidson and Rex 1980). The Jima Volcanites, reaching one thousand meters of thickness in the Omo valley, represent most of the effusives in SW Ethiopia with a wide fringe in the southern portion of the Somali plateau NW of the Amaro Mts. and south of the west–east Bonga–Goba line (Abbate and Sagri 1980). The Jima Volcanites are mainly composed of massive, white, pinkish, and gray rhyolites, comendites, and pantellerites in thick flows alternating with tuffs and subordinate basalts (Fig. 2.15). Reliable radiometric datings are quite scarce. At the Omo village, toward the top of the succession, the Jima Volcanites gave an age of 27 Ma (Merla et al. 1979). An overall age range from ca. 30 to 27 Ma can be assigned to this unit. The Wollega Basalts, resting on the basement and on the tilted Omo Basalts and Jima Volcanites, consist of 200–400 m of predominant columnar alkaline basalt flows interbedded, particularly in the upper portion, with acidic tuff and loose fluvial lacustrine deposits. Two samples gave ages of 15 and 13 Ma.

A few huge rhyolite plugs and domes (e.g. Mt.Egan, Mt. Mizan Tafari, Fig. 2.13) constitute a prominent and peculiar feature of the southern Ethiopian plateau. According to their relationships with the surrounding volcanites, they are doubtfully assigned to the Pliocene (Merla et al. 1979).

2.5.3 The Volcanites of the Somali Plateau

Above a patch of limited extent of Jima Volcanites, the succession in the Somali plateau, assigned to the Traps, consists of the Arussi and Bale Basalts of Miocene age (K/Ar ages ranging from ca. 24 to 9 Ma, Kunz et al. 1975; Morbidelli et al. 1975; Merla et al. 1979) with variable composition from transitional to alkaline basalts (Zanettin et al. 1980) and a thickness of about 3,000 m (Juch 1975). Rhyolitic intercalations are particularly abundant and thick in the middle portion of the Arussi and Bale Basalts (Juch 1975). They become the predominant component of the overlying Ghinir Unit alternating with basalts and fluviolacustrine intercalations. From the data supplied by Juch (1975), it can be inferred that the Ghinir Unit attains several hundred meters in thickness. The radiometric ages fall in the range of 6–2 Ma (Kunz et al. 1975; Morbidelli et al. 1975; Merla et al. 1979). A lacustrine unit with abundant diatomites and rare sands and conglomerates (Chorora Formation,

Sickenberg and Schoenfeld 1975) is discontinuously present beneath the Ghinir Unit at the foot of the escarpment. It is up to 200 m thick and has yielded Late Miocene mammals (Bernor et al. 2004).

A peculiar morphological feature in the Somali plateau is the huge Plio-Quaternary volcanic complex of the Bale Mts. with cones and plugs, probably connected with the Bonga–Goba line (Figs. 2.10 and 2.14). It reaches an elevation of 4,300 m (Mt. Batu, Quaternary, Merla et al. 1979), rests on the Arussi and Bale Basalts, and is covered by patches of glacial deposits.

2.5.4 The Afar Volcanites

The Afar region is a quasi-triangularly shaped depressed area at the intersection of the Red Sea, Gulf of Aden, and the Main Ethiopian Rift (MER) (Fig. 2.1). Due to its 25-million-year-long story of rifting and incipient oceanization, volcanic rocks cover wide areas of the Afar depression (Fig. 2.10).

According to Barberi et al. (1975), the Afar volcanites can be assigned to a first stage of continental rifting which lasted about twenty million years, starting from 25 Ma, and a later stage which commenced 4 Ma during which the oceanic floor in the central portion of Afar began to develop.

The older volcanites of the first stage include the Adolei Basalts, Mabla Rhyolites, and Dalha Basalts (Barberi al. 1975). They are more than 1,000 m thick and cover a time range from 26 to 6 Ma. Associated with them are alkaline and peralkaline granites aged 25–22 Ma linked to an early phase of continental breakup. The most extensive volcanic sequence connected with the second stage is the Plio-Pleistocene Afar Stratoid Series which covers about two-thirds of the Afar depression. This consists of transitional basalts, about 1,500 m thick and lies unconformably on the Dalha Basalts after a phase of magmatic quiescence. Intercalated in the top of and above the Afar Stratoid Series are the transversal volcanics and marginal rhyolitic centers (e.g., Dubbi, Ado Ale, Fig. 2.13). The axial volcanic ranges of Quaternary age are a typical morphological feature of the Afar depression from which they rise prominently up to 1,500 m. The northern range (Erta Ale) parallels the Afar axis with a NNW trend; to the south, the volcanic ranges shift gradually to WNW. They consist of fissure eruptions and shield volcanoes with basaltic flows and alkaline to peralkaline silicic rocks. Many of them have been active in historical times, and the Erta Ale volcano exhibits a spectacular lava lake.

2.5.5 The Main Ethiopian Rift Volcanites

The Main Ethiopian Rift (MER) is a NNE–SSW to N–S-trending trough 80 km wide in its central portion and 1,000 km long (Fig. 2.14). It separates the southern Ethiopian plateau to the west from the Somali plateau to the east. Northward, the MER progressively widens out into the complex Afar triple junction, while at its southern end, a 200–300-km tectonically disturbed area (Gofa basin and range, Baker et al. 1972) marks the transition to the Kenyan Gregory Rift in the Turkana depression.

The volcanic history of the MER has been dealt with in numerous papers which often take into account limited sectors of the rift. This has resulted in a proliferation of volcanic units with significant problems of correlation among the northern, central, and southern sectors.

The MER volcanic stratigraphy was summarized by Corti (2009). He envisages a lower basalt unit with trachybasalts and subordinate silicic flows from 11 to 8 Ma old followed by a widespread ignimbrite cover (e.g., Nazaret Group) ranging in age from 7 Ma in the northern sector to 2 Ma to the south and up to 700 m thick. Most of the ignimbrite layers are believed to have formed by catastrophic eruptions related to the collapse of large calderas, such as the 3.5-Ma-old Munesa caldera now buried beneath the Ziway–Shala lakes (Fig. 2.13). These two units, common to the whole MER, are followed by Late Pliocene basalts with pyroclastics fed by calderas which are limited to the northern and central sectors. The subsequent Quaternary volcanic unit, which outcrops throughout the MER, is the Wonji Group associated with the oblique Wonji fault belt (Mohr 1962). It includes basalt flows and scoria cones, and large silicic central volcanoes with calderas. These edifices and calderas rise up to 700 m above the plain (e.g., Bora Bericho, Alutu, Gademotta, Fig. 2.13), and some of them experienced phreatomagmatic activity and historical flows. They are referred to as en-echelon arranged magmatic segments connected to the Wonji fault belt (see later).

Off-axis magmatism is mainly concentrated on the Somali plateau with huge shield volcanoes of basaltic and trachytic composition and Mio-Pliocene age (Chillalo, Badda, Chike, Kecha, Figs. 2.13 and 2.16). Some of them exceed 4,000 m in elevation and have a base of 30–40 km diameter rising from the plateau level for 1,000–1,500 m. Glacial cirques and massive moraines occur at an altitude of about 4,000 m, indicating that these high mountains were glaciated during the Late Quaternary (Grove et al. 1975).

2.5.6 The Intertrappean Beds

Common intercalations in the previously described trap pile of the Ethiopian highlands as well as in the Yemen plateau are terrestrial sediments (intertrappean beds) composed of red clays, sands, diatomites, and lignite seams, generally a few tens of meters thick, but also reaching hundreds of meters (Abbate et al. 2014). Within the abrupt and steep trap

Fig. 2.16 The Main Ethiopian Rift floor with a Quaternary volcano (*hill* on the *left*) and the faulted escarpment of the Somali plateau (*light-* and *dark-colored* stripes in the *midground*) overlain by the huge Chillalo shield volcano (in the *background*)

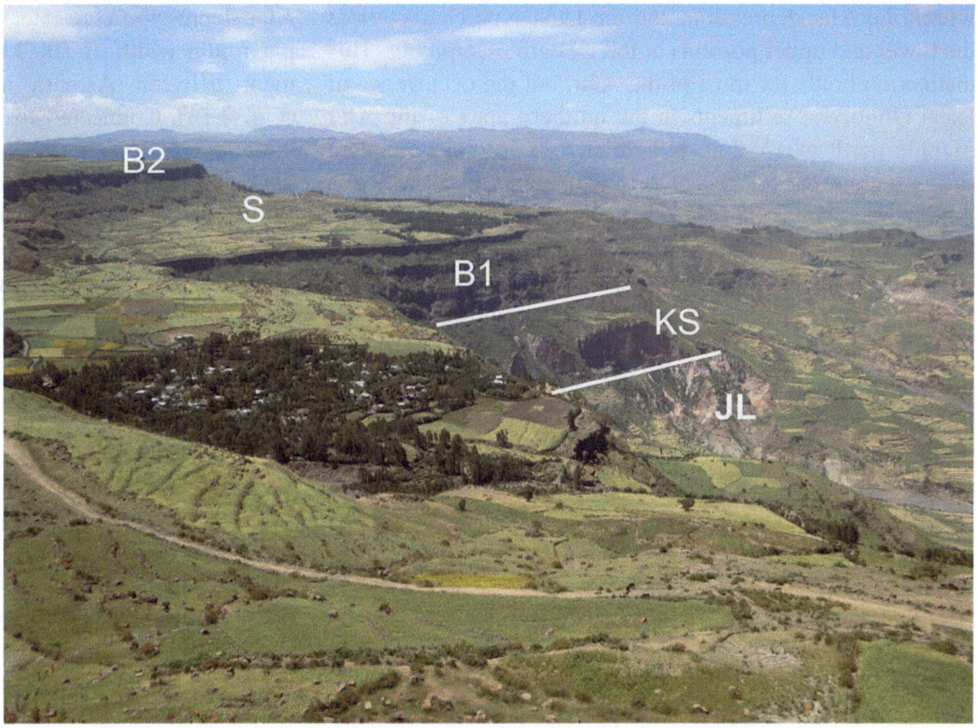

Fig. 2.17 The flat morphology of the soft intertrappean sediments (*S*) contrasting with the steep slopes cut through the trap basalts (*B1*, *B2*), the Cretaceous sandstones (*KS*), and the Jurassic limestones (*JL*). Slope of the Jema river, a Blue Nile left tributary, 100 km NNW of Addis Ababa

escarpment, these loose sediments give rise to a pronounced morphological break (Fig. 2.17). The intertrappean sediments occur in lenses which are sometimes continuous for tens of km in the middle/upper portion of the traps and span a time interval between ca. 29 and 27 Ma (Early/Late Oligocene transition). They mark a period of volcanic quiescence following a voluminous and rapid lava outpouring. The volcanic activity resumed after two/three million years but was delayed a couple of million years in the marginal areas of the trap effusion (Abbate et al. 2014). The presence of endemic proboscideans in the intertrappean sediments is particularly significant in investigating the origin of these mammals and in clarifying the relationships between African and Eurasian faunas (Abbate et al. 2012, 2014).

2.6 The Sediments in the Afar Depression, Main Ethiopian Rift, and Adjoining Areas

In addition to magmatic and pyroclastic materials, the subsiding basins of the Afar and MER host sediments which are particularly useful in reconstructing Plio-Pleistocene human evolution and climatic variations as well as the paleogeography and structural development of the region.

2.6.1 The Afar Depression

Different types of basins occur within the Afar depression and along its margins. They originated during successive phases of tectonic deformation since the Miocene and have generated a significant morphological response in the present-day topography. According to their position with respect to the Afar depression axis, we distinguish axial, peripheral, and marginal basins (Fig. 2.18). The first two types are located within the Afar depression, and the latter type characterizes the lower and upper portions of the western escarpment. This distinction holds for most of the Afar, but the occurrence of latest Oligocene sediments in the northern apex of the Afar triangle north of 13° Lat. N is related to an older incipient rifting with the development of a sedimentary basin which is not recorded southward. These older sediments, which were the first laid down in the Afar Depression, are the Red Series (Bannert et al. 1970) or Danakil Formation (Brinkmann and Kürsten 1970; Garland 1980). They are composed of violet-red to bright-red conglomerates and sands with mudstones (Fig. 2.19) locally gypsiferous, and rare freshwater gastropods-bearing limestones. Alluvial fans and high-energy streams with some swampy to lacustrine ponds were the main features of the Danakil Formation environment. Frequent basalt flows are found intercalated in this succession. Those at the base and toward the top gave K/Ar ages of 24 and 4 Ma, respectively (Bannert et al. 1970). A thickness of about 1,000 m is commonly assumed. The whitish gypsiferous Enkafala Formation, a few meters thick, unconformably overlies the Danakil Formation and marks a marine transgression in the northern Afar depression. Its marine fossils have been dated by U/Th methods at 200 to 24 ky (Lalou et al. 1970). Toward the center of the basin, the Enkafala Formation passes transitionally into the salt formation composed of halite, gypsum, potash salts, and clays (Fig. 2.20). The salts have precipitated at the surface and generate an impressive, bright-white salt plain. A thickness of 975 m has been drilled for potash exploitation, and a thickness of 2,200 m has been estimated by geophysical investigations.

Peripheral and axial basins occur south of 13° Lat. N (Figs. 2.14 and 2.18). Fluviolacustrine sediments fill peripheral basins at the foot of the western and southern Afar escarpment along the Middle Awash valley (e.g., Woramso/Mille, Hadar, Kesem/Kebena, and Chorora sub-basins). Kalb et al. (1982) include these continental sediments under the Late Miocene to Pleistocene Awash Group which is more than 1,000 m thick. Main lithological components are shales, sands, pebbles, and diatomites with intercalated ash layers and basalt flows (Fig. 2.21). The Awash Group is famed for its rich hominid content, artifacts, and mammal faunas (e.g., Hadar, Bodo, Buri, Busidima, Daka, Sidi Koma, Sagantole, Chorora, Adu Asa) (a rich bibliography on this matter can be acquired from Quade and Wynn 2008).

In central and eastern Afar, elongated graben structures (e.g., Abhe/Gob Ad, Hanle/Dobi, Tendaho/Dubti, Gaggade, Figs. 2.14 and 2.18) filled by Pleistocene fluviolacustrine deposits are related to the axial basins (Tiercelin et al. 1980). In geothermal drillings in the Tendaho basin, more than 500 m of fine-grained Pleistocene sediments have been found alternating with basalts (Battistelli et al. 2002).

The eastern margin of the Eritrean–Ethiopian plateaus is characterized by a flexure that connects the plateau to the Afar depression (see later). Such a deformation belt with a maximum width of 100 km is developed along the escarpment between Asmara and Addis Ababa for 900 km (Fig. 2.14). Connected to this flexure are a number of tectonic depressions parallel to the escarpment and mostly developed along a north–south direction ("marginal basins" Mohr 1967) (Fig. 2.18). The flexure and associated depressions form very conspicuous features due to their morphological continuity and pronounced expression.

The depressions are filled by Pleistocene fluviolacustrine sediments and volcanoclastics. Those situated close to the plateau edge with a master fault on their western flank are well developed both longitudinally and transversally (Kobo basin: 120 km by 15 km; Haik/Borkenna basin 70 km by 10 km). The sediment thickness in these basins is generally a few tens of meters. Those developed within the flexure are usually smaller and characterized by their reduced or absent sedimentary infilling. On the contrary, basins at the base of flexure host thick sedimentary successions. From north to south, they are Buia in Eritrea, and Garsat and Teru in Ethiopia. These basins are filled by up to 1,000 m of the fluviolacustrine sediments of the Early-to-Middle Pleistocene Dandiero Group (Abbate et al. 2004). The Buia section yielded a one-million-year old *Homo erectus* skull, abundant mammal faunas, and lithic industries (Abbate et al. 1998).

2.6.2 The Main Ethiopian Rift

An additional area of Late Cenozoic sediment accumulation is the MER where fluviolacustrine sediments with rare diatomites (Fig. 2.22) cover a large area. They represent the deposits laid down in a very wide lake which, in the past,

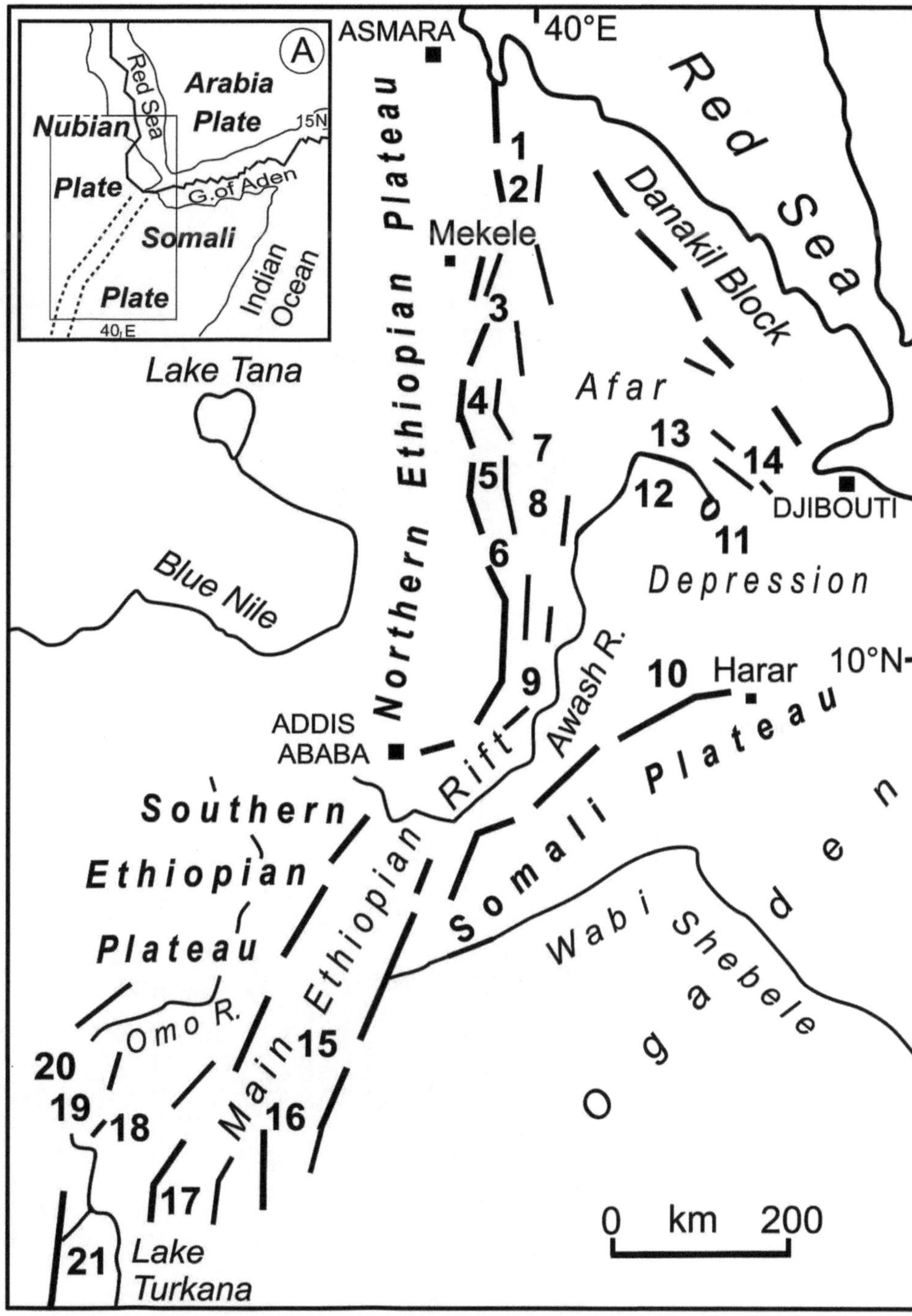

Fig. 2.18 Location map of the major Plio-Quaternary sedimentary basins (marginal, peripheral, central) according to their structural setting in the northern Ethiopia plateau escarpment, Afar depression, and Main Ethiopian Rift. Marginal basins: *1* Buia; *2* Garsat; *3* Teru; *4* Kobo; *5* Hayk; *6* Borkenna. Peripheral basins: *7* Woramso; *8* Mille; *9* Kesem-Kebena; *10* Chorora. Central basins—in Afar: *11* Abhe-Goba Ad; *12* Tendaho-Dubti; *13* Hanle-Dobi; *14* Gaggade; in the Main Ethiopian Rift: *15* Galana; *16* Konso-Gardula; *17* Chew Bahir; *18* Usno; *19* Omo; *20* Kibbish; *21* Turkana. Inset map A (*top left*) shows location of study area relative to the Nubian, Arabia, and Somali plates. Main Ethiopian Rift within *dotted lines*

occupied most of the rift floor. In the northern and central part of the MER north of the Lake Shala, only a few tens of meters of Late Pleistocene/Holocene sediments outcrop. They record at least two major phases of lake expansion separated by a prolonged period of lowstand and erosion (Fig. 2.22) (Street 1979; Le Turdu et al. 1999; Benvenuti et al. 2002). On the basis of geophysical investigations, the presence of about 600 m of sediments has been estimated by Le Turdu et al. (1999). It has been proposed that their maximum age is as old as 500 ka (Le Turdu et al. 1999).

Poorly studied similar fluviolacustrine deposits occur in the southern MER in the Abaya/Chamo Lake regions with a thickness of more than 500 m (Ebinger et al. 1993). More detailed studies have been carried out on the fluviolacustrine sediments containing *Australopithecus boisei, Homo erectus,* and abundant artifacts in the Konso/Gardula area (Katho et al. 2000). These sediments comprise the Konso Formation and consist of dark-gray clay, red-brown silt, sand, and gravel, with frequent intercalations of volcanic rocks and tephra. The Konso Formation is more than 200 m thick and was deposited between 1.9 and 1.4 Ma.

Fig. 2.19 *Red* fluviatile sandstones and mudstones of the Danakil Formation with a dark intercalation of a basalt flow, unconformably capped by coarse-grained Quaternary alluvial deposits

Fig. 2.20 Highly dissected salt crust hosting iron-rich brine deposits near Dallol, northern Afar depression. The mountains in the background are the Afar western escarpment at the edge of the northern Ethiopian plateau

The MER terminates to the south with the Chamo and Konso/Gardula basins and abuts the Sagan line (Baker et al. 1972; Abbate and Sagri 1980). Through the intermediate Chew Bahir, Usno, Omo, and Kibish graben structures, the deformation shifts to the Kenyan Turkana rift (Fig. 2.14). With the exception of the Chew Bahir, which is a symmetrical graben, all the others are half-graben with the master fault on the western side (Davidson 1983). All these basins are filled by fluviolacustrine deposits interbedded with volcanic rocks and ashes. The Omo basin, in the lower course of the Omo river, is the northern extension of the Turkana rift and is particularly well studied owing to the occurrence of abundant vertebrate

Fig. 2.21 Cyclically arranged lacustrine, lake margin, and fluviatile deposits with tuff horizons in the Hadar basin, Middle Awash, northern Afar (basin 8 in Fig. 2.18). This section is adjacent to the site where the 3.2-million-year-old partial skeleton of "Lucy" was discovered

Fig. 2.22 Late Quaternary lacustrine sediments consisting of massive diatomites (*whitish*) and thinly stratified, *light gray* volcanoclastic deposits. A deep erosional surface separates the Holocene from the Late Pleistocene successions. Central Main Ethiopian Rift. Photograph courtesy of Marco Benvenuti

fossils including hominids (e.g., Howell 1969; Arambourg 1972). It is filled by one thousand meters of sediments, and most of them are represented by the fossiliferous Plio-Pleistocene Shungura formation (deHeinzelin 1983). The development of this basin dates back to the Early Pliocene (Kidane et al. 2007; McDougall and Brown 2008).

2.7 The Morphological Responses to the Cenozoic Geodynamic Events

Three main physiographic provinces characterize the Ethiopian region: the highlands, represented by the northern and southern Ethiopian plateaus and Somali plateau, the Afar depression, and the MER (Fig. 2.1). The birth, evolution, and present-day morphology of these provinces, although not achieved at the same time, are linked to the uplift and rifting of the Afro-Arabian plate since the Oligocene.

2.7.1 The Ethiopian Plateaus and their Uplift

Using observations from along the Gulf of Aden, Dainelli (1943) assumed an Eocene age for the main uplift of the whole Horn of Africa and claimed that has occurred prior to the trap outpouring and rifting episodes. He was also the first to visualize the uplift by drawing contours of the present surface of the crystalline basement.

The issue of the exact time of uplift is still strongly debated. Mohr (1967) and Mohr and Zanettin (1988) assumed that the major uplift of the Afro-Arabian swell occurred during the Pleistocene. Alternatively, Merla et al. (1979) proposed that the domal uplift started in the Oligocene with a climax at the beginning of the Miocene at about 25 Ma, inferring this timing of events from the stratigraphy of the Neogene formations in the Somali shoulder along the Gulf of Aden.

More recently, Pik et al. (2003) suggested on the basis of (U-Th)/He thermochronometry that the Ethiopian plateau has been an elevated and stable dome since the Oligocene, with its highest region along the present-day Afar escarpment. The regional high structure was the result of the combined effects of the Afar plume impingement and associated large basalt effusions (see also Ebinger and Sleep 1998). According to Pik and coworkers, steady-state erosion commenced in the Blue Nile canyon as early as 25–29 Ma and is still active. About 20 My ago, the drift of the Arabian plate and a concomitant collapse along the western Afar margin gave rise to the Afar depression (Pik et al. 2003).

The morphotectonic history of the northern Ethiopian plateau has also been assessed by Gani et al. (2007) using the long-term incision rate of the Blue Nile catchment. Their picture proposes that starting from a broad dome with slow rate of uplift from 29 to 10 Ma (phase I), a rapid rate of increase in the uplift occurred at 10 Ma (phase II) followed by a dramatic plateau rise at 6 Ma (phase III).

An episodic succession of tectonic events is also accepted by Ismail and Abdelsalam (2012) on the basis of morphotectonic analyses of the Tekeze and Blue Nile drainage systems. The first event, characterized by a low to moderate incision rate over the entire plateau, was associated with a broad and regional uplift of the plateau after the impingement of the Afar plume at ca. 30 Ma. The second event resulted in the localized increase of the incision rate controlled by the buildup of the shield volcanoes at ca. 22 Ma along the Tekeze and Blue Nile watercourses. A particularly significant increase in the incision rate took place at ca. 11 Ma, but was limited to the eastern portions of the Tekeze and Blue Nile catchments. It was the result of the rapid uplift of the eastern margin of the plateau facing the Afar depression.

Paleobotanical data can also be used to decipher the complex history of uplift of the Ethiopian plateau. Near Lake Tana, 130 m of fluviolacustrine sediments are interbedded with lavas in the middle/upper portion of the traps. They are the famous Chilga intertrappean beds renowned for their vertebrate and flora content (among others, Unger 1866; Merla and Minucci 1938; Yemane et al. 1987; Kappelman et al. 2003; Currano et al. 2011; Abbate et al. 2014). Their Ar/Ar age of 27.4 Ma has been obtained from an intercalated ash layer (Kappelman et al. 2003). The Chilga beds with their Guineo-Congolean wet forest trees and palynoflora devoid of gymnosperms were deposited at an altitude much lower than the 1,950 m of their present-day elevation (Yemane et al.1987). On the basis of this broad indication, we tentatively assume that the Chilga beds were deposited at an altitude not higher than 900–1,000 m. Thus, the difference from the present-day elevation would be ca. 1,000 m and has to be ascribed to the uplift of this plateau segment after 27 Ma. The Chilga beds also provide constraints relevant to the height of the pre-trappean surface above the sea level. Since they lie above ca. 600 m of trap basalts and were deposited, as we assumed, at ca. 1,000 m, we deduce a modest (ca. 400 m) elevation for the pre-trappean surface. This argues against a pronounced pre-trappean regional doming.

As to the age of the uplift, the Ethiopian plateau began to rise at ca. 20 Ma according to Moucha and Forte (2011) who applied a numerical model of mantle flow to reconstruct the uplift amount and time in East Africa.

For the southern Ethiopian and Somali plateaus, detailed morphotectonic data are scanty apart from cursory mentions within general papers on the Horn of Africa.

The southern Ethiopian plateau reaches its highest elevations along the eastern margin close to the MER (over 3,000 m in the Guraghe region) and progressively slopes westward to the Sudan lowlands and southward to the Turkana depression. A low elevation peneplained pre-trappean surface, continuously covered by laterites, is suggested by Davidson and Rex (1980). Conversely, a pre-trappean doming stage is tentatively put forward by Woldegabriel et al. (1990) on the basis of a tilted Mesozoic sequence unconformably overlain by Oligocene basalts in the Kella horst (Guraghe region), but fission-track analyses in the same region indicate that updoming and concomitant denudation of the Kella horst did not begin before the Late Miocene (Abebe et al. 2010).

To these conflicting hypotheses, we add that, as in the case of the northern plateau (Chilga), palynological data are also available for the southern plateau for paleoaltitude assumptions. Samples collected by Wolela (2007) in the Oligocene intertrappean sediments in the Jimma region at altitudes between 1,700 and 2,200 m proved to lack gymnosperms. As in the northern plateau, this absence points to an altitude much lower than that of the present day and, consequently, to an uplift substantially after the accumulation of the main trap effusion.

The Somali plateau, another sector of the Ethiopian highlands, reaches elevations of more than 3,000 m along its margins facing the Afar depression and the MER. It gradually slopes down toward the southeast (Ogaden and Indian Ocean). The scarce available data on the time and amount of its uplift derive from these margins. Juch (1980), who has studied a large extent of the Somali plateau escarpment, proposes a major uplift of 1,500 m younger than 2–3 Ma. Near the margin of the plateau, this uplift was accomplished along large normal faults cutting previous flexure-like structures.

In a wider regional analysis, many authors, beginning with Pickford (1990), have pointed out that the uplift of the East Africa plateaus caused a drastic reorganization of atmospheric circulation. This induced strong hydroclimatic changes and heavy impacts on ecosystems with a trend toward more arid conditions. The first event of the East African aridification, marked by an expansion of savanna grassland replacing the wet forests, occurred around 8–10 or 13.5 Ma (Sepulchre et al. 2006; Bonnefille 2010; Wichura et al. 2010).

From the discussion presented above, we conclude that the uplift of East Africa occurred essentially during the Middle/Late Miocene after the trap accumulation. By the Early/Late Oligocene transition, these volcanic successions had produced a wide bulge, particularly pronounced in the eastern portion of the northern Ethiopian plateau. At that time, the drainage was poorly defined due to the reorganization of the African river network (Goudie 2005; Stankievicz and De Witt 2006). Beginning in the Early Miocene, shield volcanoes were superimposed on the northern Ethiopian highland and began to control the courses of the Blue Nile and Tekeze rivers with their circular bases (Fig. 2.1). The courses of these rivers became annular and stable through pronounced erosional action enhanced by the rise of the highland. Incised valleys developed in their upper reaches and have gradually evolved into the present-day canyons to 1,600 m deep. Starting from the Miocene, in addition to the updoming, the Ethiopian region experienced the Afar rifting and, successively, the development of the MER. These tectonic events affected the plateau margins and caused further uprise due to flexural deformations. These marginal uplifts have firmly established the present-day fluvial systems. In the Somali plateau, the consequence was a pronounced dendritic drainage toward the Indian Ocean with canyons in the upper reaches of the Juba and Webi Shebeli rivers (Fig. 2.1).

2.7.2 The Afar Depression and Adjacent Plateau Margins

The Afar depression is bounded to the west by the N–S-trending, 700-km-long northern Ethiopian escarpment, to the south by the W–E trending, 350-km-long Somali plateau escarpment, and to the east by the NNW–SSE trending, 700-km-long Danakil/Ali Sabieh block (Fig. 2.14). Its minimum elevation reaches 120 m below sea level in the northern sector where the axial volcanic ranges of the volcanic shield complexes dominate the landscape.

As previously discussed, the northern Ethiopian and Somali escarpments are characterized by a continuous flexure (Figs. 2.14 and 2.23) (Mohr 1962; Abbate and Sagri 1969, 1980; Justin-Visentin and Zanettin 1974; Zanettin and Justin-Visentin 1975; Morton and Black 1975; Juch 1975; Beyene and Abdelsalam 2005) with the exception of short segments north of 13°N and in the Harar-Dire Dawa area which show block tilting toward Afar (e.g., Abbate and Sagri 1980; Beyene and Abdelsalam 2005).

The onset of the escarpment formation along the western margin of Afar varies in age from north to south (Zanettin and Justin-Visentin 1975): latest Oligocene (24 Ma) in the northern sector, and between 13 and 8 Ma southward. The foot of the escarpment along the whole margin was affected during the Plio-Pleistocene by major normal faults dipping east.

The westernmost escarpment of the Somali plateau underwent a downflexing toward Afar during the Miocene with important normal faults activity in the Pliocene (Juch 1975). In the remaining portion of the Somali plateau escarpment, near Dire Dawa, block faulting becomes predominant and involves mainly Mesozoic sediments and, subordinately, trap basalts of latest Oligocene to Miocene age. Horizontal Afar Stratoid basalts of Late Miocene to Pliocene age seal the block-faulted structures (Juch 1975; Ethiopian Institute of Geological Survey, Dire Dawa sheet 1985).

Both the flexural structures and block faulting, involving a deformed belt 50–100 km wide, produced rugged slopes at the margins of the tabular Ethiopian highlands (Fig. 2.1). The deformation is most pronounced parallel to the margins with closely spaced faults and dikes. Morphological features characteristic of the edge of the northern Ethiopian plateau and associated with the marginal graben are long river reaches parallel to the margin (from north to south, Gabala in the Garsat plain, Alomata in the Kobo plain, Borkenna in the homonymous basin) locally giving rise to endorheic lakes (Ashangi, Haik-Ardibbo) (Fig. 2.13).

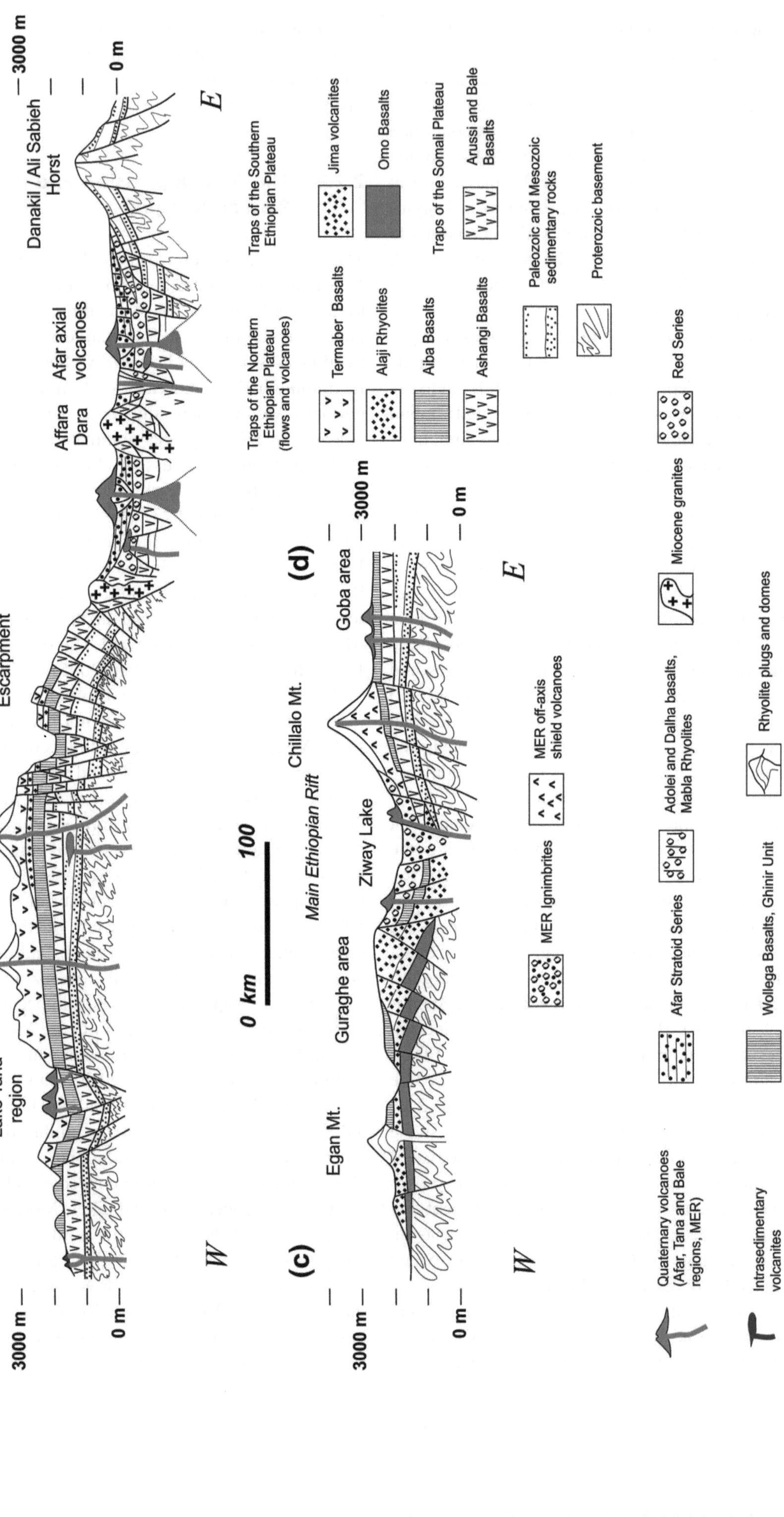

Fig. 2.23 Cross sections across northern (*A–B*) and southern (*C–D*) Ethiopia; location in Fig. 2.14. Modified after Merla et al. (1979) and Abbate and Sagri (1980)

The eastern margin of the Afar depression lies against the so-called Danakil block. This continental microplate stretches from the Gulf of Zula to the Gulf of Tadjoura and is interposed between the Afar depression and the southern Red Sea. It is composed of Proterozoic basement covered by Mesozoic sediments, trap basalts, and southward by Neogene volcanites including large edifices. The Danakil block reaches an elevation of 1,300 m in its northern portion and more than 2,000 m to the south (Sork Ale and Mussa Ale volcanoes, Fig. 2.13). No recent structural data are available for this Afar margin facing the Danakil block. According to the geological map by Brinckmann and Kürsten (1970) and in Bannert et al. (1970), flexures seem to affect most of it.

The youthful morphology of the Afar depression is substantially the result of Plio-Pleistocene events connected with the thinning of the crust and the first oceanization episodes. The magmatic and tectonic processes produced wide areas covered by the flood basalts of the Afar Stratoid Series, imposing axial shield volcanoes, and, in the central and southern sector mainly, an intricate network of fractures and faults (CNR-CNRS 1971, 1975) (Fig. 2.14).

The morphotectonic features vary from north to south. Over a distance of 200 km at the triangular apex of the Afar depression, between the Gulf of Zula and Lake Afrera, the Erta Ale axial shield volcano dominates an inhospitable, but fascinating, landscape. This segment of the Afar depression has a maximum width of ca. 100 km along 13° Lat. N and hosts at its margins the exclusive occurrences of the bright-red Miocene Danakil Formation. At its extreme north, the depression merges into the 10-km-wide Gulf of Zula bounded by Proterozoic basement rocks and terminates at the Alid axial volcanic range.

South of 13° Lat. N, the regional picture changes abruptly, probably in connection with a W/E regional tectonic alignment. The morphotectonic scenario becomes particularly complex and dominated by a series of narrow graben (often half-graben, e.g., Abbate et al. 1995; Acocella 2010) a few km wide and some tens of km long. Their sedimentary filling has been referred to as the axial basins. According to Hayward and Ebinger (1996) and Beyene and Abdelsalam (2005), three divisions can be recognized in central and southern Afar (Fig. 2.14). The east-central area is affected by a diffuse rifting, trending from NNW–SSE to W–E, and in many cases forming curvilinear structures (e,g., Immino graben). This area is bounded to the SW and south by the Tendaho-Goba Ad rift, a major structure with a length of ca. 300 km and a width up to 50 km (Acocella et al. 2008). The southeastern area is characterized by a W–E fault system parallel to the Somali plateau escarpment. In the southwestern area, the NNE–SSW fault trend of the MER, which propagates in this sector of Afar, is dominant.

Toward the west and south of this complex tectonic region, peripheral basins developed at the base of the Ethiopian and Somali escarpments in connection with intense downfaulting, subsidence, and crustal stretching.

The structural complexity of the Afar depression, particularly evident in the central and southern sectors, is the result of the interactions of geodynamic events affecting the Afro-Arabian plate since the Early Miocene. The structural history began with the anticlockwise rotation of the Danakil microplate (Baker 1970; Burek 1970; Sichler 1980) which caused it to detach from the Nubian plate with a later northward movement (Chorowicz et al. 1999; Beyene and Abdelsalam 2005). This was also facilitated by the NE drift of the Arabian plate.

By ca. 10 Ma, the Gulf of Aden propagator with its active spreading axis progressively invaded the southern Afar region between the Somali plateau and the southern portion of the Danakil block. Subsequently, the extension of the MER propagated northward into the southern Afar depression giving rise to a full-fledged triple junction (Wolfenden et al. 2004). Within the Afar east-central area, rotation of small blocks resulted in curvilinear grabens cut into the Afar Stratoid Series (e.g., Barberi and Varet 1977; Manighetti et al. 1998).

2.7.3 Morphotectonic History of the Main Ethiopian Rift

The MER is the salient morphotectonic feature that together with the southern Ethiopian and Somali highlands constitutes the spectacular scenery of the southern Ethiopia landscape. This depressed area hosts beautiful lakes, also famous for their bird life. They have a tectonic or volcanic origin and are variously colored, from the blue of Shala, the green of Bishoftu, the whitish of Abiyata, and the pink of Langano and Abaya, to the pearl gray of Chamo. All these lakes form endorheic systems connected by outlet–inlet fluvial reaches or by groundwater infiltration (Street 1979). The southern end of this endorheic system is located in the present-day ephemeral Chew Bahir lake, described at the end of nineteenth century as a wide stretch of permanent water pool. During the Late Quaternary wet periods, Lakes Abaya and Chamo were connected to Lake Turkana through the Sagan river and Lake Chew Bahir. In turn, Lake Turkana overspilled via the Akobo–Sobat rivers into the White Nile river system (Butzer et al. 1972; Street 1979). This Mediterranean connection is testified by the occurrence of the typical Nilotic Nile perch (Tilapia) in Lakes Abaya and Chamo.

The MER is confined between the uplifted shoulders of the southern Ethiopian and Somali highlands (Figs. 2.1 and 2.23). The escarpments can be very steep with a difference in altitude of 1,000–1,500 m from the plateau margin to the rift floor (e.g., Guraghe) and narrow (5–7 km) and are marked by major border normal faults. In other places (e.g., east of Langano), the shoulder is less steep and the difference is up to 1,000 m. Also the width of the escarpments varies between 10 and 20 km and is characterized by a series of steeply dipping faults with individual vertical displacement of 200–300 m, covered by a veneer of ignimbrites (Di Paola 1972).

From a morphological and geological point of view, the MER has been subdivided into three main segments: the northern, central, and southern (Mohr 1983; Woldegabriel et al. 1990; Hayward and Ebinger 1996; Bonini et al. 2005) (Fig. 2.14). The northern MER funnels from the Afar depression, where it is about 100 km wide, to the 80-km-long Dubeta Col sill (north of Ziway lake). The central MER, which is 80 km wide, includes most of the lake region and extends southward up to the W/E Goba–Bonga line (Fig. 2.14). This portion of the MER has an average elevation of 1,600 m, and the lowest altitude is at Lake Abiyata (1,580 m). At the Bonga–Goba line, the southern MER narrows up to 60 km, shifts to a N–S trend, and reaches an elevation of 2,000 m, decreasing southward to 1,000 m. From its middle portion to the south, the southern MER bifurcates into two branches (the Lake Chamo and Galana river rifts) separated by the 3,000-m-high Amaro horst ("a small Ruwenzori" according to Mohr 1967). The southern MER keeps its morphological identity until the Sagan line. To south and west of this, a wide basin and range structure connects the MER with the Kenyan rift (Fig. 2.14).

The rift floor is not uniformly flat but is occupied by recent volcanic edifices rising some hundreds of meters above the plain (e.g., Fantale, Boseti Gudda, Alutu, Tosa Sucha) and calderas (e.g., Gademota, lake Shala O'a caldera) (Fig. 2.13). Furthermore, the strongly deformed Wonji fault belt (see later) is characterized by rough and irregular morphology with narrow uplifted blocks, valleys, lava fields, spatter cones, and swampy depressions.

The origin of the MER, a continental rift with an average crustal extension rate of about 2.5 mm/a (Wolfenden et al. 2004), or 6.0 mm/a according to Corti (2009), is still a matter of debate. It has been related to pure tensional deformation by McKenzie et al. (1970), Di Paola (1972), and Le Pichon and Franchetau (1978). A sinistral shear component has been postulated by Mohr (1968), Gibson (1969), Kazmin (1980), and Boccaletti et al. (1992). According to Bonini et al. (1997) and Boccaletti et al. (1998), a sinistral oblique rifting related to an E–W extension followed a pure extension orthogonal to the rift trend, whereas Chorowicz et al. (1994) maintain that a right-lateral component of motion along the rift structure produced a NW-to-NNW-oriented extension.

The MER development was characterized by an early phase (Mio-Pliocene, Corti 2008) of activity of a series of large boundary fault-formed local asymmetric basins (e.g., Abbate and Sagri 1980; Kazmin et al. 1980; Woldegabriel et al. 1990; Boccaletti et al. 1992; Corti 2009) (Fig. 2.23). The total vertical displacement across the boundary faults reaches some thousand meters. Above the pre-Cambrian basement and Mesozoic sediments, the rift is floored by ca. 1,000 m of syn-rift Miocene to Recent volcanoclastic and sedimentary deposits (Cornwell et al. 2010).

This early phase was followed during the Pleistocene by a rift-in-rift stage, in which volcanic and tectonic activity was concentrated riftward with right-stepping en-echelon magmatic segments, the Wonji fault belt (Mohr 1962; Gibson 1969; Kazmin et al. 1980; Boccaletti et al. 1992; Chorowicz et al. 1994; Ebinger and Casey 2001; Wolfenden et al. 2004; Bonini et al. 2005; Corti 2009). In this phase, the bordering faults were no longer active and the deformation was magma assisted with diffuse diking (Ebinger and Casey 2001; Ebinger 2005). According to Bonini et al. (1997), Boccaletti et al. (1998), and Wolfenden et al. (2004), the first and second phases were related to rift orthogonal and oblique extension, respectively. However, Corti (2008) showed that both phases may have resulted from a constant post-10-Ma oblique rifting in line with early suggestions by Gibson (1969).

There is a general consensus that rifting was diachronous along the whole MER. Woldegabriel et al. (1990) recognized a 18–15-Ma rifting initiation in the southern and central MER with extension in the northern MER commencing after ca. 11 Ma (see also Ebinger and Casey 2001; Wolfenden et al. 2004). Bonini et al. (2005) propose a different picture with the onset of rifting in the southern MER between 20 and 11 Ma connected to the deformation of the Kenyan rift. At that time, no major tectonic activity was affecting the central and northern MER sectors. At the northern end of the MER, the southward rift propagation started from the Afar depression and progressively affected the northern MER in the Late Miocene (11 Ma), the central MER in the Pliocene (5.6–3 Ma), and, eventually, joined the southern MER in the Late Pliocene/Pleistocene (3–0 Ma). For the central MER, the timing is consistent with the thermochronology data provided by Abebe et al. (2010).

2.8 Conclusions

After the Neoproterozoic orogenic cycle, the cratonic history of the Ethiopian region was dominated by phases of vertical motion and rifting recorded by various erosional cycles. They are marked by regional planation surfaces (Fig. 2.6) which constitute geomorphological features useful in deciphering the different stages of the geological history of the Ethiopia (see also Coltorti et al. this volume).

The oldest and particularly significant planation surface (PS 1) results from the intense erosion which destroyed any relief of the Precambrian orogeny across the whole of East Africa. Some occurrences of more resistant granites and gneisses escaped this peneplanation and presently stand as inselbergs along the Ethiopia/Sudan boundary (Fig. 2.3).

The PS 1 records a time gap of at least one hundred million years (between the emplacement of the post-orogenic batholiths and the deposition of the Ordovician sediments), during which planation and lateritization took place.

The Triassic planation surface (PS 2) is mainly limited to northern Ethiopia where it intersects a few preserved patches of Paleozoic fluvial and glacial deposits. This surface marks a relatively short period of erosion and/or non-deposition, and in some basins, it is lacking with a continuous succession from Paleozoic to Jurassic (e.g., in the Ogaden).

The Cretaceous planation surface (PS 3) cuts across the Middle-to-Late Jurassic marine sediments and is most evident in northern Ethiopia where it represents a key marker (Coltorti et al. 2007) and is overlain by continental Cretaceous (Aptian–Albian) sandstones. The development of this surface spans a period of a few tens of million years.

After the Late Jurassic marine regression and deposition of Cretaceous continental deposits, the Ethiopian region was an exposed land affected by tectonic deformation and regional uplifts for a period of about 70 Ma. A planation surface (PS 4) developed before the trap outpouring, cutting the previous sedimentary successions, in some cases, down to the basement. It is recognizable over all the Ethiopian region and could correspond on a continental scale to the African Surface of Burke and Gunnel (2008). The topographical elevation of PS4 at the time of the traps outpouring was modest, a few hundred meters above sea level, according to the information provided by the palynoflora assemblages preserved in the intertrappean sediments of the northern plateau (Chilga, Tana lake area) as well as the southern (Jima area) Ethiopian plateaus. This is good evidence against the hypothesis of a prominent pre-trappean updoming affecting the whole Ethiopian region.

The Tana and Jima areas also provide constraints on the amount and timing of their uplift. Indications from there suggest that uplift reached a maximum of one thousand meters and was achieved after the major volcanic outpouring. Similar relationships between uplift and time were probably common to both Ethiopian plateaus. The volcanic activity, particularly prolific around 30 My, and the uplift of the Ethiopian region were essential in producing a bulge which resulted in high elevated plateaus. This new topographic barrier induced the reorganization of the eastern Africa river network with a shifting of the regional divide eastward. Rivers, originally flowing toward the Indian Ocean, were gradually captured by the Mediterranean-directed paleo-Nile system (Said 1993; Goudie 2005).

As to the role of a mantle plume in the morphogenesis of the Ethiopian region, the local impingement of a rising mantle plume has been considered as a possibility since the 1970s by Morgan (1971). Ebinger and Sleep (1998) proposed a model with a single large plume beneath the Ethiopian plateau with lateral flows exploiting preexisting lithospheric thinnings. A different reconstruction was provided by George et al. (1998) who envisage a Kenya plume initiating at 45 Ma and a 15 Ma younger Afar plume. More recent authors (e.g., Furman et al. 2006; Meshesha and Shinjo 2008) reconsidered the African superplume hypothesis of Ritsema et al. (1999) and assumed multiple plumes stemming from this large-scale thermal upwelling.

The impact of the plume (or plumes) was concomitant with the dismembering of the Nubia/Arabian continental block. During the Oligo-Miocene transition, the Danakil block began to separate from Nubia leaving behind the Afar region, and after a few million years, the Somali plate began to separate from Nubia with the development of the MER.

In addition to its salient morphological features, the Ethiopian region records some of the most important events of the Earth history since the Neoproterozoic:

(i) The aggregation of the East and West Gondwana continental blocks to form the Gondwana supercontinent and the development of the East Africa Orogen.
(ii) The long plate stillness of the southern portion of Gondwana during the Paleozoic with the erosion of the Late Proterozoic mountain chains and development of a wide planation surface.
(iii) The major, widespread, and long-lasting glaciation during the Carboniferous to Early Permian with Ethiopia located at the margin of the southern hemisphere ice sheet of Pangea.
(iv) The beginning of the breakup of Gondwana and the related Jurassic sea-level highstand resulting in the East Africa marine ingression from the Paleotethys and the India/Madagascar nascent ocean.
(v) The profuse and rapid basaltic activity which contributed to the Oligocene climatic deterioration and ensuing mass extinctions at a global scale.
(vi) The uplift of the East Africa plateaus which resulted in a drastic reorganization of atmosphere circulation and river pattern during the Late Neogene.
(vii) The development of a variegated landscape with well-defined rift valleys which prompted the birth, evolution, and radiation of the human species later impelled to colonize more suitable territories in nearby continents.

Acknowledgments Constructive commentaries and careful editorial suggestions by Frances Williams (Adelaide University, Australia) greatly improved the manuscript. The authors have benefited from helpful discussions in the field with Abebe Tsegaye (MASSA Spin off, Pisa, Italy) and Miruts Hagos (Mekele University, Ethiopia) and in the office with Marco Bonini, Giacomo Corti, and Federico Sani (Department of Earth Sciences, Florence University, Italy). Special thanks to World Geomorphological Landscapes Series Editor Piotr Migoń (University of Wrocław, Poland) for his precious editorial assistance. We are grateful to Volume Editor Paolo Billi (Ferrara University, Italy) who prompted our contribution and gave valuable suggestions.

References

Abbate E, Sagri M (1969) Dati e considerazioni sul margine orientale dell'altopiano etiopico nelle province del Tigrai e del Wollo. Boll Soc Geol It 88:489–497

Abbate E, Sagri M (1980) Volcanites of Ethiopian and Somali Plateaus and major tectonic lines. Atti Convegni Acc Lincei Roma 47:219–227

Abbate E, Passerini P, Zan L (1995) Strike-slip faults in a rift area: a transect in the Afar Triangle, East Africa. Tectonoph 241:67–97

Abbate E, Albianelli A, Azzaroli A, Benvenuti M, Tesfamariam B, Bruni P, Cipriani N, Clarke RJ, Ficcarelli G, Macchiarelli R, Napoleone G, Papini M, Rook L, Sagri M, Tecle TM, Torre D, Villa I (1998) A one-million-year-old *Homo* cranium from the Danakil (Afar) depression in Eritrea. Nature 393:458–460

Abbate E, Woldehaimanot B, Bruni P, Falorni P, Papini M, Sagri M, Girmay S, Tecle TM (2004) Geology of the Homo-bearing Pleistocene Dandiero basin (Buia region, Eritrea Danakil Depression). Riv It Paleont Strat 110:5–34

Abbate E, Bruni P, Coppa A, Aria D, Ferretti MP, Libsekal Y, Rook L, Sagri M (2012) A new Oligocene-Miocene mammal-bearing site from a sedimentary intercalation in the Trap basalts of central Eritrea. Riv It Paleont Strat 118:545–550

Abbate E, Bruni P, Ferretti MP, Delmer C, Laurenzi MA, Hagos M, Bedri O, Rook L, Sagri M, Libsekal Y (2014) The Oligocene intertrappean beds: regional distribution, depositional environments and Afro/Arabian mammal dispersal corridors. J Afr Earth Sci 99:463–489

Abebe T, Mazzarini F, Innocenti F, Manetti P (1998) The Yerer-Tullu Wellel volcanotectonic lineament: a transtensional structure in central Ethiopia and the associated magmatic activity. J Afr Earth Sci 26:135–150

Abebe T, Balestrieri ML, Bigazzi G (2010) The Central Main Ethiopian Rift is younger than 8 Ma: confirmation through apatite fission-track thermochronology. Terra Nova 22:470–475

Abul-Haggag Y (1961) A contribution to the physiography of Northern Ethiopia. University of London Athlone Press, London, 153 pp

Acocella V (2010) Coupling volcanism and tectonics along divergent plate boundaries: collapsed rifts from central Afar, Ethiopia. Geol Soc Am Bull 122:1717–1728

Acocella V, Abebe B, Korme T, Barberi F (2008) Structure of Tendaho Graben and Manda Hararo Rift: implications for the evolution of the southern Red Sea propagator in Central Afar. Tectonics 27 (TC4016). doi:10.1029/2007TC002236

Alvarez F (1540) Verdadeira Informação das Terras do Preste João das Indias. 458 pp

Andrews Deller ME (2006) Facies discrimination in laterites using Landsat Thematic Mapper, ASTER and ALI data—examples from Eritrea and Arabia. Int J Remote Sens 27:2389–2409

Arambourg G (1972) Expedition internationale de rechérche paléontologique dans la vallèe de l'Omo, Ethiopie, en 1967. Proc Pan Afr Congr Prehist 6:135–140

Asfaw B, Gilbert WH, Beyene Y, Hart WK, Renne PR, WoldeGabriel G, Vrba ES, White TD (2002) Remains of *Homo erectus* from Bouri, Middle Awash, Ethiopia. Nature 416:317–320

Baker BH (1970) The structural pattern of the Afro-Arabian rift system in relation to plate tectonics. Phil Trans Roy Soc London A267:383–391

Baker BH, Mohr PA, Williams LAJ (1972) Geology of the eastern rift system of Africa. Geol Soc Am Spec Pap 136:67 pp

Baker J, Snee L, Menzies M (1996) A brief Oligocene period of flood volcanism in Yemen: implications for the duration and rate of continental flood volcanism at the Afro-Arabian triple junction. Earth Planet Sci Lett 138:39–55

Baldacci L (1891) Osservazioni fatte nella Colonia Eritrea e pubbl. a cura del R.Ufficio Geologico con una carta. Tipografia Nazionale Roma 110 pp

Bannert D, Brinckmann J, Käding KCh, Kretsch G, Kürsten N, Mayrhofer H (1970) Zur Geologie der Danakil Senke (Nördliches Afar Gebiet), NE Aethiopien. Geol Rund 59:409–443

Barberi F, Varet J (1977) Volcanism of Afar: small-scale plate tectonics implications. Geol Soc Am Bull 88:1251–1266

Barberi F, Borsi S, Ferrara G, Marinelli G, Varet J (1970) Relations between tectonics and magmatology in the northern Danakil depression (Ethiopia). Philos Trans R Soc Lond A267:293–311

Barberi F, Tazieff H, Varet J (1972) Volcanism in the Afar depression: its tectonic and magmatic significance. Tectonoph 15:19–29

Barberi F, Ferrara G, Santacroce R, Varet J (1975) Structural evolution of the Afar triple junction. In: Pilger A, Roesler EA (eds) Afar Depression of Ethiopia. Schweizerbart, Stuttgart, pp 38–54

Barnes SU (1976) Geology and oil prospects of Somalia, East Africa. Am Assoc Pet Geol Bull 60:389–413

Battistelli A, Yiheyisb A, Calore C, Ferragina C, Abatnehb W (2002) Reservoir engineering assessment of Dubti geothermal field, Northern Tendaho Rift, Ethiopia. Geothermics 31:381–406

Beauchamp J (1977) La série sédimentaire en Ethiopie central and oriental. Ph.D. thesis, University Claude Bernard Lyon, 419 pp

Beccaluva L, Bianchini G, Natali C, Siena F (2009) Continental flood basalts and mantle plumes: a case study of the Northern Ethiopian Plateau. J Petrol 50:1377–1403

BEICIP (1985) Geological map of Ogaden and surrounding area, scale 1:1,000,000. Ministry of Mines and Energy Addis Ababa

Benvenuti M, Carnicelli S, Belluomini G, Dainelli N, Di Grazia S, Ferrari GA, Iasio C, Sagri M, Ventra D, Atnafu B, Kebede S (2002) The Ziway-Shala lake basin (main Ethiopian rift, Ethiopia): a revision of basin evolution with special reference to the Late Quaternary. J Afr Earth Sci 35:247–269

Bernor RL, Kaiser TM, Nelson SV (2004) The oldest Ethiopian Hipparion (Equinae, Perissodactyla) from Chorora: systematics, Paleodiet and Paleoclimate. Cour Forsch Inst Senckenberg 246:213–226

Beyene A, Abdelsalam MG (2005) Tectonics of the Afar depression: a review and synthesis. J Afr Earth Sci 41:41–59

Beyth M (1972) The geology of central and western Tigray. Ph.D. thesis, Reinische Friedrick-Wilhelm University of Bonn, 200 pp

Beyth M, Stern R, Matthews A (1997) Significance of high-grade metasediments from the Neoproterozoic basement of Eritrea. Precambr Res 86:45–58

Beyth M, Avigadb D, Wetzelc HU, Matthewsb A, Berhe SM (2003) Crustal exhumation and indications for Snowball Earth in the East African Orogen: north Ethiopia and east Eritrea. Precambr Res 123:187–201

Blanford WT (1869) On the geology of a portion of Abyssinia. Q J Geol Soc London 25:401–406

Blanford WT (1870) Observations on the geology and zoology of Abyssinia, made during the progress of the British expedition to that country in 1867–68. Macmillan, London, 487 pp

Boccaletti M, Getaneh A, Tortorici L (1992) The main Ethiopian Rift: an example of oblique rifting. Ann Tectonicae 6:20–25

Boccaletti M, Bonini M, Mazzuoli R, Abebe B, Piccardi L, Tortorici L (1998) Quaternary oblique extensional tectonics in the Ethiopian Rift (Horn of Africa). Tectonoph 287:97–116

Bonini M, Souriot T, Boccaletti M, Brun JP (1997) Successive orthogonal and oblique extension episodes in a rift zone: laboratory experiments with application to the Ethiopian Rift. Tectonics 16:347–362

Bonini M, Corti G, Innocenti F, Manetti P, Mazzarini F, Abebe T, Pecskay Z (2005) Evolution of the main Ethiopian Rift in the frame of Afar and Kenya rifts propagation. Tectonics 24:TC1007. doi:10.1029/2004TC001680

Bonnefille R (2010) Cenozoic vegetation, climate changes and hominid evolution in tropical Africa. Glob Planet Change 72:390–411
Bosellini A (1989) The continental margins of Somalia: their structural evolution and sequence stratigraphy. Mem Sci Geol Padova 41:373–458
Bosellini A, Russo A, Fantozzi PL, Assefa G, Solomon T (1997) The Mesozoic succession in the Mekele outlier (Tigrai province, Ethiopia). Mem Sci Geol Padova 49:95–116
Bosellini A, Russo A, Schroeder R (1999) Stratigraphic evidence for an Early Aptian sea-level fluctuation: the Graua Limestone of south-eastern Ethiopia. Creta Res 20:783–791
Bosellini A, Russo A, Assefa G (2001) The Mesozoic succession of Dire Dawa, Harar Province, Ethiopia. J Afr Earth Sci 32:403–417
Bosworth W, Huchon P, McClay K (2005) The Red Sea and Gulf of Aden Basins. J Afr Earth Sci 43:334–378
Brassier M, Geleta S (1993) A planktonic marker and Callovian-Oxfordian fragmentation of Gondwana: data from Ogaden basin, Ethiopia. Palaeogeogr Palaeoclimatol Palaeoecol 104:177–184
Brinkmann J, Kürsten M (1970) Geological sketch map of the Danakil Depression, scale 1:250,000. Bunderanstalt für Bodenforschungen, Hanover
Burek PJ (1970) Paleomagnetic evidence for an anti-clockwise rotation of the Danakil Alps, Ethiopia. Eos Trans Am Geophys Un Abs 51:271
Burke K, Gunnell Y (2008) The African erosion surface: a continental-scale synthesis of geomorphology, tectonics, and environmental change over the past 180 Million Years. Geol Soc Am Mem 201:66 pp
Bussert R, Schrank E (2007) Palynological evidence for a latest Carboniferous-Early Permian glaciations in Northern Ethiopia. J Afr Earth Sci 49:201–210
Butzer KW, Isaac GL, Richardson JL, Washbourn-Kamau C (1972) Radiocarbon dating of East Africa lake levels. Science 175:1069–1076
Chavaillon J (1971) Les habitants acheuleens de Melka-Kontoure. Proc PanAfr Congr Prehist 230:17–25
Chorowicz J, Collet B, Bonavia FF, Korme T (1994) Northwest to north-northwest extension direction in the Ethiopian Rift deduced from orientation of the extension structures and fault slip analysis. Geol Soc Am Bull 105:1560–1570
Chorowicz J, Collet B, Bonavia FF, Korme T (1999) Left-lateral strike-slip tectonics and gravity induced individualization of wide continental blocks in the western Afar margin. Eclogae Geol Helv 92:149–158
Clark JD (1985) Découvertes effectuée dans le middle Awash en 1981. L'Anthropologie 88:129–130
CNR-CNRS (1971) Geological map of the northern Afar, scale 1:500,000. Geotechnip, La Celle St. Claude
CNR-CNRS (1975) Geological map of the central and southern Afar, scale 1:500,000. Geotechnip, La Celle St. Claude
Coltorti M, Dramis F, Ollier CD (2007) Planation surfaces in northern Ethiopia. Geomorphology 89:287–296
Cornwell DG, Maguire PKH, England RW, Stuart GW (2010) Imaging detailed crustal structure and magmatic intrusion across the Ethiopian Rift using a dense linear broad band array. Geochem Geophys Geosyst 11. doi:10.1029/2009GC002637
Corti G (2008) Control of rift obliquity on the evolution and segmentation of the main Ethiopian rift. Nature Geosci 1:258–262
Corti G (2009) Continental rift evolution: from rift initiation to incipient break-up in the main Ethiopian Rift, East Africa. Earth Sci Rev 96:1–53
Coulié E, Quideleur X, Gillot PY, Courtillot V, Lefevre JC, Chiesa S (2003) Comparative K–Ar and Ar/Ar dating of Ethiopian and Yemenite Oligocene volcanism: implications for timing and duration of the Ethiopian traps. Earth Planet Sci Lett 206:477–492
Courtillot V, Féraud G, Maluski H, Vendamme D, Moreau MG, Besse J (1988) Deccan flood basalts and the Cretaceous/Tertiary boundary. Nature 333:843–846
Currano ED, Jacobs BF, Pan AD, Tabor NJ (2011) Inferring ecological disturbance in the fossil record: a case study from the late Oligocene of Ethiopia. Palaeogeogr Palaeoclimatol Palaeoecol 309:242–252
Dainelli G (1943) Geologia dell'Africa Orientale. (3 vols. text, 1 vol. maps). Reale Accademia Italia, Roma, 1916 pp
Dainelli G, Marinelli O (1912) Risultati scientifici di un viaggio nella Colonia Eritrea. Pubblicazioni Regio Istituto di Studi Superiori Pratici e di Perfezionamento Firenze, 601 pp
Davidson A (1983) Reconnaissance geology and geochemistry of parts of Illubabor Kefa, Gemu Gofa and Sidamo, Ethiopia. The Omo River Project, Ministry of Mines and Energy, Ethiopian Institute of Geological Survey Bull, 289 pp
Davidson A, Rex CD (1980) Age of volcanism and rifting in southwestern Ethiopia. Nature 283:657–658
deHeinzelin J (ed) (1983) The Omo Group. Koninklijk Muesum voor Midde-Afrika, Tervuren Belgie Annalen 85:365 pp
Di Paola GM (1972) The Ethiopian Rift Valley (between 7° 00′ and 8° 40′ lat.). Bull Volcanol 36:517–560
Dow DB, Beyth M, Hailu T (1971) Paleozoic glacial rocks recently discovered in northern Ethiopia. Geol Mag 108:53–60
Ebinger CJ (2005) Continental breakup: the East African perspective. Astronomy and Geophysics 46:2.16–2.21
Ebinger CJ, Casey M (2001) Continental breakup in magmatic provinces: an Ethiopian example. Geology 29:527–530
Ebinger CJ, Sleep NH (1998) Cenozoic magmatism through east Africa resulting from impact of a single plume. Nature 395:788–791
Ebinger CJ, Yemane T, Woldegabriel G, Aronson JL, Walter RC (1993) Late Eocene-Recent volcanism and faulting in the southern main Ethiopian rift. J Geol Soc London 150:99–108
Ethiopian Institute of Geological Survey (1985) Geological map of the Dire Dawa Sheet, scale 1:250,000, NC 37-12 Ethiopian Government Addis Ababa
Furman T, Bryce J, Rooney T, Hanan B, Yurgu G, Hayalew D (2006) Heads and tails: 30 million years of the Afar plume. In Yurgu G, Ebinger CJ, Maguire PKH (eds) The Afar volcanic province within the East Africa rift system. Geological Society Special Publications No 259, pp 95–119
Gani NDS, Gani MR, Abdelsalam MG (2007) Blue Nile incision on the Ethiopian Plateau: pulsed plateau growth, Pliocene uplift, and hominin evolution. GSA Today 17:4–11
Gani NDS, Abdelsalam MG, Gera S, Gani MR (2009) Stratigraphy and structural evolution of the Blue Nile Basin, Northwestern Ethiopia Plateau. Geol J 44:30–56
Garland CR (1980) Geology of the Adigrat Area. Geological survey of Ethiopia, Addis Ababa, 51 pp
George R, Rogers N, Kelley S (1998) Earliest magmatism in Ethiopia: evidence for two mantle plumes in one continental flood basalt province. Geology 26:923–926
Gibson IL (1969) The structure and volcanic geology of an axial portion of the main Ethiopian rift. Tectonophysics 8:561–565
Gortani M (1973) La fauna mesocretacea degli Stati di Graua. In Missione Geologica dell'AGIP nella Dancalia Meridionale e sugli Altopiani Hararini (1936–1938). Acc Naz Lincei Roma 4:3–98
Goudie AS (2005) The drainage of Africa since the Cretaceous. Geomorphology 67:437–456
Gregnanin A, Piccirillo AM (1974) Considerazioni sulle serie vulcaniche e sulla struttura dell'Altopiano Etiopico centrale. Mem Mus Tridentino Sci Nat 20:79–100
Grove AT, Street FA, Goudie AS (1975) Former lake levels and climatic changes in the rift valley of southern Ethiopia. Geogr J 141:177–202
Hagos M, Koeberl C, Kabeto K, Koller F (2010) Geochemical characteristics of the alkaline basalts and the phonolite-trachyte

plugs of the Axum area, northern Ethiopia. Austrian J Earth Sci 103:153–170

Haile-Selassie Y, Deino A, Saylor B, Umer M, Latimer B (2007) Preliminary geology and paleontology of new hominid-bearing Pliocene localities in the central Afar region of Ethiopia. Anthropol Sci 115:215–222

Haile-Selassie Y, Suwa G, White TD (2009) *Ardipithecus Kadabba*: late miocene evidence from the Middle Awash, Ethiopia. In: Haile-Selassie Y, WoldeGabriel G (eds) The middle awash series. University of California, Berkeley, pp 159–236

Hayward NJ, Ebinger CJ (1996) Variations in the along-axis segmentation of the Afar Rift system. Tectonics 15:244–257

Hofmann C, Courtillot V, Féraud G, Rochette P, Yirgu G, Ketefo E, Pik R (1997) Timing of the Ethiopian flood basalt event and implications for plume birth and global change. Nature 389:838–841

Howell FC (1969) Remains of Hominidae from Plio/Pleistocene formations in the lower Omo basin, Ethiopia. Nature 223:1234–1239

Hunegnaw A, Sage L, Gonnard R (1998) Hydrocarbon potential of the intracratonic Ogaden Basin, SE Ethiopia. J Petrol Geol 21:401–425

Ismail EH, Abdelsalam MG (2012) Morpho-tectonic analysis of the Tekeze River and the Blue Nile drainage systems on the Northwestern Plateau, Ethiopia. J Afr Earth Sci 69:34–47

Jepson DH, Athearn MJ (1964) Land and water resources of the Blue Nile Basin. Appendix II Geology US Dept Interior, Addis Ababa, p 221

Johanson DC, Taieb M (1976) Plio-Pleistocene hominid discoveries in Hadar, Ethiopia. Nature 260:293–297

Johanson DC, Splingaer M, Boaz NT (1976) Paleontological excavation in the Shungura Formation, lower Omo basin, 1969–1973. In: Coppens Y et al. (eds) Earliest man and environments in the Lake Rudolph basin: stratigraphy, paleoecology, and evolution. University Chicago Press, Chicago, pp 402–420

Johnston C (1844) Travels in southern Abyssinia through the country of Adal to the Kingdom of Shoa. Madden, London, 492 pp

Juch D (1975) Geology of the south eastern escarpment of Ethiopia between 39° and 42° long, East. In: Pilger A, Roesler EA (eds) Afar depression of Ethiopia. Schweizerbart, Stuttgart, pp 310–316

Juch D (1980) Tectonics of the southeastern escarpment of Ethiopia. Atti Convegni Acc Lincei Roma 47:407–418

Justin-Visentin E, Zanettin B (1974) Dike swarms, volcanism and tectonics of the western Afar margin along the Kombolcha-Eloa traverse (Ethiopia). Bull Volcanol 38:187–205

Kalb JE (1993) Refined stratigraphy of the hominid-bearing Awash Group, middle Awash valley, Afar depression, Ethiopia. Newsl Stratigr 29:21–62

Kalb JE, Oswald EB, Tebedge S, Mebrate A, Tola E, Peak D (1982) Geology and stratigraphy of Neogene deposits, Middle Awash valley, Ethiopia. Nature 298:17–25

Kappelmann J, Rasmussen DT, Sanders WJ, Faseha M, Bown T, Copeland P, Krabaugh J, Fleage J, Glanz M, Gordon A, Jacobs B, Maga M, Muldoon K, Pan A, Pyne L, Richmond B, Ryan T, Selffert ER, Sen S, Todd L, Wlemann MC, Winkler A (2003) Oligocene mammals from Ethiopia and faunal exchange between Afro-Arabia and Eurasia. Nature 426:549–552

Katho S, Nagaoka S, Woldegabriel G, Renne P, Snow MG, Beyene Y, Suwa G (2000) Chronostratigraphy and correlation on the Plio-Pleistocene tephra layers of the Konso formation, southern main Ethiopian Rift, Ethiopia. Quat Sci Rev 19:1305–1317

Kazmin V (1973) Geological map of Ethiopia, scale 1:2,000,000. Geological Survey of Ethiopia, Addis Ababa

Kazmin V (1980) Transform faults in the east African Rift system. Atti Convegni Acc Lincei Roma 47:65–73

Kazmin V, Shiferaw A, Balcha T (1978) The Ethiopian basement, stratigraphy and possible manner of evolution. Geol Rund 67: 531–546

Kazmin V, Seife MB, Nicoletti N, Petrucciani C (1980) Evolution of the northern part of the Ethiopian rift. Atti Convegni Acc Lincei Roma 47:275–292

Kidane T, Otofuji YI, Brown FH, Takemoto K, Eshete G (2007) Two normal paleomagnetic polarity intervals in the lower Matuyama Chron recorded in the Shungura Formation (Omo Valley, Southwest Ethiopia). Earth Planet Sci Lett 262:240–246

Kieffer B, Srndt N, Lapierre H, Bastien F, Bosh D, Pecher A, Yirgu G, Ayalew D, Weis D, Jerram DA, Keller F, Meugniot C (2004) Flood and shield basalts from Ethiopia: magmas from the African superswell. J Petrol 45:793–834

Kröner A (1985) Ophiolites and the evolution of tectonic boundaries in the late proterozoic Arabian–Nubian shields of northeast Africa and Arabia. Precambr Res 27:277–300

Kumpulainen R, Uchman A, Woldehaimanot B, Kreuser T, Ghirmay S (2006) Trace fossil evidence from the Adigrat sandstone for an Ordovician glaciation in Eritrea, NE Africa. J Afr Earth Sci 45:408–420

Kunz K, Kreuzer H, Muller P (1975) Potassium-argon age determinations of the Trap Basalts of the SE part of the Afar rift. In: Pilger A, Roesler EA (eds) Afar depression of Ethiopia. Schweizerbart, Stuttgart, pp 370–374

Lalou C, von Nguyen H, Faure H, Moreira L (1970) Datation par la method Uranium-Thorium des hautes niveaux de coraux de la dépression de l'Afar (Ethiopie). Rev Géo Phys Géol Dyn 12:3–8

Larson PJ Jr (1977) Matabaietu, an Olduwan site from the Afar, Ethiopia. Nyame Akuma 11:6–10

Le Bas MJ, Mohr PA (1968) Feldspathoid rocks from the Cainozoic volcanic province of Ethiopia. Geol Rundsch 58:273–280

Le Pichon X, Francheteau J (1978) A plate-tectonic analysis of the Red Sea-Gulf of Aden area. Tectonoph 46:369–406

Le Turdu C, Tiercelin JJ, Gibert E, Travi Y, Lezzar K, Richert J, Massault M, Gasse F, Bonnefille R, Decobert M, Gensous B, Jeudy V, Tamrat E, Mohamed MU, Martens K, Balemwal A, Chernet T, Williamson D, Taieb M (1999) The Ziway-Shala lake basin system, main Ethiopian rift: influence of volcanism, tectonics, and climatic forcing on human formation and sedimentation. Palaeogeogr Palaeoclimatol Palaeoecol 150:135–177

Leakey REF (1974) Further evidence of lower Pleistocene hominids from East Rudolf, North Kenya. Nature 248:653–656

Lewin R (1983) Fossil lucy grows younger, again. Science 219:43–44

Maguire PKH, Keller GR, Klemperer SL, Mackenzie GD, Keranen K, Harder S, O'Reilly B, Thibo H, Asfaw L, Khan MA, Amha M (2006) Crustal structure of the Northern Main Ethiopian Rift from the EAGLE controlled source survey: a snapshot of incipient lithospheric break-up. In: Yirgu G, Ebinger CJ, Maguire PKH (eds) The Afar volcanic province within the East African rift system. Geol Soc Spec Publ 259:269–291

Makris MJ, Rhim R (1991) Shear-controlled evolution of the Red Sea: pull apart model. Tectonophysics 198:441–466

Manighetti I, Tapponnier P, Gillot PY, Jacques E, Courtillot V, Armijo R, Rurgg JC, King G (1998) Propagation of rifting along the Arabia–Somalia plate boundary into Afar. J Geophys Res 103:4947–4974

Martini IP, Brookfield EM, Steven S (2001) Principles of glacial geomorphology and geology. Prentice-Hall, New Jersey, 380 pp

McDougall I, Brown FH (2008) Geochronology of the pre-KBS Tuff sequence, Omo Group, Turkana Basin. J Geol Soc London 165:549–562

McKenzie DP, Davies D, Molnar P (1970) Plate tectonics of the Red Sea and East Africa. Nature 226:243–248

Meert JG, Lieberman BS (2008) The Neoproteozoic assembly of Gondwana and its relationship to Ediacaran–Cambrian radiation. Gondwana Res 14:5–21

Merla G (1963) Missione geologica nell'Etiopia meridionale del Consiglio Nazionale delle Ricerche 1959–1960. Notizie geomorfologiche e geologiche. Giorn Geol 31:1–56

Merla G, Minucci E (1938) Missione geologica nel Tigrai. Volume primo: La serie dei terreni. Reale Accademia, Italia Roma, 363 pp
Merla G, Abbate E, Canuti P, Sagri M, Tacconi P (1973) Geological map of Ethiopia and Somalia, scale 1:2,000,000. Consiglio Nazionale delle Ricerche, Roma
Merla G, Abbate E, Azzaroli A, Bruni P, Canuti P, Fazzuoli M, Sagri M, Tacconi P (1979) A geological map of the Ethiopia and Somalia and comment with a map of major landforms (scale 1:2,000,000). Consiglio Nazionale delle Ricerche, Roma, 95 pp
Meshesha D, Shinjo R (2008) Rethinking geochemical feature of the Afar and Kenya mantle plumes and geodynamic implications. J Geoph Res 113. doi:10.1029/2007JB005549
Mohr P (1962) The geology of Ethiopia. University College of Addis Ababa Press, Ethiopia, 268 pp
Mohr P (1963) Geological map of Horn of Africa, scale 1:2,000,000. Philip and Tacey, London
Mohr P (1967) The Ethiopian rift system. Bull Geoph Obs Addis Ababa 11:1–65
Mohr P (1968) Transcurrent faulting in the Ethiopian rift system. Nature 218:938–941
Mohr P (1983) Ethiopian flood basalt province. Nature 303:577–584
Mohr P (2009) Africa beckoning. Explorers of eastern Africa, its rift valleys and geology, in the 19th and early 20th centuries. Millbrook Nova press, Tóin An Ghárrain, Corr an Dola, Co na Gaillime, 192 pp
Mohr P, Wood CA (1976) Volcano spacings and lithospheric attenuation in the Eastern Rift of Africa. Earth Planet Sci Lett 33:126–144
Mohr P, Zanettin B (1988) The Ethiopian Flood Basalt Province. In: Macdougall JD (ed) Continental flood basalts. Kluwer Academic Publisher, The Netherlands, pp 63–110
Morbidelli L, Nicoletti C, Petrucciani C, Piccirillo EM (1975) Ethiopian south-eastern plateau and related escarpment: K/Ar ages of the main volcanic events (Main Ethiopian Rift from 8° 10' to 9° 00' lat. N). In: Pilger A, Roesler EA (eds) Afar depression of Ethiopia. Schweizerbart, Stuttgart, pp 362–369
Morgan WJ (1971) Convection plumes in the lower mantle. Nature 230:42–43
Morton WH, Black R (1975) Crustal attenuation in Afar. In: Pilger A, Roesler EA (eds) Afar depression of Ethiopia. Schweizerbart, Stuttgart, pp 55–65
Moucha R, Forte AM (2011) Changes in African topography driven by mantle convection. Nature Geosci 4:707–712
Munzinger W (1864) Ostafrikanische studien. Schaffhausen, p 584
Natali C, Beccaluva L, Bianchini G, Siena F (2013) The Axum-Adwa basalt-trachyte complex: a late magmatic activity at the periphery of the Afar plume. Contr Min Petrol. doi:10.1007/s00410-013-0879-0
Overstreet WC, Stoeser DB, Overstreet EF, Goudarzi GH (1977) Tertiary laterite of the As Sarat mountains, Asir Province, Kingdom of Saudi Arabia. Min Res Bull Saudi Arabia Direct Gen Min Res 21:1–30
Peccerillo A, Donati C, Santo AP, Orlando A, Yirgu G, Ayalew D (2007) Petrogenesis of silicic peralkaline rocks in the Ethiopian rift: geochemical evidence and volcanological implications. J Afr Earth Sci 48:161–173
Pickford M (1990) Uplift of the roof of Africa and its bearing on the evolution of mankind. J Hum Evol 5:1–20
Pik RD, Coulon C, Yirgu G, Hofmann C, Ayalew D (1998) The Northwestern Ethiopian plateau flood basalts: classification and spatial distribution of magma types. J Volc Geoth Res 81:91–111
Pik R, Marty B, Carignan J, Lavé J (2003) Stability of the Upper Nile drainage network (Ethiopia) deduced from (U-Th)/He thermochronometry: implications for uplift and erosion of the Afar plume dome. Earth Plan Sci Lett 215:73–88
Prior GT (1900) On Aegirite and Riebeckite Anorthoclase rocks related to the "Grorudite-Tinguaite" series, from the neighborhood of Aodowa and Axuma, Abyssinia. Min Mag 12:255–273
Quade J, Wynn JG (Eds) (2008) The geology of early humans in the Horn of Africa. Geol Soc Am Spec Publ 446:234 pp
Ritsema J, van Heijst HJ, Woodhouse JH (1999) Complex shear wave velocity structure imaged beneath Africa and Iceland. Science 286:1925–1928
Rochette P, Tamrat E, Féraud G, Pik R, Courtillot V, Ketefo E, Coulon C, Hoffmann C, Vandamme D, Yirgu G (1998) Magnetostratigraphy and timing of the Oligocene Ethiopian traps. Earth Planet Sci Lett 164:497–510
Rogers NW (2006) Basaltic magmatism and the geodynamics of the East Africa rift system. In: Yrgu G, Ebinger CJ, Maguire PKH (eds) The Afar volcanic province within the East Africa rift system. Geological Society London Special Publication, London, 259, pp 77–93
Rüppell WE (1834) Skizze der geologischen Formation Abyssiniens. Abhandlungen Museum Senkenbergianum 1:286–288
Russo A, Getaneh A, Atnafu B (1994) Sedimentary evolution of the Abbai River (Blue Nile) Basin, Ethiopia. N J B Geol Palaeo Mh 5:291–308
Sacchi R, Alene M, Barbieri M, Conti A (2007) On the Paleozoic Tillite of the Adigrat Group (Tigrai, Ethiopia). Periodico Mineralogia 76:241–251
Said R (1993) The River Nile geology, Hydrology and Utilization. Pergamon Press, Oxford, p 320
Schilling JG (1973) Afar mantle plume: rare earth evidence. Nature 242:2–5
Sepulchre P, Ramstein G, Fluteau F, Schuster M, Tiercelin JJ, Brunet M (2006) Tectonic uplift and eastern Africa aridification. Science 313:1419–1423
Sichler B (1980) La biellette danakile. Bull Soc Geol Fr 6:925–933
Sickenberg O, Schoenfeld M (1975) The chorora formation—lower pliocene limnical sediments in the southern Afar (Ethiopia). In: Pilger A, Roesler EA (eds) Afar depression of Ethiopia. Schweizerbart, Stuttgart, pp 277–284
Stankiewicz J, de Wit MJ (2006) A proposed drainage evolution model for the Central Africa—did the Congo flow east? J Afr Earth Sci 44:75–84
Stefanini G (1933) Carta Geologica dell Eritrea, della Somalia e dell'Etiopia alla scala di:2.000.000. Istituto Geografico Militare Firenze
Stern RJ (1994) Neoproterozoic (900–550 Ma) arc assembly and continental collision in the East African Orogen. Ann Rev Earth Planet Sci 22:319–351
Street FA (1979) Late quaternary lakes in the Ziway-Shala basin, Southern Ethiopia. Ph.D. thesis, University of Cambridge, Cambridge, 457 pp
Taieb M (1974) Evolution quaternaire du basin de l'Awash (rift éthiopienne et Afar). Doctoral Univ Paris, 442 pp
Tefera M, Chernet T, Haro W (1996) Geological map of Ethiopia, scale 1:2,000,000. Ethiopian Mapping Authority, Ethiopia
Tiercelin JJ (1986) The pliocene hadar formation; Afar depression of Ethiopia. In: Frostick LE, Renaut RW (eds) Sedimentation in the African Rifts. Geological Society Special Publication No 25, pp 221–240
Tiercelin JJ, Taieb M, Faure H (1980) Continental sedimentary basins and volcano-tectonic evolution of the Afar Rift. Atti Convegni Acc Lincei Roma 47:491–504
Ukstins IA, Renne PR, Wolfenden E, Baker J, Ayalew D, Menzies M (2002) Matching conjugate volcanic rifted margins: 40/39Ar chrono-stratigraphy of pre-and syn rift bimodal flood volcanism in Ethiopia and Yemen. Earth Planet Sci Lett 198:289–306

Unger F (1866) Notiz über fossile Hölzer aus Abyssinien. Sitzungenesberichte der Kaiserlichen Akademie der Wissenshaften Math. Naturwissenshaft Klasse, Wien 54:289–296

Vinassa de Regny P (1931) La geologia delle Alpi Dancale. Boll Soc Geol It 50:1–24

Walker A, Leakey REF (1978) The hominids of East Turkana. Sci Am 239:54–66

Walter RC (1994) Age of Lucy and the First Family: Single crystal $^{40}Ar/^{39}Ar$ dating of the Denen Dora and lower Kada Hadar members of the Hadar Formation, Ethiopia. Geology 22:6–10

White RS, McKenzie DP (1989) Magmatism at rift zones: the generation of a continental margins and flood basalts. J Geoph Res 94:7685–7729

White TD, Johanson DC (1982) Pliocene hominid mandibles from the Hadar Formation, Ethiopia; 1974–1977 collections. Am J Phys Anthropol 57:501–544

White TD, Asfaw B, Beyene Y, Haile Selassie Y, Lovejoy CO, Suwa G, WoldeGabriel G (2009) *Ardipithecus ramidus* and the paleobiology of early hominids. Science 326:75–86

Wichura H, Bousquet R, Oberhäusli R, Strecker MR, Trauth MH (2010) Evidence for middle Miocene uplift of the East African Plateau. Geology 38:543–546

WoldeGabriel G, Aronson JL, Walter RC (1990) Geology, geochronology, and rift basin development in the central sector of the Main Ethiopia Rift. Geol Soc Am Bull 102:439–458

Wolela A (2007) Fossil fuel energy resources of Ethiopia: coal deposits. Int J Coal Geol 72:293–314

Wolela A (2008) Sedimentation of the Triassic–Jurassic Adigrat sandstone formation, BlueNile (Abbai) basin, Ethiopia. J Afr Earth Sci 52:30–42

Wolfenden E, Ebinger C, Yirgu G, Deino A, Ayalew D (2004) Evolution of the northern Main Ethiopian rift: birth of a triple junction. Earth Plan Sci Lett 224:213–228

Yemane K, Taieb M, Faure H (1987) Limnologic studies on an intertrappean continental deposit from the northern Ethiopian Plateau (37° 03′E, 12° 25′N). J Afr Earth Sci 6:91–101

Zanettin B, Justin-Visentin E (1973) Serie di vulcaniti etiopiche: 1—La serie dell'altopiano etiopico centro-orientale. Boll Soc Geol It 92:313–327

Zanettin B, Justin-Visentin E (1974) The volcanic succession in central Ethiopia. 2—The volcanics of western Afar and Ethiopian rift margins. Mem Ist Min Univ Padova 31:1–19

Zanettin B, Justin-Visentin E (1975) Tectonical and volcanological evolution of the western Afar margin (Ethiopia), In: Pilger A, Roesler EA (eds) Afar depression of Ethiopia. Schweizerbart, Stuttgart, pp 300–309

Zanettin B, Gregnanin A, Justin-Visentin E, Nicoletti M, Petrucciani C, Piccirillo EM, Tolomeo L (1974) Migration of the Oligocene-Miocene ignimbritic volcanism in the Central Ethiopan Plateau. NGB Geol Palaeo Mh H 9:567–574

Zanettin B, Justin Visentin E, Nicoletti M, Piccirillo EM (1980) Correlations among Ethiopian volcanic formations with special references to the chronological and stratigraphical problems of the "Trap Series". Atti Convegni Acc Lincei Roma 47:231–252

The Climate of Ethiopia

3

Massimiliano Fazzini, Carlo Bisci, and Paolo Billi

Abstract
Several papers have been published on different issues regarding the climate of Ethiopia or of some specific region. This presentation attempts to revise the knowledge of the climate of Ethiopia by means of updated, longer time series and including a larger number of meteo-stations than previous studies. Basic climatic parameters such as temperature, rainfall, relative humidity, wind, evapotranspiration, and aridity are considered and their spatial distribution is analyzed. The main results of such elaborations have been regionalized to obtain climatic maps by means of geostatistical interpolation, also taking into account topogeographic variables. These parameters were also used to update the Köppen classification of the Ethiopian territory. Climate change is a very important issue with worrying repercussions on agriculture and hence the social and economic development of the country. Trends of temperature and annual, spring, and summer rains were interpolated for the last 3–5 decades. Temperature shows a markedly increasing trend especially as regards the minimum values, whereas annual rainfalls tend to decrease with the spring, 'small rains' decreasing at faster rate.

Keywords
Climate • Ethiopia • Temperature • Precipitation • Aridity index

3.1 Introduction

Ethiopia is among the largest countries of Africa and it is characterized by a wide variety of landscapes, with marked contrasts in relief and altitudes ranging from about 155 m below sea level of Assale Lake, in the Danakil depression, to about 4,533 m a.s.l. at Ras Dejen (EMA 1988). For these reasons and given its geographic position close to the equator and the Indian Ocean, the country is subjected to large spatial variations in temperature and precipitation (see Sect. 3.3). The climate of Ethiopia is therefore mainly controlled by the seasonal migration of the *Intertropical Convergence Zone* (ITCZ) and associated atmospheric circulations as well as by the complex topography of the country. Landscapes with contrasting characteristics in terms of physiography and elevation, such as the highlands and the lowlands, experience a variety of climates from desert climate to that typical of equatorial mountains.

Climate, in turn, has many obvious implications on landforms and morphodynamic evolution of natural landscapes as much as on the living conditions of local people, in a country whose economy is heavily dependent on rain-fed agriculture (Hadgu et al. 2013).

The relevance of climate and its variability on natural resources of Ethiopia and their potential exploitation has been recognized since long by the first Western explorers visiting the country. The establishment of a few weather stations with regular observations since the beginning of the twentieth

M. Fazzini · P. Billi (✉)
Physics and Earth Sciences,
University of Ferrara, Ferrara, Italy
e-mail: bli@unife.it

M. Fazzini
e-mail: fzzmsm@unife.it

C. Bisci
School of Science and Technology,
University of Camerino, Camerino, Italy

P. Billi (ed.), *Landscapes and Landforms of Ethiopia*, World Geomorphological Landscapes,
DOI 10.1007/978-94-017-8026-1_3, © Springer Science+Business Media Dordrecht 2015

century led Ethiopia to assume a prominent position in the Horn of Africa as far as the availability of reference meteorological data is concerned, also within the operational framework of international climate organizations.

Several studies on the climate of Ethiopia have been published throughout the past two centuries (see the next sections), but only very few of them benefited from the inherent, ongoing expansion of the time series in refining the definition of the different climate types and their spatial variability. Instead, they rather focused on specific topics, with climate change and its implications becoming the most investigated theme in the last decade. The aim of this chapter is therefore to update the knowledge of the main characteristics of the climate of Ethiopia and to analyze the occurrence of climate change trends on the base of time series longer than those used in previous studies.

3.2 Previous Studies

3.2.1 Historical Data

In sub-Saharan Africa, century long time series of meteo-climatic data are scarce and sparse over large portions of territory (Houghton et al. 2001; Conway et al. 2004). Ethiopia is a large country, with very few well-documented long series of climate observations. The longest and uninterrupted climate record for the Horn of Africa is the rainfall and thermometric (but also wind, pressure, relative humidity, and sunshine) data record from Asmara which began in 1890 and from Addis Ababa since 1898. These data derived from four intermittent sources, covering different time intervals: Russian (1898–1903), Italian (1903–1904), British (1901–1904), and French (1898).

Some preliminary data on temperature, rainfall, relative humidity, and wind for Addis Ababa and Addis Alem, plus sparse climatic information for a few other sites in Ethiopia, were reported by Eredia and De Castro (1914), but the quality of these data is questionable for both the short duration and the methodology of measurements.

The study by Fantoli (1940) considered the meteo-climatic data of three thermometric stations in Ethiopia—Addis Ababa (1902–1930), Gambela (1914–1932), and Harar (1902–1918)—and other six thermometric stations, including Asmara (1890–1930), located in the bordering territories, and 18 rain gauges, some of which were located beyond modern Ethiopia. For a few of these meteorological stations, additional meteo-climatic parameters such as atmospheric pressure, relative humidity, wind speed and direction, and insolation were reported.

These data series were used by Hurst and Black (1937), Fantoli (1940, 1965), WWR (1959), and Conway et al. (2004). Though some stations have incomplete records (Lovett and Wood 1976), nonetheless, they are of great interest for a preliminary description of the general climate of Ethiopia and surrounding regions at the nineteenth–twentieth-century transition. Given their spatial distribution, these data can be considered suitable only for a climatic point characterization and as a baseline, though some stations may have changed location and instrumentation, to study climate variations, if any, but do not allow to define the climate of Ethiopia with sufficient detail.

3.2.2 More Recent Data and Studies

In Ethiopia, most of the longest rainfall and temperature records began in the 1950s and 1960s. Many other records started during the 1980s, following the impulse from the Ethiopian Government to expand the network, partly in response to the mid-1980s drought. In 1951, a small meteorological unit was formally established within the Civil Aviation Department of Ethiopia, to provide meteorological information solely for flying purposes. As the need for meteo-climatic data increased, a Meteorological Department was established in 1964, under the Civil Aviation Authority. Later, this department became the modern National Meteorological Agency (NMA), after designation as an autonomous organization on December 31, 1980 (ENRAEMED 2003).

Since 1984, FAO (1984) has been publishing mean monthly agroclimatic data, including decadal rainfall data for Ethiopia. FAO obtained these data from the National Meteorological Agency and Chernet (1982) used them for a concise description of the Main Ethiopian Rift Valley climate, whereas Gamachu (1988) based on them his study on climatic elements of mountainous regions of Ethiopia.

In 1994, NOAA–NCDC put together a huge database at global scale, the 'Global Historical Climate Network (GHCN),' which includes daily data of several meteo-stations located in Ethiopia, the source of which, however, is always the NMA.

3.2.3 The National Atlas

In 1988, the Ethiopian Mapping Authority published the first edition of the National Atlas of Ethiopia. The Atlas is likely the first publication with a comprehensive description of the climate of Ethiopia. In fact, it includes several maps depicting the spatial distribution of a few climatic parameters such as rainfall, rainfall patterns, hail, frost, temperature, sunshine hours, climate classification, wind velocity and direction and the location of the meteorological stations as well. The data to prepare these climate sections of the Atlas were provided by the Ethiopian Meteorological Agency as far as 1977, and the data analysis and summary are partly based on the studies of Gamachu (1977). Though the spatial distributions of climatic parameters for the three decades before the mid-1970s are reported, this publication still represents a reliable reference and a milestone in the history of climatological studies in Ethiopia. This paper draws on the inspiring climatic maps and chapters of the National Atlas of Ethiopia and attempts to update the analysis of climatic parameters and climate classification considering also the data of the following three decades as far as 2010.

3.2.4 Modern Studies

In the scientific literature, there are a number of papers on precipitation in Ethiopia and related problems, such as extreme rains, droughts, trend variations, and evapotranspiration, whereas the spatial and temporal distribution of temperatures have not been widely and satisfactorily analyzed. The majority of modern studies deal with limited areas and, commonly, are constrained by the availability of homogeneous time series of good quality. For these reasons, modern investigations on the general characteristics of the climate of Ethiopia are missing, with the exception of the National Meteorological Agency publications (e.g., Tadege 2001, 2007) that provide a reliable description of the climate across the whole country, though limited to the end of the second millennium and early beginning of the third. In the last decade, the NMA has also expanded the number of recording stations to a remarkable quantity and presently the entire territory of Ethiopia is densely, though heterogeneously, covered (Fig. 3.1).

In the 2007 NMA report, temperature data are considered, and for the 1951–2006 interval, increasing rates of 0.13 °C/

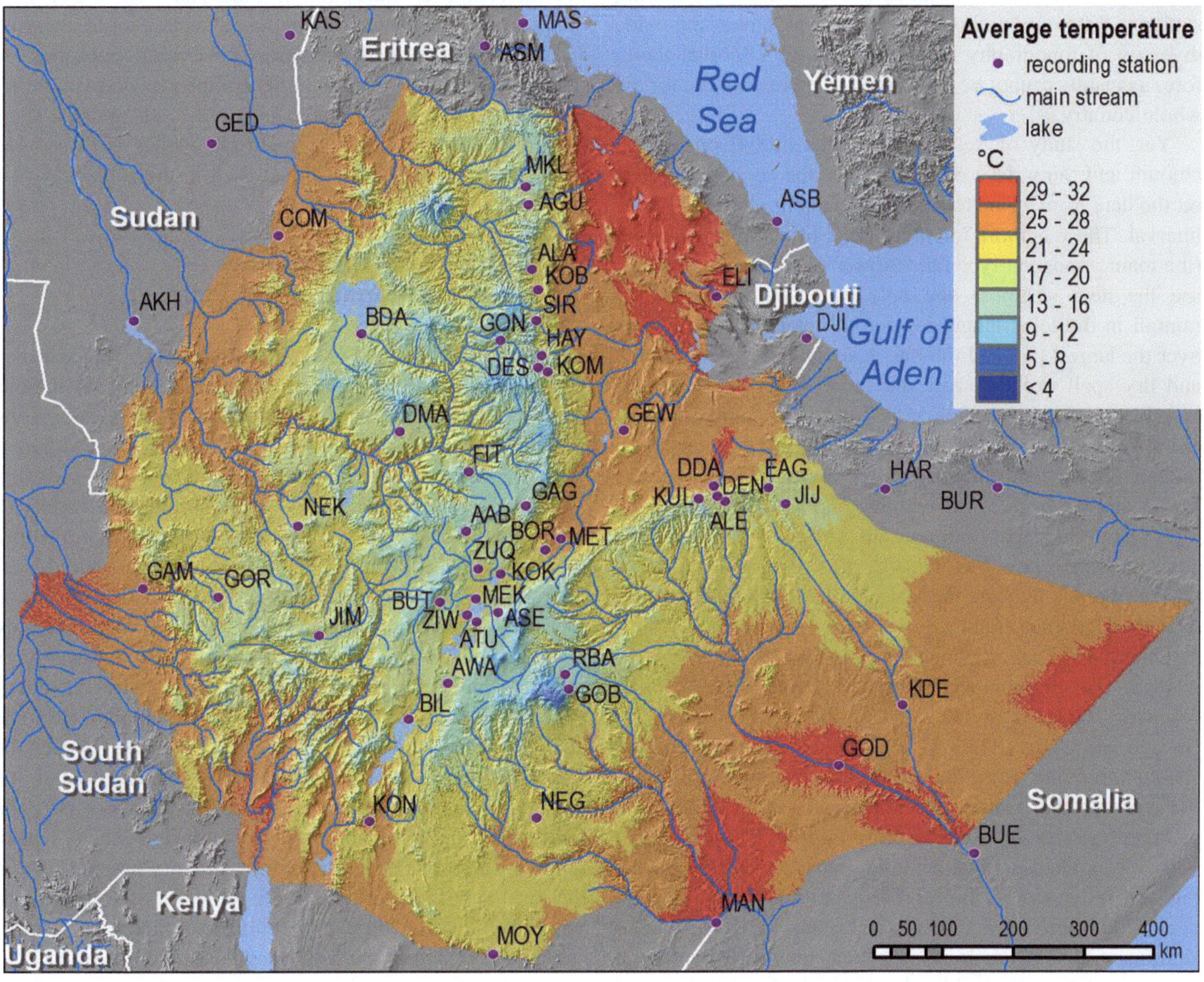

Fig. 3.1 Mean annual temperatures

decade and 0.37 °C/decade for mean annual minimum and maximum temperature, respectively, are presented. For the 1960–2006 interval, the report by McSweeney et al. (2010) indicates a mean annual temperature increase of 0.28 °C per decade and a significant increasing trend in the frequency of hot days and a decrease of cold days. These authors also found that no statistically significant trend is recognizable for mean precipitation in any season.

In a more recent study on the upper Blue Nile basin, based on statistical and geostatistical techniques and gridded data, reconstructed from NMA weather stations and meteorological satellite records, Mengistu et al. (2013) found an insignificant change in temperature across the 1981–2010 interval.

Mekasha et al. (2013) have analyzed temperature and rainfall extremes for 11 stations and found a general tendencies of increasing warm and decreasing cold extremes, whereas trends in precipitation extremes were much more variable and increasingly inconsistent among neighboring stations. This study, however, does not cover the wide diversity of topography and landscapes of Ethiopia; therefore, its results cannot be considered as representative for the whole country.

Yet, the study of Selashi and Zanke (2004) on rainfall amount and rainy days changes across the country is based on the data from 11 meteo-stations spanning the 1965–2002 interval. These authors found no trend in the annual, *kiremt* (the main, monsoon-type rainy season in the local language, see the next sections) and *belg* (the small, spring time rainfall in the local language) rainfall totals and rainy days over the largest part of Ethiopia. A study on extreme rainfall and dry spell events across the same 1965–2002 interval, based upon the same time series, has been carried out by Selashi and Camberlin (2006) who found contrasting, though weak, trends in extreme precipitation for both the *kiremt* and the *belg* in different geographic areas, showing a high variability for both rainfall seasons. These results contrast with a previous work by Easterling et al. (2000), whose findings indicate that extreme intensity is decreasing over the Ethiopian highlands.

Other papers report climatic information on specific areas or regions of Ethiopia (e.g., Conway 2000a, b; Bewket and Conway 2007; Taye and Zewdu 2012). The paper by Bewket and Conway (2007) reports about the temporal and spatial variability of rainfall for a relatively small portion of the country, the Amhara Region, and found no consistent emergent pattern or trend in daily rainfall. These authors, however, pointed out that the time span selected may have a relevant influence on the results of trend analysis. Cheung et al. (2008) arrived at similar conclusions, verifying that no significant changes or trend in annual rainfall is evident at the national or watershed level in Ethiopia. Finally, Jury and Funk (2013), using gridded data from different sources, analyzed the temperature and rainfall trends for the 1948–2006 interval and performed a simulation projecting these trends to 2050 under the assumption of a gradual doubling of greenhouses gases. For the time series bounded to 2006, these authors find that air temperatures increase at the rate of about 0.03 °C per year across most of Ethiopia, with the exceptions of the lowlands and the northern Rift Valley, whereas rainfall trends indicate a weak increase in the arid lowlands of southeastern Ethiopia.

Droughts have been studied by Shanko and Camberlin (1998)—in connection with the development of tropical cyclones—and by Viste et al. (2013). These latter authors, analyzing rainfall data for the 1971–2011 interval, found that the drought patterns are largely influenced by the variation in the seasonal precipitation cycle among different areas of Ethiopia.

Scarce information is available on other climatic parameters such as velocity and direction of winds, air humidity, barometric pressure, and evapotranspiration (Enku and Melesse 2013).

Though the climate of Ethiopia received a lot of attention from scientists and many papers have been published on different climatic topics, given its complexity and variability and the availability of longer time series, an updated summary and description of the main parameters is considered here of great help for those interested in the subject.

3.3 Summary Description of the Climate of Ethiopia

The climate of Ethiopia is mainly controlled by the seasonal migration of the ITCZ and associated atmospheric circulations (Beltrando and Camberlin 1993), the complex physiography and the marked contrast in elevation among large different areas of the country. The transition between lowlands and highlands is commonly very sharp, resulting in a variety of climates, from very arid to very humid typical of equatorial mountains, with further differentiation at local scale. Moreover, precipitation varies with latitude, decreasing from south to north, whereas the meteorological framework is deeply affected also by elevation and physiography, especially as regards temperature distribution and anemometric characteristics. Finally, the proximity of the Asiatic continent has to be considered as well. In winter, in fact, the contrast between the thermal anticyclone of western Asia and Egypt and the equatorial low pressures determines the presence of trade winds blowing from northeast to southwest. These winds, relatively cool but rather dry, control the dry period (*bega* in the local language). In spring, the influence of southwestern winds, coming from the Congo basin, determines the season of 'little rains' (*belg* in the local language) that can bring relatively abundant precipitation in the southern part of the country. In summer, the Guinean monsoon, consisting of equatorial warm and humid winds, results in bountiful rains (*kiremt* in the local language) which are also substantially influenced by the orographic diversity mentioned above.

Such a complex meteorological framework is reflected by the distribution of annual precipitation. In the Danakil depression, it is constantly less than 250 mm but can be as low as 50 mm. By contrast, on the highlands, 2,000 mm can be locally exceeded. Similar values of annual precipitation, however, are recorded also in the southwestern lowlands, likely due to a larger contribution of the spring rains.

Altitude patently determines marked annual thermal gradients with sharp transitions from zones with a desert climate—among the hottest of the planet—to tropical, high mountain climates with minimum temperatures below zero in each month and modest annual thermal amplitude.

3.4 Data Sources

As briefly introduced in the previous section, the high variability of physiography and orography of Ethiopia is reflected in the complex spatial distribution of precipitation and temperature as briefly introduced in the previous section. Having in mind such a complexity, large sets of data and long time series for the main climatic parameters are required to produce an updated description of the climate of Ethiopia and, possibly, of its variations through time. For this purpose, a large number of monthly temperature and, whenever possible, daily rain gauge data were collected from different sources including the Ethiopian National Meteorological Agency (NMA), which is the main provider of field data measurements, and other international organizations such as the Global Historical Climatology Network (GHCN) of the National Oceanic and Atmospheric Administration (NOAA) in the USA and the United Nations Food and Agricultural Organization (FAO), the data sets of which are substantially based on NMA data. The NMA produces daily rainfall data for all the many stations that are presently operative; in 22 of them also hourly precipitation is measured, whereas FAO reports rainfall data at decadal (10 days) intervals.

The temperature data of the NMA, and those of the GHCN as well, are reported as daily data for the whole operational time interval of first level meteo-stations and as monthly data for lower level gauges, whereas FAO provides only decadal data.

3.5 Data Validation and Processing

Unfortunately, the above data sets are commonly affected by long and reiterated gaps and may include entire years of measurements missing. This posed several problems in the validation of data and in the selections of uniform time series intervals and of the meteo-stations to be considered in this study.

The notable physiographic variability of the country in many cases did not allow to fill the gaps by means of ordinary correlation analysis with other stations in the vicinity, resulting in insignificant correlation coefficients (Hadgu et al. 2013). Such constraint is particularly relevant for time series and trend analysis, especially for rainfall. Notwithstanding such limitations and nonuniformity, some indications on long-term trends can be given anyway.

Jury and Funk (2013) report about spatial variability of rainfall on the basis of gridded data and interpolation. Given the low density of meteorological stations and the rugged topography of Ethiopia, this procedure does not seem to provide a realistic pattern of rainfall distribution. Additionally, the data set used by these authors includes a five stations correlation to fill data gaps, and a correlation as low as 0.2 has been accepted as significant. In the worst case, the mean values of the time series are included. This procedure seems, at least, questionable; in fact, it does not avert the data limitations and introduces arbitrary conditions that are not necessarily making the data analysis stronger or more reliable than the use of raw data.

Spatial distribution of rainfall has been investigated also by Viste et al. (2013) by using the raw data of the NMA, but their study focused mainly on monthly precipitation. Other authors such as Seleshi and Cumberlin (2006) investigated the spatial and seasonal (*bega*, *belg* and *kiremt*) rainfall extremes, but their paper is focused on 5 days consecutive rains rather than daily rainfall.

Notwithstanding the awareness of data limitations, the approach of this paper attempts to complement the analyses of seasonal and spatial variability reported in previous papers (e.g., Segele and Lamb 2005; Seleshi and Cumberlin 2006; Korecha and Barnston 2007; Viste et al. 2013).

As an alternative to the manipulation of data beyond any reasonable confidence and notwithstanding the existing gaps and limitation and the need for the time series as longest as possible, in order to attain a certain degree of coherence with the actual physical phenomena, the following procedure was used. Statistical tests were carried out in order to ascertain the quality of data and the uniformity of the temperature time series. The standard normal homogeneity test (Alexandersson 1986) and the Kolmogorov–Smirnov test were used to determine the homogeneity of records, whereas the Mann-Kendall test was used to verify the statistical significance of trends at a 0.05 level of significance.

Both the quality tests indicate that rainfall time series are generally homogeneous, whereas only four out of 18 temperature time series proved inhomogeneous. Furthermore, the variability of annual rainfall and temperatures over the study period was analyzed by calculating the coefficient of variability (CV).

In order to provide a general picture of temperatures and precipitation variability across the whole territory of Ethiopia in the last decades and in consideration of the above-mentioned quality limitation of the available time series, a simple procedure, based on the normalization of data, was used. For each station, the annual precipitation and the mean minimum and maximum temperature of each year have been normalized and the Z_i score calculated as follows:

$$Z_i = (D_i - D_m)/s_d$$

in which D_i is the parameter annual value for the ith year, D_m is the mean value of the time series, and s_d its standard deviation.

For each year of the time span considered, a representative Z score has been calculated by averaging the Z_i scores of all the stations. In this way, a new time series of Z scores is obtained. By this procedure, though not orthodox, it is possible to minimize the relative effects of gaps that are present and almost randomly distributed within the whole data set and to depict the time distribution of higher and lower than average values. Any attempt to fill the gaps through correlations with neighboring stations as suggested by Jury and Funk (2013) resulted to be unsatisfactory. On the other hand, the use of normalized data and averaged Z scores has been considered appropriate to reduce the effect of local outliers and to delineate trends that depict a general tendency for the entire country. This issue, however, would need further detailed and more sound statistical analyses that are beyond the scope of this paper.

For the main climatic parameters, the data elaboration is portrayed in colors overlying a shaded relief map. The regionalization of the data has been obtained through geostatistical interpolations considering the local elevation, latitude, distance from the sea and the largest lakes, and slope aspect for the whole area. For temperatures, the theoretical monthly solar radiation was considered as well. All the maps were interpolated and drawn using advanced GIS tools (ESRI ArcGIS–ArcInfo).

3.6 Temperatures

To analyze the spatial distribution of temperature, the monthly data from 30 meteo-stations, with continuous series or with negligible gaps throughout the World Meteorological Organisation (WMO) 1981–2010 reference interval, were selected (Table 3.1). Additional 17 meteo-stations, located in the adjoining territories of Djibouti, Eritrea, Somalia, and Sudan, were considered as well (Table 3.1) in order to account also for the conditions right beyond the Ethiopian borders and thus to improve as much as possible the reliability and quality of the analyses (Fig. 3.1).

The elevation of the stations considered varies between 10 m a.s.l. at Massawa and 2,760 m a.s.l. at Fitche. Though marked differences are observed for the average annual and monthly values in the highlands—typical of equatorial high mountains—and in the northeastern lowlands, Tigray and the Omo valley, the climate of Ethiopia is generally very pleasant and among the most equilibrate of the African continent since the temperatures are mitigated by the high average elevation. Only in the Danakil Desert areas the temperatures reach extreme values.

The lowest mean annual temperatures are recorded at elevations over 2,300–2,600 m a.s.l. (Fig. 3.1 and Table 3.1), irrespective of their geographic position. At Fitche, the highest and coldest station, the mean annual temperature is 14.2 °C, whereas at the elevation of 2,000 m it is around 18.5 °C.

In the Rift Valley, the temperatures gradually increase from south to north. Around the elevation of 1,500 m, they are slightly over 20 °C and the 25° isotherm stands around 1,000 m a.s.l. At lower elevations, the temperatures noticeably increase and peak to 30 °C in the steppe area of Gore and to higher values in the Danakil Desert.

The mean maximum temperatures are particularly pleasant at elevations above 1,500 m. At Fitche, they are 20, 23.5 °C at Addis Ababa and about 25 °C around the 2,000 m elevation. At 1,000 m, the mean values are close to 30 °C but can be higher than 35 °C in the southwestern lowlands and, mainly, in the steppe areas and the Danakil Desert (Fig. 3.2).

The mean minimum temperatures (Fig. 3.3) are rather low all across the highlands, where a few days of frost are commonly recorded every year. At Robe Bale, the station with the lowest minimum temperatures, the mean annual value is 8 °C, whereas at Addis Ababa it is 10 °C. Higher, typically tropical values of 20–25 °C are recorded at elevations below 500 m.

The lowest temperature ever recorded in the meteo-stations considered is −7 °C on the Bale Mountains (highest peak 4,377 m a.s.l.), whereas in Addis Ababa frost conditions only sporadically occur.

In Ethiopia, one of the hottest areas of the planet is found: the Danakil depression (Vinassa de Regny 1931; Fantoli 1940). Unfortunately, recent data are not available to confirm this evidence. At Dallol (130 m b.s.l.), which is considered the hottest, inhabited place of the world, Pedgley (1967) measured a mean annual temperature of 34.7 °C, mean maximum temperatures of 41.2 °C, and a peak of 45.7 °C for the 1960–1966 interval. According to Billi (personal communication), in November 2007, temperatures of 42–44 °C were commonly recorded in the Samoti plain in the Eritrean Danakil.

Table 3.1 Main temperature data

Meteo-stations	H (m a.s.l.)	T_{mean} (°C)	T_{max} (°C)	T_{min} (°C)	E_a (°C)	E_d (°C)	M_h	M_c
Addis Ababa	2,354	17.0	23.5	10.4	3.1	13.1	May	Dec
Al Damazin	470	27.5	35.1	19.8	7.2	15.3	Apr	Sep
Alamata	1,520	22.5	30.0	14.9	6.3	15.1	Jun	Jan
Alemaya	2,047	16.5	23.7	9.3	5.7	14.4	Jun	Dec
Arba Minch	1,285	23.8	30.2	17.3	2.9	12.9	Mar	Jul
Asab	14	30.4	35.4	25.4	9.5	10.0	Aug	Feb
Asela	2,430	15.4	21.9	8.9	2.7	13.0	May	Dec
Asmara	2,325	15.2	22.9	7.4	5.0	15.5	Jun	Dec
Awasa	1,750	20.0	27.3	12.6	1.8	14.7	Apr	Nov
Bahir Dar	1,770	19.1	26.9	11.2	4.6	15.7	May	Jan
Belet Uen	198	28.6	34.8	22.4	2.6	12.4	Apr	Aug
Berbera	1	30.5	35.2	25.8	13.1	9.4	Jul	Jan
Borama	1,454	21.2	27.3	15.0	4.9	12.3	Mar	Jul
Burao	970	22.8	29.3	16.3	6.2	13.0	May	Jan
Debre Marcos	2,515	16.4	22.5	10.3	3.7	12.2	Apr	Aug
Dire Dawa	1,260	25.5	31.8	19.1	7.0	12.7	Jun	Jan
Djibouti	21	29.8	33.8	25.7	11.3	8.1	Jul	Jan
Fitche	2,750	14.2	20.3	8.1	3.1	12.2	May	Nov
Galcayo	247	30.4	34.1	26.7	9.8	7.4	Jul	Feb
Gambela	484	27.6	34.8	20.4	5.3	14.4	Mar	Jun
Garowe	473	30.3	33.9	26.7	9.9	7.2	Jul	Dec
Gewane	627	21.5	24.9	18.1	8.0	13.8	Jun	Dec
Gode	295	28.9	34.7	23.1	2.8	11.6	Mar	Nov
Gonder	1,967	20.0	26.7	13.3	4.6	13.4	Apr	Aug
Gore	2,002	18.9	23.9	13.9	3.7	10.0	Mar	Jul
Hargeysa	1,347	22.3	27.4	17.1	5.0	10.3	May	Jan
Hayk	2,030	18.3	25.8	10.7	5.8	15.1	Jun	Dec
Jijiga	1,644	19.4	26.5	12.3	4.5	14.2	Aug	Jan
Jimma	1,725	19.7	27.6	11.8	3.0	15.8	Apr	Dec
Kassala	502	29.5	37.1	21.8	8.3	15.3	May	Jan
Kebry Dehar	549	26.6	33.1	20.1	2.2	13.0	Apr	Aug
Kobo	1,610	22.1	29.3	14.9	7.0	14.4	Jun	Dec
Kombolcha	1,903	19.3	26.5	12.1	5.9	14.4	Jun	Dec
Mandera	231	29.5	34.9	24.1	3.7	10.8	Mar	Jul
Massawa	10	29.4	34.5	24.3	10.1	10.2	Jul	Jan
Mekele	2,070	17.6	23.8	11.3	4.6	12.5	Jun	Dec
Metahara	947	25.4	33.3	17.4	6.3	15.9	Jun	Dec
Mogadishu	7	26.7	30.2	23.1	3.1	7.1	Apr	Aug
Moyale	1,097	22.4	27.3	17.5	5.2	9.8	Feb	Jul
Negelle	1,544	20.7	26.3	15.0	4.7	11.3	Mar	Jul
Nekemte	2,080	18.3	24.0	12.6	4.1	11.4	Mar	Jul
Robe Bale	2,480	14.8	21.6	8.0	2.2	13.6	Jun	Dec
Sirinka	2,081	18.8	24.1	13.5	6.1	10.6	May	Dec
Ziway	1,640	20.6	27.1	14	3.3	13.1	Apr	Dec

H elevation, T_{mean} mean annual temperature, T_{max} mean annual maximum temperature, T_{min} mean annual minimum temperature, E_a mean annual temperature excursion, E_d mean daily temperature excursion, M_h hottest month, and M_c coldest month

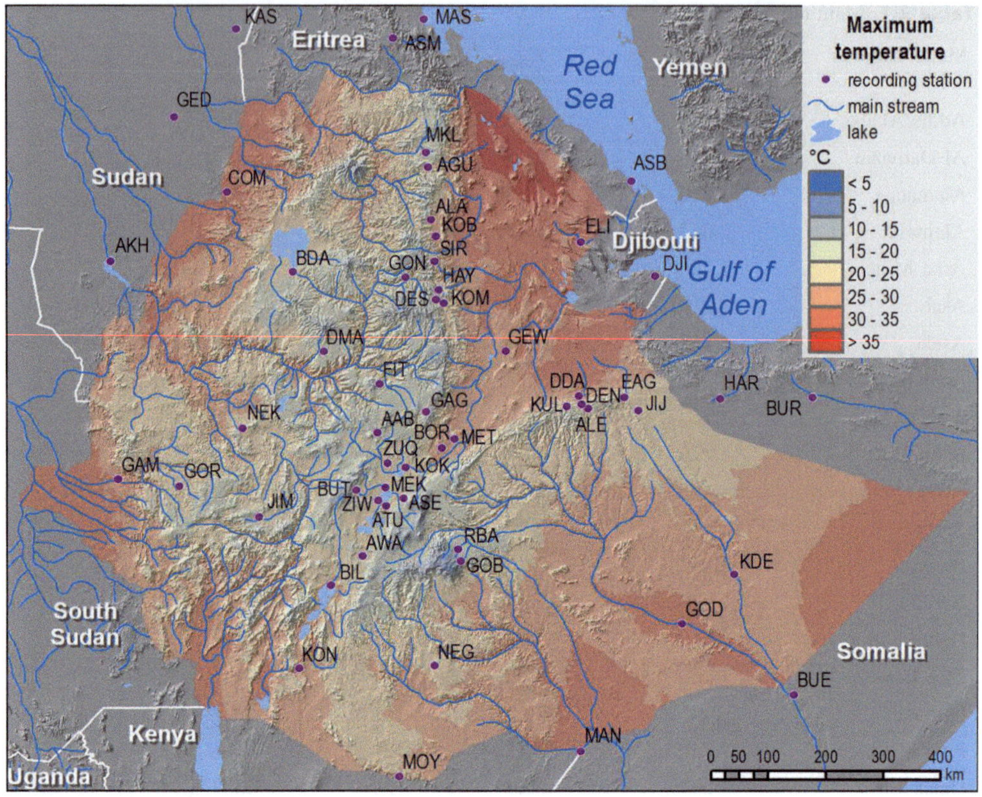

Fig. 3.2 Mean maximum annual temperatures

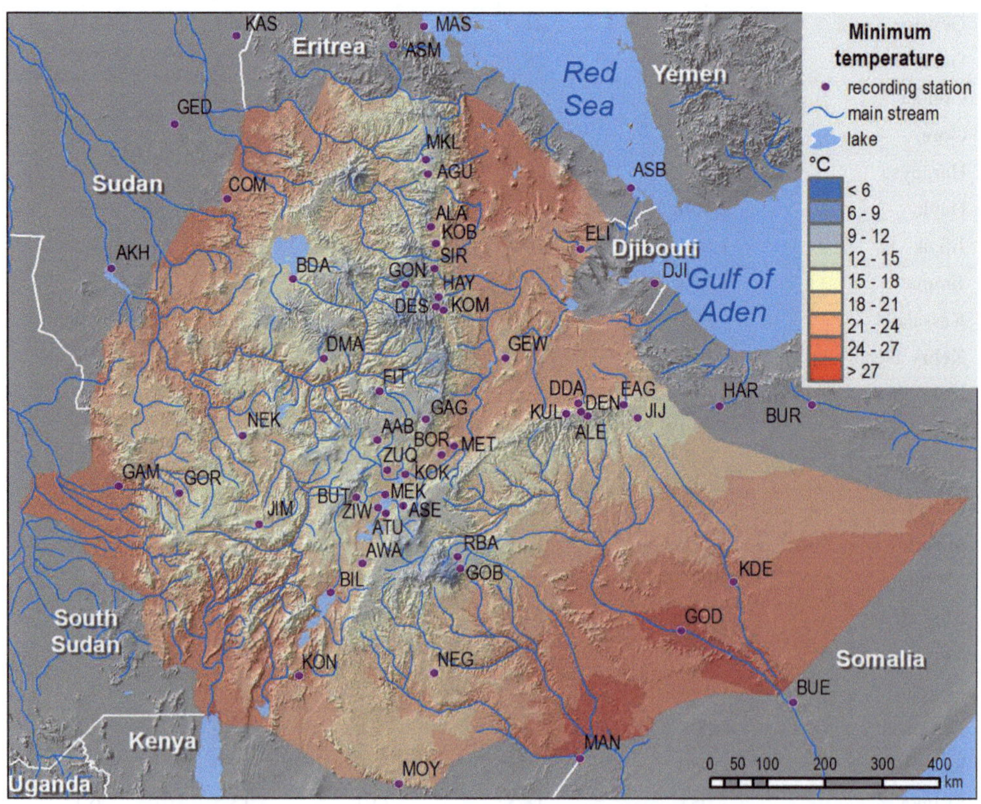

Fig. 3.3 Mean minimum annual temperatures

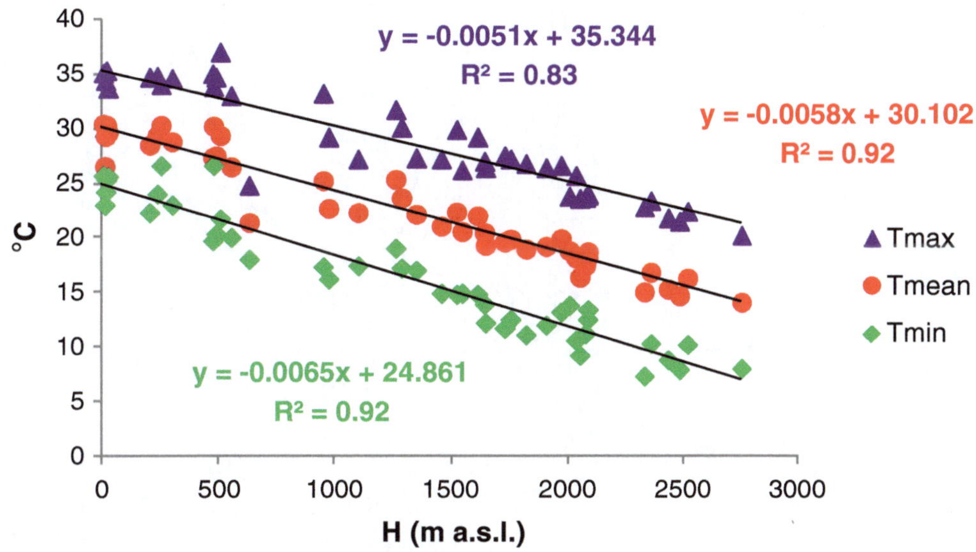

Fig. 3.4 Lapse rate for maximum mean and minimum mean annual temperatures

From the above considerations and Fig. 3.4, it is evident that elevation is a main parameter influencing temperature in Ethiopia and adjoining areas. The lapse rate is 5.8 °C every 1,000 m, and the mean annual 10 and 5 °C isotherms were estimated to be around 3,450 and 4,300 m a.s.l., respectively. The maximum and minimum temperature lapse rates are slightly different with 5.1 °C/1,000 m and 6.5 °C/1,000 m, respectively.

Considerable differences characterize the diurnal and annual temperature ranges (Fig. 3.5). The former is rather large, especially in the central-western steppe and desert area (Fig. 3.6), varying between 10 and 15 °C; the latter is much narrower, commonly less than 5 °C in the highlands areas below 2,000 m in elevation and in the tropical areas (Fig. 3.6). In fact, sunny days prevail during most of the year in almost the whole country, including the areas with a monsoon climate; the alternation of cloudy and clear sky conditions results in ample thermal differences (Fig. 3.5 and Table 3.1). By contrast, the annual thermal amplitude is limited and seems to be controlled by the intertropical circulation and the monsoon pattern rather than elevation. In the central-western highlands and in the Rift Valley, the hottest month is in spring and anticipates the summer, big monsoon rains (*kiremt*), whereas in the northern and eastern quadrants it occurs at the beginning of summer.

In the highlands, the coldest month is in the middle of winter, whereas in the southern and eastern regions, including also high elevated areas, subjected to perturbed conditions brought about by the monsoon and characterized by heavy downpours and cloudy skies, the coldest month occurs in summer.

3.7 Precipitation

To describe the rainfall characteristics of Ethiopia, the monthly data of 49 meteo-stations in Ethiopia and 9 in the adjoining territories of Eritrea, Djibouti, Somalia, and Sudan, and the daily data of 27 Ethiopian stations were used (Table 3.2). The longest time series is that of Moyale covering 73 years of almost uninterrupted daily data from 1935 to 2008, whereas the average length of the Ethiopian time series considered is 33 years (ranging between 20 and 68 years), with only four stations covering a shorter interval (10–20 years). These latter were included as well in order to attain the larger and denser distribution of data as possible for the spatial interpolations used in producing the annual rainfall distribution map (Fig. 3.7).

The annual precipitation, averaged all across the country, is 817 mm but, given the complex physiography and the different seasonal and spatial influences of the prevailing air masses and winds, a large diversity is observed among various regions. The highest annual rainfall is recorded at Gore, in the western highlands at an altitude of 2,002 m a.s.l., with 2,101 mm, whereas the lowest value in Ethiopia is found at Elidar with 153 mm. This rain gauge is located in the Afar triangle at an elevation of 423 m s.l.m., but the driest site is Asab, in Eritrea, with only 45 mm. Gore and Elidar are not the most and the least elevated stations, respectively, considered in this study; nevertheless, they share the extreme values of annual rainfall. This is reflected by the poor correlation ($R^2 = 0.42$) between precipitation and elevation. Figure 3.7 shows that higher precipitation occurs in the western part of

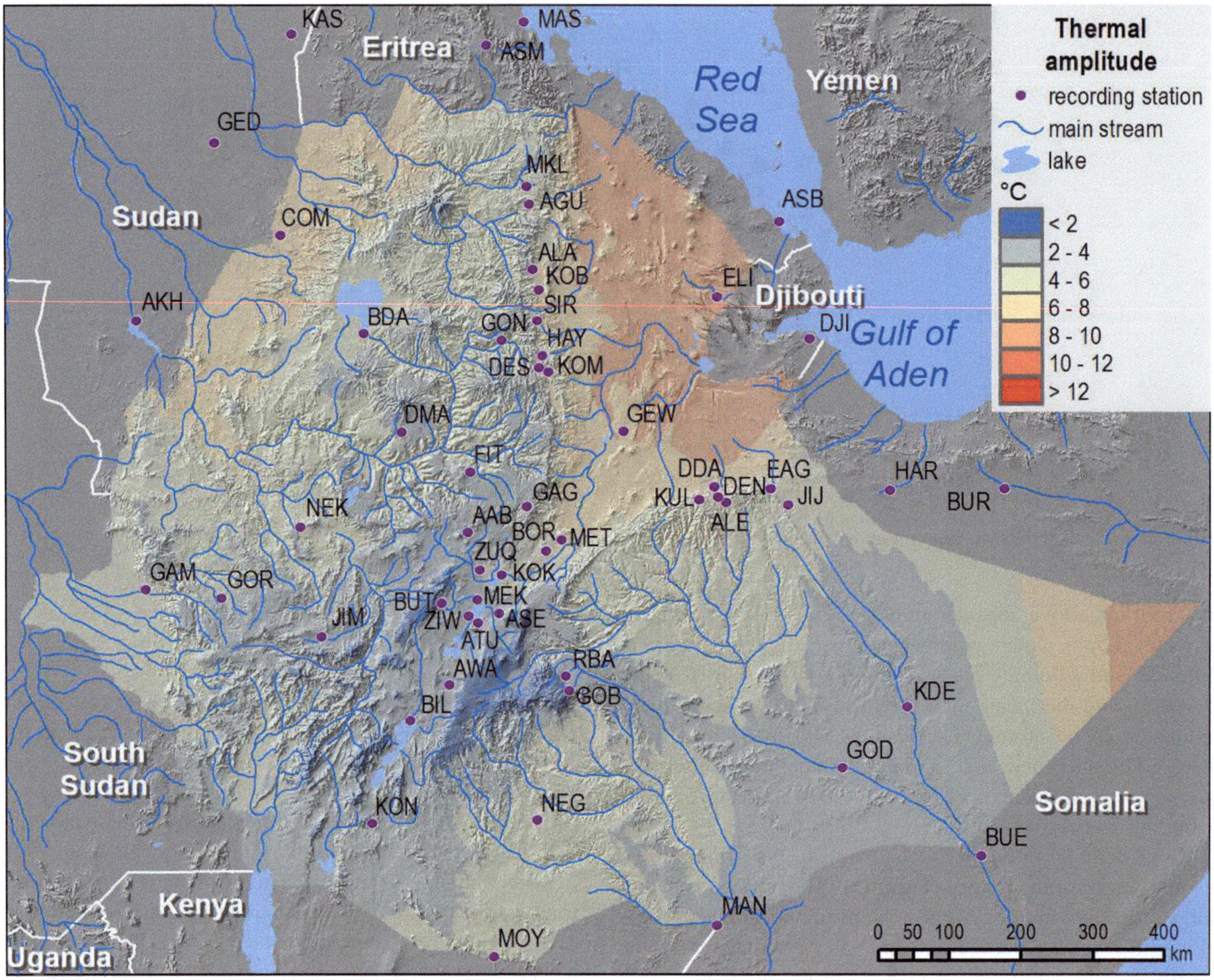

Fig. 3.5 Annual temperature range

Ethiopia and is partly irrespective of the station altitude. In Fig. 3.7 also, a westward gradient is rather evident too. Following these considerations, a physiographic factor (*F*) was developed. It is the ratio of the meteo-station elevation (*H*) to *L*, i.e., the difference between its longitude and the 34th meridian, taken as a western reference. Figure 3.8 shows that 85 % of annual precipitation variability is explained by the physiographic parameter $F = H/L$.

Since in two-thirds of the meteo-stations, the summer rains (*kiremt*) contribute more than 50 % of the annual precipitation, with peaks of 80–85 % in the north, and 87 % of the annual precipitation variability is explained by the *kiremt* (Fig. 3.9); the monsoon and its seasonal and geographic pattern appear to be the most relevant factors in controlling the spatial distribution and amount of rainfall in Ethiopia (Fig. 3.10), though in some places the *belg* precipitation (Fig. 3.11) may be higher than the *kiremt*, contributing about 50 % of the annual rainfall such as at a few stations in the south and in Ogaden. Figure 3.9 shows also that the *belg* rains have little influence on the annual precipitation as they explain only 31 % of its variability.

The spring rains are also the most unpredictable for their variation coefficient is 47 %, whereas the summer rains (CV = 33 %) result the most dependable for the farmers. Only on the coast, namely at Djibouti and Massawa, the winter rains account for most of the annual precipitation. Along the Eritrean coast and in Ogaden, the largest variability of annual precipitation is found (Aseb CV = 127 %), whereas the least variability is observed in the central highlands (Table 3.2). This observation is confirmed by the diagram of Fig. 3.12 in which the variation coefficient is well explained ($R^2 = 0.75$) by the physiographic parameter *F*.

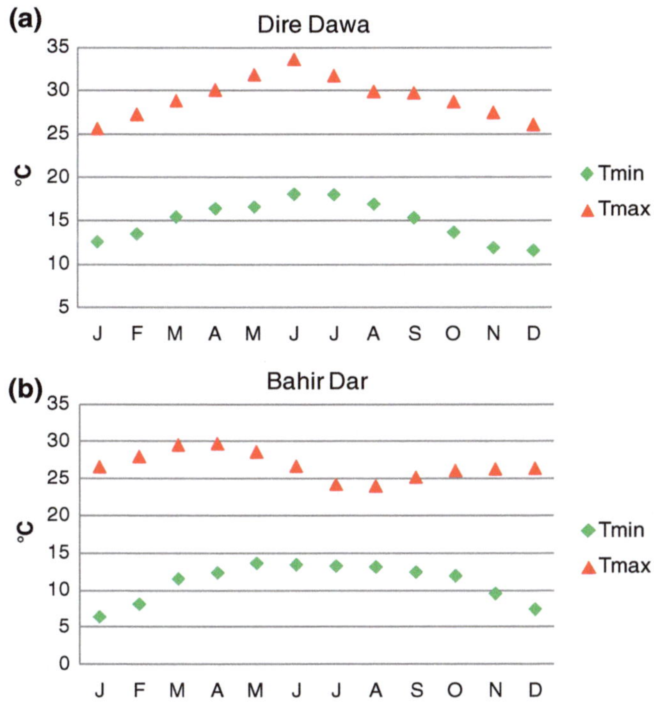

Fig. 3.6 Mean maximum and minimum monthly temperatures at Dire Dawa and Bahir Dar

Rainfall intensity was investigated using daily data recorded at 26 stations. The daily interval was selected because, unfortunately, long time series of hourly data are not available. Certainly, this is a strong constraint to sort out the regions with aggressive rains since commonly, also during the monsoon season, rainstorms are short (1–3 h) and a daily datum may be not sufficiently accurate to describe the rainfall intensity parameter, especially in relation to the activation of soil erosion processes. Nevertheless, also using daily data, some general results can be drawn. The mean maximum rainfall intensity (I_{max}) at each meteo-station was calculated by averaging the maximum values recorded in each year, whereas I_p is the highest value ever recorded by each rain gauge. Gambela is subjected to the highest rainfall intensities with an average of 89.4 mm in 24 h and a maximum value ever recorded of 181.4 mm/24 h. By contrast, the lowest values are observed in Elidar (I_{max} = 27.8 mm/24 h) and Gewane (I_p = 31.6 mm/24 h), i.e., two rain gauges located in the Afar depression characterized by desert and semidesert climate conditions. Neither physiographic factor, such as elevation or the geographic position, nor any pluviometric parameter, such as annual rainfall, is able to explain the variability of rainfall intensity, which is likely more influenced by local orographic or morphologic factors.

High intensity rains may occur in any month, but the highest frequencies are recorded in April and August, i.e., in the middle of the spring and summer rainy seasons (Fig. 3.13). Hourly data for Mekele and Ziway show that maximum intensities around 20 mm h^{-1}, with peaks of 40–50 mm h^{-1}, are rather common in these two localities.

Soil erosion is a major environmental and economic problem in Ethiopia. In order to highlight the regions affected by severe hydro-meteoric erosion conditions, the Modified Fournier Index (MFI) (Arnoldus 1980) was calculated (Table 3.2) and the spatial distribution of this parameter is mapped (Fig. 3.14). At two-thirds of the weather stations considered, rainfall erosivity is from moderate to high and at 43 % of them it is high. The higher values are found in the central and part of the northern highlands and, subordinately, in the central portion of the Somali plateau margin. These areas are classified as subjected to high and very high runoff by Berhanu et al. (2013) and high erosion rates would be expected, but the occurrence of soils with low erodibility, as reported by the USLE's factor *K* map reported by these authors, partly mitigates this threat.

3.8 Wind

In Ethiopia, the wind regime is influenced by barometric synoptical variations associated with the position of the ITCZ. Winds are not strong because the Horn of Africa is not a cyclone-genetic area and, as pointed out in Sect. 3.6, temperatures are rather uniform throughout the whole year. As a consequence, winds, though seldom absent, have low velocities, and only during deep convective processes, short but intense gusts and whirls are observed. Windiness is characterized by a large spatial variability due to the physiographic complexity which affects the prevailing seasonal air masses, especially along the Rift Valley and its margins.

From the analysis of daily and, occasionally, 6 h data from a few meteo-stations located in the highlands, in Tigray, along the Rift Valley, and in Ogaden for the 2000–2006 interval (Fig. 3.15), it is evident that in the central-northern part of the country, in a few sites of the Eritrean coast and the Sudan lowlands, winds from the northern quadrant dominate, whereas in the southern areas, winds from the second and the third quadrants prevail throughout the whole year. On the margin of the Somali Plateau, western winds prevail, whereas in the far southwest of Ethiopia southwestern winds are more common. On the highlands, the annual frequency of days with calms and light breezes amounts to 12–18 % of the total.

At seasonal level and particularly during the *kiremt*, noticeable differences from the above framework can be traced:

1. In winter, the situation described above is confirmed with winds from the first quadrant and from northwest prevailing on the central highlands and on the Somali

Table 3.2 Characteristic data of the rain gauges considered in this study

Meteo-stations	*H* (m a.s.l.)	*P* (mm) annual	CV (%) annual	*P* (mm) *belg*	*P* (mm) *kiremt*	*P* (mm) *bega*	*k/a* %	I_{max} mm/24 h	I_p mm/24 h	T_r 100 mm/24 h (years)	MFI
Adami Tulu	1,636	756.8	27	186.2	447.6	123.0	59.1	47.4	81.9	68.6	93.9
Addis Ababa	2,354	1,204.8	18	238.4	863.5	102.9	71.7				186.1
Adigudum	2,095	490.5	49	52.5	428.5	9.5	87.4				127.1
Al Damazin	470	691.0		56.0	596.0	39.0	86.3				137.3
Alamata	1,520	750.4	24	219.9	387.1	143.4	51.6	52.8	96.0	32.4	106.2
Alemaya	2,047	802.9	26	248.4	431.0	123.5	53.7	56.4	118.0	15.3	95.0
Arba Minch	1,285	931.9	17	375.9	236.6	319.4	25.4				99.7
Aseb	14	45.4	127	6.1	20.6	18.7	45.4				7.5
Asela	2,430	1,225.5	17	321.6	762.8	141.1	62.2				150.2
Asmara	2,325	512.3	34	87.8	375.7	48.8	73.3				109.7
Awasa	1,750	951.5	16	292.9	472.6	186.0	49.7				99.3
Azezo	1,966	1,153.0		146.0	914.0	93.0	79.3				220.7
Bahir Dar	1,770	1,413.2	17	105.0	1,189.8	118.4	84.2				293.7
Belet Uen	198	259.0		138.0	21.0	100.0	8.1				53.7
Bilate	1,500	744.2	23	240.3	326.9	176.9	43.9				78.9
Borciota	1,062	505.0		122.0	309.0	74.0	61.2				76.4
Burao	970	195.0		102.0	62.0	31.0	31.8				33.0
Butajira	2,000	1,118.1	19	377.7	560.6	179.8	50.1				121.1
Comar	764	920.0		90	785	45.0	85.3				180.5
Debre Marcos	2,515	1,325.5	11	208.9	970.1	146.5	73.2				204.4
Dengeco	2,111	750.5	22	249.6	429.0	71.9	57.2	53.8	98.8	21.0	103.6
Derbiga el Agemsa	1,644	678.0		239.0	340.0	99.0	50.1				80.4
Desse	2,250	1,207.1	16	244.6	789.5	173.0	65.4	54.9	94.0	47.0	189.4
Dire Dawa	1,260	638.7	27	220.2	313.6	104.9	49.1	54.8	122.3	18.1	78.3
Djibouti	21	188.0	71	41.59	17.52	128.9	9.3				24.8
Elidar	423	153.1	40	29.0	99.1	25.0	64.7	27.8	55.0		26.5
Fitche	2,750	1,125.2	12	183.0	826.0	116.2	73.4	51.4	90.9	42.8	226.2
Gambela	484	1,114.1	22	187.3	697.6	229.2	62.6	89.4	181.4		139.7
Gedaref	599	579.7	16	15.8	525.2	38.7	90.6				146.6
Gewane	627	521.9	21	135.3	268.7	117.9	51.5	29.1	31.6		71.8
Gina Ager	3,160	1,693.2	20	415.3	977.0	300.9	57.7	74.3	170.8	6.4	216.9
Goba	2,743	943.7	20	292.4	412.2	239.1	43.7				98.9
Gode	295	236.9	61	135.0	8.4	93.5	3.6	58.4	174.0	7.0	50.7
Gonder	1,967	1,131.9	18	139.4	878.4	114.1	77.6	52.4	99.1	40.9	206.7
Gore	2,002	2,101.3	23	439.7	1,261.9	399.7	60.1	62.1	107.7	24.9	247.4
Hargeysa	1,347	350.0		120.0	210.0	20.0	60.0				50.6
Hayk	2,030	1,174.0	17	280.9	717.1	176.0	61.1	62.1	132.8	16.5	176.4
Jijiga	1,644	721.3	39	249.9	329.2	142.3	45.6				75.6
Jimma	1,725	1,500.5	13	401.1	821.0	278.4	54.7	53.0	105.7	39.5	164.4
Kassala	502	278.6	43	13.0	229.0	36.6	82.2	49.0	104.9		56.2
Kebri Dehar	549	325.3	49	170.4	20.6	134.3	6.3	59.4	128.0	11.8	77.7
Kobo	1,610	725.8	22	151.5	411.8	162.5	56.7	53.2	101.5	18.3	97.7
Koka Dam	1,595	698.8	48	135.6	496.0	67.3	71.0				112.3
Kombolcha	1,903	1,043.6	15	221.3	663.1	159.2	63.5				160.9
Konso	1,053	805.6	14	358.2	153.0	294.4	19.0	55.2	96.9	28.4	86.2

(continued)

Table 3.2 (continued)

Meteo-stations	H (m a.s.l.)	P (mm) annual	CV (%) annual	P (mm) *belg*	P (mm) *kiremt*	P (mm) *bega*	*k/a* %	I_{max} mm/24 h	I_p mm/24 h	T_r 100 mm/ 24 h (years)	MFI
Kulubi	2,410	1,022.3	18	308.7	565.6	148.0	55.3	60.2	100.5	14.4	125.6
Mandera	231	286.7	56	159.4	4.5	122.8	1.6				59.6
Massawa	10	175.8	58	22.5	19.6	133.7	11.1				24.0
Mekele	2,070	598.7	27	91.3	487.7	19.7	81.5	49.4	95.5	59.6	162.6
Meki	1,660	771.2	21	185.6	482.1	103.5	62.5				103.6
Metahara	947	549.2	23	140.2	322.4	86.6	58.7	52.3	96.0	31.5	78.0
Moyale	1,097	685.0	40	352.5	63.4	269.1	9.3	68.7	167.6	6.3	104.1
Negelle	1,544	726.6	27	414.5	52.7	259.4	7.3	66.1	137.0	8.2	128.6
Nekemte	2,080	2,037.7	12	369.3	1,429.0	239.4	70.1	75.9	137.5	7.0	294.9
Robe Bale	2,480	876.3	16	281.0	380.1	215.2	43.4	43.3	112.3	8.7	92.1
Sirinka	2,081	1,128.9	14	206.1	361.3	561.5	32.0				161.8
Ziway	1,640	742.5	24	201.5	442.1	98.9	59.5				89.3
Zuquala	2,980	1,124.8	22	251.8	763.3	109.7	67.9	68.0	132.6	8.7	178.3

H elevation, *P* precipitation, *CV* variation coefficient, *k/a* contribution of the *kiremt* rainfall to the annual precipitation, I_{max} mean maximum rainfall intensity in 24 h, I_p highest rainfall intensity in 24 h ever recorded, T_r return time for a rainfall intensity of 100 mm in 24 h, and *MFI* Modified Fournier Index Arnoldus 1980)

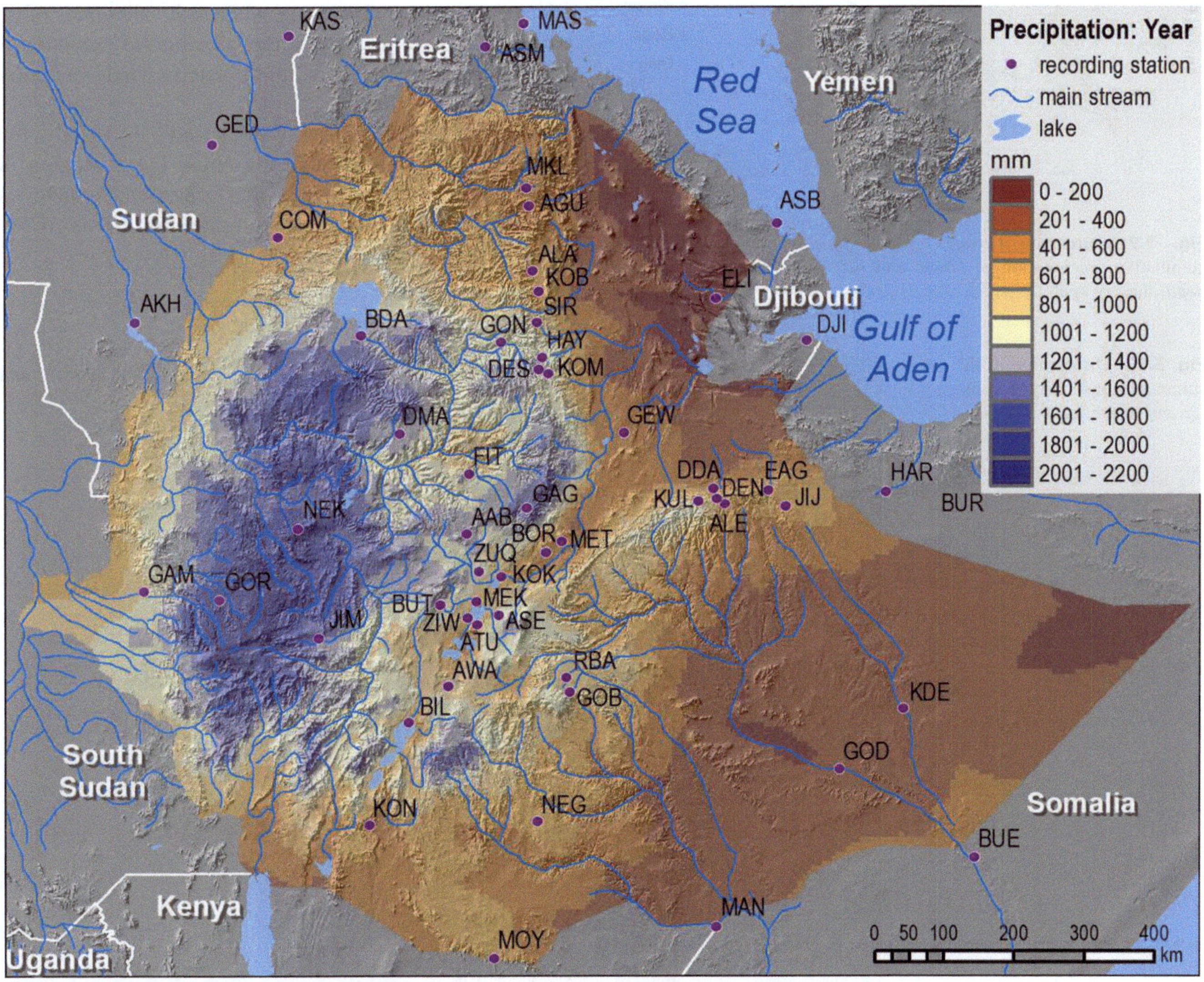

Fig. 3.7 Distribution of annual precipitation

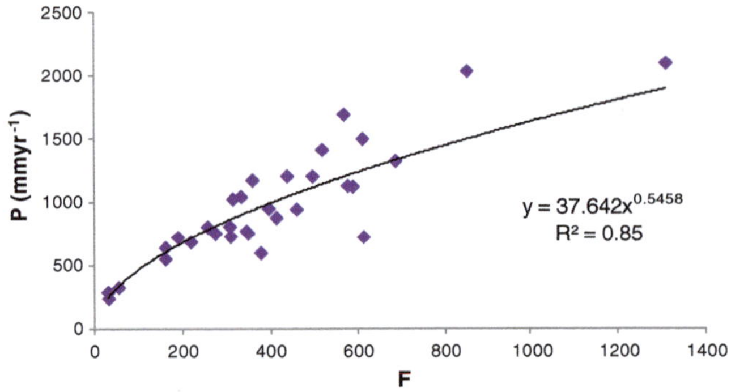

Fig. 3.8 Correlation between annual precipitation (*P*) and the physiography factor (*F*—see text for explanation)

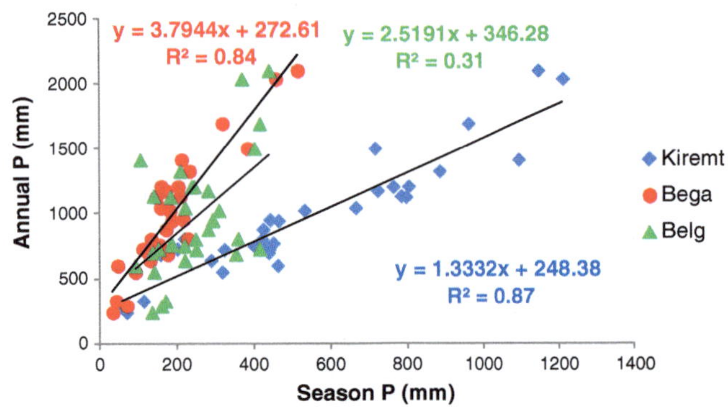

Fig. 3.9 Correlation between annual precipitation (*P*) and the cumulative precipitation of *kiremt* and *belg* main rainy seasons and *bega*, the dry spell

plateau margin, respectively. Elsewhere, southeastern and southwestern winds prevail.

2. In spring, only small changes occur. In fact, the western winds rise in latitude along the Sudanese border as far as Tana Lake, whereas in the southernmost areas they are replaced by winds from the second quadrant. On the Somali plateau margin and in Ogaden northeastern winds predominate.
3. During summer, wind circulation turns eastward on the central-northern highlands, whereas such orientation is not so evident in the area of Addis Ababa and in the Rift, and finally to northwest in Eritrea (e.g., Asmara). Southwestern winds predominate on the Sudanese lowlands and as far as Gambela where winds from east and southeast prevail. These winds are present also in the southern part of the country as far as the Somali plateau margin (e.g., Harar).
4. In autumn, the part of Ethiopia to the west of an ideal Gonder-Nekemte line and the southern sectors are subjected to western air masses, whereas in other parts of the country eastern air masses prevail with the exception of the Harar region where northwestern winds are more common.

On the highlands, mean wind speed is typically low (12 km h^{-1}) but increases in the Great Lakes Region, in the sub-desert and bush lands of Ogaden and around the border with Sudan, especially during the *kiremt* (EMA 1988).

Fig. 3.10 Distribution of the summer, monsoon-type rains (*kiremt* in the local language)

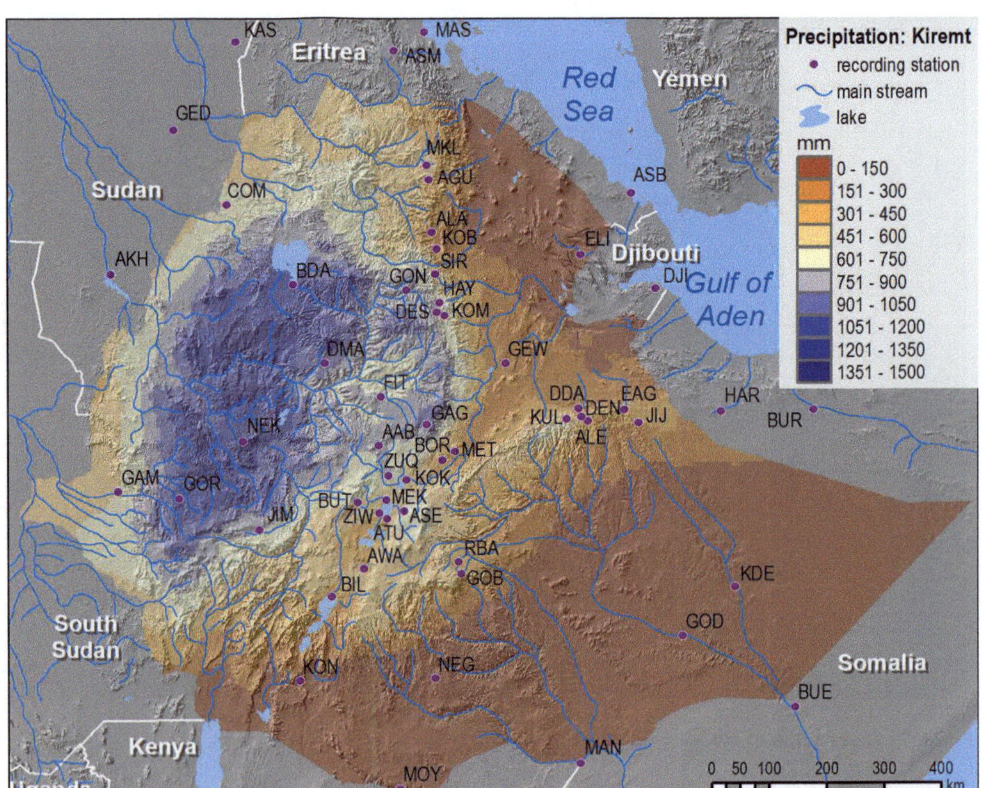

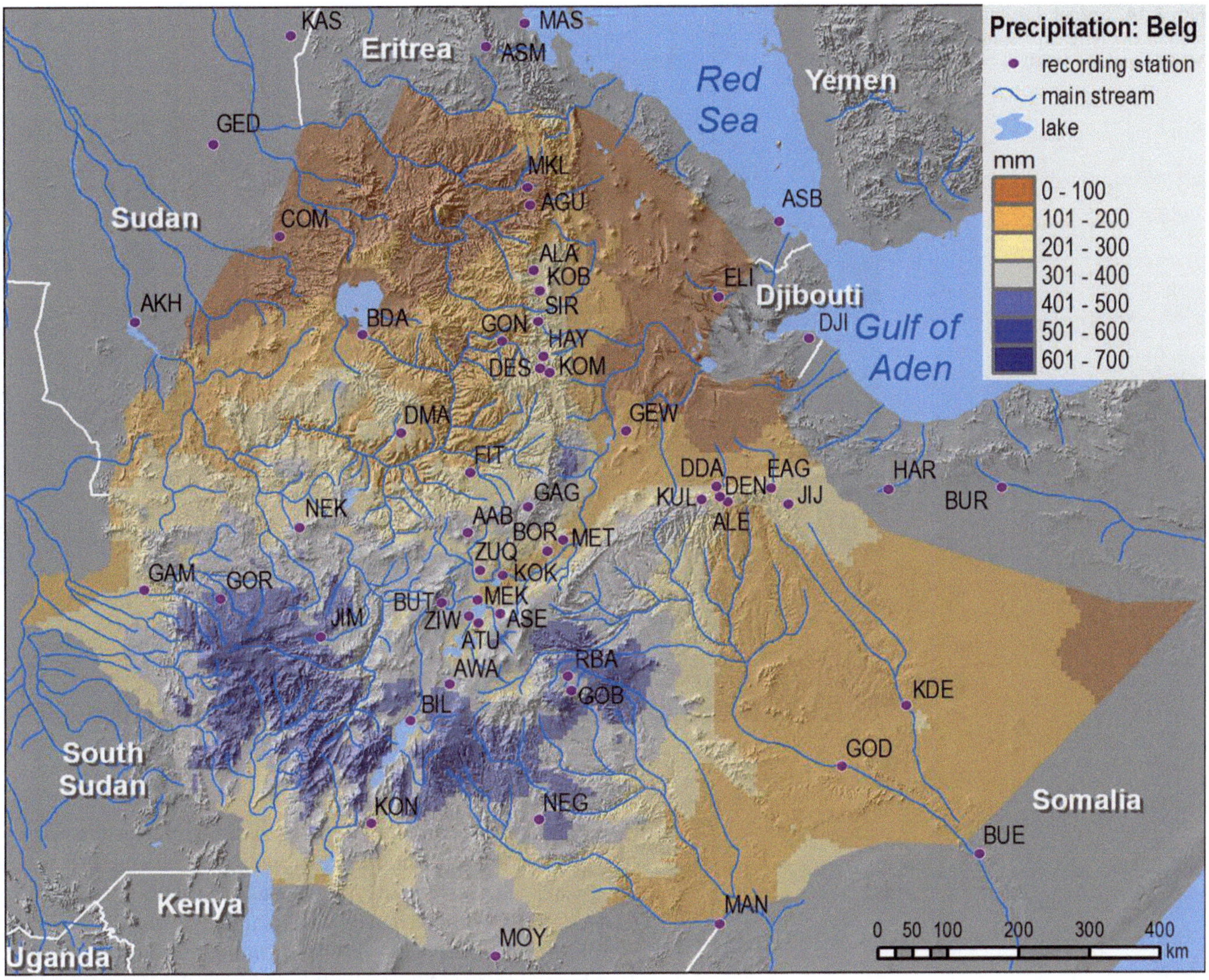

Fig. 3.11 Distribution of the little, spring rains (*belg* in the local language)

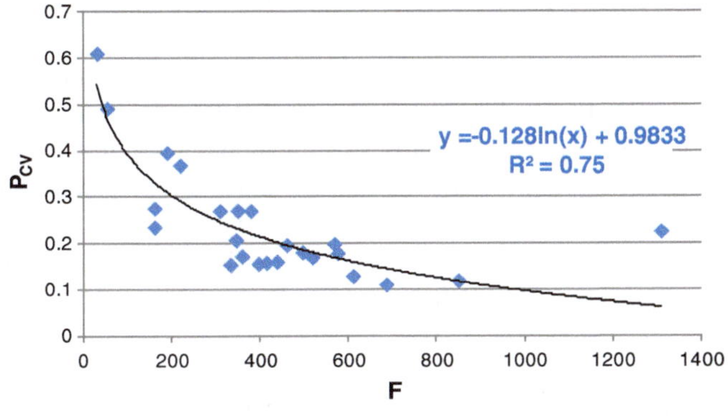

Fig. 3.12 Correlation between the annual rainfall variation coefficient (P_{CV}) and the physiographic factor (*F*—see text fro explanation)

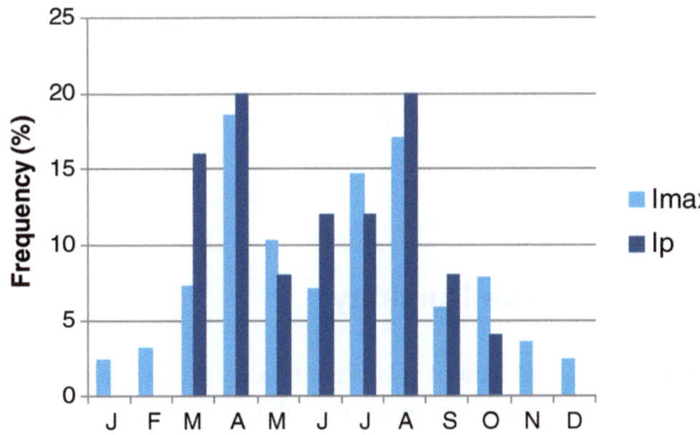

Fig. 3.13 Monthly distribution of the frequency of occurrence of the annual maximum rainfall intensity in 24 h (I_{max}) and of the highest value (*Ip*) recorded throughout the period of recording for all the meteo-stations considered in this study

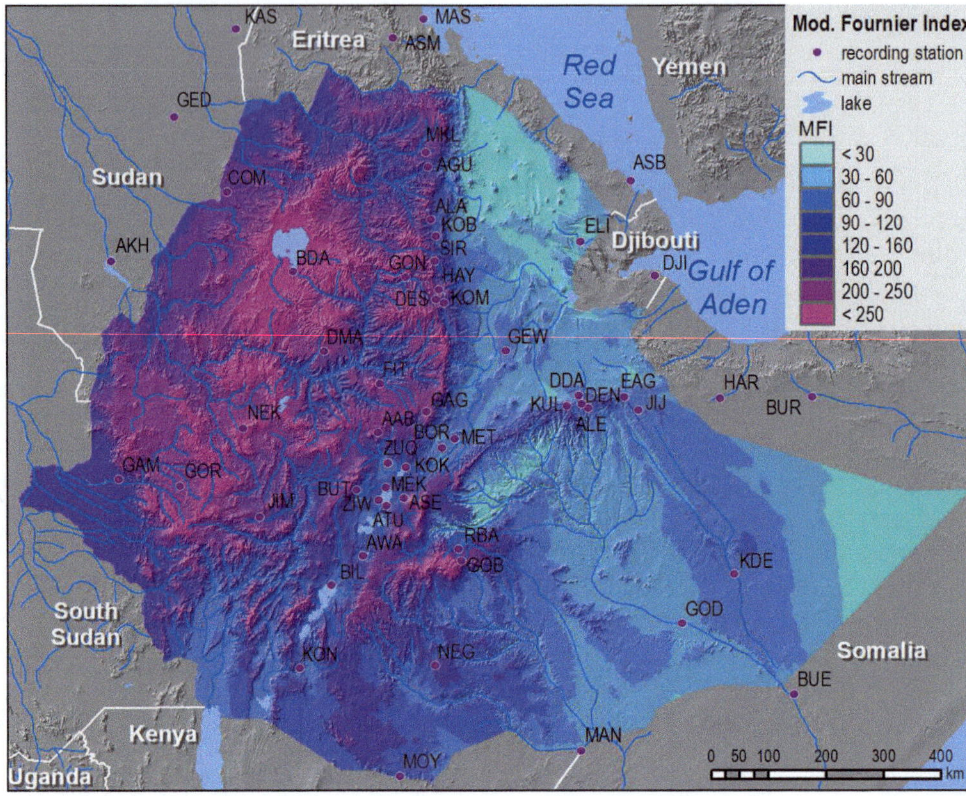

Fig. 3.14 Spatial distribution of rainfall erosivity expressed by the MFI (Arnoldus 1980)

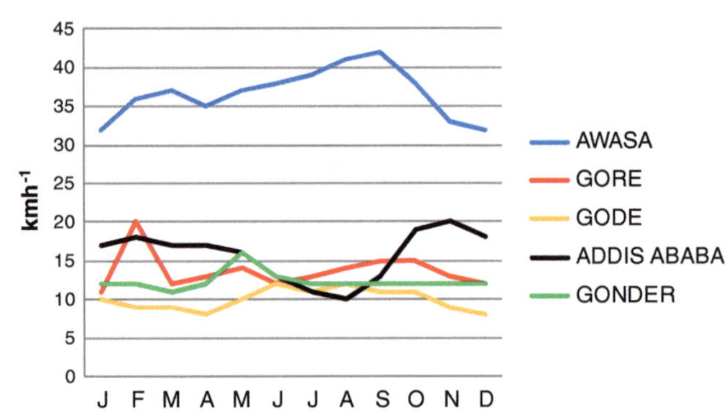

Fig. 3.15 Monthly variation of average wind speed for a few meteo-stations

3.9 Relative Humidity

Only inadequate data of relative humidity are available since few stations have been measuring this parameter in the last decades. Notwithstanding this constraint, combining the available data with the maps of the National Atlas (1988), some considerations on the spatial distribution of this parameter can be drawn.

The most part of Ethiopia—and in particular the north—is subjected to the influence of trade winds commonly rich in water vapor. Obviously, the spatial distribution of humidity is tightly correlated with precipitation. In the southwest of Ethiopia, areas with very high humidity, among the most humid in the African continent (e.g., the great equatorial lakes region), are found.

By contrast, during winter, high values are recorded in Ogaden and along the Red Sea, whereas low humidity characterizes the Sudanese border belt with mean values around 23 %.

The data recorded in 13 meteo-stations during the last decade indicate that average annual relative humidity is about 60 % on the highlands, in the Rift and along the Kenya border (Fig. 3.16), but even higher values are presumed for very high mountain areas (EMA 1988). Lower values between 40 and 50 % are instead recorded in the northernmost portion of the Rift (e.g., Dire Dawa—Fig. 3.16).

3.10 Cloudiness and Sunshine

Ethiopia is the country with the most extensive cloudiness in Africa. Unfortunately, only very few data of Fantoli (1940) are available and refer to the first decades of the twentieth

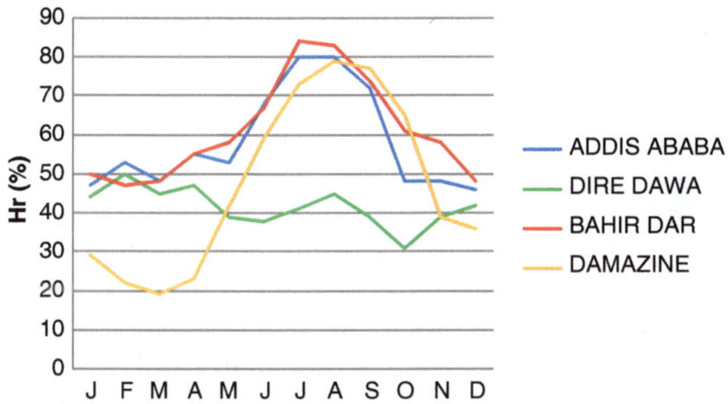

Fig. 3.16 Monthly distribution of mean relative humidity for a few representative meteo-stations

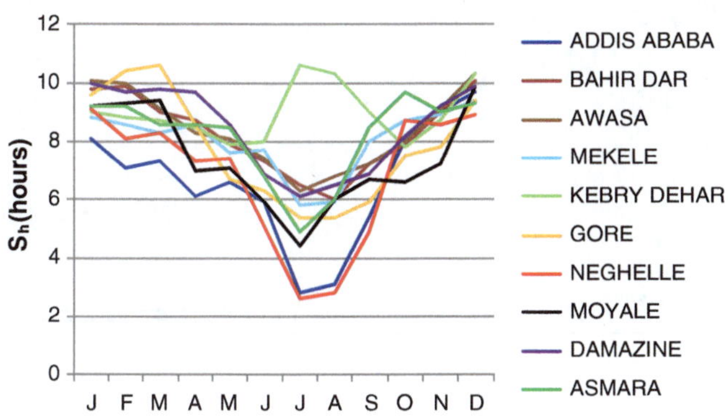

Fig. 3.17 Monthly distribution of mean sunshine hours for a few representative meteo-stations

century. According to this author, average cloudiness is 2/10 in January and 6/10 in July, whereas clear sky days are a little less frequent than mixed and cloudy ones (170 vs 195). The maximum and minimum cloudiness occurs during the rainy season and the boreal winter, respectively. In Ogaden and mainly in the Danakil depression, cloudiness is almost an exceptional phenomenon. Along the Eritrean coast and for a few kilometers inland, night advection fogs may commonly form and turn into morning drizzle, integrating the scarce precipitation of this area.

Sunshine is obviously complementary to cloudiness. Monthly data of sunshine hours (S_h) measured by nine stations in the last decade (Fig. 3.17) show that most of the highlands and the lowlands close to the Sudan border receive more than 7 h of mean daily sunshine, whereas from Tigray to the southeast of the country the respective value increases to 10 h. During autumn and winter, 300–350 h per month are commonly recorded in Tigray and Ogaden. In summer, vice versa, in the highlands between Gonder and Addis Ababa, the sunshine commonly does not exceed 50 h per month.

3.11 Evapotranspiration and Aridity

Evapotranspiration is limited by soil moisture supply, surface and atmospheric temperature and the dew point threshold (Law et al. 2002). The coupling of evapotranspiration, temperature, and precipitation is particularly pronounced in moisture-limited subtropical regions at the interface between wet (monsoonal) and dry climate regimes.

Potential evapotranspiration (PET) was calculated following the Thorntwaite's method, using the data from 29 meteo-stations for the 1981–2010 time range (Fig. 3.18). The same data were used also to reckon the aridity index (A_i) (Fig. 3.19), known also as the desertification index (UNEP 1992), as $A_i = P/\text{PET}$ in which P is annual precipitation.

Almost all the areas over 2,000 m a.s.l. are characterized by annual PET values less than 1,000 mm, whereas precipitation is typically around 1,500 mm or higher. Therefore, the highlands can be classified as humid and hyper-humid in the Gore–Jimma–Tepi triangle (Fig. 3.18). As elevation decreases, PET commonly increases as it is observed in the Rift Valley. Extreme evapotranspiration conditions are found in the Danakil depression where values of as much as 4,000 mm can be estimated for Dallol. This figure, combined with inappreciable precipitation and extremely hot temperatures, makes this area as one of the most arid on the planet.

A_i values less than 0.65 are typical of drylands (Thomas and Middleton 1994) and UNEP (1992) takes 0.65 as the threshold value to identify desertification prone areas. Large portions of Ethiopia fall in this category: the whole Ogaden, the Afar triangle, the Danakil depression, the northern portion of the Rift Valley, and the belt across the border with Sudan (Fig. 3.19).

3.12 Climate Classification

As highlighted in Sect. 3.3, the climate of Ethiopia is mainly controlled by the seasonal migration of the ITCZ and by the complex topography of the country. This combination results in a variety of climate types and climate conditions that may vary also within a short distance (Fig. 3.20).

A tight relation is evident between elevation and climate type, especially along the plateau margins. East of this margin and in the southeastern portion of the country, a predominantly arid and very hot climate (tropical *A*, Köppen classification) fades, throughout a belt of climatic transition at higher elevations, into a temperate C climate, characterized by mild temperatures, higher than 18 °C in every month, and by rainfalls with a maximum in summer (Tigray and Somali plateau margin), in spring and a secondary

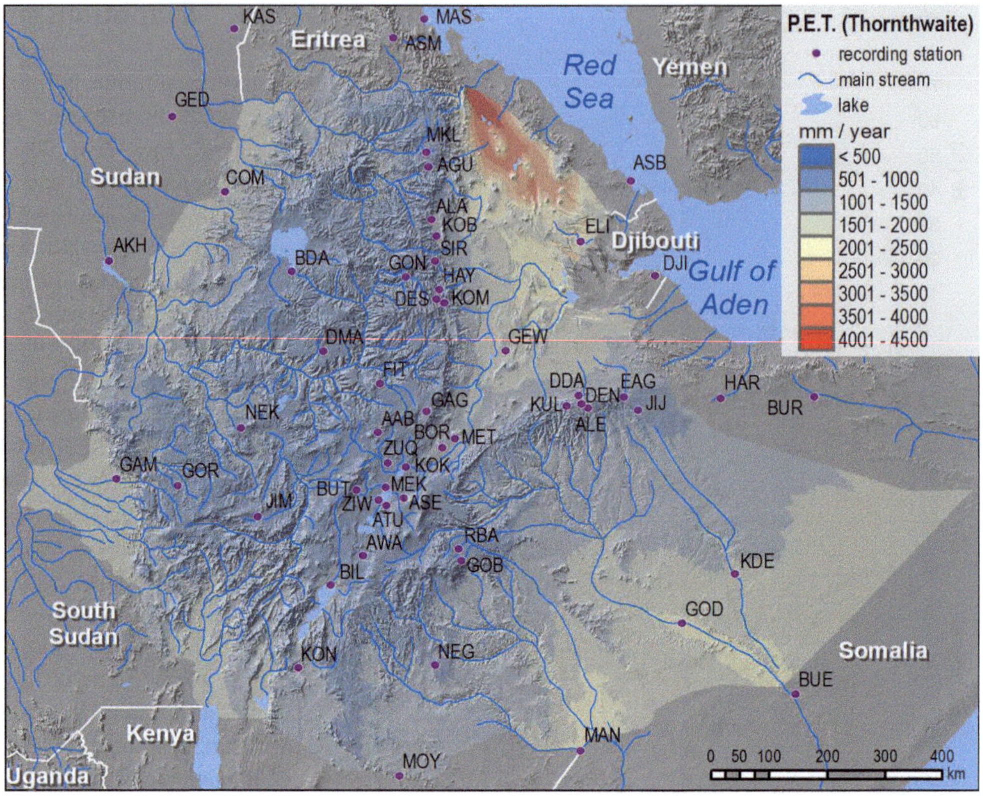

Fig. 3.18 Spatial distribution of potential evapotranspiration

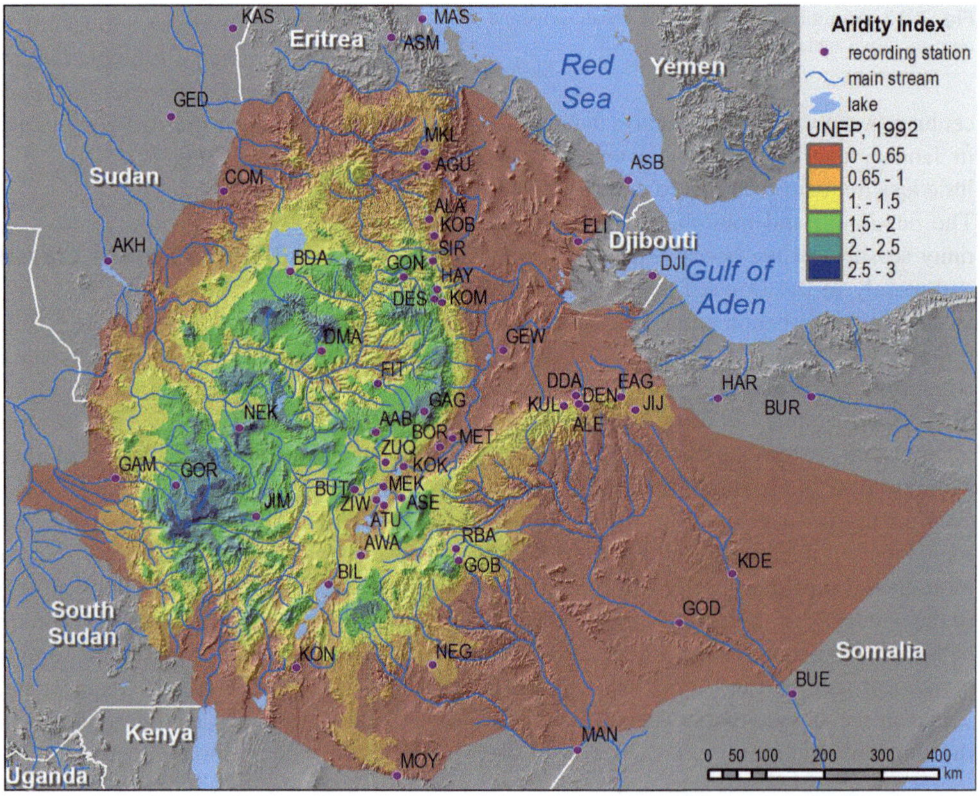

Fig. 3.19 Spatial distribution of the aridity index (UNEP 1992)

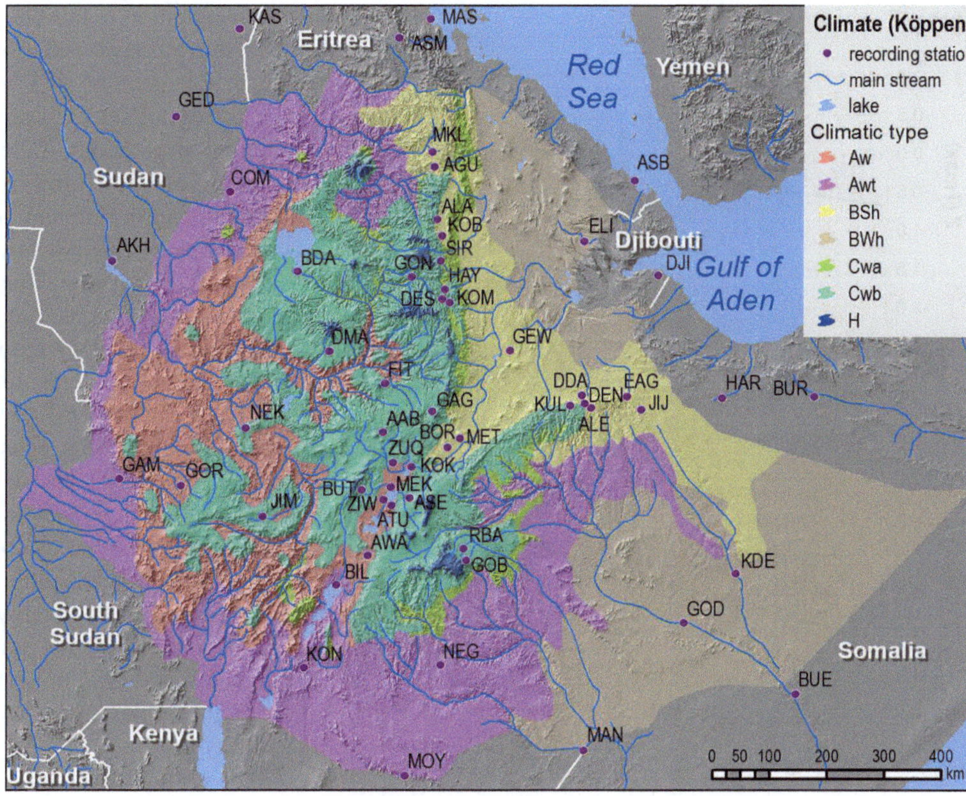

Fig. 3.20 Map of the Köppen climate classification

maximum in autumn (southern portion of the country) rather than constant precipitation throughout the four seasons (type *AM* of the southwestern areas). The thermal transition from *Af*, *Aw*, or *BSh* to the temperate *C* climate occurs at an elevation ranging between 1,800 and 2,000 m a.s.l., though in the northernmost part of the highlands the climate becomes cooler and remains arid also at high elevations (e.g. Asmara, type *BSh*). In the areas with a temperate climate, the temperature ranges are never large and the minimum mean temperatures are recorded in summer, during the big, monsoon-type rains, whereas the milder months are in spring. The prevailing climate is *Cw*, with a patent scarcity of precipitation during the winter semester (*bega*).

In the tropical climate region with a dry season in winter (*Aw*), areas characterized by the maximum temperature in spring—just before the beginning of the *kiremt*—and the minimum temperature during the perturbed phase (July or August)—climate sub-types *Awg* or *Awt*—are also found (e.g., Jimma, Gambela, and Gonder). In the Rift Valley (e.g., Awasa, Ziway), these conditions are paired by a particularly small annual range (less than 3 °C) and the climate is *Awi*.

In places of the southwestern portion of the country (e.g., Nekemte and Bonga), rainfalls are rather uniformly distributed in all months (cool variant of the tropical climate) and determine the *Cfc* climate.

Over 2,600 m a.s.l. and in a few areas of the Semen and Bale Mountains, the climate becomes relatively cold with mean monthly temperatures less than 10 °C. At elevations higher than 3,300 m a.s.l., the climate is characterized by very low temperatures and it is classified as type *H* which includes the peaks of the intertropical massifs. Here, vegetation consists mainly of alpine prairie and arboreal species is absent.

3.13 Climate Change

Climate change is one of the most investigated issues of Ethiopian climate since its rain-fed agriculture is largely dependent on the amount and regular onset of seasonal precipitation (Hadgu et al. 2013). Throughout the last century, temperature and rainfall trend lines indicate a climatic anomaly that cannot be associated with ordinary cyclic oscillations of these parameters.

In all the meteo-stations considered, an average temperature increase of 1.1 °C is recorded for the 1981–2010 interval (0.04 °C per year). The same increasing rate is observed for the mean minimum and maximum temperatures. The highest rates are recorded at Neghelle (2.7 °C in 30 years), Gonder and Robe Bale (2 °C in 30 years). Such marked positive trends do not seem to be influenced by the

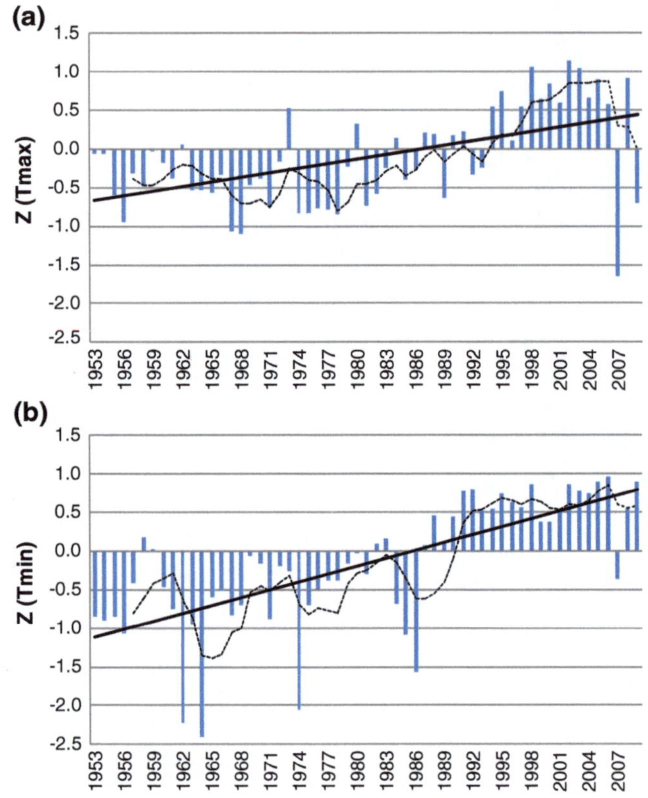

Fig. 3.21 Variation through time of normalized mean maximum (**a**) and minimum (**b**) temperatures. *Dotted line* is the 5 years moving mean

local physiography nor by the expansion of the urban areas. In fact, at Addis Ababa, the temperature change is negligible (0.15 °C/30 years) as are the modest decreasing rates measured at Mekele, Alamata, and Arba Minch (0.1–0.2 °C/30 years).

Eight meteo-stations have longer time series starting in the 1950s (Fig. 3.21). The analysis of their normalized data shows a higher increasing rate as far as the 1980s, whereas during the last 15–18 years the thermal signal becomes stable at 0.6–0.8 °C above the semi-secular average. It is also worth noticing that for these stations, the mean temperature increase is mainly due to the contribution of the minimum temperatures that increase at a rate twice as much the maximum temperatures.

Figure 3.22 shows the individual temperature changes recorded by the same eight meteo-stations for two periods: 1953–1980 and 1981–2010. At Addis Ababa, the mean temperature reported by Fantoli (1940) for three decades at the beginning of the twentieth century is the same (16.4 °C) as that of the 1953–2010 interval, but the former have a much larger daily excursion, i.e., 15.2 versus 13.2 °C.

Precipitation trends are very difficult to analyze since the time series considered are punctuated by gaps and, sometimes, the data of 3–5 years in a row are missing. As discussed in Sect. 3.5, any attempt to fill the gaps with a correlation procedure failed to give reliable results. Moreover, the available time series span different intervals which start and end in different years. Nevertheless, some general, statistically nonsignificant tendency can be depicted on the basis of 20 of the longest (29–46 years) time series selected. 12 stations out of 20 show a negative trend; however, in order to reduce the weight of the gaps scattered across the data set, the trend analysis has been restricted to the 1961–2009 interval and a representative time series has been constructed by averaging the annual precipitation at all the stations for each year. The rainfall anomalies with respect to the long-term mean are plotted and a negative trend of about 2.6 mm per year is observed (Fig. 3.23). This diagram confirms that the 1980s and the first decades of the twenty-first century were characterized by recurrent droughts that caused the spread of severe famines from 1983 to 1985, known as the Great Famine (1 mln fatalities), and another one in 2003 (Bewket and Conway 2007).

The normalized data of the *kiremt* (Fig. 3.24 a) and *belg* (Fig. 3.24 b) rains over the last five decades show that the spring rains are decreasing at a rate 1.5 times faster than the monsoon summer rains. As discussed above, the *belg* rains are less voluminous, less reliable, and poorly predictable, and though agriculture in Ethiopia rely mainly on the summer rains, the decreasing trend of the spring rains may pose severe constraints to the economic development of the country.

The 20 meteo-stations selected are too small in number to sort out any significant regional pattern of change in precipitation as neighboring stations may have contrasting trends (e.g., Gore and Jimma, the trend line angular coefficient of which is −16.2 and +1.4, respectively).

Very few stations have data suitable for an analysis of maximum rainfall intensity variation with time. However, it seems that no significant change occurred during the 1961–2009 interval.

Older data are available only for very few stations and refer mainly to the study of Fantoli (1940) who worked on data ranging approximately from 1890 to 1936, with different lengths and time intervals, but for some stations the beginning of the measurements is not indicated. Any comparison with this author's data is therefore speculative since neither information is provided about the exact location of the rain gauge nor and whether it was moved during the operation period as discussed by Conway et al. (2004).

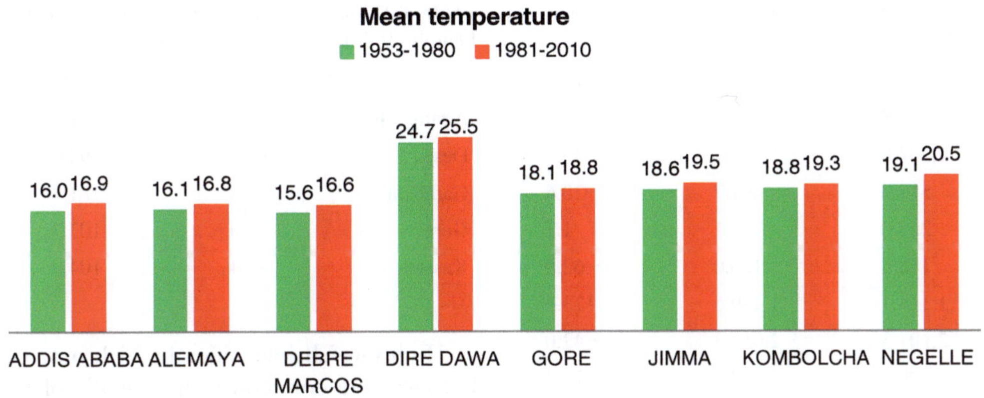

Fig. 3.22 Mean annual temperatures recorded at eight meteo-stations with long data records relative to two distinct intervals, 1953–1980 and 1981–2010

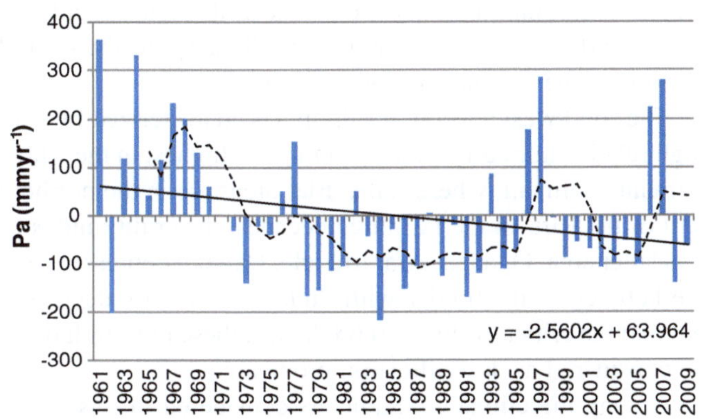

Fig. 3.23 Rainfall anomalies obtained by averaging the annual precipitation of all the meteo-stations considered (*Pa*). *Dashed line* is the 5 years moving mean

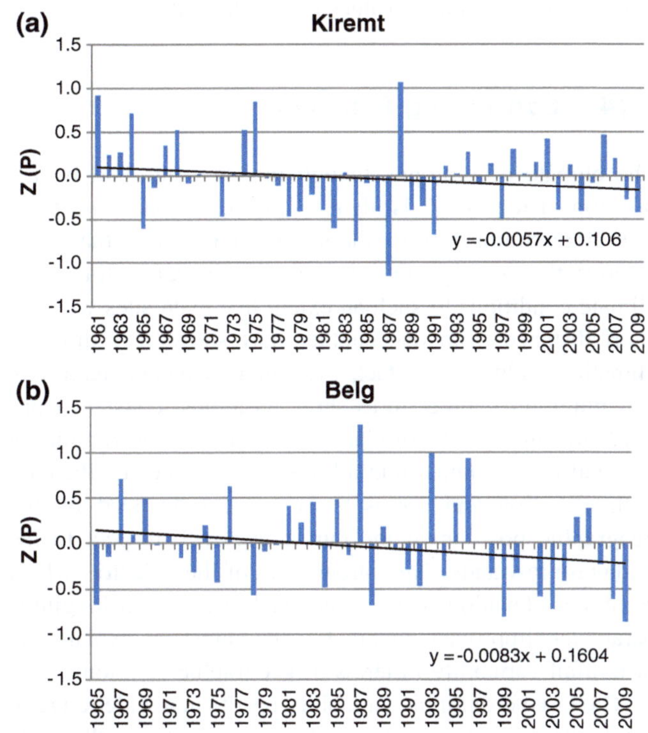

Fig. 3.24 Variation through time of normalized *kiremt* (**a**) and *belg* (**b**) precipitation

However, just for a qualitative reference, the differences in annual precipitation for the 1890–1936 and the 1960–2009 intervals recorded by ten meteo-stations are reported in Table 3.3. In six out of ten stations, a moderate to marked decrease in rainfall emerges, with an average value of −81.1 mm, and in two of them (Adami Tulu and Kassala), it is very pronounced (around −176 mm). The average rainfall increase of the other four stations is 58.4 mm, but at Gambela, a maximum difference of +175 mm is observed.

Also for the analysis of rainfall intensity change over a long time, only the data of Fantoli (1940) were used. Unfortunately, only five stations can be used for a comparison with the modern data of this study (Table 3.4). Two stations, Adami Tulu and Gore, show a modest decrease between the two periods with −5.9 and −19.3 mm in 24 h, respectively, and are paired by a small increase at Dese and Kassala, 7.0 and 15.5 mm/24 h, respectively. The most relevant change is observed at Gambela with an increase of about 100 mm in 24 h. The significance of these results, for the reasons explained above, is highly questionable and is to be taken just as a mere indication. Though Fantoli (1940) provides some data on other climate parameters such as relative humidity, evaporation and wind speed, they refer to very few stations and any comparison with modern data is totally insignificant.

Finally, within this framework of climatic change, it is also worth noticing that a few areas, such as the Rift Valley, characterized by aridity index values in the 0.65–0.8 range,

Table 3.3 Mean annual precipitation difference between the old data of Fantoli (1940) and the present study

Meteo-stations	1890–1936	1960–2009	Difference
Adami Tulu	756.8	581.1	−175.7
Addis Ababa	1,204.8	1,251.7	46.9
Asmara	512.3	491.7	−20.6
Belet Uen	259.0	260.1	1.1
Djibouti	188.0	126.0	−62.0
Gambela	1,114.1	1,289.5	175.4
Gore	2,101.3	2,059.3	−42.0
Kassala	500.0	322.6	−177.4
Massawa	175.8	186.0	10.2
Moyale	685.0	675.9	−9.1

in the last two decades, frequently experienced desertification conditions with A_i values of less than 0.5.

3.14 Concluding Remarks

The climate of Ethiopia, though mainly controlled by the position of the ITCZ, is highly variable in space and time. The large contrasts in elevation and morphology that characterize the country are very effective in controlling local climate conditions to such an extent that even sites within a short distance from one another may show very different climatic conditions. In fact, even though mean, maximum and minimum temperatures show an almost constant lapse rate, no significant correlation was found for rainfall at national level. Annual rainfall does not change significantly with elevation and seems to be more influenced by the geographic position of the rain gauge with respect to the Sudanese lowlands. The introduction of the *F* factor, which combines elevation with a specifically developed longitude parameter, improved remarkably the statistical explanation of rainfall variability in terms of geographic position.

The large orographic variability results in a wide range of temperatures that may be as low as −7 °C in the Bale Mountains and as much as 50 °C in the Danakil Desert.

The higher rainfalls are recorded in the western highlands, with 2,101 mm year^{-1} at Gore, and the lower in the Afar depression with 145 mm year^{-1}, though even lower values are expected for the desert areas of the Danakil. The summer, monsoon rains account for more than 50 % and as much as 85 % of the annual rainfall, whereas the spring rains are more variable in amount and predictability.

Table 3.4 Maximum daily rainfall comparison between the old data of Fantoli (1940) and the present study

	1890–1936	1960–2009	Difference
Adami Tulu	87.8	81.9	−5.9
Dese	87.0	94.0	7.0
Gambela	81.3	181.4	100.1
Gore	127.0	107.7	−19.3
Kassala	89.4	104.9	15.5

High rainfall intensities in 24 h may occur every month but are more common in the middle of the two main rainy seasons. The highest average and absolute rainfall intensities are recorded at Gambela (89.4 and 181.4 mm/24 h, respectively), whereas the lowest values are measured at Elidar and Gewane in the Afar triangle. The MFI shows that rainfalls are very aggressive over large part of Ethiopia; the higher values of rainfall erosivity are found in the central and part of the northern highlands and, subordinately, in the central portion of the Somali plateau margin.

The southwest portion of Ethiopia is characterized by very high relative humidity values, among the highest in the whole African continent, whereas the drier areas are in the northern part of the Rift. In the last three decades, the aridity index of this latter area, Ogaden, Afar triangle, Danakil depression and the belt across the border with Sudan, commonly decreased beyond the critical value of 0.65; hence, these regions have to be considered as desertification prone areas.

Among the climatic parameters considered in this study, long time series were available only for temperature and precipitation data. The change of both maximum and minimum temperature through the last three decades is rather evident with an average increase of 1.1 °C. The data analysis of the eight weather stations with longer time series (starting in the 1950s) shows a marked temperature increase as far as the 1980s, whereas its rate slowed down and has become almost stable in the last 15–18 years. In most of the stations considered, it is the minimum temperature, rather than the maximum temperature that contributes to the mean temperature increase. The former increases at a rate twice as much higher than the latter. For the eight stations with longer time series, the comparison between 1953–1980 and 1981–2010 intervals indicates that in the most recent decades, mean temperatures show an average increase of about 0.9 °C. By contrast, annual precipitation is characterized by a decreasing trend with an average decline of about 125 mm in 49 years, with the small spring rains decreasing at a rate 1.5 time higher than the monsoon summer rains.

References

Alexandersson H (1986) A homogeneity test applied to precipitation data. J Clim 6(6):661–675

Arnoldus HM (1980) An approximation of the rainfall factor in the universal soil loss equation. In: de Boodts M, Gabriels D (eds) Assessments of erosion. Wiley, Chichester, pp 127–132

Beltrando G, Camberlin P (1993) Interannual variability of rainfall in the eastern Horn of Africa and indicators of atmospheric circulations. Int J Climatol 13:533–546

Berhanu B, Melesse AM, Seleshi Y (2013) GIS-based hydrological zones and soil geo-database of Ethiopia. Catena 104:21–31

Bewket W, Conway D (2007) A note on the temporal and spatial variability of rainfall in the drought-prone Amhara region of Ethiopia. Int J Climatol 27:1467–1477

Chernet T (1982) Hydrogeology of the Lakes Region, Ethiopia. Memoir N.7, the Provisional Military Government of Socialist Ethiopia, Addis Ababa

Cheung WH, Senay GB, Singh A (2008) Trends and spatial distribution of annual and seasonal rainfall in Ethiopia. Int J Climatol 28:1723–1734

Conway D (2000a) Some aspects of climate variability in the north east Ethiopian Highlands—Wollo and Tigray. SINET—Ethiop J Sci 23:139–161

Conway D (2000b) Climate and hydrology of the upper blue Nile river. Geogr J 166(1):49–62

Conway D, Mould C, Bewket W (2004) Over one century of rainfall and temperature observations in Addis Ababa, Ethiopia. Int J Climatol 24:77–91

Easterling DR, Evans JL, Groisman PYa, Karl TR, Kunkel KE, Ambenje P (2000) Observed variability and trends in extreme climate events. Bull Am Meteor Soc 81:417–425

Enku T, Melesse AM (2013) A simple temperature method for the estimation of evapotranspiration. Hydrol Process. doi:10.1002/hyp.9844

ENRAEMED (2003) Ethiopian natural resources and environmental meta-database. http://clearinghouse5.fgdc.gov/enraemed/NMSA.php

Eredia F, DeCastro L (1914) Sula climatologia dell'Etiopia. Boll Real Soc Geog Ser V III(8):843–884

Ethiopian Mapping Authority (EMA) (1988) The National Atlas of Ethiopia. EMA, Addis Ababa

Fantoli A (1940) Elementi preliminari del clima dell'Etiopia. Sansoni, Firenze

Fantoli A (1965) Contributo alla climatologia dell'Etiopia. O.P.I, Roma

FAO (1984) Agroclimatological data for Africa, vol 1: Countries North of the Equator. FAO Plant Production and Protection Series No. 22, FAO, Rome

Gamachu D (1977) Aspects of climate and water budget in Ethiopia. Addis Ababa University Press, Addis Ababa

Gamachu D (1988) Some patterns of altitudinal variation of climatic elements in the mountainous regions of Ethiopia. Mount Res Dev 8 (2/3):131–138

Hadgu G, Tesfaye K, Mamo G, Kassa B (2013) Trend and variability of rainfall in Tigray, Northern Ethiopia: analysis of meteorological data and farmers' perception. Academia J Env Sci 1(8):159–171

Houghton JT, Ding Y, Griggs DJ, Noguer M, van der Linden PJ, Xiaosu D (eds) (2001) Climate change 2001: the scientific basis: contributions of working group I to the third assessment report of the intergovernmental panel on climate change. Cambridge University Press, Cambridge

Hurst H, Black P (1937) The Nile basin vol VI. Monthly and annual rainfall totals and the number of rainy days at stations in and near the Nile basin for the period ending 1937. Ministry of Public Works Press, Cairo

Jury MR, Funk CR (2013) Climatic trends over Ethiopia: regional signals and drivers. Int J Climatol 33:1924–1935

Korecha D, Barnston AG (2007) Predictability of June–September rainfall in Ethiopia. Mon Weather Rev 135(2):628–650

Law B, Falge E, Baldocchi D, Bakwin P, Berbigier P, Davis K, Dolman A, Falk M, Fuentes J, Goldstein A, Granier A, Grelle A, Hollinger D, Janssens I, Jarvis P, Jensen N, Katul G, Mahli Y, Matteucci G, Monson R, Munger W, Oechel W, Olson R, Pilegaard K, Paw UK, Thorgeirsson H, Valentini R, Verma S, Vesala T, Wilson K, Wofsy S (2002) Environmental controls over carbon dioxide and water vapor exchange of terrestrial vegetation. Agric For Meteorol 113:97–120

Lovett R, Wood CA (1976) Rainfall reliability in Ethiopia. Weather 31:417–424

McSweeney C, New M, Lizcano G (2010) UNDP climate change profile: Ethiopia. http://countryprofiles.geog.ox.ac.uk

Mekasha A, Tesfaye K, Duncan AJ (2013) Trends in daily observed temperature and precipitation extremes over three Ethiopian eco-environments. Int J Climatol. doi:10.1002/joc.3816

Mengistu D, Bewket W, Lal R (2013) Recent spatiotemporal temperature and rainfall variability and trends over the upper blue Nile river basin. Int J Climatol, Ethiopia. doi:10.1002/joc.3837

Pedgley DE (1967) Air temperature at Dallol, Ethiopia. Meteorol Mag 96(1967):265–271

Segele ZT, Lamb PJ (2005) Characterization and variability of Kiremt rainy season over Ethiopia. Meteorol Atmos Phys 89(1):153–180

Seleshi Y, Camberlin P (2006) Recent changes in dry spell and extreme rainfall events in Ethiopia. Theoret Appl Climatol 83:181–191

Seleshi Y, Zanke U (2004) Recent changes in rainfall and rainy days in Ethiopia. Int J Climatol 24:973–983

Shanko D, Camberlin P (1998) The effects of the southwest Indian Ocean tropical cyclones on Ethiopian drought. Int J Climatol 18:1373–1388

Tadege A (ed) (2001) Initial national communication of Ethiopia to the United Nations Framework Convention on Climate Change (*UNFCCC*). National Meteorological Services Agency, Addis Ababa

Tadege A (ed) (2007) Climate change national adaptation programme of action (NAPA) of Ethiopia. National Meteorological Services Agency, Addis Ababa

Taye M, Zewdu F (2012) Spatio-temporal Variability and Trend of Rainfall and Temperature in Western Amhara, Ethiopia: a GIS approach. Glob Adv Res J Geogr Reg Plann 1(4):65–82

Thomas DSG, Middleton NJ (1994) Desertification: exploding the myth. Wiley, Chichester

UNEP (1992) Status of desertification and implementation of the United Nations plan of action to combat desertification. Report of the Executive Director, United Nations Environment Programme, Nairobi

Vinassa de Regny P (1931) La geologia delle Alpi Dancale. Bollett Soc Geol Ital 50(1):1–24

Viste E, Korecha D, Sorteberg A (2013) Recent drought and precipitation tendencies in Ethiopia. Theoret Appl Climatol 112 (3–4):535–551

WWR (1959) World weather records 1941–50. US Department of Commerce, Weather Bureau, Washington, DC

Ethiopian Rivers

4

Paolo Billi, Semunesh Golla, and Dawit Tefferra

Abstract

The scientific literature about the geomorphology and hydrology of Ethiopian rivers is very poor, though large and socially important rivers have their source in this country. The largest rivers deliver their waters into the Mediterranean Sea or the Indian Ocean, whereas the most of the smaller ones have an endorheic drainage. Flow data were collected from different sources, and the longest as possible time series were considered. A new index, the runoff concentration index is used by analogy with the precipitation concentration index to analyze the monthly variation of runoff. Characteristic discharges such as bankfull discharge or longer return time discharges are calculated as well. Due to the remarkable variety of physiography of Ethiopia landscapes and recent tectonics, rivers show different drainage network and channel morphology. Their main features are described, and recent geomorphological changes are analyzed. Flow data are investigated and a number of correlations with catchment parameters, capable also to predict discharge in ungauged rivers, are reported. Runoff changes through time do not show any significant trend, whereas the sediment yield of Ethiopian river is of the same order of magnitude of much larger African rivers indicating a high soil erosion rate of Ethiopian highlands.

Keywords

Hydrography • Channel morphology • Discharge • Runoff • Sediment transport

4.1 Introduction

Rivers are among the most powerful, effective, and ubiquitous geomorphic agents, capable to shape the Earth surface and to produce a large variety of landscapes. Rivers are present and active all over the world, draining almost 70 % of the Earth land surface (Petts 1983) and encompassing almost all environments from cold polar regions to hot drylands. That is the result of the deployment of a huge amount of energy that rivers use to convey runoff, generated by excess precipitation, and sediment supplied by erosions processes on slopes. The morphological, hydrological, and hydraulic characteristics of rivers differ in space and time and local geological, climatic, and, recently, human factors make each river a unique physical system. Rivers are a resource and a hazard at the same time. Their political, social, economic, and environmental relevance are patent to everyone and both industrialized and developing countries have benefited from their sustainable management. With the recent climate changes and a growing human impact, rivers became central to many environmental problems and in many regions, the resulting increased risks, associated with extreme low or high flows, are posing serious constraints forcing land managers and decision makers to rethink the current approach to urban and rural areas planning within a short timescale.

Nowadays, the scientific literature on rivers, their characteristics, and processes is extensive and studies on river morphology, sedimentology and hydrology have been carried

P. Billi (✉)
Physics and Earth Sciences, University of Ferrara, Ferrara, Italy
e-mail: bli@unife.it

S. Golla · D. Tefferra
Ministry of Water, Irrigation and Energy, Addis Ababa, Ethiopia

P. Billi (ed.), *Landscapes and Landforms of Ethiopia*, World Geomorphological Landscapes,
DOI 10.1007/978-94-017-8026-1_4,

out in almost any region of the world (see Knighton 1993; Bridge 2003—for a comprehensive reference list). By contrast, information about the geomorphology and hydrology of Ethiopian rivers is almost nonexistent and generally limited to the main Ethiopian river, the Blue Nile (Abay in the local language) (Conway 2000; Tesemma 2010). A relatively larger number of studies were instead carried out on much smaller streams in order to investigate other important aspects such as sediment transport (Billi 2000, 2004, 2011), sediment yield (Haregeweyn 2008 and Chap. 13 of this volume; Vanmaerke et al. 2010) and channel morphology (Billi 2007, 2008). The scarcity of papers on Ethiopian rivers is due to many reasons, the most important of which is the lack of data. In fact, though the Ethiopian Ministry of Water, Irrigation and Energy has recently greatly improved the network of flow gauges across the country, the number of rivers with a long and continuous record of data is still very limited. As a consequence of that and, at the same time, aiming to attempt to supplement and to go beyond the present data gaps, most of the recent papers on Ethiopian rivers deal with hydrological modeling to predict flow regime (Collick et al. 2009; Wale et al. 2009; Legesse et al. 2010; Melesse et al. 2010; Uhlenbrook et al. 2010; Di Baldassarre et al. 2011; Gebrehiwot et al. 2011), sedimentation (Ahmed 2008; Billi and Bedri 2010; Easton et al. 2010) and sediment transport (Haregeweyn et al. 2008 and Chap. 13 of this volume; Steenhuis 2009; Vanmaercke et al. 2010; Billi 2011) or to assess the impact of climate change (Conway 2000; Kim et al. 2008; Abdo 2009; Gebremicael et al. 2013), land use change (Tamene 2006; Gumindoga 2010; Rientjes et al. 2011) and human impact (Batlthazar et al. 2013) on river hydrology and sediment yield. Though these studies may provide a useful framework to depict the main hydrological and sedimentation characteristics of the Abay/Blue Nile and other smaller rivers, still large gaps of basic knowledge on river processes in Ethiopia do exist. Some papers, in fact, are based on old data (e.g., Gebrehiwot et al. 2011), have to cope with insufficient data (e.g., Avery 2010) or test their models against very short hydrological time series.

The aim of this presentation is to provide an updated picture of the hydromorphological characteristics of the Ethiopian rivers on the base of the largest as possible data set, despite the data availability limitations reported above.

4.2 General Setting and Hydrologic Data

The hydrography of Ethiopia consists of 12 river systems and the ten larger ones are reported in Fig. 4.1. The western ones, namely Tekeze, Abay/Blue Nile, and Baro, flow into the Mediterranean Sea, whereas those of the east and Ogaden, Wabe Shebele and Genale-Dawa, flow into the Indian Ocean. The river systems in between the former, Danakil, Awash, Rift Valley, and Omo, are instead closed basin. Other small endorheic systems are found within the western main basins and are associated with the drainage into small lakes such as Ashange or structural basins such as the Kobo–Alamata basin (see Chap. 19).

Within the borders of Ethiopia are included also a small portion of the Mereb River headwaters (the small green area close to the border with Eritrea in Fig. 4.1), the headwaters of small ephemeral streams forming terminal fans in the lowland between Dikhil and Ali Sabih within the Djibouti territory and some tributaries of larger ephemeral streams, running through Djibouti and outflowing into the Indian Ocean near the town of Zeylac in Somalia (the dark yellow area close to the border with Djibouti in Fig. 4.1).

The Tekeze R. has a catchment area of 82,350 km^2 (Awulachew 2007) and takes its source from Abuna Yosef (4,190 m a.s.l.) in the portion of the northern highlands across Tigray and Amhara region that includes also the Simen Mountains with the highest peak of Ethiopia, the Ras Dashen (4,620 m a.s.l.). The river turns around the Simen Mountains, points westward to join the Atbara R. in Sudan, near the small town of Showak, and proceeds with the name of Atbara as far as its confluence with the River Nile near the town of Atbara. Main tributaries within the Ethiopian territory are the Tirari and the Zarema rivers, whereas the Angereb R. is a tributary of the Atbara (Fig. 4.2).

The Abay/Blue Nile is the largest river of Ethiopia. It is 1,610 km long and drains a watershed of about 176,000 km^2 upstream of El Diem (Kim 2008). It originates from Lake Tana that spills out by the town of Bahir Dar. The lake is fed by small rivers, the largest of which are the Gilgel Abay (catchment area is 4,100 or 5,000 km^2, according to Abdo et al. 2009 and Uhlenbrook et al. 2010, respectively, though the first figure seems to be more appropriate) from the south, the Rib (1,303 km^2) and Gumara (1,283 km^2) from the east (Wale et al. 2009; Poppe et al. 2013) and the Megetch river (514 km^2—Wale et al. 2009) from the north. According to Kebede et al. (2006), these four major rivers contribute 93 % of the inflow. From Lake Tana, the Abay/Blue Nile flows southward into a deep canyon and then makes a large turn to the west pointing to the Sudan border before which it makes another wide turn, first to the south and then to the north. The river proceeds beyond the Sudanese border to join the White Nile at Khartoum. Within Ethiopia, the largest left tributaries are the Didessa, with a watershed of about 25,800 km^2 (Sima 2011) and Dabus (21,032 km^2—World Bank 2008) rivers, whereas the largest right tributary is the Beles R. (14,200 km^2—World Bank 2008). Other two important right tributaries, the Dinder and the Rahad (14,891 and 8,269 km^2, respectively—World Bank 2008) have their headwaters in Ethiopia, but join the Blue Nile in Sudan (Fig. 4.2).

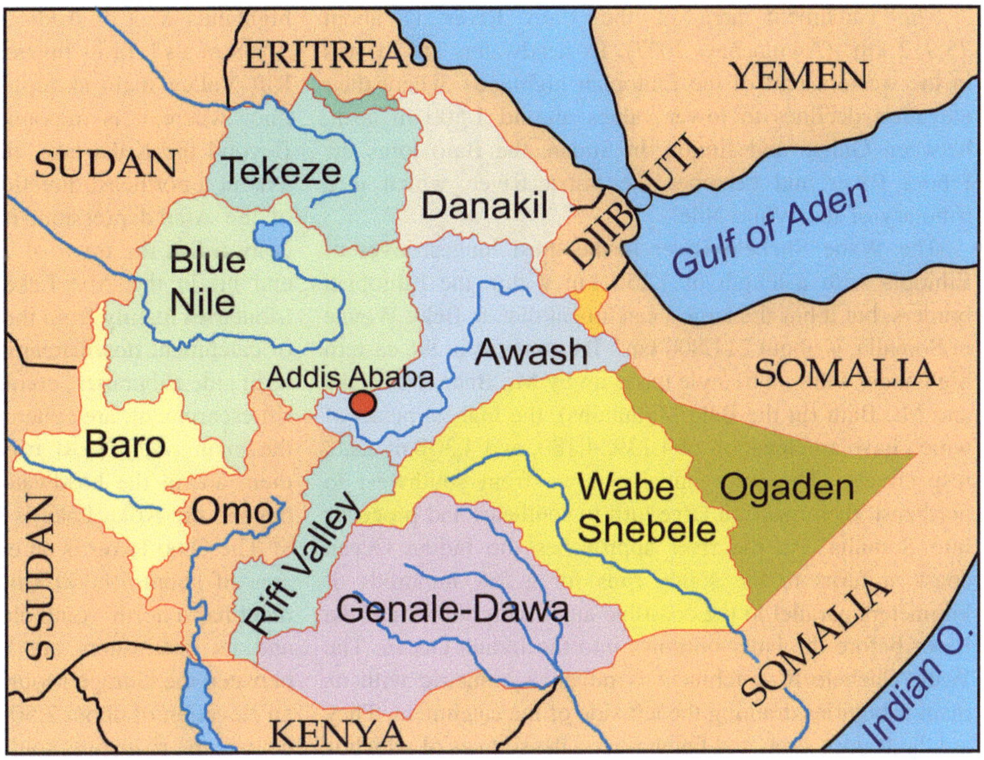

Fig. 4.1 The main hydrographic basins of Ethiopia

Fig. 4.2 The study rivers. The triangles indicate the flow gauges

The catchment area of the Baro River is about 75,912 km^2 (Awulachew 2007). Its headwaters are located in the western part of the Ethiopian highlands, where their elevation declines to lower values around 1,500 m a.s.l., between Guliso and Jimma. In Sudan, the Baro joins the Pibora River and becomes the Sobat River, which is a tributary of the White Nile.

The Wabe Shebele River is the third longest river of Ethiopia with a length of 1,150 km within the Ethiopian borders, but it has the largest catchment that, at Belet Weyne in Somalia, is about 211,800 km^2. Its source is in the eastern highlands, within a triangle made up by Mt. Bada, Mt. Kaka and Mt. Batu (in the Bale Mountains), the highest peaks of which have an elevation of 4,139, 4,180, and 4,307 m a.s.l., respectively. The river initially flows from southwest to northeast, then makes a large turn to southeast and proceeds into Somalia. As the river approaches the Indian Ocean coast, it turns to the south, runs for a few hundreds of kilometers parallel to the coastline and finally joins the Juba River before the latter outflows into the Indian Ocean. The Wabe Shebele R. catchment is rather asymmetric with its main tributaries draining the left side of the catchment. They are the Ramis, Erer, and Fafen rivers, the sources of which is the eastern highlands, from Mt. Gara Muleta (3,381 m a.s.l.) to Harar.

The Genale-Dawa is the third largest river system of Ethiopia with a catchment of about 171,000 km^2 (Awulachew 2007). It originates from the Bale Mountains and the southeastern margin of the Rift Valley. Within the Ethiopian territory, the main tributary is the Weyb River, the source of which is from Mt. Batu. This tributary joins the Genale a few kilometers before this river crosses the border with Somalia, beyond which it takes the name of Juba River and, shortly downstream of the confluence of the Wabe Shebele, outflows into the Indian Ocean.

The Ogaden river systems (Fig. 4.1) consist of parallel, relatively small ephemeral rivers that originate in the highlands south of Jijiga and from the margin of the Gulf of Aden Rift. These rivers do not have any outlet into the sea since they end up due to infiltration and evaporation around the border with Somalia, at elevations between 400 and 300 m a.s.l. and at a distance of 200–300 km from the Indian Ocean coastline.

The Danakil river systems are all ephemeral and drain endorheic basins. None of these rivers, in fact, has an outlet into the Indian Ocean. The base level of a very few of them are small lakes such as Afrera and Asele, whereas the majority of Danakil rivers forms terminal fans or distributary systems on the borders or on the floor of structural basins (see also Chap. 19).

The Awash river is the second longest river of Ethiopia (about 1,200 km) and has a catchment area of about 112,696 km^2 (Awulachew 2007). Its source is in the highlands west of Addis Ababa, where the highest peak is 3,198 m a.s.l. In its upstream reach, the river runs down the Rift Valley main escarpment and then, it reaches the Rift floor where it is impounded to form the Koka reservoir. Beyond the Koka lake, the river flows parallel to the Rift axis in a northeast direction. As the river enters the triangle of the Afar depression, it slightly turns to the north, then eastward at the town of Tendaho and finally southward to end up in the Abe Lake. The Awash River receives its tributaries mainly from the left side, especially in the portion of catchment downstream of the Koka reservoir, where the right side tributaries, coming from the eastern margin of the rift escarpment, are ephemeral and commonly unable to join the main river. Most tributaries are small. The largest of them are in the headwaters basin, whereas the larger one beyond the Koka Lake is Mille River (4,467 km^2).

The Omo River is an endorheic system with a catchment area of about 79,000 km^2 (Awulachew 2007). The catchment has a north–south elongated, almost rectangular shape, and its headwaters are located in the central highlands, between the Gurage Mountains and the town of Nekemte, at an elevation of about 2,500 m a.s.l. For the largest part of its course, the river runs southward and after 760 km enters the Turkana Lake, which is its base level, at an elevation of about 360 m a.s.l.

All the Rift Valley river systems (Fig. 4.1) drain closed basins since they outflow into the large lakes on the rift floor. The largest of these rivers are the Segen (13,612 km^2), that supplies Chew Bahir Lake in the southernmost portion of the Rift Valley, close to the Kenya border; the Bilate (5,791 km^2), the source of which is in the Gurage Mountains (Main Ethiopian Rift Valley); the Gidabo (3,473 km^2) and the Gelana (3,411 km^2) (Bilete 2009), both coming from the eastern margin of the southern Rift, with the latter three rivers entering the Abaya Lake.

4.2.1 Hydrological Data

Most of flow data were obtained from the Ministry of Water, Irrigation and Energy of Ethiopia in the form of monthly discharge data with additional information on maximum and minimum discharge. Though Ethiopia is currently committed to improve its network of flow gauges, many important rivers are still ungauged, data gaps are still present and time series are discontinuous with the exception of a few rivers. Due to high variability of catchment altitude, physiography, orientation and spatial rainfall distribution, any attempts to fill the gaps by correlation method with reference rainfall time series fail and the original data were used without any manipulation that would only artificially and unrealistically increase their validity and reliability. Additional data were taken from international organizations databases such as

UNESCO, FAO-SWALIM and RIVDIS (Vorosmarty et al. 1998).

Notwithstanding these limitations, it was possible to compile a relatively large and complete database of river flows that covers almost the entire territory of Ethiopia and includes the most important river systems. The main hydrological data of the rivers considered in this study are reported in Table 4.1. In this table, the rivers/flow gauges were divided into three categories: (1) highlands, i.e., rivers with the headwaters in the Ethiopian highlands and flowing mainly in Ethiopia, with the most downstream flow gauge in Ethiopia; (2) lake fed, i.e., those exiting from a lake/reservoir or with a flow gauge located a few kilometers downstream of it; (3) Ethiopian/Somali, i.e., those rivers with the headwaters in Ethiopia but with the most downstream reference flow gauge located a few kilometers beyond the Somali border.

A new index, the runoff concentration index (RCI) is also included. The RCI is introduced by analogy with the precipitation concentration index proposed by Oliver (1980) in order to define the monthly runoff variability within the year as follows:

$$\mathrm{RCI} = 100 \times \sum r_i^2 / R_a^2 \qquad (4.1)$$

in which r_i is the runoff volume in each individual month and R_a is the annual runoff volume. According to Oliver (1980), a PCI, and by analogy a RCI, less than 10 suggests a

Table 4.1 Main hydrological characteristics of the Ethiopian rivers considered in this study

River/gauge	A (km^2)	R_a (10^6 m^3)	R_{au} (10^3 m^3 km^{-2})	R_a CV	RCI	Q_m (m^3 s^{-1})	$Q_{1.58}$ (m^3 s^{-1})	Q_{bu} (m^3 s^{-1} km^{-2})	Q_{10} (m^3 s^{-1})	Q_{peak} (m^3 s^{-1})
Highland										
Abay/Blue Nile @ Kessie	65,784	17,136	260	0.39	22.9	543	4,548	0.069	9,373	13,681
Awash @ Hombole	7,656	1,382	180	0.29	26.9	44	357	0.047	494	497
Baro @ Gambela	23,461	11,541	492	0.13	15.0	366	1,133	0.048	1,415	1,594
Baro @ Itang	24,636	11,991	487		15.0	380				
Baro @ Masha	1,653	1,872	1,132	0.19	14.9	59	208	0.126	302	357
Dabus @ Asosa	10,139	5,043	497		16.3	160				
Didessa @ Dembi	1,806	189	105	0.19	21.8	6	31	0.017	87	190
Didessa @Arjo	9,981	3,960	397		18.5	128				
Genale @ Chenemasa	9,273	2,911	314	0.23	11.7	92	367	0.040	630	925
Gibe @ Abelti	15,746	5,742	365	0.29	18.2	182	879	0.056	1,517	1,811
Gibe @ Asendabo	2,966	1,165	393	0.22	16.8	37	156	0.053	242	269
Gidabo @ Aposto	646	240	371	0.25	10.6	8	28	0.044	55	77
Gojeb @ Shebe	3,577	1,789	500	0.20	14.7	57	239	0.067	392	559
Ketar @ Abura	3,350	389	116	0.28	18.7	12	70	0.021	153	262
Leliso @ Adaba	135	45	331	0.26	13.7	1	9	0.067	17	27
Maribo @ Adaba	185	94	508	0.26	14.4	3	16	0.086	24	28
Meki @ Meki Town	2,433	283	116	0.40	16.8	9	58	0.024	122	218
Mille @ Mille	4,467	313	70	0.61	15.8	10	291	0.065	817	1,227
Mojo @ Mojo Village	1,264	211	167	1.03	18.5	7	89	0.071	231	346
Tekeze @ Embamadrie	45,694	5,223	114	0.35	27.0	166	1,492	0.033	2,883	3,063
Wabe Shebele @ Dodolla B.dge	1,035	218	210	0.26	12.6	7	20	0.019	121	77
Weib @ Alemkerem	3,578	386	108	0.38	14.3	12	106	0.030	190	183
Lake-fed										
Abay/Blue Nile @ Bahir Dar	15,321	3,873	253	0.39	15.3		337	0.022	644	779
Awash @ Tendaho	63,485	2,178	34	0.31	10.9		430	0.007	1,065	1,687
Awash @ Melka-Sedi	21,510	1,472	68	0.27	12.1		188	0.009	373	443
Awash @ Metahara	16,417	815	50	0.58	10.9		96	0.006	245	374
Bulbula @ Kerkersitu	7,488	168	22	0.59	16.5		14	0.002	31	37
Ethiopian/Somali										
Juba @ Lugh	179,520	6,063	34	0.29	11.4					
Wabe Shebele @ Belet Weyne	211,800	2,132	10	0.32	12.1					

A catchment area; R_a annual runoff; R_{au} unit annual runoff; *CV* variation coefficient; *RCI* runoff concentration index (see text for explanation); Q_m mean discharge; $Q_{1.58}$ discharge with 1.58-years return time, equivalent to bankfull discharge; Q_{bu} unit bankfull discharge; Q_{10} discharge with 10-years return time; Q_{peak} highest discharge ever recorded

uniform distribution; a value from 10 to 15 denotes a moderately seasonal distribution; a value from 15 to 20 indicates a seasonal distribution.

$Q_{1.58}$ and Q_{10} were calculated with the Gumbel extreme values method only for river gauging stations with time series of at least 15 years, with the only exceptions of the Tekeze R. at Embamadrie (12 years) and the Baro R. at Masha (10 years).

Sediment yield and sediment transport data are even more discontinuous in space and time and were taken from the available scientific literature (see Sect. 4.5 for appropriate citations), international organizations reports (Omuto et al. 2009) and databases (e.g., FAO Land and Water Development Division—AQUASTAT).

4.3 River Geomorphology

4.3.1 Drainage Network

Ethiopia consists of a variety of landscapes resulting mainly from the combination of tectonics, volcanism, and fluvial processes. The drainage network of Ethiopian rivers is a reflection of such interactions. The northeastern plateau is deeply dissected by large perennial rivers such as the Abay/Blue Nile, the Tekeze and their main tributaries that developed feathered and dendritic networks, whereas in the southwestern plateau, sub-parallel networks prevail (e.g., Dawa, Genale, Weyb, Fafen and Webe Shebele rivers). Given the large occurrence of volcanoes all over Ethiopia, the radial network is very common as well, the Zuqwala and Fantalé volcanoes being good examples (Fig. 4.3).

The rectangular drainage network is rather common in the Rift Valley and its margins. Here, channel pattern changes, due to even very recent faulting, can be easily observed. For instance, in the eastern coastal plain of Ziway Lake, south of Ogolcho, a dry channel connects the Shetemata swamps with the Ketar River, joining it a few kilometers upstream of Ketar R. mouth into Ziway Lake (Fig. 4.4). This channel is wider than the Ketar and, though in the vicinity of the Shetemata swamps it presently flows into the swamps, in the past, it was an alternative channel of the Ketar that activated during former high stands of Shetemata. Axial faulting and likely the transformation from a permanent to a seasonal water body of the Shetemata turned the Ketar R. channel into its present position (Corti and Manetti 2012).

A more complex faulting system and tectonic deformation affected the Awash River course between Ombole and the Koka Lake in the Early Holocene. Initially, in fact, the Awash was the main river of a larger fluvial system, including the Meki and Mojo rivers, that outflowed into Ziway Lake forming a large delta, presently fed only by the Meki River. The former Awash flew southwestward, west of Ombole and Koye Lake, and received the Meki R. near Ejersa Lele, where today this latter river makes a sharp, angular turn to the right, before reaching Ziway Lake (Fig. 4.5). The former connection between the Awash and the Meki is witnessed by field geomorphological evidence of a dry, abandoned river valley and by a number of disconnected small streams, draining the Midrekebd ridge and formerly joining the Awash R. (Sagri et al. 2008). The Mojo R. joined the Awash a few kilometers upstream of the delta in Ziway Lake, as it is presently shown by the Cheleleka Lake, which maintains the former meandering pattern of the Mojo R. (Figs. 4.5 and 4.6). Local faulting and large-scale tectonic deformation then forced the Awash to flow northeastward, involving also the Mojo R. in this drainage pattern rearrangement, reaching the present setting (Sagri et al. 2008). Both the old and the present river systems show a typical rectangular drainage network largely controlled by axial and transverse fault systems as shown by the perpendicular orientation of first and fifth order rivers in the Meki catchment reported in Fig. 4.7 (Sagri et al. 2008).

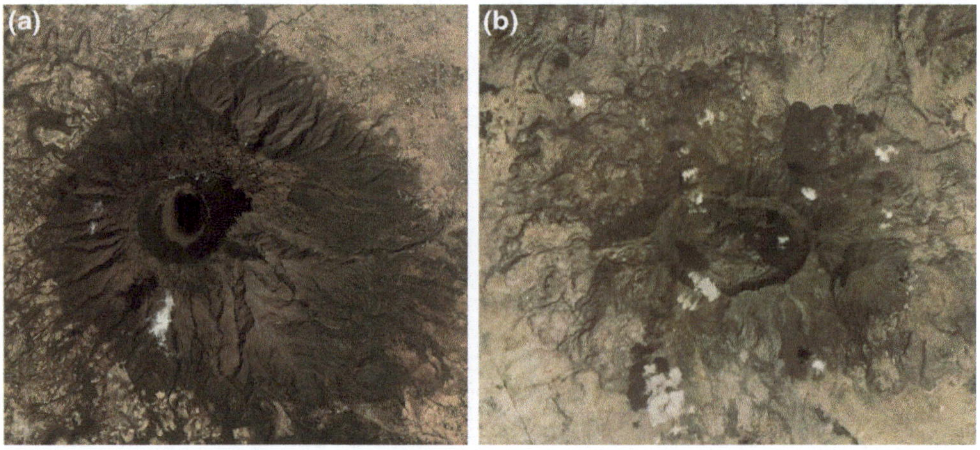

Fig. 4.3 Examples of radial river network around a volcano: **a** Zuqwala (08° 32′ 30″N–38° 51′ 20″E); **b** Fantale (08° 59′ 07″N–39° 54′ 26″E)

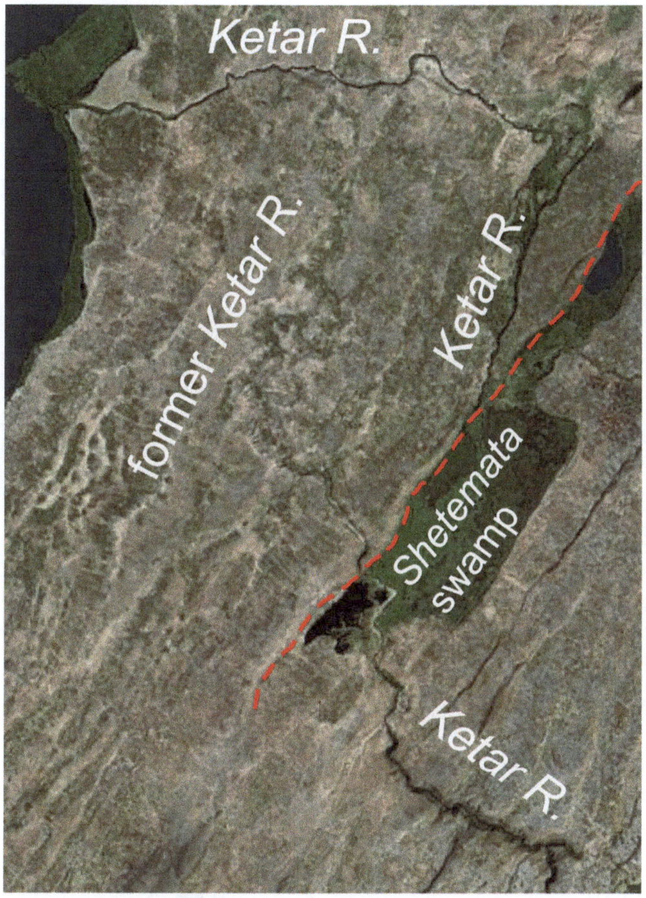

Fig. 4.4 The old channel of the Ketar River in the plain of Ogolcho (08° 00′ 00″N–38° 59′ 00″E). The river abandoned the old channel and moved into its present position is connection with the movement of the normal fault indicated by the *dashed red line* (modified from Corti and Manetti 2012)

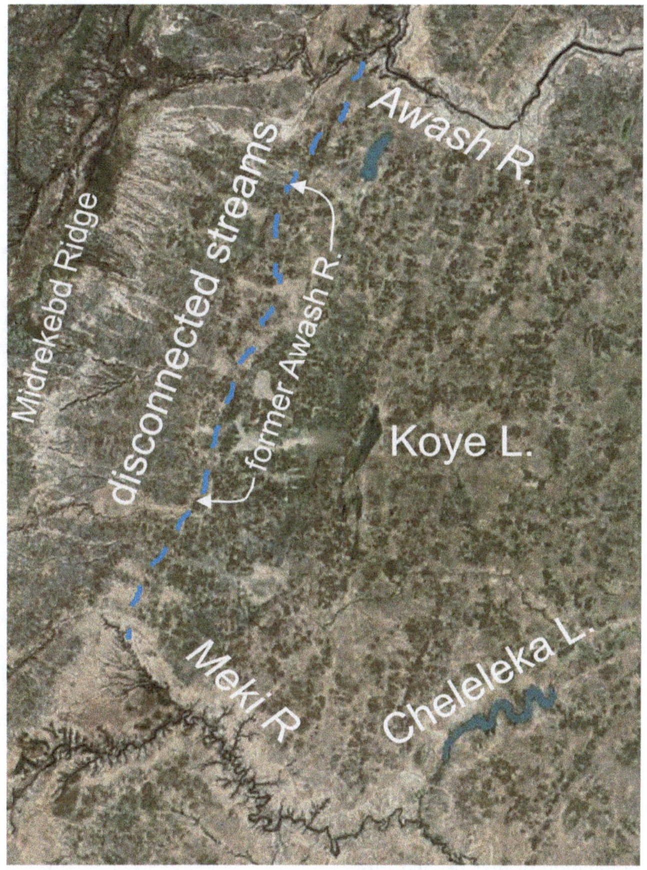

Fig. 4.5 Early Holocene Awash River channel (*dashed blue line*) (08° 19′ 54″N–38° 43′ 28″E). At this stage of the river evolution, the Awash received the Meki and Mojo rivers. Notice the presently disconnected, small tributaries of the Awash and the "meandering" Cheleleka Lake, relict of the former Mojo river (modified from Sagri et al. 2008)

Channel avulsion (Fig. 4.8) and anabranching (Fig. 4.9) are also common in low-gradient alluvial areas or close to and beyond the borders of the plateau swelling.

4.3.2 Channel Morphology

Three different types of headwaters are common in Ethiopia: (1) volcano headwaters, stretching from an individual volcano or a volcanic ridge. They are found on the flat surface of the highland plateau and in the Rift Valley and structural basins floor; (2) plateau headwaters. In this case, the headwaters divide is poorly defined since it is commonly marked by small hills or a few meters high ridge resulting from a basalt flow; (3) escarpment headwaters, developed from the margins of structural depressions such as the rift flanks or the shoulders of smaller structural basins. Given the physiography of headwaters and the recent structural setting, most of Ethiopian rivers are therefore incised into bedrock (Fig. 4.10) and a real alluvial plain is missing or it is patchy and very limited (4.11). Also on the bottom of the Rift Valley or other structural basins, rivers commonly flow entrenched into poorly consolidated volcanic ashes (Fig. 4.12) and fluvial–lacustrine deposits (Fig. 4.13). This is due to tectonics that in many parts of Ethiopia is witnessed by vertical movements, active also in very recent times. Rivers with a large alluvial plain are found only west of Gambela, in lowland areas underlain by Quaternary deposits (e.g., the Baro River), in the downstream Omo River valley and similar flat lands surrounding small lakes and swamps (e.g., the desiccated Chew Bahir Lake/swamps and the Segen River; the Awash River between Beda and Yardi lakes). Other lowlands or closed, structural basins are crossed by ephemeral streams that form distributary systems or terminal fans.

Type 1 and especially type 3 headwaters bestow a mountainous geomorphology to the landscape with very steep slopes and streambeds. In small mountain streams, all the main geomorphological features described by Montgomery

Fig. 4.6 The Cheleleka Lake (08° 12′ 57″N–38° 46′ 56″E)

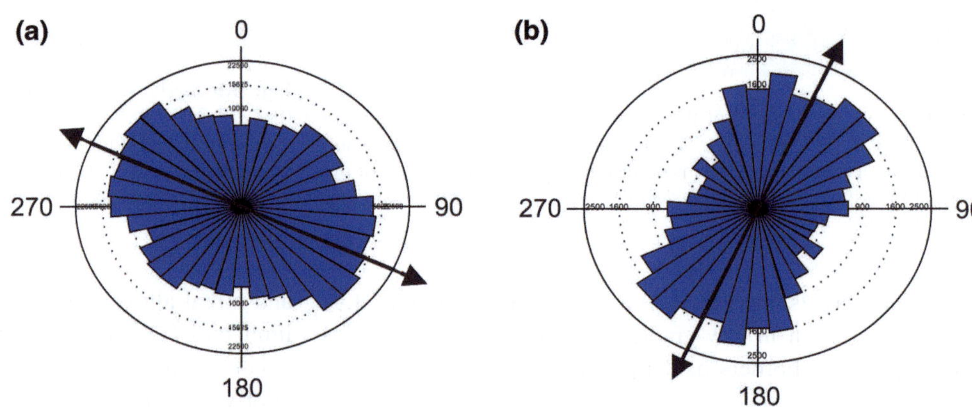

Fig. 4.7 Polar diagrams with orientation of first (**a**) and fifth (**b**) order streams in the Meki R. catchment. The larger rivers follow the main rift-oriented faults, whereas the smaller, more recent streams have a mean transverse orientation (modified from Sagri et al. 2008)

and Buffington (1997) characterizing these steep channels are present (Fig. 4.14). They include step-pool sequences (Grant et al. 1990), boulder berms (Carling 1987), glides (Bisson et al. 1982), riffle-pool (Leopold et al. 1964) and transverse ribs (Koster 1978). Commonly, the streambed of a small mountain stream takes up the whole valley bottom and the alluvial plain is missing or consists of terraced narrow strips and small patches (Fig. 4.15). The valley side slopes bound the streambed on which braided or pseudomeandering (Hickin 1972) channels develop during low flow conditions.

During floods, steep mountain streams experience high shear stresses and large boulders (0.5–2.0 m in mean diameter) can be easily entrained and transported (Fig. 4.16). Evidence of transport of such large boulders as bedload is provided by their imbricated position and the formation of huge particle clusters (Brayshaw 1984) (Fig. 4.17). Figure 4.17 also shows that large and small particles are transported en mass and cobbles of 15–20 cm move in saltation as proved by the particle embedded between the two larger boulders.

On the plateaus or on the treads of faulted steps, which are almost flat and characterized by low gradients, high sinuosity rivers are more common. Most of them have a typical meandering pattern but are commonly entrenched

Fig. 4.8 Avulsion of the Baro R. near Berhane Selam (08° 10′ 16″N–34° 07′ 00″E)

Fig. 4.9 Anabranching of the Baro R. close to the Sudan border (8° 21′ 34″N–33° 47′ 03″E)

Fig. 4.10 The bedrock channel of the Didessa R. near Barri (09° 01′ 50″N–36° 09′ 18″E)

Fig. 4.11 The Gibe River at Abelti. In this mountain reach a small alluvial plain is present on the left bank. Flow is toward the reader. (08° 13′ 48″N–37° 34′ 40″E)

within shallow bedrock banks. Gravel beds (Fig. 4.18) prevail over sand beds (Fig. 4.19).

As these low-gradient rivers exit the plateau/faulted blocks, their downstream pattern is determined by the margin morphology: (1) If it is abrupt and sharp, a waterfall may result (Fig. 4.20); (2) if it is incised, a steep mountain stream develops (Fig. 4.21).

The majority of the Ethiopian land consists of highlands. The larger rivers and their tributaries have dissected the plateau and formed deep and more or less narrow gorges and

Fig. 4.12 The flow gauge on the Ketar R. near Ogolcho (08° 01′ 58″N–39° 01′ 08″E). The streambed is incised into poorly consolidated volcanic ash

Fig. 4.13 Ephemeral stream incised into Quaternary colluvial–alluvial deposits at the margin of the Danakil depression on the way to Dallol (13° 54′ 42″N–40° 09′ 29″E). Notice the three orders of fluvial terraces on the left. Flow is away from the reader

canyons. This is by far the most common river morphology in Ethiopia, the best example of which is given by the Abay/Blue Nile canyon near Degen (Fig. 4.22). This spectacular gorge is about 1,500 m deep, with the plateau at an average elevation of 2,500 m a.s.l. and the river bed at 1,000 m a.s.l. (Ayalew and Yamagishi 2004). The river has cut through the trap basalt series and the Mesozoic sedimentary sequence down to the Adigrat Sandstones. The development of the Abay/Blue Nile canyon is associated with the fast uplift of the area that occurred during the Pliocene and the Pleistocene (McDougall et al. 1975; Yemane et al. 1985), whereas Ayalew and Yamagishi (2004) assume the valley incision was initiated in the upper 150 km of the Abay/Blue Nile River course before or a little after the beginning of the

Fig. 4.14 A typical mountain stream in the central highland. Notice the small lateral bars, a sequence of three small step-pools and a glide upstream of them. The stream is mainly incised into the bedrock, the channel is bounded by the slope foots and only small patches of flood plain are present

Quaternary. According to these authors, the phase of maximum river degradation occurred at the end of the Pleistocene or at the beginning of the Holocene and valley widening occurred as a result of landslides and retreating rock cliffs as is now evident by the remaining tabular hills and mesas. Other deep canyons are formed by the Wabe Shebele in the upper portion of the Somali plateau (see Chap. 1).

Notwithstanding the much larger size, these rivers share their main morphological characteristics with the smaller, steep mountain streams (Figs. 4.15 and 4.21). In fact, they take up the whole valley bottom, may be incised into the bedrock, or may have a small, stripped alluvium. The channel is almost straight or with a low sinuosity (mainly imposed to the river by structural factors) and punctuated by alternating lateral bars (Leopold and Wolman 1957) (Fig. 4.22). A similar river and valley morphology is also attained by large rivers where they cross a structural threshold between adjacent faulted steps, where rapids can develop (Fig. 4.23), or flow on lowlands such as the Rift Valley floor. Here, recent tectonics and, primarily, the emplacement of thick basaltic lava flows have forced the river to incise deep canyons (Fig. 4.24) and to form

Fig. 4.15 A mountain stream draining a main rift escarpment near Dese (11° 06′ 52″N–39° 39′ 30″E). The streambed takes up the whole valley bottom. Notice the steep slopes bounding the streambed, the occurrence of only a small portion of flood plain and the braided/pseudomenadering pattern of the low flow channel

Fig. 4.16 A steep mountain stream draining the deep escarpment of the Abay/Blue Nile gorge (10° 06′ 24″N–38° 10′ 30″ E). The vicinity with the headwaters favors the supply of very coarse boulders that are easily entrained given the steep bed gradient and high flow energy

spectacular waterfalls (Fig. 4.25), as is the case of the Awash River in the Awash National Park.

Most rivers, as they enter lowland areas, develop a braided stream morphology in their proximal reaches (Fig. 4.26) and then, if the gradient decreases, a typical meandering pattern (Fig. 4.27). By contrast, some of the rivers flowing on the rift bottom and entering the rift lakes may have a large portion of their channel entrenched into the Rift Valley floor due to recent change in the lake base level as for instance in the case of the Meki R. flowing into the Ziway Lake (Fig. 4.5) or the Bilate River (Fig. 4.28) which runs southward for more than 100 km parallel to the Rift axis and form a large delta into the Abaya Lake.

Dryland ephemeral streams have specific morphological characteristics that are described in more detail in Chap. 12 of this volume.

Fig. 4.17 A huge pebble cluster. Notice the small particle interlocked in between the very large, imbricated boulder in the cluster stoss side and the core boulder

Fig. 4.18 A coarse grained meandering channel on top of the central highlands plateau near Debre Birhane (09° 49′ 21″N–39° 43′ 04″E)

4.4 Discharge

The rivers considered in this study have very different catchment size and hydrologic characteristics. The Wabe Shebele river system includes both the smallest, 135 km^2, and the largest catchment, 211,800 km^2, if we consider the flow gauge at Belet Weyne, located a few kilometers beyond the border with Somalia (Table 4.1). As a consequence, mean discharge (Q_m) of Ethiopian rivers varies largely between 1 (Leliso R.) and 543 m^3 s^{-1} (Abay/Blue Nile at Kessie) and the same large variability is observed for bankfull discharge ($Q_{1.58}$), 9–4,548 m^3 s^{-1} (Leliso and Abay/Blue Nile at Kessie, respectively). Bankfull discharge per unit catchment area (Q_{bu}) ranges between 0.002 (Bulbula River) and 0.126 m^3 s^{-1} of the Baro River at Masha, where Q_{bu} is almost double that of the Abay/Blue Nile at Kessie (Table 4.1) and almost four time larger than the Tekeze River at Embamadrie. Mean, peak, and bankfull discharge are well correlated (R^2 = 0.83, 0.90 and 0.89, respectively) with catchment area (Fig. 4.29) whereas, as expected for the high climatic conditions variability, no correlation was found between $Q_{1.58}$ and Q_{bu}, nor between catchment area (A) and Q_{bu}, although highland- and lake-fed rivers can be clearly distinguished (Fig. 4.30).

The data set of this study was also used to test the application range of the empirical equation of Syvitski and Milliman (2007), developed from a globally distributed database of 488 rivers, which predicts mean discharge as a function of catchment area:

$$Q_m = 0.075 \cdot A^{0.8} \qquad (4.2)$$

Equation 4.2 proved to predict with relative accuracy (91 %) the actual mean discharge of Ethiopian rivers with the exception of the Tekeze R. at Embamadrie and the

Fig. 4.19 The Weja R., a sand bed, meandering river flowing on top of a faulted block in the Meki R. catchment near Koshe (08° 01′ 15″N–38° 31′ 07″E). Flow is away from the reader

Fig. 4.20 The waterfall of the Aleltu R. entering the deep canyon of the Sodoblè R. (Muger R. catchment) on the central highland plateau near Chancho (09° 25′ 47″N–38° 39′ 27″E)

Ethiopian/Somali rivers (Fig. 4.31). The strength of this relationship is confirmed by the Spearman's rank correlation coefficient at a 0.01 significance level. This result indicates that Eq. 4.2 can be used in Ethiopia to predict the mean discharge of ungauged rivers, provided they are not located in dryland areas. Nevertheless, a closer inspection of the diagram of Fig. 4.31 and a more detailed analysis of data show that Eq. 4.2 predictions are less accurate when mean discharge is expected to be less than 20 $m^3\ s^{-1}$. In the case of small rivers, in fact, a different equations was derived as follows:

$$Q_m = 0.1755 \cdot A^{0.51} \tag{4.3}$$

which is capable to explain 85 % of mean discharge variability.

4.4.1 Runoff Volume

Annual runoff (R_a) ranges widely among the study rivers, given their large variability in catchment area and climatic conditions, with the lowest value (45 × 10^6 m^3 $year^{-1}$) recorded in the smallest stream (the Leliso R. headwater tributary of the Wabe Shebele) and the largest (17,136 × 10^6 m^3 $year^{-1}$) measured on the Abay/Blue Nile at Kessie (Table 4.1). Annual runoff is well correlated (R^2 = 0.87) with catchment area only for the highland-fed rivers, with the datum of the Tekeze R. as an outlier

Fig. 4.21 A mountain stream draining the main Rift escarpment of the Danakil lowland (13° 51′ 53″N–39° 50′ 36″E)

Fig. 4.22 The Abay/Blue Nile Gorge near Degen (10° 4′ 28″N–38° 11′ 22″E). The river is deeply incised into the tabular bedrock (**a**), has elongated lateral bars but no alluvial plain (**b**). The streambed is bounded by foot slopes (**b**)

Fig. 4.23 The rapids of the Awash R. at Melka Konturè (08° 42′ 14″N–38° 36′ 27″E)

(Fig. 4.32). Including also the Tekeze, the correlation coefficient decreases to 0.75 whereas, if also the lake fed and the Somali rivers are included, the best fit is given by a power function ($R_a = 2.324 \cdot A^{0.7}$), the determination coefficient of which is only $R^2 = 0.63$. Even poorer correlations are obtained between unit runoff (R_{au}), i.e., the annual runoff per unit catchment area, and the watershed area though, similarly to unit bankfull discharge, the three groups of highland, lake-fed, and Ethiopian/Somali rivers can be clearly distinguished (Fig. 4.33).

In order to point out the potential water resources associated with the runoff of Ethiopian rivers and to rank them within Africa (Table 4.2), a diagram of unit annual runoff (R_{au}) versus catchment area (A) for Saharan, Sub-Saharan, and Ethiopian rivers is plotted (Fig. 4.34). This diagram shows clearly that the Ethiopian rivers, though being among those with the smaller catchment area, have a unit runoff comparable to that of other African (especially equatorial) rivers, whose catchments are as much as three orders of magnitude larger.

The variation coefficient (CV) values indicate that the least inter-annual variability occurs in the Baro R. at Mash and the Didessa R. at Dembi (CV = 0.19), whereas the largest variability is observed in the rift valley rivers with the

Fig. 4.24 The canyon of the Awash R. in the Awash National Park (08° 52′ 21″N–40° 05′ 28″E)

Fig. 4.25 The Awash R. water fall in the Awash National Park (08° 50′ 32″N–40° 00′ 42″E)

highest values recorded in Mojo R. at Mojo Village (CV = 1.03) (Table 4.1). The annual runoff of more than 50 % of the studied rivers is highly variable, with a CV ranging between 0.50 and 0.75, and about one quarter of the sample is characterized by a medium to low variability with CVs in the 0.25–0.50 range (Fig. 4.35). Therefore, though the unit annual runoff (R_{au}) is relatively high with respect to the catchment size (Fig. 4.34), the high values of the annual runoff CV may result in some constraints to the exploitation of river water for agriculture irrigation.

Fig. 4.26 A braided river a few kilometers upstream of its mouth into Abaya Lake, near Arba Minch (06° 02′ 40″N–37° 32′ 58″E)

Fig. 4.27 A meandering reach of the Awash R. upstream of the Gora marshes (08° 23′ 20″N–38° 54′ 35″E)

In order to investigate the runoff variability within a year, by analogy with the precipitation concentration index proposed by Oliver (1980) to define the monthly rainfall distribution in a year, the runoff concentration index (RCI) is used (see Eq. 4.1).

According to Oliver (1980), values of PCI (and by analogy of RCI) less than 10 indicate a uniform distribution, values from 10 to 15 denote a moderately seasonal distribution, and values from 15 to 20 are characteristic of a seasonal distribution. Values above 20 are associated with strong seasonal variations. The RCI data set obtained by Eq. 4.1 for the study rivers, although the lake-fed rivers were not considered since their runoff volume is directly controlled by the lakes supply and/or reservoir water routing programmes, shows that more than 50 % of them are characterized by a seasonal (37 %) and a markedly seasonal (15 %) distribution of runoff, whereas only for 37 % of them the seasonality is moderate (Fig. 4.36). This result is not surprising given the dominant role of the monsoon type rainfalls that are concentrated in July and August all across the largest part of Ethiopia.

Fig. 4.28 The Bilate River flowing southward entrenched into the rift valley floor. Flow is toward the reader (07° 17′ 09″N–38° 04′ 22″E)

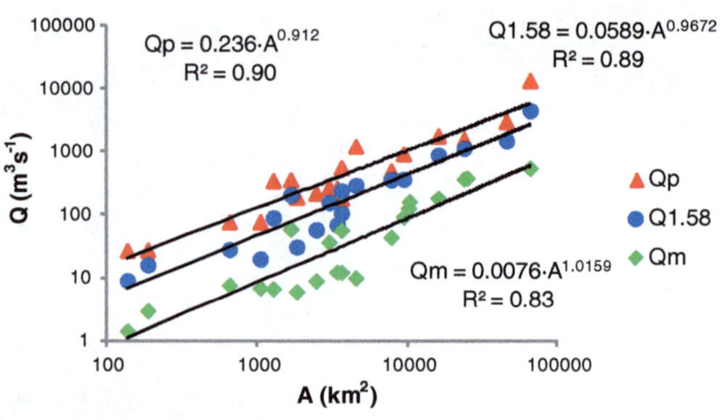

Fig. 4.29 Catchment area (A) versus characteristic discharges for the study highland-fed rivers. Q_p = maximum discharge ever recorded; $Q_{1.58}$ = discharge with 1.58-year return time equivalent to bankfull discharge; Q_m = mean discharge

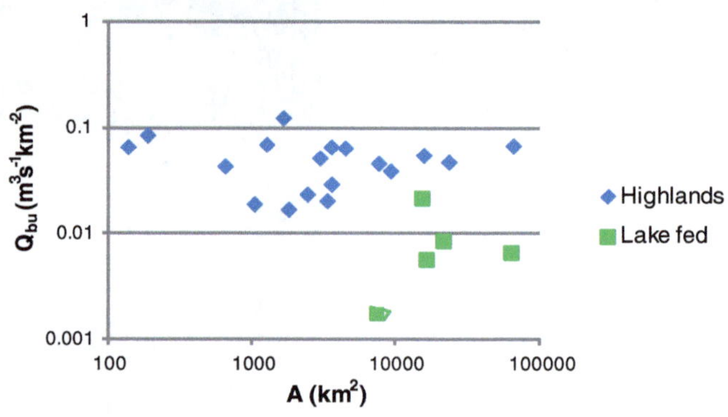

Fig. 4.30 Catchment area (A) versus unit bankfull discharge (Q_{bu}). Highland- and lake-fed rivers of this study are quite distinct. The latter have an order of magnitude smaller unit bankfull discharge

4.4.2 Floods

Though some studies have been carried out on floods in Ethiopia, most of them were focused on flood forecast and flood frequency analysis, regionalization and modeling (e.g., Demissie 2008; Baratti et al. 2012). Very few scientific papers and field studies dealing with recent decades floods and flash floods (e.g., Alemu 2009; Billi et al. 2015) are available in the international literature, though these devastating events are cited in the international news, humanitarian organizations, and local authorities reports.

According to EM-DAT (2013), from 1900 to 2013, 50 floods occurred in Ethiopia with 1,976 people killed, about 2.4 million of people affected and about 17 million US$ of damages. The worst floods occurred in 2006, with 862 fatalities and 361,600 people affected (Alemu 2009). Sena and Woldemichael (2006) report some additional data on floods in different parts of the country:

1996 displacement of 40,000 people in Wonji and Metehara due to burst of Awash River banks and overflow from the Koka reservoir;

2003 100,000 people affected by Wabe Shebele river flooding in Somali region;

2005 170 people died and 260,000 were displaced due to flooding from Wabe Shabele, Hargeysa, Bilate, Ashewa, Genale, Dawa, Fafen, Sile, and Sego rivers;

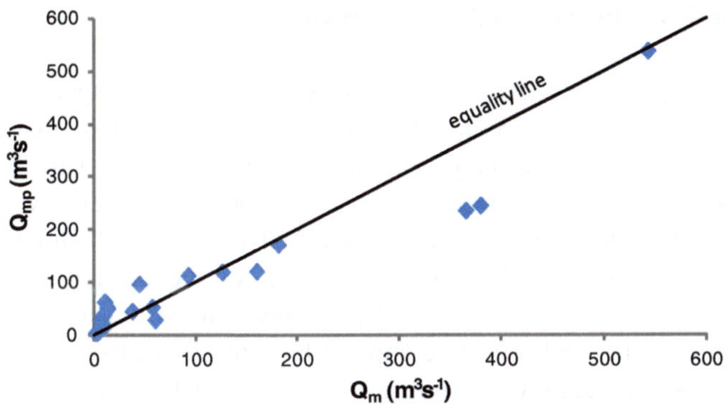

Fig. 4.31 Very good correlation (91 %) between observed (Q_m) and predicted (Q_{mp}) mean discharge by Eq. 4.2 (Syvitski and Milliman 2007)

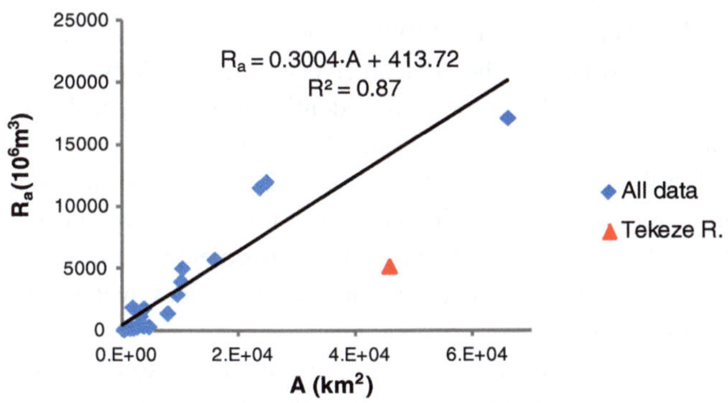

Fig. 4.32 Catchment area (A) explains 87 % of highland-fed rivers annual runoff (R_a) variability with the exception of the Tekeze R

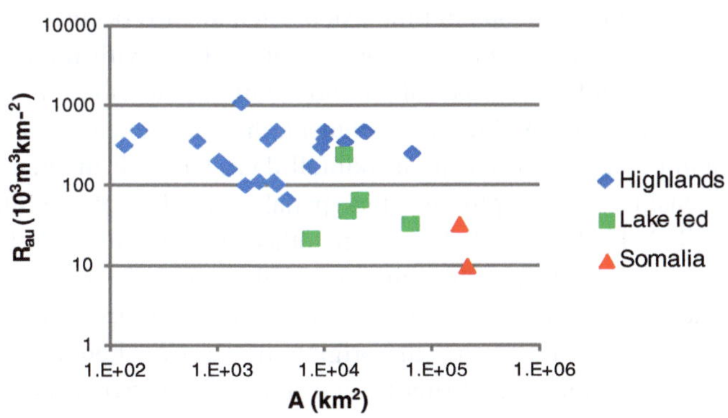

Fig. 4.33 The combination of parameters such as catchment area (A) and unit annual runoff (R_{au}) well distinguishes highland, lake-fed, and Ethiopian/Somali rivers. These latter two categories of rivers have one order of magnitude smaller unit annual runoff in spite of a catchment area that is one to two orders of magnitude larger

2006 the worst scenario of flood in Ethiopia, resulting in loss of lives, damage of property and destruction of livelihoods of tens of thousands of people in South Omo, Dire Dawa and 19 in other parts of the country

Table 4.2 Mean annual unit runoff volume (R_{au}, 10^3 m^3 km^{-2}) of Ethiopian and other African rivers

Region/country	No of data	R_{au}	Range
Algeria	6	147	2–486
Egypt	1	30	
Morocco	4	217	27–449
Tunisia	2	23	18–28
North Africa	13	104	2–486
Benin	4	98	36–152
Burkina Faso	7	68	27–174
C.A.R.	11	337	24–683
Cameroon	9	602	168–1,216
Chad	4	226	21–544
Congo	4	426	332–511
Gabon	5	765	507–1,081
Ghana	9	159	62–288
Guinea	4	1,973	314–6,509
Ivory Coast	6	375	87–653
Kenya	2	4,524	2,979–6,076
Lesotho	2	144	102–185
Liberia	2	1,457	694–2,220
Malawi	2	139	102–177
Mali	10	213	61–377
Mozambique	8	162	87–391
Niger	3	148	33–376
Rwanda	2	219	203–234
Senegal	6	140	21–311
South Africa	9	37	5–122
Sudan	4	96	27–166
Tanzania	5	157	15–419
Togo	3	277	122–420
Uganda	5	182	31–478
Zaire	1	366	
Zambia	3	125	19–186
Zimbabwe	13	79	5–146
Sub-Saharan	*143*	*354*	*5–6,509*
Ethiopia	**22**	**290**	**10–1,132**

Flood hydrological data measured in the field or inferred "post event" are substantially missing (Alemu 2009; Billi et al. 2015). The highest discharge (Q_p) recorded for the highlands fed rivers are very close to Q_{10}, i.e., the discharge with 10-years return time (Table 4.1) and Q_p is well correlated with both Q_{10} (R^2 = 0.99) and bankfull discharge ($Q_{1.58}$) (R^2 = 0.97).

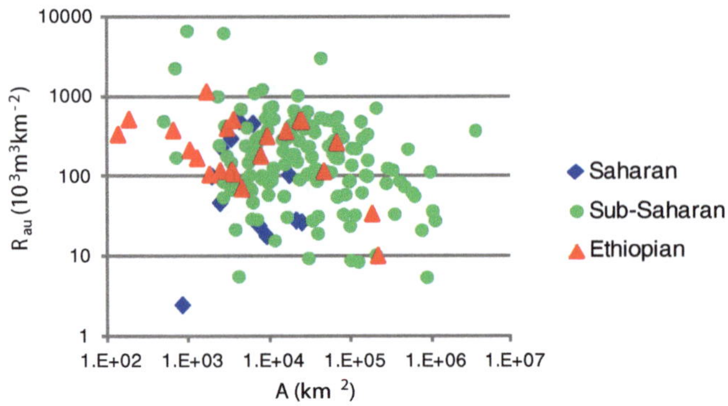

Fig. 4.34 Catchment area (A) and unit annual runoff (R_{au}) for African rivers of different regions. Sub-Saharan rivers include a number of rivers south of the Sahara desert (i.e., sub-Saharan s.s., equatorial, and southern Africa). Though the study Ethiopian rivers have a smaller catchment area, their unit annual runoff is comparable to that of much larger rivers

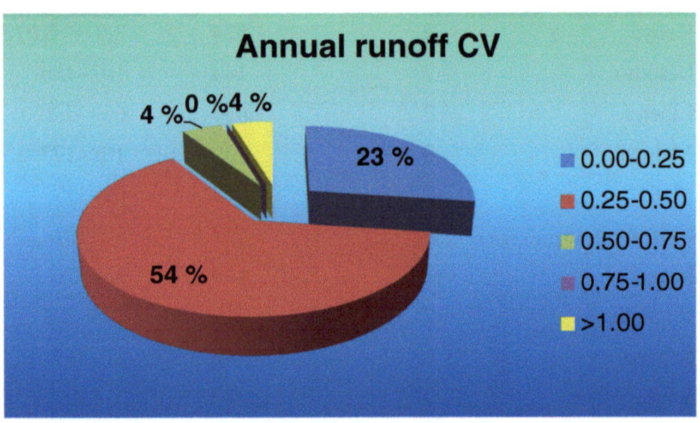

Fig. 4.35 Frequency distribution of the annual runoff variability coefficient of the Ethiopian study rivers

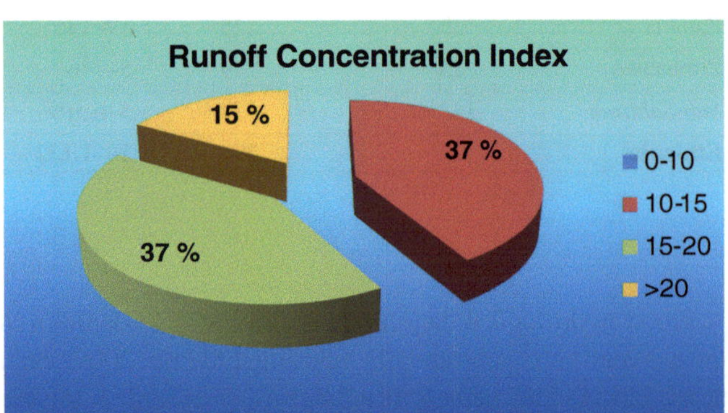

Fig. 4.36 Frequency distribution of runoff concentration index (*RCI*) of the study rivers

4.4.3 Flow Discharge Trend

Though the effects of climate change on several aspects of the physical environment have been investigated in many regards, comparatively little attention has been paid to the hydrological response of rivers throughout the last century or the last decades (e.g., EEA 2012). The effects of climate change on river discharge at global, continental and local scale have been investigated by a few authors, but most of the studies and reports compiled by international organizations are focused on future scenarios (IPCC 2007; Bates et al. 2008) rather than on the observation and interpretation of the changes which occurred in the past decades (Bates et al. 2008). This situation is well reflected in the whole African continent, including Ethiopia (Conway 2005; Bates et al. 2008).

The Hadley Center (2005) reports a general decreasing trend for runoff of the African rivers, whereas Conway et al. (2009) concluded that robust identification of hydrological change was severely constrained by data limitations for the sub-Saharan Africa and Conway (2005) found no clear indication of climate change effects on the Abay/Blue Nile River flow, because of uncertainty in projected rainfall patterns and the influence of complex land and water management structures.

Notwithstanding such a complex situation and the lack of continuous long time series, it is worth attempting to trace the runoff trends across the last four decades using the database constructed for this study. As reported early in this paper, both daily and monthly flow data of the study rivers are not homogeneous in time extension and gaps occur. Any attempt to fill the gaps by correlation analysis with rainfall amounts failed because of the non uniform distribution of the rain gauges within the catchment, the occurrence of gaps and/or long periods without rainfall data, the high diversity in basin physiography and the spatial variability of rainfall (Billi 2011; Billi et al. 2015). Any other manipulation of data is considered to produce significant alteration of the actual time series with errors the extent of which is comparable to those expected by using the original data set unaltered.

Though the annual runoff time series of the rivers under study range from 1957 to 2009, in order to work with a more uniform and consistent database, only the 1968–2004 interval was considered. For the annual runoff volume of each year, the z score is calculated as follows:

$$z = (x_i - \mu)/\sigma \quad (4.4)$$

in which for a given river x_i is the annual runoff (R_a) of the ith year, μ is the mean, and σ the standard deviation of R_a time series.

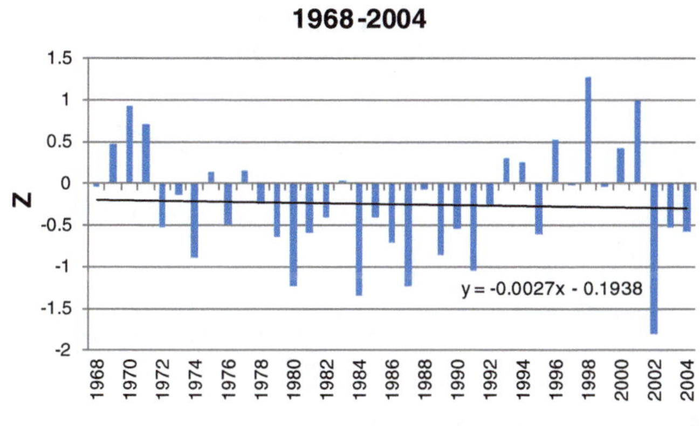

Fig. 4.37 Time series of the annual runoff *Z* score averaged over all the study rivers *z* score data. No significant change is observed for the 1968–2004 interval

For each year, a mean value of *Z* is calculated by averaging the *z* values of all rivers considered and finally, the time series of *Z* scores is obtained (Fig. 4.37). Figure 4.37 indicates that, though a minuscule decrease results from the interpolating line, in general, no clear trend is present; hence, no evident change in annual runoff volume has occurred in the Ethiopian rivers during the 36-years interval (1968–2004) considered in this study. Within this general picture, obviously, there are exceptions such as marked increasing trends for the Abay/Blue Nile at Kessie, the Genale at Chenemasa, the Ghibe at Abelti and one Ethiopian/Somali river such as the Wabe Shebele at Belet Weyne, but these are counterbalanced by less prominent decreasing trends for another Ethiopian/Somali river, the Juba at Lugh, and modest negative trends for all other rivers. The negligible decrease in river runoff is paired by a very modest (of the same order of magnitude) decreasing trend of annual rainfall across the country (see Chap. 4, this volume).

Unfortunately, runoff data of the first half of the twentieth century (1905–1959) are available only for the Baro R. at Gambela (Sutcliffe and Parks 1999). A comparison with the 1977–2003 interval shows that the annual runoff has decreased by 12 %. The reasons behind that are manifold but human impact may have played an important role as observed in other African rivers (Conway et al. 2009; Mahe et al. 2013). By analogy with this finding, it could be speculated that the runoff time series of Ethiopian rivers not affected by reservoirs and spanning the past century would probably depict a different (decreasing?) trends conversely to the general stability depicted by Conway et al. (2009) for several large rivers in the sub-Saharan Africa in the last decades. Unfortunately, the lack of old river flow data is presently a tight constraint to any adequate and reliable analysis.

4.5 Sediment Transport

Data on sediment transport and sediment yield of Ethiopian rivers are very scarce. The available data on specific sediment yield (SSY) are reported in a very few papers and in a few cases only occasional data are reported. SSY data from reservoir filling are reported by Tamene et al. (2006) and Haregeweyn et al. (2008). FAO (2000), Billi (2004, 2011), Vanmaercke et al. (2010) and Haregeweyn et al. (Chap. 13 this volume) studies are instead based on data obtained from field measurements. Discontinuous field measurements on the Awash R. at Ombole from 1986 to 1990 are reported in detail in the FRIEND/Nile report (UNESCO 2007) and let to construct the suspended sediment concentration rating curve for this station (Fig. 4.38). Guzman et al. (2013) present detailed suspended sediment field measurements for three very small catchments (area ranging from 113 to 447 ha) in the sub-humid highlands, but their rating curves are poorly significant as it would be expected in such small catchments in which local factors, such as seasonal changes in vegetation and cultivation practices, may play a relevant role and introduce a high degree of randomness in the data.

The Ethiopian rivers considered in these studies have a catchment area ranging from 0.72 km^2 of the Adihilo R. in Tigray to 62,732 km^2 of the Awash R. at its most downstream end close to Djibouti. The SSY varies widely as well, with the lowest value of 4 tkm^2 year^{-1} of the Woyb (Genale-Dawa river system) to 4,935 tkm^2 year^{-1} measured on the Adikenafiz reservoir, again in Tigray. The mean values of SSY measured for 24 reservoirs and 47 river sites of Ethiopia considered in this study, 78 river sites in northern Africa, 112 in the sub-Saharan Africa and the 668 world river data compiled by Summerfield and Hulton (1994)

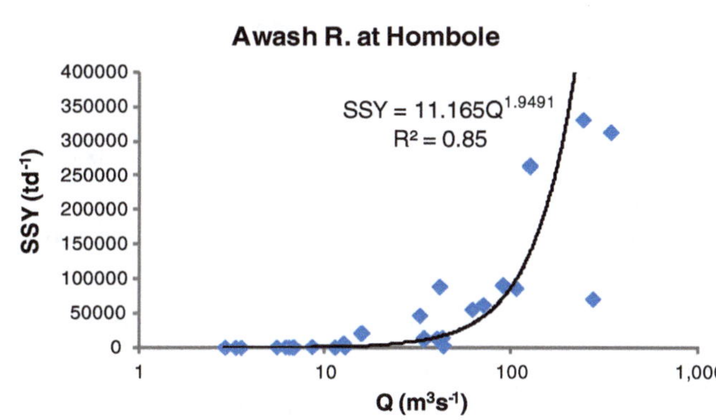

Fig. 4.38 Suspended sediment yield (*SSY*) rating curve for the Awash R. at Ombole

without considering the African rivers, are reported in Table 4.3.

The SSY data of the Ethiopian rivers, when plotted against the catchment area, do not follow the typical decaying pattern, as reported in the classic study of the global sediment yield by Milliman and Meade (1983), and no significant correlation was found. By contrast, the total sediment yield (Y) obtained from field measurements on the Ethiopian rivers varies significantly ($R^2 = 0.84$) with catchment area (Fig. 4.39) and is predicted by the following equation:

$$Y = 0.0005A + 0.1989 \quad (4.5)$$

in which Y is expressed in 10^6t year^{-1} and A is catchment area in km^2.

Table 4.3 Suspended sediment yield (SSY, tkm^{-2} year^{-1}) of Ethiopian and other African and world rivers

Region/country	No of data	Mean SSY	Range
Algeria	39	945.2	36–6,654
Morocco	36	1,173.4	100–4,620
Tunisia	1	708.0	
Egypt	2	39.5	39–40
North Africa	***78***	***1,024.3***	***36–6,654***
Cameroon	4	85.7	20–210
C.A.R.	4	5.5	3–9
Chad	4	7.0	1–15
Ghana	16	14.4	1–85
Ivory Coast	2	15.7	9–22
Kenya	14	2,136.0	4–19,520
Lesotho	19	615.9	3–2,050
Madagascar	2	2,358.0	1,586–3,130
Mali	2	27.3	15–40
Mozambique	3	44.1	17–80
Niger	1	4,000	
Nigeria	14	196.1	5–483
Senegal	1	8.0	
South Africa	2	495.0	100–890
Sudan	2	2,189.5	927–3,422
Tanzania	4	195.7	95–390
Zaire	2	14.5	11–18
Zimbabwe	16	150.6	1–39
Sub-Saharan	***112***	***558.1***	***1–19,520***
World river data[a]	***668***	***176.0***	***136,000***
Ethiopian reservoirs	24	1,428	345–4,935
Ethiopian rivers	47	670	4–3,784
Ethiopia	***71***	***935***	***4–3,784***

[a] Summerfield and Hulton (1994) data excluding the African rivers

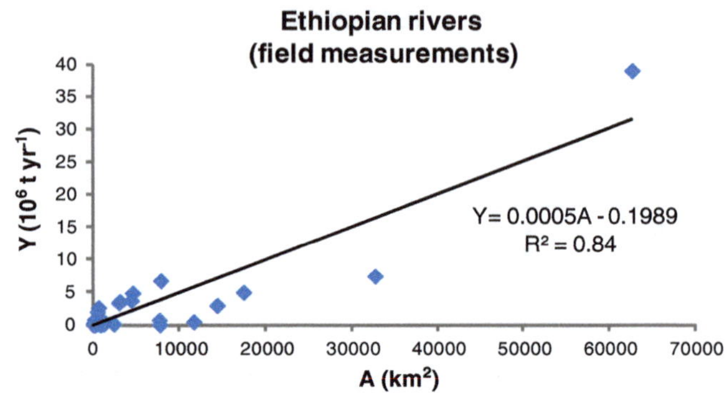

Fig. 4.39 Catchment area (A) versus sediment yield (Y)

Also Boateng et al. (2012) found a good correlation ($R^2 = 0.98$) between these two parameters for a few rivers in Ghana, but their variability is best described by an exponential function. Omuto et al. (2009) report a suspended sediment rating curve with a high correlation coefficient ($R^2 = 0.88$) for both the Juba and Wabe Shebele rivers:

$$Q_s = 0.087Q^{1.96} \quad (4.6)$$

where Q_s is suspended sediment discharge in kgs^{-1} and Q is discharge in m^3 s^{-1}. This equation was obtained by field measurements during two short sampling campaigns in 2007 and 2008 at a few stations, including also Lugh and Belet Weyne that are located just a few kilometers beyond the Somali border. For the Gereb Oda, a small ephemeral stream close to Alamata in Tigray, the field measurements of Billi (2011) indicate that suspended sediment transport is best described ($R^2 = 0.97$) by an exponential law of the following type:

$$Q_s = 108\, e^{0.12Q} \quad (4.7)$$

Also the Gereb Oda data were measured during one rainy season and for these reasons, though the correlation coefficient is high, Eqs. 4.6 and 4.7 must be considered as indicative and used with caution.

In order to compare the sediment yield of Ethiopian rivers with that of other African rivers, a diagram of SSY versus catchment area (A) is plotted (Fig. 4.40). From this figure, it is evident that though the Ethiopian rivers have a comparatively smaller catchment area, their SSY is of the same order of magnitude as other, much larger rivers of the sub-Saharan Africa. This is also the evidence of high soil erosion that is taking place in the Ethiopian highlands as reported in a number of studies (e.g., Hurni 1983; Nyssen et al. 2004, 2007; Tamene and Vlek 2008).

Bedload field measurements data for Ethiopian rivers are very scarce and limited to the Kulfo R. in the southern Rift Valley (Girma and Horlachen 2004) and to the Gereb Oda in

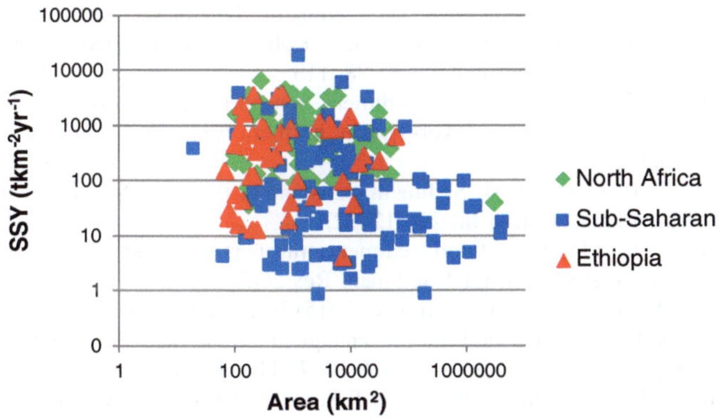

Fig. 4.40 Catchment area (*A*) versus suspended sediment yield (*SSY*) for a number of African rivers. Sub-Saharan rivers include a number of rivers south of the Sahara desert (i.e., sub-Saharan s.s., equatorial, and southern Africa). Though the Ethiopian rivers have a comparatively smaller catchment area, their SSY is of the same order of magnitude of much larger African rivers

the Kobo Basin (Tigray) (Billi 2011). In both studies, bedload transport rate was found to be mainly controlled by unit stream power (ω). In fact, the bedload rate (Q_b) of Gereb Oda (Billi 2011) is well correlated (R^2 = 0.71) with ω through the following linear equation:

$$Q_b = 0.0567\omega - 0.097 \quad (4.8)$$

whereas the study of Girma and Horlachen (2004) proved that the Bagnold's criterion (Bagnold 1980), which is based on stream power, predicts with the largest accuracy the bedload of their study river. By contrast, Bagnold equation performs poorly with Gereb Oda bedload rates, which are best predicted by the empirical power formula of Martin (2003), yet based on unit stream power. The different performance of these stream power-based equations in predicting bedload of the Kulfo and Gereb Oda rivers stands likely in the ephemeral character and the finer, sandy bed material of the latter, whereas the former is a gravelly-sand bed and an almost permanent braided river.

4.6 Conclusions

Rivers of Ethiopia present a variety of drainage networks, channel morphologies, flow discharges, and sediment transport rates, reflecting the complex combination of tectonics, volcanism, physiography and climate. Geomorphic and hydrologic data have been analyzed and led to the following conclusions:

1. In the Rift Valley, recent axial and minor transverse faults have markedly affected the river pattern of the Ketar and the Awash rivers, with evidence of older abandoned channels.
2. The physiography of headwaters, commonly located on the highland plateau or within the rift margins, and the recent structural setting provided the most of Ethiopian rivers with a pattern incised into bedrock. Wide alluvial plains are commonly missing or patchy and very limited alluvium may be present. As a consequence, larger rivers and their tributaries have dissected the plateau and formed deep and more or less narrow gorges and canyons in which the rivers show typical channel features of mountain streams. Rivers with a large alluvial plain are found only west of Gambela, in lowland areas at the far margins of the structural swell, underlain by Quaternary deposits (e.g., the Baro River), in the downstream Omo River valley and similar flat lands surrounding small lakes and swamps.
3. Flow data of the studied rivers show that discharge is well correlated with catchment area. In fact, catchment area explains 83 % of mean, 89 % of bankfull, and 90 % of peak discharge, respectively, whereas, given the high variability of climatic and physiographic conditions, no correlation was found between bankfull and unit bankfull discharge, nor between catchment area and unit bankfull discharge.
4. The equation of Syvitski and Milliman (2007) proved to predict with a high accuracy (91 %) the actual mean discharge of Ethiopian rivers, with the exception of the Tekeze River and the Somali rivers. Both this equation and those developed in this study are reliable tools to predict characteristic discharges of ungauged rivers, provided they are not located in dryland areas.
5. The unit runoff of Ethiopian rivers is comparable to that of other African rivers with a catchment area as much as three orders of magnitude larger. The flow data indicate that annual runoff volume of more than 50 % of the study rivers is highly variable and, though the unit annual runoff is relatively high with respect to the catchment size, the high values of the annual runoff variation coefficient may reduce the potential exploitation of river water for agriculture irrigation.
6. Time series of annual runoff were constructed for the 1968–2004 interval and the normalized data of the study rivers averaged. The interpolating line show a very modest decreasing trend indicating that no evident change in annual runoff volume has occurred in the Ethiopian rivers during the last decades.
7. The suspended sediment yield of the Ethiopian rivers is not correlated with the catchment area nor it decreases following a decaying pattern, as reported in the classic study of the global sediment yield by Milliman and Meade (1983). By contrast, the annual sediment volume of Ethiopian rivers, obtained from field measurements, varies significantly (R^2 = 0.84) with the catchment area.

8. Though the Ethiopian rivers have a comparatively smaller catchment areas, their suspended sediment yield is of the same order of magnitude of other, much larger rivers of Africa. This result confirms the high rate of soil erosion that is taking place in the Ethiopian highlands as reported by a number of studies (e.g., Hurni 1983; Nyssen et al. 2004, 2007; Tamene and Vlek 2008).

Acknowledgments The Ministry of Water, Irrigation and Energy of Ethiopia is acknowledged for it provided the most of the flow data used in this paper. This study was carried out with University of Ferrara 2012 and 2013 FAR funds.

References

Abdo KS, Fiseha BM, Rientjes THM, Gieske ASM, Haile AT (2009) Assessment of climate change impacts on the hydrology of Gilgel Abay catchment in Lake Tana basin, Ethiopia. Hydrol Process 23:3661–3669

Ahmed AA (2008) Sediment in the Nile River system. UNESCO-IHP-International Sediment Initiative, Khartoum

Alemu YT (2009) Socio-economic impact of flooding in Dire Dawa, Ethiopia. Unpublished MSc. thesis, International Center for Water Hazard and Risk Management, Tsukuba, Japan

Avery S (2010) Hydrological impacts of Ethiopia's Omo basin on Kenya's Lake Turkana water levels and fisheries. African Development Bank, Tunis

Awulachew SB, Yilma AD, Loulseged M, Loiskandl W, Ayana M, Alamirew T (2007) Water resources and irrigation development in Ethiopia. Working Paper 123, International Water Management Institute, Colombo

Ayalew L, Yamagishi H (2004) Slope failures in the Blue Nile basin, as seen from landscape evolution perspective. Geomorphology 57:95–116

Bagnold RA (1980) An empirical correlation of bedload transport rates in flumes and natural rivers. Proc Royal Soc (London) 372A:453–473

Balthazar V, Vanacker V, Girma A, Poesen J, Golla S (2013) Human impact on sediment fluxes within the Blue Nile and Atbara River basins. Geomorphology 181:231–241

Baratti E, Montanari A, Castellarin A, Salinas JL, Viglione A, Bezzi A (2012) Estimating the flood frequency distribution at seasonal and annual time scales. Hydrol Earth Syst Sci 16:4651–4660

Bates BC, Kundzewicz ZW, Wu S, Palutikof JP (2008) Climate Change and Water. Technical Paper of the Intergovernmental Panel on Climate Change, IPCC Secretariat, Geneva

Belete A (2009) Climate change impact on Lake Abaya water level. MSc thesis, Addis Abaab University, Addis Ababa

Billi P (2000) A study on the geomorphology and sediment yield of the Meki river (Great Lakes Region, Ethiopian Rift Valley): preliminary results. In: Lenzi MA (ed) Dynamics of water and sediments in Mountain Basins, vol 20. Quaderni di Idronomia Montana Special Issue, pp 145–157

Billi P (2004) Sediment yield of a closed river system: the Meki River, western margin of the Ethiopian rift valley. In: Proceedings of the IAHS international symposium on sediment transfer through the fluvial system. Moscow, 2–6 Aug 2004, pp 94–100

Billi P (2007) Morphology and sediment dynamics of ephemeral streams terminal reaches in the Kobo basin (northern Welo, Ethiopia). Geomorphology 85:98–113

Billi P (2008) Flash floods, sediment transport and bedforms in the ephemeral streams of Kobo basin, northern Ethiopia. CATENA 75 (1):5–17

Billi P, Bedri O (2010) Sediment transport of the Blue Nile River at Khartoum. Quatern Int 226:12–22

Billi P (2011) Flash flood sediment transport in a steep sand-bed ephemeral stream. Int J Sediment Res 26(2):193–209

Billi P, Alemu YT, Ciampalini R (2015) Increased frequency of flash floods in Dire Dawa, Ethiopia: change in rainfall intensity or human impact? Nat Hazards 75. doi:10.1007/s11069-014-1554-0

Bisson PA, Nielsen JL, Palmason RA, Grove LE (1982) A system of naming habitat types in small streams, with examples of habitat utilization by salmonids during lowstream flow. In: Armantrout NB (ed) Proceedings of a symposium on acquisition and utilization of aquatic habitat inventory information. Western Division of the American Fisheries Society, Portland, pp 62–73

Boateng I, Bray M, Hooke J (2012) Estimating the fluvial sediment input to the coastal sediment budget: a case study of Ghana. Geomorphology 138:100–110

Brayshaw AC (1984) Characteristics and origin of cluster bedforms in coarse-grained alluvial channels. In: Koster EH, Steel RJ (eds) Sedimentology of gravels and conglomerates, vol 10. Canadian Soc Petroleum Geol, Memoir, pp 77–65

Bridge JS (2003) Rivers and floodplains: forma, processes, and sedimentary record. Blackwell, Oxford

Carling PA (1987) Hydrodynamic interpretation of a boulder berm and associated debris-torrent deposits. Geomorphology 1(1):53–67

Collick AS, Easton ZM, Ashagrie T, Biruk B, Tilahun S, Adgo E, Awulachew SB, Zeleke G, Steenhuis TS (2009) A simple semi-distributed water balance model for the Ethiopian highlands. Hydrol Process 23:3718–3727

Conway D (2000) The climate and hydrology of the upper Blue Nile. Geogr J 166(1):49–62

Conway D (2005) From headwater tributaries to international river: observing and adapting to climate variability and change in the Nile basin. Global Env Change 15:99–114

Conway D, Persechino A, Ardoin-Bardin S, Hamandawana H, Dieulin C, Mahe G (2009) Rainfall and water resources variability in sub-saharan Africa during the twentieth century. J Hydrometeorol 10(1):41–59

Corti G, Manetti P (2012) Geologia e paesaggi della rift valley in Etiopia. Consiglio Nazionale delle Ricerche, Pacini, Pisa

Demissie M (2008) Regional flood frequency analysis for upper Awash sub-basin (upstream of Koka). MSc thesis, Addis Ababa Univeristy

Di Baldassarre G, Elshamy M, van Griensven A, Soliman E, Kigobe M, Ndomba P, Mutemi J, Mutua F, Moges S, Xuan Y, Solomatine D, Uhlenbrook S (2011) Future hydrology and climate in the River Nile basin: a review. Hydrol Sci J 56(2):199–211

Easton ZM, Fuka DR, White ED, Collick AS, Biruk Ashagre B, McCartney M, Awulachew SB, Ahmed AA, Steenhuis TS (2010) A multi basin SWAT model analysis of runoff and sedimentation in the Blue Nile, Ethiopia. Hydrol Earth Syst Sci 14:1827–1841

EEA (European Environment Agency) (2012) Cliamte change, impacts and vulnerability in Europe. An indicator-based report. Report 12/ 2012, EEA Copenhagen

EM-DAT (2013) The international disaster database. http://www.emdat.be/result-country-profile

FAO (2000) Global river sediment yields database. http://www.fao.org/nr/water/aquastat/sediment/index.stm

Gebrehiwot SG, Ilstedt U, Gardenas AI, Bishop K (2011) Hydrological characterization of watersheds in the Blue Nile basin, Ethiopia. Hydrol Earth Syst Sci 15:11–20

Gebremicael TG, Mohamed YA, Betrie GD, van der Zaag P, Teferi E (2013) Trend analysis of runoff and sediment fluxes in the upper Blue Nile basin: a combined analysis of statistical tests, physically-based models and landuse maps. J Hydrol 482:57–68

Girma NT, Horlachen HB (2004) Investigation of performance of sediment transport formulas in natural rivers based on measured data in Kulfo River, Southern Ethiopia. In: Proceedings of the Lake Abaya research symposium, vol 4. Addis Ababa, 10–12 Nov 2004, pp 35–42

Grant GE, Swanson FJ, Wolman MG (1990) Pattern and origin of stepped-bed morphology in high-gradient streams, western cascades, Oregon. Geol Soc Am Bull 102:340–352

Gumindoga W (2010) Hydrological impact of landuse change in the upper Gilgel Abay River basin, Ethiopia. TOPMODEL application. MSc thesis, International Institute for Geo-Information Science and Earth Observation, Enschede

Guzman CD, Tilahun SA, Zegeye AD, Steenhuis TS (2013) Suspended sediment concentration–discharge relationships in the (sub-) humid Ethiopian highlands. Hydrol Earth Syst Sci 17:1067–1077

Hadley Centre (2005) Climate change, rivers and rainfall. Met Office, Exeter

Haregeweyn N, Poesen J, Govers G, Verstraeten G, de Vente J, Deckers J, Meryersons J, Haile M (2008) Sediment yield variability in northern Ethiopia: a quantitative analysis of its controlling factors. Catena 75:65–76

Hickin ED (1972) Pseudomeanders and point dunes—flume study. Am J Sci 272:762–799

Hurni H (1983) Soil erosion and soil formation in agricultural ecosystems of Ethiopia and Northern Thailand. Mt Res Dev 3:131–142

IPCC (Intergovernmental Panel on Climate Change) (2007) Synthesis report. IPCC Secretariat, Geneva

Kebede S, Travi Y, Alemayehu T, Marc V (2006) Water balance of Lake Tana and its sensitivity to fluctuations in rainfall, Blue Nile basin. Ethiop J Hydrol 316(1–4):233–247

Kim U, Kaluarachchi JJ, Smakhtin VU (2008) Climate change impacts on hydrology and water resources of the upper Blue Nile River basin, Ethiopia. IWMI Research Report 126, International Water Management Institute, Colombo

Knighton D (1993) Fluvial forms and processes: a new perspective. Arnold, London

Koster EH (1978) Transverse ribs: their characteristics, origin and paleohydraulic significance. In: Miall AD (ed) Fluvial sedimentology, vol 5. Canadian Soc Petroleum Geol, Memoir, pp 161–186

Legesse D, Abiye TA, Vallet-Coulomb C, Abate H (2010) Streamflow sensitivity to climate and land cover changes: Meki River, Ethiopia. Hydrol Earth Syst Sci 14:2277–2287

Leopold LB, Wolman MG (1957) River channel pattern: braided meandering and straight. US Geol Surv Prof Paper 282B:39–85

Leopold LB, Wolman MG, Miller JP (1964) Fluvial processes in geomorphology. Freeman, San Francisco

Mahe G, Lienou G, Descroix L, Bamba F, Paturel JE, Laraque A, Meddi M, Habaib H, Adeaga O, Dielun C, Chahnez Kotti F, Khomsi K (2013) The rivers of Africa: witness of climate change and human impact on the environment. Hydrol Processes 27:2105–2114

Martin Y (2003) Evaluation of bed load transport formulae using field evidence from the Vedder River, British Columbia. Geomorphology 53:75–95

McDougall I, Morton WH, Williams MAJ (1975) Age and rate of denudation of Trap Series basalts at Blue Nile gorge, Ethiopia. Nature 254:207–209

Melesse A, Abtew W, Dessalegne T, Wang X (2010) Low and high flow analyses and wavelet application for characterization of the Blue Nile River system. Hydrol Process 24:241–252

Milliman JD, Meade RH (1983) World-wide delivery of river sediment to the oceans. J Geol 91:1–21

Montgomery DR, Buffington JM (1997) Channel-reach morphology in mountain drainage basins. Geol Soc Am Bull 109(5):596–611

Nyssen J, Poesen J, Moeyersons J, Deckers J, Haile M, Lang A (2004) Human impact on the environment in the Ethiopian and Eritrean highlands—a state of the art. Earth-Sci Rev 64(3–4):273–320

Nyssen J, Descheemaeker K, Haregeweyn N, Haile M, Deckers J, Poesen J (eds) (2007) Lessons learnt from 10 years research on soil erosion and soil and water conservation in Tigray. Tigray Livelihood papers no. 7, Mekelle: Zala-Daget Project, Mekelle University, K.U. Leuven, Relief Society of Tigray, Africamuseum and Tigray Bureau of Agriculture and Rural Development

Oliver JE (1980) Monthly precipitation distribution: a comparative index. Prof Geographer 32:300–309

Omuto CT, Vargas RR, Paron P (2009) Soil erosion and sedimentation modelling and monitoring framework of the areas between rivers Juba and Shabelle in southern Somalia. FAO-SWALIM Technical Report No. L-16, Nairobi

Petts GE (1983) Rivers. Butterworths, London

Poppe L, Frankl A, Poesen J, Admasua T, Dessied M, Adgoc E., Deckers J, Nyssen J (2013) Geomorphology of the Lake Tana basin, Ethiopia. J Maps 93: 431–437

Rientjes THM, Haile AT, Kebede E, Mannaerts CMM, Habib E, Steenhuis TS (2011) Changes in land cover, rainfall and stream flow in Upper Gilgel. Abbay catchment, Blue Nile basin—Ethiopia. Hydrol Earth Syst Sci 15:1979–1989

Sagri M, Bartolini C, Billi P, Ferrari G, Benvenuti M, Carnicelli S, Barbano F (2008) Latest Pleistocene and Holocene river network evolution in the Ethiopian lakes region. Geomorphology 94:79–97

Sena L, Woldemichael K (2006) Disaster prevention and preparedness. Ethiopia Public Health Training Initiative, Jimma

Sima BA (2011) Flow regime and land cover changes in the Didessa sub -basin of the Blue Nile river, south-western Ethiopia: combining empirical analysis and community perception. MSc thesis, Department of Aquatic Science and Assessment Department of Urban and Rural Development SLU, Swedish University of Agricultural Sciences, Uppsala

Steenhuis TS, Collick AS, Easton ZM, Leggesse ES, Bayabil HK, White ED, Awulachew SB, Adgo E, Ahmed AA (2009) Predicting discharge and sediment for the Abay (Blue Nile) with a simple model. Hydrol Process 23:3728–3737

Summerfield MA, Hulton NJ (1994) Natural control of fluvial denudation rates in major world drainage basins. J Geophys Res 99(B7):13871–13883

Sutcliffe SV, Parks YP (1999) The hydrology of the Nile. IASH Spec Publ No 7, Institute of Hydrology, Wallingford

Syvitski JPM, Milliman JD (2007) Geology, geography and humans battle for the dominance over the delivery of fluvial sediment to the coastal ocean. J Geol 115(1):1–19

Tamene L, Vlek PLG (2008) Soil erosion studies in Ethiopia. In Braimoth AK, Vlek PLG (eds) Land use and soil resources. Springer, New York, pp 73–100

Tamene L, Park SJ, Dikau R, Vlek PLG (2006) Analysis of factors determining sediment yield variability in the highlands of northern Ethiopia. Geomorphology 76:76–91

Tesemma ZK, Mohamed YA, Steenhuis TS (2010) Trends in rainfall and runoff in the Blue Nile basin: 1964–2003. Hydrol Process 24:3747–3758

Uhlenbrook S, Mohamed Y, Gragne AS (2010) Analyzing catchment behavior through catchment modeling in the Gilgel Abay, upper Blue Nile River basin, Ethiopia. Hydrol Earth Syst Sci 14:2153–2165

UNESCO (2007) Flow Regimes from International Experimental and Network Data (FRIEND) of the River Nile basin. FRIEND/NILE, Final Project Report, UNESCO Regional Office, Cairo

Vanmaercke M, Zenebe A, Poesen J, Nyssen J, Verstraeten G, Deckers J (2010) Sediment dynamics and the role of flash floods in sediment export from medium-sized catchments: a case study from the semi-arid tropical highlands in northern Ethiopia. J Soils Sediments 10 (4):611–627

Vorosmarty CJ, Fekete BM, Tucker BA (1998) Global River Discharge, 1807–1991, V. 1.1 (RivDIS). Data set. Available online http://www.daac.ornl.gov from Oak Ridge National Laboratory Distributed Active Archive Center, Oak Ridge

Wale A, Rientjes THM, Gieske ASM, Getachew HA (2009) Ungauged catchment contributions to Lake Tana's water balance. Hydrol Process 23:3682–3693

World Bank (2008) Tana and Beles integrated water resources development project. World Bank Report No: 43400-ET

Yemane K, Bonnefille R, Faure H (1985) Palaeoclimatic and tectonic implications of Neogene microflora from the northwest Ethiopian highlands. Nature 318:653–656

Planation Surfaces and the Long-term Geomorphological Evolution of Ethiopia

5

Mauro Coltorti, Dario Firuzabadi, Andrea Borri, Pierlorenzo Fantozzi, and Pierluigi Pieruccini

Abstract

Four major planation surfaces (PS) characterize the Ethiopian geology and landscape. They were modelled near or at sea level before the Ordovician (PS1), before the Late Triassic (PS2), before the Cenomanian (PS3), and before the Oligocene (PS4). These are unconformities in the sedimentary sequence recognizable across the entire country and the surrounding regions that due to uplift can be locally exhumed to generate wide steps in the landscape. At the top of the sequence, that also corresponds to the higher parts of the highlands, the flat depositional surface of the continental flood basalts (CFB) is preserved over large areas. However, a series of deep large palaeovalleys dissecting the CFB have been recognized on the water divides of the Afar, the Main Ethiopian Rift (MER), the Southern Ethiopian Rift (SER), and the Somalian and Sudan lowlands that are the main geomorphological features of the country. The palaeovalleys testify to a radial pattern that originated from the dome created by the emplacement of the CFB. They are easily recognizable to the south of Dire Dawa and in the northern part of the country. They were modelled over the pre-volcanic bedrock but in most of the highlands were buried under later volcanic products. The deep erosion of the dome and the later volcanic filling can explain the results of the previous thermochronological investigation. Unfortunately, the thickness of the CFB could have led to the reset of the apatite fission tracks (AFT) during the Oligocene and it is difficult to separate the effects of deep erosion from that of uplift, also considering that they can be closely related. In any case, a progressive incision of the thicker part of the dome is documented soon after the CFB deposition in the Blue Nile Gorge and during the Miocene and the Plio-Pleistocene in many other parts of the country.

Keywords

Planation surfaces • Supersequence • Plateau • Uplift • Erosion palaeovalleys • Geomorphology • Ethiopia

5.1 Introduction

Ethiopia is one of the few regions on Earth where continental break-up is ongoing, generating peculiar large-scale geomorphological features. In the southern Red Sea and the Aden Gulf, the break-up already occurred and the drifting stage is ongoing (Bosworth et al. 2005). The Afar area is the triple junction where the two more evolved branches join the third branch, the Main Ethiopian Rift (MER) (Mohr 1962, 1967; Huchon et al. 1991; Wolfenden et al. 2004) which is a major ca. NNE–SSE oriented tectonic depression that hosts a series of lakes (Zway, Langano, Abijata, Shala, Awassa, Abaya, Chamo) (Fig. 5.1). The northern sector of the MER has an almost regular width of ca. 90–100 km, but moving southward, it enlarges and the border faults split to create the Southern Ethiopian Rift (SER) or "broad rift zone" (BRZ,

M. Coltorti (✉) · D. Firuzabadi · A. Borri · P. Fantozzi · P. Pieruccini
Dipartimento di Scienze Fisiche, della Terra e dell'Ambiente, Via di Laterina, 8, 53100 Siena, Italy
e-mail: mauro.coltorti@unisi.it

P. Billi (ed.), *Landscapes and Landforms of Ethiopia*, World Geomorphological Landscapes, DOI 10.1007/978-94-017-8026-1_5, © Springer Science+Business Media Dordrecht 2015

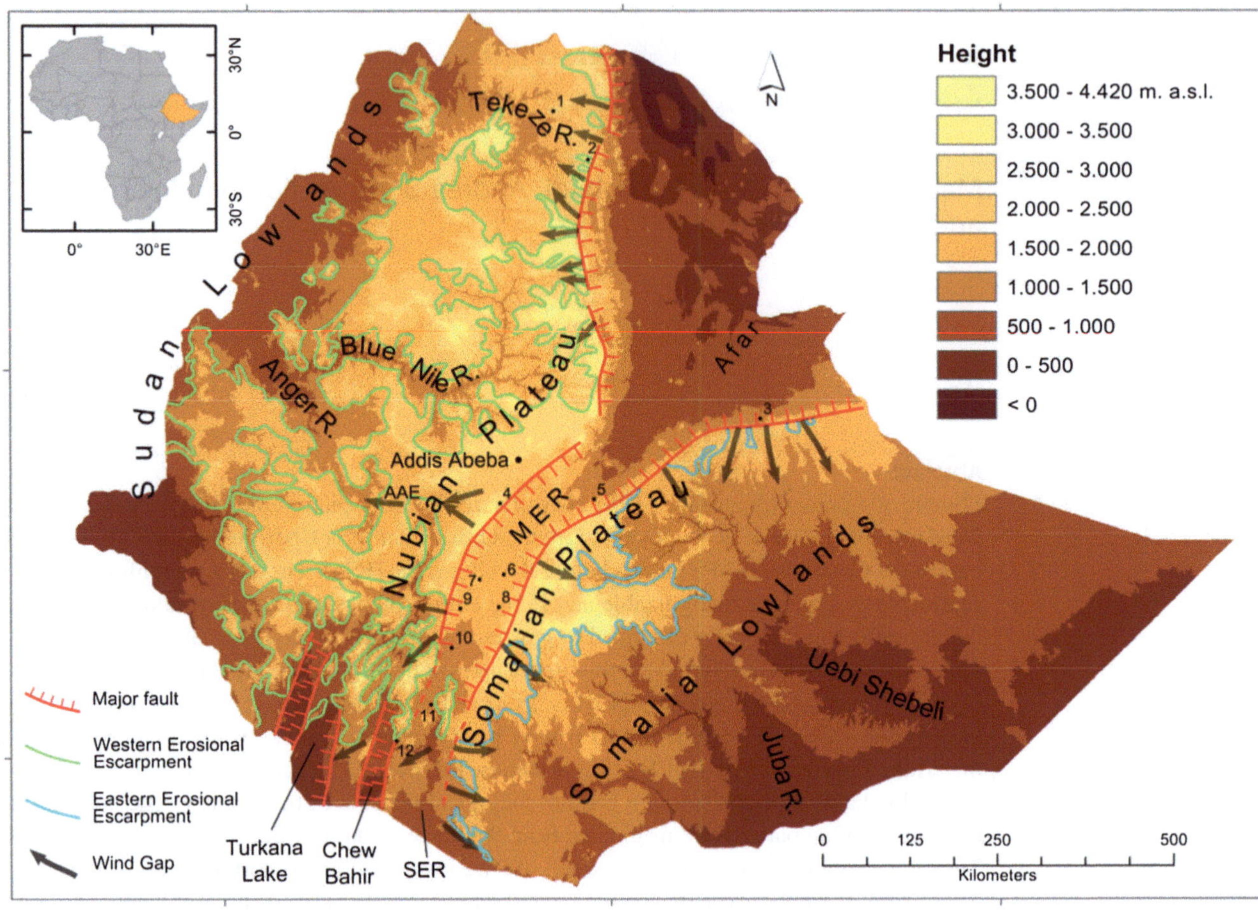

Fig. 5.1 Digital elevation model of Ethiopia (Shuttle Radar Topography Mission 2005) with the main geomorphological features. In the SER, the escarpment delimiting the horsts has been associated with the Western and Eastern Ethiopian Escarpment although they are in many cases fault escarpments deeply remodelled by erosional processes. Main localities mentioned in the text: *1* Mekellé; *2* Amba Alaji; *3* Dire Dawa; *4* Kella; *5* Chorora; *6* Ziway Lake; *7* Abijata Lake; *8* Langano Lake; *9* Shala Lake; *10* Abaja Lake; *11* Chamo Lake; *12* Turmi

Ebinger et al. 2000; Pik et al. 2008)—a series of tectonic basins, the westernmost of which continue southward generating the Turkana Lake, a part of the East African Rift System (EARS). The Afar and the MER are sinking and act as a trap for volcanic products and continental deposits. By contrast, the Nubian Plateau to the west and the Somalian Plateau to the east are rising (Mohr 1986; Faure 1975; Weissel et al. 1995; Pan et al. 2002; Pik et al. 2006; Pizzi et al. 2006). The uplift activated deep incision of the drainage network, especially to the west and east of the rift (Fig. 5.1). Incision is responsible for the typical landscape of the country, made of stepped slopes with the remains of the depositional surface of the continental flood basalts (CFB; Trap) preserved at higher elevations. The dissection allowed the exposure of a series of unconformities along the sides of the valley as well as along the margin of the rift. These correspond to buried planation surfaces that separate one depositional cycle from another. Coltorti et al. (2007) demonstrated that in northern Ethiopia, there were four major sedimentary cycles older than the CFB, bounded by four major planation surfaces that were buried and later exhumed due to uplift. A clear relationship exists between pre-CFB sedimentation and planation surfaces that have been modelled at sea level and later uplift, rifting and erosional processes. This work is aimed to show the importance, extent and significance of the planation surfaces and discuss the relationships with the CFB depositional surface, uplift, rifting and erosional processes, especially the evolution of the drainage network that has led to the present-day geomorphological setting.

5.2 The Sedimentary Sequence and the Planation/Unconformities Surfaces

The oldest rock, or "basement", is made of Proterozoic complexes that include intrusive and ultramafic rocks, highly metamorphosed rocks, schists and metasandstones

(Mohr 1962; Arkin et al. 1971; Aklilu et al. 1978; Merla et al. 1979; Kazmin et al. 1980; Tefera et al. 1990). These rock formed at a depth of over 20 km in the crust, but before the Early Palaeozoic, they were uplifted and moulded into an extensive flat planation surface, named PS1 in the northern part of the country (Coltorti et al. 2007). In fact, to the north of the Mekelle outlier, this major unconformity is buried by the Middle–Upper Ordovician continental and marine successions (Enticho Sandstone Formation; Arkin et al. 1971), including tillites (Edaga Arbi Glacials; Beyth 1972; Dow et al. 1971; Saxena and Assefa 1983; Bosellini et al. 1997). The thickness of this succession does not exceed a few hundred metres, but this likely underestimates the original thickness because a part of it was later eroded. In fact, a second major planation surface or unconformity cuts these rocks that are preserved only to the north of Mekelle and in the western side of this outlier. This is the PS2 by Coltorti et al. (2007) although firstly described as 1st erosional (peneplain) cycle by Merla and Minucci (1938). This unconformity is buried by the Adigrat Sandstones, deposited in a meandering alluvial plain (Bosellini et al. 1997), which are also known as the "Lower Sandstone" (Blanford 1869; Merla and Minucci 1938). The proximity to the sea level is indicated by the occurrence of beach deposits made of evenly laminated, well rounded and sorted quartz and gravels (Merla et al. 1979; Garland 1980; Tefera et al. 1990). The thickness varies, and as much as 700-m-thick deposits are preserved to the south-west of Mekelle (Beyth 1972). The transition from the continental/coastal sandstones to the overlying marine Antalo supersequence is reported to be gradual in Tigray and marked by 20- to 30-m-thick shales with calcarenite and sandstone intercalations (Bosellini et al. 1997). The age of this transgression is Late Callovian–Early Oxfordian. The Antalo supersequence in Tigray is made of limestone, and shales deposited during the Late Jurassic in a shallow water marine transgressive succession (Bruni and Fazzuoli 1977; Bosellini et al. 1997). Massive limestones characterize the lower part of the sequence, whereas in the upper part, shales with thin layers of coquina limestone, calcarenites and gypsum are dominant ("Agula Shales" of Merla and Minucci 1938). In northern Ethiopia, the distribution of the sedimentary facies is also affected by a series of synsedimentary normal faults oriented from WNW–ESE to E–W. The sequence and faults are cut by a major angular unconformity named PS3 by Coltorti et al. (2007) that, according to Bosellini et al. (2001), corresponds to a sedimentary hiatus and related erosion that lasted ca. 20 Ma. The unconformity is buried under the Amba Aradam Formation, also known as the "Upper Sandstones" (Shumburo 1968; Getaneh 1991; Bosellini et al. 1997, 2001), mostly made of sandstones and subordinately conglomerate and shales of fluvial origin with local aeolian sediments (Bosellini et al. 1997, 2001; Wolela 2008). The occurrence of very well sorted and rounded evenly laminated sandstones and quartz conglomerate indicates the interlayering with beach and coastal environment (i.e. the type area to the south of Mekelle). The age of the Amba Aradam Formation is attributed to the Early Cretaceous because of the age of the ca. 10-m-thick "Orbitolina Limestones" found in the Dire Dawa area (Bosellini et al. 2001). A series of red buried palaeosoils are interbedded in the lower and upper part of the succession and are also found in the other coeval formation cropping out in Horn of Africa and in Arabia (i.e. Yesomma Sandstones in Somalia and Mukalla Fm in Yemen; Bosellini et al. 2001). Although detailed studies of these soils in Ethiopia are lacking, their red colour, some clay illuviation and Fe/Mn precipitation features are the main indicators of pedogenic processes. In the palaeosoils we observed in Tigray, the thickness is limited and does not exceed few metres. The lack of a saprolite or deep weathering horizons affecting the parent material suggests short pedogenetic intervals similar to the Last Interglacial palaeosoils in Italy that evolved in ca. 20 ka (Coltorti and Pieruccini 2006). The thickness of the Amba Aradam Formation does not exceed 200 m, but it is much less in many places. Yet, it is a minimum evaluation due to the later erosive processes.

The Amba Aradam Formation contains the youngest marine deposits which are preserved on the highlands up to elevation of 2,500–2,800 m (i.e. the Amba Aradam-type area in Tigray) and constitute a topographic reference for the later uplift (Coltorti et al. 2007).

Another very flat planation surface or unconformity cuts the pre-basaltic sedimentary successions including the basement. This was firstly described by Blanford (1869) and later indicated as "pre-volcanic erosion" (Merla and Minucci 1938), "pre-trappean peneplanation" (Mohr 1962) or PS4 by Coltorti et al. (2007). The PS4 was buried under the CFB (Trap volcanics of Blanford 1870). These are basaltic and trachytes lava flows with different regional names (Arussi, Bale, Wallega, Omo, Amba Alaji, Ashangi, Blue Nile, etc.; Merla et al. 1979; Tefera et al. 1990). The lava flows originated from fissural activity and local volcanic centres that nowadays are largely eroded and only locally preserved as skeleton volcanoes. The maximum thickness of the CFB has been reported up 2,000 m. During this phase, a series of sills, dikes and laccolites were also emplaced inside the older sedimentary successions (Merla et al. 1979; Schultz et al. 2008), especially in the Antalo supersequence. Radiometric dates seem to constrain the emplacement of the CFB in the Afar area during the Oligocene, between 30.5 and 28.5 Ma (Baker et al. 1996; Hoffman et al. 1997; Rochette et al. 1998; Couliè et al. 2003), while in southern Ethiopia, it seems to have occurred around 40 Ma (Ebinger et al. 1993; George et al. 1998). The lava flows expanded radially from the highlands locally filling shallow paleovalleys such as the Juba in Somalia, over 500 km in distance from the plateau (Carmignani et al. 1983; Abdirahim et al. 1993). The continuous and slow basalt outpouring occurred over an almost flat landscape. Bohannon et al. (1989, p. 1687) after a review of the geology of the Red Sea margin stated that "the entire north-eastern part of the Afro/Arabian continent was almost flat and near sea level from Late Cretaceous to at least Early Oligocene". In fact, the basalts along the Juba River, after filling the valley outflowed along the watersheds creating wide lava fields still partially preserved in western Somalia (Abbate et al. 1994). The CFB are capped by giant ignimbrites and more acid volcanic rocks that received local names (Jimma, Amba Alaji, Magdala, etc.; Merla et al. 1979; Tefera et al. 1990; Ayalew et al. 2002). After the emplacement of major ignimbrites that were almost synchronous with the CFB, two other events occurred at 15 and 8 Ma (Ayelew et al. 2002).

In any case, an erosional surface interrupting the emplacement of the CFB was recognized in the Blue Nile Gorge area (Gregnanin and Peccirillo 1974; Merla et al. 1979) and named "Ashangi peneplain" because it cuts the Ashangi basalt (Blue Nile Basalts of Merla and Minucci 1938). This and later unconformities deeply differ from the previous planation surfaces because they seal a contrasted topography and have much more local significance. A younger phase of flood basalt emplacement (Tarmeber-Megezez Fm of Kazmin et al. 1980) occurred during the Middle Miocene (13 Ma, Mohr 1967 and Kazmin et al. 1980; 10 Ma, Chernet et al. 1998). Kazmin et al. (1980) associated the volcanic centres of these younger lava flows to the opening of the Afar depression. However, according to Chernet et al. (1998) and Wolfenden et al. (2004), the opening of the Afar was associated with the emplacement of the Lower Afar Stratoid Series (LASS). These flood basalts have been dated for between 4.5 and 5.6 Ma (Chernet et al. 1998), although an age of 0.63 Ma obtained near Dire Dawa by Audin et al. (2004), if confirmed, would cast some doubts on the previous chronological setting. In the southwestern Afar, the LASS are interlayered with the Chorora Formation, diatomite lacustrine layers bearing mammal fauna (Tiercelin et al. 1980; Juch 1980). They are a few tens of metre thick and displaced by normal faults, but apparently, there is no evidence of synsedimentary faulting. They are still preserved almost at the top of the Somalian Plateau suggesting that during their deposition, at least, this part of the Afar depression was still not yet developed. The LASS are in turn buried by ignimbrites that can be up to 300 m thick (Nazreth Group; Butajira ignimbrites and tuffs; Woldegabriel et al. 1990; Chernet et al. 1998). The Nazreth Group is dated between 6.8 and 1.8 Ma (Bigazzi et al. 1993; Chernet et al. 1998) and is interstratified with the Butajira Ignimbrites dated between 4.2 and 3 Ma (Woldegabriel et al. 1990). Almost at the same time, apparently between 5.6 and 5 Ma, a series of Rift Margin Silicic Centres erupted (Chernet et al. 1998). To the south of Addis Ababa, Pliocene volcanic activity occurred along the Ambo lineament, also known as the Addis Ababa Embayment, the westward continuation of faults delimiting the southern margin of the Afar. These lava flows belong to the Wechecha Formation and are dated between 4.3 and 3 Ma (Chernet et al. 1998), overlying the Addis Ababa basalts dated between 6.6 and 5 Ma. In the same area, a younger set of volcanoes (Bishoftu Fm) were active between 2.8 and 0.88 Ma (Chernet et al. 1998). The basalts of the Bishoftu Formation are found also on the rift floor, where the volcanic activity is included in a single unit, the Bofa Formation (EIGS 1978, 1985). In the MER, they are overlain by the volcanic Wonji Group attributed to 3.5–1.5 Ma (Kazmin et al. 1980) and more recently considered younger than 1.8 Ma (Chernet et al. 1998; Boccaletti et al. 1999; Abebe et al. 2005, 2007). Continental sedimentary succession is interlayered in these volcanic products during the Pliocene and the Pleistocene.

5.3 Methods

In order to describe the main geomorphologial features of the area and the relationships between the planation surfaces/unconformities and the geological history of Ethiopia, a general geological map of the country was derived from the available regional geological map (1:2,000,000 scale; Tefera et al. 1990). The elevation data were derived from the SRTM DEM (Shuttle Radar Topography Mission; United States Geological Survey 2005). Unfortunately, geological vector data of the entire study area were not available. Therefore, the map of Tefera et al. (1990), once georeferenced, was digitalized.

The analysis of the geological formations (i.e. Amba Aradam Fm and Antalo Fm), their origin (sedimentary, volcanic, continental, etc.) and the presence of the main unconformities/planation surfaces made possible to distinguish four supersequences (or supersynthems) (Fig. 5.2). The map of the supersequences is in WGS84 datum and UTM 38 N projection (Fig. 5.2). A series of geological cross sections (Fig. 5.3) were obtained from the SRTM DEM, available for the study area at 3-arcsec resolution (ca. 90 m) using ESRI ArcGIS 10.2 software. The sections were obtained from series of elevation points, automatically created at a distance of 100 m from each other, a short distance considering the large area investigated. A vertical exaggeration factor of 10× was adopted in order to improve the visual resolution of the topography.

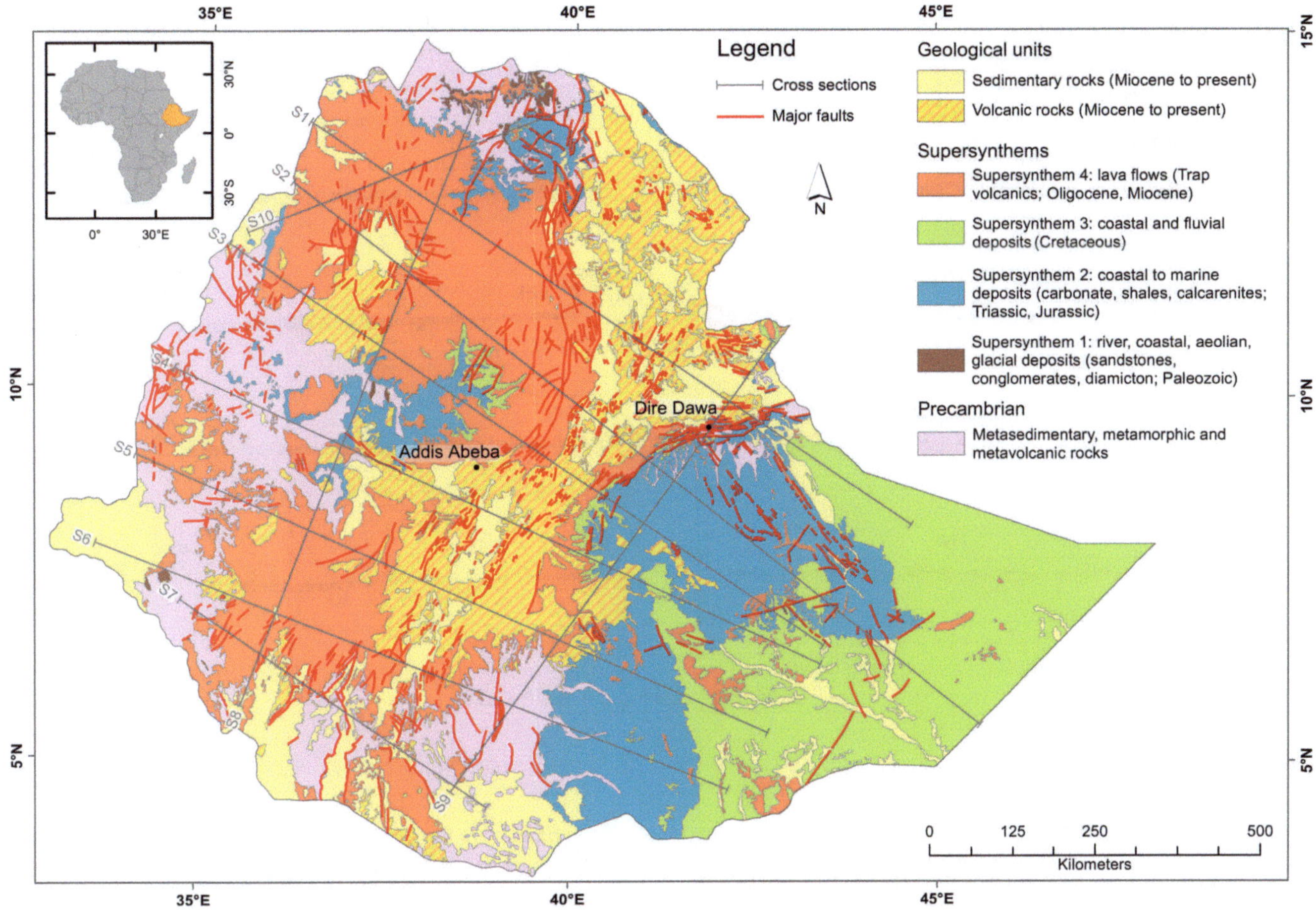

Fig. 5.2 Geological map with the Palaeozoic to Oligo-Miocene rocks grouped in supersynthems/supersequences bounded by planation/unconformity surfaces. PS1 separates the basement from the Palaeozoic rocks; PS2 separates the basement and the Palaeozoic rocks from Triassic–Jurassic rocks; PS3 separates the basement, Palaeozoic and Triassic–Jurassic rocks from Cretaceous rocks; PS4 separates the older sedimentary rocks from Oligocene–Miocene continental flood basalts. The geological map of Tefera et al. (1990) has been used to locate sedimentary sequence and their boundaries

5.4 The Planation Surfaces in Ethiopia

The planation surfaces are a major feature of the Ethiopian highlands. They are buried but well recognizable because, on both sides of the plateaus, deep valley incisions, along with rift systems tectonics, allow for the observation of the lowermost parts of the successions. The preservation of the PS1 is limited to the northern part of the country, whereas its absence to the south of Mekelle suggests that levelling of PS2 caused erosion of the Palaeozoic rocks in the rest of the country. The modelling of PS2 was likely preceded by a large-scale folding that generated wide synforms where the Palaeozoic sequences are still preserved. PS2 is a very flat surface and recognizable in most of the country because the Adigrat Sandstones lie over the Precambrian basement in the rest of the country and also in Somalia (Fantozzi et al. 2002), southern Arabia (Kohlan Formation, Simmons and Al-Thour 1994; Toland et al. 1994) and northern Kenya (Mansa Guda Formation, Bosellini et al. 2001) (Fig. 5.4). The thickness of the Adigrat Sandstones varies across the country from a few tens of metres in the eastern regions (south-east Ethiopia, Somalia and Yemen) up to 1,000 m in the Blue Nile Gorge suggesting that the sandstones covered a landscape made of "depressions and structural highs" (Bosellini et al. 2001). The difference in thickness might also be related to the occurrence of a neglected "paraconformity" located at the base of the Antalo Formation. In fact, in many places, the Antalo Formation lies directly over the Precambrian basement such as to the south-west of Mekelle (Bosellini et al. 1997), in the Danakil Alps, and in the Dire Dawa area (Bosellini et al. 2001). The end of the Adigrat Formation deposition is placed at the appearance of the lowermost layers of the Antalo Formation, but an exact age for the sandstones is unknown, and therefore, a hiatus cannot

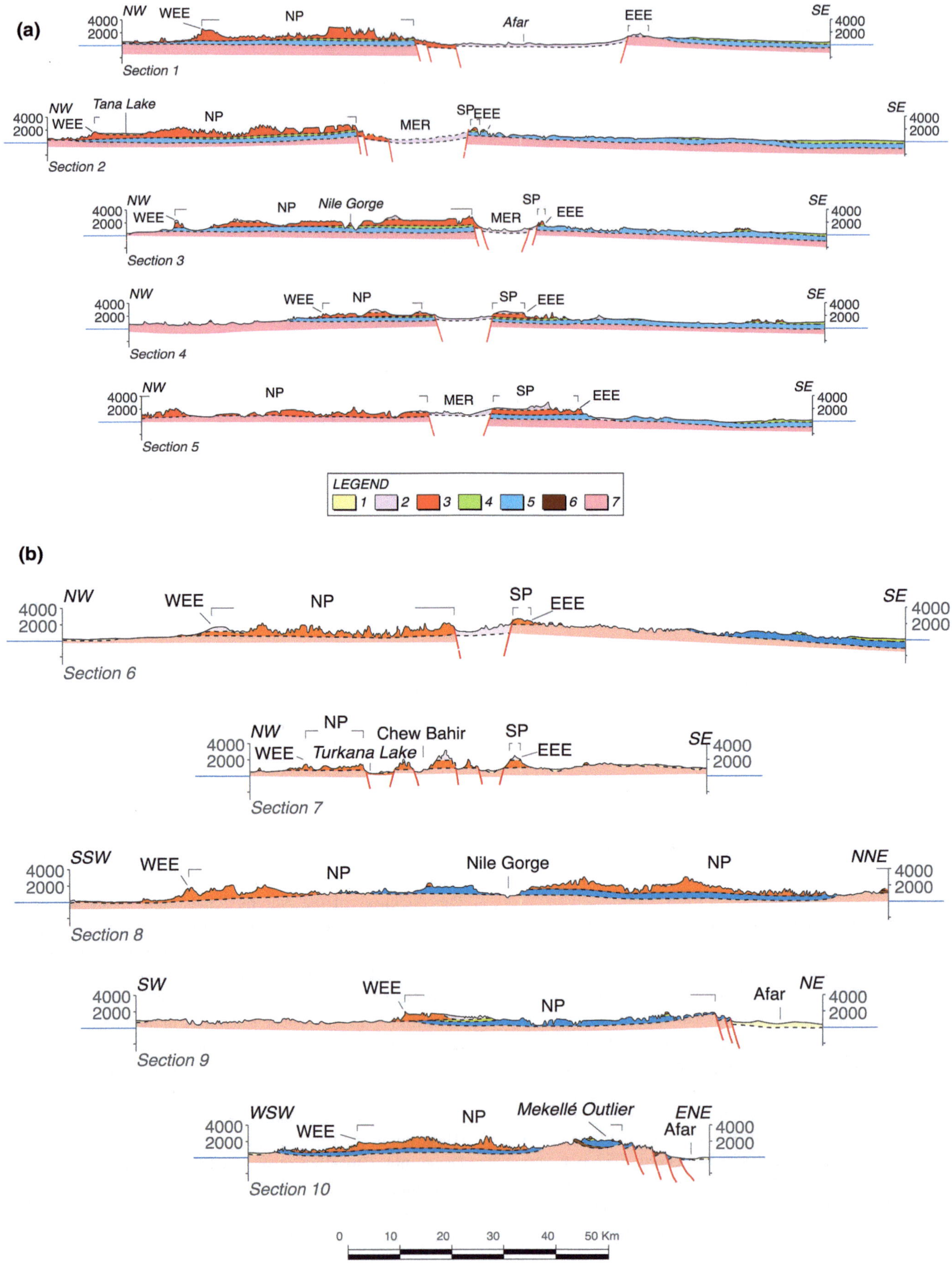
(a)
NW WEE NP Afar EEE SE
4000 2000
Section 1
NW Tana Lake WEE NP MER SP EEE SE
Section 2
NW WEE NP Nile Gorge MER SP EEE SE
Section 3
NW WEE NP SP EEE SE
Section 4
NW NP MER SP EEE SE
Section 5
LEGEND
1 2 3 4 5 6 7
(b)
NW WEE NP SP EEE SE
Section 6
NW NP Chew Bahir WEE Turkana Lake SP EEE SE
Section 7
SSW WEE NP Nile Gorge NP NNE
Section 8
SW WEE NP Afar NE
Section 9
WSW WEE NP Mekellé Outlier ENE Afar
Section 10
0 10 20 30 40 50 Km

◀**Fig. 5.3** Geological cross sections of the Ethiopian supersynthems. Location in Fig. 5.2. The absence of a supersynthem in a section testifies to efficacy of erosional processes during the modelling of different planation surfaces. The boundaries are extrapolated, and major uncertainties are due to the variation in thickness that corresponds to the different importance of erosional processes, especially where the sequence is buried under CFB. *1* sedimentary and volcanic rocks (Miocene to present); *2* volcanic rocks (supersynthem *4*; Trap, Oligocene–Miocene); *4* coastal and fluvial sediments (supersynthem *3*; Amba Aradam Formation, Cretaceous); *5* coastal to marine deposits (supersynthem *2*; Triassic–Jurassic); *6* river, coastal, aeolian and glacial sediments (supersynthem *1*; Palaeozoic); *7* metamorphic and intrusive rocks (Precambrian)

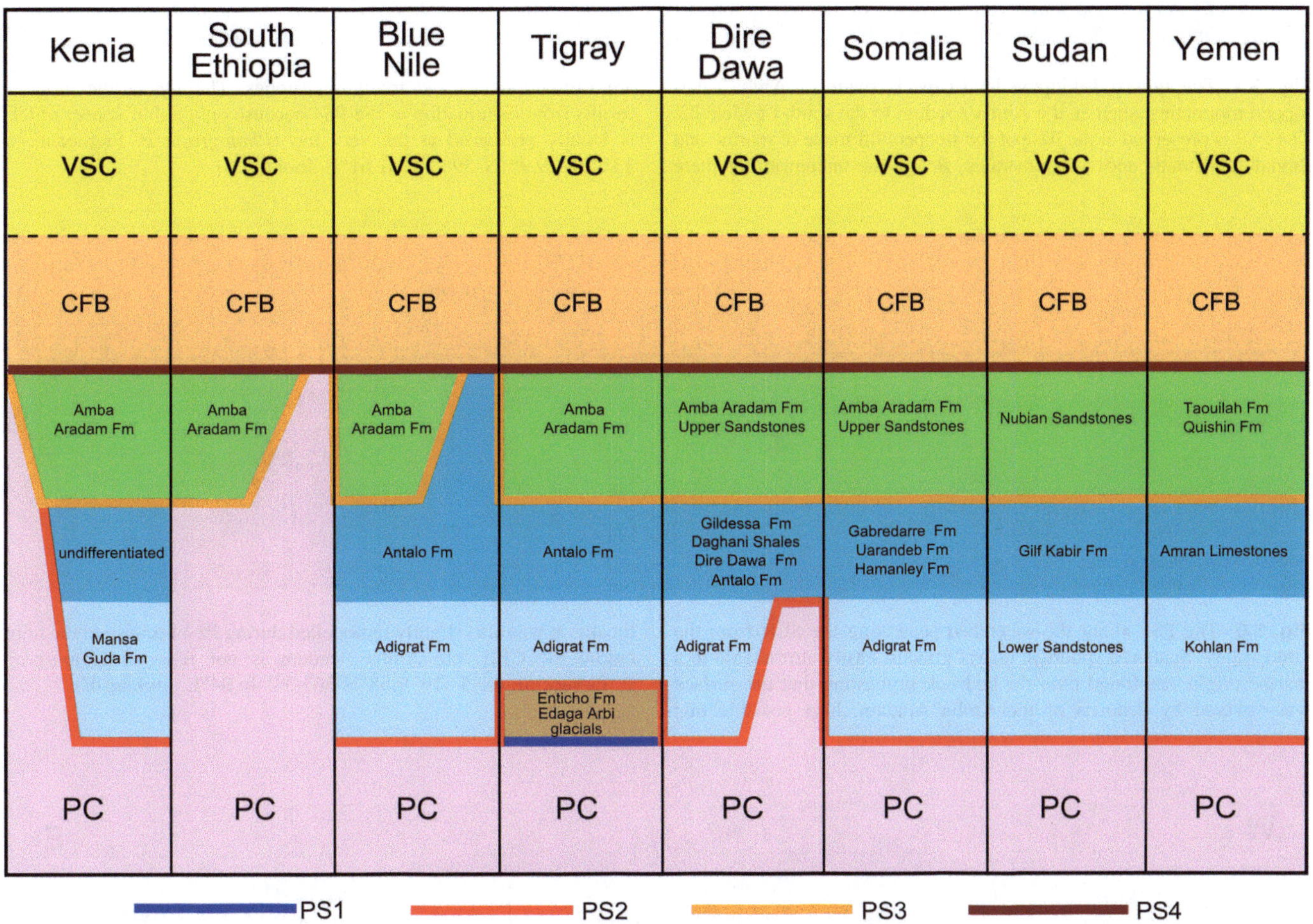

Fig. 5.4 Relationship between planation surfaces and sedimentary sequence in Ethiopia and surrounding regions. *PC* precambrian rocks; *CFB* continental flood basalts; *VSC* post-CFB volcanic and sedimentary sequence

be excluded. Sedimentary evidence of repeated fluctuations of the sea level during the early stages of the Antalo marine transgression is also reported by Bosellini et al. (1997, 1999). The occurrence of a hiatus would be also supported by a lateritic crust at the top of the Adigrat Sandstone Formation in different parts of the Mekelle outlier (Tigray) and in Eritrea (Merla and Minucci 1938; Abul-Haggag 1961; Bosellini et al. 1997). The marked difference in thickness of the Adigrat Formation might also be explained with the occurrence of tectonic movements and the formation of wide synforms and antiforms. PS3 would have erased the succession in the antiforms and preserved it in the synforms. In this case, the "structural highs" would have existed before the deposition of the younger rocks. However, in order to make it easier to read the supersequence Map and to pinpoint the major events, all the Triassic–Jurassic sequences of Ethiopia were grouped into a single Adigrat–Antalo supersequence. The Adigrat and Antalo limestones occur in the Blue Nile, buried under shales (Mugher Fm; Sagri et al. 1998; Gani et al. 2009). In the Dire Dawa area, the supersequence includes the Antalo limestones, unconformably covered by black limestones and marls (Dire Dawa Fm), marly limestones and marls with shales and gypsum (Daghani Shales), and finally by reef limestones (Gildessa Fm)

Fig. 5.5 The typical Ethiopian landscape is made of Amba—flat-topped mountains, such as the Amba Aradam to the south of Mekellè. The PS3 is preserved at the base of the steeper cliff made of marine and fluvial sandstones and conglomerates. *Below* the unconformity there are Jurassic shales and marly limestones. The *top* of the mountain results from exhumation of the PS4 because only a thin veneer of CFB is locally preserved at the very top (*Photograph* P. Pieruccini, ca. 13° 20′ 29.95″N 39° 30′ 41.61″E, looking S)

Fig. 5.6 The PS3 along the watershed separating the SER from the Omo valley. Scattered spherical quartz gravels, easily attributable to a marine origin, are found over the bedrock suggesting that the surface was covered by deposits of the Amba Aradam. It is possible that locally, it was also slightly remodelled during PS4 because to the north, below the CFB, the Amba Aradam is not reported (*Photograph* P. Pieruccini, ca. 5° 19′ 9.58″N 36° 34′ 46.04″E, looking E)

Fig. 5.7 The exhumation of the PS3 in southern Ethiopia and the erosion of a thick cover of the CFB is easily recognizable north of Turmi because the surface modelled over the basement rocks (foreground) only a few kilometres to the north is covered by over 1,000 m of CFB (background). (*Photograph* P. Pieruccini, 5° 36′ 7.92″N 36° 41′ 36.11″E, looking N)

(Bosellini et al. 2001). The westernmost outcrops of the Adigrat–Antalo supersequence in central-south Ethiopia are found in the Blue Nile Gorge sequence that represents the emersion of the Jurassic sediments from under the CFB and later volcanics that cover both sides of the central and northern MER. The Antalo limestones crop out with local facies variation in Ogaden and Somalia where they correspond to the superimposed Hamanley (Antalo limestones-equivalent), Uarandeb (Agula Shales-equivalent) and Gabredarre Formation, a lateral facies variation of the Gildessa Formation of Dire Dawa (Bosellini 1992) (Fig. 5.4). In Yemen, the equivalent of the Antalo limestones is the Amran limestones around San'a and Suqhraq in southern Yemen, also covered by shales (Bosellini et al. 2001). They were recognized also in Somalia (Fantozzi 1998; Fantozzi and Ali Kassim 2002). A tectonic phase is documented after the deposition of the Adigrat–

Fig. 5.8 The Eastern Ethiopian Escarpment (*EEE*) at the Afar Window, to the south of Debresina. The top of the escarpment is modelled over the CFB. The slope is dissected by river erosion and shaped by large-scale landslides (*Photograph* M. Coltorti, ca. 9° 50′ 0.92″N 39° 44′ 23.96″E, looking E)

Fig. 5.9 The Western Ethiopian Escarpment (*WEE*) to the north of Chenca modelled over CFB whose depositional surface is preserved at the top of the relief. The slope is affected by large-scale gravitational movements (*Photograph* M. Coltorti, ca. 6° 25′ 16.92″N, 37° 37′ 15.77″E, looking S)

Antalo supersequence that could be responsible also for the minor Jurassic unconformities described in Dire Dawa district (Russo et al. 1994, 1996; Bosellini et al. 1997).

PS3 affected the Adigrat–Antalo supersequence in most parts of the country and in the nearby regions (Bosellini et al. 1997, 2001). It is remarkable that the Antalo Formation, grouped together with the Adigrat Formation in the map (Fig. 5.2), is absent in most of the southern part of the country where only the Adigrat Formation crops out. This would suggest the existence of another unconformity between these two formations that is difficult to recognize in the field. Exhumed remains of the PS3 are rarely observed over long distances because the overlying Amba Aradam Formation is also made of strongly cemented rocks and there is no contrast to trigger processes of selective erosion and re-exhumation (Fig. 5.5).

The supersequence 3, made by the Amba Aradam Formation, crops out in large part of the country, including the north-eastern border of the Somalian Plateau (Beyth 1972; Hutchinson and Engels 1970; Assefa 1991; Russo et al. 1994; Bosellini et al. 1997; Sagri et al. 1998). It is missing in the westernmost part of the Blue Nile Gorge and in the southern part of Ethiopia. It could be inferred that it was never deposited in the former areas, but its presence on the western

Fig. 5.10 The Somalian Plateau and the Eastern Ethiopian Escarpment (*EEE*) modelled over the CFB, the Amba Aradam and the Jurassic rocks deeply dissected by fluvial erosion (*Photograph* M. Coltorti, ca. 8° 13′ 34.03″N, 39° 50′ 20.00″E, looking SE)

Fig. 5.11 The Nubian or Ethiopian Plateau to the south of Chenca (*Photograph* M. Coltorti, ca. 6° 32′ 5.25″N, 37° 37′ 53.94″E, looking S)

side of the Turkana Lake (Merla et al. 1979; Nyagah 1997) suggests its later erosion. Reworked scattered spherical quartz gravels and sands are found over the basement along the new road from Jinka to Turkana Lake, where wide and flat watersheds largely correspond to the exhumed PS3 (Figs. 5.6 and 5.7). The thin layers of sands described by Davidson (1983) along the borders of the SER underlying the CFB could also be attributed to the supersequence 3. This formation crops out also in Yemen (Taouilah Fm, Simmons and Al-Thour 1994; Quishin Fm, Toland et al. 1994) (Fig. 5.4), in Afar, in Ogaden, Somalia and north-western Kenia (Merla et al. 1979; Tefera et al. 1990), as well as in Sudan (Upper Sandstones, Whiteman 1971; Yassin et al. 1984).

The PS4 planation is responsible for the complete erosion and/or the reduced thickness of the older supersequence and the erosion of the PS that had previously affected the Palaeozoic rocks. In fact, in many parts of the Ethiopian highland as well as in northern Kenya, the CFB lay directly on the Precambrian basement (Merla et al. 1979; Tefera et al. 1990). There are no elements in Ethiopia to establish the position relative to the sea level of the CFB but, as reported in Chap. 2, Bohannon et al. (1989) suggested that the marine sequences preserved along the margins of the Red Sea were deposited over an almost flat landscape close to the sea. In fact, in Saudi Arabia, there are also marine strata interbedded within the CFB.

Fig. 5.12 The Nubian Plateau to the south of Debresina. This photograph is taken from the same locality of Fig. 5.8 (*Photograph* M. Coltorti, ca. 9° 50′ 0.92″N 39° 44′ 23.96″E) but looking W. Note the strong contrast of relief's energy

Fig. 5.13 The Somalian Plateau deeply dissected by fluvial erosion (*Photograph* M. Coltorti, ca. 8° 10′ 35.61″N, 39° 45′ 14.73″E looking S)

5.5 Large-scale Geomorphological Features

From west to east, the major geomorphological units of the country (Figs. 5.1 and 5.2) are as follows: 1. Sudan lowlands; 2. Nubian Plateau; 3. Afar; 4. MER; 5. SER basins; 6. Somalia Plateau; and 7. Somalian lowlands.

The transition from the Sudan and the Somalian lowlands to the Nubian Plateau and the Somalian Plateau, respectively, is a sharp boundary located in correspondence with the Western (WEE) and Eastern Ethiopian Escarpment (EEE) (Fig. 5.1). These escarpments can exceed 1,000 m in height, and the major jump commonly develops within a few kilometres of horizontal distance (Figs. 5.8 and 5.9). It delimits the top of the depositional surface of the CFB or the later ignimbrite and volcanic rocks (Figs. 5.10, 5.11 and 5.12), from the basement rocks exposed in front of the escarpment to the west and south-east and/or the Jurassic and later sedimentary sequence to the east. In plan view, the boundary is very irregular, with very deep indentations corresponding with the major rivers and their tributaries such as the Blue Nile or the Tekeze in the west or the Juba and Wabi Shebelle in the east (Figs. 5.1 and 5.13). This indicates a long-lasting headward retreat that eroded the original margins of the CFB. In central Somalia, lava fields associated with the CFB are preserved at over 500 km from the margin of the plateau (Abdirahim et al. 1993; Abbate et al. 1994) but have been eroded in the rest of the Somalian lowlands. The Blue Nile Gorge and the Tekeze valley on the western side are among the best examples of deep dissection involved in the retreat of the WEE.

On the other hand, the boundaries of the rift-related tectonic depressions (Afar, MER, SER, Figs. 5.14 and 5.15) are usually sharp and rectilinear and marked by well-expressed faulted escarpments (Mohr 1962, 1967, 1987; Wolfenden et al. 2004; Pizzi et al. 2006; Abebe et al. 2007). The rivers have short valleys with stepped profiles, indicating a very

Fig. 5.14 The main fault escarpment to the east of Asela. In this sector, the displacement is accommodated by dozens of faults (*Photograph* M. Coltorti, ca. 8° 2′ 1.45″N, 39° 6′ 1.71″E)

Fig. 5.15 The main fault escarpment along the southern margin of the Afar, to the south of Dire Dawa (*Photograph* M. Coltorti, ca. 9° 29′ 53.97″N, 41° 45′ 27.33″E)

recent age for rifting processes, especially in the MER and SER. The morphology at the transition from the MER to the Afar (Ebinger et al. 2000; Pik et al. 2008) is not much different but more complex because the faults of the MER cut the faults of the south-eastern corner of the Afar (Chernet et al. 1998; Wolfenden et al. 2004; Abebe et al. 2007). The southern and western borders of the Afar have a rather rectangular drainage pattern influenced by the presence of rift-in-rift basins. The rivers have stepped profiles and are much longer than in the MER but much shorter than those crossing the WEE and the EEE. In the Dire Dawa area, the margin of the Afar is characterized by a more than 20-km-long, deeply incised and terraced pediment (Pizzi et al. 2008), the distal part of which is sealed by LASS basalts dated 0.6 Ma (Audin et al. 2004).

5.6 The Evolution of the Drainage Network and the Lesson from the Dire Dawa Watershed

Our investigations on the watersheds (Sect. 1 in Fig. 5.16) that separate the Afar to the north from the Somalian lowlands to the south of Dire Dawa allowed us to recognize two different drainage patterns. To the north of the watershed, along the steep main escarpment, drainage pattern is rectangular, with long branches trending ca. E–W, parallel to the fault system bounding the Afar. Minor branches are normal to these lineaments. The southward-running streams originate from small wind gaps hosting small lakes at ca. 2,000 m asl.

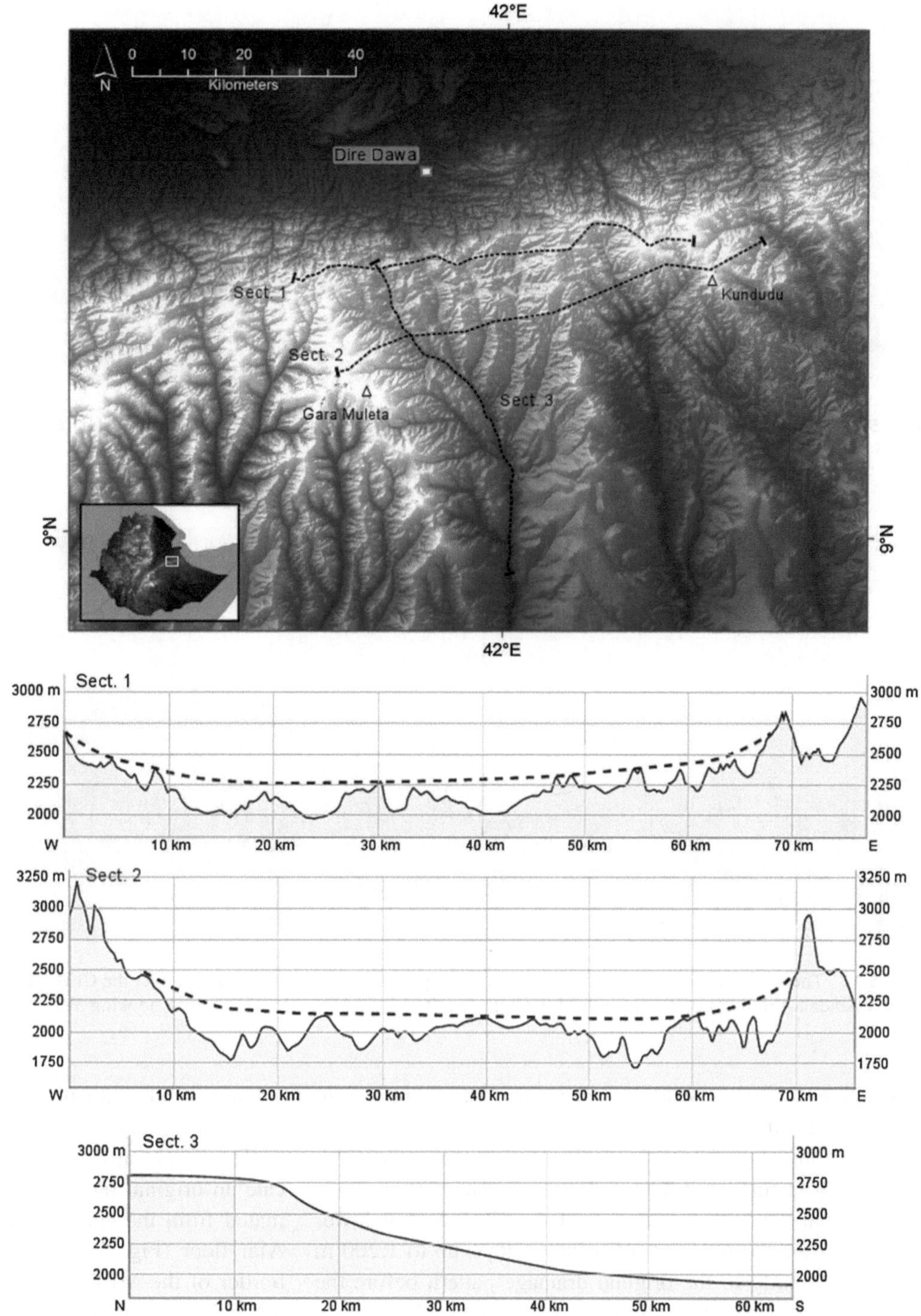

Fig. 5.16 A rectangular pattern influenced by faults is recognizable in the Dire Dawa area, to the north of the watershed, marked by Sect. 1. A straight southward-oriented drainage is recognizable to the south. The latter is associated with the continuation of a valley that was beheaded by the activation of the fault escarpment. A larger palaeovalley at higher elevation on the watershed was already modelled over the basement and the pre-Oligocene sedimentary sequence (Sect. 1). It is also well recognizable slightly to the south from Sect. 2 where the higher part of the slope is modelled over the CFB. The *flat upper part* of the longitudinal profile of the river is influenced by the original palaeovalley before beheading, while the lower tract is due to headward erosion

They slightly dissect the exhumed PS2 that contains also limited patches of the Adigrat Formation (Fig. 5.17). At ca. 2,200 m, above the smaller wind gaps, the divide shows the remnants of a much larger beheaded valleys (Fig. 5.16, Sect. 1 dotted line). The flanks of this large palaeovalley are actually the slopes modelled over the CFB whose depositional surface reaches ca. 3,500 m on the Mt Gara Mullata to the south-west (Fig. 5.18) and ca. 3,200 on the Kundudu, a residual relief, to the east (Fig. 5.16, Sect. 2). To the south of the watershed, the rivers are straight and their longitudinal profile shows a bimodal slope with the inner gradient very flat and increasing in steepness southward (Fig. 5.16, Sect. 3). The upper low-gradient profile is evidently influenced by the presence of the original paleovalley modelled before the

Fig. 5.17 The rectilinear southward-oriented valley to the south of the water divide of the Afar (*Photograph* M. Coltorti, ca. 9° 20′ 52.16″N, 41° 54′ 13.62″E looking S)

Fig. 5.18 The southern slope of the Gara Mullata, the upper part of which is modelled over the CFB. The PS4 is exhumed at the contact with the Amba Aradam Formation (*Photograph* M. Coltorti, ca. 9° 12′ 7.11″N, 41° 47′ 21.26″E looking W)

beheading, whereas the lower steeper part can be related to the recent headward erosion.

Therefore, the plateau is made of two major geomorphological features at different elevations. The higher one is the original depositional surface of the CFB, and at lower elevations are the remnants of palaeovalleys, up to 1,200 m deep, that testify to the original drainage pattern before the activation of the rifting. Similar palaeovalleys have been observed in other parts of the southern margin of the Afar, where the original watersheds are easily recognizable due to the occurrence of mesas and residual reliefs topped with the CFB at the sides of the wind gap (Figs. 5.1 and 5.16a). Palaeovalleys are also found along the margins of the Nubian Plateau and especially in northern Ethiopia. In fact, to the north of the Amba Alagi (ca. 3,500 m asl), the CFB is completely eroded and the watershed between the Nile river and the Afar is located around 2,000–2,200 m. A deep erosion is also documented in Eritrea because CFB are recorded in very limited patches (Balestrieri et al. 2005).

The wind gaps on the watersheds around the Afar indicate an original radial Palaeodrainage that could have originated from the centre of the CFB now lowered under the Afar floor (Fig. 5.16b). Palaeovalley also occur along the border of the MER, although in many parts of the plateau, the palaeomorphology is covered under younger volcanic products. Along the north-western side of the Abaja Lake, for example, the water divide with the Omo River is located at very low elevation (palaeovalley west of 9 and 10 in Fig. 5.1). Similar evidences are also found to the south of the Chamo Lake, on both sides of the MER, where the CFB are deeply dissected and the Precambrian basement crops out (paleovalleys south of 12 in Fig. 5.16). The importance of erosion of the watershed in this area comes out from the abrupt reduction in thickness and locally the absence of the CFB that, slightly to the north, on both sides of the Chamo Lake are over 2,000 m thick.

Most probably, the so-called Addis Ababa embayment (Chernet et al. 1998) corresponds to one of these palaeovalleys because there are over 1,000 m between the top of the preserved CFB to the north and the plateau to the south of Addis Ababa, where the surface is made of Plio-Pleistocene volcanics (AAE in Fig. 5.16). However, we do not exclude that the Addis Ababa embayment followed E–W faults along the so-called Yerer-Tellu Wellel lineament (Chernet et al. 1998; Tsegay et al. 1998).

5.7 Uplift and Erosion

Many authors claimed that Ethiopian relief was created rapidly around 30 Ma ago, slightly before and synchronously with the emplacement of the CFB (Baker et al. 1996; Pik et al. 2003, 2008), when the Ethiopian lithosphere moved over the Afar mantle plume (Hart et al. 1989; Schilling et al. 1992). The present elevation would be preserved by the sub-lithospheric mantle upwelling and outflow from under Afar (Wolfenden et al. 2004). However, a strong uplift in later time and especially during the Plio-Pleistocene is claimed by others (Merla et al. 1979; Juch 1980; Faure 1975; Mohr 1986; Weissel et al. 1995; Bohannon et al. 1989; Pizzi et al. 2006). The amount of long-term uplift can be roughly estimated from the mean elevation of the sedimentary succession preserved in an area. In fact, the successions are mostly made of alluvial plain and littoral deposits and the planation surfaces were levelled at or close to the sea level. This also includes the PS4 because the CFB expanded over an almost flat PS4 without filling deep valleys.

An estimation of the plateau uplift rates of 0.015 mm/year can be inferred taking into account that the Amba Aradam Formation (ca. 105 Ma) is commonly located between 2,000 and 2,500 m, a rate also suggested by McDougall et al. (1975). A minimum uplift rate of 0.066–0.083 mm/year is obtained if we consider that the area remained at or close to the sea level also during the emplacement of the CFB at ca. 30 Ma. However, values of 0.5–1 mm/year have been suggested if the movements are associated with flexural flank uplift, with a much later onset (Faure 1975; Weissel et al. 1995).

The data obtained from titanite (U-Th)/He method (Pik et al. 2003) that considered a partial retention zone (PRZ) at a depth 4–5 km with a gradient of 25–30 °C/km and a surface temperature of 20 °C, and apatite (U-TH)/He method (PRZ at 2 km with the same gradient) suggest that a sedimentary cover with a thickness exceeding 2,000 m was never deposited after 40 Ma in the Blue Nile Gorge and the northern part of the country. However, the youngest titanite (U-TH)/He age of 213 Ma found at the bottom of the Blue Nile gorge while the overlying rocks have ages of 400–520 Ma has two possible explanations (Pik et al. 2003): 1. the erosion of thicker CFB and 2. the existence of a thicker Mesozoic sedimentary cover before the emplacement of the CFB. The last explanation seems more reliable because of the occurrence of the planation surfaces and a thickness of the Mesozoic sediments exceeding 2,000 m in the northern part of the country (Getaneh 1991). An enhanced heating of the crust during the emplacement of the CFB such as suggested by Mock et al. (1999) is not consistent with the apatite (U-Th)/He as old as 45–107 Ma (Pik et al. 2003). On the other hand, the younger apatite (U-Th)/He ages of ca. 40 Ma found at 1,100 m below the base of the CFB in the Blue Nile Gorge is still older than the emplacement of the CFB and thus indicates that the CFB never exceeded 900 m, a thickness that would led to the reset of the apatite (U-Th)/He system. A sedimentary cover of at least 900 m (2,000 m of the apatite/He PRZ—1,100 m of the thickness of the sequence above the sample) was therefore eroded before 30 Ma (end of the modelling of the PS4). Pik et al. (2003) inferred, on the base of thermal modelling, that in the Blue Nile Gorge area, the incision of the plateau started soon after the emplacement of the CFB, 29–25 Ma ago, while in the Anger River catchment, a minor tributary, 10–5 Ma ago. However, if this occurred in an already uplifted area or this was the consequence of the dissection of the dome generated by the accumulation of the CFB is difficult to establish.

(U/Th)/He thermochronometry and apatite fission tracks (AFT) dating of the western rift escarpment have been performed in Eritrea, west of Asmara, but with controversial results (Abbate et al. 2002; Ghebreab et al. 2002; Balestrieri et al. 2005; Drury et al. 1994, 2006). Ghebreab et al. (2002) did not find any correlation between age and elevation, with younger ages also at the edge of the escarpment suggesting that tectonic denudation was the major agent at work. AFT ages across normal faults in the middle part of the escarpment indicate an old age of cooling between 23 and 16 Ma, followed by a thermal resetting event during Late Miocene and Early Pliocene (9–3 Ma) that the authors attributed to the rising of hot waters along faults and fractures. On the other hand, Balestrieri et al. (2005) found ages older than 33 Ma moving towards the plateau except for two ages found at the base of the escarpment that gave ca. 10 and 16 Ma, respectively. According to these results, the plateau surface was not buried anymore under 2,000 m of sediments since 200 Ma and in many cases also 400 Ma (close to Asmara). The younger ages, but always older than 30 Ma, found along the escarpment seem to testify that an important erosion preceded the formation of PS4, as in the Blue Nile Gorge (Pik et al. 2003). Along the escarpment, Balestrieri et al. (2005) found that AFT dating would indicate a rapid cooling during the Early–Middle Miocene (20 Ma) and suggest a post-breakup erosion of more than 3.5–4 km on top of the escarpment.

However, slightly to the east, the upper member of the Dogali Formation, dated between 15 and 6 Ma, testifies to marine deposition and hence the existence of the Red Sea graben.

The occurrence of deep palaeovalley affecting the CFB could easily explain the AFT date from the Kella horst, located on the western side of the MER, 100 km to the south of Addis Ababa (Di Paola et al. 1993; Tsegay et al. 2010). In fact, in this area, a thickness between 1,500 and 2,300 m of rocks was eroded after 7–6 Ma.

U-Th/He ages between 34 and 49 Ma have been observed in the upper part of the scarp delimiting the Hamar Range from the Chew Bahir in the SER (Pik et al. 2008). Ages between 19 and 12 Ma are found at the bottom of the escarpment. Therefore, erosion was active before the formation of PS4, but afterwards, since 19–12 Ma, a thickness of ca. 2,000 m was eroded, not necessarily due to fault activity. The later event led to the complete erosion of the over 2,000-m-thick CFB from the Hamar Range. Thus, the fact that the CFB are missing in many parts of the plateaux and that their present-day borders do not coincide with their original border suggests that downwearing due to weathering and progressive removal of weathering products from the top of the CFB was coupled with much faster erosion along the major palaeoriver systems

5.8 Conclusion

The planation surfaces of Ethiopia are unconformities between different depositional sequences recognizable in the entire country and the surrounding regions. The country was planated at sea level at least four times before the emplacement of the CFB, ca. 30 Ma. A ravinement surface generated by marine erosion during a major transgression, that is a plain of marine erosion (Davis 1899; Johnson 1930; Coltorti et al. 2007), is the best explanation for the modelling of these surfaces. In fact, except PS4, the planation surfaces are covered by marine and fluvial sediments. Alternative hypothesis of etchplanation and pediplanation (King 1975) has been already ruled out by Coltorti et al. (2007). In fact, the lacking of deep weathering profiles below the planation surfaces rules out etchplanation and the very flat nature of the surfaces rules out pediplanation. The latter hypothesis would also require very long periods of arid conditions and tectonic stability, a combination that it was also difficult to achieve. Moreover, there is no evidence that before or at the onset of the CFB deposition, the area was elevated. This had already been stated by Bohannon et al. (1989) after a review of the geology of the north-eastern Africa and the Arabian Peninsula. These authors also described sites on the eastern side of the Red Sea where marine deposits interlayer with the CFB. However, the deposition of the volcanic products that are 2,000 m in places, also if occurred over a flat surface, created reliefs. AFT confirm that at the periphery of the dome, where CFB thickness was reduced, there was no total reset of the system during the Oligocene. In the Blue Nile Gorge, the tracks were annealed due to the burial under the CFB, and later, as early as 29–25 Ma, rocks started to cool again because valley erosion was already at work (Pik et al. 2003). Miocene erosion is documented in the SER (Pik et al. 2008), where over 2,000 m of CFB have been eroded, as well as along some minor tributaries of the Nile (Pik et al. 2003). These authors associated the evidence in the SER to the activation of faults, but a down-cutting along lines of weakness in the basement, later reactivated as normal faults, would have a similar effect. A Miocene incision is documented by AFT and U-Th/He analysis also along the south-western borders of the Red Sea and the north-western side of the Afar in Eritrea (Ghebreab et al. 2002; Balestrieri et al. 2005), although the different results leave the interpretation of subsequent drainage evolution controversial. In any case, the upper member of the Dogali Fm in Eritrea lying at the edge of the Red Sea testifies to the opening of this sector at ca. 15 Ma. In our opinion, there is still not enough evidence to establish whether the nearby plateau was already elevated at that time or the uplift, as we also support, was triggered as a consequence of flexural flank uplift (Faure 1975; Mohr 1986; Weissel et al. 1995).

The recognition of beheaded palaeovalleys on the watersheds of the Afar cut over 1,200 m below the depositional surface of the CFB allows us to establish the main location of the original radial pattern. No chronological constraints are available for the beheading of the valleys and therefore the activation of the faults in Afar. However, the recognition of hanging palaeovalleys that in the western side of the Afar and the MER are commonly buried under volcanic products allows also to establish that the plateaux are complex geomorphological features that bear witness of the oldest evolution of the drainage network. A large drainage system possibly occupied the depression, known as the Addis Ababa Embayment, located to the south of the capital area (Fig. 5.19). If the Kella horst belongs to this drainage system and the complete erosion of the CFB is attributed to a westward drainage system, the beheading in this area would have occurred after 3 Ma that is the age of the basalt filling the palaeovalleys (Tsegay et al. 2010). This should also be the lower limit for the activation of border faults of the central MER.

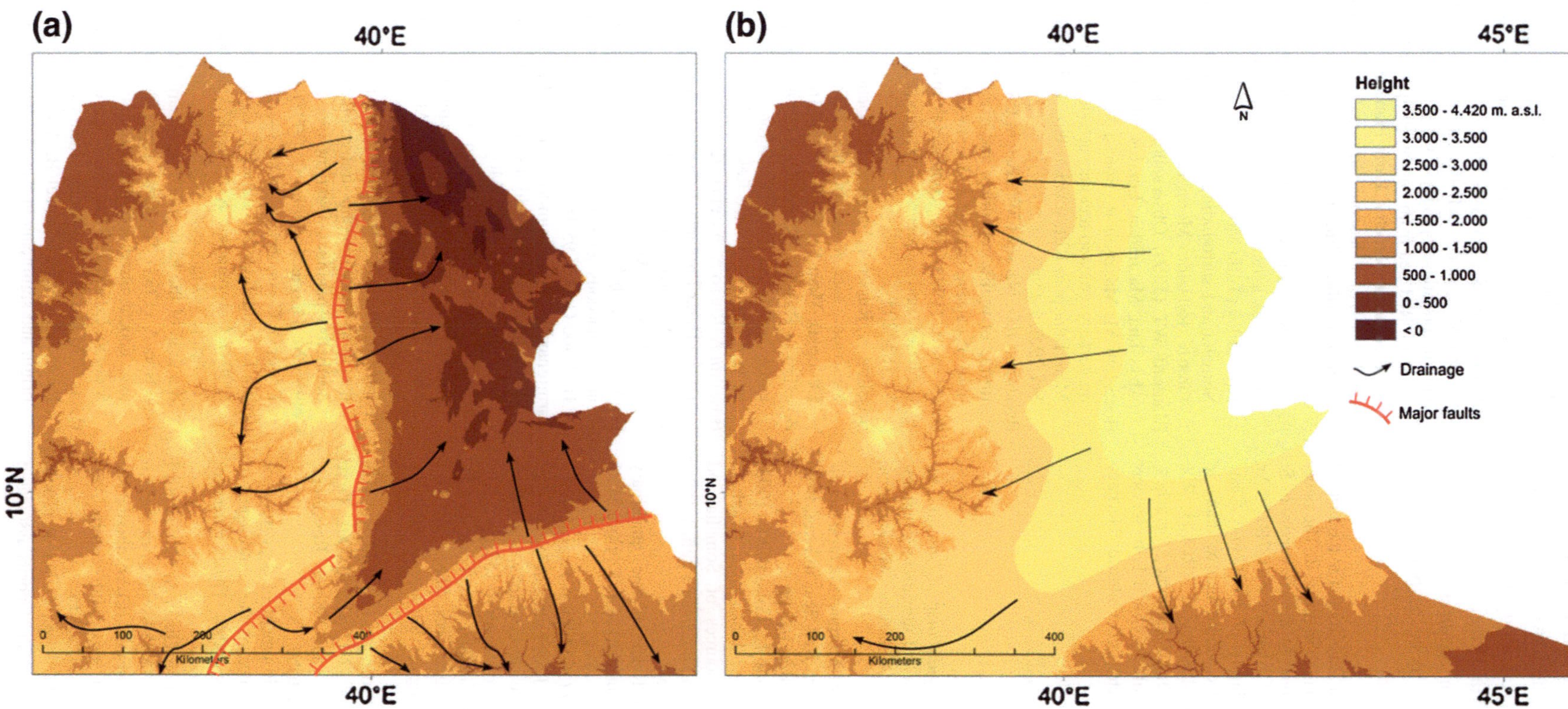

Fig. 5.19 **a** Present-day drainage network with location of major wind gaps corresponding to palaeovalleys hanging on the watershed of the Afar and northern MER; **b** Reconstruction of the original drainage resulting from the dissection of the dome generated by the CFB. The Addis Ababa Embayment (*AAE* in Fig. 5.1) corresponds to a palaeovalleys now diverted towards the Omo River Valley but was originally oriented to the Nile basin. Note that out of the Afar, we used the present-day topography that is the result of Miocene and later dissection. The reconstructed dome has a topography at the same level or slightly higher than the highest relief of the plateau

Pediplanation worked during the arid phases of the Quaternary at the feet of the Western and Eastern Escarpment (WEE and EEE), and a pediplain over 20 km long, nowadays terraced, has been observed along the southern margin of the Afar prior to 0.6 Ma (Pizzi et al. 2008). However, at a larger scale, the major features of the Afar margins, the WEE and the EEE, are to be associated with headward river erosion.

Acknowledgments We thank Maria Laura Balestrieri for the critical review of the paper and the useful suggestions to improve it. The paper was also realized thanks to the support received by the University of St Petersburg, FL, USA, NSF Project Award Abstract #1027607 "An Ethnoarchaeological and Archaeological Study of the Gamo Caste System in south-western Ethiopia". We also would like to thank in Addis Ababa Dr. Yonas Desta, General Manager of ARCCH; Wz. Mamito, Director at the National Museum of Ethiopia; Getachew Senishaw, Director of Cultural Heritage Collection and Lab Service; and Desalegn Abebaw Andualem, Director of Cultural Heritage Research.

References

Abbate E, Sagri M, Sassi FP (1994) Geological map of Somalia. In: Geology and mineral resources of Somalia and surrounding regions, Istituto Agronomico Oltremare Firenze, Relazione Monografica, Firenze

Abbate E, Balestrieri ML, Bigazzi G (2002) Morphostructural development of the Eritrean rift flank (southern Red Sea) inferred from apatite fission track analysis. J Geophys Res 107 pp

Abdirahim MM, Ali Kassim M, Carmignani L, Coltorti M (1993) Geomorphological evolution of the upper Jubba river (southern Somalia). In: Geology and mineral resources of Somalia and surrounding regions, Istituto Agronomico Oltremare Firenze, Relazione Monografica 113:241–250 (Firenze)

Abebe B, Coltorti M, Pizzi A (2005) Rates of late quaternary deformation along the Wonji fault belt in the Lakes Region, Main Ethiopian Rift. Rend Boll Soc Geol It 1:41–43

Abebe B, Acocella V, Korme T, Ayalew D (2007) Quaternary faulting and volcanism in the Main Ethiopian Rift. J Afr Earth Sci 48:115–124

Abul-Haggag Y (1961) A contribution to the physiography of northern Ethiopia. The Athlone Press, University of London, 153 pp

Aklilu A, Amenti A, Beyth M, Dow DB, Garland CR, Temesgen H, Hailu T (1978) Geological map of Adigrat area, Sheet ND 37-7 (1:250.000). Ministry of Mines, Geological Survey of Ethiopia, Addis Ababa

Arkin Y, Beyth M, Dow DB, Levitte M, Temesgen H, Hailu T (1971) Geological map of Mekele, Sheet ND 37-11 (1:250.000), Tigre Province. Ministry of Mines, Energy and Water Resources, Geological Survey of Ethiopia, Addis Ababa

Assefa G (1991) Lithostratigraphy and environment of deposition of the late Jurassic-early Cretaceous sequence of the central part of northwestern Plateau, Ethiopia. Neues Jahrb Geol Paleontol Abhandunglen 182(3):255–284

Audin L, Quidelleur X, Coulie E, Courtillot V, Gilder S, Manighetti I (2004) Paleomagnetism and K-Ar and 40Ar/39Ar ages in the Ali Sabieh area (Republic of Djibouti and Ethiopia). Constrains on the mechanism of Aden ridge propagation into south-eastern Afar during last 10 M year. Geophys J Int 158:327–345

Ayalew D, Barbey P, Marty B, Reisberg M, Yirgu G, Pik R (2002) Source, genesis, and timing of giant ignimbrite deposits associated with Ethiopian continental flood basalt. Geoch Cosmochim Acta 66 (8):1429–1448

Baker J, Snee L, Menzies M (1996) A brief Oligocene period of flood volcanism in Yemen: implications for the duration and rate of continental flood volcanism at the Afro-Arabian triple junction. Earth Planet Sci Lett 138:39–55

Balestrieri ML, Stuart FM, Persano C, Abbate E, Bigazzi G (2005) Geomorphic development of the escarpment of the Eritrean margin, southern Red Sea from combined apatite fission tracks and (U/Th)/He thermochronometry. Earth Planet Sci Lett 231:97–110

Beyth M (1972) Paleozoic-Mesozoic sedimentary basin of Mekele Outlier, northern Ethiopia. Am Ass Petrol Geol Bull 56(12):2426–2439

Bigazzi B, Bonadonna FP, Di Paola GM, Giuliani A (1993) K-Ar and fission track ages of the last volcano tectonic phase in the Ethiopian Rift Valley (Tullu Moye area). In: Geology and mineral resources of Somalia and surrounding regions. Istituto Agronomico Oltremare, Firenze, Relazioni Monografie 113:311–322

Blanford WT (1869) Observations on the geology and zoology of Abyssinia. Macmillan, London, 407 pp

Boccaletti M, Mazzuoli R, Bonini M, Trua T, Abebe B (1999) Plio-quaternary volcano-tectonic activity in the northern sector of the Main Ethiopian Rift: relationships with oblique rifting. J Afr Earth Sci 29:679–698

Bohannon RG, Naeser CW, Dwight LS, Zimmermann RA (1989) The timing of uplift, volcanism and rifting peripheral to the Red Sea: a case of passive rifting? J Geophys Res 94:1683–1939

Bosellini A (1992) The continental margin of Somalia. Structural evolution and sequence stratigraphy. Am Ass Petrol Geol Mem 53:185–205

Bosellini A, Russo A, Fantozzi PL, Getaneh A, Tadesse S (1997) The Mesozoic succession of the Mekele outlier (Tigre province, Ethiopia). Mem Sci Geol 49:95–116

Bosellini A, Russo A, Schroeder R (1999) Stratigraphic evidence for an early Aptian sea-level fluctuation: the Graua Limestone of south-eastern Ethiopia. Cretac Res 20:783–791

Bosellini A, Russo A, Assefa G (2001) The Mesozoic succession of Dire Dawa, Harar Province, Ethiopia. J Afr Earth Sci 32(3):403–417

Bosworth W, Huchon P, McClay K (2005) The Red Sea and Gulf of Aden Basin. J Afr Earth Sci 43:334–378

Bruni P, Fazzuoli M (1977) Sedimentological observations on Jurassic and Cretaceous sequences in northern Somalia. Preliminary Report. Boll Soc Geol It 95:1571–1588

Carmignani L, Ali Kassim M, Fantozzi PL (1983) Carta Fotogeologica della Regione di Gedo (Somali sud-occidentale). Quaderni di Geologia della Somalia, Mogadiscio

Chernet T, Hart WK, Aronson JL, Walter RC (1998) New age constraints on the timing of volcanism and tectonism in the northern Main Ethiopia Rift-southern Afar transition zone (Ethiopia). J Vulcanol Geoth Res 80:267–280

Coltorti M, Pieruccini P (2006) The last interglacial pedocomplexes in the litho- and morpho-stratigraphical framework of the central-northern Apennines. Quat Int 156–157:118–132

Coltorti M, Dramis F, Ollier C (2007) Planation surfaces in northern Ethiopia. Geomorphology 89(3–4):287–296

Coulié E, Quidelleur X, Gillot PY, Courtillot V, Lefèvre JC, Chiesa S (2003) Comparative K-Ar and Ar/Ar dating of Ethiopian and Yemenite Oligocene volcanism: implications for timing and duration of the Ethiopian Traps. Earth Planet Sci Lett 206:477–492

Davidson A (1983) The Omo River Project: reconnaissance geology and geochemistry of part of Ilubabor, Kefa, Gemu Gofa and Sidamo, Ethiop Eth Inst Geol Surveys Bull 2:89 pp

Davis WM (1899) The geographical cycle. Geogr J 14:481–504

Di Paola GM, Seife-Michael B, Arno V (1993) The Kella Horst: its origin and significant in crustal attenuation and magmatic processes

in the Ethiopian Rift Valley. In: Geology and Mineral resources of Somalia and surrounding regions, Istitituto Agronomico Oltremare Firenze, Relazione Monografica 113:311–323 (Firenze)
Dow DB, Beyth M, Hailu T (1971) Paleozoic rocks recently discovered in northern Ethiopia. Geol Mag 198(1):53–59
Drury SA, Kelley SP, Berhe SM, Collier REL, Abraha M (1994) Structures related to Red Sea evolution in northern Eritrea. Tectonics 13:1371–1380
Drury SA, Ghebreab W, Deller AME, Talbot CJ, Berhe SM (2006) A comment on "geomorphic development of the escarpment of the Eritrean margin, southern Red Sea from combined apatite fission-track and (U–Th)/He thermochronometry". Earth Planet Sci Lett 242:28–432
Ebinger CJ, Yemane T, Woldegabriel G, Aronson JL, Walter RC (1993) Late eocene-recent volcanism and faulting in the southern main Ethiopian rift. J Geol Soc London 150:99–108
Ebinger CJ, Yemane T, Harding DJ, Tesfaye S, Kelley S, Rex DC (2000) Rift deflection, migration, and propagation: linkage of the Ethiopian and eastern rifts, Africa. Geol Soc Am Bull 112:163–176
EIGS (Ethiopian Institute of Geological Surveys), Ministry of Mines and Energy (1978) Geological map of the Nazret Sheet _NC37-15., scale 1:250,000
EIGS (Ethiopian Institute of Geological Surveys), Ministry of Mines and Energy (1985) Geological Map of the DireDawa Sheet _NC37-12., scale 1:250,000
Fantozzi PL (1998) Transition from continental to oceanic rifting in the gulf of Aden: structural evidence from field mapping in Somalia and Yemen. Tectonophysics 259:285–311
Fantozzi PL, Ali Kassim M (2002) Geological mapping in northeastern Somalia (Midjiurtinia region): field evidence of the structural and paleogeographic evolution of the northern margin of the Somalia plateau. J Afr Earth Sci 34:25–53
Faure H (1975) Recent crustal movements along the Red Sea Gulf of Aden coasts in Afar (Ethiopia and TFAI). Tectonophysics 29:479–486
Gani NDS, Abdelsalam MG, Gera S, Gani R (2009) Stratigraphic and structural evolution of the Blue Nile Basin, northwestern Ethiopian plateau. Geol J 44:30–56
Garland CR (1980) Geology of the Adigrat Area. Ministry of Mines, Energy and Water Resources Geological Survey of Ethiopia, Memoir 1:51 pp
George R, Rogers N, Kelley S (1998) Earliest magmatism in Ethiopia: evidence for two mantle plumes in one food basalt province. Geology 26:923–926
Getaneh A (1991) Lithostratigraphy and environment of deposition of the Late Jurassic-Early Cretaceous sequence of the central part of northwestern Plateau, Ethiopia. Neues Jahrb Geol Paleontol Abhandunglen 182:255–284
Ghebreab W, Carter A, Hurford AJ, Talbot CJ (2002) Constraints for timing of extensional tectonics in the western margin of the Red Sea in Eritrea. Earth Planet Sci Lett 200:107–119
Gregnanin A, Peccirillo EM (1974) Considerazione sulle serie vulcaniche e sulle strutture dell'Altopiano Etiopico Centrale. Mem Mus Tridentino Sci Nat 20:79–100
Hart WK, Woldegabriel G, Walter RC, Mertzman SA (1989) Basaltic volcanism in Ethiopia: constraints on continental rifting and mantle interactions. J Geophys Res 94:7731–7748
Hofmann C, Courtillot V, Féraud G, Rochette P, Yirgu G, Ketefo E, Pik R (1997) Timing of the Ethiopian flood basalt event and implications for plume birth and global change. Nature 389:838–841
Huchon P, Jestin F, Cantagrel JM, Gaulier JM, Al Khirbash S, Gafaneh A (1991) Extensional deformations in Yemen since Oligocene and the Africa-Arabia-Somalia triple junction. Ann Tectonicae 2:141–162
Hutchinson RW, Engels GG (1970) Tectonic significance of regional geology and evaporate lithofacies in northeastern Ethiopia. Phil Trans Roy Soc London A267:313–329
Johnson DW (1930) Planes of lateral corrasion. Science 73:174–177
Juch D (1980) Tectonics of the southeastern escarpment of Afar, Ethiopia. Accad Naz Lincei 47:407–418 (Rome)
Kazmin V, Berhe SF, Nicoletti M, Petrucciani C (1980) Evolution of the northern part of the Ethiopian rift. Atti Convegni Lincei 47:275–292
King LC (1975) Planation surfaces upon highlands. Z Geomorph NF 20 (2):133–148
McDougall I, Morton WH, Williams MAJ (1975) Age and rates of denudation of trap series basalt at Blue Nile Gorge, Ethiopia. Nature 254:207–208
Merla G, Minucci E (1938) Missione geologica nel Tigrai, vol. 1 "La serie dei terreni". Rend Reale Acc Italia, Centro Studi per l'Africa Orientale Italiana 3:362 pp
Merla G, Abbate E, Azzaroli A, Bruni P, Canuti P, Fazzuoli M, Sagri M, Tacconi P (1979) A geological map of Ethiopia and Somalia (1:2.000.000 scale) and comments with map of major landforms. CNR, University of Florence, Centro Stampa Firenze, 95 pp
Mock C, Arnaud NO, Cantagrel J-M, Yirgu G (1999) 40Ar/39Ar thermochronology of the Ethiopian and Yemeni basements: reheating related to the Afar plume? Tectonophysics 314:351–372
Mohr PA (1962) The geology of Ethiopia. Addis Ababa University Press, Addis Ababa, 268 pp
Mohr PA (1967) The Ethiopian rift system. Bull Geophys Observatory 11:1–65 (Addis Ababa)
Mohr P (1986) Sequential aspects of tectonic evolution of Ethiopia. Mem Soc Geol Ital 447–461
Mohr P (1987) Pattern of faulting in the Ethiopian rift valley. Tectonophysics 143:169–179
Nyagah K (1997) Late Paleozoic through Mesozoic cyclostratigraphy, sedimentology and depositional history of the Mandera Basin, Northeast Kenya and implication for hydrocarbon exploration. National Oil Corporation, Kenya, 27 pp (unpublished manuscript)
Pan M, Sjoberg LE, Asenjo E, Alemu A, Asfaw LM (2002) An analysis of the Ethiopian Rift Valley GPS campaigns in 1994 and 1999. J Geodyn 33:333–343
Pik R, Marty B, Carignan J, Lavé J (2003) Stability of the Upper Nile drainage network (Ethiopia) deduced from (U-Th)/He thermochronometry: implications for uplift and erosion of the Afar plume dome. Earth Planet Sci Lett 215:73–88
Pik R, Marty B, Carignan J, Yirgu G, Ayalew T (2008) Timing of east African Rift development in southern Ethiopia: implication for mantle plume activity and evolution of topography. Geology 36:167–170
Pizzi A, Coltorti M, Bekele A, Disperati L, Sacchi G, Salvini R (2006) The Wonji fault belt (Main Ethiopian Rift, Ethiopia): structural and geomorphological constraints and GPS monitoring. In: Yurgu G, Ebinger CJ, Macguire PKH (eds) The Afar Volcanic Province within the east African Rift System. Geol Soc, 259:191–207 (special publications)
Pizzi A, Pomposo G, Abebe B, Coltorti M (2008) Stile deformativo e reticolo idrografico lungo la scarpata del Plateau Somalo al margine meridionale dell'Afar (Etiopia). Rend Online Soc Geol It 1:137–139
Rochette P, Tamrat E, Féraud G, Pik R, Courtillot V, Ketefo E, Coulon C, Hoffmann C, Vandamme D, Yirgu G (1998) Magnetostratigraphy and timing of the Oligocene Ethiopian traps. Earth Planet Sci Lett 164:497–510
Russo A, Assefa G, Atnafu B (1994) Sedimentary evolution of the Abay River (Blue Nile) Basin, Ethiopia. Neues Jb Geol Palaontol Monatsh 5:291–308
Russo A, Fantozzi PL, Solomon T, Getaneh A, Neri C, Russo F, Asfwossen A, Peccerillo A, Valera P (1996) Geological map of the Mekele outlier (western sheet). Italian Cooperation—Addis Ababa University, Addis Ababa
Sagri M, Abbate E, Azzaroli A, Balestrieri ML, Benvenuti M, Bruni P, Fazzuoli M, Ficcarelli G, Marcucci M, Papini M, Pavia G, Reale V,

Rook L, Tecle TM (1998) New data on the Jurassic and Neogene sedimentation in the Danakil Horst and northern Afar Depression, Eritrea. In: Crasquin-Soleau S, Barrier E (eds) Perythetis Memoir 3: Stratigraphy and evolution of peri-Tethian platforms. Mem Mus Nat Hist Nat Paris, 177:193–214

Saxena GN, Assefa G (1983) New evidence on the age of glacial rocks of northern Ethiopia. Geol Mag 120(6):549–554

Schilling JG, Kingsley RH, Hannan BB, McCully BL (1992) Nd–Sr–Pb isotopic variations along the Gulf of Aden: evidence for Afar mantle plume-continental lithosphere interaction. J Geophys Res 97:10927–10966

Schultz RA, Mège D, Diot H (2008) Emplacement conditions of igneous dikes in Ethiopian traps. J Volcanol Geoth Res 178:683–692

Shumburu MM (1968) The Amba Aradom formation (formerly the upper sandstone). Mobil Petroleum Ethiopia Inc., Asmara, 15 pp (unpublished report)

Simmons MD, Al-Thour K (1994) Micropaleontological biozonation of the Amran Series (Jurassic) in the Sana'a Region, Yemen Republic. In: Simmons MD (ed) Micropaleontology and hydrocarbon exploration in the middle east. Chapman and Hall, London, pp 43–79

Tefera M, Chernet T, Haro W (1990) Geological Map of Ethiopia, scale 1:2.000.000. Geol Surv Ethiopia, Addis Ababa, Ethiopia

Tiercelin JJ, Taieb M, Faure H (1980) Continental sedimentary basins and volcano-tectonic evolution of the Afar Rift. Accad Naz Lincei 47:491–504 (Rome)

Toland C, Simmons MD, Walkden GM (1994) A new sequence stratigraphic reference section for the Upper Jurassic of Yemen. In: Al-Husseini MI (ed) Geo 94 The middle east Petroleum Geosciences, 2, Gulf Petrol Ink Bahrain, pp 891–899

Tsegay AT, Mazzarini F, Innocenti F, Manetti P (1998) The Yerer-Tullu Wellelvolcanotectonic lineament: a transtensional structure in central Ethiopia and the associated magmatic activity. J Afr Earth Sci 26:135–150

Tsegay A, Balestrieri ML, Bigazzi G (2010) The central main Ethiopian rift is younger than 8 Ma: confirmation through apatite fission-track thermochronology. Terranova 22(6):470–476

Weissel JK, Malinverno A, Harding DJ (1995) Erosional development of the Ethiopian plateau of Northeast Africa from fractal analysis of topography. In: Barton CC, Pointe PRL (eds) Fractals in petroleum geology and Earth processes. Plenum Press, New York, pp 127–142

Whiteman AJ (1971) The geology of the Sudan Republic. Clarendon Press, Oxford, 290 pp

Woldegabriel G, Aronson JL, Walter RC (1990) Geology, geochronology, and rift basin development in the central sector of the main Ethiopian rift. Geol Soc Am Bull 102:439–458

Wolela A (2008) Sedimentation of the Triassic-Jurassic Adigrat Sandstone Formation, Blue Nile (Abay) Basin, Ethiopia. J Afr Earth Sci 52:30–42

Wolfenden E, Ebinger C, Yirgu G, Deino A, Ayalew D (2004) Evolution of the northern Main Ethiopian rift: birth of a triple junction. Earth Plan Sci Lett 224:213–228

Yassin AA, Khalil FA, El Shafie AG (1984) Explanatory note of the geological map at the scale of 1:2,000,000 of the Democratic Republic of the Sudan, 63 pp

Part II

Local Studies

Paleoglaciated Landscapes in Simen and Other High-Mountain Areas of Ethiopia

6

Hans Hurni

Abstract

In the present-day Ethiopia, glaciated landscapes do not exist, but paleoglaciated landscapes have been documented on a few mountain tops, which have altitudes higher than about 4,350 m asl in northern Ethiopia (Simen Mountains) and about 4,100 m asl in southern Ethiopia (Arsi and Bale Mountains). Glaciers were associated with the Late Pleistocene cold stages and reached as far down as 3,760 m asl in northern and 3,200 m asl in southern Ethiopia. Bale Mountains had the most extensive Late Pleistocene glaciation, covering over 190 km^2, followed by Arsi Mountains (about 85 km^2). In Simen, the Late Pleistocene glaciers covered merely 13 km^2. In addition, paleo-periglacial slope deposits are found on all above-mentioned paleoglaciated mountains and in further mountain systems which did not host glaciers. This allows the reconstruction of the Late Pleistocene paleoclimate as being about 8 °C colder than at present (2014), much more dry, and probably without monsoon, at least in northern Ethiopia. Most probably in the Early Holocene, the re-emergence of monsoonal rains led to a strong erosion phase, which was followed by an extended stable phase with soil formation, building up about 70-cm-deep A-horizons (Andosol) on the paleo-periglacial slope deposits. These soils have been heavily degraded due to human-induced soil erosion up to about 3800 m asl since agriculture started several decades to millennia ago.

Keywords

Late Pleistocene glaciation • Paleo-periglacial slope deposits • Paleoclimate • Early Holocene erosion • Holocene soil formation • Human-induced soil degradation

6.1 Introduction

Glaciated landscapes are extremely rare in the contemporary Africa and non-existent in Ethiopia, as her mountain tops are too low to exceed a permanent present-day snowline. The highest mountain, Ras Dejen, reaches 4,540 m asl, while the current snowline is estimated to be at about 4,900 m asl (2014) and is expected to rise even higher in future decades due to global warming. Although not all authors agree that there was no recent glaciation in Ethiopia, this would have implied colder temperatures by at least 8 °C on average. There are, however, signs of paleoglaciated landscapes on some of Ethiopia's highest mountains, featuring moraines and periglacial slope deposits. The most prominent of such landscapes are found in the Bale Mountains (Messerli et al. 1977; Osmaston et al. 2005), Mount Badda in Arsi (Potter 1976), and in the Simen Mountains (Hurni 1981). Detecting former moraines is quite demanding, while the paleo-periglacial slope deposits can be seen on virtually all mountain slopes above about 3,500 m asl in northern Ethiopia, and above about 3,200 m asl in southern Ethiopia. There are differences in altitude of these lower limits around each mountain system, as northerly-facing slopes show the cold-climate features more prominently than the southerly-facing ones.

H. Hurni (✉)
University of Bern, Bern, Switzerland
e-mail: hans.hurni@cde.unibe.ch

P. Billi (ed.), *Landscapes and Landforms of Ethiopia*, World Geomorphological Landscapes,
DOI 10.1007/978-94-017-8026-1_6, © Springer Science+Business Media Dordrecht 2015

6.2 Geographical Setting and Literature

Figure 6.1 presents an overview of Ethiopian mountain tops above 4,000 m asl. The minimum height of 4,000 m asl is taken because mountain tops with lower altitudes do generally not show paleoglaciation features. In addition, the extent of the Ethiopian Highlands is also given, according to its agroecological belts (from *Kolla, Weyna Dega* and *Dega* to *Wurch*, cf. Hurni 1998).

In the second half of the last century, numerous investigations and observations were made on paleofeatures in these mountain systems. For Bale, Messerli et al. (1977) presented an abstract of their observations, while for Arsi, Potter (1976) described paleoglaciation on Mount Badda. For the Ethiopian lowlands, Gasse et al. (1980) studied Quaternary landforms and paleoclimates, parallel to many other authors. More recent studies on paleoglaciation features in high-mountain environments in Ethiopia include Osmaston et al. (2005), Zech et al. (2005), and Umer et al. (2004), including an excellent overview of Quarternay glacation in Africa by Mark and Osmaston (2008).

In Simen, features from a paleoglacial period were observed by Minucci (1938), Nilsson (1949), Hoevermann (1954), Buedel (1954), Werdecker (1955), Mohr (1962), Hastenrath (1974, 1977), Messerli (1975), Williams et al. (1978), and the present author (Hurni 1981, 1982, 1986). Geologically, the Simen Mountains are composed of Trap basalts that overlie Mesozoic sandstones, which in turn rest on the Precambrian basement (Kazmin 1973). The mountain tops above 3,000 m

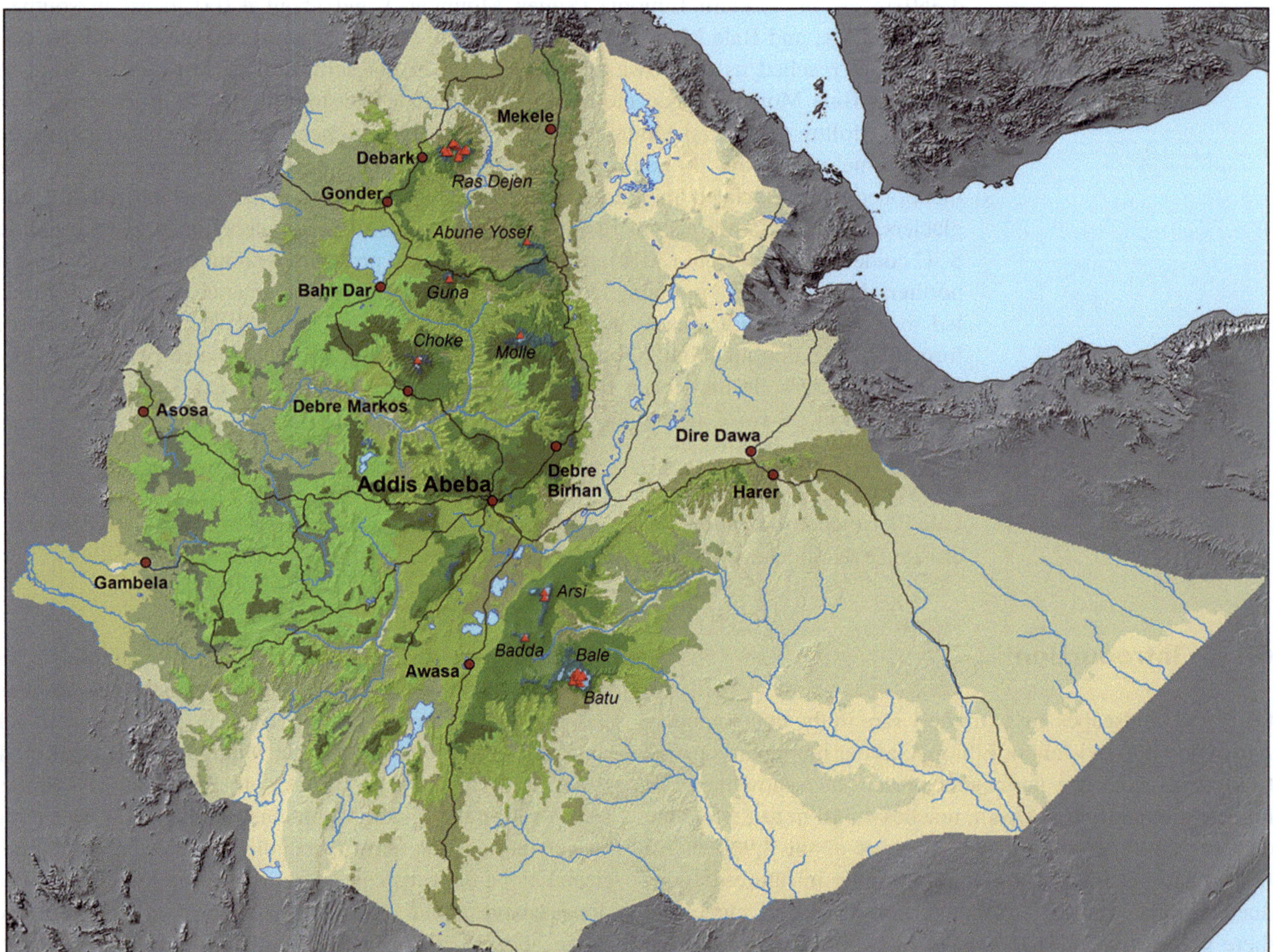

Fig. 6.1 Overview of mountain tops higher than 4,000 m asl in the Ethiopian Highlands, all with paleo-periglaciation and some with paleoglatiation features. The country is coloured according to agroecological belts: semi-arid lowlands are *yellow* to *brownish* (*Berha*, i.e. below about 1,000 m asl); sub-humid to humid highlands are olive to light green (*Kolla* and *Weyna Dega*, from about 1,000 m to about 2,500 m asl), and humid highlands are *darker green* to *bluish* (*Dega* to *Wurch*, from about 2,500 m up to 4,540 m asl). Size of Ethiopia: 1,106,000 km^2; highlands: 575,000 km^2. *Source* from Hurni (1998)

asl in Ethiopia are all remnants of ancient volcanoes, consisting of Termaber basalts and belonging to the Shield group, being porphyritic, with tuffs and paleosols in between, connected to volcanic centres and dating from the Miocene to the Pliocene (Merla et al. 1979). The present-day climate of Simen is characterised by a unimodal rainy season from May to October, about 1,500 mm of annual rainfall, and about 8.5 °C average annual temperature at 3,600 m asl (adapted from Hurni 1982). Other mountain tops show similar characteristics, although with some deviation in annual rainfall.

6.3 Paleoglacial Landforms in Simen

6.3.1 Moraines

The experienced eye will fairly easily detect former moraines in Simen, the more since the newly built roads to the towns of Mekane Birhan in Janamora and to Dilyibza in Beyeda are passing quite near to some of them at an altitude of about 4,200 m asl, namely on Mount Bwahit (4,437 m asl), Mount Mesarerya (4,360 m asl), and Mount Ras Dejen (4,540 m asl). Moraines, according to Hurni (1986), are rubble ridges that "consist of firmly compacted angular blocks, which are not sorted, not regulated in certain directions, and embedded in a yellowish-brown matrix of clayey silt." Such ridges are found above 3,780 m to above 4,320 m asl, mainly on north-facing slopes (from west to north-east), while only one is found on a south-facing slope, which is next to the top of Mount Kidis Yared (4,460 m asl), at an elevation of 4,400 m asl (Hurni 1981). Above these ridges, one can detect the catchment areas for accumulation of ice, which usually have a concave form (cirque) excavated in basalt rock.

Figure 6.2 gives a section from the map 1:100,000 by Hurni (1981) with all moraines he detected in Simen during fieldwork in 1974–1977, and Fig. 6.3 is a photograph of the largest moraine found in the westerly catchment of Ras Dejen, below a cirque that steeply descends into the Mesheha Valley.

Figure 6.4 presents an analysis of all paleoglacial moraines observed in Simen by Hurni (1982) and includes the most important characteristics and parameters, such as the paleosnowline and the lower limit of the paleo-periglacial slope deposits.

The moraines in the Ras Dejen area are the longest, with ridges between 70 and 550 m, and their depth varies between 5 and 150 m. Normally, they are singular, but in a few locations, a second lateral moraine can be found behind the

Fig. 6.2 Mountain tops in Simen with paleoglaciation features of the last cold period. Features presented are moraines (*full red*) with ice flow direction (*black arrows*); periglacial slope deposits (*pale red*), and snowline (*red dotted*). East–West distance is about 36.5 km. *Source* section of map by Hurni (1981)

Fig. 6.3 Panorama of the largest (*middle*) moraine of the last cold period in the Simen Mountains, western escarpment below Ras Dejen (4,540 m asl), seen from an altitude of about 4,200 m asl. The main glacier was found towards the photographer, while a smaller glacier was situated behind the moraine, having itself a small moraine on its right side (H. Hurni, 13 April 2013)

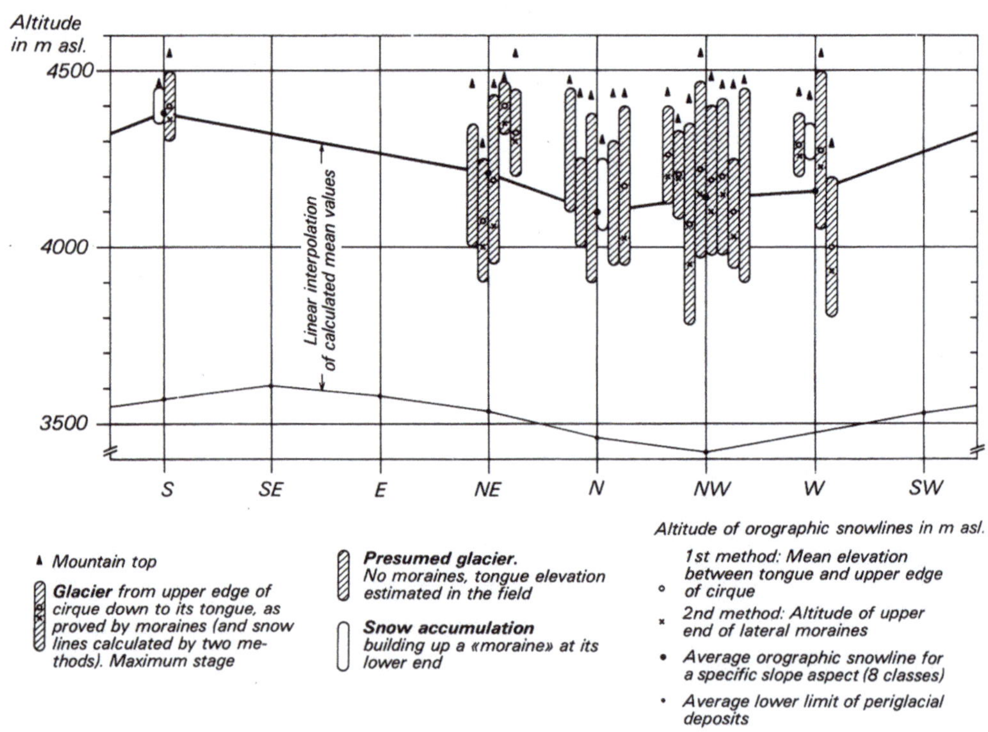

Fig. 6.4 Topographic analysis of all mapped paleoglaciers and Late Pleistocene altitudinal belts in Simen, including the last cold period snowline as reconstructed by moraines and cirques using two methods. The correlation of snowline with slope aspect is evident, as is the lower limit of the periglacial belt being parallel to the snowline. *Source* Hurni (1986)

first one. In the Bwahit/Mesarerya area, moraine ridges are between 70 and 250 m long and up to 15 m thick, hence considerably smaller. Most moraines represent lateral moraines, often with an indication of a terminal stage. The respective cirques of the paleoglaciers are quite evident, particularly within Ras Dejen north-west-facing slopes, where rock faces and signs of overdeepening are obvious.

6.3.2 Slope Deposits

Paleo-periglacial deposits are widespread throughout the Simen Mountains. These features can be best observed in the present-day Afro-alpine grasslands between 3,600 m and 4,000 m asl, such as in the upper Jinbar Valley in the centre of the park, on the western slopes of Bwahit and Mesarerya

Fig. 6.5 Paleo-periglacial deposits on a northerly-facing slope in the Upper Jinbar Valley, 3,500–4,000 m asl. The deposits are up to 15 m thick and were eroded in the Early Holocene, before soil formation built up about 70 cm of Andosol A-horizon over the entire landscape (H. Hurni, September 1974)

Fig. 6.6 Andosol soil profile as prevalent on all mountains of the Ethiopian Highlands between about 3,200 m asl and about 4,000 m asl, here near Gich Camp at 3,600 m asl in Simen. Note the *black* A-horizon overlaying a compacted B-horizon, which represents paleo-periglacial slope deposits, with the Early Holocene erosion surface in between. (H. Hurni, August 1974)

mountains, and on the south-eastern slopes of Ras Dejen (Fig. 6.2). Their altitudinal occurrence starts below the paleo-snowline, which according to the occurrence of the moraines was determined by Hurni (1981) at about 4,250 m asl, and extends as far down as 3,600–3,400 m asl according to slope aspect, similar to the moraines. Slope deposits, according to Hurni (1986), are "rubble on trough-shape slopes (which) consists of firmly compacted angular blocks, which are not sorted, but regulated along or across the fall-line, and embedded in a yellowish-brown matrix of clayey silt" (Fig. 6.5). This regulated formation of stones along the fall-line was interpreted as cold-age periglacial solifluction processes.

The thickness of the slope deposits increases downslope and can be up to 15 m thick. In some of the steep tributary

Fig. 6.7 Heavily degraded landscape in the Simen Mountains due to century-old agricultural land use without soil conservation. Looking from 3,300 m asl in a northerly direction towards the Upper Jinbar Valley (H. Hurni, 7 February 2013)

valleys to Mesheha Valley, rubble deposits are also found on the valley floors and as far down as 3,000 m, and they can be up to 40 m thick (Hurni 1981). Their association with the same formation period, however, could not be clarified during the fieldwork, and it may be that they belong to an earlier Pleistocene glaciation period.

6.3.3 Erosion of Slope Deposits

The paleo-periglacial slope deposits have been incised by gullies, which can be up to 15 m deep (Fig. 6.5). This erosion took place before a 70-cm-thick Andosol A-horizon was formed on top of them. According to Hurni (1986), "the erosion phase must have taken place immediately after the last cold period, before the upward movement of vegetation regrowth took place in those belts. The climate must have been much wetter, and possibly warmer. In the climatic history of Ethiopia, the erosion phase can be inserted into the phase of northward movement of the monsoons around 12,000–10,000 BP."

6.3.4 Soils

In all of Simen between about 3,000–4,200 m asl, the Andosol soil type is prevalent. These soils have a dark brown to black A-horizon, which on average is about 70 cm thick (Fig. 6.6). According to Frei (1978), this horizon is rich in humus (up to 30 % organic matter) and contains ashes from volcanic eruptions and Saharan dust, and it is of more or less similar depth throughout the area. The formation of such deep soils indicates a long-lasting soil formation period without much disturbance; hence, slopes must have been densely covered by vegetation with grasses and trees during their formation period.

6.3.5 Land and Soil Degradation

As in all of Ethiopia, the highland parts of Simen have been settled by farmers already about 2–3 thousand years ago. The "people of Samen" were mentioned by the Greeks about 2,000 years ago (Kirwan 1972). Such agricultural activity required deforestation, which is a form of land degradation, because no grain will grow above the tree line, which in Simen is currently at about 3,800–4,000 m asl. With the subsequent ploughing of steep slopes, soil erosion started to remove the top parts of the Andosols, at a rate of about 100 tonnes per ha per year or more during cultivation. Within a relatively short period of time, the topsoils were thus washed downstream. This process may have been slowed during regular fallowing periods, which were applied traditionally. Nevertheless, over the centuries, soil degradation became devastating, as the present-day landscape of Simen shows. Soil and water conservation measures were only introduced in Ethiopia as of the mid-1970s onwards (cf. Hurni et al. 2008) (Fig. 6.7).

6.4 Evolution of the Paleoglaciated Landscape in the Simen Mountains

The features of paleoglaciated landscape as described in Sect. 6.3 imply a sequence of processes that logically condition each other and help interpreting them in a chronological order and building a chronology. During the glaciation period, the existence of glaciers can be inferred from the presence of moraines, and their association with the paleo-periglacial belt appears logic, as no slope deposits were found inside the paleoglaciated areas. The paleoglaciers are situated about 800 m lower than present-day glaciers would be situated, although the latter are not existent due to insufficient altitudes of the mountain tops. The climate at that time, according to Hurni (1986), appears to have been about 7 °C cooler, depleted in precipitation and run-off, with a tendency towards winter (November–March) precipitation, and only occasional Summer (June–August) clouding with reduced or missing monsoons. Today, in 2014, this would be about 8 °C due to global warming since 1975. The glaciers were thus not historical, as 14 °C dating inside moor deposits on the ground moraines confirmed, showing an oldest date of 4,120 ± 90 BP (sample B-3043 in Hurni 1982). The paleoclimate during the glaciation period is most likely associated with the Late Pleistocene, between may be 20,000 and 12,000 BP (cf. Gasse et al. 1980).

The post-glaciation period, as inferred from the eroded slope deposits, started with intensive natural erosion, when the monsoons reached far into the Sahara. This phase ended when the temperature increased and vegetation cover moved upslope to the present levels (or above). With vegetation, soil formation started to build the observed deep A-horizons covering the slope deposits, the gullies eroded into them, and all other landscape elements between 3,000 and 4,000 m asl. This long and stable period must be associated with the Holocene period, lasting about 8,000 years, and soil formation has continued up to the present day, at least in places where no cultivation and associated land and soil degradation has taken place.

Processes leading to the present-day land degradation started about 2,000–3,000 years ago with widespread deforestation, cultivation of the Andosol soils, and human-induced soil erosion on slopes without soil protection. Soil degradation is widespread and still ongoing, to the extent that human livelihoods are threatened and subsistence agricultural practices are no longer possible for many farming families. Insufficient crop yields and the need to supply food aid for a considerable portion of the year have become a common practice, while the population in Simen increased by a factor of 8 since the 1950s (cf. Hurni and Ludi 2000).

6.5 Other Paleoglaciated Mountain Tops in Ethiopia

Bale Mountains A short study by Messerli et al. (1977) in April 1976 showed that the Bale Mountains, which are characterised by an extended high plateau above 4,000 m asl and peaks up to 4,377 m asl, were glaciated similar to the Simen Mountains, but at a much larger scale, with an ice shield on the Sanetti Plateau at around 4,100 m asl and individual glaciers extending into the valleys on all sides. In the Tegona Valley (near the town of Goba), the valley glacier was found to have had a thickness of 250 m and reaching down to an altitude of 3,200 m asl. The Bale Mountains thus probably had the largest glaciers in Ethiopia in the Late Pleistocene, with a glaciated area of around 190 km^2 according to Osmaston et al. (2005). Zech et al. (2005) were able to date lacustrine sediments formed after deglaciation on the Sanetti Plateau at 15,500–14,000 years BP. The other two features observed in Simen, the paleo-periglacial slope deposits and the post-glacial erosion, were also observed in Bale, but not studied in detail. Land and soil degradation, finally, was much less advanced than in the Simen Mountains.

Arsi Mountains Mount Badda (4,170 m asl) in the Arsi Region was studied by Potter (1976), who found evidence of Late Pleistocene glaciation down to an elevation of 3,650 m asl in the west-facing valleys, with an extent of 140 km^2, which is probably an overestimation. Periglacial features were not reported. According to Umer et al. (2004), "there are large, clear terminal moraines on Mount Bada at 3,200–3,700 m with the glaciated area estimated to be 85 km^2 with equilibrium line altitudes (ELAs) of 3,700 m as1 on the east and 3,900 m as1 on the west. This appears to represent the LGM," the Last Glacial Maximum.

Paleo-periglacial features on other mountains in Ethiopia Surveys on different mountain tops in Ethiopia were carried out between 1983 and 1986 and reported by Hurni (1986). None of the mountains showed paleoglaciation; however, paleo-periglacial slope deposits, post-glacial erosion, and subsequent Holocene soil formation with Andosols were observed on all mountains, followed by human-induced land and soil degradation. Mountains visited included Mount Choke in Gojam (4,052 m asl), Mount Guna in South Gonder (4,135 m asl), and Mount Molle (or Amba Farit) in Western Wello (4,247 m asl). Mount Abune Yosef (4,284 m asl) in

Northern Wello near the town of Lalibela has not been visited so far, but since the mountain top does not extend much into the paleosnowline of the Late Pleistocene (at 4,250 m asl in Simen), paleoglaciation there is not probable either.

References

Mark BG, Osmaston HA (2008) Quaternary glaciation in Africa: key chronologies and climatic implications. J Quat Sci 23:589–608. ISSN 0267-8179

Buedel J (1954) Klima-morphologische Arbeiten in Aethiopien im Fruehjahr 1953. Erdkunde VIII 2:139–156

Frei E (1978) Andepts in some high mountains of East Africa. Geoderma 21:119–131

Gasse F (1975) L'evolution des lacs de l'Afar central (Ethiopie et T.F. A.I.) du Plio-Pleistocene à l'Actuel, vol 3. These Université Paris VI, 568 pp

Gasse F, Rognon P, Street FA (1980) Quaternary history of the Afar and Ethiopian rift lakes. The Sahara and the Nile. Balkema, Rotterdam, pp 361–499

Hastenrath S (1974) Glaziale und Periglaziale Formbildungen in Hoch-Semyen, Nord-Aethiopien. Erdkunde 28:176–186

Hastenrath S (1977) Pleistocene mountain glaciation in Ethiopia. J Glaciol 18(79):309–313

Hoevermann J (1954) Ueber glaziale und "periglaziale" Erscheinungen in Erithrea und Nordabessinien. Veroeffentlichungen der Akademie fuer Raumforschung und Landesplanung. Abhandlung Band 28:87–111

Hurni H (1981) Simen mountains—Ethiopia: palaeoclimate of the last cold period (late wurm). Paleoecology Afr 13:127–137

Hurni H (1982) Simen Mountains—Ethiopia, vol II (in German): climate and the dynamics of altitudinal belts from the last cold period to the present day (Part II in co-authorship with Peter Staehli). Geogr Bernensia G 13, 196 pp

Hurni H (1986) Late quaternary in Simen and other mountains in Ethiopia. Mahaney WC, Balkema AA (eds) Quaternary and environmental research on East African mountains, Rotterdam, pp 105–120

Hurni H (1998) Agroecological belts of Ethiopia. Explanatory notes on three maps at a scale of 1:1,000,000. Soil conservation research programme, research report 43. Addis Abeba and Bern: Centre for Development and Environment (CDE). 30 pp. with map, scale 1:5 million

Hurni H, Ludi E (2000) Reconciling conservation with sustainable development. A participatory study inside and around the Simen Mountains National Park, Ethiopia. Produced with the assistance of an interdisciplinary group of contributors. Bern: Centre for Development and Environment. ISBN 3-906151-44-1, 476 pp

Hurni H, Debele B, Zeleke G, Ludi E, Abate S, Bantider A (2008) Land degradation and sustainable land management in the highlands of Ethiopia. In: Research for development. A synthesis of NCCR North–South research, 2001–2008. Perspectives of the Swiss National Centre of Competence in Research (NCCR) North–South, University of Bern, vol 4. Geographica Bernensia, Bern

Kazmin V (1973) Geological map of Ethiopia, 1:2 million. Geological survey of Ethiopia, Ministry of Mines and Energy, Addis Abeba

Kirwan LP (1972) The Christian topography and the kingdom of Axum. Geogr J 138:166–177

Merla G, Abbate E, Azzaroli A, Bruni P, Canuti P, Fazzuoli M, Sagri M, Tacconi P (1979) A geological map of Ethiopia and Somalia (1973), scale 1:2 million. Department of Geology, University of Florence, 95 pp

Messerli B, Hurni H, Kienholz H, Winiger M (1977) Bale mountains: Largest Pleistocene mountain glacier system of Ethiopia. INQUA Abstracts, Birmingham

Minucci E (1938) Ricerche geologiche nella regione del Semien. In: Ace R (ed) vol. I, Roma: 37–46

Mohr PA (1962) General report on an expedition to the Semien mountains. Bull Geo Obs Addis Abeba 6:155–167

Umer M, Kebede S, Osmaston H (2004) Quaternary glacial activity on the Ethiopian mountains. Dev Quat Sci Part C 2:171–174

Nilsson E (1949) Ancient changes of climate in British East Africa and Abyssinia. Geogr. Annaler 22:1–79

Osmaston HA, Mitchell WA, Osmaston JAN (2005) Quaternary glaciation of the Bale Mountains, Ethiopia. J Quat Sci 20:593–606

Potter EC (1976) Pleistocene glaciation in Ethiopia: New evidence. reprint: series 299, Center for Astrophysics, Cambridge, MA, 7 pp

Werdecker J (1955) Beobachtungen in den Hochlaendern Aethiopiens auf einer Forschungsreise 1953/54. Erdkunde IX: 395–317

Williams MAJ, Street FA, Dakin FM (1978) Fossil periglacial deposits in the Semien highlands, Ethiopia. Erdkunde 32:46–49

Zech R, Manhart A, Glaser B, Solomon, D, Zech W (2005) A high-resolution lateglacial climate record from Lake Sediments in the Ethiopian Bale Mountains. American Geophysical Union, Fall Meeting 2005, abstract #PP21A-1554

Geomorphology of the Archaeological Area of Aksum

7

Giovanni Ferrari, Rossano Ciampalini, Paolo Billi, and Piotr Migon

Abstract

Aksum is one of the most important archaeological and historical towns in Ethiopia. The archaeological area stretches on a plateau ranging from 2,250 to 2,460 m, and the general landscape consists of a few dome-shaped hills standing on a plateau crossed by small ephemeral streams. The main geomorphic features are the result of the emplacement of basalt flows; the intrusion of syenite sub-volcanic plugs; and domes, tectonics, weathering and its interaction with various surface erosion processes. In such a context, between 700 BC and 800 AD, the rise and the decay of the Aksumite Kingdom, one of the most known civilisations of East Africa, took place. The core of the archaeological area of Aksum, with tall carved obelisks, is a UNESCO World Heritage listed property. The integration of archaeological data with the analysis of soil conservation measures and agricultural practices, adopted till present, allowed to develop a plough marks-based methodology to assess the historical soil erosion rate and, hence, to understand historical landscape evolution within a man-controlled environment. The occurrence of archaeological evidence in such a characteristic geomorphological context provides an excellent opportunity to study and understand interaction between geomorphic and anthropic processes.

Keywords

Structural landforms • Geoarchaeology • Soil erosion • Aksum • Ethiopia

7.1 Introduction

Northern Ethiopia is home to one of the most fascinating ancient civilisations in the sub-Saharan Africa—the Aksumite Kingdom. The now rather inconspicuous town of Aksum (or Axum) in the north of Tigray Province used to be the administrative centre of a flourishing state for more than a millennium, since c. 400 BC to around 800 AD when it began to decline and eventually collapsed. However, a number of monuments and archaeological sites survived until today, including the famous stelae carved out of local rocks and erected to commemorate Aksumite rulers. A measure of the worldwide cultural significance of the area is its inscription on the World Heritage List in 1980. This diverse material legacy of the Aksumite civilisation, along with an opportunity to assess long-term interactions between soils, land use and soil erosion, has made Aksum a place where geoarchaeological approach to the environment can be developed with a particular success. However, the broader geomorphological context of the former Aksumite capital is no less interesting. From a relatively flat surface, a number of flat-top domes rise (Natali et al. 2013), whose associations provide insights into structural geomorphology

G. Ferrari
Di.P.S.A, Università di Firenze,
Piazzale Cascine 18, 50144 Florence, Italy

R. Ciampalini (✉)
INRA-US InfoSol, Centre de recherche Val de Loire, 2163
Avenue de la Pomme de Pin, 45075 Orléans, France
e-mail: rossano.ciampalini@gmail.com

P. Billi
Physics and Earth Sciences, University of Ferrara,
Via Saragat 1, 44122 Ferrara, Italy

P. Migon
Department of Geography and Regional,
Development University of Wroclaw, Wroclaw, Poland

P. Billi (ed.), *Landscapes and Landforms of Ethiopia*, World Geomorphological Landscapes,
DOI 10.1007/978-94-017-8026-1_7,

of syenite intrusive complexes and weathering and denudation patterns in basement rocks, including characteristic residual landforms. Both these topics and their mutual interactions will be presented in this chapter.

7.2 Geographical Setting of Aksum

The town of Aksum is located in the north of Ethiopia, at roughly 14°N and 39°E, and is one of the most important archaeological and historical cities of Ethiopia. The surrounding area stretches on a plateau ranging from 2,250 to 2,460 m a.s.l., and the general physiography consists of a few scattered hills with steep to almost vertical slopes standing on a plateau crossed by small ephemeral streams (Figs. 7.1 and 7.2).

Annual rainfall ranges from 600 to 800 mm and is characterised by high interannual variability. Precipitation is concentrated in the main, monsoon-type rainy season, from mid-June to mid-September (about 60–70 % of the total) and, at a much lesser extent, in the unpredictable "small" rainy season from mid-March to mid-May, originated by the development of cyclonic cells. Mean annual temperature is around 18 °C, with low seasonal variability, but large diurnal range. The vegetation is typically a highland sparse woodland, dominated by *Eucalyptus brevifolia* and *Euphorbia*

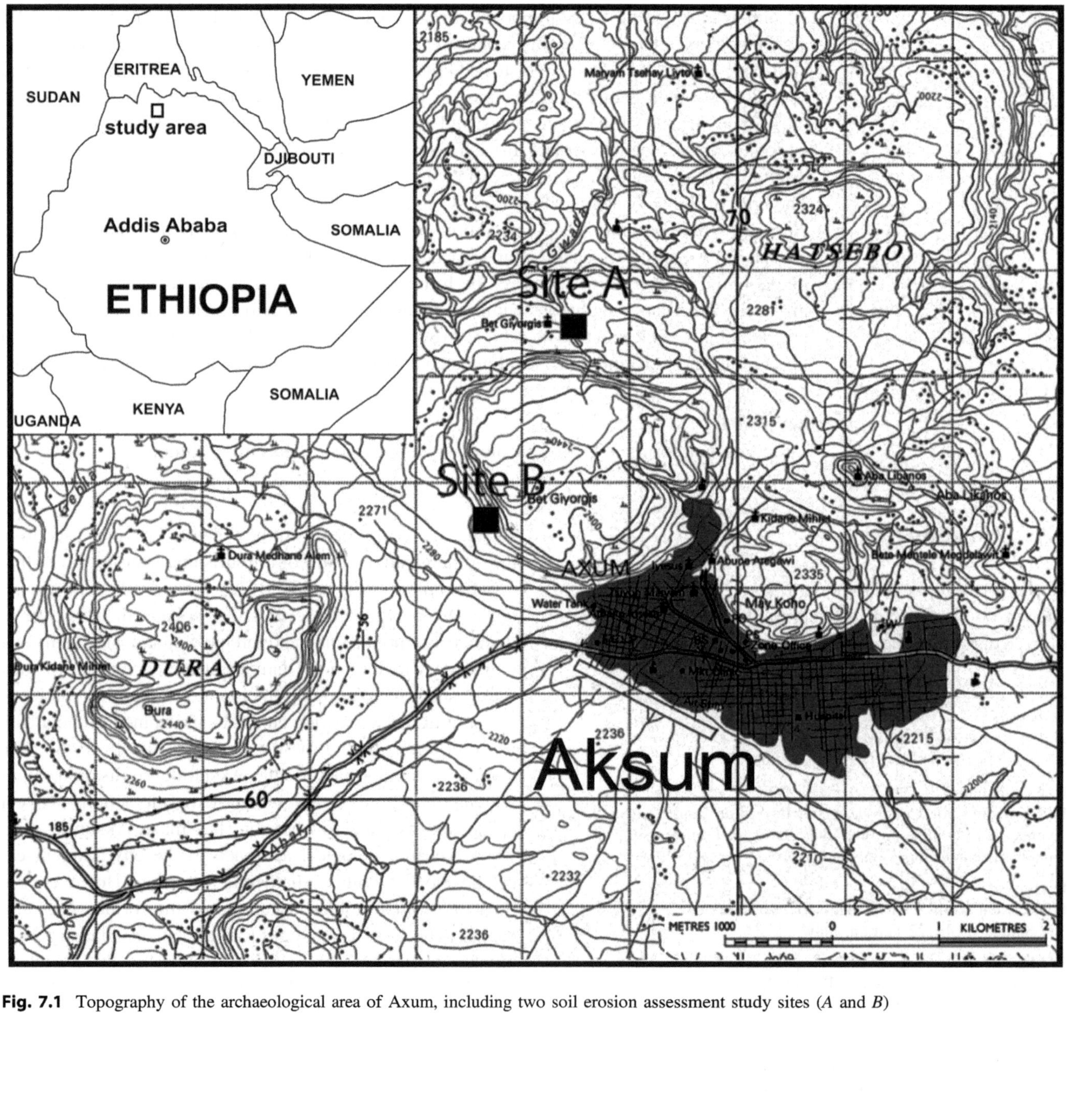

Fig. 7.1 Topography of the archaeological area of Axum, including two soil erosion assessment study sites (*A* and *B*)

Fig. 7.2 Aerial photograph of the area showing the flat-top of Bet Giyorgis hill

candelabrys, though much of the area is covered only by sparse shrubs and grasses and many slopes are rocky or almost bare. The town of Aksum itself has population of c. 45,000, and its downtown area is located right at the foot of two prominent hills of Bet Giyorgis in the west and May Koho in the east (Fig. 7.1).

7.3 Geology

The geology of the area is complex and includes Proterozoic basement of crystalline rocks (mainly granites and schist), sedimentary rocks of Mesozoic age (mainly sandstones), overlain by the Ethiopian Plateau flood basalts, emplaced between 31 and 29 Ma) (Hofmann et al. 1997), the later phase basaltic rocks (17–15 Ma) and the post-trap, Oligocene to Miocene trachytic lavas and syenite sub-volcanic intrusions in the form of plugs and domes (Hagos et al. 2010; Natali et al. 2013). These silica-poor volcanic to hypabyssal rocks belong to the Aksum–Adwa phonolite–trachyte volcanic field that is a part of a larger regional magmatic district that extends for ca. 150 km from west of the town of Aksum to the north–south belt of Adwa and covers an area of about 800 km^2 (Zanettin et al. 2006; Hagos et al. 2010; Natali et al. 2013). Magmatic rocks and the eruptive centres are placed along E–W and NE–SW volcano-tectonic trends, which conform to NE–SW striking shear zones affecting the Pan-African basement of the area (Natali et al. 2013). The chronological relationships between basalts and felsic rocks are not everywhere clear. Locally, basalts seem to overlie felsic rocks, and in some places, e.g. west of Aksum, basaltic dikes clearly cross-cut a felsic intrusion. Even though the absolute age of this complex is unknown, compositional similarities suggest a link with basalt–trachyte series from the shield volcanoes overlying the Northern Ethiopian continental flood basalts, which have been dated as Miocene in age (Kieffer et al. 2004). Analogous magmatic associations have been observed in the neighbouring Eritrean site of Senafe, where radiometric dating indicates ages from 21 to 23 Ma (Zanettin et al. 2006).

The archaeological area around Aksum is dominated by the nepheline syenites domes of Bet Giyorgis and Gobo Dura, dated for between 19 and 15 Ma (Natali et al. 2013), and the dome-shaped trachytic plug hills of May Koho, likely emplaced concomitantly (Bianchini, personal communication). These domes stand out from a relatively flat area underlain by trap basalts. On the western and eastern sides of Bet Giyorgis, small outcrops of lacustrine deposits occur. They are tilted and in places hydrothermally affected by the emplacement of the syenite plug. In the study area, these sedimentary rocks are commonly associated with the Enticho Sandstones of Carboniferous to Permian age (Beyth 1972; Bussert and Schrank 2007; Bussert 2010; Sacchi et al. 2007) and, though it was not possible to establish a clear stratigraphy, they can be considered as a laterally interfingering glacigenic facies unit of the Edaga Arbi Formation (Bussert and Schrank 2007). Both formations are unconformably overlain by the Adigrat Sandstones of Mesozoic age, mostly of continental origin, as shown by the occurrence of palaeosols, fluvial point bar

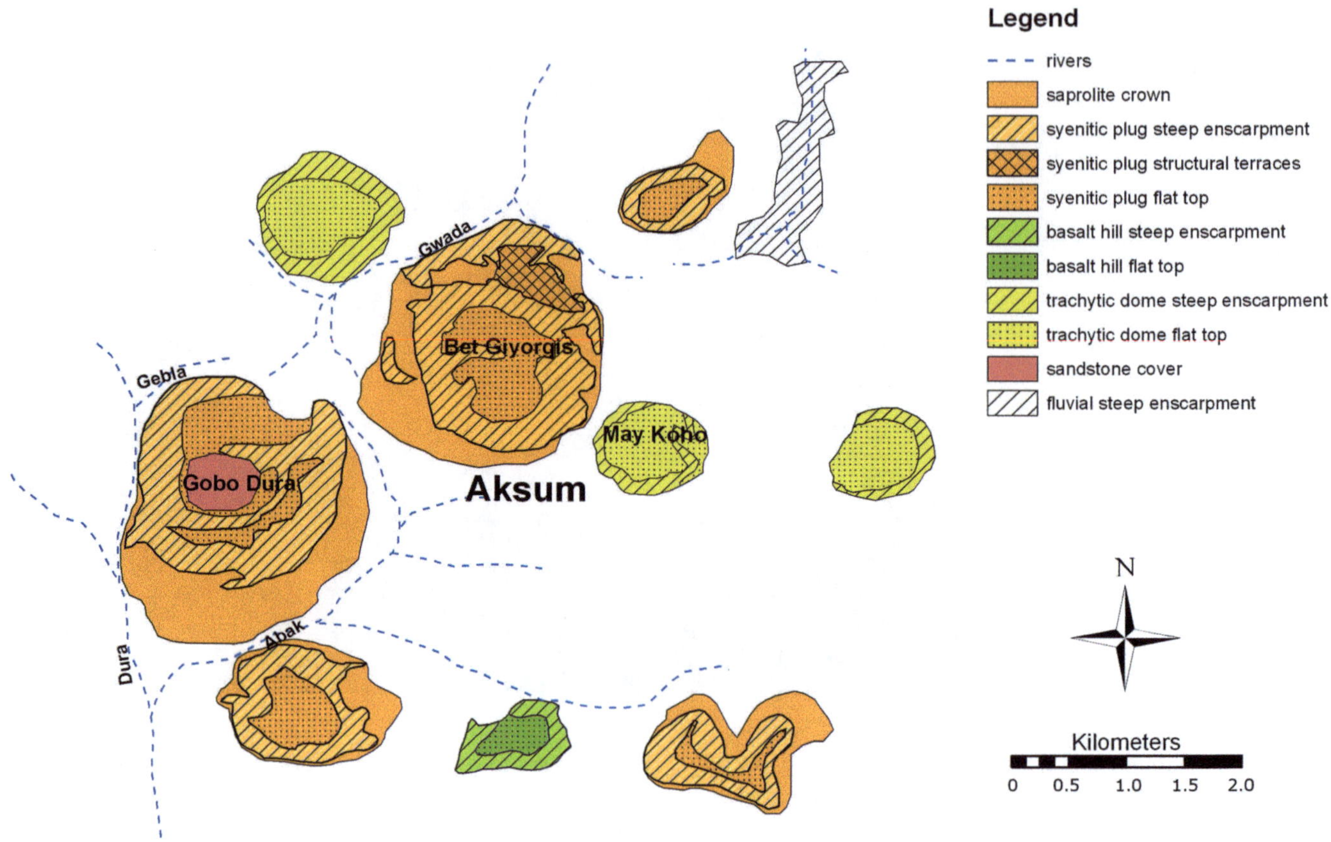

Fig. 7.3 Geomorphological sketch map of the study area

sequences and fossil wood (Beyth 1972; Merla et al. 1979; Bosellini et al. 1997). Blocks of sandstone and fine conglomerate, probably belonging to the Adigrat Sandstone unit, are also found on the top of Bet Giyorgis and on the northern side of Gobo Dura.

7.4 Landforms and Landscapes

The geomorphological landscape of the archaeological area of Aksum is characterised by a flat surface at an elevation around 2,200 m a.s.l., above which a few individual, dome-shaped sub-volcanic plugs stand (Fig. 7.3). The long-term weathering and denudation enriched the morphology of these plugs, leading to the origin of a number of landforms associated with the typical geomorphological evolution of crystalline rocks such as rock cliffs, castellated tors, block fields, boulder slopes, and mid-slope benches. In parallel, the intervening flat areas between the domes have been shaped by a variety of erosional and depositional processes. Thus, the typical landform association in the broader Aksum area consists of (i) domes with distinctive subordinate elements such as flat-tops, steep or stepped hillslopes, and boulder talus; (ii) an etchplain developed in the felsic rocks around the hills and connected with the lower plain; (iii) fluvial valleys incised into the etchplain; and (iv) lower clayey alluvial plain.

7.4.1 Main Domes and Related Forms

Most of the domes protrude abruptly from the surrounding flat area and rise above it by 200–250 m, emerging as the most evident geomorphological forms of the region. However, in contrast to many igneous rock domes known from elsewhere in Africa and beyond (see Thomas 1994), the domes around Aksum are distinguished by their remarkably flat-tops surrounded by steep slopes, not uncommonly cliff lines, which grade into boulder-covered hillslopes further below.

The flat-tops may attain considerable dimensions (Fig. 7.2). At Bet Giyorgis, the mountain-top surface is nearly circular, with a diameter of 2 km, enough to accommodate a small village and extensive agricultural area. An efficient drainage and soil types with less developed vertic properties than on a floodplain have provided an ideal support for cropping since the first stages of the Aksumite Kingdom.

Outcrops of massive syenite in the form of cliffs, castellated spurs (Fig. 7.4) and isolated tors (Fig. 7.5) are present on

Fig. 7.4 Rock cliffs and castellated towers above a boulder-mantled hillslope of the syenite plug of Gobo Dura

Fig. 7.5 Structural bench with emerging syenite tors

the flanks of the hills. They are best developed on their northwestern sides, where clusters of towers rise abruptly from the surrounding etchplain and connect two main relief elements, namely the etchplain and the dome tops. Otherwise, hillslopes are less steep (6–45 %) and consist of a mosaic of bare rock outcrops, exposed regolith with rinded corestones (Fig. 7.6), boulder blankets and patches of soil cover. While bedrock outcrops are more frequent in steeper section, regolith- and soil-covered surfaces are dominant. These steep hillslopes are affected by three main geomorphic processes: (1) deep bedrock weathering leading to the formation of grus-type regolith with corestones; (2) gravity-driven downhill movement of boulders (corestones); and (3) creeping and soil displacement due to gravity and water erosion.

The main feature of the lower sections of the syenite plugs is benches, giving the slopes a step-like appearance.

Fig. 7.6 Grus-type regolith in syenite with corestones surrounded by weathering rinds

Fig. 7.7 A step at the foot hill of Bet Giyorgis

Fig. 7.8 Fluvial valley incised into basaltic substrate the north-west of Bet Giyorgis

They are large flat areas, located at the foot of the escarpments, with the inner part of the treads deeply weathered and their edges made of exposed rocks, commonly crowned with tors. The origin of benches is not entirely clear. It is likely that they are structure-controlled features, formed due to selective weathering of marginal parts of the plugs. More susceptible compartments decompose at a faster rate and turn into regolith, but barriers made of more resistant, and hence, less weathered parts of the intrusion prevent regolith evacuation and sustain the bench. Once exposed, differential rates of weathering between exposed and buried rock surfaces seem to play a major role in maintaining characteristic footslope morphology (Wahrhaftig 1965) (Fig. 7.7).

7.4.2 Etchplain

In tropical and subtropical landscapes, etchplanation is the name given to a specific pathway of geomorphic evolution in which deep chemical weathering produces a thick saprolite that is progressively removed by erosion (Thomas 1989, 1994). Circulating groundwaters lower the weathering front

Fig. 7.9 Floating syenite boulders in the lower plain, with flat-topped plugs in the background

at various rates according to the different density of joints in the substrate and lithology-controlled susceptibility to weathering. Such etching process led to progressive differentiation of relief features, including the formation of inselbergs, steps, scarps and flat valleys.

In the case of Bet Giyorgis hill, the structure of the plug, with the outer parts apparently more densely jointed, favours such differentiation. In fact, the high rate of weathering in the foothill settings (sapping—Thomas 1974) and the adjacent areas results in the lowering and planation of the lower surfaces, whereas the main elements of hilly relief are preserved. Along and around the plug hill base, syenite rocks undergo intense chemical weathering and the soluble products are removed by the percolating waters.

7.4.3 Incised Valleys

Along the north-western margin of the Bet Giyorgis hill, the deeply weathered etchplain surface is considerably incised by a few ephemeral streams. Most of the valleys in the area seem to have formed before the emplacement of basalts associated with the second phase of the Miocene flood basalt emission (Natali et al. 2013). In fact, basalt flows appear to have filled most of the palaeovalleys between the plug hills and were incised again in more recent times to achieve the modern valley and river channel morphology (Fig. 7.8). This phase of erosion resulted in deeply incised box-like streams with sub-vertical bedrock banks 3–6 m high and 5–20 m wide (Fig. 7.8). The beds of these streams may consist of metric and sub-metric boulders resulting from weathering and mass movement processes active on the hill slopes.

7.4.4 Lower Plains

The flat areas around the main domes have been formed by different processes: etchplanation, flood basalt emplacement and sedimentation of ephemeral river floods (Fig. 7.9). Long-term deposition and pedogenesis produced a deep soil cover where vertisols are the dominant type. The lower plains have been widely used for cropping and cereal production since the antiquity and, because of the vertic properties of the soil, are considered of inferior quality by the local farmers. Notwithstanding the low gradient (0–3 %), along almost the whole lower plain soil conservation practices such as continuous terracing with stone bounds as high as to 0.5 m are common. This setting allows to control the erosive energy of the monsoon intense rainfalls and to prevent the cultivated fields from rill development.

The lower plains, especially those south-west of the Bet Giyorgis dome, are punctuated by scattered boulders of metric size (Fig. 7.9). Field evidence indicates that their presence on top of the lower plain surface at a far distance from the plug hillslopes can be accounted for by a sequence of processes, involving detachment by weathering and gravity displacements in which wetting and drying of the soil may have played a significant role (Cook 1993 in Poesen and Lavee 1994), allowing for sliding of boulders on saturated muddy substrate.

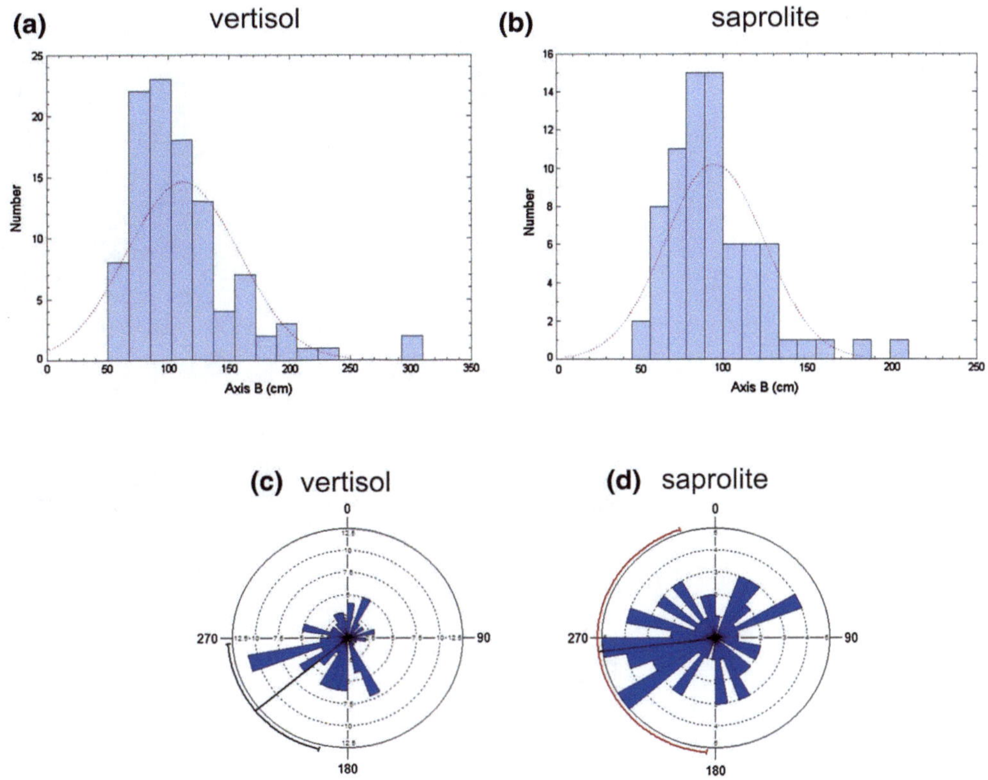

Fig. 7.10 Frequency distribution and main direction, respectively, of the intermediate *b* axis of the boulders in a site west of the Bet Giyorgis: **a** vertisol (n = 104) (*valley*); **b** saprolite substrate (n = 74) (*mid-slope*); **c** vertisol (n = 104) (*valley*); and **d** saprolite substrate (n = 74) (*mid-slope*)

In Fig. 7.10, the frequency distribution of the intermediate, *b* axis of the boulders and their orientation are reported, respectively. Two groups of boulders were considered, i.e. those floating on the top of saprolite at the plug hill footslope, ideally representing the baseline of the boulder movement, and those resting on the vertisols in the distal slope area. The results show the mean value of the *b* axis of 111.6 and 94.7 cm for boulders located on the vertisol and the saprolite, respectively. The mean *b* axis orientations for the boulders on these two different substrates are 230.6° and 263.3°, respectively, i.e. almost consistent with the main slope direction (220°). The consistency is more evident for boulders on top of the vertisols and indicates that they slid down "floating" on top of the ground surface and keeping their intermediate axis almost parallel to the main slope dip. This pattern is observed when particles are moved solely by gravity (Dumas and Raffy 1993), and not perpendicular to the bed gradient as it is by contrast observed when a particle is moved by the drag of a fluid (particle imbrication) (Briggs 1977).

7.5 Soils and Soil Erosion Assessment

7.5.1 Soils

On the flat-top of Bet Giyorgis, an association of truncated luvisols and shallow soils with vertic properties is present. On the plug foothill slopes, truncated palaeosoils and, in places, deep saprolites buried by colluvial processes and truncated luvisols ("*makaeo*" in the local language) are found. The low-gradient slopes around the hills are strongly weathered, and relict soils, probably nitisols ("*keyah*"), with argillic horizon and peds with shiny faces, are preserved under boulder cover. Soils with vertic features ("*bakahel*") occupy the central parts of the benches within plug hillslopes, while vertisols ("*walka*") (Nyssen et al. 2008; Schmid et al. 2008; French et al. 2009) are typically found on basalts, siltstones and alluvial plains.

Table 7.1 Chronology of the Aksumite Kingdom (Fattovich 1997a; Phillipson 2000)

Historical period	Year
Pre-axumite	700–400 BC
Proto-axumite	400–40 BC
Early axumite	40 BC–150 AD
Classic axumite	150–350 AD
Middle axumite	350–550 AD
Late axumite	550–800 AD
Post-axumite	800–1100 AD

Fig. 7.11 Typical hillslope terrace made for cultivation purposes

7.5.2 Soil Use and Agriculture

The land use of the area is primarily agriculture, and the agricultural landscape consists mainly of level bench terraces with stone bunds decreasing in height according to the slope gradient. Stony bunds to control and reduce soil erosion were particularly implemented during the DERG (*Coordinating Committee of the Armed Forces, Police, and Territorial Army*; 1974–1987) period within the food for work programme. They are present on almost any slope, with the exception of the steepest ones. Structural steps, such as those on the northern side of Bet Giyorgis, are used as natural agricultural terraces bounded by natural rock outcrops (kopjes) or man-made stone mounds. Large grazing areas, without any conservation practices, are found on gently sloping areas covered by soils with vertic properties, whereas large arable fields on deep vertisols are in the topographically lower parts of the study area. All crops are rain-fed and include teff (*Eragrostis tef*), barley (*Hordeum vulgare*), wheat (*Triticum aethiopicum*) and millet (*Eleusine coracana*) (D'Andrea et al. 2011). Some forest exclosures are present near residential and religious areas or in marginal rocky grounds. On the steeper slopes, a sparse eucalyptus forest alternates with bedrock exposed by erosion processes and fluvial dissection.

The hilly land around the city of Aksum has been extensively utilised for agriculture purposes since the antiquity and early traces of agricultural land use date back to the first millennium BC (Butzer 1981; Fattovich 1997a, b; Fattovich et al. 2000). According to a few authors (e.g. Bard et al. 2000; Sernicola 2008), archaeological evidence suggests that the northern slope of Bet Giyorgis has been uninterruptedly cultivated since the Early Aksumite time (40 BC–150 AD) (Table 7.1). The urban and rural populations reached their maximum during the Classic Aksumite period (150–350 AD), implying an expansion of land used for agricultural purposes (Sernicola 2008; Sulas et al. 2009; Sernicola and Phillipson 2011). Since at several sites, artefacts dating back to the different periods are found in place, it is reasonable to postulate that these locations have been continuously cropped at least since the Early Aksumite period.

In Ethiopia, relevant agricultural techniques and implements seem to have been mainly imported from South Arabia, probably before the Aksumite period (2,600–2,200 BP; Butzer 1981), though stone terracing is a very old approach to land management since it started probably earlier than 2,500 years ago (Michels 2005). In the Aksum area, the oldest terraced surfaces are estimated to date back to pre-Aksumite period (700–400 BC) (Fattovich et al. 2000). This conservation technique regards the slope as a whole, like in the Middle East, and considers the upper section not as an enlargement of the lower cultivated area under human pressure but as an integrated part of the system, serving different purposes (Hallsworth 1987). Thus, the slope is managed in the following manner: upslope terraces are mainly used for grazing and water harvesting, while at

Fig. 7.12 The "Maresha" ard plough

Fig. 7.13 Plough marks features on a boulder and various intensities of patinas indicating the history of exposure (after Ciampalini et al. 2012, modified)

downslope, they are cropped (Vogel 1988) (Fig. 7.11). The terraced surfaces in the higher portion of slopes do not show any sign of ploughing, confirming that this area was used specifically for water harvesting and grazing rather than for cropping. In this context, agricultural practices were and still are based on the intensive use of the *maresha*, a traditional ard plough (Gebregziabher et al. 2006; Nyssen et al. 2000, 2011). The *maresha* seems to have been introduced in Ethiopia between 1000 and 400 BC by Semitic tribes from Yemen, or even before by Cushitic-speaking peoples from

Table 7.2 Median values of the estimated erosion (t ha^{-1} $year^{-1}$) for ancient (*Site A* and *B*) and recent soil erosion (*Site B*) at two study sites

Sites	Proto-axumite (since 2400 year BP)	Early axumite (since 2040 year BP)	Classic axumite (since 1850 year BP)	Middle axumite (since 1650 year BP)	Recent erosion (last 35 years)
A	2.6	3.1	3.4	–	–
B	–	2.5	2.8	3.1	65.8

Fig. 7.14 Geoarchaeological legacy in the Axum area. **a** The broken Great Stele in the archaeological park in the town; **b** ancient quarry at Gobo Dura, with an unfinished obelisk; **c** rock carving of Lioness at Gobo Dura

north-eastern Sudan. The plough is a pointed, steel-tipped tine fixed to a pole (Fig. 7.12). The soil, therefore, is not turned over, but incised and pushed aside by two lateral wings; a handle allows the implement to be driven and lifted when an obstacle is encountered.

7.5.3 Soil Erosion Assessment

In the study area, soils are characterised by a large content of rock fragments. Nepheline syenite boulders and core stones with weathering rinds lie on thin, buried red clayey palaeosols or protrude from a saprolite mantle. These boulders are scattered across gentle slopes at the base of the domes and used to build the bounds of the cropped terraces. During the last two millennia, the terraces have been ploughed with the use of an ard plough pulled by one or two oxen. Mechanical impact of the plough on the boulders is therefore unavoidable, and they are scratched on the top, if embedded in soil at a shallower depth than tillage depth (8–16 cm), or on the sides, if they are partially exposed. As soon as the farmer hears and feels the scraping, he lifts the implement, but a deep mark on the rock surface is already made. Plough marks are found at different heights on the rocks faces, thus providing valuable information on the thickness of soil loss (Fig. 7.13). In association with other rock weathering features such as patinas, etches, pitting and lichen cover, plough marks can be used to reconstruct the position of old soil surfaces, and by identifying their occurrence, it is possible to recognise ancient cultivated lands that are presently abandoned, degraded or strongly dissected (Ciampalini et al. 2008).

The spatial relationship between plough marks and rock surface orientation gives information on the ploughing direction, whereas the relative height from the top of the marks belt to the present ground surface, plus the ploughing depth, i.e. additional 15–18 cm, is assumed as the actual soil loss. By this method, developed by Ciampalini et al. (2008), it is possible to assess soil erosion rate throughout long-time intervals, provided reliable dating of the cultivated surface (e.g. by archaeological evidence, C^{14} and thermoluminescence) is available. For instance, in a study plot of the Ma' Qono area, at the north-west slope of Bet Giyorgis, archaeological evidence in the form of old ruined building, artefacts, pottery, iron slags and pollen analysis indicates this site has been occupied and cultivated since the Proto-Aksumite times (about 2,400 BP) (Fattovich et al. 2000).

Similarly, two sites represented by two series of agricultural terraces with plough marks features still well preserved have been dated and investigated (Fig. 7.1). These are the site A, where the agricultural exploitation has been estimated to start from Proto- to Classic Aksumite (2400–1850 year BP) and the site B, at which cultivation started from Early Aksumite to Middle Aksumite (2040–1650 year BP). The latter site is particularly interesting as a major land use change occurred in 1975 when cultivation was abandoned and terrace maintenance was halted. This change produced a significant degradation of the terraces by accelerated erosion. The long-term soil loss calculated at these two sites ranges from 2.6 to 3.4 t ha^{-1} $year^{-1}$ and for the site B, an erosion rate ranging from 2.5 to 3.1 t ha^{-1} $year^{-1}$ (Table 7.2). These values can be contrasted with the 20-fold increase in erosion rate at the site B after the recent abandonment. The median value of soil loss over a period of 31 years (1975–2006) is 65.8 t ha^{-1} $year^{-1}$ (Ciampalini et al. 2008, 2012).

7.6 Stone Industry and Aksum Stelae

The presentation of the geomorphology of the Aksum area would be incomplete if the geomorphic impact of quarrying and stone dressing is ignored. The archaeological area of Aksum is famous for the stone-carved monuments—the stelae (obelisks). The most imposing ones are displayed in the Northern Stelae Field, at the outskirts of the town and at the foot of Bet Giyorgis hill. The highest standing intact obelisk, dedicated to the King Ezana (reigned 330–356 AD), is 24 m high, but the most impressive of all is the lying and broken, 33-m-long Great Stele (Fig. 7.14a). In 2008, another obelisk, the Rome Stele, broken into pieces in antiquity and taken to Italy in 1930s, was reinstated on the site. All are made of local syenite and elaborately decorated. Many more obelisks can be found at this and other archaeological sites in the vicinity. Further, stone monuments of geoarchaeological interest are numerous megalithic tombs, including a highly complex underground Mausoleum erected from huge syenite slabs in a purposefully made trench.

Stone for obelisks was sought and quarried on the slopes of the Gobo Dura hill, about 3 km west of the town. Apparently, they were the best source of rock solid enough to withstand cutting, transport and final dressing, pointing to an excellent indigenous knowledge of syenite weathering patterns, rock properties and rock behaviour. Until today, remnants of ancient quarries can be seen at Gobo Dura, including unfinished stelae and stones abandoned during transport (Fig. 7.14b). The slopes of Gobo Dura feature further examples of interactions between rocks, landforms and people. One of these is the 2-m-long image of a lioness carved on the site of huge monolithic syenite boulder, linked with local legends. Chiselling the animal outline required removal of the thin outer layer of rock, resulting in the currently observed colour difference between the red iron-rich crust developed over long time and the carving itself which lacks the crust (Fig. 7.14c).

Finally, we have to consider the amount of stone blocks that have been carved from the crystalline rocks of the plugs throughout the past centuries to present. These stones have been used as construction material to build the majority of the ancient buildings.

7.7 Conclusions

The close vicinity of Aksum belongs to the most attractive geomorphological landscapes of Ethiopia and offers a singular combination of natural and anthropogenic features. The geomorphological evolution of the area is controlled by a complex combination of local tectonics, lithological diversity, climate and surface processes. The main geomorphic features of the area represent a classical, well-expressed example of volcanic and sub-volcanic intrusions in the form of lava flows, plugs and domes. Contemporary surface processes include deep weathering of syenite rocks, disintegration of rock cliffs along the rims of the flat-top domes, continuing development of boulder talus, regolith and soil erosion, as well as fluvial incision and valley widening. The development of the Aksumite Kingdom provides an additional value to such particular geomorphological context and favourable conditions to investigate about landscape evolution and geomorphic processes during the last two millennia. Historical erosion rates were calculated since the beginning of the Aksumite Kingdom owing to the concurrence of a few main factors: the plough marks, the presence of artefacts and other archaeological evidence, and the persistence of the cropping practices in the same plots for about two millennia. There is little doubt that while the archaeological area of Aksum has been inscribed on the UNESCO World Heritage List as a cultural property, its geomorphological features are also of major interest and international significance.

References

Bard KA, Coltorti M, Di Blasi MC, Dramis F, Fattovich R (2000) Environmental history of Tigray (northern Ethiopia) in the middle and late holocene: a preliminary outline. Afr Archaeol Rev 17:65–85

Beyth M (1972) The geology of central and western Tigray. Ph.D thesis, Rheinische Friedrich-Wilhelms Universität, Bonn, W. Germany, 200 pp

Bosellini A, Russo A, Fantozzi PL, Assefa G, Tadesse S (1997) The Mesozoic succession of the Mekelle Outlier (Tigrai Province, Ethiopia). Mem Sci Geol 49:95–116

Briggs D (1977) Sediments. Butterwoths, London

Bussert R (2010) Exhumed erosional landforms of the late palaeozoic glaciation in northern Ethiopia: indicators of ice-flow direction, palaeolandscape and regional ice dynamics. Gondwana Res 18:356–369

Bussert R, Schrank E (2007) Palynological evidences for a latest Carboniferous-Early Permian glaciation in Northern Ethiopia. J Afr Earth Sci 49:201–210

Butzer KW (1981) Rise and fall of Axum, Ethiopia: a geo-archaeological interpretation. Am Antiq 46:472–495

Ciampalini R, Billi P, Ferrari G, Borselli L (2008) Plough marks as a tool to assess soil erosion rates: a case study in Axum (Ethiopia). Catena 75:18–27

Ciampalini R, Billi P, Ferrari G, Borselli L, Follain S (2012) Soil erosion induced by land use changes as determined by plough marks and field evidence in the Aksum area (Ethiopia). Agric Ecosyst Environ 146:197–208

D'Andrea AC, Richards MP, Pavlish LA, Wood S, Manzo A, Wolde-Kiros HS (2011) Stable isotopic analysis of human and animal diets from two pre-Aksumite/proto-Aksumite archaeological sites in northern Ethiopia. J Archaeol Sci 38(2):367–374

Dumas B, Raffy J (1993) The study of stone trails with painted tracers on unstable slopes in the Southern French Alps: preliminary results. Geog Fis Din Quat 16(1):37–45

Fattovich R (1997a) The peopling of the Tigray Plateau in ancient and medieval times (ca. 4000 B.C.–A.D. 1500): evidences and synthesis. In: Bard KA (ed) Environmental history of human ecology on Northern Ethiopia in the late holocene. Istituto Universitario Orientale, Napoli, pp 81–105

Fattovich R (1997b) Archaeology and historical dynamics: the case of Bieta Giyorgis (Aksum), Ethiopia. Annali dell'Istituto Universitario Orientale 57:46–79

Fattovich R, Bard KA, Detrassi L, Pisano V (2000) The Axum archaeological area: a preliminary assessment. Ist. Univ. Orientale, Dip. Di Studi e Ricerche su Africa e Paesi Arabi. Working paper 1

French C, Sulas F, Madella M (2009) New geoarchaeological investigations of the valley systems in the Aksum area of Northern Ethiopia. Catena 78:218–233

Gebregziabher S, Mouazen AM, Van Brussel H, Ramon H, Nyssen J, Verplancke H, Behailu M, Deckersf J, De Baerdemaeker J (2006) Animal drawn tillage: the Ethiopian and plough, maresha: a review. Soil Tillage Res 89:129–143

Hagos M, Koeberl C, Kabeto K, Koller F (2010) Geochemical characteristics of the alkaline basalts and the phonolite—trachyte plugs of the Axum area, Northern Ethiopia. Austrian J Earth Sci 103 (2):153–170

Hallsworth EG (1987) Anatomy, physiology and psychology of soil erosion. Wiley, New York

Hofmann C, Courtillot V, Feraud G, Rochette P, Yirgu G, Ketefo E, Pik R (1997) Timing of the Ethiopian flood basalt event and implications for plume birth and environmental change. Nature 389:838–841

Kieffer B, Arndt N, Lapierre H, Bastien F, Bosch D, Pecher A, Yirgu G, Ayalew D, Weis D, Jerram DA, Keller F, Meugniot C (2004) Flood and shield basalts from Ethiopia: magmas from the African Superswell. J Petrol 45:793–834

Merla G, Abbate E, Azzaroli A, Bruni P, Canuti P, Fazzuoli M, Sagri M, Tacconi P (1979) A geological map of Ethiopia and Somalia. Consiglio Nazionale delle Ricerche, Firenze

Michels JW (2005) Changing settlement patterns in the Aksum-Yeha Region of Ethiopia: 700 BC–AD 850. Cambridge Monographs in African Archaeology. British Archaeological Reports (BAR), Oxford

Natali C, Beccaluva L, Bianchini G, Siena F (2013) The Axum-Adwa basalt–trachyte complex: a late magmatic activity at the periphery of the Afar plume. Contrib Miner Petrol 166:351–370

Nyssen J, Poesen J, Haile M, Moeyersons J, Deckers J (2000) Tillage erosion on slopes with soil conservation structures in the Ethiopian highlands. Soil Tillage Res 57:115–127

Nyssen J, Naudts J, De Geyndt K, Haile M, Poesen J, Moeyersons J, Deckers J (2008) Soils and land use in the Tigray highlands (Northern Ethiopia). Land Degrad Dev 19:257–274

Nyssen J, Govaerts B, Araya T, Cornelis WM, Bauer H, Haile M, Sayre K, Deckers J (2011) The use of the marasha ard plough for conservation agriculture in Northern Ethiopia. Agron Sustain Dev 31:287–297

Phillipson DW (2000) Archaeology at Aksum, Ethiopia, 1993-7. DW Phillipson Eds, London, p. XII + 538

Poesen J, Lavee H (1994) Rock fragments in top soil: significance and processes. Catena 23:1–28

Sacchi R, Alene M, Barbieri M, Conti A (2007) On the Paleozoic Tillite of the Adigrat Group (Tigrai, Ethiopia). Periodico di Mineralogia 76(2–3):241–251

Schmid T, Koch M, DiBlasi M, Hagos M (2008) Spatial and spectral analysis of soil surface properties for an archaeological area in Aksum, Ethiopia, applying high and medium resolution data. Catena 75:93–101

Sernicola L (2008) Il modello d'insediamento sull'altopiano tigrino (Etiopia settentrionale/Eritrea centrale) in epoca Pre-Aksumita e Aksumita (ca. 700 a.C.–800 d.C.). Un contributo da Aksum. Unpublished PhD dissertation, Department of African and Arabian Studies, University of Naples, Italy

Sernicola L, Phillipson L (2011) Aksum's regional trade: new evidence from archaeological survey. Azania: Archaeol Res Afr 46(2):190–204

Sulas F, Madella M, French CAI (2009) State formation and water resources management in the Horn of Africa: the Aksumite Kingdom of the Northern Ethiopian highlands. World Archaeol 41:2–15

Thomas MF (1974) Tropical Geomorphology. A study of weathering and landform development in warm climates. The MacMillan Press ltd, London and Basingstoke

Thomas MF (1989) The role of etch processes in landform development. I. Etching concepts and their application. Zeitschrift für Geomorphologie NF 33:129–142

Thomas MF (1994) Geomorphology in the tropics: a study of weathering and denudation in low latitudes. Wiley, Chichester

Vogel H (1988) Impoundment-type bench terracing with underground conduits in Jabal Haraz, Yemen Arab Republic. Transactions Institute of British Geographers, N.S. 13: 29–38

Wahrhaftig C (1965) Stepped topography of the South Sierra Nevada, California. Geo Soc Am Bull 76:1165–1190

Zanettin B, Bellieni G, Justin Visentin E (2006) Stratigraphy and evolution of the trachy-rhyolitic volcanism of the Senafe area (East Eritrean Plateau). J Afr Earth Sci 45:478–488

Geomorphology of the Adwa District

8

Maria J. Machado

Abstract

This chapter describes the geologic and geomorphic characteristics of the Adwa district and provides a reconstruction of the major phases of the long-term landscape evolution from the Late Proterozoic to the Quaternary. Two landscape features of the Adwa district exemplify the control of tectonic uplift and erosion, often in a feedback loop: the plateau landscape and river incision. The plateau landscape developed upon metamorphic, sedimentary and volcanic materials, with major contacts marked by planation surfaces correlative with major regional- and sometimes continental-scale unconformities. In the southern Adwa district, the most prominent landscape feature is a regionally extensive erosion surface bevelled across the Precambrian basement and later buried by Palaeozoic and Mesozoic sedimentary rocks. In the northern district, the flat-top horizontal structural plateaux are formed by the Mid-Cenozoic upper basaltic Trap sequence, intruded by trachyte and phonolite plugs originated during the Miocene and Pliocene volcanic activity, and a lower plateau developed over the laterised Adigrat Sandstone Formation. The post-Pliocene landscape evolution has led to exhumation of the former erosion surfaces and dissection of the drainage network. These stream channels have cut large amphitheatre headwaters and flow southwards to join the regional base level of the Weri River, a contributor to the Nile basin. The Quaternary sedimentary record in the Adwa area is composed by consolidated carbonate Late Pleistocene rocks (travertines), unconsolidated Pleistocene/Holocene alluvial fan deposits, Holocene fluvial deposits (river terraces and valley fill deposits) and historical to present-day colluvial deposits.

Keywords

Northern Ethiopia • Plateau landscape • Erosion surfaces • Valley fill deposits

8.1 Introduction

The Adwa *woreda* (Amharic word for a district level division) is located in the north-west part of the extensive and elevated Ethiopian Plateau, largely above 1,500 m a.s.l., and is built of metamorphic, sedimentary and volcanic materials. Geographically, it corresponds to the central zone (Maákelay zone) of the northern province of Tigray. The Adwa region shows a complex assemblage of dynamic landscapes, each preserving a unique history of landscape evolution of the area.

The general landscape traits of the Adwa area reflect the relentless and unbalanced competition through time between rock basement uplift, climate and the resistance of exposed rocks against weathering and erosion. The geomorphology, as in the entire Ethiopian Highlands, has an evident lithological and structural control showing a heterogeneous mosaic preserving land surfaces and landforms of relative antiquity. This structural landscape is characterised in the north by a high surface with a low-gradient landscape at a prominent elevation of 2,200 m, which is cut by large

M.J. Machado (✉)
Departamento de Geología, Museo Nacional de Ciencias Naturales, CSIC, Calle Serrano, 115bis, 28006 Madrid, Spain
e-mail: machado@mncn.csic.es

P. Billi (ed.), *Landscapes and Landforms of Ethiopia*, World Geomorphological Landscapes,
DOI 10.1007/978-94-017-8026-1_8, © Springer Science+Business Media Dordrecht 2015

amphitheatre headwaters that results from an evident scarp retreat. Southwards, the incised valleys running north–south cut folded and faulted metamorphic rocks (volcanic and sedimentary) of the Neoproterozoic basement. The stream channels trend southwards to join the regional base level of the Weri River at an elevation of ~1,380–1,350 m. A remarkable feature in the geomorphology of the Adwa district is the presence of so-called inherited landscapes (e.g. denudation surfaces, volcanic calderas and regoliths), which are components of different past geodynamic and climatic settings. The most prominent is a regionally extensive erosion surface bevelled across the Precambrian basement and later buried by Mesozoic sedimentary rocks. Cenozoic rock uplift and denudation of rocks overlying the unconformity have resulted in exhumation of these ancient landforms, revealing new clues to the region's long-term landscape evolution.

The resulting physiography, therefore, consists of a heterogeneous array of geomorphic provinces, each featuring a distinctive landform assemblage that preserves unique evidence which can be used to reconstruct the history of landscape evolution. During the last decades, there has been a growing interest in physical, environmental and palaeoenvironmental studies in northern Ethiopia (e.g. Virgo and Munro 1978; Butzer 1981; Machado et al. 1998, 2001; Dramis et al. 2003). Most of these works put emphasis on the contemporary processes of environmental degradation and soil erosion (e.g. Hurni 1983; Billi and Dramis 2002; Frankl et al. 2012) and the societal causes and consequences of environmental degradation (e.g. Hurni 1986; Nyssen et al. 2001, 2004; Feoli et al. 2002). However, there is a lack of basic understanding of the regional land systems and their connection through the long-term landscape evolution. The comprehensive and novel work of Coltorti et al. (2007) on the planation surfaces in Ethiopia constitutes an example of such studies. Regional studies are the basis for spatial interpolation of process-oriented studies, assessment of natural resources and environmental planning. The present chapter describes the geologic and geomorphic characteristics of the Adwa district and provides a reconstruction of major phases of long-term landscape evolution, from ancient landforms up to most recent geomorphic features linked to Quaternary climatic changes and human environmental history.

8.1.1 Geologic Setting

The oldest rock outcrops in the Adwa area, Neoproterozoic in age, belongs to the Arabian–Nubian Shield, and corresponds to the Tsaliet Group (lower) and Tambien Group (Beyth 1972; Tadesse et al. 1999). The foliated metavolcanic and greenschist facies rocks of the Tsaliet Group are overlain unconformably (Beyth 1972) by the metasedimentary rocks of the Tambien Group, that in the case of the Adwa area is formed by four of the six formations identified in the northern highlands (Tadesse et al. 1999; Miller et al. 2009; Roberts et al. 2012): the Weri Slate formation (lower), the Assem Limestone, the Tsedia Slate and the May Kinetal Limestone formation. This sequence is exposed within the May Kinetal synclinorium. Thrust faults trending NE–SW and N–S striking faults (Fig. 8.1) affecting the Neoproterozoic basement rocks (Tsaliet and Tambien groups) resulted from regional deformation phases (Alene et al. 2006).

Towards the south of Adwa region, near May Kinetal, an intrusive body can be found. Avigad et al. (2007) dated this granitoid body at ~612.3 ± 57 Ma, correlating it with another post-tectonic intrusive body in a nearby area next to the southern Eritrean border, the Mareb granite (Tadesse et al. 1997, 2000; Avigad et al. 2003; Miller et al. 2003; Asrat et al. 2004).

A major continental-scale geomorphic event occurred between 640 and 545 Ma (Beyth and Heimann 1999; Beyth et al. 2003), and the shaping capacity of this erosive event is still evident in Adwa landscape (see Sect. 8.2). This rapid exhumation of the Neoproterozoic bedrock, estimated for up to ~2 km/Ma (Ghebreab 1999), led to the origin of a levelled and peneplained surface, not only in northern Ethiopia, but on a continental-scale, extending from north Africa towards Oman (Sandler et al. 2012). Beyth et al. (2003) suggest extensive erosion to be associated with post-glacial erosion. Palaeoedaphic units, whose characteristics resembling modern oxisols, dated as sub-Cambrian, were described from several weathering profiles of the Arabian–Nubian Shield (Sandler et al. 2012), in NE Africa and southern Israel, indicating a change of climate in the Early Cambrian, towards warm and humid conditions. This could have enhanced fluvial as well as subaerial erosion processes, allowing for the formation of extensive, regional-scale platforms of clastic sediments.

Lying unconformably on the top of this surface, the Ordovician Enticho Formation (Merla and Minucci 1938; Beyth 1972; Merla et al. 1979) is formed by a subhorizontal white, often calcareous, medium- to coarse-grained sandstone, cross-bedded at the base. The upper unit of this formation interfingers southwards in the Adwa area with the Edaga Arbi glacial deposits (Dow et al. 1971). These deposits grade upwards into a massive grey shale (Indaba T'shama). Remnants of the basal unit lie unconformably on the top of Late Neoproterozoic/Early Cambrian peneplain, or on the top of yellow-weathered sediments (north of Golongolo). Remnants of this glacial basal unit, reworked by agriculture activities in the top 70 cm, contain gravels and some erratic submetric blocks. Recent palaeomagnetic studies in northern Ethiopia conducted by Kidane et al. (2013) indicate a Late Carboniferous to Early Permian age of the Edaga Arbi glacial sediments.

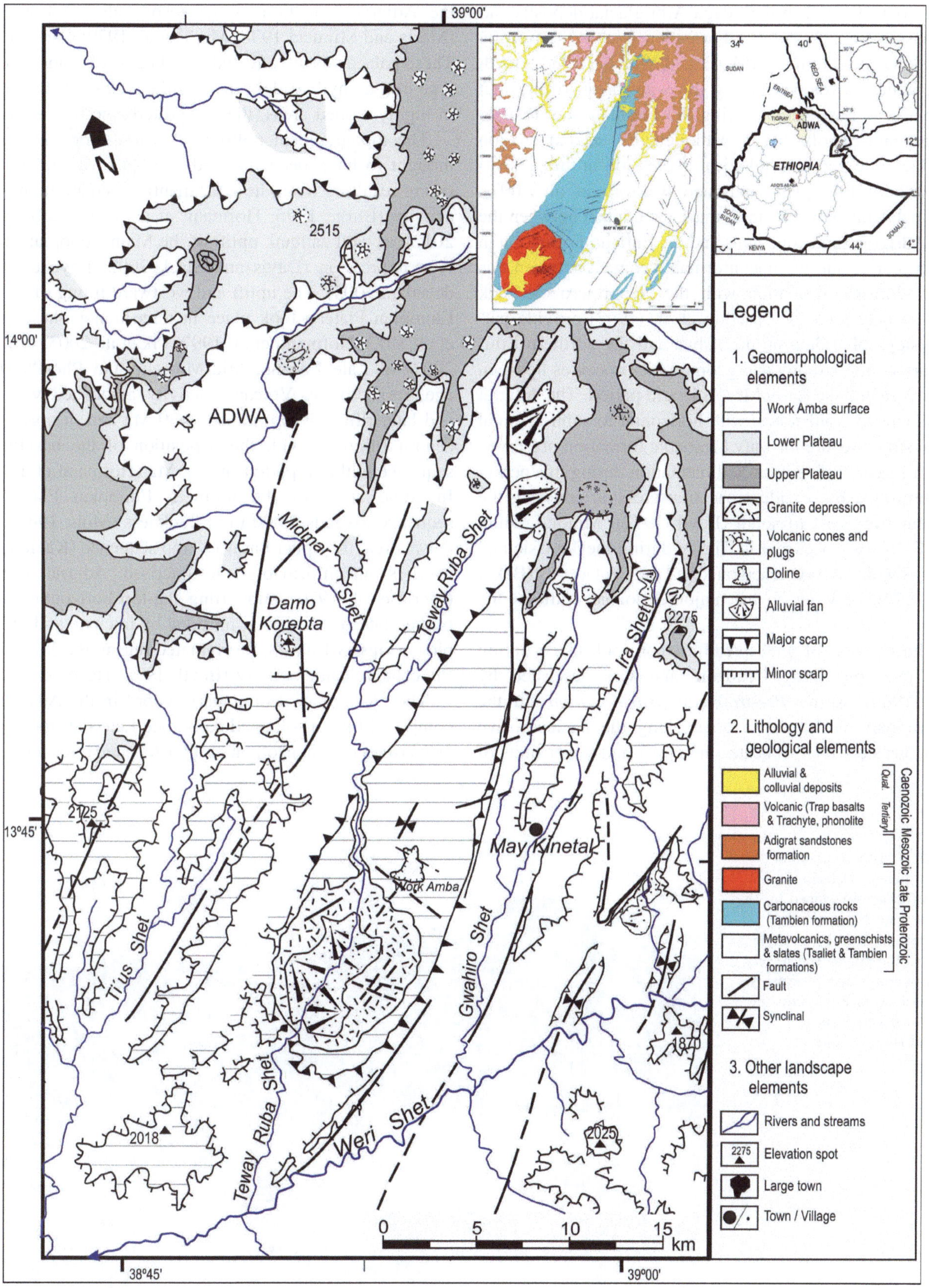

Fig. 8.1 Geomorphology of the Adwa district area, northern Ethiopia

The contact between the Edaga Arbi glacial deposits and the Triassic–Jurassic Adigrat Formation is characterised by an erosional discontinuity (Merla and Minucci 1938; Beyth 1972), which developed during the widespread rifting that took place in East Africa during the Permian and Triassic (Ebinger and Scholz 2012). Although this second erosion phase has little geomorphological expression in the present-day northern Ethiopian landscape (Coltorti et al. 2007), metric palaeoincision channels mark the contact between the lower Adrigrat Sandstone units and the Enticho Formation in the Adwa area, and the unconformable contact between this Triassic–Jurassic formation with the Neoproterozoic rock basement described by Bosellini et al. (1997) provide palaeophysiographic data on the Adwa and the northern Ethiopia surface topography and geomorphic processes between the Permian and the Early Mesozoic arid period. The Adigrat red siliciclastic sandstones, with horizontal to subhorizontal general structure, are the only Mesozoic formation preserved in the Adwa area. The basin sedimentation during this period was controlled by a widespread transgression covering the continent from east to south (Beyth 1972). Minor faulting along E–W axes occurred in Tigray during the Jurassic and the faults were later on reactivated in Late Cretaceous (Baker et al. 1972), previous to the major Cenozoic uplift of the area.

A third period of widespread erosion took place during Late Cretaceous–Early Cenozoic and was described by Mohr (1962) as the *Pre-trappean peneplanation*. In the northern part of the Adwa area, along the palaeoerosive terrain surface which truncates the upper and lower units of the Adigrat formation, a palaeosol (laterite) was formed (Merla and Minucci 1938; Merla et al. 1979; Garland 1980). These palaeosols are affected by a heavy separation of plinth type iron oxihydroxides affecting the Adigrat sediments, and in highly eroded areas, the Neoproterozoic rock basement.

The topographical uplift of this non-orogenic region is thought to have been initiated ~30–40 Ma and linked to rifting resulting from the Afar mantle circulation and plume activity (Burke 1996; Hofmann et al. 1997; Roberts et al. 2012), as well as flank uplift of the Main Ethiopian Rift and Afar depression (Davis and Slack 2002; Beyene and Abdelsalam 2005). The uplift and westward tilting of the rising Ethiopian Plateau took place in three major phases (Baker et al. 1972; Hofmann et al. 1997; Pik et al. 2003: Gani et al. 2007) of Late Eocene, Mid-Miocene and Plio-Pleistocene age, respectively. Volcanic activity associated with uplift and later rifting was initiated ~30 Ma ago in northern and central Ethiopia, with the deposition of the basaltic Trap sequence within a period of 1–2 Ma (Hofmann et al. 1997). In some areas of the northern Ethiopian Plateau, this sequence exceeds 2,000 m in thickness (Mohr 1962). In the Adwa area, the upper basaltic Trap sequence (Kazmin 1972) overlies unconformably the laterised Adigrat formation (Merla et al. 1979), supporting flat-top horizontal structure of the plateaux, however, intruded by trachyte and phonolite plugs (Figs. 8.1 and 8.2) originated during the Miocene and Pliocene volcanic activity (Beyth 1972; Hagos et al. 2010).

The Quaternary sedimentary record in the Adwa area is composed by consolidated carbonate Late Pleistocene rocks (travertines), unconsolidated Pleistocene/Holocene alluvial

Fig. 8.2 Flat-top horizontal structure plateau (laterised Adigrat formation) topped by an upper plateau (basaltic Trap sequence), intruded by trachyte and phonolite plugs. The lower plateau overlays unconformably the erosion surface that cuts the Neoproterozoic bedrock basement (light colours)

fan deposits, Holocene fluvial deposits (river terraces and infilled valley deposits) and historical to present-day colluvial deposits.

8.2 Geomorphological Features Associated with Long-Term Landscape Evolution

Two millennia of written archives document the political and socio-economic history, as well as the distinctive legacy of the Adwa district and of Ethiopia in general, as the dissemination core of important worldwide food cultivars (Kirwan 1972; Butzer 1981; Pankhurst 1985). Yet, not until the early sixteenth century, with the first comprehensive written description made by Francisco Alvares in 1520 (particularly focused on human settlements, climate, rivers, routes, society and natural resources) were the main physiographic elements of the Adwa region and of the Ethiopian Highlands described (Alvares 1540; Ross 1922a, b). Elements such as incised rivers and the prominent hills (corresponding to the phonolite and trachyte plugs) that emerge from extensive, high-altitude levelled surfaces were depicted for the Adwa and Axum areas.

Not until the late nineteenth century was there an attempt to explain the origin and formation of the Adwa district landforms: a stepped, low-gradient high surfaces, with large amphitheatres in the area of scarp retreat and incised valleys trending southwards, cutting the faulted rock basement. Blandford (1870) not only attempted a general geologic description of most of the Ethiopian territory, but was also trying to explain the origin of landforms, which he attributed to fluvial and soil erosion. Based on field observation of fine sediment transport on slopes and by rivers during floods carried out during the main rainy season, he also suggested the role of concentrated rainfall in the detachment of soil/rock particles.

During the 1930s, extensive studies were conducted in the area by Merla and Minucci (1938) within a broad research and cartographic work, focused not only on geology but also on the main physiographic characteristics of a vast area of Tigray. The work of Merla and Minucci has indisputably set the basis for all subsequent studies on regional geomorphology in the Ethiopian Highlands. Particular interest was put on the plateau geomorphology. Merla and Minucci identified three main erosion surfaces, corresponding to three topographical-levelled surfaces (suggested at the time to be end-Mesozoic, sub-Miocene and end-Tertiary in age) and tentatively correlated with the three major East African erosion surfaces. These erosion surfaces would be linked with three main phases of regional uplift (Merla and Minucci 1938; Merla et al. 1979; Mohr 1962).

The comprehensive work of Coltorti et al. (2007) on the planation surfaces of northern Ethiopia marks a rebirth of regional geomorphology and of the studies on the genetic mechanisms of the long-term geomorphological evolution of the area.

Although landscape shaping by erosion is an inexorable ongoing process in the Adwa area, the main geomorphic features give evidence of different relationships in the past between the two main geomorphic driving forces imprinted in the geological history of the area: tectonic uplift and erosion.

Two landscape features of the Adwa district exemplify the control of tectonic uplift and erosion (often in a feedback loop): the plateau landscape and river incision.

8.2.1 The Plateau Landscape

Past, as well as contemporary travellers heading to Adwa, either from the nearby Ethiopian civilisation pillars of Axum (east) and Yeha (north), or from further east (Adigrat) was impressed by a singular landscape of high, stepped surfaces, above 2,200 m a.s.l., intruded by a myriad of steep slope hills, corresponding to columnar basalt necks and chimneys as well as to trachyte and phonolite plugs that can rise as much as 300 m above the summit surface (Fig. 8.3). These slightly undulating or flat-top plateaux where, throughout the central and northern Ethiopian Highlands, villages and fortresses would be located due to its difficult access (Abdul-Haggag 1961; Uhlig et al. 2003) are called locally *Ambas* (in Tigrinyan and Amharic languages).

In the Adwa district, two types of flat-top surfaces can be recognised, taking into account its underlying structure, and rock erodibility resistance. Some plateaux are underlain by sedimentary and volcanic Mesozoic and Cenozoic strata, whilst other planation surfaces cut across the metasedimentary and metavolcanic rocks of the Neoproterozoic basement. The flattened surfaces of this landscape correspond to regional-scale erosion surfaces (Merla and Minucci 1938; Beyth 1972; Coltorti et al. 2007) and can be correlated, from the stratigraphic point of view, with major regional-, and sometimes continental-scale (Sandler et al. 2012) unconformities.

8.2.1.1 The Plateaux

The lower plateau ($\sim$2,200–2,300 m a.s.l.) is built on a laterised unit that tops the nearly horizontally bedded Adigrat Sandstone Formation. The laterite was formed on a palaeosurface that truncates the upper Adigrat formation. This surface corresponds to the second oldest regional-scale planation surface, described by Coltorti et al. (2007) as the Late Triassic planation.

Fig. 8.3 Columnar basalt necks and trachyte and phonolite plugs intruding the basaltic trap flows

The upper plateau (>2,300 m a.s.l.) is built on horizontal thick upper basaltic layers of the Trap sequence, whose deposition followed the major three-phase uplift and rifting initiated ~40–30 Ma. The soils of the plateau, developed over a detrital layer, are of the Ap-Bw type (Inceptisols) or have a marked vertic character. This detrital layer, which contains fine material from the weathering of basalt rocks and trachyte gravels from later, Plio-Pleistocene volcanic activity, would correspond to a Quaternary period of subaerial erosion processes.

Differences in erodibility between the lower, laterised Adigrat Sandstone and the upper basaltic Trap layer, explain the steep-walled cliffs of the Lower Plateau and faster retreat of the slope walls of the Upper Plateau.

Cliff retreat, like the drainage system draining to the Weri River (see Sect. 8.2.3), follows the NNE–SSW structures (striking faults affecting the Neoproterozoic basement) and could have been triggered by the inception of the contemporary drainage network in the area, initiated after the last (Plio-Pleistocene) phase of regional uplift and westward tilting of the Ethiopian Plateau.

Active slope erosion processes and landforms, such as gullies, debris flows and rock falls in the laterised Adigrat sandstone unit, alluvial fans, past large landslide scars and extensive alluvial Quaternary fans at the headwaters of the Teway Ruba shet (stream or river in the local language), indicate considerable slope erosion during the Quaternary.

Debris-covered relict slopes observed in the road that connects Adwa and Indaba T'shama (Fig. 8.4) indicate that slope retreat was not continuous in time. Tripartite slopes or talus flatirons are common in semiarid and arid environments and are thought to be an indicator of climatic (Gerson 1982; Sancho et al. 1988) or anthropic changes (Everard 1964; Gutiérrez and Peña 1998), or may be the result of substantial gully erosion at the talus base (Boroda et al. 2013). The sequence of three talus flatirons could suggests that during the Early–Middle Pleistocene, up to three phases of wetter/drier climatic condition (accumulation/erosion activity) may have existed, when plateau cliff retreat would have been more active than the one observed during the Holocene.

8.2.1.2 The Erosion Surfaces

The best preserved evidence of the continental-scale rapid exhumation of the Neoproterozoic basement is located southwards of Adwa. This exhumation was estimated at ~2 km/Ma (Ghebreab 1999) and resulted in the origin of a levelled and peneplained surface. Remnants of this surface constitute the upper summit interfluvial surfaces (~1,900–2,150 m a.s.l.) of the *shet's* draining towards the Weri River (Fig. 8.1). The preservation of this surface, defined by Coltorti et al. (2007) in northern Ethiopia as the PS1 (Pre-Ordovician planation surface), shows a strong lithological and structural control.

In the Adwa area, it is best preserved in Work Amba (Figs. 8.1 and 8.5) cut across the carbonaceous rocks of the Tambien formation. This planation surface was eventually covered by Late Palaeozoic and Mesozoic sedimentary sequences (Enticho, Edaga Arbi and Adigrat formations) and Tertiary basalts (Trap series) rocks. After the later Plio-Pleistocene uplift, this surface was exhumed as the cover rocks were eroded and subsequently affected by erosion,

Fig. 8.4 Tripartite slopes in the Adwa area

Fig. 8.5 Work Amba erosive surface cutting the carbonaceous rocks of the Tambien formation

resulting in a polygenetic surface. Quaternary denudation processes allowed for the development of karst landforms and well as the developed *terra rossa* soils that will be described in Sect. 8.3.1.

8.2.2 Quaternary Alluvial Terraces

In the Adwa region, the fluvial headwaters are located in the plateaux (on volcanic and sandstone rocks) and rivers flow southwards to the Weri River cutting the metamorphic rocks of the Neoproterozoic basement. Three major streams cut the Adwa region, namely the Teway Ruba Shet (1,101 km^2 in catchment area), the Kinetal Shet (262 km^2) and the Ira Shet (212 km^2), whose direction is largely controlled by the major structures trending NNW–SSE. In the southern part, the Weri River flows ENE–WSW cutting into the Precambrian structures and carving a spectacular gorge with a meandering pathway. Between September and June, the river discharge, including also the Weri, is greatly reduced, and rivers may even dry up. This river system is dissecting the Plio-Pleistocene volcanic materials, indicating that the current drainage pattern of the area was developed only after the most recent uplift and tilting of the Ethiopian Plateau (Adamson and Williams 1987; Gani et al. 2007). Furthermore, Said (1993) indicated that prior to the setting of the lower Trap series, the stream network must have been directed eastwards, towards the Red Sea, since there is no indication of the existence of any significant drainage to the Nile. The post-Pliocene landscape evolution has led to the exhumation of the polygenic Work Amba erosion surface and fluvial dissection. The most complete sequence of staircase alluvial fill terraces was found next to the bridge crossing the old Adwa–Mekele road at the Weri River (Fig. 8.6) The terrace levels can be divided into three groups (Table 8.1): (a) Weri group or older alluvium (T0 to T2); (b) Daba Tadis group (T3 to T6) and (c) the present floodplain (T7). This sequence of fill terraces is not continuous along the Weri River valley and varies in terms of their number and height above the current river bed. The youngest terraces show a higher preservation and can be traced all along the Weri valley.

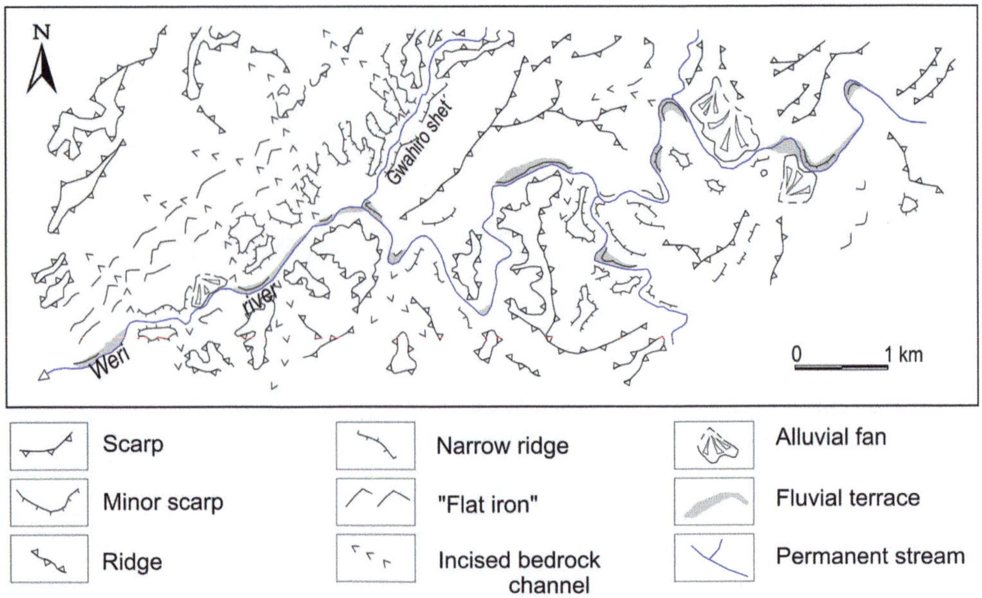

Fig. 8.6 Fill and strath terraces location along the Weri River between the Gwahiro and the Ts'edyia shet's reaches

Table 8.1 Topographical and sedimentological characteristics of fill and strath terraces of the Weri River between the Gwahiro and Ts'ediya reaches

Level		Alt. (m)	Grain size (cm) modal	D_{90}	Lithology	Other
	T7	+1	5–10	90	Volcanic and metavolcanic (95 %)	2 m in thickness
			14–32		Quartz (5 %)	
DABA TADIS group	T6	+3	1–3	19	Volcanic and metavolcanic (80 %)	90-cm-thick deposit with silty sand matrix (5YR). At top, 1.3-m-thick, medium- to coarse-laminated sand well sorted. Matrix colour 10YR5.5/3.5
			5–7		Quartz (15 %)	
					Schists (5 %)	
	T5	+10	1.5–5	16	Volcanic and metavolcanic (80 %)	4.2 m in thickness. Matrix colour 5YR
			6–7		Quartz (15 %)	
					Schist (5 %)	
	T4	+13/14	1–4	28	Volcanic and metavolcanic (85 %)	3.5 m in thickness
			7–12		Quartz (12 %)	
					Schist < 3 %	
	T3	+20/21	1.5–5	28	Volcanic and metavolcanic (85 %)	D_{99} boulder long axis of 62 cm
			8–15		Quartz (10 %)	
					Schist (5 %)	
WERI group	T2	+24	1.5–7	56	Metavolcanics (75 %)	<1 m in thickness
			9–14		Schist (20 %)	
					Quartz (5 %)	
	T1	+31	1.5–7	35	Metavolcanics (70 %)	3 m in thickness. D_{99} boulder long axis of 1.3 m
			12–15		Schist (20 %)	
					Quartz (10 %)	
	TO	+35/40				Strath terrace cutting Neoproterozoic bedrock

The alluvial deposits of the Weri group terraces (older alluvium) consist of gravels and boulders with metavolcanic (70 %), schist (20 %) and quartz (10 %) lithologies. The fill terraces of the Daba Tadis group (younger alluvium) show gravels and boulders with the dominant lithology of metavolcanic and volcanic (85 %) and quartz (10–15 %), and occasional schist (<5 %). The grain size of the coarser fraction (D_{90}) for the Weri group terraces (56–35 cm) is larger than those in the Daba Tadis terraces (16–28 cm) which, adopting the equation of Williams (1983) for critical stream power, implies a decrease of flow energy by a factor 2.5. The alluvium of the current floodplain consists of even higher proportion of metavolcanic and volcanic gravels (95 %) and quartz (5 %) and lacks schist gravels. In the floodplain alluvium, the size of D_{90} is similar to the younger Daba Tadis terrace group indicating a similar flow competence. The chronology of these stream terraces is unknown, although we can speculate that the Weri group terraces may correspond to the Early to Middle Pleistocene and the Daba Tadis group to the Late Pleistocene and Holocene.

8.3 Geomorphological Features Associated with Quaternary Surface Processes

The Plio-Pleistocene regional uplift and westward tilting have enhanced denudation processes that exhumed ancient landforms and deepened river valleys connected to the Weri River. The most characteristic landforms resulting from Quaternary surface processes are those produced by weathering of limestones (karst and travertines), granites (tors and weathering pits—gnammas), as well as alluvial deposits accumulated on valley flat bottoms.

8.3.1 Karstic Landforms and Travertines

As previously indicated, the Work Amba erosion surface is an exhumed planation surface that cuts the Neoproterozoic carbonated Tambien Group formations and the May Kinetal granitoid intrusion. It resulted from multiple erosion phases during the Early Palaeozoic and Mesozoic and was covered, locally, by Cenozoic rocks. As a consequence of the regional uplift that took place ~30–40 Ma ago, the erosion of cover rocks exhumed the old peneplain and led to the development of a new pediment surface. The latter is slightly tilted southwestwards and was likely covered by sediments during the Pliocene–Pleistocene, although currently the only indirect evidence is the exposed relict landforms typical of a mantled karst. Soils display an Ap-Bt-R or Bt-R profile type are reddish in colour (particularly the Bt horizon) with shallow intermittent or broken diagnostic horizons of lithic nature. Soil loss due to surface processes is active even in gently slopping areas, and northwards, closer to the plateau foot area, A-R-type profiles were identified. The area closer to the contact with the depression carved into the May Kinetal granitoid body presents the thickest soil depth (>1 m, vs. <0.75 m of soil depth closer to the plateau contact area), developed over the May Kinetal limestones.

Doline-shaped depressions were identified on the Work Amba surface (Fig. 8.1) with ones most noticeable in size present in the vicinity of the faulted granitoid contact area. A major accumulation of fine subaerial dissolution material, fine sand detrital limestone material and sediments resulting from the erosion of the surrounding *terra rossa* have evolved here under conditions different from the rest of the surrounding area. Soils developed in these sediment-filled karstic depressions are rich in smectite clays, responsible for the presence of common soil cracks during the dry season. As in the case of the Vertisols developed over the weathered volcanic materials of the Trap volcanic plateaux, soils in these depressions are characterised by pH ranging between 7 and 8, with a good supply of bases and water storage capacity which favour an intense agricultural use.

Other assemblages from epikarst processes due to dissolution beneath a soil and vegetation cover are hohlkarren microforms (according to Bögli 1978 classification). The hohlkarren micromorphology is characterised by groove and pitted morphologies formed by corrosion of water which percolates from pedogenetic cover and is enriched with biogenic CO_2 (Sweeting 1972). Other karren types are also present on bare rock surfaces, such as rillenkarren (Fig. 8.7) characterised by the presence of small-scale linear solution flutes on steeply sloping limestone surfaces, resulting from a combination of chemical (dissolution), biological and physical (raindrop impact) processes. Their trace disappears at the soil-ground level. At some karren outcrops, the current ground surface is ~20–30 cm below the termini of the rillenkarren bottom level, indicating a period of soil erosion following the karren development, which implies that these dissolution processes are not active in the present time.

The development of travertines in the eastern rim of the Work Amba surface contributes to the evidence of past active karst processes. The travertines show a cascade morphology (Pentecost and Viles 1994; Pentecost 2005) formed at spring outlets with vertical walls along resistant carbonated rock layers of the Tambien formation. The accretionary nature of cascades indicates a slow moving flow that prevents them from being eroded. The wall travertines are relict, although lower level springs are still an important source of water supply for the local population. Sedimentologically, they are composed by porous calcic carbonate indicating an in situ cemented plant material with a highly porous framework which can be classified as phytoherm framestone facies type according to Pedley's (1990)

Fig. 8.7 Dissolution morphologies (*rillenkarren*) on May Kinetal Limestone rock surfaces

classification. The travertines yielded a radiocarbon date of 34,500 ± 1,100 years BP (author's data). Lake carbonaceous material from a southern Egypt area was found to yield a radiocarbon dating of 36,620 ± 1,100 years BP (Brookes 1993). Brookes (1993) suggested that these were formed under much wetter conditions and from an increase in rainfall in the Ethiopian Highlands between 20 and 40 kyears. This wetter condition period would correlate in time with the May Kinetal travertines.

8.3.2 Landform and Processes on Granite Bedrock

Other singular landforms are tors and gnammas produced on granite rocks. In the study area, tors of about 10 m high are mainly located near the contact between the limestone and the granites, where other types of contact metamorphic rocks (hornstones) were formed (Fig. 8.8). During the cooling phase of the granite emplacement, a well-developed network of transversal vertical cracks was produced facilitating subsequent hydrothermal circulation and mineral precipitation (Twidale 1982). Other sets of fractures were formed as the granite pluton was subject to pressure release close to the surface and these are subhorizontal joins following the shape of the land surface. Subsequent weathering along these joints and fractures has resulted in the formation of mushroom-shaped tors characteristic for the May Kinetal area. In addition, on horizontal granites surface there are ubiquitous gnammas, developed as a result of chemical processes exacerbated by wetting and drying processes during the rainy season.

8.3.3 Flat Infilled Valleys

Holocene alluvial deposits in the valley bottoms (Fig. 8.9) provide a unique opportunity for palaeoenvironmental reconstructions of the area considering both the human disturbance and other external environmental stresses such as climatic variability. Machado et al. (1998) studied the valley infill stratigraphy within two drainage basins, the Teway Ruba (Adwa and Wechi sites) and the Gwahiro rivers (May Kinetal site). This study provided a high-resolution record of the main landscape stability and instability stages during the past 4,000 year: (1) three major wetter periods (ca. 4,000–3,500 year B.P., 2,500–1,500 year B.P. and 1,000–960 year B.P.), during which soils were formed and (2) two degradation episodes (ca. 3,500–2,500 year B.P. and 1,500–1,000 year B.P.), during which there was an increase of sediment yield from the slopes into the valleys. For the past 1,000 years, and in particular since the early seventeenth century, stratigraphic records together with historic chronicles suggest an increasing aridity. Although some of these stages seem to have affected the area of Axum and the Tigray province in particular, other stages show a good regional correlation not only with palaeoenvironmental proxy data from different areas within Ethiopia, but also from other eastern and north-eastern African regions.

The oldest preserved stability phase is represented by the buried Vertisol at the Adwa site (unit 3; Machado et al. 1998) which was correlated with one from the confluence of the Midmar and the Teway Ruba rivers and dated to 3,510 ± 80 ^{14}C year B.P. This date is consistent with a Vertisol described by Semmel (1971) in the Ethiopian

Fig. 8.8 Tors generated by granite weathering along joints and fractures (May Kinetal intrusive granitoid)

Fig. 8.9 Holocene sedimentary valley fill sequence in the May Kinetal site

Central Highlands and dated between 3,865 ± 105 ^{14}C year B.P. and 3,670 ± 105 ^{14}C year B.P. The existence of wetter conditions within Tigray and the Ethiopian Highlands can also be inferred from the Nile river-level records registered at the Roda Nilometer. Between 1840 and 1770 BC, there are records of high waters (Butzer 1981) preceded by a period of 35 years with frequent floods (Said 1993). An increase in sediment yield from the slopes into the valleys is shown in a colluvial unit (unit 4) of the Adwa Site (Machado et al. 1998). The probably reworked orange pottery fragments collected at the base of the unit (3,747 ± 328 year B. P. TL), and the flecks of charcoal at the top (2,230 ± 30 ^{14}C year B.P.) suggest that this drier phase may have persisted over a long time span.

A second wet phase is reflected in the formation of a Vertisol (Soil II, Machado et al. 1998), which was developed between 2,470 ± 30 ^{14}C year B.P. and 1,681 ± 169 year B. P. (TL). This phase may be related to other Vertisols formed around 2000 ^{14}C year B.P. in Sudan (Blokhuis 1993) and is consistent with a lacustrine phase of lake Hanlé-Dobi between 2,500 and 1,500 ^{14}C year B.P. (Gasse et al. 1980). A pollen study in the Lake Turkana (Kenya) by Umer et al. (1995) also detected higher values of arboreal pollen between 2,400 and 1,900 ^{14}C year B.P. In the Weri River (catchment area over 4,300 km^2), the analysis of slack-water flood deposits shows at least four major extreme floods, with the uppermost flood unit yielding an age of 1,790 ± 60 ^{14}C year B.P. Hydroclimatically, these flood events are related to anomalous rainfalls in both amount and intensity during the rainy season, rather than exceptional heavy storms over a small part of the basin. The soil formation and the occurrence of extreme events can be interpreted as resulting from an increase in rainfall. Historically, this period coincides with the Early Axumite period (Butzer 1981; Sutton 1989). At the archaeological site of the Staele Park, Axum, Butzer (1981) described colluvial and reworked cultural debris of Middle to Late Axumite age (fourth to eighth century AD) and interpreted this as resulting from slope instability due to over-intensive land use. A clear degradational stage was also found in the valley infill deposits at the study area (May Kinetal site), with evidence of lateral debris (Machado et al. 1998, unit 4: 1,490 ± 60 ^{14}C year B.P.) due to intensive slope erosion processes. A period of landscape recovery and stability, probably the last major one, enabled the formation of a well-developed Vertisol (Soil III: 980 ± 50 ^{14}C year B. P. and 963 ± 60 ^{14}C year B.P.; Machado et al. 1998). In a sequence of four flood deposit units in the Weri River, two individual events were dated with charcoal and cultural charcoal yielding ages of 970 ± 60 ^{14}C year B.P. and 980 ± 50 ^{14}C year B.P.

For the past 1,000 years, the valley infill deposits show two major changes in the character of the sequences: (a) an anomalous increase in the intensity and length of aggradation phases and (b) shorter periods of soil formation giving rise to poorly developed soils. The aggradational sequence for the past 1,000 years is up to 13 m in thickness, in contrast to only 4 m of aggradation for the previous 3,000 years. These deposits differ from the previous aggradational phases in their fluvial character, with gravel bars and sand sheets. These can be interpreted not only as a result of an increase in fluvial competence (flash floods), transporting large amounts of sediments through the valley systems, but may also reflect severe degradation of slopes producing high sediment yield. An example of this phenomenon has been reported by Alvares (1540) on 30 April AD 1520. During his travel from Massawa to Axum, Alvares witnessed a flash flood during which the previously dry streams (arroyos), reached a high-water level in only 2 h.

Two buried soils (Soils IV and V; Machado et al. 1998) were developed in fine texture sediments over a time span shorter than 100 years. In Soil V, a charcoal sample collected at the base of the sediment yielded an age of 480 ± 30 ^{14}C year B.P., whereas the overlaying unit contains charcoal with a radiocarbon age of 380 ± 30 ^{14}C year B.P. These poorly developed soils suggest that the stability of the system is irregular and short lived, either as a result of lower soil moisture conditions or most probably due to intensive land use.

A general trend towards drier climatic conditions since the middle sixteenth century can be inferred from the chronology of drought and famine (Pankhurst 1985; Webb and von Braun 1994) and landscape stability recorded in the alluvial fills. The major change in the environmental conditions of the area may have been triggered by the drought of early seventeenth century. Drier weather conditions during the seventeenth and eighteenth centuries were registered not only in East Africa, but also in the entire east–west extent of the Sahel zone (Nicholson 1980). This drought period correlates with the accumulation of ca. 150-cm-thick gravelly colluvial deposit which, together with the pottery fragments, contains also numerous boulders with marks attributed to iron-tipped plough, indicating a strong human intervention. This evidence points out to progressive occupation of marginal lands and steep slopes, for both cultivation and grazing.

The present degradation period, already perceptible in the aerial photographs from 1963, was accelerated through a combination of increased variability in annual rainfall, and an increase in human pressure on the land. The active gravel and boulder bars, deposited within a braided system, reveal the present fluvial dynamics of the area, related to the ephemeral nature of river discharges and frequent flash floods. The sharp-peaked and short-lived hydrographs may also be related to the presence of a thin patchy soil and scarce vegetation on slopes. A minimum average erosion rate of 21 tons/ha/year for the period from AD 1985 to 1993 has been estimated for a 6.7 km^2 watershed located in the northern part of the study area by Machado et al. (1995, 2001), based on the filling of an irrigation dam.

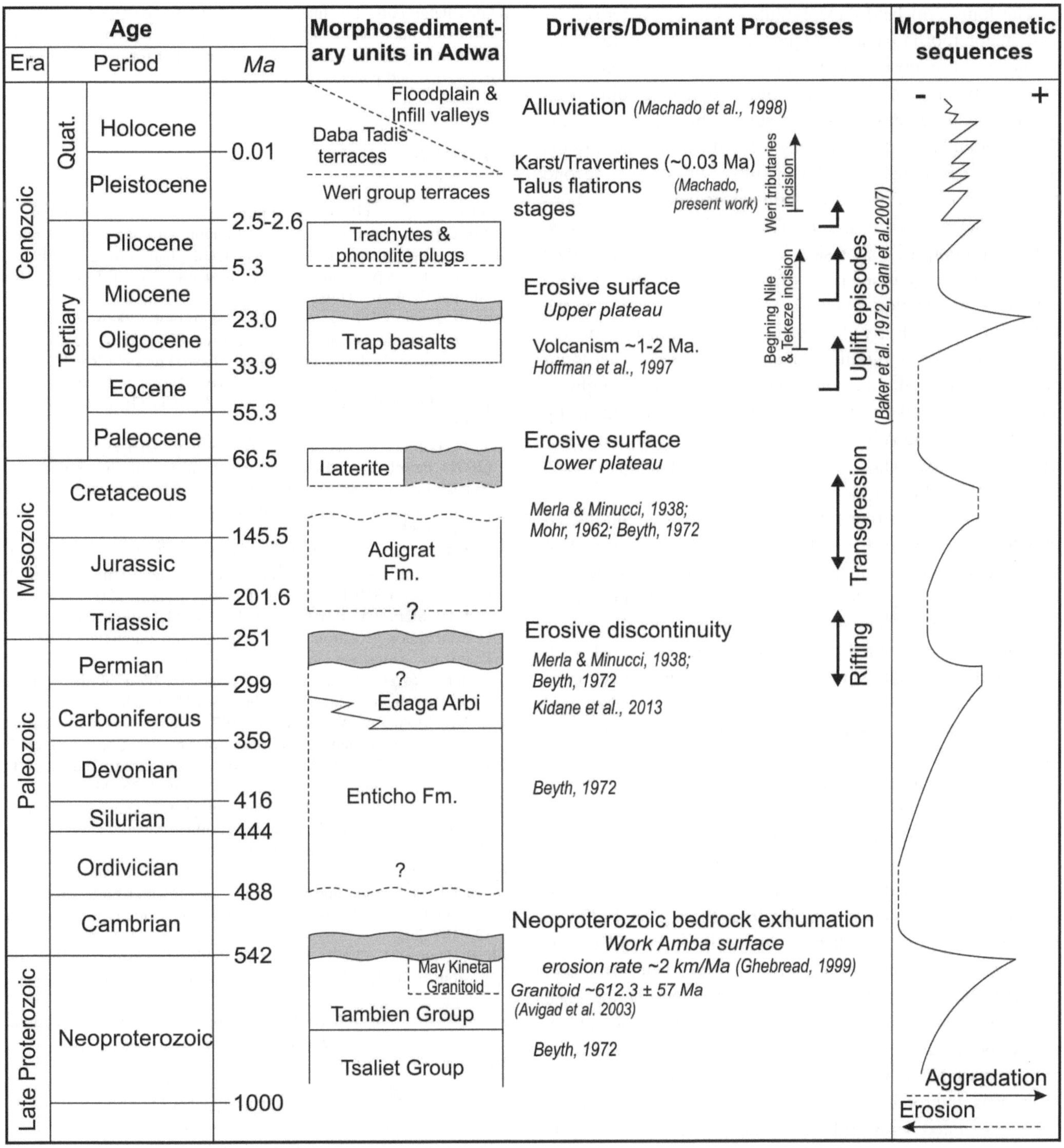

Fig. 8.10 Landscape evolution in the Adwa district area since Late Proterozoic: drivers and major morphogenetic sequences

8.4 Conclusions

The landscape of the Adwa region has an evident lithological and structural control and shows a heterogeneous mosaic of ancient land surfaces and landforms preserving a unique landscape history of the Ethiopian Highlands. The region was characterised by well-marked periods of basement uplift alternating with long-periods of tectonic stability and regional erosion with formation of pedimentation surfaces (Fig. 8.10). The climate control is more evident during the stability periods, leading to the formation of palaeosols, rock weathering, travertine formation and alluviation on floodplains and infill valleys.

The long-term landscape evolution was characterised by four major morphogenetic sequences since the Late Proterozoic era, composed by cycles of aggradation sequences followed by erosion phases. The oldest sequence culminated in a regionally extensive erosion surface bevelled across the Precambrian basement, with an excellent development in the southern Adwa district (Work Amba surface). A second phase of erosion provides the contact between the Carboniferous glacial deposits and the Triassic–Jurassic sandstones, but its geomorphic expression is limited in both the Adwa area and in many other parts of northern Ethiopia. A third extensive aggradation–planation cycle took place during the Late Cretaceous-Early Cenozoic and culminated in the development of a plinthic palaeosol (laterite). The topographical uplift of this non-orogenic region is thought to have been initiated ~30–40 Ma, causally linked with rifting resulting from the Afar mantle circulation and plume activity, within three major phases of volcanic activity of Late Eocene, Mid-Miocene, and Plio-Pleistocene age.

During the Quaternary, the river system was entrenched into the Plio-Pleistocene volcanic materials indicating that the current drainage pattern of the area draining to the Nile was developed only after the most recent uplift and tilting of the Ethiopian Plateau. The Quaternary morphogenetic sequences are controlled by changes in the base level, with a dominant climatic signal. Changes in base level controlled the terrace formation in the Weri River valley and scarp retreat with evidence of multiple phases of talus flatirons at the slopes. The most complete sequence of staircase alluvial fill terraces shows up to seven terrace treads (Weri group and Daba Tadis group), composed by volcanic, metavolcanic, schist and quartz. The grain size of the coarser fraction (D_{90}) decreases in the youngest terraces indicating a decrease of flow energy by a factor of 2.5. Surface processes in relation with the regional climate favoured limestone weathering during more humid conditions, leading to the development of karst landforms and travertine accumulation at spring outlets. During the Late Pleistocene and Holocene, alluvial units were deposited in valley bottoms and these provide a detailed record of climate variability and human disturbance. Landscape stability was interpreted during three wetter periods, having been associated with soil development (dated ca. 4,000–3,500 year B.P., 2,500–1,500 year B.P., and 1,000–960 year B.P.). Subsequent degradation episodes (ca. 3500–2,500 year B.P. and 1,500–1,000 year B.P.) resulted in an increase of sediment yield from the slopes into the valleys. For the past 1,000 years, and in particular since the early seventeenth century, stratigraphic records together with historic chronicles suggest an increasing aridity.

Acknowledgments I would like to express my very great appreciation to Professor Paolo Billi, not only for his valuable constructive suggestions during the planning of this chapter, but for his willingness to give his time and patience so generously.

References

Abdul-Haggag Y (1961) A contribution to the physiography of Northern Ethiopia. University of London, Athlone Press, London

Adamson D, Williams MAJ (1987) Geological setting of Pliocene rifting and deposition in the Afar depression of Ethiopia. Jour Human Evol 16:597–610

Alene M, Jenkin GRT, Leng MJ, Darbyshire FDP (2006) The Tambien group, Ethiopia: an early Cryogenian (ca. 800–735 Ma) Neoproterozoic sequence in the Arabian-Nubian shield. Precambr Res 147:79–99

Alvares F (1540) Verdadeira informação das Terras do Prestes João das Indias. Original re-edition of 1889, Agência Geral das Colónias, Lisboa

Asrat A, Barbery P, Ludden JN, Reisberg L, Gleizes G, Ayalew P (2004) Petrology and isotope geochemistry of the Pan African Negash Pluton, Northern Ethiopia: Mafic-felsic magma interactions during the construction of shallow-level calc-alkaline plutons. Jour Petrol 45:1147–1175

Avigad D, Kolodner K, McWilliams M, Persing H, Weissbrod T (2003) Origin of northern Gondwana Cambrian sandstone revealed by detrital zircon SHRIMP dating. Geology 31:227–230

Avigad D, Stern RJ, Beyth M, Miller N, McWilliams MO (2007) Detrital zircon geochronology of Cryogenian diamictites and Lower Paleozoic sandstone in Ethiopia (Tigrai): age constraints on Neoproterozoic glaciation and crustal evolution of the southern Arabian-Nubian shield. Precambr Res 154:88–106

Baker BH, Mohr PA, Williams LAJ (1972) Geology of the eastern rift system of Africa. Geological Society of America, Special paper 136

Beyth M (1972) The Paleozoic-mesozoic sedimentary basin of the Mekele outlier-northern Ethiopia. AAPG 56:2426–2439

Beyth M, Heimann A (1999) The youngest igneous event in the crystalline basement of the Arabian-Nubian Shield, Timna Igneous Complex. Isr. J. Earth Sci. 48:113–120

Beyth M, Avigad D, Wetzel HU, Matthews A, Berhe SM (2003) Crustal exhumation and indications for snowball Earth in the East African Orogen: north Ethiopia and east Eritrea. Precambr Res 123:187–201

Billi P, Dramis F (2002) Geomorphological investigation on gully erosion in the Rift Valley and the northern highlands of Ethiopia. Catena 50:353–368

Beyene A, Abdelsalam MG (2005) Tectonics of the Afar depression: a review and synthesis. J Afr Earth Stud 41:41–59

Blandford WT (1870) Observations on the geology and zoology of Abyssinia, made during the progress of the British expedition to that country in 1867–1868. Macmillan, London

Blokhuis WA (1993) Vertisols in the central clay plain of the Sudan. Wageningen Agricultural University, Wageningen

Bögli A (1978) Karsthydrographie und physische Spelëologie: Springer Verlag, Berlin

Boroda R, Matmon A, Amit R, Haviv I, Porat N, Team ASTER, Rood D, Eyal Y, Enzel Y (2013) Long-term talus flatirons formation in the hyperarid northeastern Negev, Israel. Quat Res 79(2):256–267

Bosellini A, Russo A, Fantozzi PL, Assefa G, Tadesse S (1997) The Mesozoic succession of the Mekelle outlier (Tigre Province, Ethiopia). Memore di Scienze Geologiche 49:95–116

Brookes IA (1993) Geomorphology and Quaternary geology of the Dakhla oasis region, Egypt. Quat Sci Rev 12:529–552

Burke K (1996) The African plate. S Afr J Geol 99:339–409

Butzer KW (1981) Rise and fall of Axum, Ethiopia: a geo-archaeological interpretation. Am Antiq 46(3):471–495

Coltorti M, Dramis F, Ollier CD (2007) Planation surfaces in Northern Ethiopia. Geomorphology 89:287–296

Davis PM, Slack PD (2002) The uppermost mantle beneath the Kenya dome and relation to melting, rifting and uplift in East Africa. Geophys Res Lett 29(7):21-1–21-4. doi:10.1029/2001GL013676

Dow DB, Beyth M, Hailu T (1971) Palaeozoic glacial rocks recently observed in northern Ethiopia. Geol Mag 108:53–60

Dramis F, Umer M, Calderoni G, Haile M (2003) Holocene climate phases from buried soils in Tigray (northern Ethiopia): comparison with the lake level fluctuations in the Main Ethiopian Rift. Quatern Res 60:274–283

Everard CE (1964) Climatic change and man as factors in the evolution of slopes. Geogr J 130(1):65–69

Ebinger C, Scholz CA (2012) Continental rift basins: An East African perspective. In C Busby A Azor (eds) Tectonics of Sedimentary Basins: Recent Advances. John Wiley & Sons, Chichester, 183–208

Feoli E, Gallizia L, Woldu Z (2002) Evaluation of environmental degradation in northern Ethiopia using GIS to integrate vegetation, geomorphological, erosion and socio-economic factors. Agric Ecosyst Environ 91(1–3):313–325

Frankl A, Poesen J, De Dapper M, Deckers J, Haile M, Nyssen J (2012) Gully head retreat rates in the semiarid highlands of North Ethiopia. Geomorphology 173–174:185–195

Gani NDS, Gani MR, Abdelsalam MG (2007) Blue Nile incision on the Ethiopian Plateau: pulsed plateau growth, pliocene uplift, and hominin evolution. GSA Today 17(9):4–11. doi:10.1130/GSAT01709A.1

Garland CR (1980) Geology of the Adigrat area. Memor No. 1. Ministry of Mines and Energy, Addis Ababa

Gasse F, Rognon P, Street FA (1980) Quaternary history of the Afar and Ethiopian rift lakes. In: Williams MAJ, Faure H (eds) The Sahara and the Nile. AA Balkema, Rotterdam, pp 361–400

Gerson R (1982) Talus relict in deserts: a key to major climatic fluctuations. Isr J Earth Sci 31:123–132

Ghebread W (1999) Tectono-metamorphic history of the Neoproterozoic rocks in eastern Eritrea. Precambr Res 98:83–105

Gutiérrez Elorza M, Peña JL (1998) Geomorphology and Late Holocene climatic change in northeastern Spain. Geomorphology 23:205–217

Hagos M, Koeberl C, Kabeto K, Koller F (2010) Geochemical characteristics of the alkaline basalts and phonolite-trachyte plugs of the Axum area, northern Ethiopia. Austrian J Earth Sci 103(2): 153–170

Hofmann C, Courtillot V, Feraud G, Rochette P, Yirgu G, Ketefo E, Pik R (1997) Timing of the Ethiopian flood basalt event and implications of Plume birth and global change. Nature 389:838–841

Hurni H (1983) Soil erosion and soil formation in agricultural ecosystems: Ethiopia and northern Thailand. Mt Res Dev 3 (2):131–142

Hurni H (1986) Soil conservation in Ethiopia. Community forests and Soil Conservation Development Department Ministry of Agriculture, Addis Ababa

Kazmin V (1972) The Geology of Ethiopia. Ethiopian Institute of Geological Surveys. Note N° 821-051-12

Kidane T, Bachtadse V, Alene M, Kirscher K (2013) Palaeomagnetism of Palaeozoic glacial sediments of Northern Ethiopia: a contribution towards African Permian palaeogeography. Geophys J Int, First online October 3, 2013; doi:10.1093/gjl/ggt336

Kirwan LP (1972) The Christian topography and the kingdom of Axum. Geogr J 138(2):166–177

Machado MJ, Pérez-González A, Benito G (1995) Reconstrución de ambientes naturales y antrópicos en la región de Axum (Tigray, Norte de Etiopía). In Aleixandre T and Pérez-González A (eds) Reconstrucción de paleoambientes y cambios climáticos durante el Cuaternário. CSIC Monograph 3:163–174

Machado MJ, Pérez-González A, Benito G (1998) Palaeoenvironmental changes during the last 4000 years in the Tigray, northern Ethiopia. Quatern Res 48:312–321

Machado MJ, Pérez-González A, Benito G (2001) The role of geomorphology and geoarchaeology in understanding land degradation. In: Feoli F, Pottier D and Woldu Z (eds) Sustainable development of dryland areas of East Africa. EC Directorate General XII, Science, Research and Development, pp 63–67

Merla G, Minucci E (1938) Missione geológica del Tigray. Rendiconti della Reale Academia d'Italia. Centro Studi per l'Africa Orientale Italiana 3:1–362

Merla G, Abbate E, Azzaroli A, Bruni P, Canuti P, Fazzuoli M, Sauri M, Tacconi P (1979) A Geological Map of Ethiopia and Somalia (1973), 1;2,000,000 and Comment. Consiglio Nazionale delle Ricerche. Univeristy of Florence, Italy

Mohr PA (1962) The Geology of Ethiopia. University Collection Addis Ababa Press, Ethiopia

Miller NR, Alene M, Sacchi R, Stern R, Conti A, Kröner A, Zuppi G (2003) Significance of the Tambien Group (Tigre. N. Ethiopia) for Snowball Earth Events in the Arabian–Nubian Shield. Precambrian Research 121:263–283

Miller NR, Stern RJ, Avigad D, Beyth M, Schilman B (2009) Cryogenian slate-carbonate sequences of the Tambien Group, Northern Ethiopia. Precambr Res 170:129–156

Nicholson SE (1980) Saharan climates in historic times. In: Williams MAJ, Faure H (eds) The Sahara and the Nile. AA Balkema, Rotterdam, pp 173–200

Nyssen J, Haile M, Poesen J, Deckers J, Moeyersons J (2001) Removal of rock fragments and its effect on soil loss and crop yield, Tigray, Ethiopia. Soil Use Manag 17:179–187

Nyssen J, Poesen J, Moeyersons J, Deckers J, Haile M, Lang A (2004) Human impact on the environment in the Ethiopian and Eritrean highlands—a state of the art. Earth-Sci Rev 64:273–320

Pankhurst R (1985) The History of Famine and Epidemics in Ethiopia Prior to the Twentieth Century. Relief and Rehabilitation Commission, Addis Ababa

Pedley HM (1990) Classification and environmental models of cool freshwater tufas. Sed Geol 68:143–154

Pentecost A, Viles H (1994) A review and reassessment of travertine classification. Géog Phys Quatern 48(3):305–314

Pentecost A (2005) Travertine. Springer, Berlin

Pik R, Marty B, Carignan J, Lave J (2003) Stability of the upper Nile drainage network (Ethiopia) deduces from (U-Th)/he thermochronometry: implication of uplift and erosion of the Afar plume dome. Earth and Planetary Science Letters 215:73–88

Roberts EM, Stevens NJ, O'Connor PMO, Dirks PHGM, Gottfried D, Clydes WC (2012) Initiation of the western branch of the East African Rift coeval with the eastern branch. Nat Geosci 5. doi:10.1038/NGEO1432

Ross ED (1922a) Early travellers in Abyssinia: part I. J R Afr Soc 21 (84):268–278

Ross ED (1922b) Early travellers in Abyssinia: part II. J R Afr Soc 22 (85):5–16

Said R (1993) The River Nile. Geology, hydrology and utilization. Pergamon Press, Oxford

Sancho C, Gutiérrez M, Peña JL (1988) A quantitative approach to scarp retreat starting from triangular slope facets, central Ebro Basin, Spain. Catena supplement 13:139–146

Sandler A, Teutsch N, Avigad D (2012) Sub-Cambrian pedogenesis recorded in weathering profiles of the Arabian Nubian shield. Sedimentology 59:1305–1320

Semmel A (1971) Zur junquartaren klima und Reliefentwicklung in der Danakilwuste und ihren westlichen Randgebieten. Erdkunde 25:199–209

Sutton JEG (1989) The history and chronology. In: Munro Hay SC (ed) Excavations at Aksum: memoirs of the British Institute in Eastern Africa, No. 10, pp 7–26

Sweeting MM (1972) Karst Landforms. MacMillan, London

Tadesse T, Suzuki K, Hoshino M (1997) Chemical Th–U total Pb isochron age of zircon from the Mareb Granite in northern Ethiopia. J Earth Planet Sci Negoya Univ 44:21–27

Tadesse T, Hoshino M, Sawada Y (1999) Geochemistry of low-grade metavolcanic rocks from the Pan-African of the Axum Area, northern Ethiopia. Precambr Res 99:101–124

Tadesse T, Hoshino M, Suzuki K, Iisumi S (2000) Sm-Nd, Rb-Sr, and Th-U-Pb zircon ages of syn- and post-tectonic granitoids from the Axum area of northern Ethiopia. J Afr Earth Sci 30:313–327

Twidale CR (1982) Granite Landforms. Elsevier, Amsterdam

Uhlig S, Yiman B, Crummey D, Goldenberg G, Marrasini P, Aregay MW, Wagner E (2003) Encyclopaedia Aethiopica, vol. 1: A-C, Harrassowitz Verlag, Wiesbaden

Umer M, Bonnefille R, Johnson TC (1995) Pollen and isotopic records in late Holocene sediments from Lake Turkana, Kenya. Palaeogeography Palaeoclimatoogy Palaeoecology 119:371–383

Virgo KJ, Munro RN (1978) Soil and erosion features on the Central Plateau region of Tigrai, Ethiopia. Geoderma 20:131–157

Webb P, von Braun J (1994) Famine and food security in Ethiopia: lessons for Africa. Wiley, Chichester

Williams GP (1983) Paleohydrological methods and some examples from Swedish fluvial environment, I—cobble and boulder deposits. Geogr Ann 65 A:227–243

The Amba Landscape of the Ethiopian Highlands, Shaped by Rockfall

9

J. Nyssen, J. Moeyersons, J. Deckers, Mitiku Haile, and J. Poesen

Abstract

Cliff retreat occurs on the *ambas* or structurally determined stepped mountains of the northern Ethiopian highlands. This chapter describes the rock fragment detachment from cliffs by rockfall, quantifies its annual rate and identifies factors controlling rock fragment movement on the scree slopes. It further presents a conceptual model explaining rock fragment cover at the soil surface in these landscapes. In the May Zegzeg catchment (Dogu'a Tembien district, Tigray), rockfall from cliffs and rock fragment movement on debris slopes by run-off and livestock trampling were monitored over a 4-year period (1998–2001). Rockfall and rock fragment transport mainly induced by livestock trampling appear to be important geomorphic processes. Along a 1500 m long section of the Amba Aradam sandstone cliff, at least 80 t of rocks are detached yearly and fall over a mean vertical distance of 24 m resulting in a mean annual cliff retreat rate of 0.37 mm year^{-1}. Yearly unit rock fragment transport rates on scree slopes ranged between 23.1 and 37.9 kg m^{-1} year^{-1}. This process is virtually stopped when exclosures are established. A conceptual model indicates that besides rockfall from cliffs and argillipedoturbation, all factors and processes of rock fragment redistribution in the study area are of anthropogenic origin.

Keywords

Debris slope • Livestock trampling • Rock fragment redistribution • Subhorizontal structural relief

9.1 Introduction

The subhorizontal geological formations of the Ethiopian highlands have been epirogenically uplifted over the last 25 million years. This has led to important incision and cliff retreat. Taking into account the high rates of other mass movements (landsliding, sheet and rill erosion), fallen material is rapidly removed and cliff retreat processes are maintained. The process immediately following rockfall which occurs downslope is the further removal of rock fragments by rolling, mainly induced by animal trampling.

Previous studies in the northern Ethiopian highlands (Nyssen et al. 2000, 2002a; Moeyersons et al. 2006a) have also analysed the origin of the extensive rock fragment covers. The rock fragment cover (R_C) of Vertisols and soils with vertic properties is clearly a result of swell–shrink action in these soils. The rapid appearance of new rock fragments at the surface after field clearing has been related to the active polygonal structures of the Vertisols. Rock fragments, appearing yearly at the surface of Vertisols, have been shown to belong lithologically to deposits underlying the Vertisols. Given this stratigraphical situation and the activity of Vertisols in the study area, the rock fragment

J. Nyssen (✉)
Department of Geography, Ghent University, Krijgslaan 281 (S8), 9000 Ghent, Belgium
e-mail: jan.nyssen@ugent.be

J. Moeyersons
Royal Museum for Central Africa, 3080 Tervuren, Belgium

J. Deckers · J. Poesen
Department of Earth and Environmental Sciences, KU Leuven, Celestijnenlaan 200E, 3001 Heverlee, Belgium

M. Haile
Department of Land Resource Management and Environmental Protection, Mekelle University, P.O. Box 231, Mekelle, Ethiopia

P. Billi (ed.), *Landscapes and Landforms of Ethiopia*, World Geomorphological Landscapes,
DOI 10.1007/978-94-017-8026-1_9, © Springer Science+Business Media Dordrecht 2015

covers are thought to be squeezed up as a consequence of argillipedoturbation accompanying swell–shrink cycles in the vertic horizon.

Another study in the northern Ethiopian highlands (Nyssen et al. 2002b) showed that the balance between lateral and vertical movements of rock fragments controls the spatial distribution of rock fragment cover. Vertical supply of rock fragments to the soil surface is caused by (1) tillage-induced kinetic sieving, bringing preferentially large rock fragments (>7.5 cm) to the surface, even in the case of continuous fine sediment deposition, (2) argillipedoturbation in Vertisol areas and (3) selective run-off erosion and the development of erosion pavements. With respect to the lateral displacement processes, one can distinguish between (1) lateral transport over the soil surface by trampling and concentrated overland flow, especially on steep slopes and (2) rockfall from the cliffs. The latter two processes are studied in detail in this chapter, based on observations in the northern Ethiopian highlands.

Gardner (1970), Lee et al. (1994), Govers and Poesen (1998) and Oostwoud Wijdenes et al. (2001) showed that the above two processes play a significant role in slope development of some mountain areas. Ayalew and Yamagishi (2004) also insisted on the importance of rockfall, concurrently with landsliding in shaping of the Blue Nile gorge. Therefore, it was expected that these processes partly explain the presence of rock-fragment-rich layers in Skeletic Regosols on debris slopes reported from the study area (Nyssen et al. 2008). Hence, in this chapter, based on an earlier publication (Nyssen et al. 2006), we will (1) introduce the nature of stepped topography present so widely in northern Ethiopia, (2) quantify the annual rock fragment transport rate caused by these processes in the northern Ethiopian highlands, (3) analyse the factors controlling rock fragment movement on scree slopes and (4) develop a conceptual model explaining R_C at the soil surfaces based on major controlling factors.

9.2 The *Amba* Landscape

9.2.1 Geomorphic Context

In Ethiopia, most of peneplained Palaeozoic and Mesozoic sedimentary rocks have been concealed by Tertiary basaltic flows (Mohr 1963; Merla et al. 1979; Coltorti et al. 2007). The dome-like uplift of the Arabo-Ethiopian region started during the Oligocene and had two periods of intense tectonic activity: in the Miocene, about 25 million years ago and in the Plio-Pleistocene (Williams and Williams 1980). The elevation above sea level of the base of the basalt—about 500 m in 25 million years in the southernmost areas of Ethiopia and up to 2,500 m in the north—shows the importance of this uplift. All the rivers are deeply incised (Adamson and Williams 1980), and various lithologies are exposed giving rise to a typical structural subhorizontal relief with tabular, stepped landforms.

Major geological formations outcropping in the region comprise Precambrian metamorphic rocks at the base, the only formations that are strongly folded and faulted and that were subsequently truncated by erosion. Next, the subhorizontal formations comprise the Palaeozoic Enticho sandstone (of fluvioglacial origin) and the Edaga Arbi tillites (Bussert and Schrank 2007). The Mesozoicum is represented by the lower transgressional Adigrat sandstone (particularly cliff-forming—Fig. 9.1), overlain by alternating hard and soft Antalo limestone layers, some 400 m thick, and by Amba Aradam sandstone (Hutchinson and Engels 1970) (Fig. 9.2). Two series of Tertiary lava flows, separated by silicified lacustrine deposits (Merla and Minucci 1938; Arkin et al. 1971; Merla et al. 1979) cap these Mesozoic sedimentary rocks. As a consequence of the vertical succession of numerous subhorizontal sedimentary rock formations as well as Tertiary sills and lava flows, a subhorizontal structural landscape of scarps and dips (Young 1987; Young and Wray 2000) has come into existence. This stepped topography, resulting from the variable hardness of the different geological formations, is locally described as *amba* landscape.

9.2.2 Study Area

For this study, the May Zegzeg catchment (Dogu'a Tembien district), a 199-ha subcatchment of Geba and Tekezze (Fig. 9.3), situated at 2,280–2,650 m a.s.l., was selected as a representative catchment for the northern Ethiopian highlands. The subhorizontal geological formations in the catchment comprise layers of the Mesozoic Antalo limestone and Amba Aradam sandstone in the lower parts, and Tertiary basalt flows (traps) with silicified interbedded lake deposits in the upper parts. Quaternary deposits, consisting of alluvium, colluvium and tufa, are also found. The study area comprises a typical red-black soil catena (Driesen and Dudal 1991) on basalt and Calcisols and Calcaric Regosols at the foot of the limestone cliff (Nyssen et al. 2008).

The main rainy season (>80 % of total rainfall) extends from June to September, but is preceded by 3 months of dispersed small rains. Average annual rainfall is 750 mm (Nyssen et al. 2005). Intense rains falling on bare soils, which have already lost most of their natural vegetation by century-long action of human society, cause severe soil erosion. Erosive rains and the predominance of steep slopes induce a natural vulnerability of the study area to soil erosion, despite overall low soil erodibility due to high clay contents and high rock fragment content. Daily air–temperature variations are large (range of more than 20 °C, with 5 and 28 °C as extreme range values) during the dry season, without, however, dropping below freezing point.

Fig. 9.1 The Tsaliet valley draining the Tembien highlands in Tigray towards the north, seen from Ba'ati Woyane (13.670704°N, 39.162169°E). At the lower position, in stratigraphical order, are the folded Precambrian rocks, visible at the far end, that result in an undulating topography (*1*). All subsequent subhorizontal formations are part of the *amba* landscape. From bottom to top in the opposite valley flank: Edaga Arbi tillites, a relatively soft rock that is easily eroded away (*2*); subvertical, high cliffs in Adigrat sandstone (*3*); alternating hard and soft rock layers of Antalo limestone (*4*); again a subvertical cliff of Amba Aradam sandstone (*5*); and finally the typical trap basalt (*6*) that caps the sedimentary transgression–regression series

After deforestation, which took place over the last 4000 years (Hurni 1985; Moeyersons et al. 2006b), topsoil and subsoil were removed in many places, predominantly by water and tillage erosion. In remnant forests of the study area, thick Phaeozems are found which are totally absent in nearby deforested areas in a comparable geomorphic context. Presently, there is an active policy to reforest steep slopes, which is, however, not expected to completely restore the original climax vegetation and soils (Descheemaeker et al. 2006).

Field observations indicate that two geomorphic processes occur on steep slope sections of the study area: (1) rock fall at the end of the rainy season, inferred from the presence of fresh sand- and limestone blocks, up to 3 m across, on debris slopes below cliffs; (2) rock fragment transport triggered by livestock trampling, inferred from the noise of rolling and falling rock fragments, when livestock, especially goats but also cattle and sheep, grazes on steep slopes.

9.3 Materials and Methods

9.3.1 Rockfall Monitoring

Rockfall was monitored along the 60-m-high and 1,500-m-long Amba Aradam sandstone and limestone cliff line (elevation: around 2,500 m a.s.l.), which forms an amphitheatre-shaped slope section in the middle of the study catchment, and the 30-m-high and 1,327-m-long basalt cliff line (elevation: around 2,600 m a.s.l.) in the upper part of the catchment (Fig. 9.3).

From 1998 to 2001, the slopes below both cliffs were inspected twice a year (at the end of the rainy season and in the middle of the dry season) in order to record rockfall events as indicated by (1) straight downslope scars of damaged vegetation along the path of the rockfall; (2) bright colour of freshly fallen and broken rocks; (3) up to 10 cm deep impacts of falling rocks in cropland soil; and (4) information provided by shepherds about the moment, conditions and magnitude of the event. For each major rockfall event (Fig. 9.4), the volume of the fallen rock fragments was measured with a metre stick and horizontal and vertical displacement distances with a theodolite. A weighed average displacement distance was calculated, taking into account the volumes of transported rock fragments (Fig. 9.5).

9.3.2 Livestock Trampling Monitoring

Medium-term monitoring of individual rock fragment movement along steep slope sections, especially by livestock trampling, was conducted on three slope sections

Fig. 9.2 A typical *amba* landscape reflects the presence of subhorizontal lithological structure and rapid incision. At Guyeha, near the study area, a prominent Amba Aradam sandstone cliff occurs (total height is ca. 40 m). People on the road for scale (*Photo* J Nyssen)

representative for the selected catchment and its surroundings (Table 9.1; Fig. 9.4). Sites 1 and 2 (Fig. 9.4) were situated in an intensely grazed rangeland. Site 3 was located in a 5-year-old exclosure, where no livestock is allowed to enter, but where people come once a year to cut grass, especially for roofing. This last site, like the rangelands, bears some rills generated by run-off, which overtops the upslope cliff.

Since painted rock fragments risked to be picked up by shepherds, limestone rock fragments were used as tracers in these basalt and sandstone environments. Their three main diameters were measured and the flatness index (*FI*; Cailleux 1945) was calculated as:

$$\mathrm{FI} = (d_1 + d_2)/(2d_3) \tag{9.1}$$

where

d_1 longest diameter;
d_3 shortest diameter, perpendicular to d_1;
d_2 intermediate diameter, perpendicular to d_1 and d_3.

Although the aim was to use tracers similar to the rock fragments naturally present on the slope, it should be noted that, at site 1, the basalt rock fragments were somewhat flatter compared to the tracers (Table 9.1). The tracers were installed in July 1999. At each site, a 20–50 m long line along the contour was materialised by a rope, following straight sections between large rocks, on which markings were painted. Rock fragments on the soil surface, or embedded for less than half their volume, which were crossed by this line, were removed and replaced by tracers of similar shape and size (1–4 tracers per m). In March 2001 (20 months later), the ropes were installed again (Fig. 9.6), tracers were recovered, and if they had moved from their original position, the shortest distance (which was along the steepest slope) to the rope was measured. In rangeland, recovery rates were relatively low (56–67 %). That is probably due to (1) recent rockfall deposits which in places visibly covered the original soil surface and (2) the shepherds might have picked up some of the tracers by curiosity,

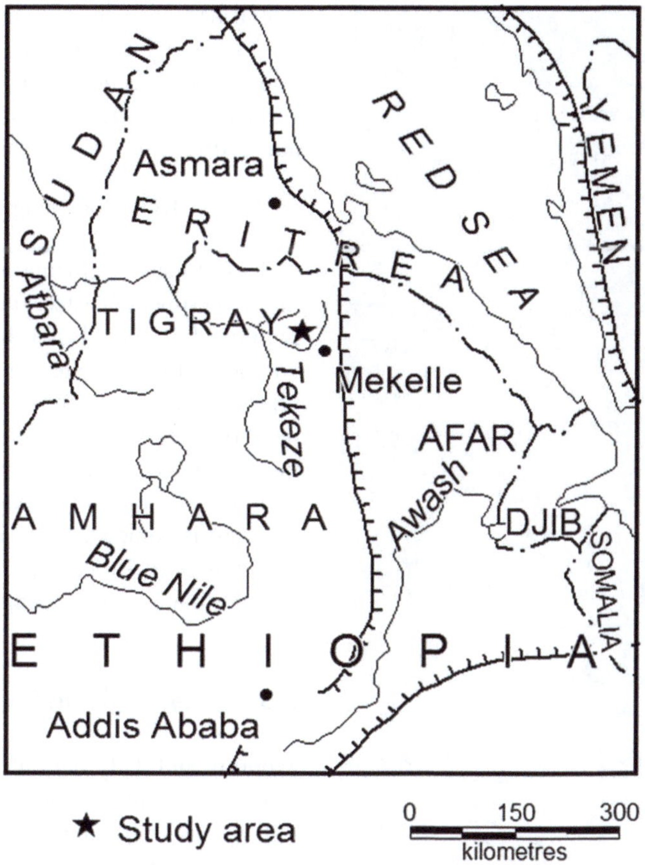

Fig. 9.3 Location of the May Zegzeg catchment in the northern Ethiopian highlands (after Nyssen et al. 2006)

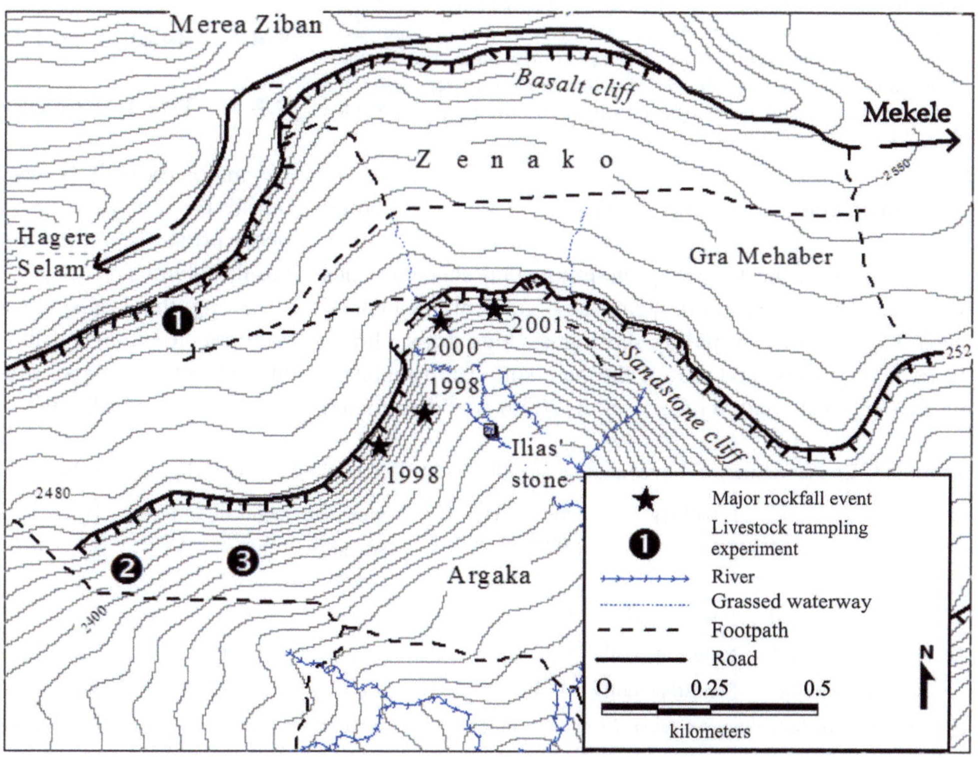

Fig. 9.4 Location map of major rockfall events (1998–2001) and of livestock trampling monitoring sites within the May Zegzeg catchment (after Nyssen et al. 2006)

Fig. 9.5 Three months after the occurrence of a rockfall event (August 1998) on the sandstone cliff. Fresh rock fragments (*arrows*) are easily recognisable by their bright colour. The *upper arrow* indicates the place of origin of these rocks on the cliff. Despite recovering vegetation, the rockfall path can still be recognised on the backslope (after Nyssen et al. 2006)

since their colour contrasted with that of the surrounding area.

At each site, the following environmental characteristics were measured or assessed: mean slope gradient, mean areal percentage of soil covered by short grass, long grass, shrubs or rock fragments, and that of the bare soil. The roughness index (*RI*; Oostwoud Wijdenes et al. 2000) was determined 4–8 times at each site, using a 200-cm long chain with 3 × 1.5 cm links, which was placed on the surface in a downslope direction following all irregularities of the soil surface, while the shortest distance between its beginning and end (D_s, in cm) was measured:

$$\text{RI} = 200\ (\text{length of chain in cm}) - D_s \quad (9.2)$$

Unfortunately, grazing intensity could not be measured at the experimental sites, and official data, organised per municipality, could not be used because (1) they were computed for large areas and (2) access to rangeland is also open to livestock from Hagere Selam town and other neighbouring villages. In a qualitative way, it can be stated that the much degraded rangeland near Harena Village (site 2) is most intensively grazed, especially by goats. Grazing pressure is somewhat less in the Zenako rangeland (site 1), which is used by cattle, sheep and goats. In the exclosure (site 3), there is no grazing.

9.3.3 Rainfall Analysis

Monthly and daily rainfall data of the nearby Hagere Selam station were analysed to assess the representativeness of the rain events triggering rockfall. Antecedent rain depth since the start of the rainy season in the years with observed rockfall was also compared to rain depths of a 20-year long series.

9.4 Results and Discussion

9.4.1 Yearly Rockfall

No major rockfall events were observed on the basalt cliff during the four years of observation. Here, rockfall seems to occur primarily in the form of toppling of parts of individual basalt columns.

Along the sandstone and limestone cliffs, four events were recorded, always in August, when the soils around and in between the rocks are saturated by water (Table 9.2). Some young shepherds from the area (Fig. 9.5) observed one

Table 9.1 Characteristics of the monitoring sites for livestock trampling and experimental conditions (after Nyssen et al. 2006)

Site	Lithology	Mean slope gradient (m m^{-1})	Rock fragments at surface			Tracers				
			Mean cover (%)	d_2	FI	d_2	FI	Installed	Recovered	Recovery rate (%)
1. Zenako rangeland	Basaltic coll	0.55	62	6.2	2.60	5.8	1.84	64	43	67
2. Harena rangeland	Sandstone coll	0.72	30	5.0	2.07	5.7	1.88	81	45	56
3. Harena exclosure	Sandstone coll	0.85	14	6.0	2.07	4.6	1.88	60	53	88

d_2 mean intermediate diameter (cm); *FI* flatness index. They are based on measurements of 30–40 randomly selected rock fragments/tracers

Fig. 9.6 Harena monitoring sites #2 and #3 for individual rock fragment movement (July 1999). *2* rangeland; *3* exclosure. Between both sites is a cropland on an ancient debris flow deposit (after Nyssen et al. 2006)

Table 9.2 Rockfall events at the sandstone and limestone cliffs (1998–2001) (after Nyssen et al. 2006)

Approx. date	Lithology	Total volume (m^3)	Rock density (kg m^{-3})	Total mass (10^3 kg)	Horizontal displacement (m)			Vertical displacement (m)		
					Min	Max	Weighted average	Min	Max	Weighted average
August 1998	Sandstone	10.49	2,400	25	4.0	150.9	80.8	1.1	77.9	46.9
August 1998	Limestone	13.36	2,535	34	9.2	35.3	14.0	8.5	24.0	11.6
August 2000	Sandstone	70.76	2,400	170	5.0	36.2	20.8	1.5	24.0	17.6
August 2001	Sandstone	37.77	2,400	91	14.0	66.5	32.5	11.2	33.1	20.2
Mean (1998–2001)		33.10		80			37.0			24.1

major rockfall event: *We saw one big rock rolling from the top of the cliff; it broke into pieces when rolling along the slope. Some large rocks reached the cropland and they broke into pieces at different sites. It was during daytime, in August 1998. It was raining. We were not very far with our cattle. The villagers came quickly when they heard the noise, because they were afraid for us. Fortunately nobody was hurt. Rockfall with big noise like this is exceptional.*

All rockfall events in 1998, 2000 and 2001 occurred at the peak of the rainy season. Unlike temperate mountain areas, where rockfall is strongly correlated with temperature variations (Perret et al. 2006), rainfall is the main triggering factor for rockfall in the study area. Rainfall analysis shows that the maximum daily rain depths in August during the study period correspond to 20-year average values of around 40 mm day^{-1}. Extreme daily rainfall events, such as those occurring in 1975 (67 mm) and 1980 (66 mm), were not observed during the study period.

Antecedent rain depth since the beginning of the rainy season and rain depth for August showed average conditions during three years, but in 1998, they had the second highest value in the 20-year series. Hence, it is anticipated that rockfall events reported here are representative for both average and extreme rainfall conditions of the region.

The mean yearly rockfall of 80 t, observed along a 1,500-m long section (parallel to the contour) of the Amba Aradam sandstone cliff over an average vertical distance of 24 m (Table 9.2), should be considered as a minimum, since we did neither account for many small events involving only a few rock fragments of some kg, nor for possible extreme rockfall events related to exceptionally high daily rain. An extrapolation of the observed rockfall rates to the 1,500-m-long and 60-m-high cliff indicates that the cliff would retreat by at least 0.37 mm year^{-1} or 3.7 cm century^{-1}. One of the few studies on sandstone cliff retreat in (sub)tropical regions (Young and Wray 2000) reported geological scarp retreat

Fig. 9.7 Tracers (*white limestone*) displaced over a distance of 6 m in 20 months (Harena rangeland, March 2001). The rope at the back shows the original tracer position (after Nyssen et al. 2006)

Table 9.3 Tracer displacement in 20 months, perpendicular to the contour (after Nyssen et al. 2006)

Site	Percentage of tracers moved	Displacement distance (m) of moved tracers		Mean displacement distance (m) of all tracers
		Mean[a]	Standard deviation	
1. Zenako range	72	1.74*	±2.37	1.25
2. Harena range	95	2.31*	±1.97	1.80
3. Harena exclosure	66	0.90#	±1.00	0.59

[a] Different symbols indicate significantly different values ($\alpha = 0.1$) based on unpaired Student's t-test

rates ranging between 1.5 and 2.5 cm century^{-1}. On the Colorado Plateau, scarp retreat rate averages 1.6 cm century^{-1} (Young 1985). These mean long-term values are lower though of a similar order of magnitude as our observed short-term cliff retreat rates.

9.4.2 Rock Fragment Movement over Debris Slopes

Twenty months after their placement, most tracers had left their original position (Fig. 9.7). In the two rangelands, 72 and 95 % of the tracers had moved, but it appeared that also in the exclosure, 66 % of the tracers received an impulse which was strong enough to initiate their movement (Table 9.3). Besides exceptional illegal livestock grazing, such impulses can also be caused by rock fragments falling and rolling from the cliff, by wild animals such as hare, jackal, hyena, caracal and porcupine (Yami et al. 2007), by run-off produced above the cliff, by forest guard and by people who come occasionally in the area, especially to harvest grass. Mean displacement distance of the transported tracers is, however, significantly smaller in the exclosure (0.9 m) than in the rangelands (1.74 and 2.31 m) (Table 9.3). In particular, the difference in tracer displacement distance between the rangeland and the exclosure on the Harena scree slope is highly significant ($\alpha = 0.001$). The percentage of tracers moved as well as the mean distances of tracer movement increase with increasing grazing pressure (Harena rangeland > Zenako rangeland > Harena exclosure).

Looking at other environmental characteristics of the three sites, further reasons for the differences in displacement distances become clear. If mean surface roughness is similar at the study sites (roughness index between 34 and 46), vegetation cover, and especially long grass cover, seems to be the primary explanatory factor for the observation that tracers moved over much smaller distances in the exclosure. Major explanatory factor for smaller tracer displacement distances in Zenako, compared to Harena, is the difference in area of smooth surface in the latter (53 %) as compared to the former (20 %).

A yearly unit rock fragment transport rate can be calculated, similar to the unit soil transport rate for contour ploughing (Poesen et al. 1997):

Table 9.4 Calculation of the yearly unit rock fragment transport rate (Qs) for the three monitored sites (t = 1.67 year) (after Nyssen et al. 2006)

Site	Mean displacement distance (m) along the slope of all tracers	Slope gradient (m m^{-1})	δ (m)	d_3 (m)	R_C (%)	ρ (kg m^{-3})	Qs (kg m^{-1} year^{-1})
1. Zenako range	1.25	0.55	1.10	0.034	62	2734	37.9
2. Harena range	1.80	0.72	1.46	0.035	30	2511	23.1
3. Harena exclosure	0.59	0.85	0.45	0.035	14	2511	3.3

δ horizontal component of the net mean downslope displacement distance of all tracers (including those that did not move), in the direction of the steepest slope; d_3 shortest stone diameter perpendicular to the longest diameter; R_C rock fragment cover; ρ rock density

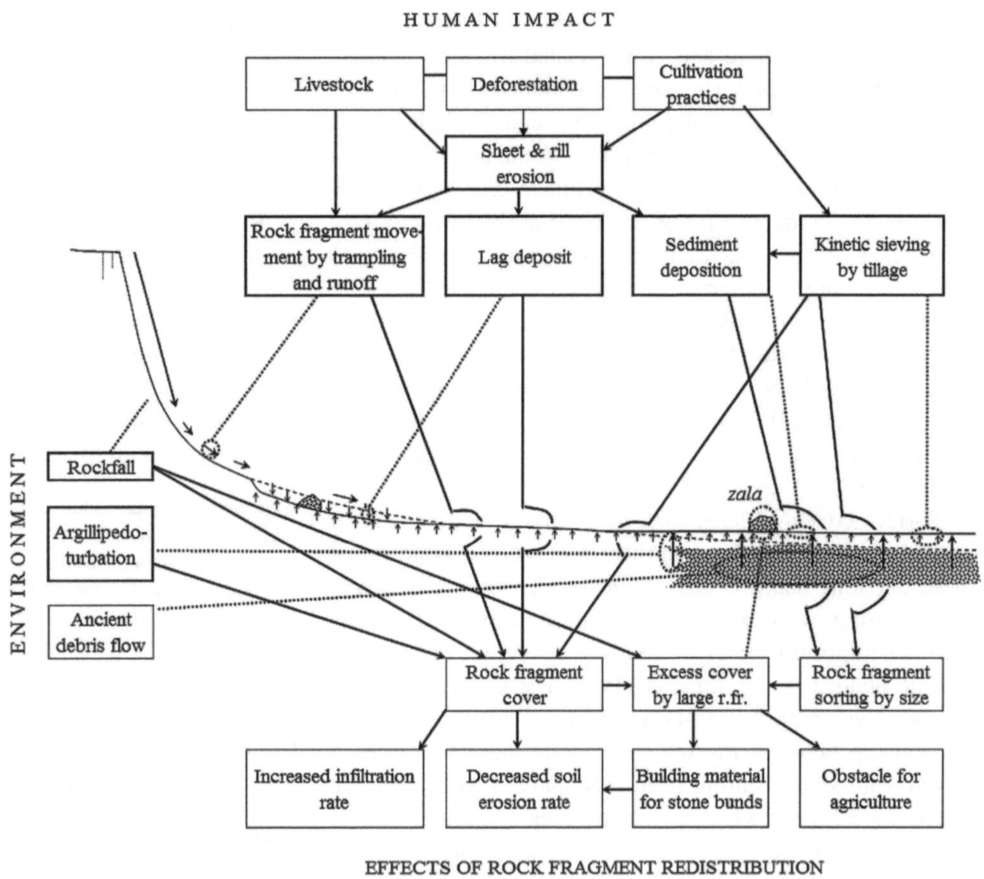

Fig. 9.8 Major factors in rock fragment redistribution, characterising the present day landscape and agricultural system. *Dotted lines* refer to rock fragment movements, schematically represented by *short arrows* on the figure, and *solid lines* indicate relations between factors and processes (*bold frames*) (after Nyssen et al. 2006)

$$Qs = \delta \cdot d_3 \cdot R_C \cdot \rho \cdot t^{-1} \quad (9.3)$$

where Qs = yearly unit rock fragment transport rate (kg m^{-1} year^{-1}); δ = horizontal component of the net mean downslope displacement distance of all tracers (including those that did not move), in the direction of the steepest slope (m); ρ = rock density (kg m^{-3}); and t = period over which tracer displacement was monitored (year).

The yearly unit rock fragment transport rate is larger in the Zenako rangeland (site 1) than in Harena (site 2) (Table 9.4), which is due to a high rock fragment content (R_C).

9.4.3 Cliffs and Scree Slopes as Part of Catenas

The current understanding of the detachment processes on cliffs and transport processes over scree slopes can now be integrated with results of previous studies to analyse rock fragment redistribution along catenas on stepped mountains in the northern Ethiopian highlands (Fig. 9.8).

The yearly unit rock fragment transport rate for cliffs (≥53 kg m^{-1} year^{-1}) is of the same order of magnitude as that for rangeland (38 and 23 kg m^{-1} year^{-1}), but significantly larger than the transport rate measured in an exclosure (3 kg m^{-1} year^{-1}). Corresponding rock fragment transport coefficients (K) for rangeland (32–69 kg m^{-1} year^{-1}) are much larger than K for densely vegetated exclosures (3.9 kg m^{-1} year^{-1}). This indicates that rocks fallen from cliffs into rangeland are transported downslope mainly by livestock trampling. By contrast, rockfall into exclosures is largely stored at the upper part of the escarpment face. Rock fragments found in remnant forests on steep slopes are also generally located at the foot of the cliffs. These forests act as "protection forests" against rockfall, similar to those found in the Alps (Stoffel et al. 2005).

Downslope from cliffs and scree slopes, the overall R_C at the soil surface on the catena developed on basalt is large (55–85 %) everywhere and is not related to slope gradient but to the location of mass movement bodies, as indicated by Nyssen et al. (2002c). In the limestone area, R_C is larger on the steeper areas close to the cliff. Due to the combination of kinetic sieving as a consequence of tillage (Oostwoud et al. 1997) and slow deposition of fine earth at the lower side of the catenas, mean rock fragment size significantly increases with decreasing slope gradient (Nyssen et al. 2002b).

The vertical processes supplying rock fragments to the soil surface include (1) tillage-induced kinetic sieving, (2) selective run-off erosion and the development of erosion pavements and (3) vertic movements (argillipedoturbation; Poesen and Lavee 1994; Nyssen et al. 2002a; Moeyersons et al. 2006a).

Surface covers of large rock fragments sometimes hinder agriculture, and therefore, farmers remove and concentrate these on stone heaps (*zala*). However, dense rock fragment covers bring also some advantage to agriculture. They increase infiltration rates, decrease evaporation and protect topsoil against water erosion (Poesen and Lavee 1994). Indeed, Nyssen et al. (2001) reported a significant negative relationship between R_C and soil loss in the study area. A recommendation resulting from this study is to rely on the following farmers' wisdom: smaller rock fragments should never be removed from the field surface, but a limited number of larger rock fragments can be removed in order to increase crop yield.

9.5 Conclusions

Rockfall from cliffs and rock fragment transport on debris slopes under rangeland, mainly by livestock trampling, appear to be important geomorphic processes in the northern Ethiopian highlands. Along a 1,500 m long section (parallel to the contour) of the Amba Aradam sandstone cliff, at least 80 t of rock are detached and fall annually over a mean vertical distance of 24 m, resulting in a mean annual cliff retreat rate of 0.37 mm year^{-1}. The hard rock of the Amba Aradam sandstone formation and especially its ferruginous cap provide structural control on the geomorphological evolution. Our observed sandstone cliff retreat rates are of the same order of magnitude as long-term sandstone cliff retreat rates reported from other (sub)tropical regions, indicating that the anthropogenic impact on the process of rock fragment detachment from cliffs is minimal.

Yearly unit rock fragment transport rates (Qs), mainly induced by animal trampling, were 37.9 kg m^{-1} year^{-1} in rangeland on basalt (slope gradient S = 0.55 m m^{-1}) and 23.1 kg m^{-1} year^{-1} in rangeland on sandstone colluvium (S = 0.72 m m^{-1}). Similar to sheet and rill erosion and gullying (Descheemaeker et al. 2006), this process is also virtually stopped after exclosures are established with Qs equalling only 3.9 kg m^{-1} year^{-1} on a 0.85 m m^{-1} slope. The importance of rock fragment movement on debris slopes is positively correlated with grazing pressure and the areal percentage of smooth surface, and inversely with the percentage of long grass cover.

Rock fragment redistribution on the lower part of the catena is controlled by the occurrence of ancient debris flow bodies as well as current processes such as kinetic sieving by tillage, argillipedoturbation, development of erosion pavements by water erosion, and manual removal by farmers. Besides rockfall from cliffs and argillipedoturbation, all the discussed factors and processes of rockfall redistribution in the study area are of anthropogenic origin (Fig. 9.8).

Acknowledgments This study was conducted in the framework of research programme G006598.N funded by the Fund for Scientific Research—Flanders, Belgium and of the Zala-Daget project (VLIR-UOS, Belgium). JN was affiliated with KU Leuven while the research was carried out. Thanks go to Berhanu Gebremedhin Abay for assistance with all the fieldwork. Local farmers and shepherds shared their knowledge with us. The local Agricultural Office, REST (Relief Society of Tigray) branch, and the authorities of the concerned villages and district facilitated the research. Conceptual discussions with Andrew Goudie, Robert Wray and David Alexander are gratefully acknowledged.

References

Adamson D, Williams F (1980) Structural geology, tectonics and the control of drainage in the Nile basin. In: Williams M, Faure H (eds) The Sahara and the Nile. Quaternary environments and prehistoric occupation in northern Africa, Balkema, Rotterdam, pp 225–252

Arkin Y, Beyth M, Dow DB, Levitte D, Temesgen Haile TH (1971) Geological map of mekele sheet area ND 37-11, Tigre Province, 1:250.000, Imperial Ethiopian Government, Ministry of Mines, Geological Survey, Addis Ababa

Ayalew L, Yamagishi H (2004) Slope failures in the Blue Nile basin, as seen from landscape evolution perspective. Geomorphology 57:95–116

Bussert R, Schrank E (2007) Palynological evidences for a latest carboniferous-early permian glaciation in Northern Ethiopia. J Afr Earth Sci 49:201–210

Cailleux A (1945) Distinction des galets marins et fluviatiles. Bulletin de la Société géologique de France 15:375–404

Coltorti M, Dramis F, Ollier CD (2007) Planation surfaces in Northern Ethiopia. Geomorphology 89:287–296

Descheemaeker K, Nyssen J, Rossi J, Poesen J, Haile M, Moeyersons J, Deckers J (2006) Sediment deposition and pedogenesis in exclosures in the Tigray highlands, Ethiopia. Geoderma 132:291–314

Driesen P, Dudal R (1991) The major soils of the world. Wageningen Agricultural University and KU Leuven, The Netherlands

Gardner J (1970) Rockfall: a geomorphic process in high mountain terrain. Albertan Geogr 6:15–20

Govers G, Poesen J (1998) Field experiments on the transport of rock fragments by animal trampling on scree slopes. Geomorphology 23:193–203

Hurni H (1985) Erosion—productivity—conservation systems in Ethiopia. In: Proceedings of 4th international conference on soil conservation, Maracay, Venezuela, pp 654–674, 3–9 Nov 1985

Hutchinson RW, Engels GG (1970) Tectonic significance of regional geology and evaporite lithofacies in northeastern Ethiopia. Phil Trans Roy Soc Lond A Math Phys Sci 267:313–330

Lee FT, Odum JK, Lee JD (1994) Rockfalls and debris avalanches in the Smugglers Notch area, Vermont. US Geol Surv Bull 2075:1–33

Merla G, Minucci E (1938) Missione geologica nel Tigrai. Volume primo: la serie dei terreni. Reale Accademia d'Italia, Centro Studi per l'AOI, Roma

Merla G, Abbate E, Azzaroli A, Bruni P, Canuti P, Fazzuoli M, Sagri M, Tacconi P (1979) A geological map of Ethiopia and Somalia (1973) 1:2.000.000 and comment. University of Florence, Italy

Moeyersons J, Nyssen J, Poesen J, Deckers J, Haile M (2006a) On the origin of rock fragment mulches on Vertisols: a case study from the Ethiopian Highlands. Geomorphology 76(3–4):411–429

Moeyersons J, Nyssen J, Poesen J, Deckers J, Haile M, Lang A (2006b) New insights in timing and steering of environmental changes in Northern Ethiopia since Late-Pleistocene times: a view from the Highlands. Palaeogeogr Palaeoclimatol Palaeoecol 230:165–181

Mohr P (1963) The geology of Ethiopia. University College of Addis Ababa Press, 270 p

Nyssen J, Moeyersons J, Deckers J, Haile M, Poesen J (2000) Vertic movements and the developments of stone covers and gullies, Tigray highlands, Ethiopia. Zeitschrift für Geomorphologie NF 44:145–164

Nyssen J, Haile M, Poesen J, Deckers J, Moeyersons J (2001) Removal of rock fragments and its effect on soil loss and crop yield Tigray Ethiopia. Soil Use Manag 171:179–187

Nyssen J, Moeyersons J, Poesen J, Haile M, Deckers J (2002a) Argillipedo-turbation and the development of rock fragment covers on vertisols in the Ethiopian highlands. Belgeo 2:183–193

Nyssen J, Poesen J, Moeyersons J, Lavrysen E, Haile M, Deckers J (2002b) Spatial distribution of rock fragments in cultivated soils in Northern Ethiopia as affected by lateral and vertical displacement processes. Geomorphology 43:1–16

Nyssen J, Moeyersons J, Poesen J, Deckers J, Haile M (2002c) The environmental significance of the remobilisation of ancient mass movements in the Atbara-Tekeze headwaters, Northern Ethiopia. Geomorphology 49:303–322

Nyssen J, Vandenreyken H, Poesen J, Moeyersons J, Deckers J, Haile M, Salles C, Govers G (2005) Rainfall erosivity and variability in the Northern Ethiopian highlands. J Hydrol 311:172–187

Nyssen J, Poesen J, Moeyersons J, Deckers J, Haile M (2006) Processes and rates of rock fragment displacement on cliffs and scree slopes in an amba landscape, Ethiopia. Geomorphology 81:265–275

Nyssen J, Naudts J, De Geyndt K, Haile M, Poesen J, Moeyersons J, Deckers J (2008) Soils and land use in the Tigray highlands (Northern Ethiopia). Land Degrad Dev 19:257–274

Oostwoud Wijdenes D, Poesen J, Vandekerckhove L, De Luna E (1997) Chiselling effects on the vertical distribution of rock fragments in the tilled layer of a Mediterranean soil. Soil Tillage Res 44:55–66

Oostwoud Wijdenes D, Poesen J, Vandekerckhove L, Kosmas C (2000) The use of marked rock fragments as tracers to assess rock fragments transported by sheep trampling on Lesvos, Greece. In: Foster I (ed) Tracers in geomorphology. Wiley, Chichester, pp 201–220

Oostwoud Wijdenes D, Poesen J, Vandekerckhove L, Kosmas C (2001) Measurements at one-year interval of rock-fragment fluxes by sheep trampling on degraded rocky slopes in the Mediterranean. Zeitschrift für Geomorphologie NF 45:477–500

Perret S, Stoffel M, Kienholz H (2006) Spatial and temporal rockfall activity in a forest stand in the Swiss Prealps—A dendrogeomorphological case study. Geomorphology 74:219–231

Poesen J, Lavee H (1994) Rock fragments in top soils: significance and processes. Catena 23:1–28

Poesen J, Van Wesemael B, Govers G, Martinez-Fernandez J, Desmet P, Vandaele K, Quine T, Degraer G (1997) Patterns of rock fragment cover generated by tillage erosion. Geomorphology 18:183–197

Stoffel M, Schneuwly D, Bollschweiler M, Lièvre I, Delaloye R, Myint M, Monbaron M (2005) Analyzing rockfall activity (1600–2002) in a protection forest—a case study using dendrogeomorphology. Geomorphology 68:224–241

Williams M, Williams F (1980) Evolution of the Nile basin. In: Williams M, Faure H (eds) The Sahara and the Nile. Quaternary environments and prehistoric occupation in Northern Africa, Balkema, Rotterdam, pp 207–224

Yami M, Gebrehiwot K, Moe S, Mekuria W (2007) Impact of area enclosures on density and diversity of large wild mammals: the case of May Ba'ati, Douga Tembien Woreda, Central Tigray, Ethiopia. East Afr J Sci 1:1–14

Young RA (1985) Geomorphic evolution of the Colorado plateau margin in west-central Arizona: a tectonic model to distinguish between the causes of rapid symmetrical scarp retreat and scarp dissection. In: Hack JT, Morisawa M (eds) Tectonic geomorphology. Allen & Unwin, London, pp 261–278

Young RW (1987) Sandstone landforms of the tropical East Kimberley region, Northwestern Australia. J Geol 95:205–218

Young RW, Wray R (2000) Contribution to the theory of scarpland development from observations in Central Queensland, Australia. J Geol 108:705–719

Gully Development in the Tigray Highlands 10

A. Frankl, J. Poesen, J. Moeyersons, and J. Nyssen

Abstract

In the Tigray highlands, gully development is linked to poverty-driven unsustainable use of the land in a vulnerable semi-arid and mountainous environment, where intense rainfalls challenge the physical integrity of the landscape. Over the last two centuries, three major phases in the hydrological regime of the region could be distinguished. In the first phase, between 1868 (or earlier) and ca. 1965, the relatively stable gully channels showed an oversized morphology inherited from a previous period when external forcing of environmental conditions caused significant channel development. In the second phase (ca. 1965–ca. 2000), increased aridity and a continued vegetation clearance accelerated dynamics of the gully system. A sharp increase in gully headcut retreat rates, network densities and volumes could be quantified for that period. With the widespread implementation of soil and water conservation measures, erosion rates decreased, which announced the start of the third hydrogeomorphic phase since ca. 2000. In 2010, about one-fourth of the gully channels were stabilized. These hydrogeomorphic developments correspond to a gully *cut*-and-*fill* cycle in the second half of the twentieth century and suggest that a pre-1868 cut cycle took place.

Keywords

Gully • Headcut retreat • Repeat photography • Vertisol

10.1 Introduction

In dryland environments, water availability and biomass production are often restricted and confined to a short rainy season. As a result, the carrying capacity of the ecosystem is rapidly exceeded by human exploitation of natural resources, especially in sub-Saharan countries like Ethiopia, with fast demographic expansion and deficient exploitation techniques (Kassas 1995). Furthermore, the resilience of the land is often reduced by recurring droughts and severe land degradation, which threatens sustainable development in these fragile environments.

The Tigray Rural Development Study (TRDS; HTS 1976), which investigated the state of the environment in the 1970s, concluded that land degradation was severe in the Tigray highlands and that natural resources were put to their limits (Virgo and Munro 1978). Almost a decade later, the devastating effects of drought and desertification were brought to a global audience with the drought that stroke the region in 1984–1985. In a context of civil war, region-wide crop failures led to massive famine and starvation. As shown on historical photographs (Fig. 10.1), severe gully erosion destroyed valuable land, jeopardizing in situ and downstream agricultural production, increasing the costs of transport and infrastructure construction and producing flash floods of polluted water which threatened human health.

A. Frankl (✉) · J. Nyssen
Department of Geography, Ghent University, 9000 Ghent, Belgium
e-mail: amaury.frankl@ugent.be

J. Poesen
Department of Earth and Environmental Sciences, KU Leuven, 3001 Heverlee, Belgium

J. Moeyersons
Royal Museum for Central Africa, 3080 Tervuren, Belgium

P. Billi (ed.), *Landscapes and Landforms of Ethiopia*, World Geomorphological Landscapes,
DOI 10.1007/978-94-017-8026-1_10, © Springer Science+Business Media Dordrecht 2015

Fig. 10.1 Active gully channel networks in a barren landscape of Tigray highland (*Photographs* by Jean Poesen—*top*, and Piotr Migoń—*middle* and *bottom*)

Several studies reported significant gully erosion in Ethiopia and especially addressed the Tigray highlands (Fig. 10.1; Virgo and Munro 1978; Berakhi and Brancaccio 1993; Nyssen et al. 2000, 2002, 2004, 2006; Billi and Dramis 2003; Munro et al. 2008; Reubens et al. 2009; Frankl et al. 2011, 2012, 2013b, c, d). Among the first, Virgo and Munro (1978) assessed the status of soils and landscapes in northern Ethiopia, documenting numerous cases of severe gully erosion. The importance of gully erosion in Vertisols, which are susceptible to soil piping, was discussed by Nyssen et al. (2000). Berakhi and Brancaccio (1993) and Nyssen et al. (2006) studied the effect of road building on gully erosion risk, and Nyssen et al. (2004) investigated the usefulness of check dams to control gullies. Nyssen et al. (2006) analysed the development of four gully systems in the Tigray highlands by developing a field method which is based on how local people remember the evolution of specific gullies. A similar approach was used by Moges and Holden (2008) who studied gully development in southern Ethiopia. A small gully network in eastern Ethiopia was studied by Daba et al. (2003), using a time series of digital elevation models derived from small-scale aerial photographs. An analysis of the development of gully headcuts, their cross sections, networks and volumes at regional scale in Tigray highlands was carried out by Frankl et al. (2011, 2012, 2013b, c, d). By using repeated terrestrial photography, aerial photographs and satellite images, these authors could define distinct phases in gully erosion development since the late nineteenth century. Nigussie Haregeweyn et al. (2005, 2008) linked the presence of gullies to the rate of reservoir sedimentation (catchment sediment yield).

10.2 Gullies and Gully Erosion

In the Oxford Dictionary of Earth Sciences (2008), a gully is defined as a "feature of water erosion that develops from the run-off of a violent torrent that bites deeply into topsoil and soft sediments". Gully erosion can thus be associated with the rapid incision of valley sides or valley floors by the erosive action of flash floods. This is also apparent from the definition of Poesen et al. (2003) who proposed that gully erosion is an "erosion process whereby runoff water accumulates and often recurs in narrow channels and, over short periods, removes the soil from this narrow area to considerable depths", a definition that is widely used in scientific literature. Gully erosion is thus the result of disruptions in a previous stage of equilibrium that cause changes in the run-off volumes and sediment delivered to a certain place in the landscape (Graf 1988; Knighton 1998). Causes of incision by concentrated overland flow are numerous and can be grouped into six categories (Schumm 2005): geologic, geomorphic, climatic, hydrologic, animal and human causes. Examples that are most relevant for the case of the Tigray highlands are uplift, base level lowering, mass movements, increased aridity or humidity, increased mean run-off discharge and/or peak discharge, decreased sediment load, animal grazing and tracking, flow diversion, urbanization and infrastructure construction. As the geomorphic development of gullies is related to alternating conditions, they are also referred to as non-regime channels (Schumm 2005).

In order to efficiently transfer water and sediment via run-off downslope, the shape and size of gullies mainly adjust to peak flow discharge properties (Knighton 1998). Channel shape and size may be spatially and temporally very variable, depending on local environmental characteristics (e.g. soil, vegetation), and the time that is needed to accommodate changes in run-off discharge and sediment load, i.e. reaction and relaxation times (Knighton 1998).

Frankl et al. (2013c) characterized gully cross-sectional characteristics in the Tigray highlands by surveying 811 gully cross sections in various environments, reflecting contrasts in lithology, topography, soil, land use and climate. Gully top width (TW) varied between 0.35 and 31.90 m, with a median of 6.34 m. Gully depth (D) varied between 0.20 and 12.77 m, with a median of 2.15 m, and bottom width (BW) ranged between 0.10 and 19.50 m, with a median of 3.00 m. The median cross-sectional area (CSA) was 10.1 m^2 and ranged between 0.15 and 236.5 m^2. As the boxplots show (Fig. 10.2a), the distributions are right-skewed and the variability of the observations, as indicated by interquartile range, is higher for *TW* (5.20) and *BW* (2.70) than for *D* (1.79). The median *TW–D* ratio was 2.7, while the median *BW–TW* ratio was 0.5. Note that for *TW*/*D* and *BW*/*TW*, median and mean do not differ much as the distributions are nearly normal. As shown in Fig. 10.2b, plotting *D* over *TW* shows wide scatter around a linear relation purged through the origin (0, 0).

Catchment area is the most important factor explaining the variability in *TW*, *D* and *CSA*, as it reflects the volume of water that drains to a certain point in the landscape. However, the large variability in soil management and environmental characteristics causes the gully shape and size to be locally very variable. As shown by Frankl et al. (2013c), besides the catchment area, lithology and the presence of check dams are the most important explanatory factors to account for gully morphology in the Tigray highlands. Gully cross sections in shale-derived deposits are on average 36.7 % larger than those in deposits derived from volcanics. This is mainly related to the higher erodibility of shale-derived materials and to the presence of travertine dams. Degradation and breaching of the latter

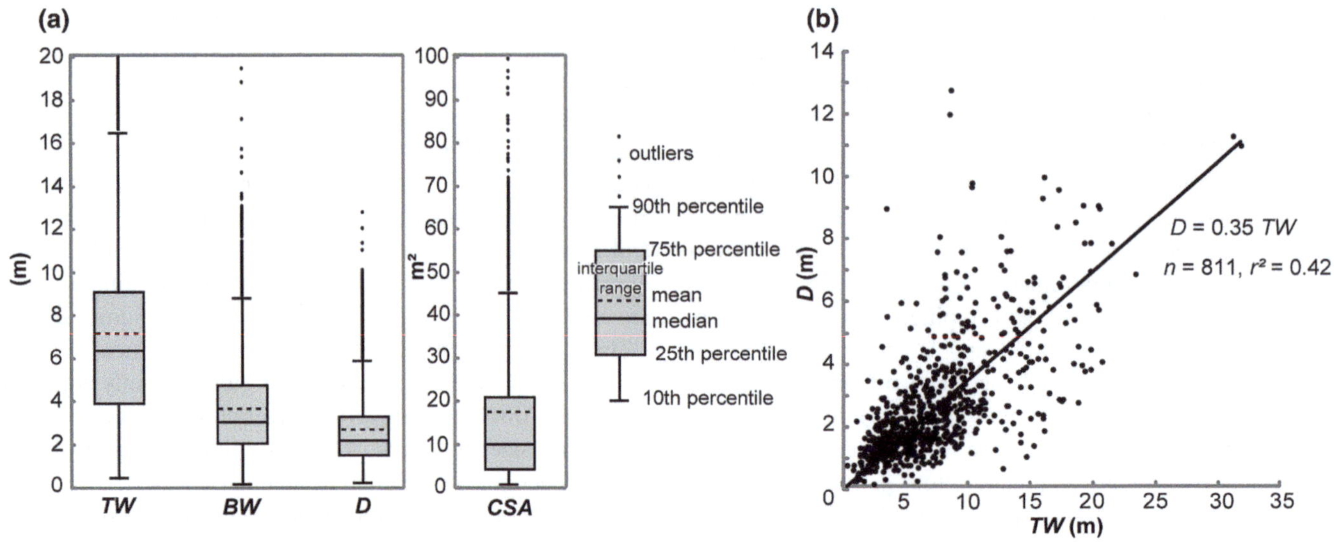

Fig. 10.2 Cross-sectional characteristics of gullies in Tigray highlands (Frankl et al. 2013c). **a** Boxplots for gully top width (*TW*), bottom width (*BW*), depth (*D*) and cross-sectional area (*CSA*) for 811 sections. Outliers larger than 20 m and 100 m^2 are not displayed, **b** plotting gully depth (*D*) over gully top width (*TW*) shows a linear relation. Based on Frankl et al. (2013c)

Fig. 10.3 The deeply incised travertine dam at May Mekdan and gully network upslope (*Photograph* of 2010 by Amaury Frankl; 13.58°N, 39.56°E)

(Moeyersons et al. 2006; Fig. 10.3) triggers regressive erosion and exposure of thick, fine-textured soils locked upslope of the dams. Check dams, which are successfully implemented in low-active gully sections, cause the gully depth to decrease by circa one-third. This makes moderately active gully cross sections with check dams on average 33.5 % smaller than those of very active gullies without check dams.

10.3 Evidence of Past Dynamics

10.3.1 Early Evidence of Gullying

The oldest terrestrial photographs known for the highlands of Tigray are those taken during the British military expedition to former Abyssinia in 1867–1868, with the objective to

Fig. 10.4 The gully draining the valley in 1939 (*left*) is low-active (smooth cross section, vegetated) and has been partially converted to cropland (*zoom*). Although wood harvesting and agricultural exploitation in 1939 severely affected the environment, shrub cover on the arable land (*white arrow*) and on the surrounding hills (*black arrow*) of this small catchment draining to Lake Ashenge is still relatively high. By 2009, the gully had extended downstream by 294 m (*right*). It is freshly incised (=high-active section) and remains active despite the recent soil and water conservation efforts. Original photograph (*left*): Maugini (Istituto Agronomico per l'Oltremare, Firenze, I). Repeat photograph by Amaury Frankl 12.56°N, 39.25°E. Based on Frankl et al. (2011)

release a number of Europeans who had been imprisoned by Emperor Tewodros (Sharf et al. 2003). Such historical photographs are often the only records available on the environmental status for the period 1868–1963 (Fig. 10.4), after which aerial photographs were produced for the area (Nyssen et al. 2010). Aerial photographs taken during the Italian occupation of Ethiopia in 1935–1941 have recently been discovered but were not yet analysed. In order to compare the previous and current situations, historical terrestrial photographs were repeated in the field according to the technique of repeat photography by Hall (2001). Qualitative and quantitative analyses of time-lapsed photographs revealed that in the late nineteenth and early twentieth centuries, gullies were common features of the northern Ethiopian landscape (Frankl et al. 2011). However, most gullies seen on historical photographs from that period show smooth and vegetated cross sections (Fig. 10.4). As their size and morphology suggest, the gullies were not in equilibrium with the prevailing conditions but were rather inherited from a previous period when external forcing of environmental conditions caused significant geomorphic change (Nyssen et al. 2009; Frankl et al. 2011). Most probably, such ancient gully incision cycles occurred during the frequent droughts of the nineteenth century, as reported by Pankhurst (1995), leading to famine and great mortality. These calamities would have increased the environmental vulnerability to human interference and, hence, also the run-off response of the land.

10.3.1.1 1960s–1990s Cut Cycle

The study of time-lapsed terrestrial photographs and aerial photographs indicates that the recent gully incision phase in Tigray highlands started around 1965, with a marked increase in gully drainage densities and volumes (Frankl et al. 2013b). In 1963–1965, gully drainage density (D_{total}) and area-specific gully volume (V_{a}) were still relatively low, i.e. 1.86 km km^{-2} and 32.23 × 10^{3} m^{3} km^{-2}, respectively (Fig. 10.5a–c). 48 % of the gully network was high-active. From a largely low-dynamic gully system in the 1960s, network expansion and increased erosion rates in the 1980s and 1990s caused the drainage density and volume to peak in 1994. D_{total} was then 2.52 km km^{-2} and V_{a} 60 × 10^{3} m^{3} km^{-2}, with 93 % of the gully network being highly active (Fig. 10.5a–c). This corresponds to a soil losses by gully erosion (*SLg*) of 17.6 ton ha^{-1} year^{-1} over the period 1963–1994. The terrestrial photographs from that period show gullies which were very active, having clear-cut walls and transporting considerable amounts of debris (Fig. 10.1). The average incision rate of gully cross sections was 0.04 m year^{-1}, and maximum incision rates of 0.13 m year^{-1} were observed in Vertisol areas (Frankl et al. 2011). At the upper gully margins, headcuts incised upslope, while at the lower ends, debris fans were deposited and incised subsequently. Long- to medium-term linear headcut retreat rates (R_{l}) were on average 3.8 ± 4.7 m year^{-1} (Frankl et al. 2012). This gully *cut* cycle, that started in the mid-1960s, lasted until ca. 2000, after which a new gully *fill* cycle was initiated.

10.4 Rehabilitation in the Twenty First Century?

As a result of huge efforts in environmental rehabilitation undertaken since the 1970s, denudation rates decreased in the highlands of Tigray—although remained fairly high in absolute terms—reducing the importance of gully erosion. Mean run-off discharge as well as sediment load and flash

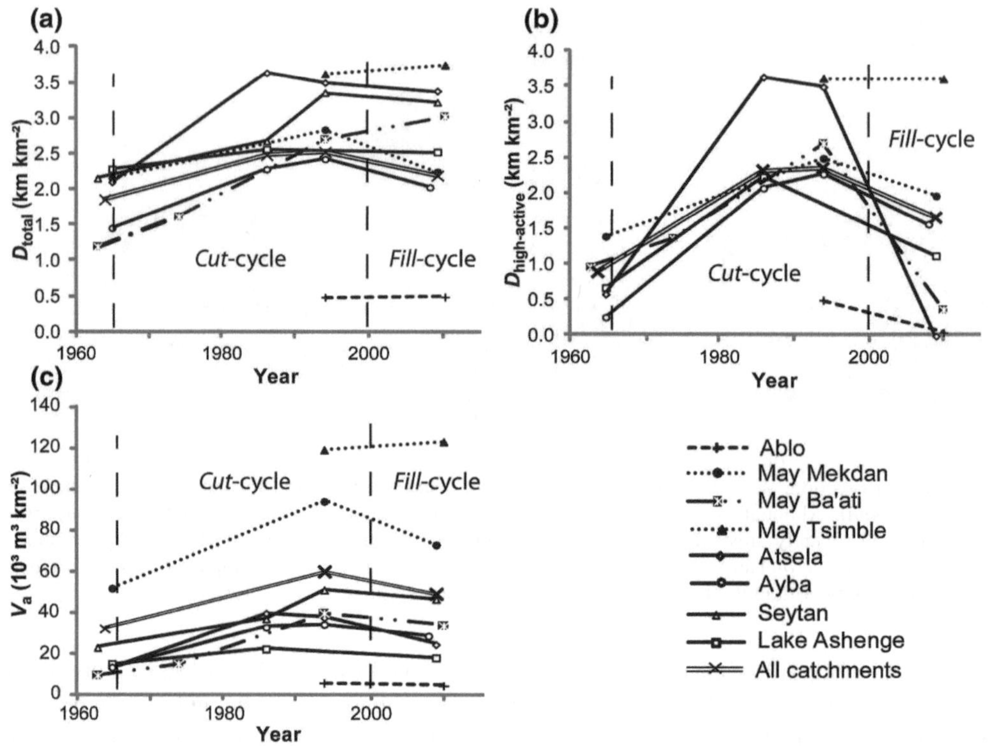

Fig. 10.5 Gully *cut*-and-*fill* cycle during the period 1963–2010 as expressed by trends in total drainage density (D_{total}) (**a**), drainage density of the high-active gullies ($D_{high\text{-}active}$) (**b**) and area-specific volume development (V_a) (**c**). Based on Frankl et al. (2013b)

flood peaks decreased (cf. Moeyersons 1989, 1990), causing gullies to become less active and to fill in partially, especially when check dams were installed (Frankl et al. 2013b; Fig. 10.6a). When proper land management was applied, gullies could even be transformed into a green linear oasis which contributes to the ecological restoration of the degraded area. Moreover, their resilience against the effects of drought or land use changes on run-off response of the land increased. Lower ends of gully channels became non-active and even migrated upslope.

This gully *fill* cycle is well evidenced by the analysis of aerial photographs and more recent satellite images (Frankl et al. 2013b). This analysis indicates that average D_{total} and V_a decreased to 2.20 km km^{-2} and 48.96 × 10^3 m^3 km^{-2}, respectively (Fig. 10.5a–c). Even more important is the decrease in average $D_{high\text{-}active}$ to 1.65 km km^{-2}, indicating that 25 % of the gully network had become low-active. These findings are also supported by repeat-photography studies, indicating that in 2006–2009, about 23 % (n = 8) of the channel cross sections were stabilized and had cross-sectional characteristics similar to those of the pre-1963 gullies (Frankl et al. 2011). Among the other 31 gullies and river sections studied from repeat photography, 44 % were highly active, whereas 23 % were in a transitional stage. Strong channel degradation can be the result of a clear-water effect, which causes the gullies in the valley bottoms to incise. This can be facilitated by the presence of thick, erodible alluvio-colluvial deposits in valley bottom position. However, strong degradation can also be the result of local factors, such as different slope/channel coupling, and connectivity or catchment scale, with large catchments having longer reaction times than small catchments. The reduced gully erosion rate is also apparent from the headcut retreat rates studied by Frankl et al. (2012). Present-day linear headcut retreat rates (R_l) are much smaller than those typical for the medium to longterm, with an average R_l of 0.34 ± 0.49 m year^{-1}. However, gullying in Vertisols remains very active as gully development is largely controlled by soil piping. In Vertisols, present-day headcut retreat rates up to 1.93 m year^{-1} were recorded and large gabion (wire net structures, filled with rocks) check dams, which are cost- and labour-intensive, were sometimes bypassed in one rainy season, forcing the gully to expand laterally into the adjacent land (Fig. 10.6b).

10.5 Factors Controlling Gully Erosion

10.5.1 Hydrogeomorphic Phases and the Role of Land Use and Precipitation

From ca. 1868 to 1965, in a first hydrogeomorphic phase, gullies were mostly low-active, displaying smooth (vegetated) cross sections. This indicates that environmental

Fig. 10.6 Effectiveness of check dams as a measure to control gully erosion. **a** Siltation behind gabion check dams caused the gully channel to fill in by approximately one-third of its depth in a catchment where slope run-off response decreased thanks to the implementation of soil and water conservation measures. Notice also on the left gully bank soil pits that are dug to plant trees, **b** at another location, the presence of a Vertisol lens in the lower soil profile (*white arrow*) caused the gabion check dam to be bypassed in one rainy season, forcing the gully flow to erode the adjacent land. Photographs taken in 2009 (**a**) and 2011 (**b**) by Amaury Frankl 13.65°N, 39.21°E

vulnerability did not yet reach a critical point for large-scale channel expansion and degradation to occur. After 1965, a marked transition from low- to high-active gullies took place in a second hydrogeomorphic phase. This is most probably related to arid pulses that occurred in the 1970s and 1980s. A similar phenomenon was observed in Senegal (Poesen et al. 2003). Such phases alter biomass production and increase human pressure on land and vegetation. In order to secure food production, farmers were forced to cultivate steeper land and overgrazing removed most of vegetation from the hillslopes. Analyses of region-wide land use and cover on the basis of Landsat imagery by de Mûelenaere et al. (2013) in the 1970s and 1980s confirmed that in 1984/1986, the area covered by bare soil was extensive and that the area covered by cropland peaked. From the analysis of land use and land cover on old terrestrial photographs, Meire et al. (2013) also indicated a minimum in vegetation cover in the period 1940s–1990s. Frankl et al. (2013a) showed that the length of the crop growth period decreases with increasing drought in the Tigray highlands, making croplands very vulnerable to high-intensity rainfall during the summer rainy season.

Since ca. 2000, the large-scale implementation of soil and water conservation measures started to yield positive effects on the environmental rehabilitation and on the stabilization of gullies. Several studies indeed indicate that vegetation cover and land management strongly improved in recent decades (e.g. Gebremedhin et al. 2004; Munro et al. 2008; Alemayehu et al. 2009; Mekuria et al. 2009; Nyssen et al. 2009; de Mûelenaere et al. 2013; Meire et al. 2013). As a reaction to severe land degradation that stroke northern Ethiopia in the 1970s and 1980s, environmental rehabilitation programs were launched with the aim of increasing land resilience to the effects of droughts. Biophysical conservation measures that were implemented include the following: (1) the establishment of exclosures in critical steep-sloped zones (Descheemaeker et al. 2006); (2) the introduction of stone bunds (Nyssen et al. 2008) and soil trenches; and (3) the construction of check dams in gullies (Fig. 10.6a; Nyssen et al. 2004). At present, exclosures cover 10–15 % of the land surface and stone bunds are found at an average density of 57 km km^{-2} (Schumacher 2012). This led to greening of the landscape in which the surface covered by bushland, forest or *Eucalyptus* plantation strongly increased. This greening is partly the result of the introduction of *Eucalyptus* trees to support the growing need of construction wood in decades where population is remarkably increasing. At a national level, population size almost doubled, from 40 million in 1980 (Maddison 2006) to 88.4 million in 2013 (FAOSTAT 2013). In Tigray, population size increased from 3.1 in 1994 to 4.3 million in 2007, representing 6 % of the total Ethiopian population (CSA 2008), whereas population density increased from 63 to 86 persons km^{-2}.

10.5.2 Check Dams as Soil and Water Conservation

In addition to soil and water conservation measures implemented in the gully catchments (e.g. stone bunds, soil trenches), the construction of loose rock or gabion check dams in gullies is a widely used conservation measure in the highlands of Tigray. Technically, the aim is to transform the long, uniform and steep longitudinal profile of the gully bed into successive, nearly horizontal steps (Ayres and Scoates 1939; Hurni 1986) by trapping sediment behind the dams.

The success of their implementation relies on the availability of loose rocks in the immediate surroundings and the free labour programs, making such dams installation community-based and cost-effective (Gebremedhin and Swinton 2003). However, check dams commonly collapse. As indicated by Nyssen et al. (2004), 39 % of loose rock dams collapse after two years which is a phenomenon strongly linked to the drainage area and slope gradient of the soil surface next to the gully, the product of these factors being a measure of run-off energy. In Vertisol areas, where soil piping occurs, flow bypassing of check dams is commonly observed and can even result in the collapse of large gabion check dams during a single rainy season (Nyssen et al. 2004; Frankl et al. 2012; Fig. 10.6b). It is therefore very important that the technical instructions (BoANR 1997; Hurni 1986; Nyssen et al. 2004) are followed when implementing new gully control structures and that maintenance is organized regularly. Moreover, Vertisol areas with piping require specific measures to prevent flow bypassing. More research is also necessary on the extent to which check dams delay run-off response and enhance infiltration.

10.5.3 Direct Human Intervention

By engineering the landscape through road and drainage canal construction, humans may cause important modifications to natural run-off pathways. As shown by Nyssen et al. (2002), a 6.5-km-long new road segment resulted in the development of 16 new gullies (total volume of 10,034 m^3) and the stabilization of five small gullies (with a total volume of 100 m^3). In these newly developed gullies, linear headcut retreat rates (R_l) were up to 10 times greater than those recorded in the gullies with no changes in their catchment area, with average R_l for the period 1994–2010 being 21.3 m $year^{-1}$. Distances between culverts are commonly large, and therefore, sites that received run-off from a relatively small drainage area before road construction may experience important increases in run-off because of the catchment area increase. In order to avoid new gully heads developing after building mountain roads, appropriate engineering works (e.g. flow energy dissipaters and/or splitters) should be undertaken during road construction.

10.5.4 Solutions for Vertisols

Soil piping is recognized as an important reason for the development of gullies (Valentin et al. 2005). In Vertisols, this is related to their periodic shrinking and swelling. In a dry Vertisol, shrinking results in the development of wide cracks, which can be as much as two metres deep. Run-off water subsequently infiltrates into the subsoil (bypass flow) and drains underground. Intense subsurface erosion of the dry, dispersive clays results in the development of soil pipes, which, once collapsed, may turn into gullies (Fig. 10.6b). When a Vertisol gets wet, swelling causes cracks to close and the Vertisol becomes almost impermeable. Consequently, run-off production is very high, and large run-off volumes drain through pipes to the gully heads (Nyssen et al. 2000; Frankl et al. 2012). Reducing gully expansion in Vertisols calls for specific measures to reduce the rates of soil piping. As proposed by Frankl et al. (2012), introducing a subsurface geomembrane dam at gully heads can increase the water storage upslope of the dam and thus reduce soil piping. This has been successfully demonstrated at May Ba'ati village (13.65°N, 39.21°E).

10.6 Conclusions

Fast land degradation may occur when improper land management is applied. Most dramatic is the development of extensive and deep gully networks, which in the highlands of Tigray produce large volumes of sediment that are transported through the gully and (ephemeral) river systems since the 1960s. However, local communities have proven that this trend can be reversed. At a regional scale, since ca. 2000, gully networks are increasingly being stabilized and the landscape is re-greening. These results have to be understood within a socio-economic context of strong population growth and a low-level technological development, where most people rely on land resources for their livelihood and where the fragility of the country's economy is frequently emphasized, for example when climatic shocks such as drought cause severe food shortages and famine. Socio-economic developments and their relation to land degradation should therefore be monitored closely. With a population size which is likely to double by 2050, Ethiopia faces immense challenges. The key is to rehabilitate the land, as a resource base for food security and ecosystem services, and to strengthen and diversify the rural economy in order to make local communities less dependent on land resources. Such challenges are embraced by many local, national and international programmes and should remain high on the agenda.

References

Alemayehu F, Taha N, Nyssen J, Girma A, Zenebe A, Behailu M, Deckers J, Poesen J (2009) The impacts of watershed management on land use and land cover dynamics in Eastern Tigray (Ethiopia). Resour Conserv Recy 53:192–198

Ayres Q, Scoates D (1939) Land drainage and reclamation. McGraw-Hill, New York

Billi P, Dramis F (2003) Geomorphological investigation on gully erosion in the Rift Valley and the Northern highlands of Ethiopia. Catena 50:353–368

BoANR (1997) Water and soil conservation, forestry development. Manual for agricultural cadres. Lekatit 1989 (Eth. Cal.) (in Tigrinya). Bureau of Agriculture and Natural Resources, Mekelle

CSA (2008) Summary and statistical report of the 2007 population and housing census. Population size by age and sex. Central Statistical Agency of Ethiopia, Addis Abbaba

Daba S, Rieger W, Strauss P (2003) Assessment of gully erosion in eastern Ethiopia using photogammetric techniques. Catena 50:273–291

de Mûelenaere S, Frankl A, Haile M, Poesen J, Deckers J, Munro RN, Veraverbeke S, Nyssen J (2013) Historical landscape photographs for calibration of LANDSAT land use/cover in the Northern Ethiopian Highlands. Land Degrad Dev 25:319–335

Descheemaeker K, Nyssen J, Rossi J, Poesen J, Haile M, Moeyersons J, Deckers J (2006) Sediment deposition and pedogenesis in exclosures in the Tigray highlands, Ethiopia. Geoderma 132:291–314

FAOSTAT (2013) Statistical division of the UN food and agriculture organization. http://faostat.fao.org/

Frankl A, Nyssen J, De Dapper M, Haile M, Billi P, Munro RN, Deckers J, Poesen J (2011) Linking long-term gully and river channel dynamics to environmental change using repeat photography (Northern Ethiopia). Geomorphology 129:238–251

Frankl A, Poesen J, Deckers J, Haile M, Nyssen J (2012) Gully head retreat rates in the semiarid highlands of North Ethiopia. Geomorphology 173–174:185–195

Frankl A, Jacob M, Haile M, Poesen J, Deckers J, Nyssen J (2013a) The effect of rainfall on the spatio-temporal variability of cropping systems and duration of the crop cover in the Northern Ethiopian highlands. Soil Use Manag 29:374–383

Frankl A, Poesen J, Haile M, Deckers J, Nyssen J (2013b) Quantifying long-term changes in gully networks and volumes in dryland environments: the case of Northern Ethiopia. Geomorphology 201:254–263

Frankl A, Poesen J, Scholiers N, Jacob M, Haile M, Deckers J, Nyssen J (2013c) Factors controlling the morphology and volume (V)—length (L) relations of permanent gullies in the Northern Ethiopian highlands. Earth Surf Process Land 38:1672–1684

Frankl A, Zwertvaegher A, Poesen J, Nyssen J (2013d) Transferring google earth observations to GIS-software: example from gully erosion study. Int J Digital Earth 6:196–201

Gebremedhin B, Swinton SM (2003) Investment in soil conservation in northern Ethiopia: the role of land tenure security and public programs. Agric Econ 29:69–84

Gebremedhin B, Pender J, Tesfay G (2004) Collective action for grazing land management in crop-livestock mixed systems in the highlands of northern Ethiopia. Agricul Syst 82:273–290

Hall F (2001) Ground-based photographic monitoring. Gen Tech Rep PNW-GTR-503. U.S. Department of Agriculture, Forest Service, Pacific Northwest Research Station, Portland

HTS (1976) Tigrai Rural Development Study (TRDS). Hunting Technical Services Ltd. Government of Ethiopia and UK Ministry of Overseas Development. Hunting Technical Services, Borehamwood

Hurni H (1986) Guidelines for development agents on soil conservation in Ethiopia. Soil conservation research project. Community Forests and Soil Conservation Development Department, Ministry of Agriculture, Addis Ababa Ethiopia

Graf WL (1988) Fluvial processes in dryland environments. The Blackburn Press, Caldwell

Kassas M (1995) Desertification: a general review. J Arid Environ 30:115–128

Knighton D (1998) Fluvial forms and processes—a new perspective. Hodder Education, London

Maddison A (2006) The world economy. OECD, Development Centre Studies, Paris

Meire E, Frankl A, De Wulf A, Haile M, Decker J, Nyssen J (2013) Land use and cover dynamics in Africa since the nineteenth century: warped terrestrial photographs of North Ethiopia. Reg Environ Change 13:717–737

Mekuria W, Veldkamp E, Haile M, Gebrehiwot K, Muys B, Nyssen J (2009) Effectiveness of exclosures to control soil erosion and local community perception on soil erosion in Tigray, Ethiopia. Afr J Agric Res 4:365–377

Moeyersons J (1989) La nature de l'érosion des versants au Rwanda. Annales, Kon. Mus. Mid. Afr., Tervuren 19, Tervuren

Moeyersons J (1990) Soil loss by rainwash: a case study from Rwanda. Z Geomorph 34:385–408

Moeyersons J, Nyssen J, Poesen J, Deckers J, Haile M (2006) Age and backfill/overfill stratigraphy of two tufa dams, Tigray highlands, Ethiopia, evidence for Late Pleistocene and Holocene wet conditions. Palaeogeogr Palaeoclimatol Palaeoecol 230:165–181

Moges A, Holden NM (2008) Estimating the rate and consequences of gully development, a case study of Umbulo catchment in southern Ethiopia. Land Degrad Dev 19:574–586

Munro RN, Deckers J, Grove AT, Haile M, Poesen J, Nyssen J (2008) Soil and erosion features of the Central Plateau region of Tigrai—learning from photo monitoring with 30 years interval. Catena 75:55–64

Haregeweyn Nigussie, Poesen J, Nyssen J, Verstraeten G, de Vente J, Govers G, Deckers S, Moeyersons J (2005) Specific sediment yield in Tigray-Northern Ethiopia: assessment and semi-quantitative modelling. Geomorphology 69:315–331

Haregeweyn Nigussie, Poesen J, Nyssen J, Govers G, Verstraeten G, de Vente J, Deckers J, Moeyersons J, Haile M (2008) Sediment yield variability in Northern Ethiopia: a quantitative analysis of its controlling factors. Catena 75(1):65–76

Nyssen J, Moeyersons J, Deckers J, Haile M, Poesen J (2000) Vertic movements and the development of stone covers and gullies, Tigray Highlands. Ethiopia Z Geomorph 44:145–164

Nyssen J, Poesen J, Luyten E, Veyret-Picot M, Deckers J, Haile M, Govers G (2002) Impact of road building on gully erosion risk: a case study from the Northern Ethiopian highlands. Earth Surf Process Land 27:1267–1283

Nyssen J, Veyret-Picot M, Poesen J, Moeyersons J, Haile M, Deckers J, Govers G (2004) The effectiveness of loose rock check dams for gully control in Tigray, Northern Ethiopia. Soil Use Manage 20:55–64

Nyssen J, Poesen J, Veyret-Picot M, Moeyersons J, Haile M, Deckers J, Dewit J, Naudts J, Teka K, Govers G (2006) Assessment of gully erosion rates through interviews and measurements, a case study from Northern Ethiopia. Earth Surf Process Land 31:167–185

Nyssen J, Poesen J, Moeyersons J, Haile M, Deckers J (2008) Dynamics of soil erosion rates and controlling factors in the Northern Ethiopian highlands—towards a sediment budget. Earth Surf Process Land 33:695–711

Nyssen J, Haile M, Naudts J, Munro N, Poesen J, Moeyersons J, Frankl A, Deckers J, Pankhurst R (2009) Desertification? Northern Ethiopia re-photographed after 140 years. Sc Total Environ 407:2749–2755

Nyssen J, Frankl A, Munro RN, Billi P, Haile M (2010) Digital photographic archives for environmental and historical studies: an example from Ethiopia. Scott Geogr J 126:185–207

Berakhi Ogbaghebriel, Brancaccio L (1993) Some reflections on the origin and land use of pediments on the Ethiopian highlands. Geografia Fisica e Dinamica del Quaternario 16:101–106

Pankhurst R (1995) The history of deforestation and afforestation in Ethiopia prior to world war I. Northeast Afr Stud 2:33–119

Poesen J, Nachtergaele J, Verstraeten G (2003) Gully erosion and environmental change, importance and research needs. Catena 50:91–133

Reubens B, Poesen J, Nyssen J, Leduc Y, Zenebe A, Tewoldeberhan S, Bauer H, Gebrehiwot K, Deckers J, Muys B (2009) Establishment and management of woody seedlings in gullies in a semi-arid environment (Tigray, Ethiopia). Plant Soil 324:131–156

Schumacher M (2012) Recent trends in gully erosion as evidenced by repeat photography around Hagere Selam (Northern Ethiopia). Unpublished Master Thesis. Institute for Geography, Technical University of Dresden, Dresden

Schumm SA (2005) River variability and complexity. Cambridge University Press, Cambridge

Sharf FA, Northrup D, Pankhurst R (2003) Abyssinia, 1867–1868: artists on campaign: watercolors and drawings from the British expedition under Sir Robert Napier. Tsehai Publishers, Addis Ababa

Valentin C, Poesen J, Li Y (2005) Gully erosion, impacts, factors and control. Catena 63:132–153

Virgo KJ, Munro RN (1978) Soil and erosion features of the Central Plateau region of Tigrai, Ethiopia. Geoderma 20:131–157

Author Biographies

Amaury Frankl Amaury Frankl (1982, Ghent) specialized in hydrogeomorphology at the Department of Geography of Ghent University (Belgium). His PhD. research was on gully development and its spatiotemporal variability since the late nineteenth century in the highlands of Tigray.

Jean Poesen (1954, Fataki, D.R.C.), a full professor and head of the Division of Geography at KU Leuven, specialized in physical geography with a focus on soil erosion, desertification, soil and water conservation. He initiated and collaborated in several research projects in Ethiopia.

Jan Moeyersons (1947, Willebroek) is an expert on gullying and landslides in tropical Africa and worked from many years at the Royal Museum for Central Africa (Belgium).

Jan Nyssen (1957, Sint-Martens-Voeren) has been based for many years at Mekelle University (Ethiopia) where he was research coordinator for various university cooperation programmes between Belgium and Ethiopia. Nowadays, he is a full professor at the Department of Geography of Ghent University (Belgium) where he leads the Physical Geography Research Group. He initiated and collaborated in several research projects in Ethiopia.

Tufa Dams in Tigray (Northern Ethiopia) as Late Pleistocene—Holocene Climate Proxies

11

Francesco Dramis and Giandomenico Fubelli

Abstract

The geomorphological–stratigraphic study of the backfill deposits of two tufa dams, Mai Makden and Tsabati Mariam, supported by ^{14}C dates, has provided a detailed description of the environmental changes occurred during the Holocene. In particular, the evolution of the investigated tufa dams points out the occurrence of some century-scale stages of tufa deposition (10.9–9.3; 8.4–7.4; 6.6–5.5 kyrs BP) interrupted by intervals characterized by lower or absent deposition and dam incision (9.3–8.4; 7.4–6.6 kyrs BP). Since 5.5–2.5 kyrs BP, the deposition rates of tufa progressively declined until stream erosion incised the dams down to the underlying bedrock. The sequence of tufa events in Tigray seems to be parallel to the record of lake-level fluctuations in the Ethiopian Rift with the high stands corresponding to the main deposition stages of tufa and the low stands corresponding to the non-deposition/erosion intervals. Quite interesting, in this context, is the progressive lowering of the lake levels after ca. 5.5 kyrs cal BP. The tufa record of Tigray also shows some correlation with the cooling–warming stages of the Mediterranean Sea and, more in general, with the main peaks of the Holocene global temperatures. The above relationships emphasize the primary role of rainfall and air temperature as factors controlling the deposition of tufa. This does not exclude human impact as an additional factor for the deposition of tufa.

Keywords

Tufa • Holocene • Paleoclimate • Ethiopia • East Africa

11.1 Introduction

Remains of tufa dams of different ages are commonly found across rivers in limestone areas (Ford and Pedley 1996). Behind such natural dams, swampy–lacustrine sedimentary sequences, consisting of clayey deposits with peaty levels and alternations of phytoclastic travertine, travertine sand layers and buried soils, may be locally observed (Pentecost 2005). The growth of tufa dams occurs where the deposition rate of calcium carbonate from flowing water is high enough to balance the stream flow erosion (Fubelli et al. 2013).

There is general agreement in referring the aggradation/degradation phases of tufa dams during geological times to climate controls. Warm/wet climates are believed to favour tufa aggradation because of: (1) higher concentration of biogenic CO_2 in soil layers, resulting in higher rates of limestone dissolution (Atkinson 1977; Brook et al. 1983); (2) higher air temperature at the springs favouring water outgassing (Gullentops and Mullenders 1972); (3) faster development of aquatic plants and related absorption of CO_2 for photosynthesis (Pedley 1990). Conversely, cold/dry climates are considered to be less favourable for $CaCO_3$ precipitation because of reduced biological activity of soils,

F. Dramis (✉) · G. Fubelli
Department of Sciences, Roma Tre University, Rome, Italy
e-mail: dramis@uniroma3.it

P. Billi (ed.), *Landscapes and Landforms of Ethiopia*, World Geomorphological Landscapes,
DOI 10.1007/978-94-017-8026-1_11, © Springer Science+Business Media Dordrecht 2015

lower air temperatures at the springs and lesser development of aquatic plants (Pentecost 2005).

An additional model to explain the increase/decrease of tufa deposition rates (Dramis et al. 1999) makes reference to variations of thermal gradient between the ground surface and the deep limestone aquifers induced by climate changes. Because of the extremely low thermal conductivity of the rocks (Vasseur et al. 1983) and the related slow penetration of thermal changes into the ground, differences of temperature up to several degrees and reversed thermal gradients may be produced between the ground surface and the underlying bedrock over timescales ranging from years to thousands of years in relation to thermal change magnitude and the aquifer depth (Williams and Smith 1989). With climatic changes to warmer conditions, water percolating through progressively colder layers in the vadose belt as well as in the phreatic zone undergoes progressive enrichment in dissolved $CaCO_3$ (Thrailkill 1968; Atkinson 1977). At the emergence, because of higher surface temperatures, the spring water becomes over-saturated with $CaCO_3$, thus inducing precipitation of tufa at waterfalls, knick points or on the river bed itself. Tufa aggradation may continue for a long time, even if with progressively lower deposition rates, till the exhaustion of the thermal disturbance in the ground. Opposite effects, such as deposition of carbonate in the upper bedrock and emergence of spring waters under-saturated with $CaCO_3$, should be expected with the onset of a climatic change to cold conditions.

Taking into account the dropping of CO_2 concentrations in soils due to man-made deforestation and other anthropogenic effects, such as changes in stream hydrology and increase of water turbidity, several authors (e.g. Nicod 1986; Goudie et al. 1993) have stressed the effects of human impact to explain the Late Holocene decline of tufa deposition observed in several localities of Europe and North Africa. In any case, the aggradation of tufa dams in the paleoclimatic record indicates a climatic regime characterized by warm–wet conditions with dense vegetation and well-developed soils. Higher rates should be expected in connection with cold to warm climate changes. On the contrary, shifting of climate towards drier or cooler conditions would induce decreasing deposition rates and the incision of dams by streams. At a more local scale, the same effects may be caused by man-made deforestation and related soil erosion on slopes.

11.2 Tufa Dams in Tigray

Due to the widespread outcrop area of Jurassic limestone (Katzmin 1972), the highlands of eastern Tigray (northern Ethiopia) are characterized by the occurrence of a large number of tufa dam remnants (Fig. 11.1), the age of which ranges from Middle Pleistocene to Holocene (Chernet and Eshete 1982; Moeyersons et al. 2006).

Some of these dams have a spectacular appearance from the geomorphological point of view. This is the case of the Romanat dam (Fig. 11.2), dated at the Late Pleistocene (Moeyersons et al. 2006), and the Mai Makden dam (Fig. 11.3) whose backfill has been the subject of study in this chapter. The Romanat dam (called Romanoff waterfall in Google Earth) shows an impressive scenery characterized by a 50-m-high tufa wall overlying an extensively jointed dolerite dyke and a great pool with several potholes in the Antalo Limestone (Mohr 1962). The jump, due the different erodibility between the two lithologies favoured the formation of the tufa dam at this point. The Mai Makden dam (Fig. 11.4) is built up over horizontally layered limestone (Antalo Limestone; Mohr 1962) and consists of at least three different units of phytohermal travertine (Pedley 1990) separated by coarse gravels and weathered material, attaining a total height of ca. 20 m. It is presently incised by the Mai Makden River that forms a narrow gorge. Below the dam, there are slots and potholes carved in the limestone bedrock by the waterfall that exists downstream from the dam incision. The backfill extends upstream for more than 1 km as far as Mai Makden Village where another dam is present.

The present climate of the area, located at the extreme northern border of the ITCZ migration, at a mean elevation of more than 2,000 m a.s.l., is characterized by annual air temperatures ranging between 15 and 25 °C and an annual rainfall ranging from 700 to over 1,200 mm, mostly concentrated in the summer (Griffiths 1962; CHP 4, this volume). The impact of human activity on natural vegetation has been particularly effective in the area since the second millennium BC (Butzer 1981; Fattovich 1990; Bard et al 2000). Though the present climatic conditions would allow Juniperus and Olea forest and mixed deciduous Juniperus woodland (Ethiopian Mapping Authority 1988), arboreal vegetation is extremely reduced, exception made for a few trees clustered around churches and monasteries and in the less accessible areas.

Two stratigraphic sequences of tufa-dammed swampy–lacustrine–alluvial deposits were investigated in Tigray (Brancaccio et al. 1997; Berakhi et al. 1998; Dramis et al. 2003) in detail. The first one is located downstream of the Mai Makden village, north of Mekelle, whereas the second crops out near the church of Tsabati Mariam, about 12 km east of Agula (Fig. 11.5). In both cases, bedrock consists of horizontally layered limestone (Antalo limestone, Merla and Minucci 1938). Both the Mai Maikden and the Tsabati Mariam streams, which deposited the tufa dams and subsequently incised them, are fed by spring waters (Chernet and Eshete 1982).

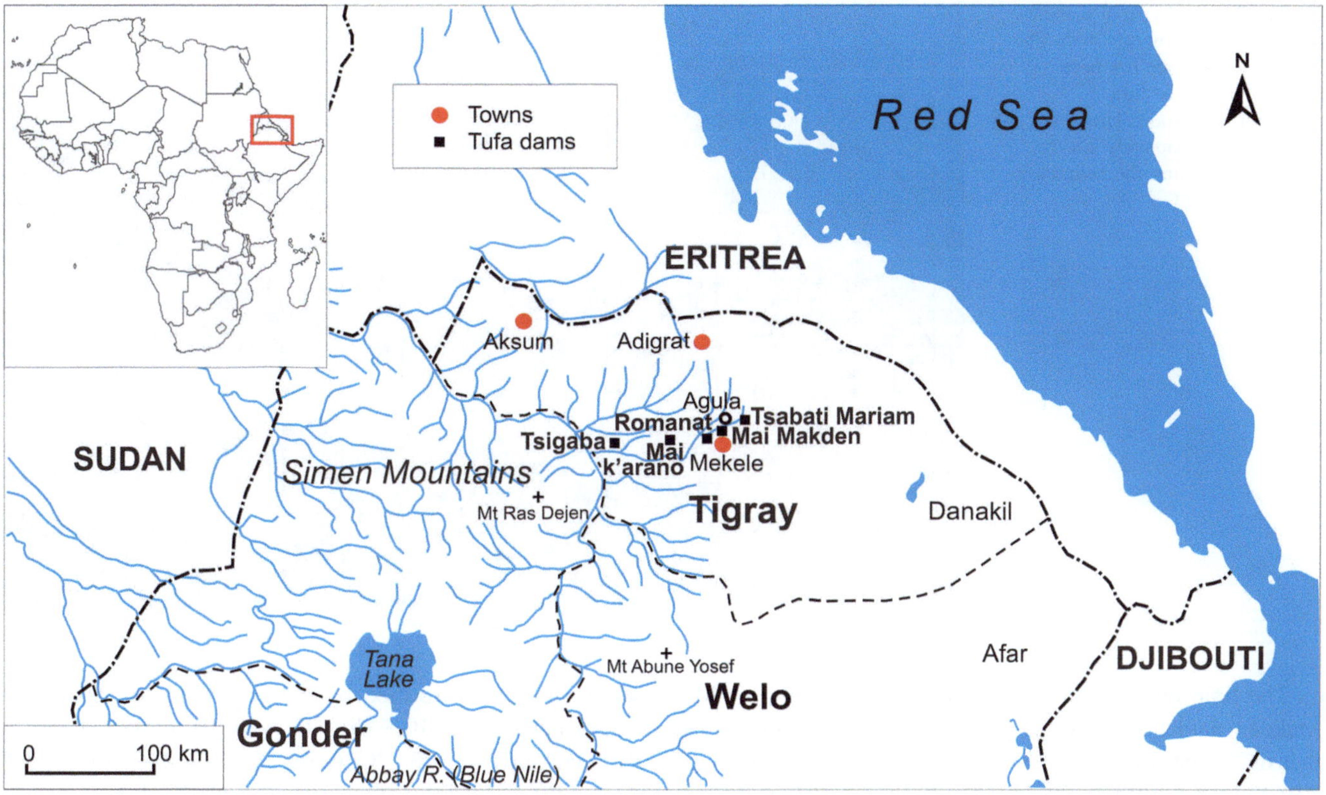

Fig. 11.1 Location map

Fig. 11.2 Picturesque tufa dam at Romanat, north-west of Mekele. The tufa wall overlies a dolerite dyke; a large pool is present at the base of the waterfall. Above the dam remnants of an indigenous forest can be seen

The temporal evolution (nine ^{14}C dates) of the Tsabati Mariam dam backfill (Fig. 11.6) spans from Early to Late Holocene: 9,510 ± 100 ^{14}C years BP (11,046–10,675 cal BP) to 2,380 ± 50 ^{14}C years BP (2,592–2,367 cal BP). The latest date refers to the uppermost organic-rich layer underlying a thin layer of alluvial gravels.

Fig. 11.3 A view from distance of the Mai Makden tufa dam. The alluvial top surface has been formed by the last tufa aggradation episode. The river incision has completely cut the tufa dam, reaching the bedrock

The time interval (fifteen ^{14}C dates) covered by the Mai Makden dam backfill (Figs. 11.7 and 11.8) is shorter than that of Tsabati Mariam (7,630 ± 80 to 3,450 ± 50 ^{14}C years BP) since the previous backfill has been completely eroded. However, the stratigraphic sequence here is much better exposed. Organic-rich levels and peaty layers from the lower part of the backfill swampy sequence were dated between 7,630 ± 80 ^{14}C years BP (8,521–8,382 cal BP) and 4,710 ± 70 ^{14}C years BP (5,552–5,355 cal BP). The latter reading constrains the last date of the swampy environment. A noteworthy desiccation event occurred during the swampy sedimentation as testified by the occurrence of desiccation cracks between 6,510 ± 70 ^{14}C years BP (7,485–7,346 cal BP) and 5,610 ± 70 ^{14}C years BP (6,472–6,333 cal BP) (Fig. 11.9). The following evolution of the dam is characterized by alternating phases of erosion and aggradation. The topmost dated level, a buried Vertisol underlying a layer of travertine sand, about 1.5 m below the summit surface of the backfill sedimentary sequence, yielded an age of 3,450 ± 50 ^{14}C years BP (3,808–3,656 cal BP). Subsequently, the tufa dam and the whole backfill were completely incised by the Mai Makden stream.

Fig. 11.4 The impressive front of the Mai Makden tufa dam. Three different levels of tufa (*T*) separated by two fluvial/debris flow events (*F*) are visible

Fig. 11.5 The Tsabati Mariam tufa dam

U/Th dates from other tufa dams of Tigray are reported by Moeyersons et al. (2006): 261.9 + 61.3/−38.8 kyrs BP; 44.4 ± 1.0 kyrs BP; 31.8 + 1.1/−1.2 kyrs BP; 28.7 + 1.4/−1.5 kyrs BP; 22.5 + 0.4/−0.5 kyrs BP; 15.8 ± 1.1 kyrs BP; and 14.1 + 0.6/−0.5 kyrs BP. However, according to these authors, only the last two dates, from the Mai K'arano and Tsigaba dams, respectively, may be considered sufficiently reliable. Moreover, a date of 3,090 ± 30 ^{14}C years BP (3,355–3,276 cal BP) provided by a charcoal sample included in top layer of the Tsigaba dam backfill constrains the end of tufa deposition.

11.3 Tufa Deposition in Neighbouring Countries

Moeyersons et al. (2006) also report U/Th and ^{14}C dates of tufa deposits measured by Brandt and Brook (1984) (U/Th: between 7,600 and 5,000, 9,700 and 9,500, 11,800 years BP) and Voight et al. (1990) (11,535 ± 105 ^{14}C years BP—13,565–13,287 cal BP; 10,800 ± 80 ^{14}C years BP—12,867–12,705 cal BP; 9,105 ± 105 ^{14}C years BP—10,427–10,186 cal BP) in northern Somalia.

Fig. 11.6 The Tsabati Mariam tufa dam backfill

Fig. 11.7 The Mai Makdem tufa dam backfill showing a sequence of swampy, lacustrine and alluvial levels

Holocene tufa dates are reported also from Sudan by Szabo et al. (1995) (9.5 ± 0.2, 7.4 ± 0.2, 5.7 ± 0.2 U/Th kyrs BP) and Swezey (2001) (5,395 ± 70 ^{14}C years BP—6,266–6,064 cal BP). Comparable dates (10,220 ± 40 ^{14}C years BP—12,060–11,804 cal BP to 4,380 ± 70 ^{14}C years BP—5,175–4,894 cal BP) were obtained in southern Yemen by Sander (2006). Finally, a very recent date (2,450 ± 110 ^{14}C years BP—2,680–2,392 cal BP) of tufa deposition, well corresponding with that obtained from the Tsabati Mariam backfill sequence, is reported by Swezey (2001) from Tchad.

Fig. 11.8 Toppling affecting the Mai Makdem tufa dam backfill

Fig. 11.9 Desiccation cracks in the lower part of the Mai Makden tufa dam backfill

More to the north, in the Saharan Libya (Fezzan, Tadrat Acacus Mts.), a well-studied detailed sequence of Early Holocene tufa deposits, spanning in age from 10,030 ± 85 to 7,731 ± 54 U/Th years BP, was obtained by Cremaschi et al. (2010).

11.4 Paleoclimatic Significance of Tufa Dam Aggradation/Erosion Phases

From the chronostratigraphic record of Mai Makden and Tsabati Mariam dam backfills, supplemented with the dates provided by Moeyersons et al. (2006) from other tufa dams of Tigray (Table 11.1), it emerges that tufa aggradation started in the Late Glacial ($\sim$15.8 U/Th kyrs BP) and followed until ca. 2,380 ± 50 ^{14}C years BP (2,450–2,345 cal BP).

However, the deposition of tufa was not continuous, as indicated by the occurrence of some century-scale main intervals of less or absent deposition and dam incision, also testified by the presence of gravel levels and buried stream channels dated for 14.1–10.9; 9.3–8.4; 7.4–6.6 kyrs BP. Since 5.5–3.7 kyrs BP, the geomorphological–stratigraphic record of tufa dams in Tigray indicates a progressive decline of calcium carbonate deposition rates, likely induced by the general trend of climate to drier conditions (Umer et al. 2004).

By comparing the sequence of lake-level fluctuations in the Rift (Gillespie et al. 1983; Street-Perrott and Perrott 1990; Gasse and Van Campo 1994; Gasse 2000; Umer et al. 2004) with that of tufa aggradation/erosion in Tigray (Dramis et al. 2003), it results that during the Early Holocene both areas experienced climate conditions definitely wetter than present, associated with a significant northward shift of the summer ITCZ, increase of solar insolation and a definitely stronger south-west Indian monsoon (Overpeck et al. 1996; Glennie 1998; Fleitmann et al. 2003a, b). As a consequence, between ca. 10 and 7.5 kyrs BP, the currently hyperarid regions such as the Arabic Peninsula (Sander 2006) and the Saharan Libya (Petit-Maire and Guo 1996; Cremaschi et al. 2010) were characterized by long-lasting rains and widespread tufa deposition.

In the following times, the record of lake levels of the Ethiopian Rift (Fig. 11.10) indicates some main episodes of drier climate (CHP 17, this volume) that seem to be consistent with the main phases of less (or absent) deposition or erosion of tufa dams (Dramis et al. 2003). Quite interesting in this context, for its rough correspondence with the most recent date of tufa in Tigray (and Tchad), is the progressive lowering of the rift lake levels after ca. 4700 ^{14}C years BP, only interrupted by a short-lived high stand interval between ca. 2500 and 1500 ^{14}C years BP.

The Late Pleistocene–Holocene sequence of aggradation phases in the tufa dams in Tigray (in particular, those between 15.8 and 14.1, 10.9 and 9.3, 8.4 and 7.4, 6.6–5.5 kyrs) seems to find some correlation also with the climate warming stages recorded from the Mediterranean Sea floor sediments (Rohling et al. 2002) and, more generally, with the main peaks of the Holocene global temperature curves (Mayewski et al. 2004; Rimbu et al. 2004; Dragons Flight 2012).

Table 11.1 Tufa dams dates and environmental changes in East Africa and the Mediterranean

Tigray Mai Makden dam	Tigray Tsabati Mariam dam (cal BP)	Tigray Mai K'Arano and Tsigaba dams	Somalia (cal BP)	Rift lakes level	Mediterranean sea
		~15.8 U/Th BP		?	Warming
		~14.1 U/Th BP			
				Low	
			13,565–13,287	High	
			12,867–12,705		
				Low	Cooling
	11,046–10,675		10,427–10,186	High	Warming
	10,730–10,483				
	10,215–9,965				
	9,408–9,166				
Hyatus (ca. 0.8 kyrs)				Low	Cooling
8,521–8,382 cal BP	8,496–8,346			High	
Hyatus? (ca. 0.3 kyrs–8.2 event?)				Low	
8,247–8,041 cal BP	7,960–7,854			High	Warming
8,012–7,867 cal BP	7,598–7,490				
8,016–7,788 cal BP					
7,612–7,337 cal BP					
7,557–7,451 cal BP					
7,485–7,346 cal BP					
Hyatus (ca 0.8 kyrs)				Low	Cooling
6,472–6,333 cal BP	6,686–6,519			High	Warming
6,423–6,292 cal BP	5,582–5,383				
6,234–6,035 cal BP					
6,159–5,962 cal BP					
6,012–5,792 cal BP					
5,926–5,775 cal BP					
5,552–5,355 cal BP					
Hyatus (ca. 1.7 kyrs)				Low	Cooling
3,808–3,656 cal BP	2,592–2,367	3,355–3,276 cal BP		High	Warming

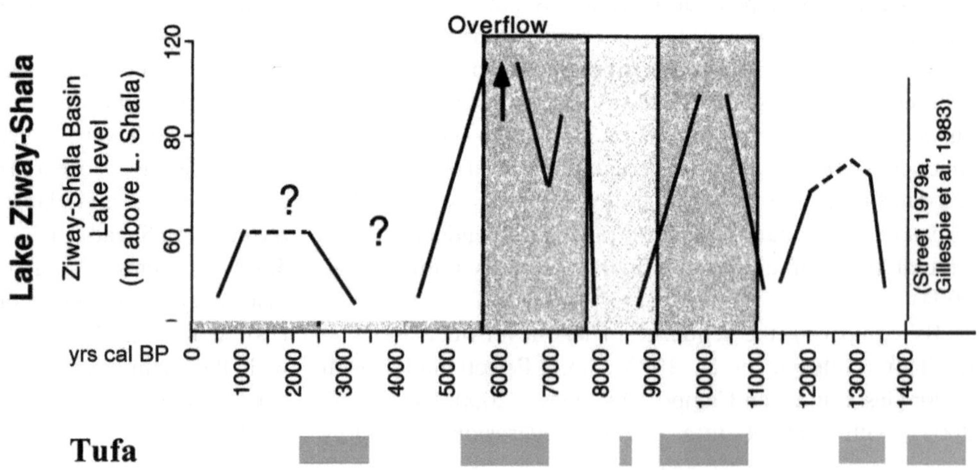

Fig. 11.10 Late Pleistocene–Holocene fluctuation of the Ethiopian Rift lakes compared with the aggradation phases of tufa dams in Tigray

In this context, it is interesting to notice that Holocene aggradation/erosion sequences of tufa dams, definitely comparable with those investigated in Tigray, have been reported from Central Italy (Fubelli et al. 2013; Dramis et al. 2014) and other parts of south-central Europe (Weisrock 1986; Vaudour 1994; Žák et al. 2002; Hlaváč et al. 2003; Lespez et al. 2005; Ollivier et al. 2005; Wehrli et al. 2010). From the above considerations, it is possible to establish the following list of events in East-Central Africa and the Mediterranean area:

(a) tufa deposits formed between ~15.8 U/Th kyrs BP and 12,867–12,705 cal BP (Somalia; Voight et al. 1990) in the Late Glacial period, characterized by climate warming in the Mediterranean Sea; a low-stand stage in the rift lakes after ca. 14 kyrs BP could correspond to a short dry interval (Older Dryas?) not recorded in the Mediterranean Sea;
(b) the 12,867–12,705 cal BP to 11,046–10,675 cal BP interval of no deposition seems to fit well with the dry–cold Younger Dryas;
(c) the 11,046–10,675 to 9,408–9,166 cal BP phase of tufa dam aggradation coincides with the rapid 'postglacial' rise of sea temperature recorded at planetary level;
(d) the 9,408–9,166 to 8,521–8,382 cal BP interval of no deposition seems to correspond to a cooling period of the Mediterranean Sea;
(e) the tufa aggradation intervals recorded in Tigray between 8,521 and 8,382 cal BP and 7,485 and 7,346 cal BP coincide with a period characterized by an increased wetness in Africa (Petit-Maire and Guo 1996) and warm sea temperature in the Mediterranean area; in this context, a short (8,496–8,346 to 8,247–8,041 cal BP) no-deposition interval in the Tigray sequence, even if not well expressed, could correspond to the 8.2 kyrs BP cooling event (Alley et al. 1997) registered in the tufa record of Fezzan (8,619 ± 57 to 8,197 ± 56 U/Th years BP in Fezzan; Cremaschi et al. 2010) and in the Mediterranean Sea;
(f) the 7,485–7,346 to 6,472–6,333 cal BP interval of low deposition rates of tufa coincides with a new cooling stage of the Mediterranean Sea and a trend to increasing aridity in the Sahara (Petit-Maire and Guo 1996); this interval is pointed out by the occurrence of well developed desiccation cracks in the swampy sequence at the base of the Mai Makden dam backfill;
(g) unlike Saharan Libya and Arabic peninsula, between 6,472–6,333 and 5,552–5,355 cal BP, the monsoon continued to keep the Tigray highlands humid enough (Arz et al. 2003; Fleitmann et al. 2003a, b) to allow for tufa deposition, at least in the more elevated areas;
(h) the progressive decline of tufa deposition with alternating short-lived aggradation/erosion events recorded at Mai Makden between 5,552–5,355 and 3,808–3,656 cal BP seems to coincide with a progressive decrease of humidity in the Sahara (Petit-Maire and Guo 1996) paired by a cooling trend of the Mediterranean Sea;
(i) the 2,592–2,367 cal BP depositional event, likely induced by a short-lived rise of humidity, as that registered by the last high stand of the southern Ethiopia lakes, marks the end of tufa deposition.

The relationships between climate changes and the evolution of tufa dams (Table 11.1) definitely confirm the influence of water availability and temperature on the rates of tufa deposition. Specifically, the rapid growth of tufa dams observed in both East Africa and Europe with changes from cold to warm temperatures, as at the transition from phases (b) to (c) or (d) to (e), and the fall of tufa deposition rates at warm to cold transition, seem to provide some support to the surface/ground temperature model proposed by Dramis et al. (1999). The exhaustion of the postglacial ground/surface thermal disequilibrium, coupled with the progressive aridification of climate in both East Africa and southern Europe (Petit-Maire and Guo 1996; Giraudi et al. 2011; Zanchetta et al. 2012), might also explain the overall decline of tufa deposition rates and the ultimate incision of tufa dams by rivers down to the underlying bedrock, observed in Ethiopia, in Central Italy (Dramis et al. 2003, 2014; Fubelli et al. 2013) and in different parts of Europe as well (Goudie et al. 1993).

11.5 Conclusions

The sequence of Late Pleistocene–Holocene aggradation/erosion phases of tufa dams in northern Ethiopia, as compared with the level fluctuations of the rift lakes and the temperature changes at the planetary level, seems to confirm the crucial control of water availability and surface temperature on the rates of calcium carbonate deposition. The rapid growth of tufa dams observed with cold to warm climate changes seems to provide some support to the surface/ground temperature model proposed by Dramis et al. (1999). In this context, the exhaustion of the postglacial ground/surface thermal disequilibrium, coupled with the progressive aridification of climate, might explain the overall decline of tufa deposition rates and the ultimate incision of dams by streams.

Some tufa dams in Tigray can be easily visited. The impressive Mai Makden dam is located next to the road connecting Mekele and Adigrat, only c. 15 km north of the former and well visible from it. The Romanat dam, c. 20 km to the NW of Mekele, is also easily accessible, whilst its geomorphological significance is enhanced by the surrounding scenery and associated bedrock channel landforms. Above the dam and the incision, one can see the remnants of an indigenous forest in the region and a monastery.

Acknowledgments Marek Kasprzak is acknowledged for improving Fig. 11.1.

References

Alley RB, Mayewski PA, Sowers T, Stuiver M, Taylor KC, Clark PU (1997) A prominent widespread event 8,200 years ago. Geology 25:483–486

Arz HW, Lamy F, Päzold J, Müller PJ, Prins M (2003) Mediterranean moisture source for an early-holocene humid period in the northern Red Sea. Science 300:118–121

Atkinson TC (1977) Carbon dioxide in the atmosphere of the unsaturated zone: an important control of hardness in limestones. J Hydrol 35:111–123

Bard KA, Coltorti M, DiBlasi MC, Dramis F, Fattovich R (2000) The environmental history of Tigray (Northern Ethiopia) in the middle and late holocene: a preliminary outline. Afr Archaeol Rev 17 (2):65–86

Berakhi O, Brancaccio L, Calderoni G, Coltorti M, Dramis F, Umer M (1998) The May Makden sedimentary sequence: a reference point for the environmental evolution of the Highlands of Northern Ethiopia. Geomorphology 23(2–4):127–138

Brancaccio L, Calderoni G, Coltorti M, Dramis F, Ogbaghebriel B (1997) Phases of soil erosion during Holocene in the Highlands of Western Tigray (Northern Ethiopia): a preliminary report. In: Bard KA (ed) The environmental history and human ecology of Northern Ethiopia in the Late Holocene. Studi Africanistici, ser. Etiopica 5. Istituto Universitario Orientale, Napoli, pp 29–44

Brandt SA, Brook GA (1984) Archaeological and paleoenvironmental research in Northern Somalia. Curr Anthropol 25(1):119–121

Brook GA, Folkoff ME, Box EO (1983) A world model of soil carbon dioxide. Earth Surf Proc Land 8:79–88

Butzer KW (1981) Rise and fall of Axum, Ethiopia: a geoarcheological interpretation. Am Antiq 46:471–495

Chernet T, Eshete G (1982) Hydrogeology of the Mekele area (ND37-11). Ministry of Mines and Energy, Addis Ababa

Cremaschi M, Zerboni A, Spötl C, Felletti F (2010) The calcareous tufa in the Tadrart Acacus Mts. (SW Fezzan, Libya). An early Holocene palaeoclimate archive in the central Sahara. Palaeogeogr Palaeocl 287:81–94

Dragons Flight (2012) Holocene temperature variations. http://commons.wikimedia.org/wiki/File:Holocene_Temperature_Variations.png

Dramis F, Materazzi M, Cilla G (1999) Influence of climatic changes on freshwater travertine deposition: a new hypothesis. Phys Chem Earth PT A 24(10):893–897

Dramis F, Umer M, Calderoni G, Haile M (2003) Holocene climate phases from buried soils in Tigray (northern Ethiopia): comparison with lake level fluctuations in the main Ethiopian Rift. Quat Res 60:274–283

Dramis F, Fubelli G, Calderoni G, Esu D (2014) Holocene aggradation/degradation phases of tufa dams in northern Ethiopia and central Italy: a palaeoclimatic comparison between East Africa and Mediterranean Europe. Z Geomorphol. doi:10.1127/0372-8854/2014/0140

Ethiopian Mapping Authority (1988) National atlas of Ethiopia. Addis Ababa

Fattovich R (1990) Remarks on the Pre-Axumite period in Northern Ethiopia. J Ethiop Stud 23:1–3

Fleitmann D, Burns SJ, Mudelsee M, Neff U, Kramers J, Mangini A, Matter A (2003a) Holocene forcing of the Indian monsoon recorded in a stalagmite from Southern Oman. Science 300:1737–1739

Fleitmann D, Burns SJ, Neff U, Mangini A, Matter A (2003b) Changing moisture sources over the last 330,000 years in Northern Oman from fluid-inclusion evidence in speleothems. Quat Res 60:223–232

Ford TD, Pedley HM (1996) A review of tufa and travertine deposits of the world. Earth-Sci Rev 41(3–4):117–175

Gasse F (2000) Hydrological changes in the African tropics since the last Glacial maximum. In: Alverson KD, Oldfield F, Bradley RS (eds) Past global changes and their significance for the future. Quaternary Sci Rev 19: 189–211

Gasse F, Van Campo E (1994) Abrupt post-glacial climate events in West Asia and North Africa monsoon domains. Earth Planet Sci Lett 126:435–445

Gillespie R, Street-Perrott FA, Switsur R (1983) Post-Glacial arid episodes in Ethiopia have implication for climate prediction. Nature 306:680–683

Giraudi C, Magny M, Zanchetta G, Drysdale RN (2011) The Holocene climatic evolution of Mediterranean Italy: a review of the continental geologic data. Holocene 21(1):105–115

Glennie KW (1998) The desert of SE Arabia: a product of climatic change. In: Alsharhan AS, Glennie KW, Whittle GL, Kendall CGStC (eds) Quaternary deserts and climatic change. Balkema Rotterdam, pp 109–116

Goudie AS, Viles HA, Pentecost A (1993) The late-holocene tufa decline in Europe. Holocene 3(2):181–186

Griffiths JF (1962) The climate of Africa. In: Russell EW (ed) The natural resources of East Africa. Hawkins, Nairobi, pp 77–87

Gullentops F, Mullenders W (1972) Age et formation de dépôts de tuf calcaire Holocène en Belgique. In: Macar P, Pissart A (eds) Processus périglaciaires étudiés sur le terrain, Comptes Rendus du Symposium International de Géomorphologie, Liège-Caen 1971, 67:113–135

Hlaváč J, Kadlec J, Žák K, Hercman H (2003) Deposition and destruction of holocene calcareous tufa cascades in the Bohemian Karst (Czech Republic). Prace Geogr 189:225–253

Katzmin V (1972) Geological map of Ethiopia. EIGS, Addis Ababa

Lespez L, Clet-Pellerin M, Limondin-Lozouet N, Pastre J-F, Fontugne M (2005) Discontinuités longitudinals des dynamiques sédimentaires holocènes dans si petit vallèes de l'Ouest du Bassin Parisien, l'exemple de la Mue (Basse-Normandie). Quaternaire 16(4):273–298

Mayewski PA, Rohling EE, Stager JC, Karlén W, Maasch KA, Meeker LD, Meyerson EA, Gasse F, van Kreveld S, Holmgren K, Lee-Thorp J, Rosqvist G, Rack F, Staubwasser M, Schneider RR, Steig EJ (2004) Holocene climate variability. Quat Res 62:243–255

Merla G, Minucci E (1938) Missione geologica nel Tigrai. 1-La serie dei terreni. Reale Accademia Lincei, Centro Studi per l'Africa Orientale Italiana

Moeyersons J, Nyssen J, Poesen J, Deckers J, Haile M (2006) Age and backfill/overfill stratigraphy of two tufa dams, Tigray Highlands, Ethiopia: evidence for late pleistocene and holocene wet conditions. Palaeogeogr Palaeocl 230:165–181

Mohr PA (1962) The geology of Ethiopia. Addis Ababa University Press, Addis Ababa

Nicod J (1986) Facteurs physico-chimiques de l'accumulation des formations travertineuses. Mediterranée 57 (1–2):161–164

Ollivier V, Guendon J-L, Ali A, Poiron P, Ambert P (2005) Évolution postglaciaire des environments travertineux provençaux et alpins: noveau cadre chronologique, faciès et dynamiques morphosédimentaires. Quaternaire 17(2):51–67

Overpeck J, Anderson D, Trumbore S, Prell W (1996) The southwest Indian monsoon over the last 18,000 years. Clim Dyn 12:213–225

Pedley HM (1990) Classification and environmental models of cool freshwater tufas. Sediment Geol 68:143–154

Pentecost A (2005) Travertine. Springer, Berlin

Petit-Maire N, Guo Z (1996) Mis en evidence de variations climatiques Holocenes rapides, en phase dans les deserts actuels de Chine et de Nord de l'Afrique. CR Acad Sci SII 322:847–851

Rimbu N, Lohmann G, Lorenz SJ, Kim JH, Schneider RR (2004) Holocene climate variability as derived from alkenone sea surface temperature and coupled ocean-atmosphere model experiments. Clim Dyn 23:215–227

Rohling EJ, Casford JSL, Abu-Zied R, Cooke S, Mercone D, Thomson J, Croudace I, Jorissen FJ, Brinkhuis H, Kallmeyer J, Wefer G (2002) Rapid holocene climate changes in the Eastern Mediterranean. In: Hassan F (ed) Droughts, food and culture: ecological change and food security in Africa's later prehistory. Plenum Press, NY, pp 35–47

Sander KM (2006) Holocene climate and hydrologic changes recorded in tufa and lacustrine deposits in Southern Yemen. Graduate School Theses and Dissertations. http://scholarcommons.usf.edu/etd/2691

Fubelli G, Dramis F, Calderoni G, Cilla G, Materazzi M, Mazzini I, Soligo, M (2013) Holocene aggradation/erosion of a tufa dam at Triponzo (Central Italy). Geogr Fis Din Quat 36. doi:10.4461/GFDQ.2013.36.0

Street-Perrott FA, Perrott RA (1990) Abrupt climate fluctuations in the tropics: the influence of Atlantic Ocean circulation. Nature 343:607–612

Swezey C (2001) Eolian sediment responses to late quaternary climate changes: temporal and spatial patterns in the Sahara. Palaeogeogr Palaeocol 167:119–155

Szabo I, Haynes CV, Maxwell TA (1995) Ages of quaternary pluvial episodes determined by uranium-series and radiocarbon dating of lacustrine deposits of Eastern Sahara. Palaeogeogr Palaeocol 113:227–242

Thrailkill J (1968) Chemical and hydrologic factors in the excavation of limestone caves. Geol Soc Am Bull 79:19–46

Umer M, Legesse D, Gasse F, Bonnefille R, Lamb H, Leng MJ (2004) Late quaternary climate changes in the Horn of Africa. In: Batterbee RW, Gasse F, Stickley CE (eds) Past climate variability through Europe and Africa. Springer, Dordrecht, pp 159–180

Vasseur G, Bernard PH, Van de Meulebrouck J, Kast Y, Jolivet J (1983) Holocene paleotemperatures deduced from borehole temperature data. Palaeogeogr Palaeocol 43:237–259

Vaudour J (1994) Évolution holocène des travertins de vallée dans le Midi méditéranéen français. Géogr Phys Quat 48(3):315–326

Voight B, Gabriel B, Lassonczyk B, Ghod M (1990) Quaternary events at the Horn of Africa. Berlin Geowiss Abh, RA 120:679–694

Wehrli M, Mitchell EAD, van der Knaap WO, Ammann B, Tinner W (2010) Effects of climatic change and bog development on Holocene tufa formation in the Lorze Valley (central Switzerland). Holocene 20(3):325–336

Weisrock A (1986) Variations climatiques et périods de sédimentation carbonatée à l'holocène – L'age des dépôts. Mediterraneée 57(1–2):165–167

Williams PJ, Smith NW (1989) The frozen earth. Fundamentals of geocryology. Cambridge University Press, Cambridge

Žák K, Ložek V, Kadlec J, Hladíková J, Cílek V (2002) Climate-induced changes in holocene calcareous tufa formations, Bohemian Karst, Czech Republic. Quat Int 91:137–152

Zanchetta G, Giraudi C, Sulpizio R, Magny M, Drysdale RN, Sadori L (2012) Constraining the onset of the holocene "Neoglacial" over the central Italy using tephra layers. Quat Res 78(2):236–247

Geomorphology of Ephemeral Streams in the Kobo Basin

12

Paolo Billi

Abstract

The structural basin of Kobo (northern Ethiopia) is characterised by a semi-arid, monsoon-affected climate, and it is drained by ephemeral streams. The basin is physiographycally asymmetric with the highest mountains along the western margin where the largest rivers originate. The river morphology well matches the ideal model proposed by Schumm (The fluvial system. Wiley, New York, 1977), consisting of the headwater, the main trunk channel and the distributary system which represents the river terminus where the whole of the water flow vanishes due to infiltration and large amounts of sediment are deposited. The main geomorphic characteristics of the main stem and the distributary systems are described. The study river streambed is flat, devoid of any bedforms and horizontal and planar lamination is by far the most common sedimentary structure. Though in a few study reaches the Froude number calculated for bankfull discharge results steadily around one, i.e. in agreement with the extensive occurrence of the upper plane bed, the occurrence of outsized particles standing on or protruding from the streambed, travelling for long distances on the bed surface and showing no flow perturbation in the fine sediment in their vicinity, is interpreted by a new model of division association capable of explaining the typical division association observed in the study river deposits. It is characterised by large boulders rooted in a core coarse layer and the ubiquitous occurrence of horizontal lamination is interpreted in terms of vertical distribution of shear stress, hyperconcentrated flow and traction carpet processes.

Keywords

Ephemeral streams • Drylands • Distributary systems • Horizontal lamination • Bedload sheets

12.1 Introduction

Ephemeral streams are a main geomorphologic feature in many drylands of the planet. In the last decades, these rivers have attracted the attention of geomorphologists and sedimentologists for the flash flood hazards connected with their impulsive nature (Garcia 1995; Lin 1999) and for providing a modern equivalent of depositional characteristics and diagnostic tools to interpret old arid river deposits as potential reservoirs in oil and gas field exploration (North and Taylor 1996). Ephemeral streams are generated by erratic, infrequent, high-intensity rainfalls of short duration or concentrated in a short rainy season. River flow is therefore intermittent and flashy and the streambed is dry for the largest part of the year.

Dryland rivers have distinctive features, such as the lack of well-defined channels, high channel-width-to-depth ratio, downstream channel narrowing, the predominance of horizontally layered deposits, subtle stratification, the lack of scour and fill sequence (Reid and Frostick 1997; Billi 2008) and the occurrence of crescent scours (Leopold et al. 1966; Karcz 1968; Picard and High 1973; Billi and Tacconi 1985),

P. Billi (✉)
Physics and Earth Sciences, University of Ferrara,
Ferrara, Italy
e-mail: bli@unife.it

P. Billi (ed.), *Landscapes and Landforms of Ethiopia*, World Geomorphological Landscapes,
DOI 10.1007/978-94-017-8026-1_12, © Springer Science+Business Media Dordrecht 2015

which are not found in humid environments (Knighton and Nanson 1997). Moreover, many of the concepts and models to describe the flow/sediment interaction, which are so relevant in controlling river channel morphology, are derived mainly from studies on perennial rivers and proved difficult to apply to ephemeral streams.

The current knowledge on many issues of dryland river processes, form and sediment has been summarised by Tooth (2000a), but as pointed out by this authors, some important aspects are still unclear and need to be studied in more detail.

The geomorphology and deposits of ephemeral streams have been studied mostly along their main stem and, with the exception of a few papers (e.g. Billi and Tacconi 1985; Abdullatif 1989; Tooth 1999; Tooth 2000b), scarce information is available in the literature about the relationship of channel morphology, sediment transport modes and sedimentary structures in terminal splays and distributary systems, though they may play the most important depositional role, especially in closed basin filling.

In the structural basin of Kobo–Alamata in northern Ethiopia, all the rivers draining the western margin end up in the basin floor where they form distributary systems, with only one exception, the Golina River, which passes beyond the eastern margin through a deep gorge and enters the Danakil depression. All these rivers are definitely ephemeral, and direct inspection of their main geomorphological and sedimentological characteristics is facilitated by their dry nature most of the time.

Though our knowledge on ephemeral rivers and their distributary systems has substantially increased in the last decades, current conceptual morphodynamics models are still incomplete and derived from a few examples in a few geographical settings. Though sand-dominated distributary systems share only some features with terminal fans and floodouts, they have not been described in detail, in spite of the patency of their role as probably the more common inland river depositional system in arid and semi-arid areas. The main aim of this presentation is therefore to contribute to fill this gap and to show yet another fascinating side of Ethiopian geomorphology and landscape.

12.2 General Setting of the Study Area

The Kobo–Alamata basin is an intermontane, structural basin stretching across the border between Welo and Tigray in northern Ethiopia (Fig. 12.1). It has an elongated, rectangular

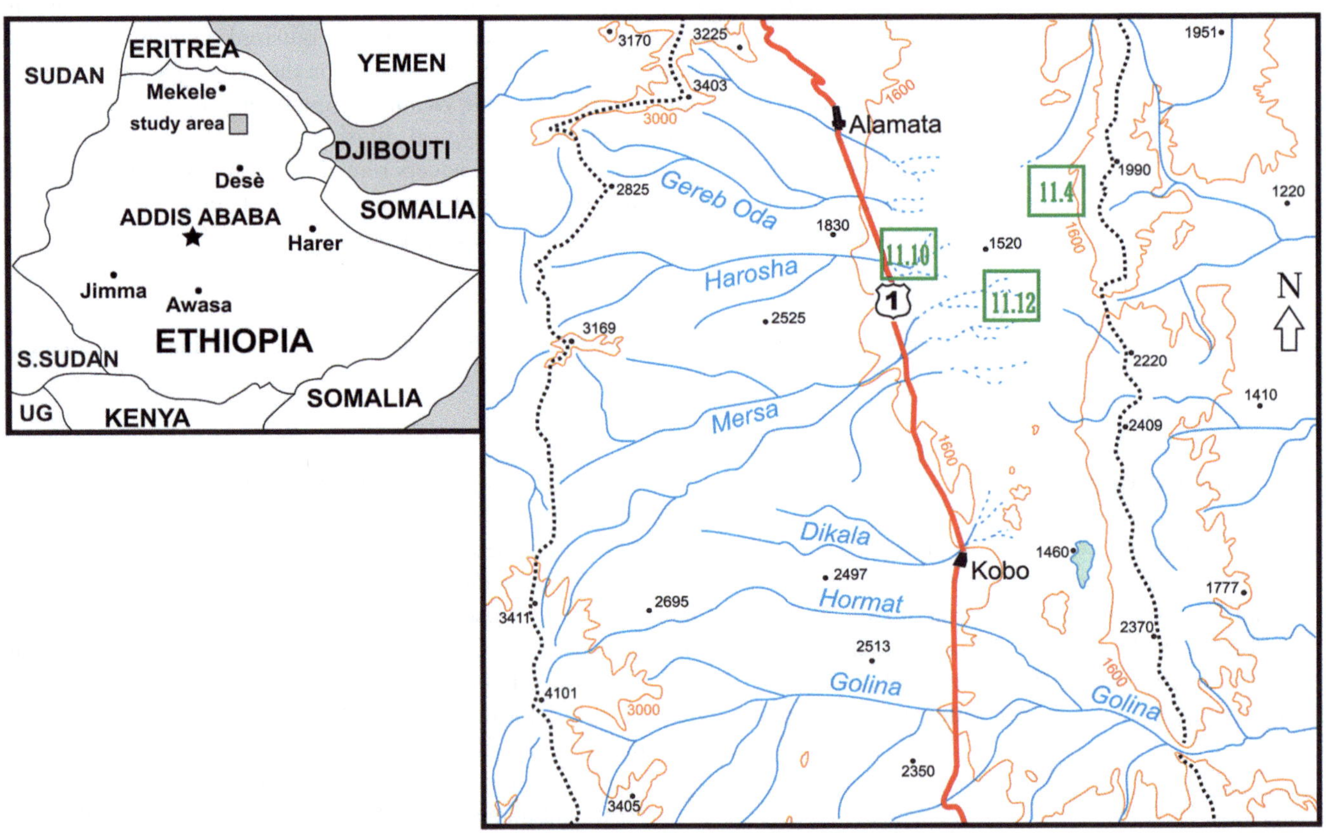

Fig. 12.1 Study area location map. *Dotted line* basin watershed divide; *red line*, Addis Ababa—Mekele road; *green squares* and *numbers* refer to the location of the satellite images of the corresponding figure

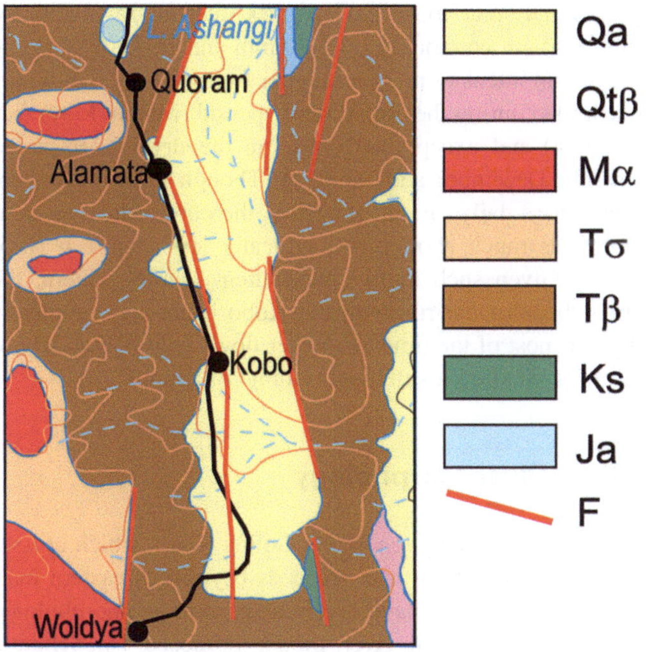

Fig. 12.2 Geology of the study area. Stratigraphic units: *Ja* Antalo Limestone (Callovian to Kimmeridgian). Neritic limestones and marls; *Ks* Amba Aradam Sandstones (Upper Jurassic to Lower Cretaceous). Fluvial/littoral variegated, quartzose sandstone; *Tβ* Oligocene to Miocene Ashenge Basalts. Olivine bsalts alternating with tuffs; *Tσ* Amba Aiba basalts (Oligocene to Miocene). Flood basalts with rare tuffs; *Mα* Amba Alaji Rhyolites (Miocene). Rhyolitic ignimbrites and tuffs, trachyrhyolites and flood basalts; *Qtβ* Miocene to Pleistocene fissural basaltic lavas, subordinately, ignimbrites and rhyolites; *Qa* Quaternary alluvial deposits, locally terraced. Symbols: *F* fault line (modified from Merla et al. 1979)

shape, reflecting its structural origin associated with the development of the main rifting system of the area. North–south-oriented master faults, in fact, bound the margins of the basin and separate it from the main Ethiopian plateau to the west, whereas the eastern ridge marks the ultimate structural relief before the Danakil depression to the east.

The basin is filled with Quaternary alluvial deposits, whereas both the eastern and western margins consist of Ashenge Basalts (basalt flows with scarce tuffs) overtopped, on the western margin, by Amba Alaji Rhyolites making up the highest peaks of the area. In the very south-east and north-east margins, small portions of Mesozoic sedimentary rock formations (Antalo Limestones and Amba Aradam Sandstones) crop out (Merla et al. 1979) (Fig. 12.2).

The lowest elevation of the basin floor is around 1,374 m a.s.l., and it is found in a restricted area east of Kobo, taken up by a seasonal marsh (Fig. 12.3). The highest peaks are in the western divide at elevations of 3,400–4,100 m a.s.l., whereas the eastern margin is less elevated with the highest peak at 2,409 m a.s.l. (Fig. 12.1). As a consequence of such physiographic asymmetry, the largest rivers flow from the western margin to the centre of the basin where they form terminal distributary channels and dry up. By contrast, the eastern margin is drained by very short and steep streams that form alluvial fans as they enter the basin floor and small distributary systems beyond them (Fig. 12.4). Only one river (the Golina R.) crosses the Kobo basin floor and overpasses the eastern margin through a deep gorge (Agamsa gorge) (Fig. 12.1). Beyond it, the river runs for about 100 km into

Fig. 12.3 The basin floor with the town of kobo on the *left* and the least elevated area (1,374 m a.s.l.) taken up by the seasonal marsh close to the eastern margin

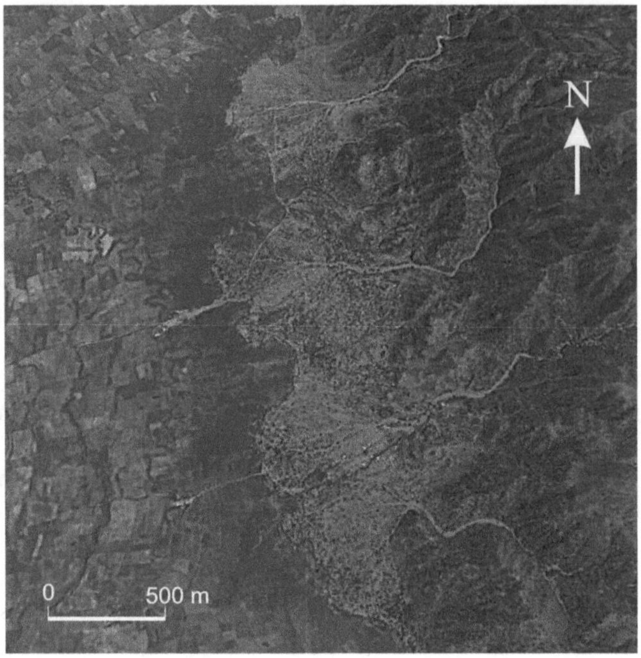

Fig. 12.4 The alluvial fans built by steep and short streams on the eastern margin of the basin

the Danakil lowland where it joins the Awira River in the Hor Mat plain at an elevation of about 700 m a.s.l. The Awira R. proceeds for further 100 km and fades out in a large distributary system in the Gerule Plain at an elevation of about 350 m a.s.l.

Unfortunately, no river flow data are available for any of the rivers draining the basin, whereas two meteorological stations, namely Kobo and Alamata, are located on the basin floor, near the western margin (Fig. 12.1). These two stations are close to each other (only 30 km apart), at about the same elevation and in a similar physiographic position within the basin; hence, their climatic data are very similar. Mean maximum and minimum temperatures are highest in June, about 34 and 18 °C, respectively, in coincidence with high sun and clear sky conditions, whereas the lowest temperatures are recorded in January, about 27 and 12 °C, respectively.

Mean annual rainfall ranges between 726 and 768 mm for Kobo and Alamata, respectively, whereas potential evapotranspiration, calculated by Thornthwaite's method, is 931 mm year^{-1}. Though the annual amount is relatively large, precipitation consists mainly of few, very intense rainstorms that are more frequent during the *kiremt* (the summer, monsoon-type big rains—see Chap. 4) and may occur as isolated events during the dry spell. The annual rainfall values are typical of sub-humid areas, and apparently, a permanent baseflow would be expected in the majority of rivers. Actually, about 50 % of the annual rain falls during July and August as typical monsoon downpours (see Chap. 4), but precipitation intensity is high the whole year-round since mean daily values (the only available for this area) are commonly over 20 mm and peaks of absolute maximum intensity in 24 h ranging from 50 to 100 mm may occur also during the 'small rains' period (March–May) (see Chap. 4) and exceptionally also in the dry winter period (mainly December and January). Though these values are reported as daily rainfall intensity, the core of heavier rainstorms is much shorter and typically less than 1 h (Billi 2011). Given such highly intermittent rainfall pattern, the streambeds of the rivers in the Kobo basin are completely dry for most of the time and water flows only in response to rainfalls yielding a sufficient volume of run-off.

12.3 River Morphology

With the only exception of the Golina River, which crosses the Kobo basin floor and proceeds beyond its eastern margin, all other rivers have a similar general geomorphology. Similar to the ideal river model proposed by Schumm (1977), they consist of three portions: (1) the headwater; (2) the main trunk channel and (3) the distributary system. The headwater reaches are typically cut into bedrock or the streambed takes up the bottom of a narrow valley filled with coarse-grained alluvium but devoid of a true alluvial plain. The steepest, upstream reaches share the same morphological and sedimentological features observed in the mountain streams of humid and sub-humid areas (see Montgomery and Buffington 1997, for a review). In the headwater lower reaches, the streambeds take up the whole valley bottom and acquire a sinuous pattern, likely influenced by local geological structures. In the western margin, the transition from the headwater to the basin floor is commonly marked by a narrow gorge (Fig. 12.5), downstream of which the main river trunk originates. Here, unlike what one would expect, no modern alluvial fan is found. This is a distinctive feature of the Kobo basin since all the rivers entering the basin floor from the highlands to the west do not develop an alluvial fan. By contrast, in spite of the less elevated basin divide and much smaller headwater catchments, the majority of the rivers coming from the eastern side show well-developed alluvial fans (Fig. 12.4). Such a geomorphic difference between the two basin sides stands probably in the basin floor being slightly tilted to the east (Fig. 12.3). The longitudinal profile of the main western rivers, in fact, does not show any discontinuity or abrupt change in gradient. This implies that, with time, the river channels have adjusted to a progressively eastward-sloping floor. Accurate field inspections indicate that, in some places, such river response is also witnessed by a chain of two or three old distributary systems, each of them overstepped by the river to form another distributary system and prograding towards the opposite side of the basin.

Fig. 12.5 The narrow gorge at the transition from the headwater to the main stem of the Dikala River

12.3.1 The Main Stem

The main stem is found on the basin floor, and it is typically straight with an overall morphology very similar to that of the arroyos of the American authors (e.g. Leopold et al. 1966; Graf 1988) (Fig. 12.6). The channel, in fact, is cut into the Quaternary basin fill deposits and its cross section is rectangular with almost vertical banks, 4–5 m high in the upstream reaches and decreasing to about one metre at the first bifurcation of the distributary system. Irrespective of river size, channel width expands rapidly downstream of the headwater gorge and, though local broadenings due to particularly extensive bank collapse may occur, it tends to reach a maximum at the distance of 2–4 km upstream of the distributary system (Fig. 12.7). Typically, the main stem is joined by tributaries only in its upstream portion where the headwater catchment of the larger rivers exceeds the range of 50–100 km^2 indicated by Wolman and Gerson (1978) as the threshold at which channel width becomes constant. However, due to the flashy nature of floods and the long

Fig. 12.6 The straight, main stem of the Dikala River upstream of the town of Kobo

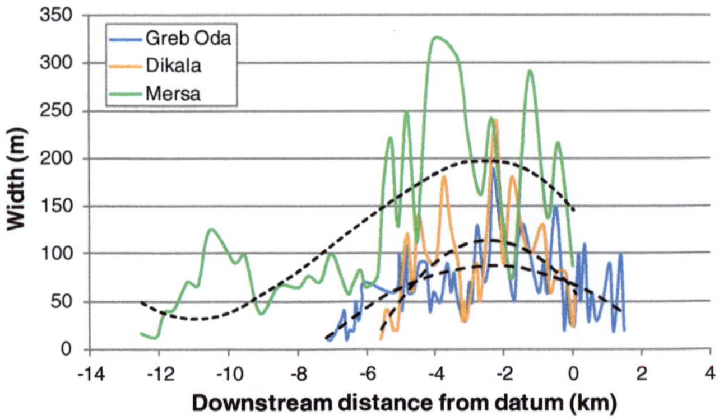

Fig. 12.7 Downstream variation of main stem channel width for, Dikala and Mersa rivers. The datum coincides with the bridge on the Kobo-Alamata road. The *dashed lines* are third (Mersa) and second grade (Gereb Oda and Dikala), respectively, polynomial interpolation *lines*

Table 12.1 Main geomorphic characteristics of the study rivers

Reach type/river	Gradient	D_{50} (mm)	w/d
Main stem	0.0141–0.0445	0.15–2.70	43–418
Distributary	0.0109–0.0272	0.72–1.35	36–325
Golina river	0.0187	46.1–93.5	ca 300

D_{50} is the particle size for which 50 % of the distribution is finer; *w/d* is width/depth ratio

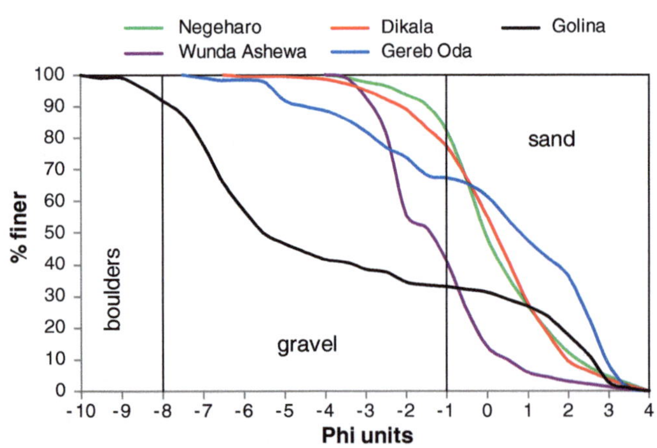

Fig. 12.8 Bed material grain-size distribution curves of a few, representative ephemeral streams of the Kobo basin

return time of bankfull flows, which seems to characterise ephemeral streams (Williams 1978; Graf 1988), rivers are unable to effectively rework the channel morphology imprinted by larger floods and oversized channel result. Given the very high intensity of rainfall, run-off from the basin floor sides may enter directly the river channel, thus contributing to increase discharge and width as far as an appreciable volume of water is lost by infiltration through the very permeable sediment of the streambed and the banks, resulting in a decrease of channel width (Dunkerley 1992; Tooth 1999, 2000b).

The data reported in Fig. 12.7 are from rivers draining the western margin, and a marked relative minimum of channel width can be observed to precede the maximum width. This narrowing is found at the transition between the river expansion beyond the headwater gorge and the most upstream development of alluvial terraces and can be interpreted as a knick point associated with the streambed incision in the basin fill deposits following the eastward tilting of the basin floor. Such a structural stetting is likely responsible for the steep gradient of the trunk river bed (Table 12.1) (Billi 2007) that is by far beyond the critical slope for sandbed streams, as are most in the study area, and close to that of the Golina, which is instead a boulder bed river (Fig. 12.8). From Table 12.1, it can be noticed that the main stems, the distributary channels and the Golina River have very similar geomorphic and sediment characteristics, with the sole exception of the Golina bed material that is much coarser.

12.3.2 The Distributary System

Though the overall morphology of the study area distributary systems is rather similar, some differences can be observed. They include the horizontal angle of the distributary system apex, the morphology and style of anabranching. Larger streams split into larger distributary systems and the average channel width of the trunk stream (W) measured in a 1–2-km-long reach upstream of the distributary apex, explains 91 % of the apex angle (α) variability (Fig. 12.9). The total mean width of the first two or three anabranches rooted in the distributary apex (W_a) was found to be constantly larger than W, irrespective of the trunk river size and the anabranch number. W_a, in fact, ranges between 1.5 and 3.3 W. This

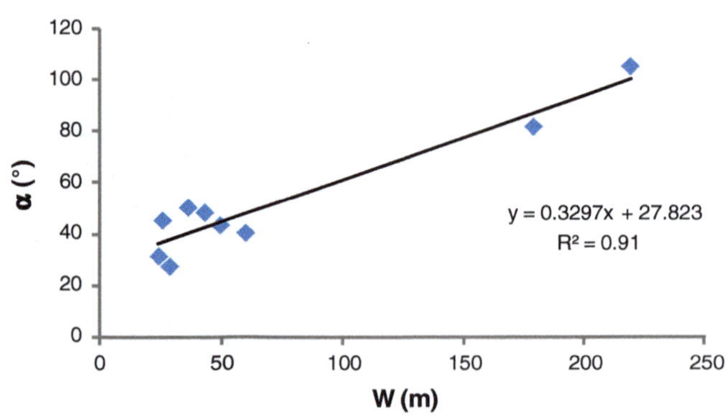

Fig. 12.9 Correlation between the average channel width in the reach upstream of the distributary system and the angle of the distributary system apex

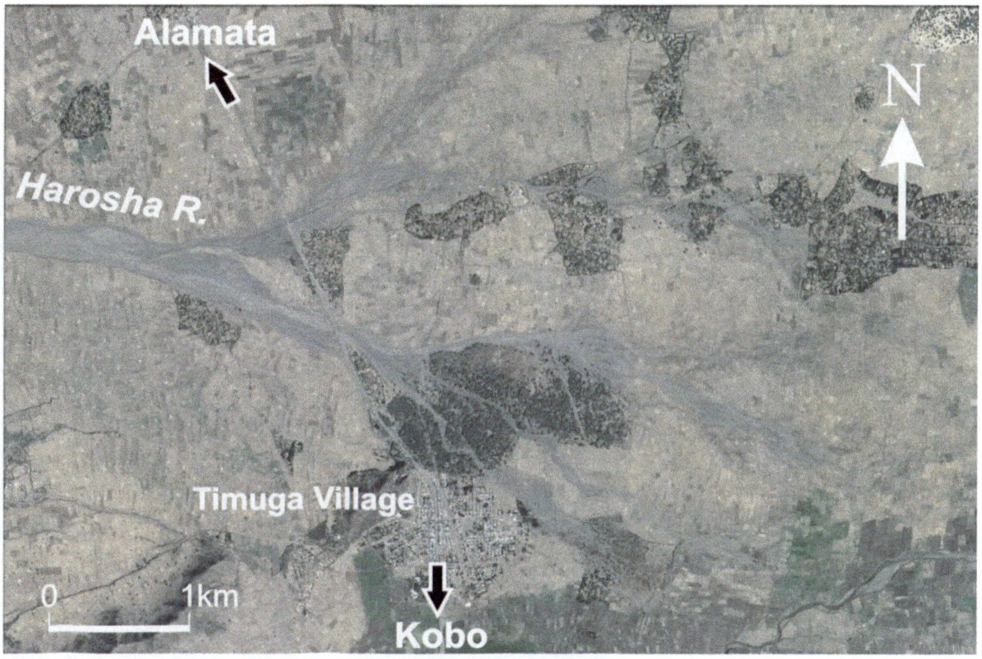

Fig. 12.10 The distributary system of the Harosha River. The distributary anabranches have comparable width, may form secondary, smaller distributary systems or can proceed as individual, narrow channels with characteristics similar to the floodouts of Tooth (1999)

implies that, though all the anabranches are active the year-round, during individual, ordinary (less than bankfull) floods, the largest proportion of the flow is conveyed only by one anabranching channel. This observation is consistent with the theoretical predictions of Bolla Pittaluga et al. (2003) and the findings of Mosley (1983) who pointed out that in braided rivers the main channel carries generally a high percentage of the total discharge, ranging from 65 to 85 %. A predictive model for the equilibrium configuration of a simple 'Y-shaped' bifurcation in which an upstream channel splits into two equivalent branches has been proposed by Wang et al. (1995). The model of Bolla Pittaluga et al. (2003), though conceived to account for the formation of braid bars in gravel-bed rivers, can be applied also to bifurcations of distributary systems. The model, in fact, depicts a condition of unbalanced equilibrium configurations, even in the case of perfectly symmetric geometry, which develops for low values of the sediment mobility, as it is the case of ephemeral streams. These rivers are commonly characterised by a high concentration of bedload transport and high sedimentation rates, associated with the flashy nature of floods and high infiltration rates (Billi 2008). High rates of sedimentation may also result in a difference of bed elevation at the upstream inlet of the anabranches and in large values of the width/depth ratio, i.e. two typical features of ephemeral streams (Table 12.1) (Billi 2007).

Some distributary channels run for a few kilometres as an individual channel and fade away like the floodouts described by Tooth (1999). Others may split again into two secondary anabranches in a tree-like structure or form a secondary, terminal splay at a certain distance from the distributary system main apex (Fig. 12.10). Though Billi (2007) addressed such distributary styles, the factors controlling them are not yet clear.

The distributary channels are slightly incised in the valley fill alluvium, and their bank height decreases from 1 to 2 m upstream to zero in the distal reaches. According to Billi (2007), the distributary systems of the Kobo basin consist commonly of five reach units, with specific morphological and sedimentological characteristics. From upstream to downstream, they are (Fig. 12.11) (1) the main feeder

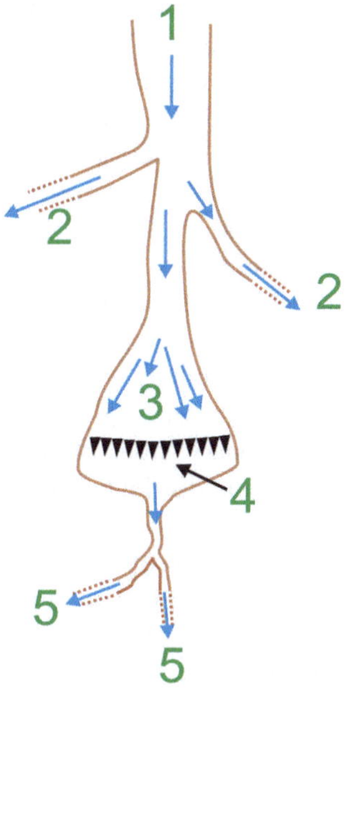

Fig. 12.11 Sketch of the distributary system reach units model: (*1*) main feeder channel; (*2*) primary distributary channel; (*3*) flow expansion reach; (*4*) accretionary slip face; (*5*) run-out distal reach

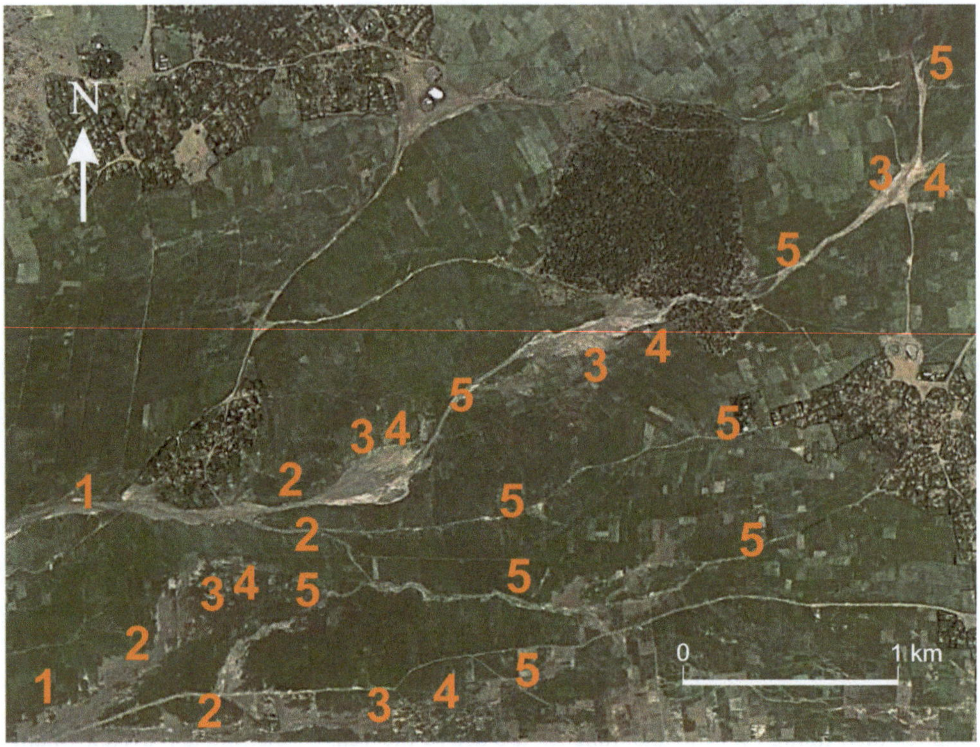

Fig. 12.12 Application of the reach units classification scheme to the distributary system of Gerday-Weja River. The *numbers* refer to the model units of Fig. 12.11

channel; (2) the primary distributary reach; (3) the flow expansion reach; (4) the accretionary front; and (5) the run-out channel.

The feeder channel is the main river stem that has been described in Sect. 12.3.1 and represents the dynamic axis of the river system (Kelly and Olsen 1993). In their upstream portion, the primary distributary reaches have an arroyo-like morphology similar to that of the main stem though bank height is smaller. Right downstream of the bifurcation, primary distributary channels tend to have comparable widths (Fig. 12.10). These channels may proceed as floodouts (sensu Tooth 1999, i.e. similar to the run-out channel of the model presented here) or, on their turn, may bifurcate into narrower and shallower distributary reaches. Further downstream, bank height decreases to a few centimetres and, in practice, flow becomes unconfined forming the flow expansion reach (Fig. 12.12). Here, the largest proportion of flow infiltrates; hence, bedload sediment came to rest and deposition takes place forming a flat, wedge-shaped body with an accretionary front. Such bed material accumulation is washed out by receding flood flows or subsequent shallow flows removing the finer particles. This process makes the accretionary front as the coarsest bed material unit of the whole distributary system. The flow infiltrating through the accretionary front may re-emerge downstream as clean water that recovers some excess shear stress capable of entraining the fine bed/soil particles of the distal areas and to incise the narrow run-out channel (Fig. 12.12). These distal channels are very narrow (1–3 m) and shallow (a few centimetres) and their discharge is less than one-tenth of the main stem flow (Billi 2007). They are typically the longest reaches within the distributary system and maintain their size as long as the flow completely vanishes due to infiltration, splays out because of lack of confinement or vegetation resistance or evaporates.

12.4 Sediment and Bedforms

Bed material of the Kobo basin ephemeral streams consists mainly of medium to coarse sand and, subordinately, of fine gravelly sand with pebbles and even large boulders scattered on a flat bed devoid of any bedforms or depositional bars. Actually, the streambed appears as made of shallow rhomboid bars, the maximum width of which is in the order of 0.2–0.05 channel width. These apparent bars are poorly developed, flat topped, commonly less than 50 cm thick and lack accretionary fronts or other progradational features, suggesting they do not result from depositional processes but are produced by streambed dissection processes during the receding or shallower flood flows (Billi 2007, 2008). Smaller-scale, 40–50-mm-thick and 0.5–2.0-m-wide, overlapping leaf-shaped bars/bedforms are a common streambed feature. Billi (2011) observed that the thickness of these bedforms coincides with that of the bedload flux measured in the field on the Gereb Oda River (Fig. 12.1) during short return time floods. This finding is consistent with the results obtained by the relation developed by Karim and Kennedy

Fig. 12.13 A large boulder resting on a thin, reversely graded, fine gravel division. No clear evidence of both crescent scour and wake deposit upstream and downstream, respectively, of the boulder can be perceived. The spatula is about 25 cm long

(1983) to predict the thickness of sandy bedload sheets, which resulted to be in the 8–65 mm range. Though more field and laboratory data and observations are needed, the leaf-shaped bedforms seem to be the product of moving bed material layers, the thickness (and likely size) of which varies with flow discharge. This process was postulated by Billi (2008) to account for the predominance of horizontal lamination and the lack of small-scale bedforms, like ripples and dunes, on the streambed of the study area ephemeral streams.

The classical flume experiments of Simons and Richardson (1966) have shown that plane bed develops at the transition from lower to upper-flow regime with Froude numbers around unity. For two ephemeral streams of the Kobo basin, Billi (2008) calculated that at bankfull flow Froude numbers are very close to one. In the flume experiments of Alexander et al. (2001), antidunes were generated under a supercritical flow and, as discharge was reduced, antidunes decreased in length and height and the bed became nominally plane with millimetre-high asymmetrical bedwaves as those described by Best and Bridge (1992).

In the study rivers, natural weavy laminae are not so evident as in the flume experiment. Their concavity is barely detectable in the field and they are characterised by upward coarsening instead of normal grading as observed in the laboratory (Billi 2008). Notwithstanding such differences, it can be accepted that horizontal laminae are generated under an upper-flow regime at the transition from plane bed to antidune, but their extensive and ubiquitous occurrence as the dominant bedform and sedimentary structure in the study rivers and the interaction with large boulders without any evident flow disturbance around them (Fig. 12.13) requires some complementary considerations and explanations. In fact, commonly, large boulders appear as floating on top of massive fine gravel layers and in places, such as the main stem and the first anabranches of the Dikala River (Fig. 12.1), there is evidence of their movement on the streambed surface for a distance as far as 2 km (Fig. 12.14). Also Hassan et al. (1999) found that coarse gravel particles move during the transition regime in their study of a sandy ephemeral stream in Botswana, and field inspection after floods revealed a flat bed with no bedforms. In the Kobo basin study reaches, Froude numbers around one are typical at bankfull discharge (Billi 2008) and, though no field data of the hydraulic characteristics of the flood vanishing flows is available, the lack of sedimentary structure and bedforms, such as low flow regime ripples and dunes, cannot be accounted for by a gradually changing flow. In fact, shallow receding flows with transport capacity enough to rework the bedforms should be able to develop also horseshoe vortices on the upstream side of virtually still boulders standing on the streambed, to carve crescent scour holes and deposit fine wakes (Picard and High 1973; Billi 1988). On the streambed of the study reaches, evidence of a late filling of the scour pockets is very uncommon and present only upstream of very large boulders (Fig. 12.15). This suggests that both the finer material and the boulders are transported and deposited together and the development of turbulent eddies, tumbling, divergent and convergent flows around the large particles is inhibited.

From Figs. 12.13, 12.14 and 12.15, a tight connection between the larger boulders and the much finer, horizontally laminated sediment is evident. Billi (2008) has proposed a

Fig. 12.14 White boulders scattered on the primary distributary channel upstream of Gara Lencha hill near Kobo. These large boulders were entrained from a bank protection wall at the road bridge of Kobo and travelled a few hundreds of metres downstream on top of the streambed

Fig. 12.15 A large boulder rooted in a reversely graded, fine gravel layer on the streambed of the Wunda Ashewa River (few kilometres South of Kobo). In this case a small scour pocket filling with fine sediment is present upstream of the boulder. The folded stick ruler is about 22 cm long

simple model, inspired by the bipartite model of Sohn (1997), to explain the association of particles with such diverse size and the ubiquitous occurrence of horizontal lamination. The bipartite model of Sohn (1997) implies that grain segregation is not possible in the frictional zone because particle concentration is very high and/or the occurrence of elutriation induced by pore fluid escaping. According to these considerations, reverse grading should not occur in the lower part of the moving sediment, but in the Kobo basin ephemeral streams deposits, the basal division is commonly reversely graded (Figs. 12.15 and 12.16) and only occasionally it is massive or normally graded. Though this seems to reject the hypothesis of a basal frictional region and the occurrence of a bipartite flow in the study dry rivers, the occurrence of individual large boulders with basal reverse grading can be accounted for by buoyancy generated by vertical stratification and dispersive pressure as predicted for the collision region by Sohn's model (Fig. 12.13 and 12.15).

The inspection of isolated, large particles standing on the bed surface reveals that they rest on and/or are associated with a coarser (commonly fine to medium gravel) division (Fig. 12.15) and the horizontally laminated and massive sand in their vicinity is typically undisturbed. This makes up the coarse core division of the typical internal bed arrangement observed in the study river deposits, consisting, from bottom to top, of the following four main divisions (Fig. 12.17): (1) the basal reversely graded or massive, fine-grained division;

Fig. 12.16 Application of the division classification model to a stratigraphic log of a modern streambed cut bank of the Mersa River. Horizontally laminated and massive sand in the vicinity of larger particles show no flow perturbation features such as scour holes or lee side fine-grained wake. Division notation: *Cc* coarse core division; *Lh* horizontal lamination; *MD* mud drape; *RG* reverse grading. The stick ruler for scale is about 90 cm long

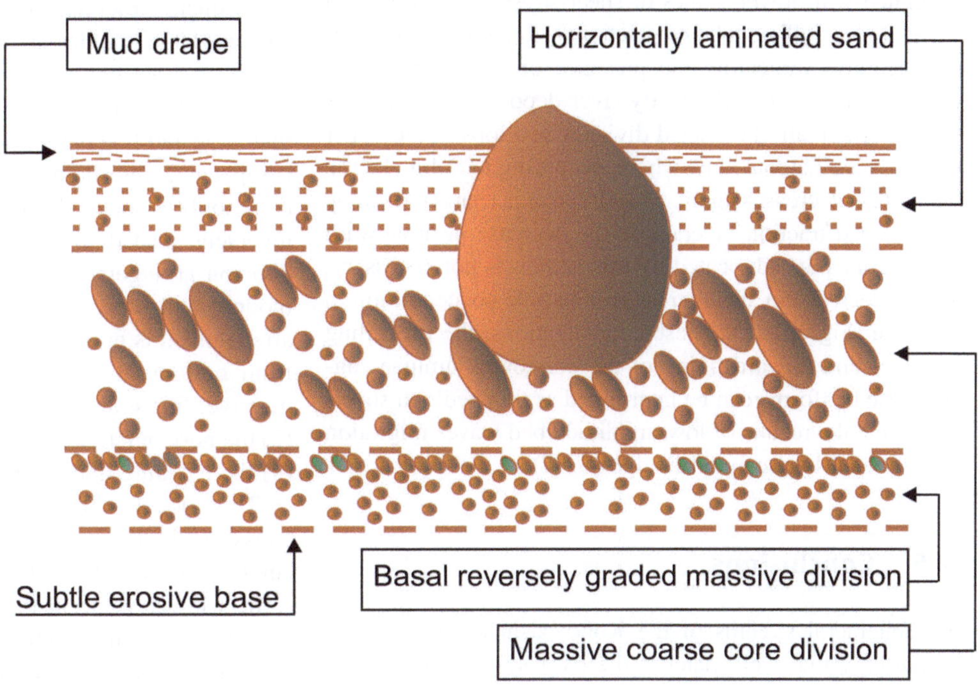

Fig. 12.17 Sketch model of the typical division arrangement within an individual layer (from Billi 2008—modified)

(2) the core coarse division; (3) the horizontally laminated sandy and grainy division; and (4) the receding flood flow mud or, subordinately, sandy mud drape. An example of application of such a scheme to channel deposits is reported in Fig. 12.16. In places, the typical division arrangement is not present with its complete sequence as it may be beheaded by subsequent flood pulses and/or the following flood and its basal division may be lacking for the onset of a surge flood that may have deeply reworked it.

This division association model seems to conform very well to the vertical shear stress distribution postulated by Billi (2008) within the collisional zone (Fig. 12.18) rather

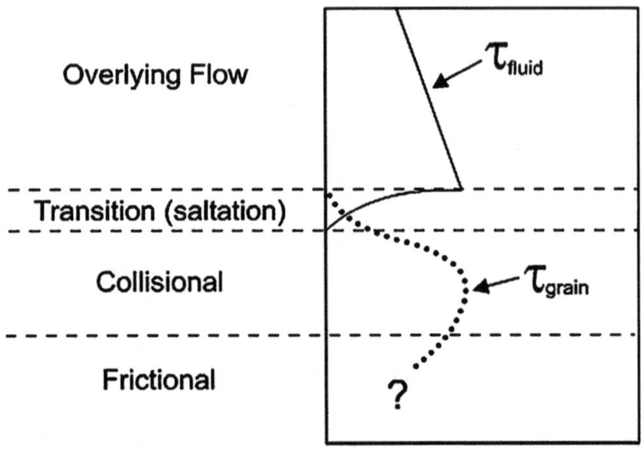

Fig. 12.18 Vertical distribution of shear stress within hyperconcentrated flow (modified from Sohn 1997)

than to a variation in the grain size of the supplied sediment as proposed by Sohn (1997). In fact, the coarse core division can be associated with the highest shear stress of the collision zone. The lower massive or reversely graded division is associated with lower values of shear stress at the transition between the collisional and frictional zones (though sediment features indicating the presence of a frictional zone were not observed in the study river deposits). Finally, the upper, horizontally laminated division develops in the upper part of the collisional zone where, according to Sohn (1997), shear stress decreases again. This division is characterised by thin bedload traction layers, and the decrease in downward flux of grains, as advocated by this author, is not necessary, given a decrease in shear stress and particle concentration in this zone (Fig. 12.18). These considerations, the prevailing geometry and upward fining of the horizontal laminae suggest these bedforms can be interpreted as thin bedload sheets rather than the results of low-amplitude bed waves migration as postulated by Best and Bridge (1992).

12.5 Conclusions

The ephemeral streams of the Kobo basin have a general morphology that well matches the river system model proposed by Schumm (1977). The trunk river is cut into the Quaternary valley fill deposits, has a straight pattern and it is characterised by large width-to-depth ratios (as much as 400). Channel width tends to increase to a maximum shortly upstream of the first bifurcation, giving way to the distributary system, and then, it decreases due to water infiltration and splitting into downstream anabranches. Tectonics is likely responsible for the lack of alluvial fans on the basin western margin, and the steep streambed gradient, ranging from 0.01 to 0.04, which is unusually higher than the critical slope expected for sandbed rivers.

The distributary systems on the floor of the Kobo basin show peculiar geomorphic characteristics, different from those described by Tooth (2000a, b) for his Australian floodouts and those of the terminal fans studied by Parkash et al. (1983) and Kelly and Olsen (1993). Field investigations by Billi (2007) revealed that the former are made of five reach units, with different morphological, sedimentological and hydrological characteristics: (1) the main feeder channel; (2) the primary distributary reach; (3) the flow expansion reach; (4) the accretionary front; and (5) the run-out channel.

The streambed of all the study rivers is flat, devoid of any bedform and punctuated by individual very large particles. In places, rhomboid bars are present, but they are only apparent longitudinal bars since no downstream grain size sorting nor accretionary fronts (sensu Bluck 1979) were observed on their surface and in cut banks. They are supposed to be formed by dissection processes active during the receding flood flows.

In both the main stem and the distributary channels, 99 % of sediment structures consist of horizontal, planar bedding, whereas large boulders, scattered on the bed surface, show no upstream crescent scour or downstream fine wake (Picard and High 1973; Billi and Tacconi 1985). This seems to indicate that large and fine particles are deposited at very similar flow competence thresholds. Moreover, the large boulders are commonly associated with or rooted into a coarse-grained, reversely graded division which is the core of a typical division arrangement making up the horizontal beds in the modern and older alluvium. A model to describe such an arrangement is presented. It consists of four divisions, which from bottom to top are (1) the basal reversely graded or massive, fine-grained division; (2) the core coarse division; (3) the horizontally laminated sandy and grainy division; and (4) the receding flood flow mud and sandy mud drape.

This model is based on the hypothesis of vertical shear stress distribution postulated by Sohn (1997) for hyperconcentrated flows. Shear stress, in fact, is predicted to increase from a lower value in the basal massive or reverse-graded division to a maximum coinciding with the coarse core division and to decrease towards the horizontally laminated division on the top layer. By this model, it is possible to account for the characteristics and origin of horizontal laminae that are the most common sedimentary feature of ephemeral streams. In the study rivers, the position of horizontal laminae in the division association model and their upward fining fabric indicates these bedforms are to be interpreted as generated by thin bedload sheets rather than by the migration of low-amplitude bed waves as postulated by Best and Bridge (1992).

The flow energy at bankfull discharge was calculated by Billi (2008) as being capable of entraining the large boulders punctuating the study river streambed, but the lack of flow disturbance around them and the erosionless transition of the coarse core division, in which they are rooted, to the lower and overlaying massive divisions suggest that both large and small particles are subjected to equal mobility conditions leading to the conclusion that, though buoyancy forces may be active on boulders, en masse bed material transport is expected to be the prevailing bedload transport process.

Acknowledgments Research funded by the Italian Ministry of Education, University and Research, PRIN Project 2003 n. 2003040420_003 and by National Geographic Society, Committee for Research and Exploration, grant n. 8400-08.

References

Abdullatif OM (1989) Channel-fill and sheet-flood facies sequence in the ephemeral terminal River Gash, Kassala. Sudan Sedim Geol 63:171–184

Alexander J, Bridge JS, Cheel RJ, Leclair SF (2001) Bedforms and associated sedimentary structures formed under supercritical water flows over aggrading sand beds. Sedimentology 48:133–152

Best J, Bridge J (1992) The morphology and dynamics of low amplitude bedwaves upon upper stage plane beds and the preservation of planar laminae. Sedimentology 39:737–752

Billi P (1988) Forme di fondo grossolane. Giorn Geol 50:15–30

Billi P (2007) Morphology and sediment dynamics of ephemeral stream terminal distributary systems in the Kobo basin (northern Welo, Ethiopia). Geomorphology 85:98–113

Billi P (2008) Flash floods, sediment transport and bedforms in the ephemeral streams of Kobo basin, northern Ethiopia. Catena 75 (1):5–17

Billi P (2011) Flash flood sediment transport in a steep sand-bed ephemeral stream. Int J Sedim Res 26(2):193–209

Billi P, Tacconi P (1985) Morfologia sedimentologia e dinamica fluviale di due corsi d'acqua effimeri della Somalia settentrionale. Quad di Geologia della Somalia 8:43–86

Bolla Pittaluga M, Repetto R, Tubino M (2003) Channel bifurcation in braided rivers: equilibrium configurations and stability. Water Resour Res 39(3):1046–1059

Bluck BJ (1979) Structure of corse-grained braided stream alluvium. Trans Roy Soc Edinb 70:182–221

Dunkerley DL (1992) Channel geometry, bed material and inferred flow conditions in ephemeral stream systems, Barrier Range, western N.S.W. Australia. Hydrol Process 6:417–433

Garcia CC (1995) Torrential flow frequency and morphological adjustment of ephemeral channels in South-East Spain. In: Hicki EJ (ed) River Geomorphol. Wiley, Chichester, pp 169–192

Graf WL (1988) Fluvial processes in dryland rivers. Springer, Berlin

Hassan M, Shick A, Shaw P (1999) The transport of gravel in an ephemeral sandbed river. Earth Surf Proc and Landforms 17:711–722

Karcz I (1968) Fluviatile obstacle marks from wadis in the Negev (southern Israel). J Sedim Petrol 38:1000–1012

Karim MF, Kennedy JF (1983) Computer-based predictors for sediment discharge and friction factor of alluvial streams. In: Proceedings of second international symposium on river sedimentation, Nanjing, China, Paper A18

Kelly SB, Olsen H (1993) Terminal fans—a review with reference to Devonian examples. Sed Geol 85:339–374

Knighton AD, Nanson GC (1997) Distinctiveness, diversity and uniqueness in arid zone river systems. In: Thomas DSG (ed) Arid zone geomorphology: process, form and change in drylands. Wiley, Chichester, pp 185–203

Leopold LB, Emmet WW, Myrick RM (1966) Channel and hillslope processes in a semiarid area New Mexico. USGS Professional Paper 352-G

Lin X (1999) Flash floods in arid and semiarid zones. UNESCO-IHP, Paris

Merla G, Abbate E, Azzaroli A, Bruni P, Canuti P, Fazzuoli M, Sagri M, Tacconi P (1979) A geological map of Ethiopia and Somalia. Consiglio Nazionale delle Ricerche, Firenze

Montgomery DR, Buffington JM (1997) Channel-reach morphology in mountain drainage basins. Geol Soc Am Bull 109(5):596–611

Mosley MP (1983) Response of braided rivers to changing discharge. J Hydrol NZ 22(1):18–67

North CP, Taylor KS (1996) Ephemeral-fluvial deposits; integrated outcrop and simulation studies reveal complexity. AAPG Bull 80 (6):811–830

Parkash B, Awasthi AK, Gohain K (1983) Lithofacies of the Markanda terminal fan, Kurukshetra district, Haryana, India. In: Collinson JD, Lewin J (eds) Modern and Ancient Fluvial Systems. Spec Publ Int Ass Sediment 6:337–344

Picard MD, High LR Jr (1973) Sedimentary structures of ephemeral streams. Developments in sedimentology 17, Elsevier, Amsterdam

Reid I, Frostick LE (1997) Channel form, flows and sediments in deserts. In: Thomas DSG (ed) Arid zone geomorphology: processes, form and change in drylands. Wiley, Chichester, pp 205–229

Schumm SA (1977) The fluvial system. Wiley, New York

Simons DB, Richardson EV (1966) Resistance to flow in alluvial channels. USGS Professional Paper 422 J

Sohn YK (1997) On traction-carpet sedimentation. J Sedim Res 67 (3):502–509

Tooth S (1999) Floodouts in central Australia. In: Miller AJ, Gupta A (eds) Varietis of fluvial form. Wiley, Chichester, pp 219–247

Tooth S (2000a) Process, form and change in dryland rivers: a review of recent research. Earth-Sci Rev 51:67–107

Tooth S (2000b) Downstream changes in dryland river channels: the Northern Plains of arid central Australia. Geomorphology 34:33–54

Wang ZB, Fokkink RJ, De Vries M, Langerak A (1995) Stability of river bifurcations in 1D morphodynamics models. J Hydraul Res 33 (6):739–750

Williams GP (1978) Bankfull discharge of rivers. Water Resour Res 14:1141–1158

Wolman MG, Gerson R (1978) Relative scales of time and effectiveness of climate in watershed geomorphology. Earth Surf Proc Land 3:189–208

Sediment Yield Variability at Various Spatial Scales and Its Hydrological and Geomorphological Impacts on Dam-catchments in the Ethiopian Highlands

13

Nigussie Haregeweyn, Atsushi Tsunekawa, Jean Poesen, Mitsuru Tsubo, Jan Nyssen, Matthias Vanmaercke, Amanuel Zenebe, Derege T. Meshesha, and Enyew Adgo

Abstract

This chapter analyzes the spatial variability, impacts, and factors of sediment yield (SY) and reservoir sedimentation rates (SRs) in the upper Blue Nile River Basin. SY data collected using reservoir sediment surveys or runoff and suspended sediment concentration (SSC) measurements at river gauging stations in the framework of different programs were compiled, screened, and used for our analysis. A large spatial variation in area-specific SY (SSY), ranging between 4 and 4,935 t km^{-2} year^{-1}, was observed among catchments. This variation is attributed to both human and environmental factors. The high SY values have drastic consequences for the life expectancy of many reservoirs in the Ethiopian highlands: 50 % of reservoirs risk losing their economic life within half of the design period. Moreover, the high trapping efficiency of the reservoirs for flow and sediment led to selective deposition of sediment fractions within the reservoir and channel stabilization and vegetation regrowth in the downstream river reaches. Unfortunately, the availability and reliability of SY data for this region is poor by international standards. The SY assessment initiatives taken through institutional collaboration projects is a good start; however, such projects have limited capacity and a short life span, so they cannot produce a sustainable solution for this important data gap. Hence, concerted efforts on the maintenance and monitoring of existing gauging stations on top of establishing new ones are needed to better understand the different eco-hydrological environments in the basin.

N. Haregeweyn (✉) · A. Tsunekawa · D.T. Meshesha
Arid Land Research Center, Tottori University, 1390 Hamasaka, Tottori 680-0001, Japan
e-mail: nigussie_haregeweyn@yahoo.com

A. Tsunekawa
e-mail: tsunekawa@alrc.tottori-u.ac.jp

D.T. Meshesha
e-mail: deremesh@yahoo.com

N. Haregeweyn · A. Zenebe
Department of Land Resources Management and Environmental Protection, Mekelle University, Mekelle, Ethiopia
e-mail: amanuelza@yahoo.com

J. Poesen · M. Vanmaercke
Department of Earth and Environmental Sciences, Physical and Regional Geography Research Group, KU Leuven, Heverlee, Belgium
e-mail: jean.poesen@ees.kuleuven.be

M. Vanmaercke
e-mail: Matthias.Vanmaercke@ees.kuleuven.be

M. Tsubo
Institute for Soil, Climate and Water, Agricultural Research Council, Pretoria, South Africa
e-mail: TsuboM@arc.agric.za

J. Nyssen
Geography Department, Gent University, Ghent, Belgium
e-mail: jan.nyssen@ugent.be

E. Adgo
Bahir Dar University, Bahir Dar, Ethiopia
e-mail: enyewadgo@gmail.com

P. Billi (ed.), *Landscapes and Landforms of Ethiopia*, World Geomorphological Landscapes,
DOI 10.1007/978-94-017-8026-1_13,

Keywords
Sediment yield • Reservoir sedimentation • Channel stabilization • Blue Nile River

13.1 Introduction

Sediment yield (SY) is the net result of soil erosion and sediment deposition processes. It is defined as the total sediment outflow from a catchment, measurable at a point of a reference and for a specified period of time (Vanoni 1975). It can be expressed in absolute terms (e.g., t year^{-1}) or in area-specific terms (specific sediment yield—SSY) (e.g., t km^{-2} year^{-1}).

Accurate estimation of SY is important for a number of applications such as the identification of net erosion (denudation) rates within catchments (Walling 1994), the design of planned hydraulic structures so that the actual lifetime of a reservoir or pond can meet the requirements, delineation of target areas which need soil conservation measures (Van Rompaey et al. 2005), and also better understanding of the impacts of past land-use or climatic changes (e.g., Ambers 2001; Vanmaercke et al. 2011). However, as in many developing countries, data on catchment SY for Ethiopia are not only limited in number, but they are also often not very reliable. The lack of a sufficient local database on SY and adoptable SY models have been a problem for reservoir designers, who have used various approaches to address the sedimentation problem in planning new reservoirs (Haregeweyn et al. 2006). This has resulted in a risky or uneconomical design of dams.

In this chapter, we present an overview of SY monitoring programs and SY variability at various spatial scales in Ethiopia. First, an overview of SY monitoring programs and studies on SY variability carried out by the Ethiopian Ministry of Water Resources (MoWR) as well as by own established stations is discussed. Second, rates of reservoir sedimentation for 14 dam catchments and the SY variability at various spatial scales and its impact on reservoirs are discussed.

13.2 SY Monitoring Program by the Ethiopian Ministry of Water Resources

The Ethiopian MoWR is the main responsible body for identifying, establishing, and managing stations designed for measuring runoff and sediment discharge from major river basins and their tributaries in Ethiopia (Fig. 13.1). At those stations, the MoWR monitors water depth, stream velocity, runoff discharge, and sediment concentration into two ways. The first one is using direct (manual) methods such as staff gauge readings and periodic manual runoff and suspended sediment concentration (SSC) sampling, while the second one is by indirect methods such as automatic data loggers/sensors for measuring water depths. Then, for both cases, stage-flow and flow-sediment concentration rating curves are established for each station in order to calculate SY or SSY for the corresponding catchment.

In total, there are 409 river gauging stations spread over the 13 major river basins of the country (Fig. 13.1). However, only 28 of these stations have relatively well-recorded data on sediment concentration and runoff discharge (Redeco 2002). The same study estimated suspended sediment transport of selected rivers under the project "Assessment and monitoring of erosion and sedimentation problems in Ethiopia" based on the existing MoWR data for Ethiopian rivers, resulting in a database with SY and runoff estimations for 230 stations, distributed over seven main river basins of the country.

However, the data of these stations result from a large number of single samplings taken for different purposes, randomly in time and space. The majority of the data at these stations consist of only few observations made over years or even decades. In addition, it was found that in most cases, the available hydrological information together with the data on the suspended sediment loads was also restricted to a limited number of instant observations taken together with the suspended sediment measurements. Due to these weak points of the dataset, it was not possible to consider the entire data for the evaluation process (Redeco 2002). Hence, only about 28 stations which have records covering wet, dry, and transition seasons and that produce a sufficient number of observations were used for the analysis (Table 13.1). The catchments have a drainage area ranging between 69 and 11,690 km^2 and a median of 248 km^2.

To estimate the sediment concentration at a given moment and location, a two-parameter regression approach was adopted:

$$\mathrm{Log(SSC)} = \mathrm{Log}\, a + b(\mathrm{Log}\, Q) \qquad (13.1)$$

where SSC is suspended sediment concentration (mg l^{-1}) and Q is river discharge (l s^{-1}).

The performance of the fitted model for each gauging station varies significantly with the coefficient of determination (R^2) ranging between 0.01 at Aleta Wondo Kolla and 0.67 at Gacheb, with a median value of 0.17 (Table 13.1). Hence,

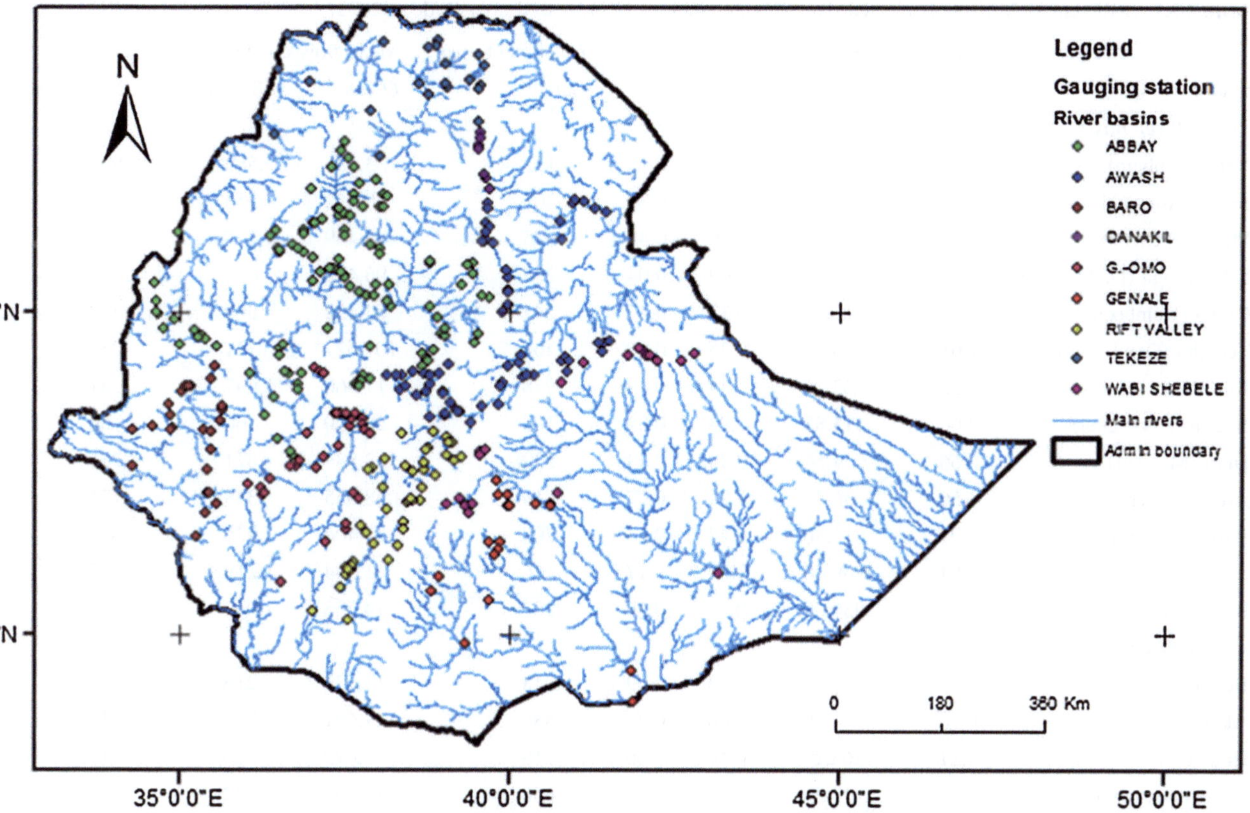

Fig. 13.1 River gauging stations for the main Ethiopian River Basins owned and run by the Ethiopian Ministry of Water Resources (MoWR database); (n = 409)

much of the SSC variability remained unexplained by flow variation; despite this, the SSY for each station (Table 13.1) was calculated on the base of these weak regression models. Therefore, the data from MoWR gauging stations need to be supplemented with site-specific measurements for detailed water resources planning and research purposes.

The lack of reliable hydrological data has been raised in several forums in Ethiopia. For example, Humphreys et al. (1997) stressed that the planning and design of the Tekeze medium-scale hydropower dam in Tigray, covering a drainage area of 30,000 km^2, has been constrained by scarcity of local hydrological data and lack of appropriate models, which forced them to depend on regional-scale data, resulting in risky or uneconomical design of the dam. Humphreys et al. (1997) recommended that accurate and regular measurements of flow discharge and sediment transport should be implemented in the future.

The overall quality of the MoWR database is also demonstrated by a comparison of runoff discharge data from two adjacent gauging stations separated by an air distance of about 100 km on the Blue Nile River, but on the both sides of the Ethiopia–Sudan border (compare Figs. 13.2 and 13.3). The data from the El Diem station in Sudan are superior in terms of length of measuring period, continuity, and consistency of records over those collected at the Ethiopian Shelkole station. The data from Shelkole station are marred with missing data throughout the period 1961–1978, while no single observation was recorded during the period 1979–1998.

Furthermore, a trend comparison between the daily observations at El Diem and Shelkole indicates an important association between both stations. River flow rate at El Diem explained about 61 % the flow rate at Shelkole station (Eq. 13.2 and Fig. 13.4).

$$Q_{\text{Shelkole}} = 0.72\, Q_{\text{Eldiem}} + 420\ \left(R^2 = 0.61; n = 198\right) \quad (13.2)$$

where Q_{Shelkole} and Q_{Eldiem} are Blue Nile River flow rates (m^3 s^{-1}) at Shelkole station in Ethiopia and at El Diem station in Sudan, respectively.

13.3 Other Initiatives to Address the Sediment Yield Data Gap in the Ethiopian Highlands

Apart from the measurements of the MoWR discussed above, there have been a few initiatives undertaken to address the lack of reliable SY data. One notable example of

Table 13.1 Sediment yield data of selected river gauging stations (reprocessed after Redeco 2002)

	Regression model: (Eq. 13.1)				Catchment area	Mean flow conditions		
Station name	Log *a*	*a*	*b*	R^2	(km^2)	*Q* ($l\ s^{-1}$)	SSC (mg^{-1})	SSY ($t\ year^{-1}\ km^{-2}$)
Abbay River Basin								
Yeda (Ambera)	2.57	369.66	0.19	0.12	125	33,136	2,589	902
Jedeb (Amanuel)	1.23	16.83	0.45	0.30	305	80,852	2,753	959
Muga (Bichena)	1.25	17.84	0.40	0.43	375	99,409	1,745	608
L-Chemoga (D. Markos)	2.56	362.24	0.09	0.04	364	96,493	1,010	352
Gudla (Dembecha)	1.22	16.62	0.37	0.47	242	64,152	968	337
Temcha (Dembecha)	1.61	40.33	0.25	0.16	406	107,627	721	251
Birr (Jiga)	1.35	22.28	0.44	0.42	978	17,860	1,697	41
Buchiksi (Metekel)	2.06	115.88	0.03	0.04	106	28,100	160	56
Dura (Metekel)	0.66	4.54	0.44	0.44	539	142,884	821	286
Ardie (Metekel)	1.56	36.37	0.21	0.07	219	58,055	348	121
Azuari (Motta)	1.07	11.86	0.47	0.33	209	55,404	1,936	674
Teme (Motta)	1.84	69.60	0.27	0.17	156	41,434	1,219	425
Fittom (Tilli)	0.90	7,930.00	0.44	0.56	282	74,755	1,042	363
Awash River Basin								
Berga (Addis Alem)	1.73	54.20	0.15	0.09	248	11,633	216	13
Akaki	0.30	1.99	0.60	0.24	884	17,691	718	19
Holeta	1.81	64.27	0.16	0.06	119	5,679	250	16
Awash (Hombole)	0.55	3.56	0.62	0.40	7,656	112,355	5,042	97
Awash (Methara)	0.57	3.71	0.52	0.16	1,641	10,239	431	4
Mojo (Mojo)	2.56	365.60	0.31	0.12	1,264	13,891	6,902	100
Wonji (Wonji)	0.80	6.37	0.53	0.17	11,690	111,793	2,914	37
Baro Akobo River Basin								
Gacheb (Mizan Teferi)	1.36	22.65	0.16	0.69	79	12,325	100	21
Genale Dawa River Basin								
Woyib (Agarfa)	1.77	58.21	0.04	0.03	7,719	270,011	90	4
Omo-Ghibe River Basin								
Sokjie (Areka)	1.10	12.65	0.47	0.18	103	23,602	1,495	450
Sheta (Bonga)	0.51	3.24	0.45	0.69	191	43,768	397	120
Amara (Sheboka)	1.57	36.90	0.27	0.20	69	15,811	497	150
Rift Valley								
Aleta Wondo Kolla	2.34	220.80	−0.09	0.01	206	21,162	95	13
Bedessa (Dilla)	1.78	59.98	0.14	0.02	81	8,321	208	28
Gelana (Yirgalem)	1.17	14.86	0.32	0.08	141	14,485	322	44

Q river flow rate; SSC suspended sediment concentration; SSY specific sediment yield

such effort is the research collaboration between Belgian Universities and Ethiopian Universities on some selected tributaries of the Tekeze and Abay (upper Blue Nile River) basins. Following the launching of different collaboration projects, more accurate SY and/or runoff discharge data for 24 catchments ranging from 1 to 5,000 km^2 have become available based on reservoir sediment surveys, as well as on runoff and SSC-measuring campaigns at gauging stations. These initiatives are discussed below.

13.3.1 Sediment Yield Study Based on Reservoir Sediment Survey Programs

The Zala Daget Research Project (a Flemish Inter University Council Own Initiative Cooperation Project between the Catholic University of Leuven, the Belgian Royal Museum for Central Africa, the Ethiopian Mekelle University and the Relief Society of Tigray) had been undertaking research

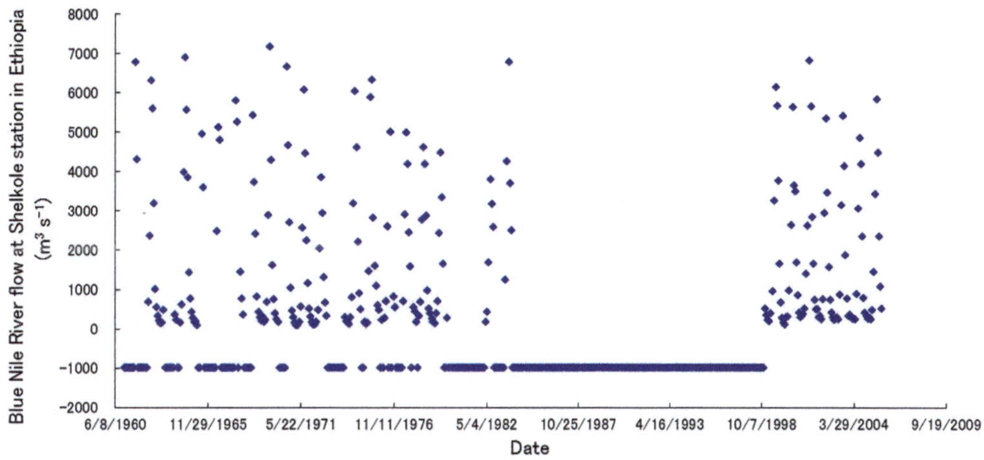

Fig. 13.2 Scatter plot showing river flow rate data at Shelkole station on the Ethiopian Blue Nile River. Missing values are indicated with the value −999

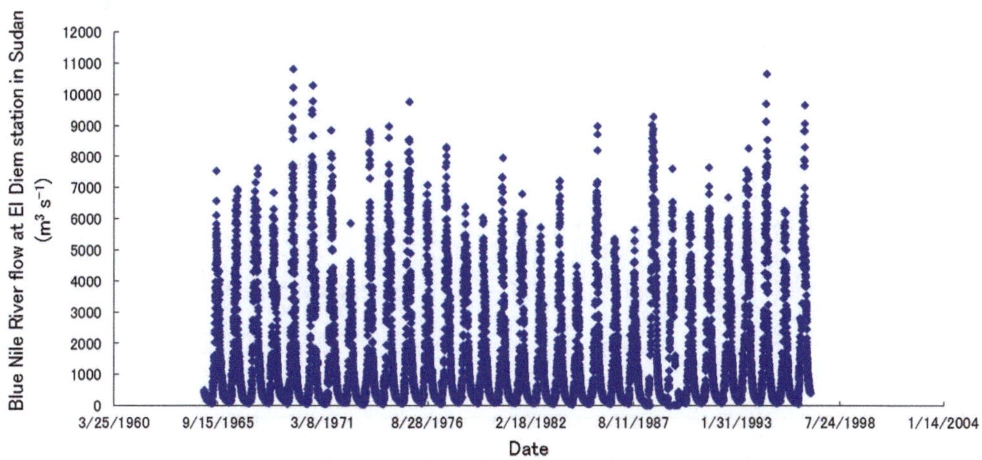

Fig. 13.3 Scatter plot showing river flow rate data at El Diem station on the Sudanese Blue Nile River

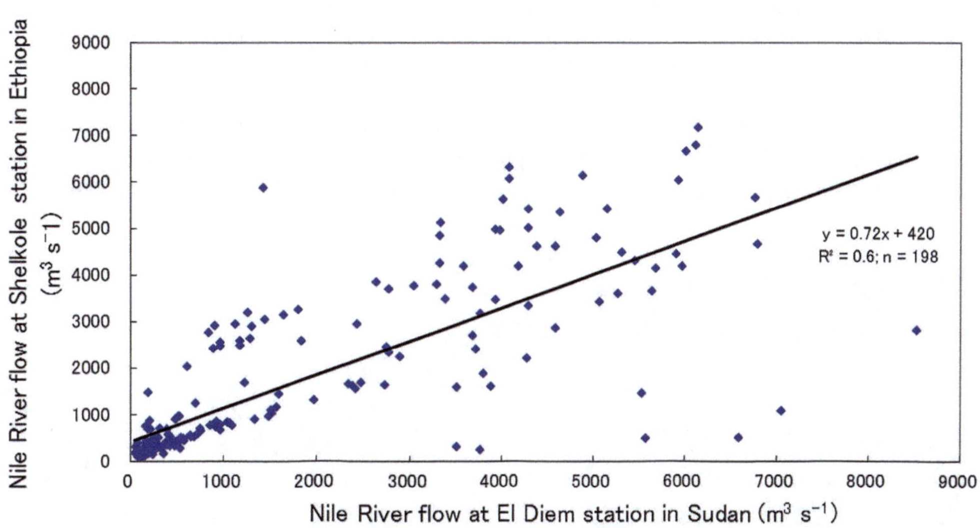

Fig. 13.4 Blue Nile River flow rate at El Diem station (Sudan) versus Shelkole station (Ethiopia) during the period 1965–2005

under the theme "Fighting desertification in the Tigray Highlands, Ethiopia: lessons to be learnt from successes and failures of soil erosion control measures" during the period 2001–2007. Part of this project consisted of assessing and modeling of reservoir sedimentation rates (SRs) and their impacts in the north-Ethiopian highlands. For this purpose, a detailed survey of deposited sediment in reservoirs was carried out for 13 micro-dam catchments located in a radius of 120 km from Mekelle, Tigray's regional capital (Fig. 13.5). The 13 reservoirs were created by constructing earth embankments to harvest seasonal runoff. The stored runoff is then used later in the season for supplementing the rain (when dry spells occur) and/or for full irrigation during the dry part of the year.

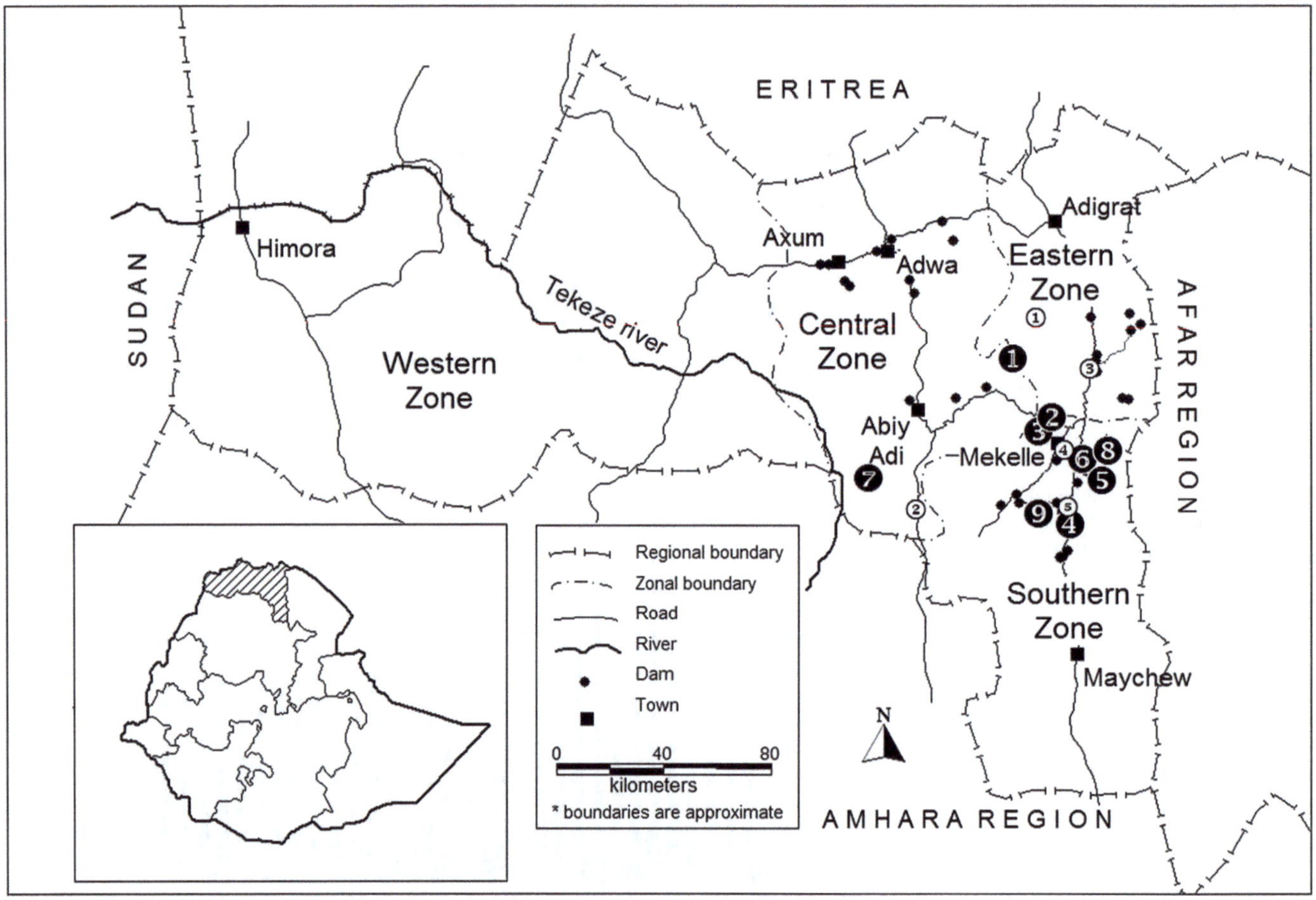

Fig. 13.5 Location of the study area (Tigray) in northern Ethiopia and the studied reservoirs (modified after Haregeweyn et al. 2013). *Dots* represent reservoirs visited, *numbers with black fill* represent reservoirs for which sediment yield was measured: *1* Gindae, *2* Gereb Shegel, *3* Sewhimeda, *4* Gereb Segen, Grashitu, Mejae, Maideli, Gum Selasa, *5* Adiakor, *6* Adihilo, *7* Agushella, *8* Endazoey, *9* Adikenafiz. *Numbers without fill* represent rainfall stations of the Ethiopian Meteorological Authority: *1* Hawzen, *2* Yechila, *3* Wukro, *4* Mekelle airport, *5* Adigudom

The sediment thickness was measured using meter sticks by opening spatially distributed 15–39 profile pits in the reservoir. The deposited sediment boundary, location of sediment profile pits, and topography of the current reservoir bottom were mapped using a theodolite (Fig. 13.6). A digital elevation model (DEM) with a resolution of 1 m for the

Fig. 13.6 Part of the layout of the 26 profile pits at Gereb Shegel reservoir where sediment deposits are up to 3 m thick near the dam (*left*) and details of topographic mapping and sediment profile pit layout in Adihilo reservoir (*right*): (*1*) theodolite readings, (*2*) sediment boundary, (*3*) profile pits, and (*4*) dam (modified after Haregeweyn et al. 2005)

sediment thickness was created from theodolite points, taking measured sediment thickness as *z* value, from which sediment volume was extracted in IDRISI® environment. More details of the reservoir survey and computation of SY values can be found in Haregeweyn et al. (2006).

A similar reservoir sediment survey initiative was reported in Haregeweyn et al. (2012) who analyzed the SR and potential mitigation strategies for the Angereb catchment in NW Ethiopia. In this case, the sediment volume was estimated by subtracting the successive reservoir's DEMs generated based on point (x, y, z) data collected by bathymetric surveys conducted in 2005 and 2007 (Fig. 13.7).

13.3.2 Sediment Yield Study Based on Hydrometric Stations Monitoring Programs

The "Research and Capacity Building for Sustainable Development in the Drylands of Ethiopia"—a collaboration project between Mekelle University and Flemish Universities funded by the VLIR Institutional University Cooperation (IUC) Program (2003–2012)—has conducted studies on various themes, including land resources in the Geba catchment, an upper tributary of the Tekeze River (Fig. 13.8). Within another VLIR-Own initiative project with Bahir Dar University in NW Ethiopia under the theme "Water and sediment budgets of Lake Tana for optimization of land management and water allocation (2010–2015)," SSC and runoff have been monitored in 16 hydrometric stations. However, at this stage only SY data for Geba catchment (Vanmaercke et al. 2010) are available for this discussion.

In the Geba catchment, runoff and suspended sediment discharges were monitored at ten gauging stations (Fig. 13.8). The fieldwork was carried out during the rainy seasons (July–September) of 2004–2007. A digital pressure transducer (TD-diver) was installed at each station, recording the (water + atmospheric) pressure every 10 min. Barometric pressure transducers were installed to allow correction for the atmospheric pressure of the TD-divers in these stations and their nearby stations.

In 2004, measurements were carried out two to three times per week, whereas in 2005–2007 they were made at least daily and also when a major change in flow depth was observed. More details on the instrumentation, monitoring, and sampling of flow and sediment and calculations of SY can be found in Vanmaercke et al. (2010) whereas flow discharge data are reported by Zenebe et al. (2013).

13.3.3 SY Variability at Various Spatial Scales and Its Controlling Factors

The sediment survey on 14 reservoirs, including also Angereb in the northwest Ethiopia, shows that there is a large spatial variation in SSY among catchments: from 189 to 2,927 t km^{-2} year^{-1}, with an average SSY of 1,013 (±735) t km^{-2} year^{-1} (Fig. 13.9). Haregeweyn et al. (2008) reported that spatial variability of SSY for these catchments is positively related to average catchment slope (Av_slope)

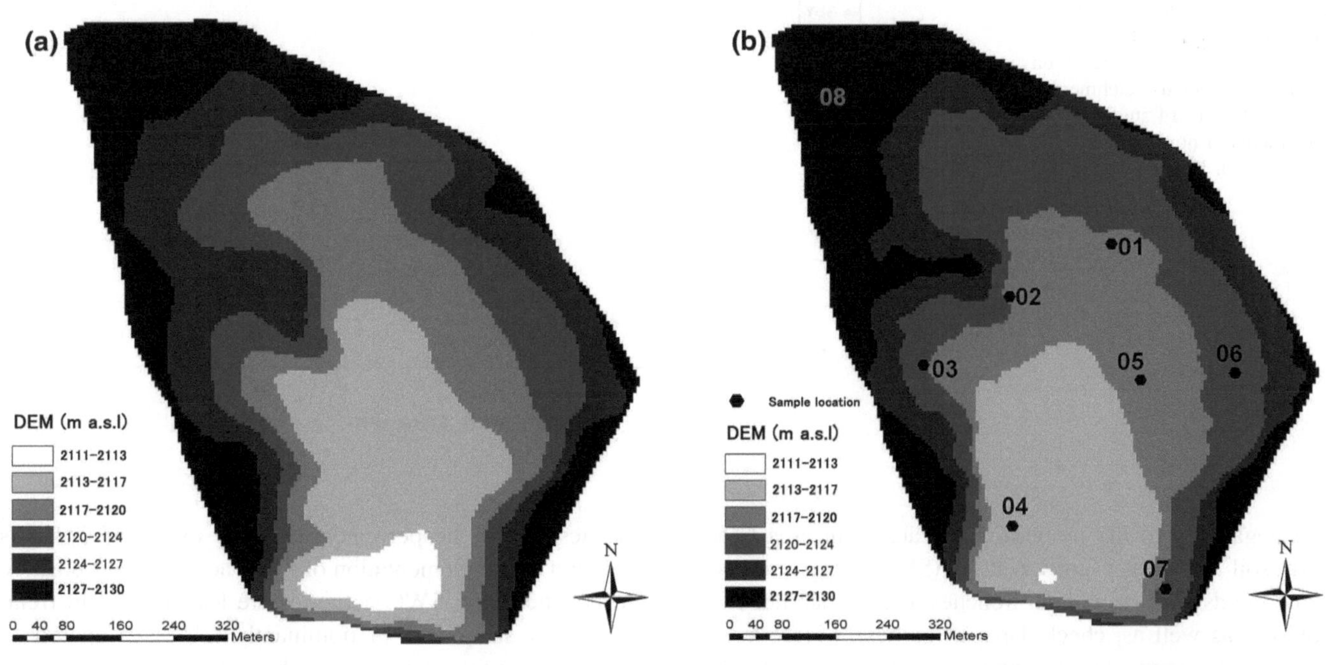

Fig. 13.7 Angereb reservoir's bottom surface DEMs (m a.s.l) created from data obtained by echo-sounder bathymetric surveys conducted in 2005 (**a**) and 2007 (**b**). The IDs on the 2007 map represent dry sediment bulk density sampling locations (Haregeweyn et al. 2012)

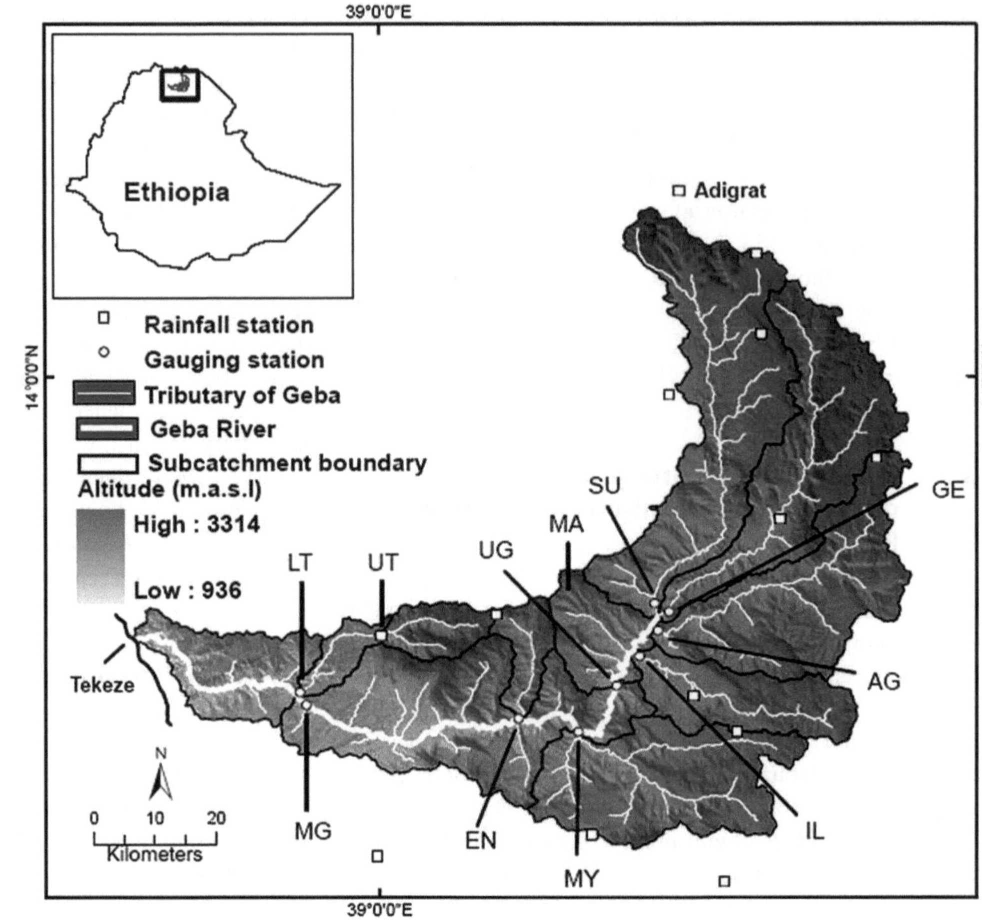

Fig. 13.8 Location of the Geba catchment and the 10 monitored subcatchments: Suluh (SU), Genfel (GE), Agula (AG), Ilala (IL), Upper Geba (UG), May Gabat (MY), Endaselassie (EN), Middle Geba (MG), Upper Tankwa (UT), Lower Tankwa (LT) (modified after Vanmaercke et al. 2010)

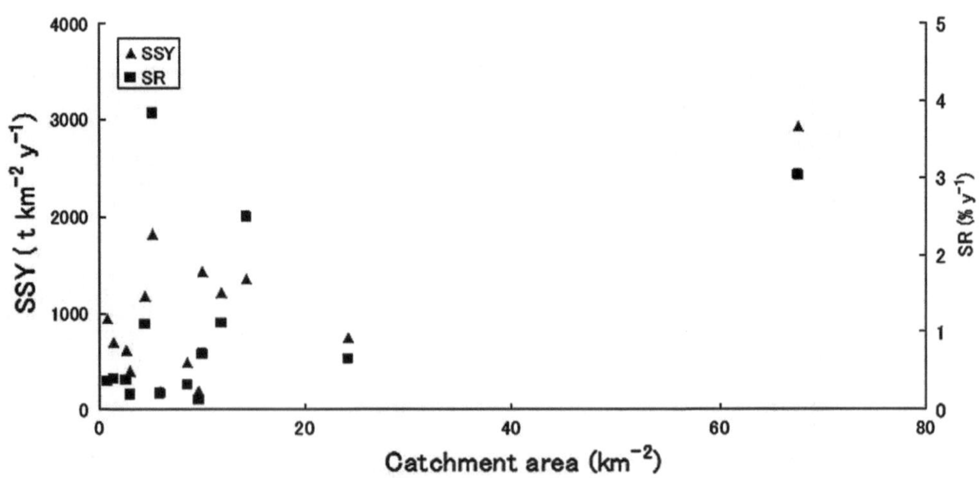

Fig. 13.9 Relationship between area-specific sediment yield (SSY) and sedimentation rate (SR, expressed as % of reservoir storage loss) versus catchment area (A) for the 14 studied dam catchments in northern and northwestern Ethiopia

and negatively to the proportion of catchment area treated with soil and water conservation (SWC) structures (i.e., stone bunds with or without trenches aligned parallel to the contour, as well as, check dams in gullies). Moreover, the same study showed that the extent of cultivated land (CUL) was inversely related to both slope and SWC practices which indicates that in steeper catchments the extent of CULs is small and the implementation of SWC activities has positive effects. Therefore, SWC activities are less applied in (relatively) less steep and in dominantly CULs compared to steeper catchments. This implies that catchments with a high proportion of CUL are prone to erosion processes.

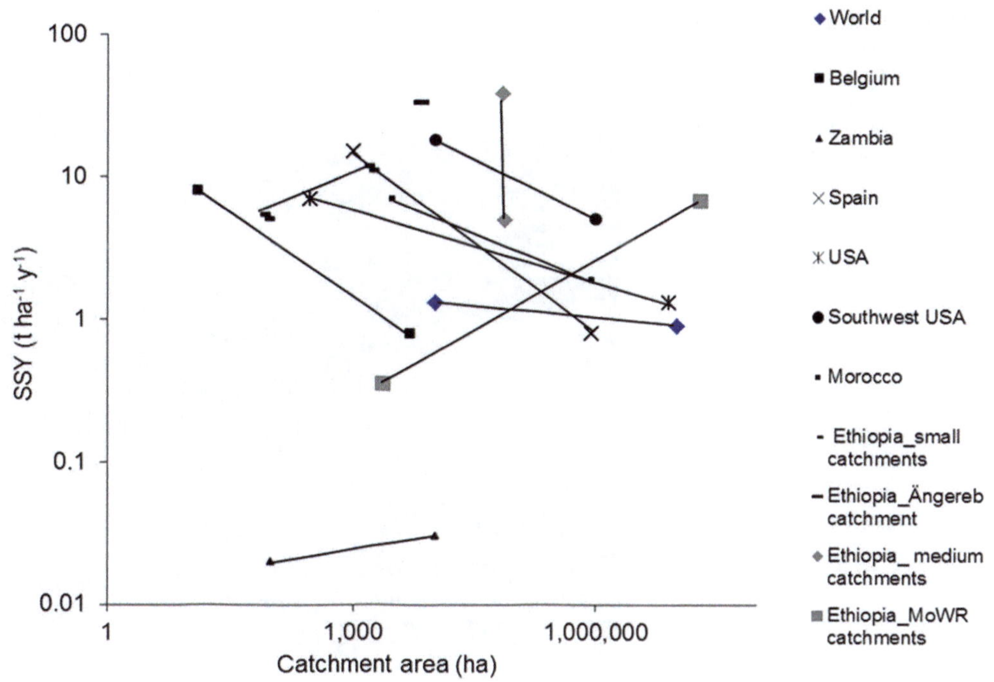

Fig. 13.10 Area-specific sediment yield (SSY) in relation to catchment area for Ethiopia and for some selected countries around the world (modified after Haregeweyn et al. 2012)

Similarly, SSY of medium-sized catchments were found to vary significantly between 497 and 6,543 t km^{-2} year^{-1} (Fig. 13.10). Most of the sediment export occurs during a few short but intense flash floods; hence, daily SY is characterized by a very large temporal variability. There are also large variations in sediment concentration and grain-size distribution which indicate changes in sediment supply during the rainy season due to the depletion of readily available sediments and the development of a vegetation cover (Vanmaercke et al. 2010). The latter study also reported that annual SY is mainly controlled by the occurrence and magnitude of one or a few flash floods that are difficult to predict since these floods mainly depend on local rainfall patterns.

A recent regional SY study in the Blue Nile and Atbara River systems by Balthazar et al. (2013) reported an important spatial variability of SSY ranging between 4 and 4,935 t km^{-2} year^{-1}. The same study reported that 41 % of the observed variation in SSY can be explained by surface vegetation cover (expressed as a percentage of poorly vegetated areas), rainfall intensity (expressed as the Modified Precipitation Index—Fournier 1960), mean annual air temperature, and the human footprint index.

Overall, measured SSY of medium-sized catchments are larger than estimations from smaller catchments (Fig. 13.10). To some extent, this can be explained by the geomorphic characteristics of the Ethiopian highlands. The highlands are generally deeply incised by the larger river systems which give rise to a sharp increase in altitudinal differences as catchment area increases. Moreover, close to the river channels, local slopes are often very steep and sediment storage on the valley bottom alluvial plain, when present, is definitely limited by its very small size. This steeper topography counteracts the fact that the probability for sediment deposition increases with increasing sediment transport distance, leading to no significant decrease in SSY (Vanmaercke et al. 2010).

The magnitude and spatial variation in SSY for Ethiopian catchments is large by regional and global standards (Fig. 13.10), which implies that land degradation and in particular soil erosion and its associated on-site and off-site impacts are severe in the Ethiopian highlands.

13.4 Impacts of Sediment Yield

13.4.1 Reservoir Sedimentation

As a result of high SYs discussed above, the SR is very high in many Ethiopian reservoirs. As much as 50 % of the reservoirs (n = 14) were found affected by extreme siltation problems so that they would lose economic life within half of the design period (Fig. 13.9). According to Haregeweyn et al. (2006), the rapid siltation is associated with (1) poor planning of the reservoirs for the expected SY during the design phase, which in turn is related to lack of a sufficient database, appropriate methodologies to predict SY, lack of skilled and experienced persons for designing dams and reservoirs and (2) poor implementation of reservoir sediment and catchment management strategies.

Fig. 13.11 Evidence of seepage water downstream of Mai Mungude dam and its effect on vegetation establishment and stabilization of river channel in Tigray, Northern Ethiopia

Reservoir sedimentation problems in northern Ethiopia are relatively severe compared to other parts of Ethiopia, the East and southeast Africa region. Haregeweyn et al. (2006) reported annual SR values of 0.18–4 % for 13 reservoirs in northern Ethiopia. Similar studies carried out by DFID (2004) reported SR ranging between 1–3 % in Zimbabwe and 1–4 % in Tanzania. DFID (2004) also reported that about 15 % of the surveyed dams were or will be filled with sediment in less than the design period. For comparison, the world average annual SR for large dams equals 0.5–1 % (WCD 2000).

13.4.2 Impacts of Reservoir Dams on the Local Ecology and River Geomorphology

Apart from the rapid sedimentation discussed above, most of the studied reservoirs had also problems of insufficient inflow and excessive seepage (Haregeweyn et al. 2006) that have influence on the local river geomorphology, surface flow and/or groundwater recharge, and vegetation establishment.

Geomorphological adjustments upstream of a dam are primarily concerned with sedimentation. This sedimentation processes are strongly related to the texture of the sediment and hence its dry bulk density. A spatial correlation between sediment dry bulk density and distance from the inlet of the reservoir to dam body (see Fig. 13.6) for 13 reservoirs in Tigray, Northern Ethiopia, yielded a strong inverse relationship ($r^2 = 0.92$) between the two with a mean dry bulk density of 1.21 t m^{-3} and standard deviation of 0.11 t m^{-3} (Haregeweyn et al. 2006). This implies that coarser fractions deposit first and the finer particles are carried farther into the dam body and remain in suspension before being deposited. This can be explained by the impact of the reservoirs to create flow velocities approaching zero thus attainment of still-water conditions (Morris and Fan 1998).

Most of the studied reservoirs are over designed for the available flow and hence consequently have sediment/flow trapping efficiency close to 100 % (Haregeweyn et al. 2006, 2012). As a result, neither sediment nor flow is supplied to the river channel located downstream of the dam. So there is no clear sign of either aggradation or degradation. Moreover, because of excessive seepage from the reservoirs as a result of highly pervious limestone–shale–marl-intercalated geological formation (Berhane et al. 2013), the immediate downstream valley of the dams has shown increased level of ground water level that led to the emergence of new springs and improvement of vegetation cover (Fig. 13.11). This combined effect of reduced erosion and vegetation establishment resulted in the stabilization of river channels downstream (Fig. 13.12).

13.5 Conclusions

SY variability in the Ethiopian highlands is very large due to the large heterogeneity of biophysical and human factors. This study also revealed that the rate of reservoir sedimentation is very high: 50 % of the reservoirs were found under extreme siltation problem so that they would lose their economic life within half of the design period. The rapid siltation is associated with poor planning and poor

Fig. 13.12 The downstream impacts of reservoirs on sediment load and river morphology are evidenced at the confluence of Illala River and its tributary Mai-Egam in Tigray, northern Ethiopia. The Illala River is relatively stabilized and has permanent base flow, even in the dry season, which is related to the presence of a cluster of five reservoirs located upstream. In contrast, Mai-Egam that is joining from the opposite slope carries significant sediment including bed load though it has a much smaller catchment, but no reservoirs upstream. *Photo* Jan Nyssen®

implementation of the reservoir sediment and catchment management interventions.

The availability of reliable, spatially distributed SY data reflecting the different environments remains the major bottleneck to understand the dynamics and drivers of SY variability in the region. The SY database is poor by international standards. The SY assessment initiatives taken through institutional collaboration projects is a good start; however, such projects have limited capacity and a short life span, so they cannot produce a sustainable solution for this important data gap. Therefore, the MoWR should concentrate efforts on the maintenance and monitoring of existing gauging stations on top of establishing new ones needed to reflect the different eco-hydrological environments in Ethiopia.

References

Ambers KR (2001) Using the sediment record in a western Oregon flood control reservoir to assess the influence of storm history and logging on sediment yield. J Hydrol 244:181–200

Balthazar V, Vanacker V, Girma A, Poesen J, Golla G (2013) Human impact on sediment fluxes within the Blue Nile and Atbara River basins. Geomorphology 180–181:231–241

Berhane G, Martens K, Al Farrah N, Walraevens K (2013) Water leakage investigation of micro-dam reservoirs in Mesozoic sedimentary sequences in Northern Ethiopia. J Afr Earth Sci 79:98–110

DFID (Department for International Development) (2004) Guidelines for predicting and minimizing sedimentation in small dams. Department for International Development, HR Wallingford Ltd., Wallingford

Fournier F (1960) Climat et érosion: la relation entre l'érosion dusol par l'eau et les précipitations atmosphériques. Presses Universitaires de France, France, p 203

Haregeweyn N, Poesen J, Nyssen J, Verstraeten G, de Vente J, Govers G, Deckers J, Moeyersons J (2005) Specific sediment yield in Tigray-Northern Ethiopia: assessment and semi-quantitative modelling. Geomorphology 69:315–331

Haregeweyn N, Poesen J, Nyssen J, De Wit J, Haile H, Govers G, Deckers J (2006) Reservoirs in Tigray: characteristics and sediment deposition problems. Land Degrad Dev 17:211–230

Haregeweyn N, Poesen J, Nyssen J, Govers G, Verstraeten G, De Vente J, Deckers J, Moeyersons J, Haile M (2008) Sediment yield variability Northern Ethiopia: a quantitative analysis of its controlling factors. Catena 75:65–76

Haregeweyn N, Melesse B, Tsunekawa A, Tsubo M, Meshesha D, Balana BB (2012) Sedimentation and its mitigating strategies: a case study of Angereb reservoir, northwestern Ethiopia. J Soils Sediment 12:291–305

Haregeweyn N, Poesen J, Govers G, Verstraeten G, de Vente J, Nyssen J, Deckers S, Moeyersons J (2013) Evaluation and adaptation of a spatially-distributed erosion and sediment yield model in Northern Ethiopia. Land Degrad Dev 24:188–204

Humphreys H, Bellier C, Kennedy R, Dokin K, (1997) Tekeze medium hydropower report, Annex B-hydrology report. Ministry Water Resour Addis Ababa, p 264

Morris G, Fan J (1998) Reservoir sedimentation handbook design and management of dams reservoirs and catchments for sustainable use. McGraw-Hill, New York

Redeco (2002) Assessment and monitoring of erosion and sedimentation problems in Ethiopia. Final report, Redeco consulting, GmbH, Hydrology Studies Department, Ministry of Water Resources, Addis Ababa, Ethiopia

Vanmaercke M, Zenebe A, Poesen J, Nyssen J, Verstraeten G, Deckers J (2010) Sediment dynamics and the role of flash floods in sediment export from medium-sized catchments: a case study from the semi-

arid tropical highlands in northern Ethiopia. J Soils Sediments 10:611–627
Vanmaercke M, Poesen J, Verstraeten G, Maetens W, de Vente J (2011) Sediment yield as a desertification risk indicator. Sci Total Environ 409:1715–1725
Vanoni VA (1975) Factors determining bed forms of alluvial streams. J Hydraul Div-ASCE 101(11):1435–1440
Van Rompaey AJJ, Bazzoffi P, Jones RJA, Montanarella L (2005) Modeling sediment yields in Italian catchments. Geomorphology 65:157–169
Walling DE (1994) Measuring sediment yield from river basins. In: Lal R (ed) Soil erosion research methods, 2nd edn. Soil and Water Conservation Society, Ankeny, pp 39–80
WCD (World commission of Dams) (2000) Dams and development, a new framework for decision making, Report of the World Commission of Dams. Earth-scan publications, London
Zenebe M, Vanmaercke M, Poesen J, Verstraeten G, Haregeweyn N, Haile M, Amare K, Deckers J, Nyssen J (2013) Spatial and temporal variability of river flows in the degraded semi-arid tropical mountains of northern Ethiopia. Zeitschrift für Geomorphologie 57(2):143–169

Climatic and Hydrologic Changes in Northern Ethiopia in the last 3,500 Years: Evidence from the Geomorphic, Stratigraphic, and Geochemical Archives of Hayk Lake

14

Massimiliano Ghinassi, Marco Benvenuti, Filippo D'Oriano, and Marialelena Fedi

Abstract

Lake Hayk (23 km^2 in surface) is located in northern Ethiopia, at an altitude of around 2,000 m a.s.l. The Late Holocene sedimentary successions preserved along the lake margins provide a detailed archive of the climate history of East Africa over the past 3,500 years. These successions are represented by colluvial/stromatolitic deposits and palustrine, deltaic, and fluvial sediments along the northern and southern coasts, respectively. During the last 3,500 years, the lake was characterized by three highstand phases, which occurred at about 3,250–3,000, 2,600–950, and 650 cal years BP—modern age (160 years BP non-calibrated). These stages are recorded by aggradation of siliciclastic deposits both along the southern and northern lake margins, whereas isotope data suggest that most of the stromatolites grew during phases of intense evaporation and lacustrine contraction. The good correlation between the Hayk Lake record and those of other late Holocene lakes of East Africa lakes allows a wide-scale generalization of late Holocene climate variability.

Keywords

Late Holocene • Climate • Isotope record • Stromatolites • East Africa • Ethiopia

14.1 Introduction

Palaeoclimatic proxies from East African lakes (Noren et al. 2002; Viles and Goudie 2003; Mayewski et al. 2004) highlighted the complexity of regional climatic fluctuations (Halfman et al. 1994; Cohen et al. 1997; Verschuren et al. 2000; Diaz et al. 2011; Xoplaki et al. 2011) which occurred mainly after the mid-Holocene humid period (Gasse 2000, 2001). During the last millennium, this marked climate variability was mainly controlled by a complex interaction between ocean–atmosphere circulation determining climate gradients over the equatorial Africa (Russel et al. 2007). The Lake Naivasha record in Kenya, representing one of the most complete climate proxies of this time interval, shows that the Little Ice Age in East Africa was characterized by humid conditions interrupted by three main droughts (Verschuren et al. 2000), which had a remarkable impact on social and economic development.

The coastal landforms of Hayk Lake (Fig. 14.1) in Northern Ethiopia and their clastic deposits and biogenic carbonates (i.e., stromatolites) are described and interpreted in detail for their geomorphological, stratigraphic and sedimentological significance. Lacustrine stromatolites are common features of the East Africa Rift lakes, representing an archive suitable for isotopic dating, reconstruction of former lake levels and paleolimnological conditions

M. Ghinassi (✉)
Dipartimento di Geoscienze, Università di Padova, V. Gradenigo 6, 35121 Padua, Italy
e-mail: massimiliano.ghinassi@unipd.it

M. Benvenuti
Dipartimento di scienze della Terra, Università di Firenze, V. G. La Pira 4, 50121 Florence, Italy

F. D'Oriano
GEOPHI srl, Area della Ricerca CNR Via Gobetti 101, 40129 Bologna, Italy

M. Fedi
INFN Sezione di Firenze, via Bruno Rossi 1, 50019 Sesto Fiorentino, FI, Italy

P. Billi (ed.), *Landscapes and Landforms of Ethiopia*, World Geomorphological Landscapes, DOI 10.1007/978-94-017-8026-1_14,

Fig. 14.1 Panorama of the Lake Hayk from the Northern coast

(Casanova 1986, 1994). This paper summarizes the results of a previous study (Ghinassi et al. 2012) and aims at extending the knowledge on Late Holocene climatic fluctuations in East Africa.

14.2 Geological and Geomorphological Settings

Lake Hayk is located on the eastern margin of the Ethiopian Highland (Fig. 14.2a) at about 2,000 m a.s.l. Bedrock is represented by the ~1,000-m-thick Termaber Basalt (Merla et al. 1973; Zanettin and Justin-Visentin 1973) that accumulated during the stage of the Ethiopian plateau shield volcanism (Miocene to Early Pliocene). The lake occupies a small fault-controlled depression bounded by N–S trending faults which crosscut two minor set of faults with NE–SW and NNW–SSE trends (Fig. 14.2c).

Hayk Lake is a hydrologically closed basin with a surface area of about 23 km^2 (Fig. 14.2c). The northern lake margin is characterized by steep, rocky slopes, whereas the southern margin hosts a wide coastal plain, which is drained by the only perennial inflow to lake, the Ankarka River (Fig. 14.2b). Since 1938, the maximum depth of the lake, occurring in the north-western sector, has decreased from 88.8 m (Baxter and Golobitsh 1970) to 81 m (Lamb et al. 2007). Occasionally, following very intense precipitation, the nearby Ardibo Lake (Fig. 14.2b) can overflow into the Hayk Lake expanding its surface catchment by about three times. The region has a subhumid, tropical climate with an average annual rainfall and a mean annual temperature of 1,158 mm and 18 °C, respectively (Demille 2000). Lamb et al. (2007) calculated a water balance pointing to a groundwater outflow of 1.4×10^6 m^3 $year^{-1}$ (3 % of the total) and evaporative losses of 33×10^6 m^3 $year^{-1}$ (65 % of the total).

14.3 Sedimentary Succession and Chronology

Sedimentary succession of the northern (Uarababo) and southern (Ankarka River) margins formed in two different morphological and depositional settings. The Ankarka River succession (Fig. 14.3) accumulated in a low-gradient setting and is characterized by swamp and fluvio-deltaic deposits, whereas the Uarababo succession (Fig. 14.4) formed along steep coasts and is made of colluvial fan deposits and stromatolitic biostromes. Radiocarbon ages obtained from different depositional units are summarized in Table 14.1.

14.3.1 Fluvio-deltaic Deposits of the Southern Coast: The Ankarka River Succession

The Ankarka River succession (Fig. 14.3), up to 7 m thick, is made of fluvial and coastal deposits organized into four main, unconformity-bounded sedimentary units (Ak1–Ak4).

- Unit Ak1 is up to 3 m thick, made of swampy coastal mud (Fig. 14.3d), sandwiched between two vertisols, which are labeled as Vertisol A and B, respectively (Fig. 14.3b). Coastal muds are laminated and show an alternation between millimeter-thick dark and light laminae. The light laminae are composed of authigenic

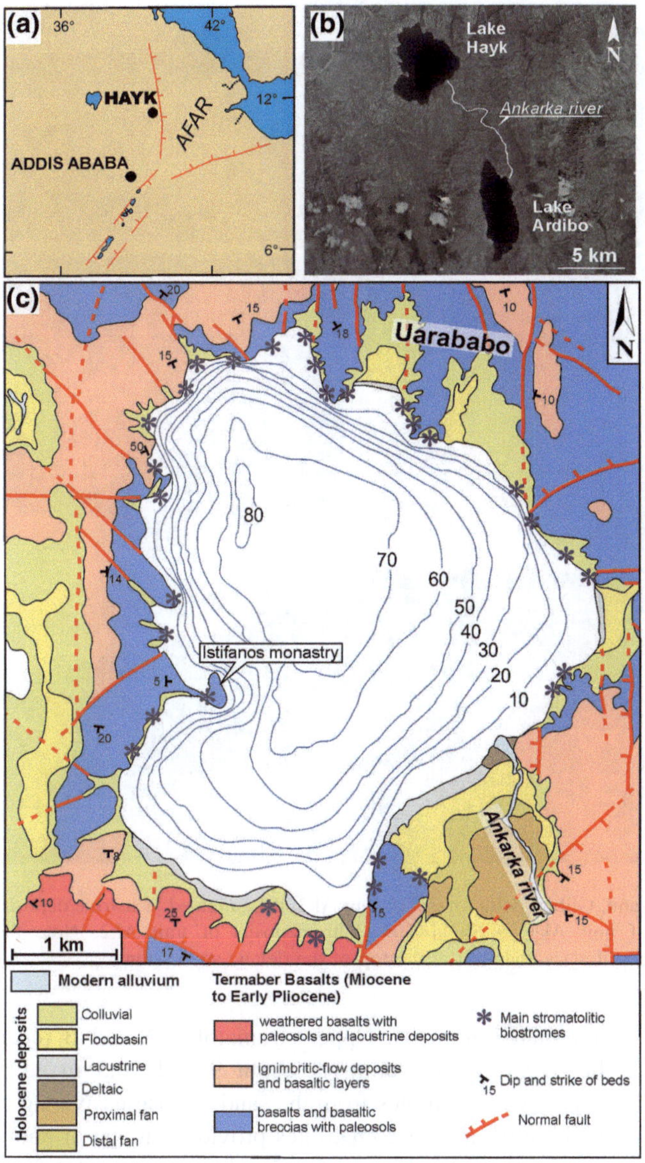

Fig. 14.2 **a** Geographic location of Lake Hayk. **b** Satellite image (Google™Earth) of Lake Hayk and Lake Ardibo. **c** Geological sketch map of the Hayk Basin. Isobaths in meters are from Demlie et al. (2007)

carbonate and shell debris, whereas the dark laminae are organic rich and contain abundant remains of palustrine vegetation. Coastal muds decrease in thickness landward, and their thinnest part is located about 5 m a.l.l. (above lake level). Radiocarbon dating provided an age of about 3,200 cal year BP for these muds (Fig. 14.5).

- Unit Ak2 consists of fluvial sand passing lakeward into shoalwater delta deposits. Fluvial sands are medium-fine grained, stacked in decimeter-thick, tabular, and gently inclined beds, locally interbedded with mud. Deltaic deposits are made of meter- to decimeter-thick, sandy and muddy tabular beds (lobes) overlain by distributary channel sands (Fig. 14.3e). Distributary channel deposits occur at an elevation of about 8.5 m a.l.l. On the whole, these deposits yielded radiocarbon ages ranging between 1,200 and 1,000 cal year BP (Fig. 14.5).
- Unit Ak3 is about 6 m thick and consists of fluvial gravels passing lakeward into their deltaic equivalent. Fluvial gravels range from planar- to cross-stratified and form amalgamated channelized units up to 1 m thick. Fluvial deposits are organized into four subunits bounded by unconformity surfaces of lower rank (Ak3a–Ak3d in Fig. 14.3b). These surfaces are characterized by a marked erosional relief (Fig. 14.3b) and are associated with repeated phases of fluvial incision. Subunits Ak3b, Ak3c, and Ak3d are dated about 450 cal year BP, 400 cal year BP, and modern age (160 year BP non-calibrated), respectively (Fig. 14.5). Deltaic deposits are poorly exposed and consist of high-angle, clinostratified gravel and sand (Fig. 14.3c) accumulated in a Gilbert-type delta front. The lack of continuous outcrop prevents the fine-tuning between the deltaic gravels and subunits Ak3a–d. The fluvial surface F3 marks the top of subunit Ak3d (see Sect. 14.3.3) and occurs about 9–10 m a.l.l.
- Unit Ak4 consists of fluvial deposits perched along the flanks of the modern Ankarka River valley (Fig. 14.3a) occurring as two subunits (Ak4a–b) capped by fluvial surfaces F1–F2 (see Sect. 14.3.3).

14.3.2 Colluvial and Stromatolitic Deposits of the Northern Coast: The Uarababo Succession

The Uarababo succession, about 10 m thick, is made of colluvial and stromatolitic deposits (Fig. 14.4) subdivided into four unconformity-bounded sedimentary units (Ub1–Ub4).

- Unit Ub1, at least 2 m thick, is made of colluvial gravels draped by 2–10-cm-thick stromatolites (Fig. 14.4e, f). The stromatolites (hereafter referred to as the "lower stromatolite") standing at a maximum elevation of 5–6 m a.l.l., developed on colluvial gravels and, locally, on a discontinuous, 0.5–2-cm-thick, sandy transgressive lag deposit containing gastropod shells (*Melanoides* sp.). Charcoal from the uppermost part of the colluvial gravel provided a radiocarbon age of about 3,500 cal year BP (Fig. 14.5).
- Unit Ub2 is up to 6 m thick and consists of colluvial fan delta gravels (Fig. 14.4c, e) with well-developed topset (i.e., subaerial part of the fan delta) and foreset (i.e., subaqueous part of the fan delta) deposits. Topsets dip about 35° lakeward and are made of debris flow and debris fall gravels with minor muddy intercalations. Foreset deposits show a similar dip and are made of debris flow, debris fall, and density current deposits ranging in grain size from sand to gravel. The topset–foreset

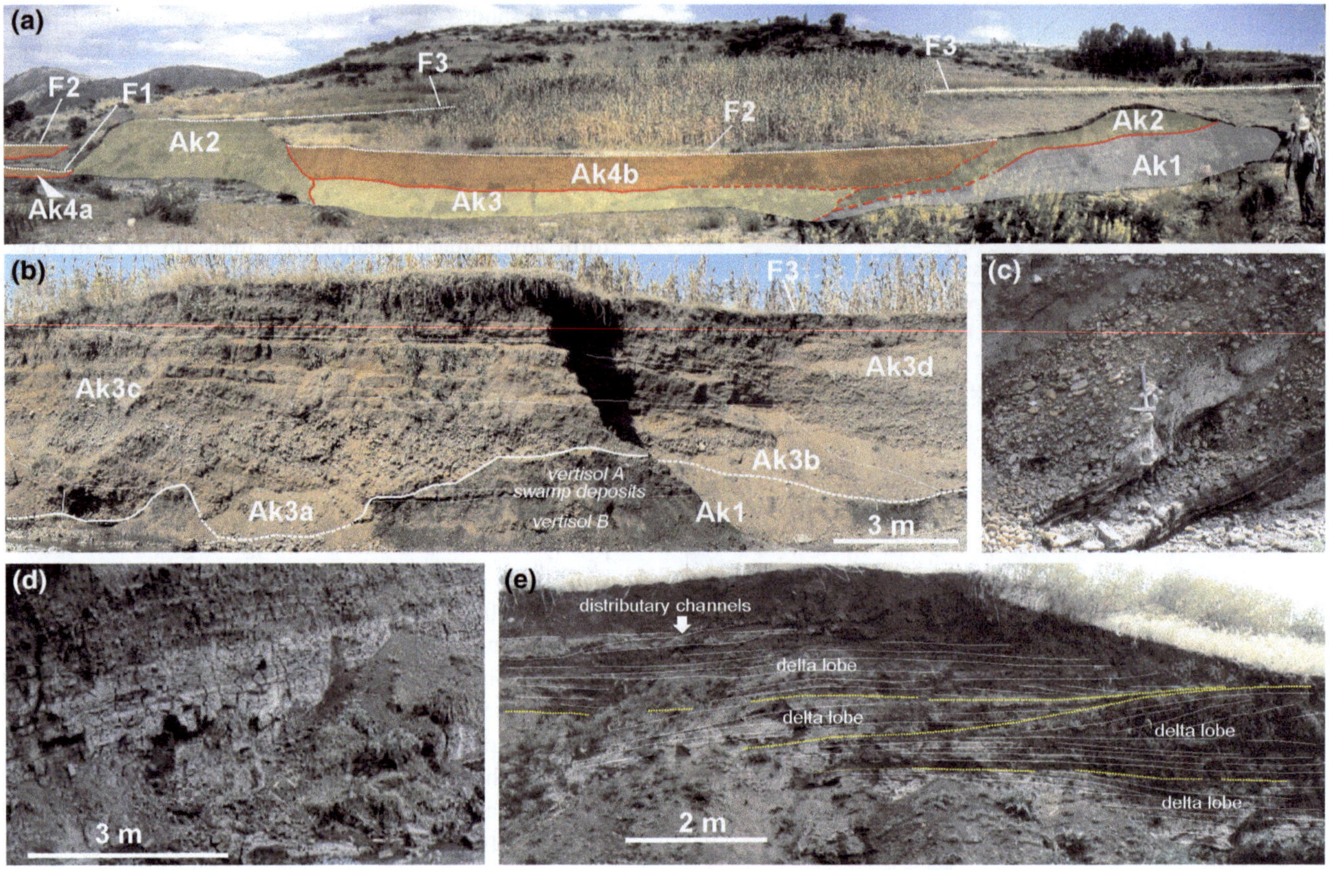

Fig. 14.3 Ankarka River sedimentary succession. **a** The northernmost outcrop of the Ankarka succession. **b** Unconformity-bounded subunits Ak3a–d erosionally overlying the basal muddy Ak1 unit. **c** Gravelly sand, Gilbert delta foreset of unit Ak3. **d** Laminated lacustrine deposits of unit Ak1. **e** Shoalwater delta deposits of unit Ak2. Note the distributary channels at the top of the deltaic succession

transition deposits are commonly wave-worked and occur about 8.5 m a.l.l. A number of samples from unit Ub2 indicate continuous deposition from about 2,500–1,650 cal year BP (Fig. 14.5).

- Unit Ub3 is up to 2 m thick and is made of a stromatolitic carbonates ("upper stromatolite") with intervening colluvial gravels. The latter allow to distinguish an upper stromatolite A from an upper stromatolite B (Fig. 14.4b). Lakeward, the upper stromatolite B rests unconformably on upper stromatolite A forming a 30–40-cm-thick biostrome (Fig. 14.4a, c, d) due to pinching-out of the colluvial deposits. A transgressive sandy layer (1–5 cm thick) rich in gastropod shells occurs at the base of upper stromatolite A. The upper stromatolite A underwent subaerial exposure as indicated by the brownish color (Fig. 14.4d) due to organic matter and mud infiltration and by dissolution cavities and then filled by well-sorted sand and gastropod shells during the following transgression. Upper stromatolite B is light yellow and contains two main veneers of fine- to medium-grained sand and shell fragments. Upper stromatolite A and B reach the maximum elevation at about 9 and 10.5 m a.l.l., respectively. Samples from the sandy transgressive lag at the base of upper stromatolites provided an age of about 650 cal year BP (Fig. 14.5). Consistently, samples from the base of upper stromatolite A and top of upper stromatolite B suggest that they developed between about 750 cal year BP and the modern age (Fig. 14.5).
- Unit Ub4 includes four subunits (Ub4a–Ub4d) of wave-worked gravelly sands topped by as many lacustrine terrace surfaces (Fig. 14.6a, b).

14.3.3 Coastal Morphological Features Related to the Stratigraphic Architecture

Hayk Lake coastal areas are characterized, particularly along the northern margin, by laterally continuous, flat to gently inclined (2° to 3° lakeward) surfaces ranging from 0.8 to

Fig. 14.4 Uarababo sedimentary succession. **a** Biostrome formed by upper stromatolites of unit Ub3. Hammer for scale is 33 cm. **b** Upper stromatolite of unit Ub3 consisting of two different portions (labeled as upper strom. A and upper strom.B) separated by subaerial colluvial gravels. **c** Biostromes of unit Ub1 and Ub3 separated by colluvial deposits of unit Ub2. Biostromes of unit Ub1 consists of upper stromatolite B resting on upper stromatolite A without any interbedded colluvial gravels. **d** Slab of upper stromatolite (unit Ub3). The *white line* marks the surface of subaerial exposure separating upper stromatolite A from B. **e** Erosion surface separating colluvial fan delta deposits of unit Ub1 from those of units Ub2. Lower stromatolites occur at the top of unit Ub1. **f** Slab of the lower stromatolite capping unit Ub1. Knife for scale is 11 cm

19 m a.l.l. (Fig. 14.6a, b, d). Five clusters of values can be determined by plotting the elevation of these surfaces above the modern lake level (Fig. 14.6d). The first three clusters (1.8–2, 3, and 5–6 m a.l.l.) are related to benches built of wave-worked gravelly sand of unit Ub4, and therefore, they can be labeled as coastal terraces (L1, L2, and L3; Fig. 14.5). The fourth cluster (about 8.5 m a.l.l.) coincides with the brink point (i.e., topset/foreset transition) of Ub2 colluvial fan deltas (Fig. 14.5) and, therefore, marks the final stage of Ub2 colluvial fan delta progradation. The fifth cluster (about 9–10 m a.l.l.) is linked with a surface capping wave-worked deposits of unit Ub4 and, similarly to the first three clusters, is interpreted as a lacustrine terrace (L4). This surface also fits with the maximum elevation reached by the upper stromatolite B and with the surface capping fluvial deposits of unit Ak3d in the Ankarka area (Fig. 14.5).

Fluvial terraces (surfaces F1–F3) occur in the distal part of the Ankarka River valley (Figs. 14.3a and 14.6c) at the top of subunits Ak4a (F1), Ak4b (F2), and Ak3d (F3). In the Ankarka delta area, terraces F1, F2, and F3 make lateral transition into lacustrine terrace surfaces L2, L3, and L4, respectively (Fig. 14.5). Lacustrine terrace surface L1 has no fluvial equivalent.

14.4 Isotopic Data from Stromatolites

Oxygen and carbon stable isotope analyses were carried out on both upper stromatolites and the lower stromatolite from the northern margin of the lake (Fig. 14.7a). A 3.5-cm-thick sample from the lower stromatolite (Fig. 14.7c) provided nine sequential samples for isotope analysis. A specimen of 17.5 cm thick of the upper stromatolite B directly over the upper stromatolite A (unit Ub3) was finely sampled (Fig. 14.7b). 26 samples from upper stromatolite A and 30 from upper stromatolite B were analyzed, yielding a curve showing frequent oscillations of the stable isotopic composition. Since Hayk Lake oxygen isotope values are more

Table 14.1 Chronological data from the Ankarka River and Uarababo district successions. (A) C14 datings on charcoal samples from the Ankarka succession; (B) C14 datings on charcoal samples from the Uarababo succession; (C) 14C data of stromatolites samples (measured radiocarbon ages, reservoir-corrected ages, and calibrated ages—at 95.4 % confidence level—are reported. Details about the correction applied for the reservoir age effect can be found in Ghinassi et al. (2012)

A	Sample	Lab.code	14C age (years BP)	Cal age (68.2 %) (Cal years BP)	Cal age (95.4 %) (Cal years BP)
Ak3d	HK05.4.3a	BETA209892	160 ± 40	Modern	Modern
Ak3d	HK05.4.3b	LABEC 14Fi 1503 and 1505	170 + 36	Modern	Modern
Ak3c	HK05.4.2	LABEC 14Fi0139 and 0144	280 ± 30	430–290	450–280
Ak3b	HK05.4.1	LABEC 14Fi0138 and 01143	350 + 50	480–320	450–310
Ak2	HK05.24.1	BETA203648	1,070 + 40	1,050–930	1,060–930
	HK18.2	LABEC 14Fi0401 and 0402	1,160 ± 30	1,170–1,000	1,170–980
	HK2	BETA-190827	1,170 + 60	1,180–990	1,250–950
Ak1	HK3	BETA-190829	3,070 + 60	3,360–3,220	3,390–3,090
B	**Sample**	**Lab.code**	**14C age (years BP)**	**Cal age (68.2 %) (Cal years BP)**	**Cal age (95.4 %) (Cal years BP)**
Ub3	HK05.28.1-10	LABEC 14F0568 and 0575	600 ± 45	645–550	660–535
	HK05.28.1	LABEC 14Fi1497 and 1504	830 ± 30	765–695	790–680
Ub2	HK 33.a1	LABFC 14Fi0968 and 0963	1,810 + 40	1,815–1,710	1,860–1,620
	HK05.33.b1	LABEC 14F0392	1,750 ± 40	1,715–1,610	1,810–1,550
	HK31.c	LABEC 14F0400 and 0405	2,220 ± 30	2,310–2,158	2,330–2,150
	HK31.b	LABEC 14Fi0406 and 0409	2,340 ± 60	2,470–2,210	2,700–1,740
	HK31.a1	BETA209891	2,550 ± 40	2,750–2,510	2,750–2,470
Ub1	HK05.30.1	BETA203650	3,550 ± 40	3,890–3,730	3,960–3,700
C	**Lab.cod**	**Measured 14C age (years BP)**	**Reservoir effect-corrected 14C age (years BP)**	**Cal age (95.4 %) (Cal years BP)**	**Cal age (68.2 %) (Cal years BP)**
STR_B2	14Fi1628	600 ± 75	185 ± 100	Modern	Modern
STR_B1	14Fi1623, 14Fi1625	1,450 ± 45	880 ± 80	930–680	910–730

sensitive to an evaporation-induced isotopic enrichment (Talbot 1990; Li and Ku 1997; Yuan et al. 2011) rather than groundwater inflow (Lamb et al. 2007), recorded high values of $\delta^{18}O$ are interpreted as indicating surface water evaporation during episodes of reduced rainfall.

In the lower stromatolites, $\delta^{18}O$ ranges between +1.68 and +5.05 ‰ and shows an overall increase from the base to the top of the sampled biostrome (Fig. 14.7c), indicating evaporative enrichment. $\delta^{13}C$ ranges from +0.67 to +4.65 ‰ and shows a good covariance with $\delta^{18}O$, suggesting a balance of the dissolved inorganic carbonate in the lake water with atmospheric CO_2 (Li and Ku 1997), hinting to a closed-lake condition during stromatolite formation (Talbot 1990).

In the upper stromatolite A, $\delta^{18}O$ ranges between of +2.99 and +6.24 ‰ and, notwithstanding a minor decrease within the first 15 mm, it shows an overall increasing trend. $\delta^{13}C$ ranges between +3.71 and +7.51 ‰ and shows a good covariance with $\delta^{18}O$ (Fig. 14.7b). This isotope record suggests that the upper stromatolite A grew in a closed lake during a stage of progressive evaporative enrichment. In the upper stromatolite B, $\delta^{18}O$ varies from a minimum of +6.06 ‰ to a maximum of +7.75 ‰ and, within an overall rising trend of $\delta^{18}O$, a short stage of slight increase can be detected. $\delta^{13}C$ fluctuates between +5.08 and +7.39 ‰, and the similar pattern between $\delta^{18}O$ and $\delta^{13}C$ (Fig. 14.7b) still indicates a closed lake.

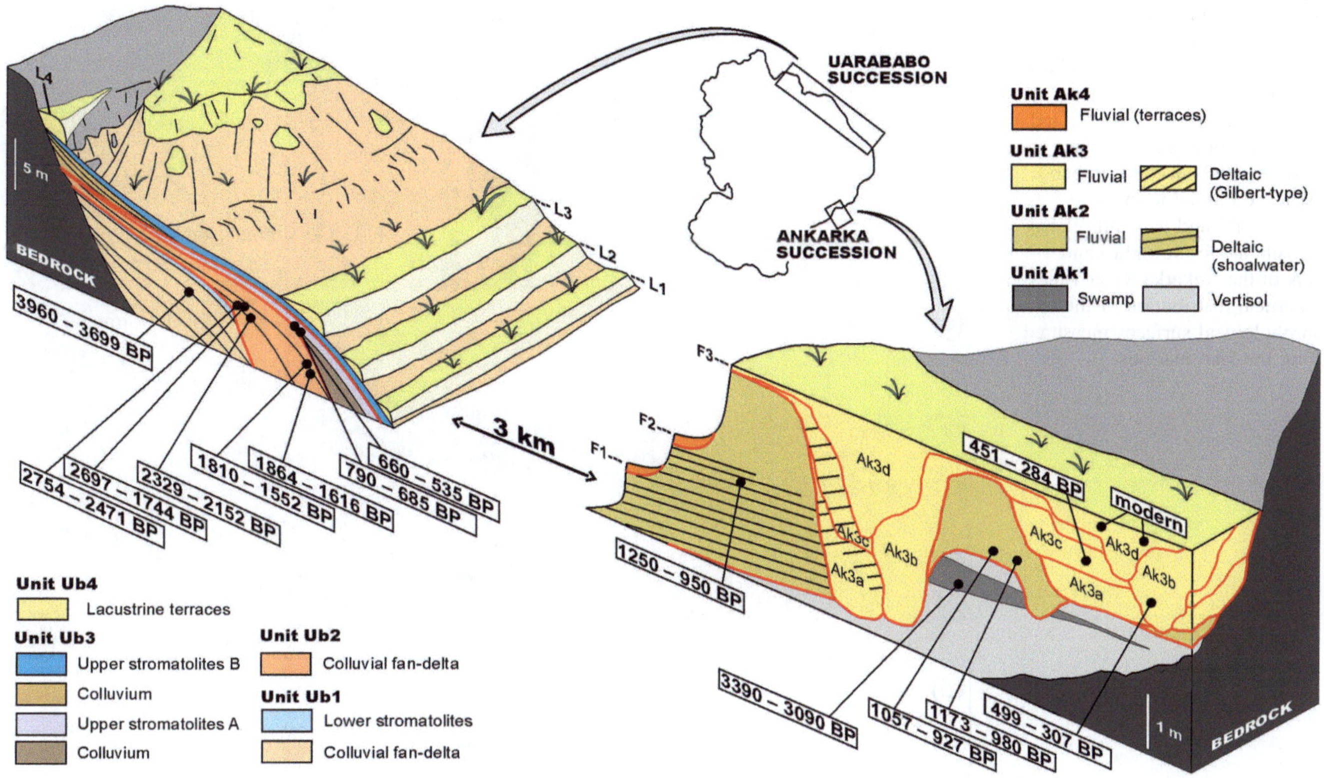

Fig. 14.5 Schematic N–S cross section across the Uarababo and Ankarka sedimentary successions. Depositional environment and age of different units are shown

The overall evaporative enrichment in $\delta^{18}O$ documented by the upper stromatolites well fits the data obtained by Lamb et al. (2007) from coeval deep-lacustrine deposits cored in the SW part of the lake.

14.5 Discussion: Lacustrine Oscillations and Comparison with Other Climate-Proxy Records

The oldest part of the Hayk succession documents a lacustrine lowstand, which was associated with colluvial sedimentation and soil formation along the northern and southern coasts, respectively. The first transgression occurred at about 3,250 cal year BP (Fig. 14.8a) and resulted in a lake level rising up to about 5 m above the modern level. In the Uarababo area, this transgression called a halt to colluvial sedimentation and caused the growth of the lower stromatolite (Ub1), whereas in the Ankarka River zone, it resulted in the establishment of a coastal marsh (Ak1). The evaporative enrichment of ^{18}O recorded in the lower stromatolites suggests that following this transgression, their development progressed under increasing evaporation conditions (i.e., dryer climate). This trend then forced subsequent lake-level fall, culminating in the subaerial exposure of the biostromes and the development of Vertisol A in the Ankarka area.

On the whole, Ub1-Ak1 units record a climatically forced regressive–transgressive cycle that finds a regional correlation with other lacustrine archives. The Hayk lowstand recorded by the Ub1 (colluvium) and Ak1 (Vertisol B) deposits is older than 3,250 cal year BP and can be tentatively correlated with a widespread dry episode centered on 4,200–4,000 cal year BP (Barker et al. 2004). In East Africa, the same event caused the regression (Fig. 14. 8b) of Lakes Bosumtwi (Talbot and Johannessen 1992) and Ziway-Shala (Gasse and Street 1978; Benvenuti et al. 2002), the isolation of Lake Albert (Talbot and Brendeland 2001), and a decline in discharge of the White Nile (Hassan 1997). The following highstand, at about 3,200 cal year BP (Fig. 14.8a), is again recorded also (Fig. 14.8b) in Lakes Bosumtwi (Talbot and Johannessen 1992), Abe (Gasse and Street 1978) and Ziway-Shala (Gillespie et al. 1983).

The lake-level fall, subsequent to this first stage and possibly greater than 5 m, was followed by renewed lake flooding between 2,600 and 950 cal year BP. During this interval, the lake level rose up to about 8.5 m a.l.l. (Fig. 14.8a), providing the space for accumulation of colluvial fan deltas (Ub2) and deltaic deposits (Ak2) in the

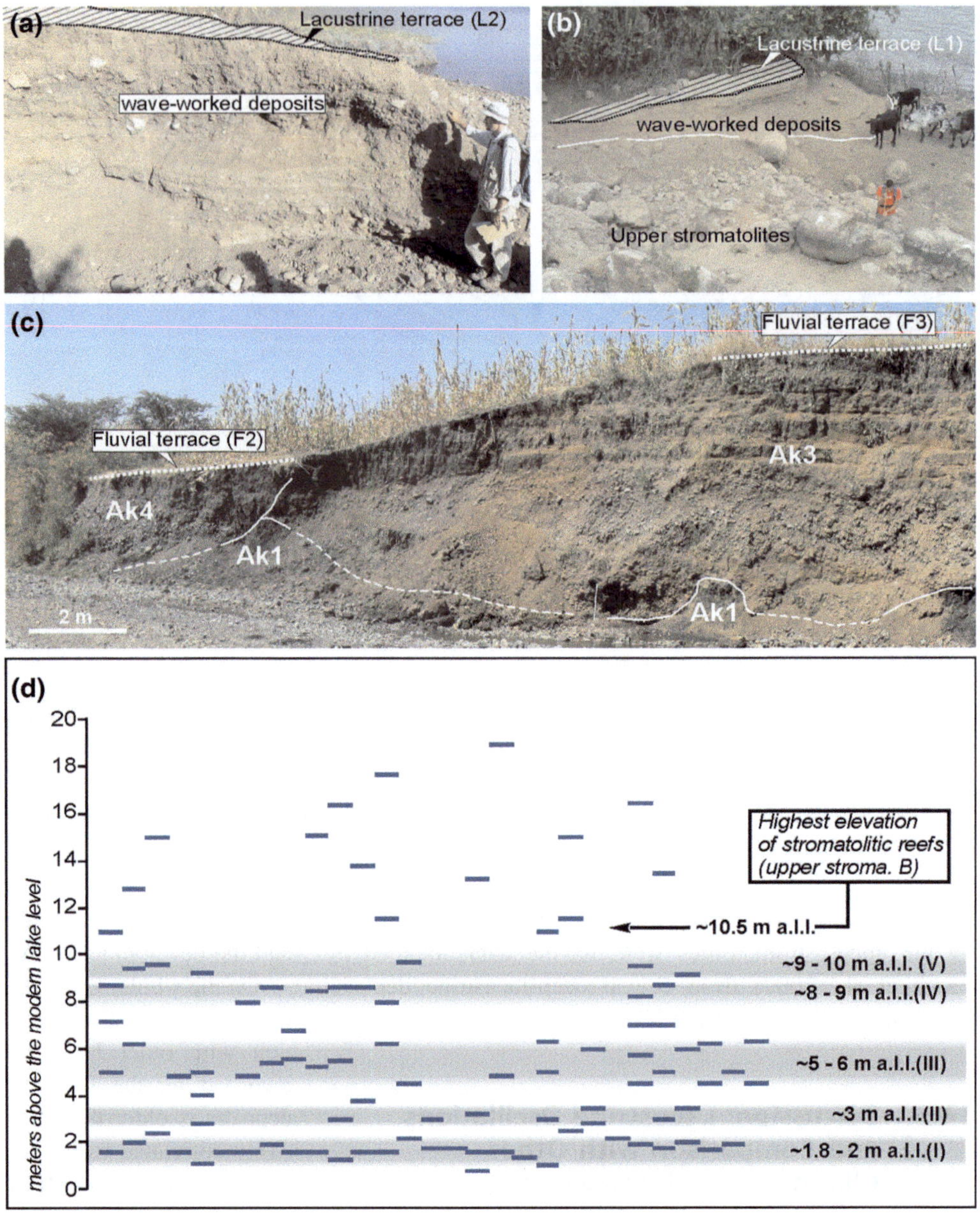

Fig. 14.6 **a** Lacustrine terraces L2 and related deposits, occurring at about 3 m a.l.l. in the Uarababo district. **b** Lacustrine terraces L1, occurring at about 2 m a.l.l. and overlying upper stromatolite deposits (Ub3) in the Uarababo district. **c** Fluvial terraces (surfaces F2 and F2) and related deposits along the right-hand flank of the Ankarka River valley. **d** Vertical distribution of the morphological surfaces measured along the lake margins

Uarababo and Ankarka areas, respectively. During the same time span, also lakes Victoria (Stager et al. 2003), Abe (Gasse and Street 1978), Bosumtwi (Talbot and Johannessen 1992), Ziway-Shala (Gillespie et al. 1983), and Tanganyika (Cohen et al. 1997) experienced high levels, hinting to a regional hydrologic signature of a moist phase (Fig. 14.8b).

A further lake-level fall occurred between 950 and 600 cal year BP, with a subsequent shoreline regression bringing the level at about 8–9 m lower than the previous highstand (Fig. 14.8a). The third lacustrine transgression started at 600 cal year BP with a level rise of about 9–10 m a.l.l. (Fig. 14.8a), which determined the fluvio-deltaic aggradation (Ak3) in the Ankarka area, whereas the sandy transgressive lag at the base of the upper stromatolite (Ub3) formed in the Uarababo area. Similarly to the previous events, this regressive–transgressive cycle has regional equivalents too. The Hayk regression at 950 cal year BP is matched (Fig. 14.8b) in lakes Tanganyika (Cohen et al. 1997), Naivasha (Verschuren et al. 2000), and Victoria (Stager et al. 2005) and represents the establishment of dry climate which affected East Africa during the European Medieval Climatic Anomaly.

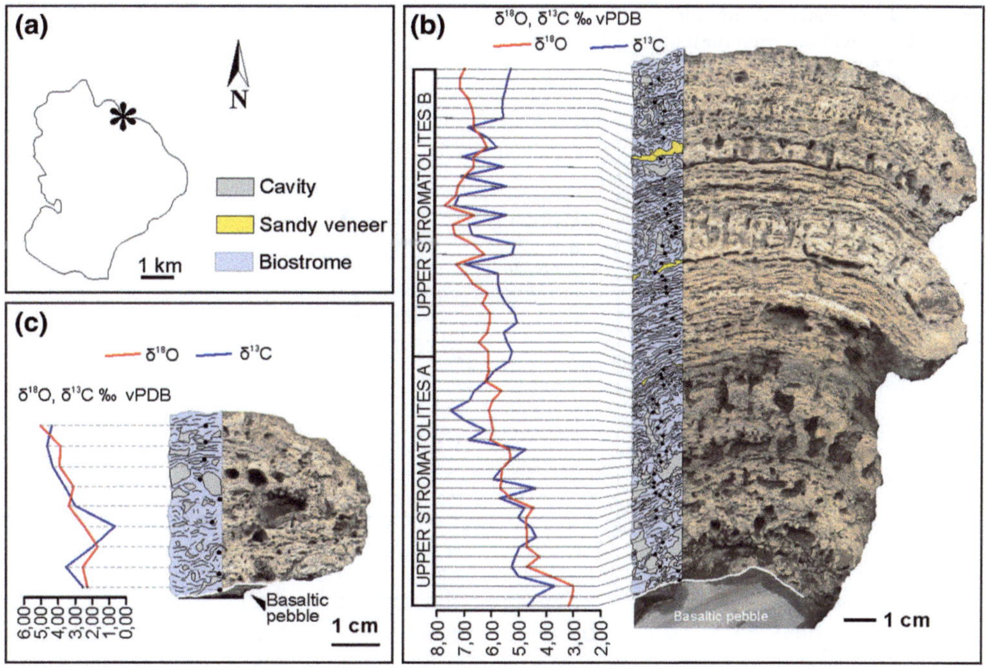

Fig. 14.7 **a** Location of the samples showed in *inset* C and D. **b** Variations of $\delta^{18}O$ and $\delta^{13}C$ in the stromatolites of unit Ub3 (upper stromatolites). Note the two main sandy veneers interbedded within the biostrome. **c** Changes of $\delta^{18}O$ and $\delta^{13}C$ in the lowers stromatolites (unit Ub1)

The last lacustrine rise started about 600 cal year BP and established highstand conditions which persisted until present (160 year BP not calibrated). This highstand phase was punctuated by three main regressive episodes, which are recorded in the Ankarka area (Ak3a–d) and in the sedimentological and isotope signatures of the upper stromatolite. Though not directly dated, the subunit Ak3a is considered to have been formed during lake-level rise, synchronously with the formation of the sandy transgressive lag and the onset of upper stromatolite A growth along the northern margin (Fig. 14.9). The upper stromatolite A started to develop during the lake-level rise as indicated by the $\delta^{18}O$ trend of the three basal samples (Fig. 14.7b). Nevertheless, its successive growth occurred under a progressive enrichment in $\delta^{18}O$, likely related to a lake level lowering under dryer climate conditions (Fig. 14.7b). The lake shrinkage culminated with the subaerial exposure of the upper stromatolites A and accumulation of colluvial deposits. In the meantime, the subunit Ak3a in the Ankarka area was undergoing erosion. Subunits Ak3b, Ak3c, and Ak3d document three highstands separated by two episodes of lake contraction between 450 cal year BP and modern age (160 year BP not calibrated). These oscillations are not evidenced in the isotope proxy of the upper stromatolites B, likely because they were too rapid to be recorded within the overall trend of $\delta^{18}O$ evaporative enrichment. Nevertheless, the signature of these oscillations may be represented by the two sandy veneers within the upper stromatolites B (Fig. 14.7b). During these two short-lived lowstands, the upper stromatolite B may have been weakly reworked by wave winnowing witnessed by the thin sandy horizons.

At a regional scale, this last phase of Lake Hayk development corresponds to the Little Ice Age as it is synchronous (Fig. 14.8b) with a similar evolution observed in Turkana (Halfman et al. 1994), Abe (Gasse and Street 1978), Tanganyika (Cohen et al. 1997), Naivasha (Verschuren et al. 2000), and Victoria (Stager et al. 2005) lakes. The three short-lived episodes of Lake Hayk level fall may find a correspondence with the main droughts (Wamara, Nyarubanga, and Lapanarat-Mahlatule) identified by Verschuren et al. (2000) in Lake Naivasha (Fig. 14.9) and by Lamb et al. (2007) in a lacustrine sediment core collected in the bottom of Hayk Lake (Fig. 14.9).

The progressive lacustrine contraction, which has led to the modern lake, is geochemically recorded in the uppermost part of the upper stromatolite B (i.e., further increase in $\delta^{18}O$ values) and in the development of lacustrine (L1–3) and fluvial (F1–3) terraces. This post-Little Ice Age lake shrinking, together with the records from lakes Naivasha (Verschuren et al. 2000) and Turkana (Halfman et al. 1994), highlights that a progressively warmer and dryer conditions have been established over East Africa in recent times.

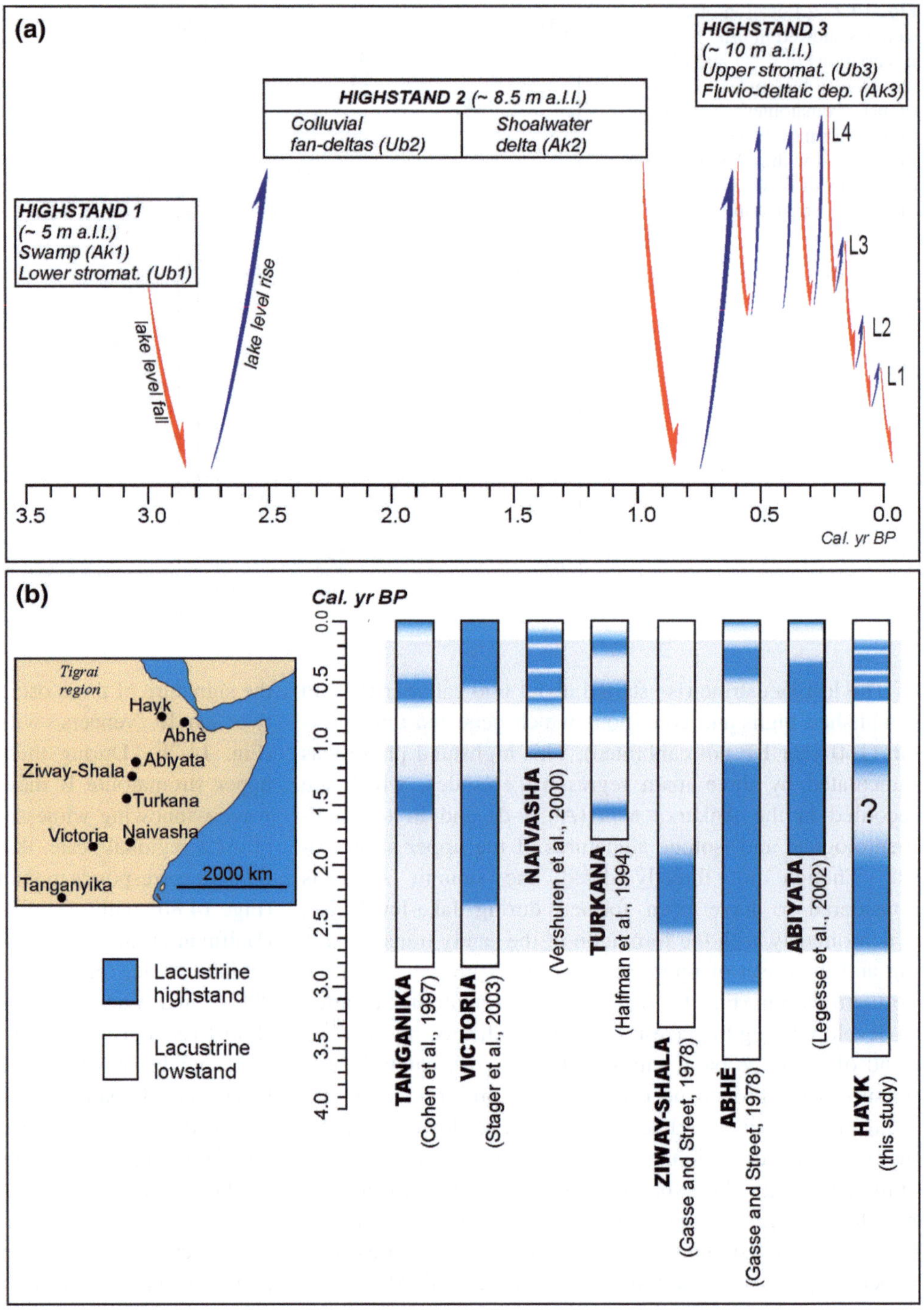

Fig. 14.8 **a** Relative changes in Lake Hayk level occurred during the past 3,500 years. **b** Comparison between Lake Hayk record and proxies from Late Holocene lacustrine successions of East Africa

14.6 Conclusions

The Hayk Lake succession documents the main climate changes occurred in the central Ethiopian highlands over the past 3,500 cal year BP. The main episodes of increase in rainfall which caused lacustrine rises are dated at about 3,250–3,000, 2,600–950, and 650 to modern cal year BP, respectively. The correlation between such lacustrine highstands and those occurred in some of the most studied East African lakes (e.g., Naivasha, Tanganika, Victoria, Turkana, Abe, Ziway-Shala, Bosumtwi, and Abyiata) supports the hypothesis of their climatic origin. Furthermore, the high-frequency fluctuations documented during the late highstand phase in Lake Hayk can be correlated with three main drought phases (Wamara, Nyarubanga, and Lapanarat-Mahlatule), which were previously identified by Verschuren et al. (2000) in Lake Naivasha.

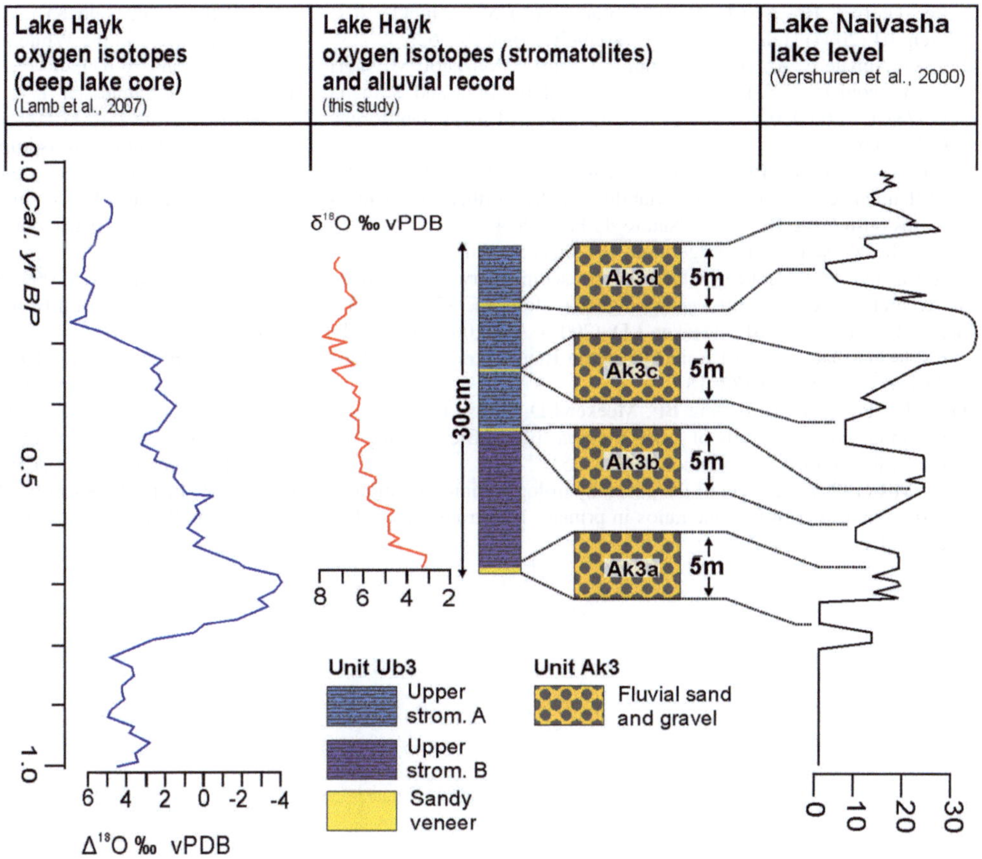

Fig. 14.9 Comparison between lacustrine oscillation documented by units Ak3 and Ub3 (*central columns*), Lake Naivasha documentation (Verschuren et al. 2000; *right-hand column*) and isotope record by Lamb et al. (2007; *left-hand column*)

References

Barker PA, Talbot MR, Street-Perrott FA, Marret F, Scourse J, Odada EO (2004)Late Quaternary climatic variability in intertropical Africa. In: Battarbee RW,Gasse F, Stickley CE (eds.) Past climate variability through Europe and Africa.Springer, Dordrecht, pp 117–138

Baxter RM, Golobitsh DL (1970) A note on the limnology of Lake Hayq, Ethiopia. Limnol Oceanogr 15:144–148

Benvenuti M, Carnicelli S, Belluomini G, Dainelli N, Di Grazia S, Ferrari GA, Iasio C, Sagri M, Ventra D, Balemwald, A, Seifu K (2002) The Ziway-Shala lake basin (Main Ethiopian Rift, Ethiopia): a revision of basin evolution with special reference to the Late Quaternary. J Afr Earth Sciences 35:247–269

Casanova J (1986) East African Rift stromatolites. Geol Soc Spec Publ London 25:201–210

Casanova J (1994) Stromatolites from the East African Rift: a synopsis. Phanerozoic stromatolites II. Springer, Berlin, pp 193–226

Cohen AS, Talbot MR, Awramik SM, Dettman DL, Abell P (1997) Lake level and paleoenvironmental history of Lake Tanganyika, Africa, as inferred from late Holocene and modern stromatolites. Geol Soc Am Bull 109:444–460

Demille M (2000) Hydrology, hydrogeology and hydrochemistry of lake Haiq-Hardibo sytem. Unpublished MSc thesis, Addis Ababa University, Ethiopia

Demlie M, Ayenewb T, Wohnlicha S (2007) Comprehensive hydrological and hydrogeological study of topographically closed lakes in highland Ethiopia: The case of Hayq and Ardibo. J Hydrol 339: 145–158

Díaz HF, Trigo RM, Hughes MK, Mann ME, Xoplaki E, Barriopedro D (2011) Spatial and temporal characteristics of climate in medieval times revisited. B Am Meteorol Soc 92:1487–1500

Gasse F (2000) Hydrological changes in the African tropics since the last glacialmaximum.Quat Sci Rev 19:189–211

Gasse F (2001) Hydrological changes in Africa. Scince 292:2259–2260

Gasse F, Street FA (1978) Late Quaternary lake-level fluctuations and environments of the northern rift valley and Afar region (Ethiopia and Djibouti). Palaeogeogr Palaeocl 24:279–325

Ghinassi M, D'Oriano F, Benvenuti M, Awramik S, Bartolini C, Fedi M, Ferrari G, Papini M, Sagri M, Talbot M (2012) Shoreline fluctuations of Lake Hayk (northern Ethiopia) during the last 3,500 years: geomorphological, sedimentary, and isotope records. Palaeogeogr Palaeocl 365, 366:209–226

Gillespie R, Street-Perrott FA, Switsur R (1983) Post-glacial arid episodes in Ethiopia have implications for climate prediction. Nature (London) 306:680–682

Halfman JD, Johnson TC, Finney BP (1994) New AMS dates, stratigraphic correlations and decadal climatic cycles for the past 4 ka at Lake Turkana, Kenya. Palaeogeogr Palaeocl 111:83–89

Hassan FA (1997) Holocene paleoclimates of Africa. Af Archaeol Rev 14:213–230

Lamb HF, Leng MJ, Telford RJ, Ayenew T, Umer M (2007) Oxygen and carbon isotope composition of authigenic carbonate from an Ethiopian lake: a climate record of the last 2,000 years. The Holocene 17:515–524

Li HC, Ku TL (1997) δC–δ^{18}O covariance as a paleohydrological indicator for closed-basin lakes. Palaeogeogr Palaeocl 133:69–80

Mayewski PA, Rohling EE, Stager JC, Karle'n W, Maasch KA, Meeker LD, Meyerson EA, Gasse F, Van Kreveld S, Holmgren K,

Lee-Thorp J, Rosqvist G, Rack F, Staubwasser M, Schneider RR, Steig EJ (2004) Holocene climate variability. Quat Res 62:243–255
Merla G, Abbate E, Azzaroli A, Bruni P, Canuti P, Mazzuoli M, Sagri M, Tacconi P (1973) A geological map of Ethiopia and Somalia (1973) 1:2,000,000 and comment with a map of major landforms. CNR, Italy
Noren AJ, Bierman PR, Steig EJ, Lini A, Southon JA (2002) Millennial-scale storminess variability in the northeastern United States during the Holocene. Nature 419:821–824
Russell JM, Verschuren D, Eggermont H (2007) Spatial complexity of 'Little Ice Age' climate in East Africa: sedimentary records from two crater lake basins in western Uganda. Holocene 17:183–193
Stager JC, Cumming BF, Meeker LD (2003) A 10,000 year high-resolution diatom record from Pilkington Bay, Lake Victoria, East Africa. Quat Res 59:172–181
Stager JC, Ryves D, Cumming BF, Meeker LD, Beer J (2005) Solar variability and the levels of Lake Victoria, East Africa, during the last millennium. J Paleolimnol 33:243–251
Talbot MR (1990) A review of the palaeohydrological interpretation of carbon and oxygen isotopic ratios in primary lacustrine carbonates. Chem Geol 80:261–279
Talbot MR, Brendeland KI (2001) Strontium isotopes as palaeohydrological tracers in the White Nile headwaters lakes, East Africa. OOS Transactions, AGU Fall Meeting Supplement 82, Absytacy PP21C-05
Talbot MR, Johannessen T (1992) A high resolution palaeoclimatic record for the last 27,500 years in tropical West Africa from the carbon and nitrogen isotopic composition of lacustrine organic matter. Earth Planet Sci Lett 110:23–37
Verschuren D, Laird KR, Cumming BF (2000) Rainfall and drought in equatorial East Africa during the past 1,100 years. Nature 403: 410–413
Viles HA, Goudie AS (2003) Interannual, decadal and multidecadal scale climatic variability and geomorphology. Earth Sci Rev 61: 105–131
Xoplaki E, Fleitmann D, Diaz H, von Gunten L, Kiefer T (2011) Medieval climate anomaly. Pages News 19:40 pp
Yuan F, Sheng Y, Yao T, Fan C, Li J, Zhao H, Lei Y (2011) Evaporative enrichment of oxygen-18 and deuterium in lake waters on the Tibetan Plateau. J Paleolimnol 46:291–307
Zanettin B, Justin-Visentin E (1973) Serie di vulcaniti etiopiche. 1- La serie dell'altipiano etiopico centro-orientale. Boll Soc Geol It 92:313–327

Rift-Related Morphology of the Afar Depression

15

Giacomo Corti, Ian D. Bastow, Derek Keir, Carolina Pagli, and Elizabeth Baker

Abstract

The Afar Depression is a subaerial triple junction between the Nubian, Somalian and Arabian Plates, the only place where the final stages of continental break-up can be observed on-land. In spite of the region being hot and inhospitable, scientists have carried out fundamental work in this unique geological setting over the last few decades. We have long-known that rifting began on large-scale border faults that now bound the Afar Depression but what role magma played in the development of this incipient ocean basin was not clear. However, in recent years, it has been revealed that repeated dike intrusions together with normal faulting accommodate extension producing a landscape dominated by spectacular fresh fault scarps and active volcanic edifices that have been created during episodic tectonic, volcano-tectonic and purely volcanic events. Observations from Ethiopia have fundamentally changed the way we think about continental break-up. The challenge now is to take what we have learned and apply it to the geological record of the rifted margins elsewhere on Earth.

Keywords

Afar • Rifting • Normal faulting • Diking • Graben • Volcano

15.1 Introduction

The Afar Depression is a triangular-shaped area of rifting at the triple junction between the Nubian, Somalian and Arabian plates (Figs. 15.1 and 15.2). Covering an area of ~200,000 km^2 Afar is up to ~300 km wide in the south and ~600 km long from south to north. Elevation drops uniformly from ~1,000 m in the south-west to below sea level in the north (Danakil depression) and in the east, where the shores of Lake Asal, fluctuating at around 155 m below sea level, represent the lowest subaerial point of the African continent. Superimposed on this topography are young volcanic landforms that only exceptionally stand 500 m above the rift floor. The Afar lowland hosts one of the most hostile environments on Earth. Maximum temperatures are, even in the coldest months, well above 30 °C and can exceed 50 °C during the summer wet season. Dallol, at the northern tip of Afar, has the highest average annual temperature for an inhabited location: 34 °C between 1960 and 1966. Rainfall is rare, averaging typically less than 200 mm per year. The Afar Depression is an endorheic basin: the main waterflow into the area is the Awash River, which flows north-eastward through southern Afar where it ends in a chain of interconnected lakes, the last of which is Lake Abhe. These saline lakes contain almost the only water in the region.

G. Corti (✉)
Consiglio Nazionale delle Ricerche, Istituto di Geoscienze e Georisorse, Via G. La Pira, 4, 50121 Florence, Italy
e-mail: giacomo.corti@unifi.it

I.D. Bastow
Department of Earth Science and Engineering, Imperial College London, South Kensington Campus, London SW7 2AZ, UK

D. Keir
National Oceanography Centre Southampton, University of Southampton, Southampton SO14 3ZH, UK

C. Pagli
School of Geography, Earth and Environmental Sciences, Plymouth University, Plymouth PL4 8AA, UK

C. Pagli
Dipartimento di Scienze della Terra, Università di Pisa, Via Santa Maria, 53, 56126 Pisa, Italia

E. Baker
Shell, Houston, TX 77079-1197, USA

P. Billi (ed.), *Landscapes and Landforms of Ethiopia*, World Geomorphological Landscapes,
DOI 10.1007/978-94-017-8026-1_15, © Springer Science+Business Media Dordrecht 2015

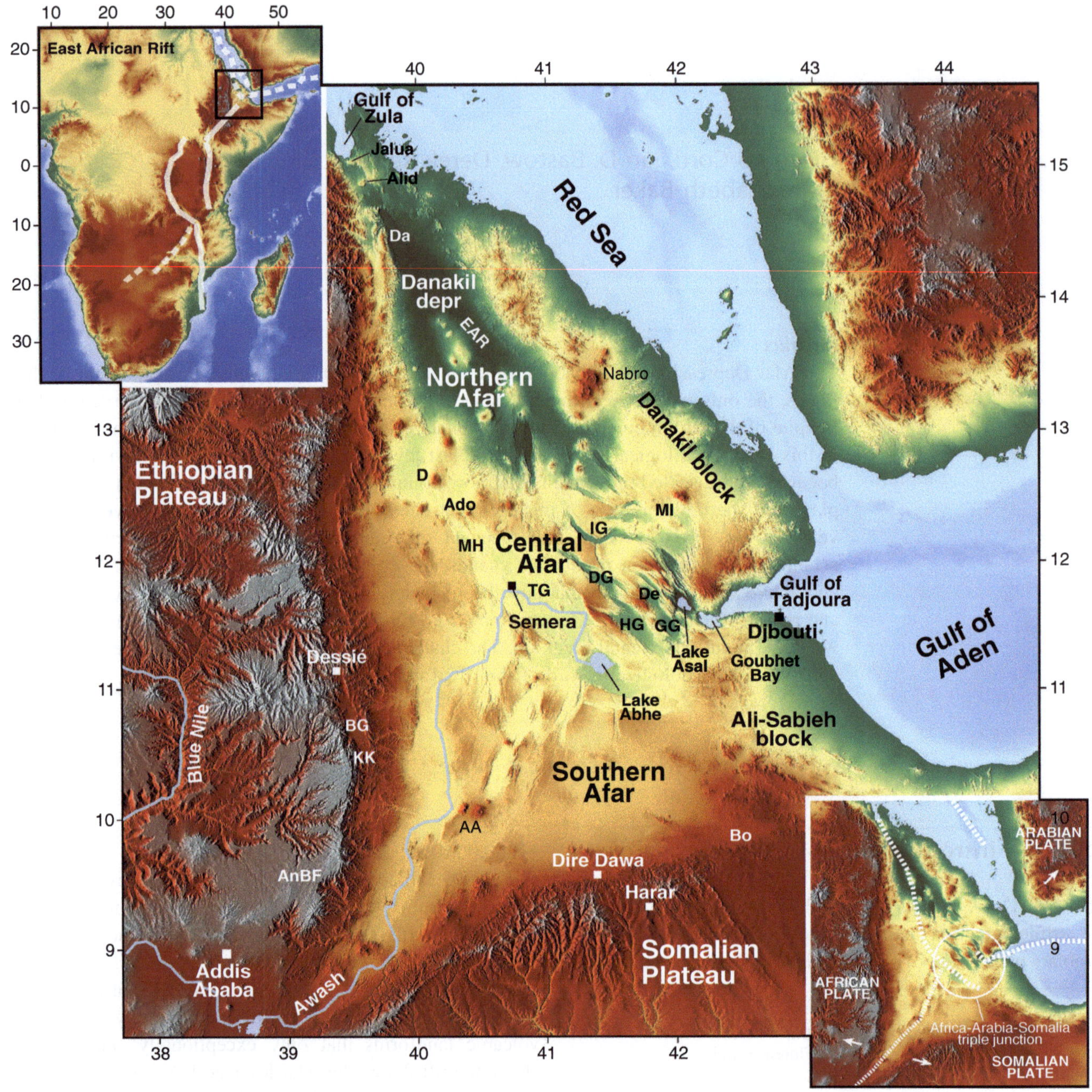

Fig. 15.1 Digital elevation model of the Afar Depression and surrounding areas (data from the Shuttle Radar Topography Mission, SRTM; resolution 90 m). *Inset on the top left* shows the extent of the system of rift valley composing the East African Rift. *Inset on the bottom right* shows the plate kinematic setting of the region; *dashed lines* indicate plate boundaries and *white arrows* illustrate plate kinematics. *Black* and *white squares* indicate major towns. *AA* Aleyu–Amoissa; *Ado* Ado Ale; *AnBF* Ankober border fault; *BG* Borkena graben; *Bo* Borama; *D* Dabbahu; *Da* Dallol; *De* Der'Ela graben; *DG* Dobi graben; *EAR* Erta Ale range; *GG* Gaddale graben; *HG* Hanle graben; *IG* Immino graben; *MH* Manda–Hararo rift; *MI* Manda-Inakir rift; *TG* Tendaho graben

The Afar Depression is separated from the Ethiopian and Somalian plateaus by major fault escarpments (Figs. 15.1 and 15.3). Separation from the Red Sea and Gulf of Aden is defined less dramatically by the low relief of the Danakil block (also called Danakil Alps) and the even more subdued hills of the Ali-Sabieh block. To the north, Afar narrows in the Danakil depression between the Danakil block and the Ethiopian escarpment, which connects the Afar lowland to the Gulf of Zula. The Gulf of Tadjoura separates the Ali-Sabieh and Danakil blocks, connecting Afar to the Gulf of Aden. However, as is the case everywhere in Afar, young volcanic centres prevent marine waters from flooding the region.

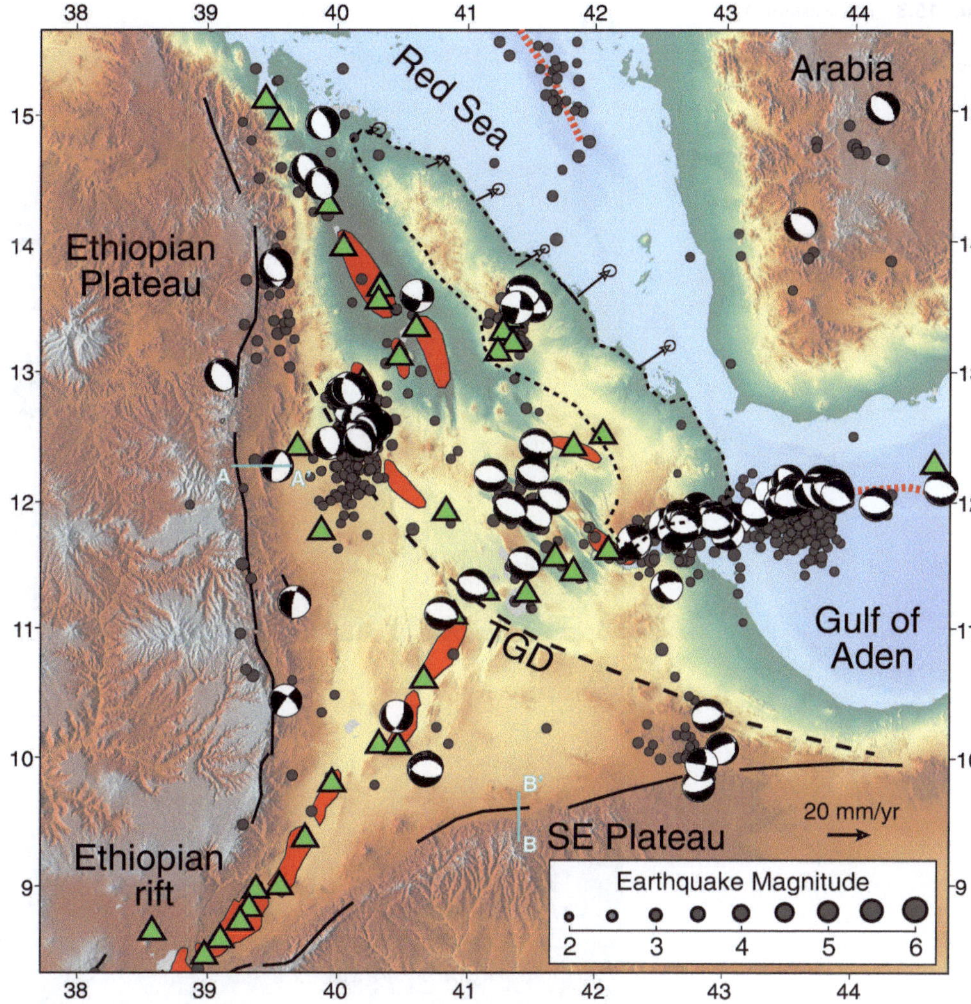

Fig. 15.2 Tectonic setting of the Afar Depression (modified after Keir et al. 2011). *Solid black lines* show Oligocene–Miocene border faults of the Red Sea, Gulf of Aden and East African rifts. *Red segments* show the Quaternary–Recent subaerial rift axes, and *green triangles* show Holocene volcanoes. *Dashed lines* show the Tendaho–Goba'ad Discontinuity (TGD). *Gray circles* show large earthquakes during 1973–2012 sourced from the National Earthquake Information Centre (NEIC) catalogue. Earthquake focal mechanisms are from the Global Centroid Moment Tensor (CMT) catalogue. The *black dashed line* defines the boundaries of the Danakil microplate. The *arrows* show the motion of the Danakil microplate (McClusky et al. 2010). *Cyan lines* with letters indicate the traces of cross sections illustrated in Fig. 15.3

In the following sections, we describe the different physiographic provinces of Afar focusing on the morphologic expression of the main tectonic–magmatic processes related to continental rifting and break-up.

15.2 Western and Southern Rift Escarpments

The Ethiopian, Somalian and Yemen plateaus are part of the so-called African Superswell, a wide region of anomalously high topography comprising the East African, Arabian and southern African Plateaus as well as a bathymetric swell in the south-eastern Atlantic Ocean basin (Nyblade and Robinson 1994). This anomalous topography results from strong uplift during the Cenozoic, with up to ~2 km rock uplift since ~30 Ma (e.g. Pik et al. 2003). Both plateaus are capped with ~2 km of Oligocene–Miocene thick sequences of flood basalts and rhyolites, with interbedded sedimentary sequences (Hoffman et al. 1997). Together with the anomalous uplift, this voluminous volcanism has been related to the presence of one or more 'traditional' narrow (e.g. Ebinger and Sleep 1998; Rogers et al. 2000) or, more likely, a broad low wave speed thermal anomaly beneath Ethiopia (e.g. Benoit et al. 2006; Furman et al. 2006; Bastow et al. 2008; Ritsema et al. 2011; Hansen et al. 2012; Rooney et al. 2012a, b, 2013).

The elevated plateaus are abruptly separated from the Afar Depression by discontinuous boundary faults that give rise to major fault escarpments that are the continuation of the margins of the Red Sea and Gulf of Aden (Wolfenden et al. 2005; Figs. 15.3 and 15.4). The faults are normally ~60 km long, widely spaced and characterised by large vertical offsets (>1 km), accommodating a drop in elevation from 2,000 to 3,000 m at the ridgeline of the escarpments to less than 1,000 m in the marginal areas of the Afar Depression (Wolfenden et al. 2005). Geochronological constraints suggest that the marginal faults were activated at ~26–31 Ma on the western margin of Afar (i.e. Red Sea rift; e.g. Ayalew et al. 2006; Wolfenden et al. 2005) and likely at

Fig. 15.3 The western Afar margin south of Dese (**a**) and in the Ankober region (**b**) (*Photos* D. Keir)

~33–34 Ma on the southern margin following rift activation all along the Gulf of Aden (e.g. Leroy et al. 2012). Their activity has been interpreted to have decreased with time starting from the Late Miocene, as extension-related deformation progressively migrated from the rift margins towards the centre of the rift depression, reflecting a process of strain localisation as rifting proceeds to continental break-up (e.g. Wolfenden et al. 2005). The crest of the escarpment is a major drainage divide. To the west, the majority of rivers feed the Nile watershed and provide the majority of fertile sediments so important for agriculture in Sudan and Egypt. To the east, the largely ephemeral rivers plummet down the steep topography of the rift margin, causing severe denudation creating badlands and hummocky terrain that averages 40 km in width between the floor of Afar and the crest of the present escarpments (e.g. Mohr 1978).

Different rift architecture and morphology characterise the western (Ethiopian) and southern (Somalian) escarpments (e.g. Beyene and Abdelsalam 2005). The Ethiopian escarpment is characterised by an overall rift-ward tilting (Fig. 15.4), which has been interpreted as resulting from a first episode of down-warping of the Afar Depression followed by subsequent marginal faulting (Zanettin and Justin-Visentin 1975; Kazmin et al. 1980). This tilting is more

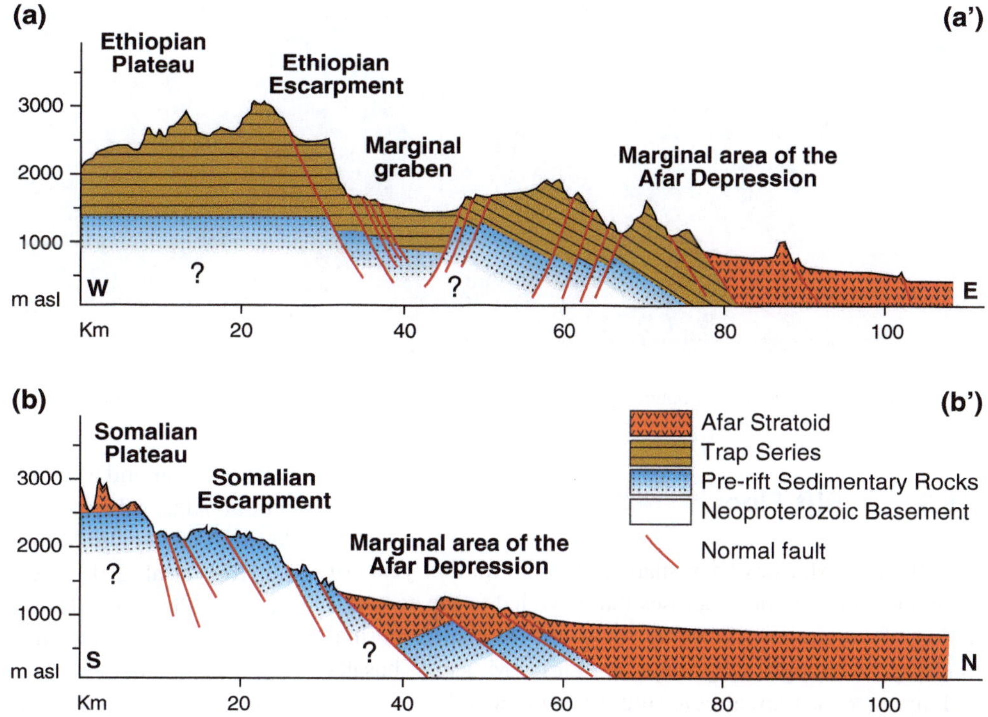

Fig. 15.4 Schematic geological cross sections across the western and southern Afar margins (after Beyene and Abdelsalam 2005). The traces of the cross sections are shown in Fig. 15.2

evident in the easternmost faulted blocks, where it gives rise to hogback-like structures; in the internal part of the margin, tilting is less developed as it is overprinted by development of marginal grabens (Figs. 15.4 and 15.5). These grabens, typically 10–20 km wide and some tens of km long, are filled with sediments of at least Pliocene–Quaternary age (Kazmin 1972; Chorowicz et al. 1999); they represent secondary features associated with major boundary fault systems (e.g. Corti 2012). In plan view, their en-echelon arrangement likely reflects a oblique component of extensional deformation during margin development, as also suggested by geological analysis (Chorowicz et al. 1999). The southern termination of the Ethiopian escarpment is at the Ankober border fault system, a major fault system that characterises the structurally complex 'corner' between the NE trending MER and the N–S trending Red Sea rift (Fig. 15.2; Wolfenden et al. 2004). Although the activity of the marginal faults is believed to have decreased with time as a result of migration of deformation towards the rift axis, the prominent morphology of the major normal faults (with clear morphotectonic indicators of recent activity, such as fresh scarps and triangular facets; Fig. 15.5; Chorowicz et al. 1999) and significant seismic activity testify an ongoing tectonic activity of the major fault escarpments. The most important earthquakes in the region struck the Kara Kore area, south of Dese, where a seismic sequence during May–September 1961 produced two main shocks with $M_L > 6.4$, seven > 5.0 and more than 3,500 with $M_L > 3.5$ (e.g. Gouin 1979). These events produced heavy alterations in the landscape, with numerous landslides and a 12–15 km-long piedmont scarp in unconsolidated materials along the escarpment of the Borkena graben with vertical differential displacement up to 2 m, depth of 5–7 m and width at the surface over 1 m (Gouin 1979; Fubelli and Dramis 2011). Significant seismic activity has been also recorded by the EAGLE network of seismic stations in the period from October 2001 to January 2003 along the complex Ankober border fault system (Keir et al. 2006a, b).

The southern (Somalian) escarpment differs from the western margin of Afar in that it is characterised by a lack of marginal basins and faulted blocks tilting away from the rift centre and a domino-style faulting that gives rise to minor half grabens only (Morton and Black 1975; Pizzi and Pontarelli 2007) (Fig. 15.4). The margin is marked by isolated volcanic centres aligned along the main structures; these volcanic centres are more common in the south-western portion of the margin, close to the MER (Beyene and Abdelsalam 2005). The presence of numerous elbows and wind gaps on the main river courses indicate a strong structural control on the local drainage network, with movements on faults inducing a large number of river captures (Acocella et al. 2011). Different from what is observed at the western escarpment, preliminary geological analysis in the Harar–Dire Dawa area suggests that tectonic activity along the Somalian escarpment decreased significantly in the Middle–Late Pleistocene (Pizzi and Pontarelli 2007). However, relatively minor clusters of earthquake are relatively common at the intersection between the southern margin of Afar and the eastern margin of the MER, and also near Borama on the Ethiopia–Somalia border (Keir et al. 2006a, b).

Fig. 15.5 Major normal fault bounding a marginal graben along the Ethiopian escarpment north of Dese (*Photo* M. Benvenuti)

15.3 Rift Floor

The floor of Afar has been shaped by ~30 million years of volcanic and tectonic processes that have led to the accumulation of voluminous volcanic rocks and to intense normal faulting. Although Oligocene to Early Miocene basalts belonging to the trap series are preserved in limited areas at the margins of the depression (e.g. Beyene and Abdelsalam 2005), the majority of volcanic rocks are Upper Miocene–Recent in age and formed after the young rift subsided between the marginal faults. Volcanism is dominated by basaltic activity, whose most important manifestation is the Afar Stratoid Series, a >55,000 km^2-wide and >1,500 m-thick sequence consisting of basaltic lava flows and minor rhyolitic ignimbrites erupted in the Pliocene–Pleistocene (~4–0.5 Ma) mainly from volcanic fissures (Barberi et al. 1975; Lahitte et al. 2003; Acocella 2010). Emplacement of this basaltic series was characterised by the highest eruption rates (>5,000 km^3/Ma) ever known for plate divergent margins and is instead more typical of a large igneous provinces (Acocella 2010). The most recent, Quaternary–Recent volcanoes are both composite stratovolcanoes, composed of interbedded rhyolites, ignimbrites, andesites and basalt flows (e.g. Dabbahu), and also shield volcanoes of mainly basalts (e.g. Erta Ale range). Both types of volcano are the source of observed fields of basaltic scoria cones, and fissure fed basaltic flows. The E–NE trending transverse volcanic alignments and trachytic–rhyolitic central volcanoes also characterise the Afar margins (e.g. Barberi and Varet 1977).

The significant volcanic activity is intimately related to extension-related faulting and magma intrusion, which since Quaternary–Recent times has been focused in axial magmatic segments (Figs. 15.2 and 15.6; e.g. Barberi and Varet 1977; Manighetti et al. 2001; Keir et al. 2013). The current faulting patterns are commonly quite complex and result from interaction between the subaerial Red Sea and Aden rifts and the northern termination of the Ethiopian rift (Figs. 15.1, 15.2 and 15.6). The Red Sea includes the Danakil depression in northern Afar and Manda–Hararo rift and Tendaho Graben in central Afar, whereas the subaerial Aden rift includes the Asal Ghoubbet and Manda–Inakir rifts to the east (Fig. 15.6; e.g. Keir et al. 2013). A series of relatively narrow grabens (such as the Dobi, Hanle and Gaddale grabens) in central Afar transfer strain from the Red Sea rift to the Gulf of Aden (e.g. Manighetti et al. 2001). The Ethiopian Rift includes the systems of NE trending grabens and volcanic alignments that result from Nubia–Somalia rifting south of the Tendaho–Goba'ad discontinuity (e.g. Tesfaye et al. 2003).

These different sectors of the Afar Depression (northern central and southern) show significant tectonic and structural differences, and hence are described separately in the following sections.

15.4 Northern Afar

The morphology of northern Afar is dominated by the Danakil Depression, a ~200–220 km-long, relatively narrow (50–150 km-wide) basin bound to the west by a prominent fault escarpment forming >2 km of relief at the western Afar margin and to the east by a less prominent series of faults forming a 500–1,000 m elevation rift margin against the Danakil microplate (Fig. 15.7). Despite being onland, the basin floor is generally 50–100 m below sea level with near-surface geology dominated by the a suite of marine evaporite deposits (e.g. calcite, gypsum, halite and sylvite). At least 2 km salt formed in the Danakil depression when it was a submarine arm of the Red Sea during the Miocene. The upper part of the series deposited during repeated marine incursions in the Pleistocene, when lava flows dammed the narrow neck between the northern end of the Danakil Horst and the western highlands and the sea water trapped in the depression soon evaporated. These incursions, the last of which occurred ~30 ka (e.g. Barberi and Varet 1970; Bonatti et al. 1971; Hutchinson and Engels 1972; Talbot 2008), left a classic pattern of marine reefs perched at altitudes of −30 to +90 m along the former

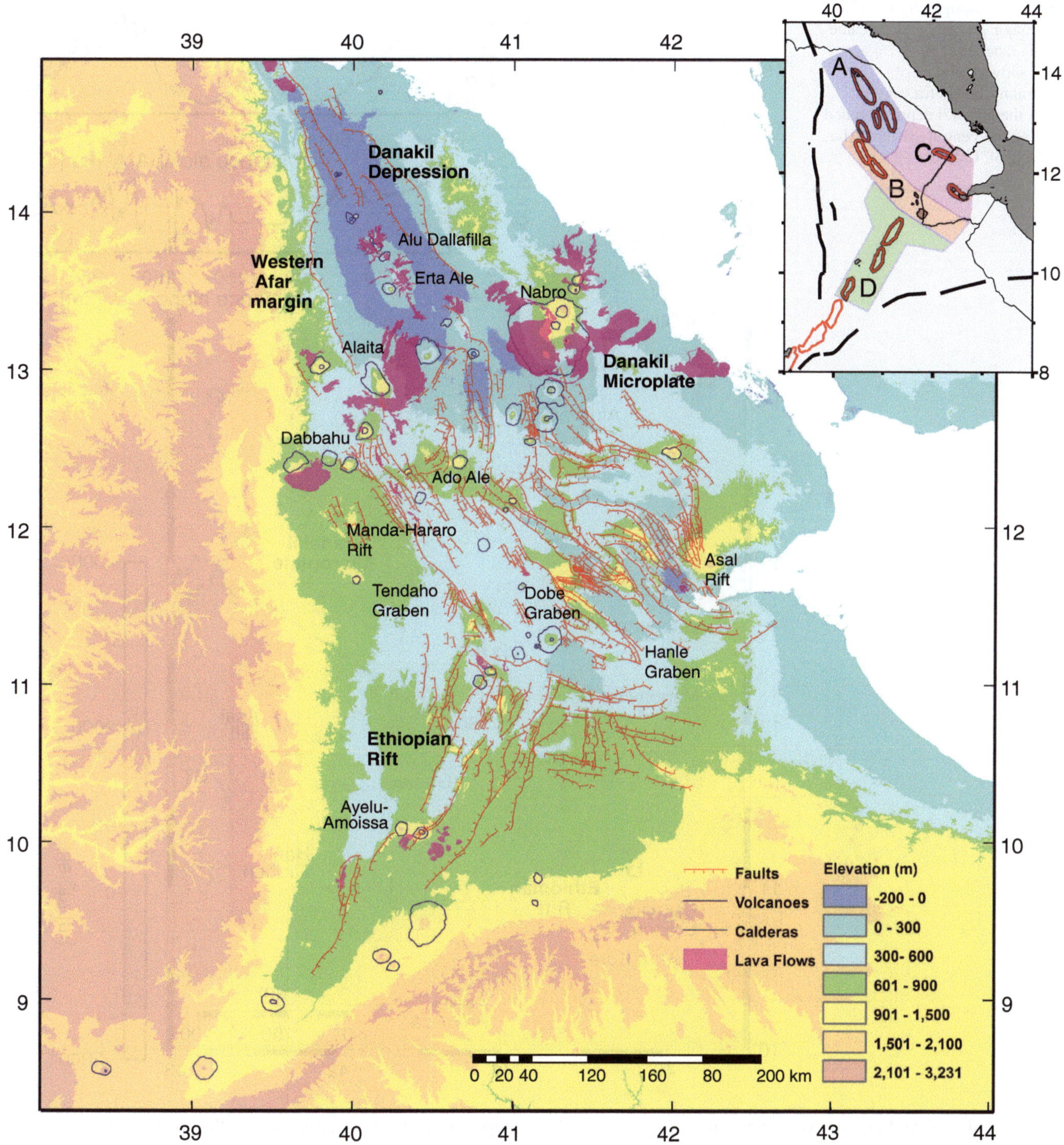

Fig. 15.6 Structural map of the Afar Depression (modified after Keir et al. 2013). Major Pliocene–recent faults affecting the rift floor are marked *red*. Recent basaltic lava flows are shaded *light purple*. The *inset* delineates the major rift system forming the Afar triple junction. The subaerial Red Sea rift is included in areas A (northern Afar) and B (western Afar). The subaerial Aden rift is included in region C (central, eastern Afar), and the northern Main Ethiopian rift is area D (southern Afar)

shoreline around the margins of the depression. The basin also contains non-marine evaporites from ongoing evaporation of seasonal terminal saline lakes and the geothermal pools (Talbot 2008).

The axis of the Danakil depression is bisected by a linear, ~NNW striking, 10–20-km-wide axial volcanic range (Erta Ale range) which is the locus of ongoing volcanism and magmatism (e.g. Field et al. 2012; Pagli et al. 2012). This

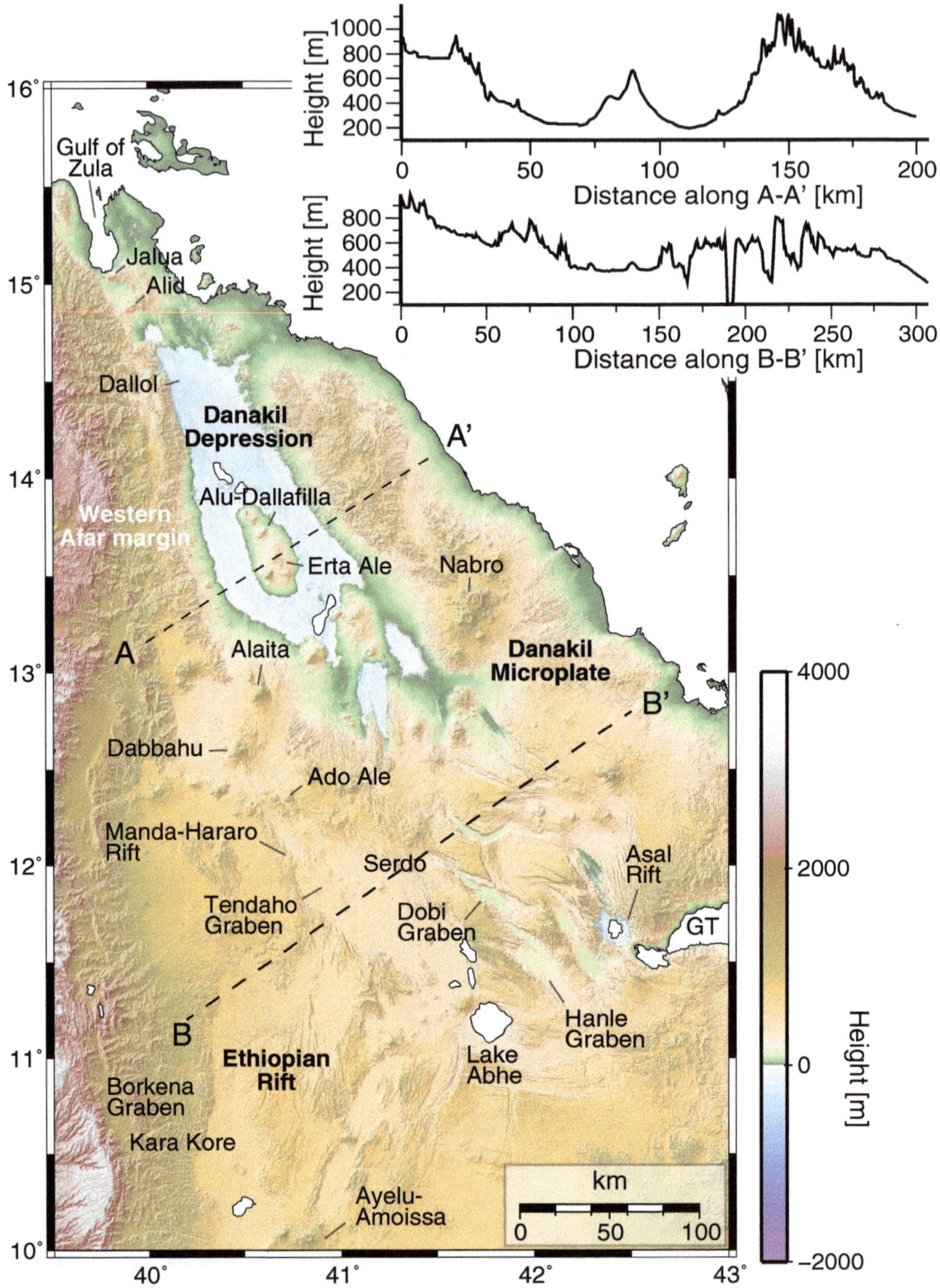

Fig. 15.7 Topography of Afar with major grabens and volcanic centres labelled. *Upper right panel* shows samples of topography on profiles A–A′ across the Danakil depression and B–B′ across central–eastern Afar. *GT* Gulf of Tadjoura

range, unlike elsewhere in Afar, does not form a heavily faulted graben, but a composite system of different volcanic centres with elevation of up >500 m above sea level (Fig. 15.8; e.g. Barberi and Varet 1970). The volcanic rocks are largely dominated by Quaternary–Recent a'a and pahoehoe basalt flows (Thurmond et al. 2006), which create the classic shield volcano-like morphology with gentle slopes. A limited number of silicic centres create more prominent morphology, such as the steep-sided conical stratovolcano of Dalafilla that rises 300 m above surrounding lava fields SE of Alu volcano. The Erta Ale range is markedly elliptical in shape, reflecting a control on the pathways of magma migration and eruption exerted by axial, rift-parallel fissures (Fig. 15.8). Basalt flows are indeed fed from a combination of ~NNW-striking fissures and from edifices such as the ~NNW elongate main caldera hosting Erta Ale lava lake (e.g. Acocella 2006). The ~1,700 × 700 m-wide, main caldera hosts a large northern crater and a smaller central crater (Fig. 15.9; Acocella 2006). The central crater hosts a small pit crater floored by the lava lake in which lava elevation changes through time. It reached its highest level in February 2010 when lava filled the pit crater; by November 2010,

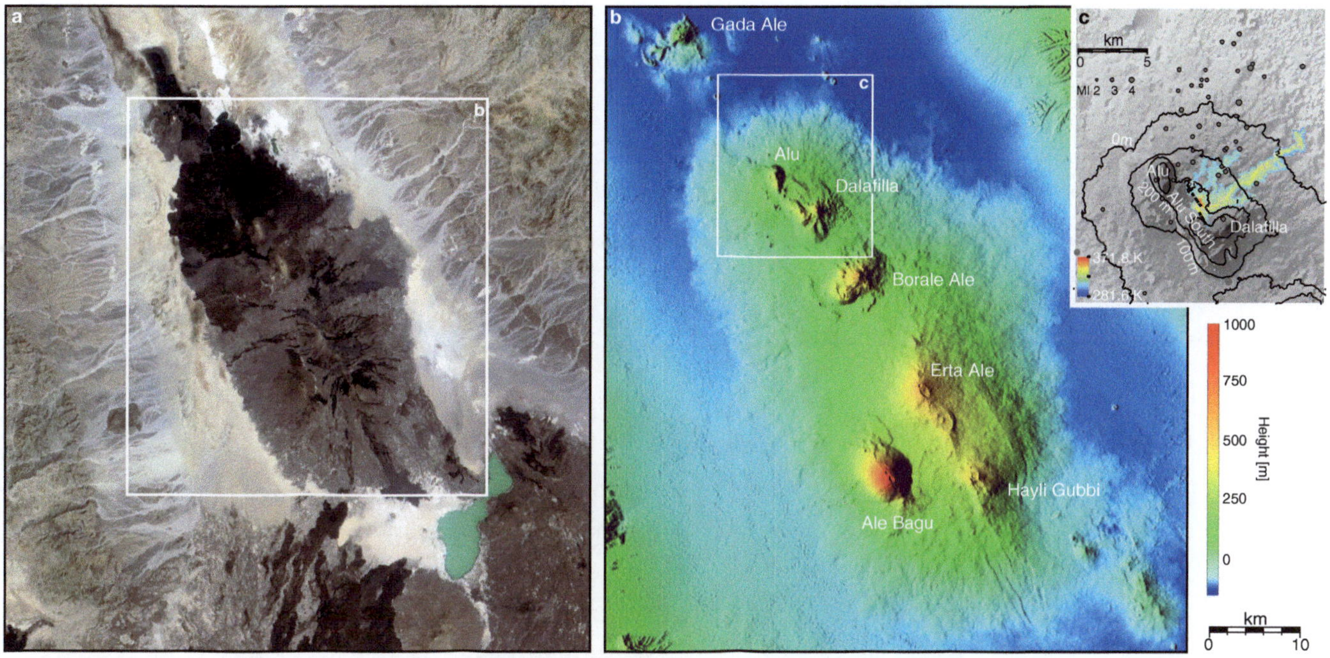

Fig. 15.8 Satellite image (*Source* Google Earth) (**a**) and digital elevation model (SRTM data) (**b**) of the Erta Ale range (**c**) The November 2008 eruption in the Alu-Dalafilla volcanic centre. Dashed lines indicate the two en-echelon volcanic fissures, and the coloured area is the surface temperature of the erupted lava flow (from Pagli et al. 2012)

Fig. 15.9 Aerial view of the elongate caldera of the Erta Ale with the large northern and the smaller central pit craters (*Photo* E. Baker)

basalt overflowed into the main crater (Field et al. 2012). The larger northern crater also hosted a lava lake until early 1975. The lava lakes have been persistently active for at least 100 years (Waltham 2010). The geometry of the main caldera, the characteristics of erupted lavas as well as the modelling of the satellite data of recent eruptions all indicate that the volcanic activity in the Erta Ale range is likely fed from long-lived shallow (uppermost 2 km) pockets of magma (e.g. Pagli et al. 2012; Keir et al. 2013). The importance of ~NNW-striking fissures in building the volcanic topography of the Erta Ale range is testified by the November 2008 eruption in the Alu-Dalafilla volcanic centre, where two en-echelon fissures opened south-east of Alu and erupted lava that flowed towards the east and north of Dalafilla (Fig. 15.8; Pagli et al. 2012). InSAR data show that the eruption was accompanied by up to ~1.9 m of subsidence at Alu. The deformation was caused by ~1 km-wide, ~10 km-long deflating sill beneath the volcanic system.

Fig. 15.10 Multicoloured salt deposits, with small salt hornitos, at Dallol (*Photo* D. Keir)

The axial lava flows of the Erta Ale range, which cover an order of magnitude more surface area than similar age basalts elsewhere in Afar (Bastow and Keir 2011), typically flow away into the lower lying, evaporite-rich basin, creating a basin stratigraphy of thinly interbedded basalts and evaporites (Talbot 2008), up to ~5 km-thick (e.g. Makris and Ginzburg 1987).

The Erta Ale range drops in elevation to the north with the Dallol volcanic centre, one of the lowest subaerial volcanic vents globally. Dallol is world renowned for multi-coloured salt and hydrothermal deposits, hot mineral springs, fumaroles and geysers (Fig. 15.10) which form as a consequence of a complex interaction of solution and recrystallisation processes driven by hydrothermal waters and rapid evaporation. Dallol has been the site of a significant phreatic eruption in 1926, which created a crater ~30 m in diameter (Siebert et al. 2010). Recently, satellite remote sensing and seismicity analysis captured the intrusion of ~0.06 km^3 dike accompanied by a M_w 5.5 earthquake and associated fault slip along the western flank of the rift during October 2004 (Nobile et al. 2012). The intrusion was fed by a previously unidentified shallow magma reservoir beneath Dallol. The volcano lacks volcanic rocks at the surface because they are covered by salt. The magmatic and tectonic activity also indicates that the Erta Ale range extends up to Dallol as a continuous and uniformly trending rift branch. Activity at the Alid and Jalua, north of Dallol, has not been reported but the spacing between magmatic centres suggests that the Erta Ale Ridge further extends to the Gulf of Zula on the Eritrean coast.

15.4.1 Central Afar

Central Afar is the complex region of mechanical interaction between the on-land southern continuation of the Red Sea Rift and the subaerial western continuation of the Aden Rift (e.g. Manighetti et al. 2001). The southern extent of the Red Sea Rift includes the Manda–Hararo Rift and the Tendaho Graben, in the west, whereas the Gulf of Aden extends into Afar in the Asal-Ghoubbet and Manda–Inakir rifts, in the east (Figs. 15.1 and 15.7). A complex architecture made of a series of relatively narrow, non-volcanic grabens (such as the Dobi, Hanle and Guma grabens) characterise the area between the Red Sea and Gulf of Aden systems, transferring extension between the two rift arms (e.g. Manighetti et al. 2001). These three different subsectors are discussed in the following sections.

15.4.1.1 Manda–Hararo Rift and Tendaho Graben

The Manda–Hararo Rift and the Tendaho Graben form a series of NNW–SSE striking, 30–60 km-wide, ~80 km-long basins bordered by 200–300–m-high escarpments (Fig. 15.7). These generally lack a clear single border fault but are characterised by numerous closely spaced relatively small-offset faults that displace the Stratoid Series and define several rigid blocks tilted toward the rift axis (Figs. 15.11 and 15.12; e.g. Acocella et al. 2008). This inward-dipping Stratoid blocks arranged in a domino configuration inside the margins result in an overall syncline-like structure of the basins (e.g. the central part of the Tendaho Graben; Acocella

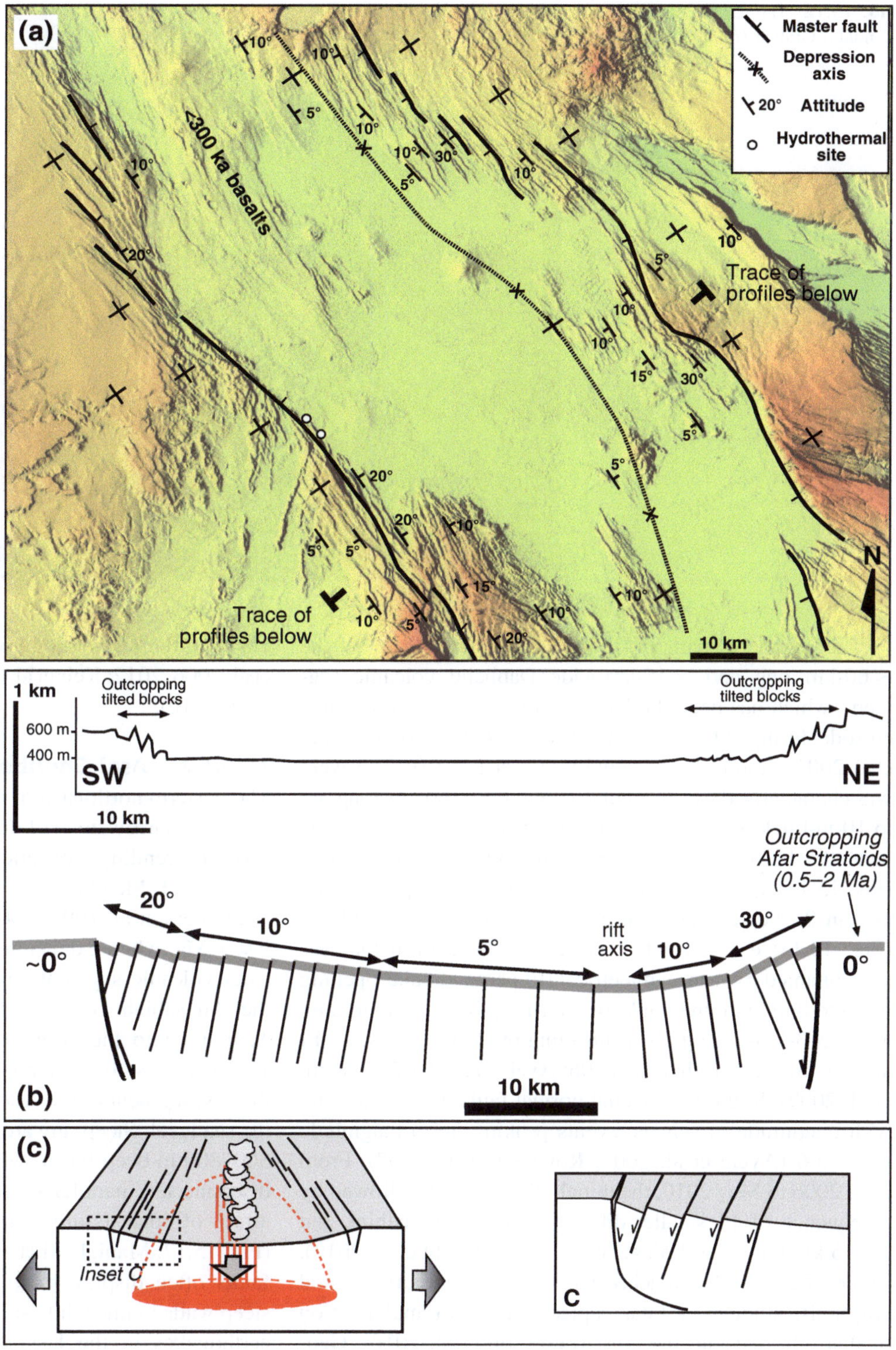

Fig. 15.11 Structure and mechanism of formation of the Tendaho Graben (modified from Acocella 2010). **a** Structure of the central part of the graben. **b** *above* topographic section along the profile indicated in **a**; *below* simplified structure (not to scale) of the Tendaho Graben in cross section; *double arrows* report extent of block domains with similar tilt. Note the domino-like structures, with blocks characterised by a uniform tilt angle separated by faults. **c** Mechanism of formation of the graben as a consequence of repeated magma withdrawal from a rift-parallel elongated reservoir

et al. 2008), a geometry that has been related to rift collapse induced by magma withdrawal during the emplacement of the Stratoid sequence (Fig. 15.11; Acocella 2010). The basins are filled with >1 km of lacustrine and fluvial deposits as well as by Quaternary–Recent volcanic rocks, the distribution and age of which suggest that faulting and volcanism have progressively localised through the Quaternary to a ~10 km-wide central axis where the youngest (0–0.2 Ma old) fissural basalts crop out (Lahitte et al. 2003; Acocella et al. 2008; Acocella 2010). At the southern tip of the Tendaho Graben, Lake Abhe is characterised by hundreds of travertine towers, giving rise to several linear chains of travertine chimneys. Each tower is up to 60 m tall, formed by geothermal activity (e.g. Waltham 2010; Hussein et al. 2013).

Fig. 15.12 The eastern side of Tendaho Graben (*Photo* D. Keir)

The northern part of the Manda–Hararo rift is the ~60-km-long and ~15-km-wide Dabbahu volcanic segment, which has been the locus of a major volcano-tectonic episode during 2005–2010 (e.g. Wright et al. 2006; Rowland et al. 2007; Grandin et al. 2009; Ebinger et al. 2010). The event was characterised by the emplacement of a ~60 km-long, up to 10-m-thick basaltic dike into the upper 10 km of the crust beneath the magmatic segment during September 2005 (Fig. 15.13), accompanied by a minor eruption of pumice and ash on 26 September from the Da'Ure vent near Dabbahu (Fig. 15.14; e.g. Wright et al. 2006). Dike intrusion was accompanied by near-symmetrical rift perpendicular opening of up to 8 m, with the flanks of the rift uplifted by up to 2 m, and a 2–3-km-wide graben subsiding by 2–3 m (Figs. 15.14 and 15.15; e.g. Wright et al. 2006; Ayele et al. 2007; Grandin et al. 2009). Slip on individual normal faults was up to 3 m, with magnitude of seismic events peaking up to magnitude M_L = 5.6 (Ayele et al. 2007; Rowland et al. 2007). From June 2006 to May 2010, the initial dike has been followed by a sequence of 13 smaller dikes typically 1–3 m thick and 10–15 km long (e.g. Ebinger et al. 2010; Wright et al. 2012). Three of these dikes reached the surface as basaltic fissural eruptions, although the total erupted volume is a small fraction of that intruded into the crust (Ferguson et al. 2010). Geophysical data of this suggest that the majority of the intrusions were fed laterally from a ~10-km-deep magma reservoir beneath the Ado'Ale volcanic complex located at the centre of the segment (Ayele et al. 2009; Grandin et al. 2009). These episodic events of normal faulting and axial graben subsidence induced by lateral dike intrusion from a segment-centred magma reservoir are modulated by cyclic variations in magma supply (Medynski et al. 2013) and function to maintain the morphology of volcanic segments (e.g. Wright et al. 2006, 2012; Keir et al. 2009; Ebinger et al. 2010; Grandin et al. 2010).

15.4.1.2 Asal-Ghoubbet and Manda–Inakir Rifts

The eastern–north-eastern edge of central Afar is characterised by a narrow zone of dense faulting that includes the two NW-SE trending volcanic rifts of Asal-Ghoubbet, to the south, and Manda-Inakir, to the north (Fig. 15.16; e.g. Manighetti et al. 1998, 2001). These two rifts formed in the last 1 Ma, when extension in the Gulf of Aden localised to axial volcanic segments breaching the Danakil Block from the Ali-Sabieh Block and proceeding along the Gulf of Tadjoura into the Afar Depression (e.g. Manighetti et al. 1998). The ~40-km-long, ~15-km-wide Asal-Ghoubbet is the first segment of the Aden ridge to rise significantly above the sea level. Steep, inward dipping normal faults delimit the 300–800-m-deep basin, which is subdivided in at least two disconnected, parallel subrifts; the rift floor is cut by dense arrays of open fissures roughly parallel to the normal faults (Fig. 15.17; Manighetti et al. 1998). Both fissures and normal faults often display sharp and youthful morphology, with steep walls in fine white silts, suggesting development during earthquakes in the last century (Manighetti et al. 1998). Deformation recorded by means of geodetical and geophysical data also shows the ongoing activity of the majority of the faults (e.g. Doubre et al. 2007). Notably, the north-western subrift has faults with sharper and fresher scarps, and wider and denser fissures; this suggests that this subrift is younger than the one to the south-east and that deformation has migrated to the north-west with time (Manighetti et al. 1998).

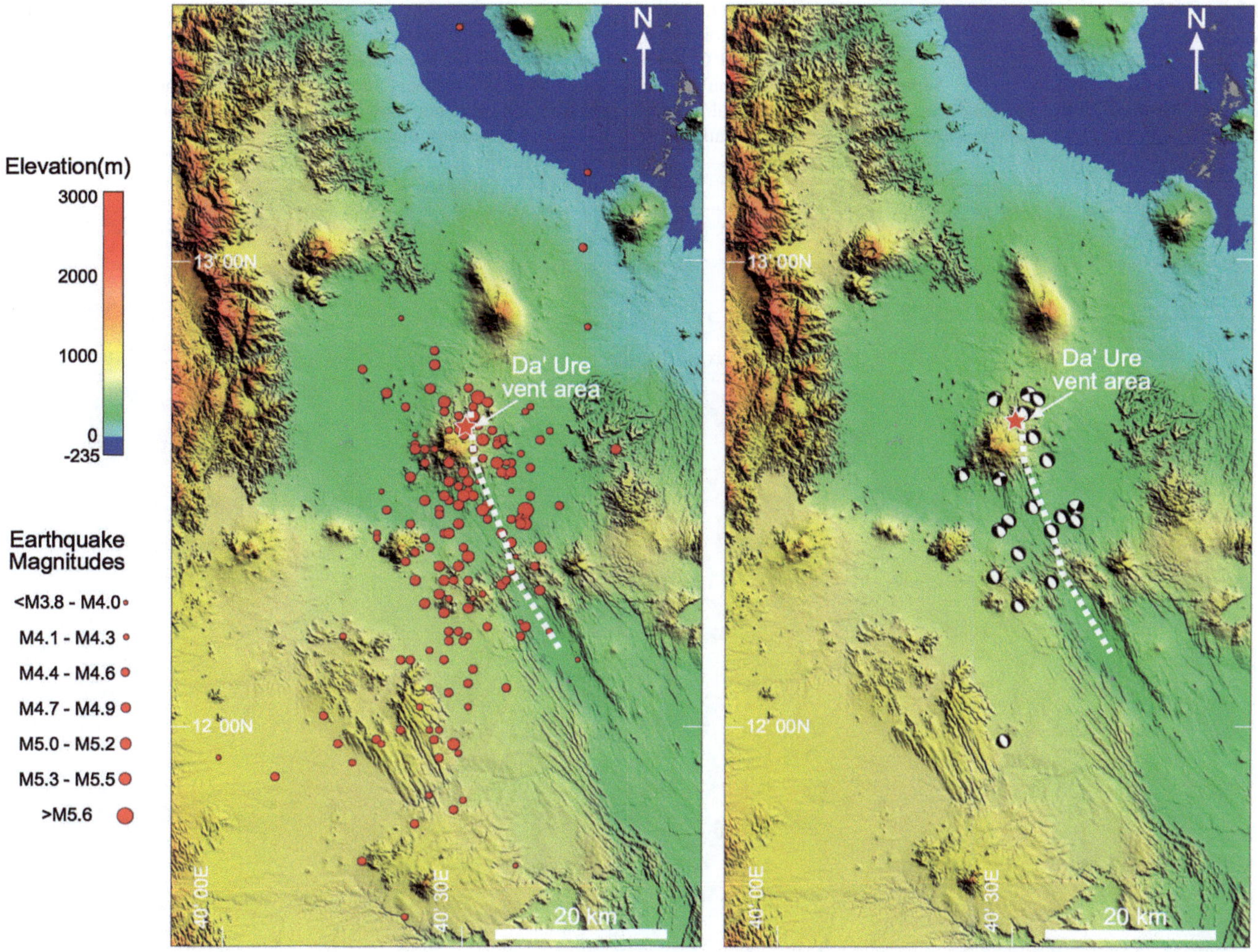
Elevation(m)
3000
2000
1000
0
-235
Earthquake Magnitudes
<M3.8 - M4.0
M4.1 - M4.3
M4.4 - M4.6
M4.7 - M4.9
M5.0 - M5.2
M5.3 - M5.5
>M5.6
N
13°00N
12°00N
40°00E
40°30E
Da' Ure vent area
20 km
N
13°00N
12°00N
40°00E
40°30E
Da' Ure vent area
20 km

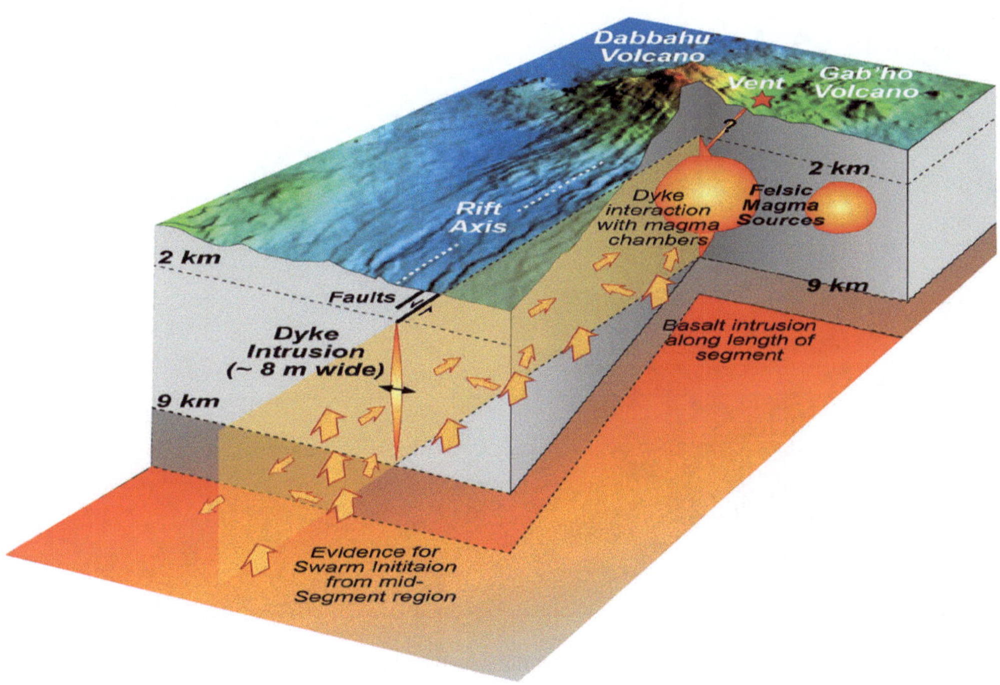
Dabbahu Volcano
Vent
Gab'ho Volcano
?
2 km
Felsic Magma Sources
Dyke interaction with magma chambers
Rift Axis
2 km
9 km
Faults
Dyke Intrusion (~ 8 m wide)
9 km
Basalt intrusion along length of segment
Evidence for Swarm Inititaion from mid-Segment region

Fig. 15.13 The *top left* and *right panels*, respectively, show earthquake locations reported by the National Earthquake Information Center (NEIC) and focal mechanism reported by the Global Centoid Moment Tensor Catalog, for the September 2005 Dabbahu rifting episode. Normal fault earthquakes were primarily located along the Dabbahu segment. The *lower panel* is a cartoon sketch illustrating that the fault slip causing graben subsidence is a relatively shallow phenomenon occurring mainly in the upper few kilometres of the Earth. The fault slip is induced by magma intrusion sourced from upper crustal magma chambers such as Dabbahu, Gabho and the Ado'Ale volcanic complex (not shown). Magma intrusion occured mainly at 2–10 km depth beneath the rift

Fig. 15.14 *Top* The Da'Ure vent near Dabbahu formed during a minor eruption of pumice and ash on 26 September 2005. *Bottom* Normal fault formed during the volcano-tectonic event (*Photos* E. Baker)

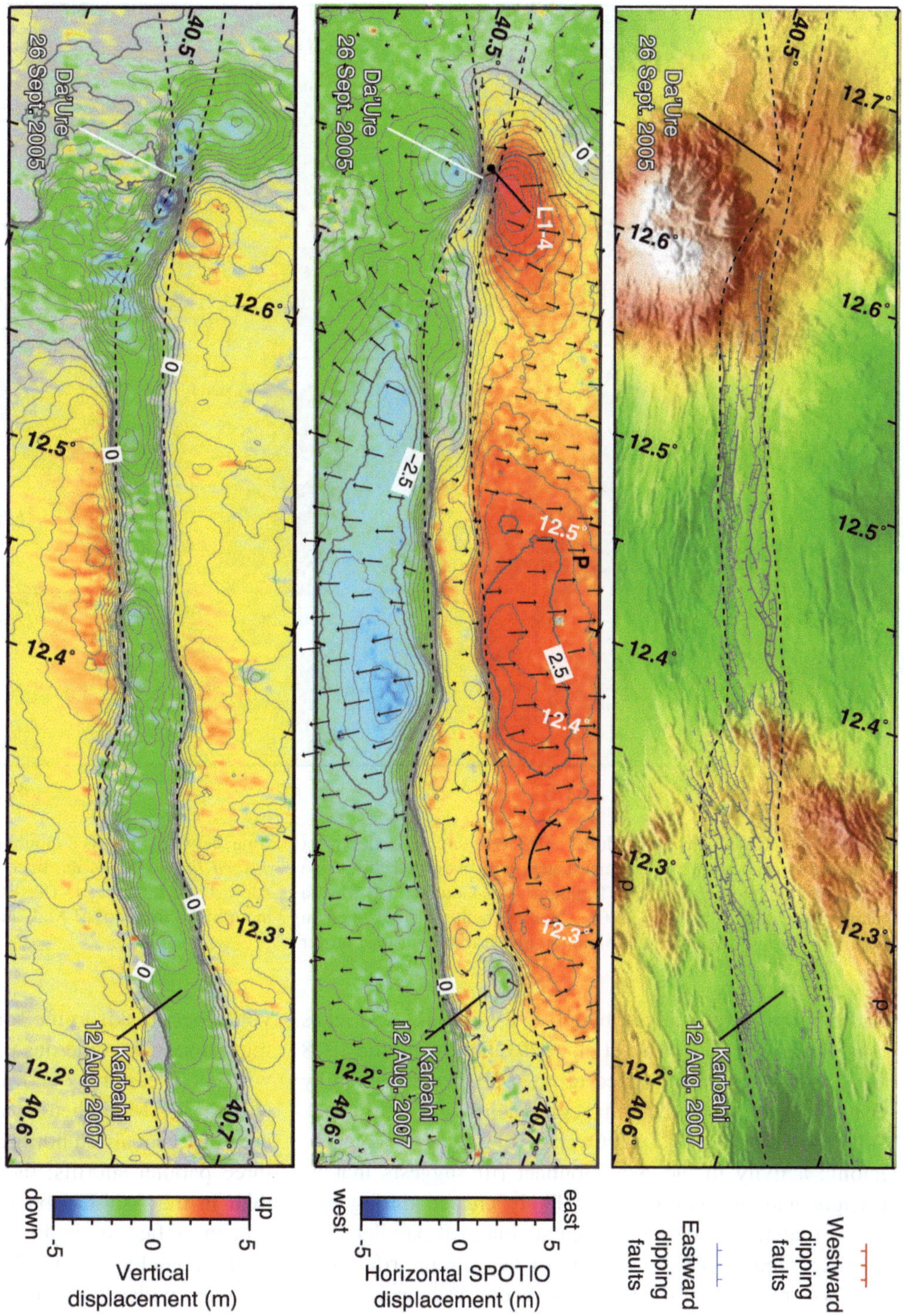

Fig. 15.15 Vertical and horizontal displacements (*left and central panels*) following the 2005 Dabbahu rifting episode and traces of the faults that were activated during the event (*right panel*). Modified after Grandin et al. (2009)

Active volcano-tectonic activity in the area is testified by a rifting event in November 1978, characterised by a major earthquake ($M_b = 5.3$) near the rift axis close to the coast of the Ghoubbet Bay and a seismic swarm that lasted for 2 months (Ruegg et al. 1979). This earthquake produced ~2 m of extension and up to ~70 cm subsidence, accommodated by the opening of several fissures and the reactivation of normal fault scarps on the rift axis; slip on individual faults varied between 0.15 and 0.5 m inside a 3-km-wide zone along 10 km of the exposed area between Goubhet and Asal (Ruegg et al. 1979). This deformation was associated with basalt flows from numerous fissures and vents at the newly formed volcanic centre of Ardoukôba, located north-west of a large volcano (Fieale) in the centre of the Asal-Goubhet rift (Figs. 15.16 and 15.17). Mechanical modelling of the deformation suggests that the rifting episode

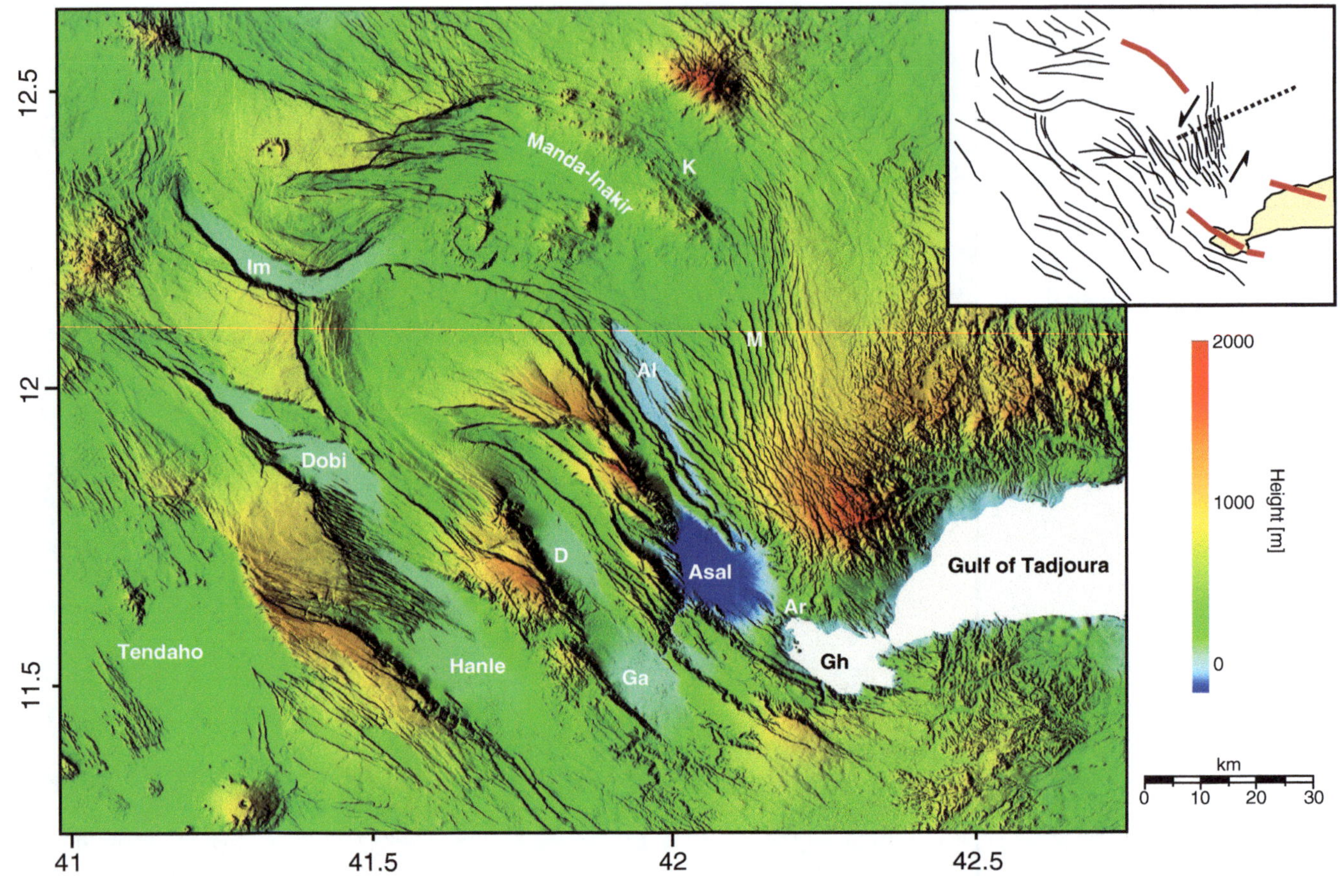

Fig. 15.16 Digital Elevation Model (SRTM data) of easternmost central Afar. *Inset* shows a tectonic sketch, with indicated the active volcano-tectonic rift segments of Tadjoura, Asal-Goubhet and Manda Inakir (after Manighetti et al. 2001). Normal faults are indicated in *black*; *arrows* indicate the relative motion of the Mak'arrasou transfer zone between the Asal-Goubhet and Manda Inakir segments; *black dashed line* indicates the trace of the profile shown in Fig. 15.18. *Al* Alol; *Ar* Ardoukoba; *D* Der'Ela; *Ga* Gaddale; *Gh* Goubhet; *K* Kammourta; *Im* Immino; *M* Mak'arrasou oblique transfer zone

resulted from the opening of two 4- to 8-km-long axial dikes at 4–5 km depth (Tarantola et al. 1979). Following the 1978 event, dike inflation continued up to 1985, accompanied by seismicity and fault activity that continues today (Cattin et al. 2005; Doubre et al. 2007). Overall, the style of volcano-tectonic activity in the Asal-Goubhet rift suggests that its structure and morphology were likely created during episodes of lateral migration of magma along rift-parallel shallow dikes, fed by a magma chamber beneath the rift centre (e.g. Doubre et al. 2007), as is observed at Dabbahu.

To the north–north-west, the Asal-Ghoubbet rift is connected to the Manda-Inakir rift, a volcanic range characterised by Pleistocene–Holocene lava flows and associated normal faults (Fig. 15.16; Manighetti et al. 1998). Connection between the two rifts occurs through the Mak'arrasou oblique transfer zone, which is composed of a network of closely spaced extensional faults and tilted blocks dipping towards Afar and giving rise to a large-scale flexure (Fig. 15.18; Le Gall et al. 2011). Volcanic products in the Manda-Inakir rift are associated to two large shield volcanoes: Inakir to the south-east, an elongated NW trending dome with numerous parasitic spatter cones, and Manda, with more easterly trend and hosting the most recent lavas (Fig. 15.19; Manighetti et al. 1998). Swarms of normal faults are mostly localised north-east of the volcanic zone and form three parallel subrifts; their topography is more attenuated than in the Asal-Goubhet rift suggesting a younger development of faulting in the area (Manighetti et al. 1998). Some of the bounding normal faults exhibit extremely fresh morphology, with light-coloured scarplets, suggesting they have likely ruptured in the last few hundred years (Tapponnier et al. 1990). As in the Asal-Goubhet rift, several open fissures (commonly more than 10 m deep and several meters wide) affect the rift floor. These fissures are often sharp and steep and with very fresh morphology suggesting very recent activity; some of them cut Late Pleistocene basalt flows in the rift floor. Most of youngest faults and fissures are concentrated in the north-western subrift, suggesting that—likewise the Asal-Goubhet rift—deformation propagated from south-east to north-west (Manighetti et al. 1998). Historical

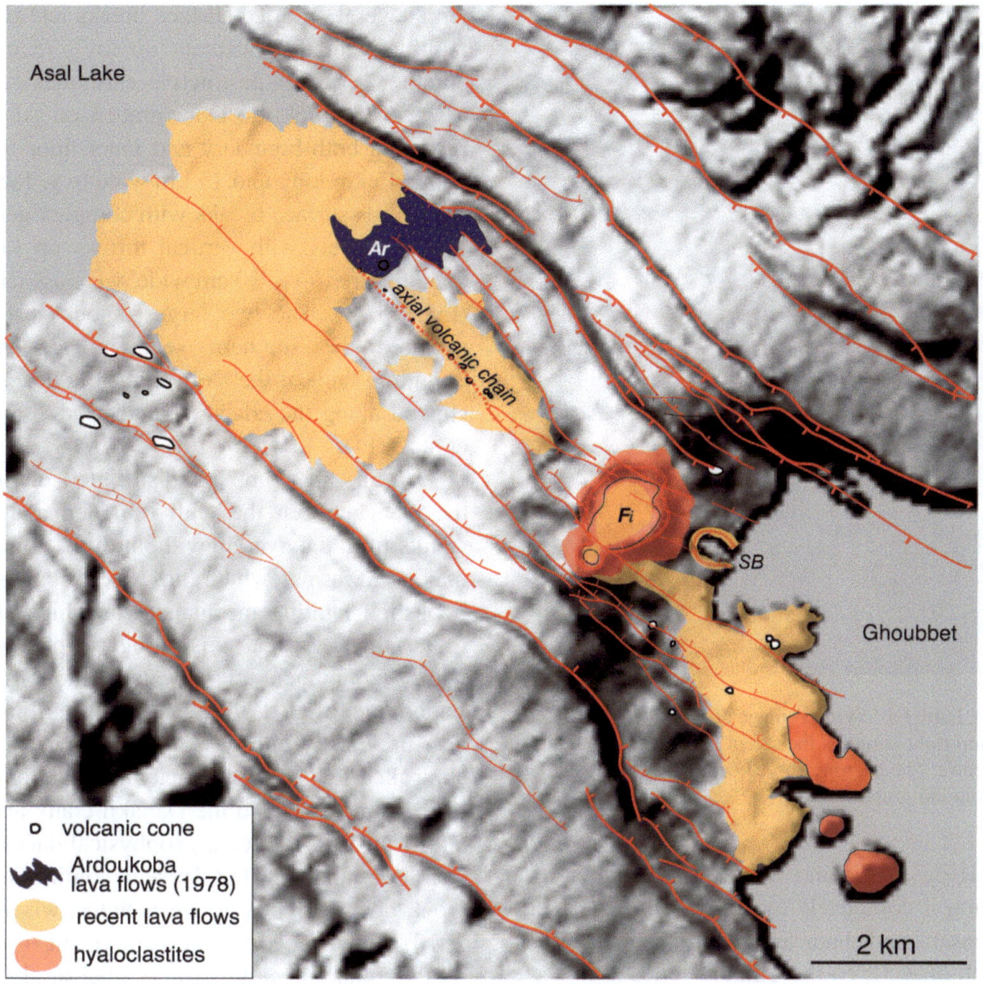

Fig. 15.17 Tectonic and simplified geological map of the subaerial section of the Asal-Goubhet Rift superimposed onto a SRTM digital elevation model (after Doubre et al. 2007). *Ar* Ardoukoba; *Fi* Fieale caldera; *SB* Shark Bay caldera

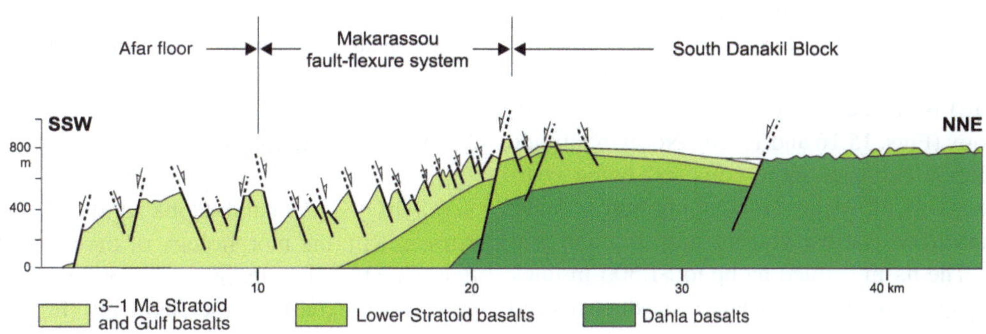

Fig. 15.18 Structural cross section of the central part of the Makarassou fault system, at the eastern margin of the Afar Depression (after Le Gall et al. 2011). Vertical exaggeration = 4. Trace of the profile shown in Fig. 15.16. Note the network of closely spaced extensional faults and tilted blocks giving rise to hogback-like structures and a large-scale flexure towards Afar

volcanic activity was reported in the area, with the last eruption occurring in 1928 or 1929 at the south-eastern end of the Manda-Inakir rift and producing the Kammourta cinder cone and a lava flow (Audin et al. 1990).

15.4.1.3 Basins Between the Red Sea and Gulf of Aden Systems

In the central portion of Afar, between the Gulf of Aden and Red Sea systems, deformation is spread over an area

Fig. 15.19 NASA Earth Observatory image of central Afar. The *black area* close to the top right corner is the recent basaltic lava field of the Manda volcanic centre. The *narrow*, curved depression in the centre of the image is the Immino graben (IG)

~120 × 100 km (Figs. 15.16 and 15.19; e.g. Manighetti et al. 2001). This region is completely overlain by the thick series of the Stratoid basalts and characterised by a lack of younger volcanism. South of ~12°N, the Stratoid series is dissected by a series of NW trending major normal fault systems with high (100–1,000 m), steep (>70°) cumulative escarpments, and numerous smaller, subparallel normal faults in between. The major border faults trend NW and define 10–20-km-wide and 30–50-km-long subparallel basins (e.g. Dobi, Hanle, Der'Ela, Gaddale), with intervening narrow horst blocks, so that the whole area is characterised by an overall horst-and-graben morphology (Figs. 15.16 and 15.20). North of latitude ~12°N, deformation becomes more complex, being characterised by a network of major crosscutting active faults giving rise to curvilinear narrow grabens (e.g. Immino graben; Fig. 15.19). The basin is filled by up to >1,500-m-thick sediments, mostly Quaternary–Recent lacustrine deposits and alluvial fans deposited during wet periods at the base of large fault escarpments (Gasse 1991). Most of the normal faults affecting the area are characterised by a fresh morphology, with a scarplet at the base of the cumulative escarpment offsetting Late Quaternary alluvial fans, thus attesting their recent/ongoing activity (Fig. 15.21; Manighetti et al. 2001). Several destructive earthquakes with recorded surface deformation have affected the area. A major earthquake affected the area of Serdo, between the Tendaho and Dobi grabens, in March–April 1969 (magnitude up to M_s = 6.3), with several reported surface breaks and oblique, left-lateral faulting with displacement up to ~75 cm (Gouin 1979). Another series of destructive earthquakes, ten with 5.5 ≤ M_s ≤ 6.3, struck the Dobi graben on August 20–21 1989, rupturing both boundary and inner floor faults over an area about 45 km long and 15 km wide (e.g. Jacques et al. 2011). Numerous surface breaks with complex geometry, including fresh scarplets with vertical throws up to 30-cm-high and open fissures up to 30 cm wide, were associated to the events (Fig. 15.22). The two largest shocks (M_s = 6.2 and 6.3) ruptured the southern segments of the south-western bounding fault of the Dobi graben; a dozen other faults also slipped along the edges of, and inside, the graben (Jacques et al. 2011).

Overall, the complex, distributed deformation affecting the area between the Gulf of Aden and Red Sea system results from the overlap without direct connection of the two major rift arms, which has been suggested to induce a bookshelf faulting mechanism in which small rigid blocks rotate about vertical axis along rift-parallel faults that slip with a component of left-lateral motion (Fig. 15.23; e.g. Tapponnier et al. 1990). The proposed block rotations were also previously cited as evidence that extension in Afar is controlled by the anticlockwise rotation of the Danakil block, a model called the Danakil crank-arm model (Souriot and Brun 1992). Recent geophysical data have been used to test these models with GPS data confirming that the Danakil block does currently rotate anticlockwise (McClusky et al. 2010). However, the large component of strike slip motion on the faults in central Afar is unclear since the majority of earthquake focal mechanisms show normal faulting (e.g. Keir et al. 2013), and the deformation field constrained using InSAR and GPS also shows rift perpendicular extension in the overlapping/en-echelon basins in central Afar with no major shear component of motion (Pagli et al. 2014).

15.4.2 Southern Afar

South of the Tendaho–Goba'ad discontinuity in the East African Rift, the morphology of the rift floor of the MER is dominated by the presence of ~30 km-wide, ~60 km-long en-echelon axial volcanic segments (Figs. 15.2 and 15.24). These are characterised by strong association of active and recent volcanoes, aligned monogenetic cones and fissures, and numerous normal faults striking roughly perpendicular to the direction of extension (e.g. Keir et al. 2013). The faults of the axial volcanic segments are normally short, closely spaced and display relatively small throws (<100 m), giving rise in places to a graben-in-graben morphology (Fig. 15.24) (Soliva and Schulz (2008). These faults are characterised by steep (subvertical) scarps and are commonly en-echelon and linear or curved in plan view over

Fig. 15.20 *Top panel* The dry alluvial plains bounded by major normal fault escarpments of the Dobi graben. *Inset* shows a satellite image of the region. *Bottom panel* A salt pan in the Dobi Graben, where locals pump water onto the flat basin floor and then wait for it to evaporate. Major boundary faults in the background (*Photos* E. Baker and D. Keir)

distances of up to a few tens of kilometres, thus delimiting several fault-bounded blocks. Associated with the faults are open fissures with or without vertical displacement, splay patterns and complex rhomb-shaped structures (e.g. Mohr 1987; Boccaletti et al. 1998; Williams et al. 2004). Structural and geomorphological analyses suggest a migration of deformation toward the axis of the volcanic segments, with inner normal faults displacing recent basalt flows erupted from rift-parallel fissures and eroded more external faults (Hayward and Ebinger 1996). The axial basins are filled with both volcanic products and sequences of Pliocene–Recent alluvial and lacustrine sediments, with rivers eroding the uplifted fault footwalls and producing depositional fans in the basins (Hayward and Ebinger 1996).

The dense fault swarms developed in the last 2 Ma (e.g. Ebinger and Casey 2001) and geodetic data support that they accommodate ~ 80 % of the present-day strain (Bilham et al. 1999). Gravity and seismic imaging show extensive

Fig. 15.21 A major boundary fault of the Dobi graben displacing the Afar Stratoid series, with the thick succession of basalt flows visible on the fault footwall (*Photo* E. Baker). Note the presence of a fresh, light-coloured scarplet at the base of the cumulative escarpment

Fig. 15.22 Ruptures associated with the 1989 Dobi seismic sequence (after Jacques et al. 2011)

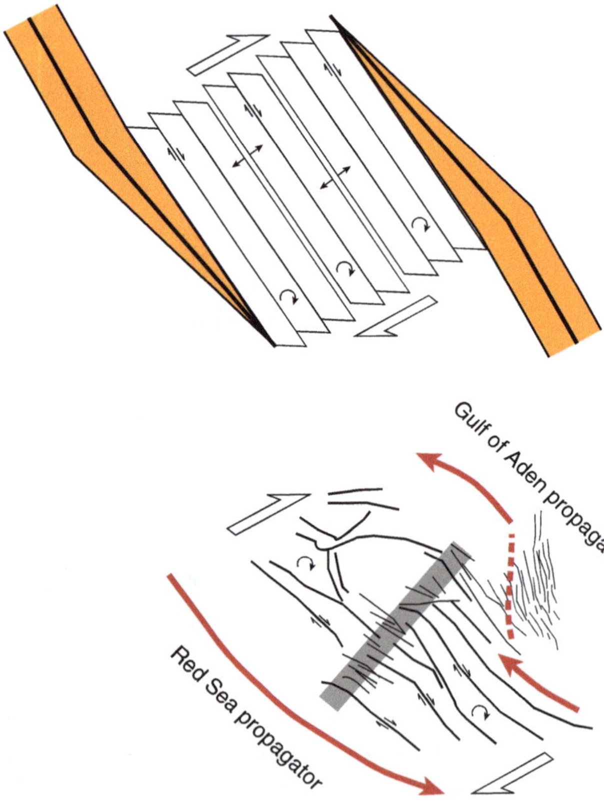

Fig. 15.23 *Top panel* Model of bookshelf faulting, in which rift propagation in opposite directions induces small-rigid blocks to rotate about vertical axis along rift-parallel faults that slip with a left-lateral component. *Bottom panel* Model of bookshelf faulting applied to central Afar; the *grey line* indicates a hypothesised major transversal structure (after Manighetti et al. 2001)

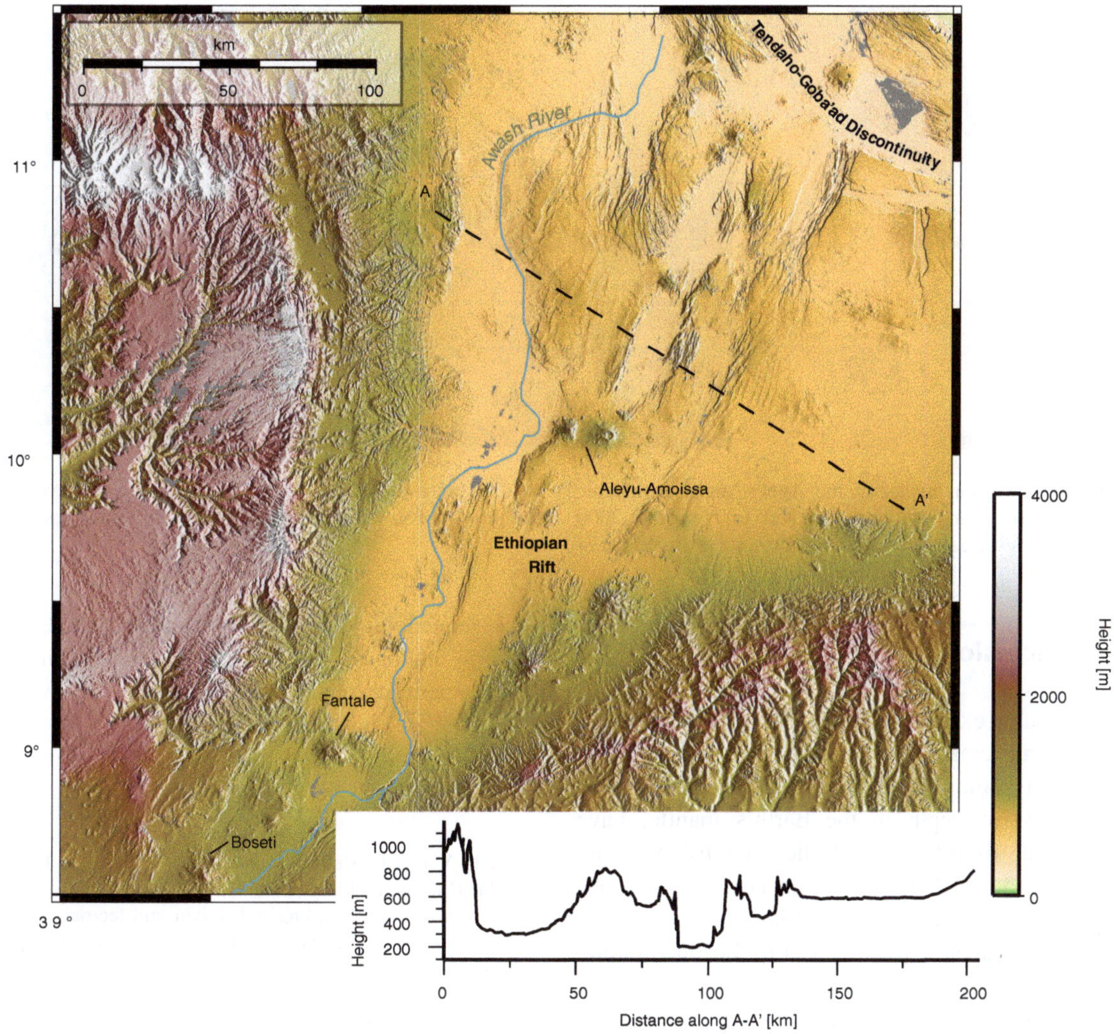

Fig. 15.24 Topography of southern Afar with major volcanic centres labelled. *Bottom right panel* shows a sample of topography on profile A–A′ across the southern Afar floor

magma intrusion throughout the lithosphere beneath magmatic segments (e.g. Bastow et al. 2011), with extension facilitated and accommodated by a combination of magmatic intrusion, diking and faulting (e.g. Keir et al. 2006b). According to Kurz et al. (2007), the architecture and morphology of the magmatic segments are symmetric around central acidic strato-volcanoes, where the deformation pattern is characterised by small faults and a low fault density. The number of faults increases longitudinally away from the volcanic centres, such that maximal fault densities and the longest faults are observed at tip domains (Kurz et al. 2007). This reflects an along-axis change from magmatic deformation accommodated by diking at segment centres to mostly brittle deformation (faulting) at the tip domains.

The NE–NNE striking magmatic segments related to the Nubia–Somalia opening overprint previous NW–NNW striking faults, suggesting that extension in the northern MER postdates extension from initial opening of the southern Red Sea and Gulf of Aden (e.g. Tesfaye et al. 2003; Wolfenden et al. 2004). However, the NW–NNW trending structures still accommodate some volcano-tectonic activity, such as the intrusion of a ~N120°E striking, ~6-km-long dike during May 2000 (Fig. 15.25; Keir et al. 2011). The intrusion occurred beneath the ~WNW striking Ayelu–Amoissa chain of aligned volcanic centres that dominate the morphology of the right stepping MER offset at ~10°N (Fig. 15.25), but did not result in either volcanic eruptions or significant surface deformation (Keir et al. 2011).

Fig. 15.25 Digital elevation model of the Ayelu-Amoissa volcanic system. *Coloured circles* are epicentres related to the May 2000 intrusion event; the *black dashed line* indicates the surface projection of the dike (after Keir et al. 2011). *Photo* showing the Ayelu (*on the right*) and the Amoissa (*the lower one on the left*) volcanoes (*Photo* D. Keir)

15.5 Conclusions

Afar is a spectacular example of how rifting at a complex triple junction can create and continuously modify the morphology of a region at different scales. Large-scale processes, initiated at depth in the Earth's mantle, have shaped the overall morphology of the area by inducing plateau uplift (up to $\sim$2 km) and emplacement of voluminous flood basalt sequences at $\sim$30 Ma. The subsequent subsidence of the Afar floor with respect to the surrounding elevated areas during initial Arabia–Africa rifting has created large fault escarpments that abruptly separate the rift floor from the plateaus. Although not characterised by significant current deformation and in places highly eroded, these escarpments still dominate the regional morphology.

The floor of Afar is primarily shaped by the tectonic–magmatic processes that developed since extension and became focused in a narrower zone of faults and volcanoes within the depression. In these rifts, continuing extension is accommodated by magma emplacement and normal faulting, which gives rise to a landscape dominated by spectacular fresh fault scarps and active volcanic edifices created during episodic tectonic, volcano-tectonic and purely volcanic events.

This morphology, and its along-axis variations, reflects the last phases of the protracted process leading to plate separation and the birth of a new oceanic basin. As break-up progresses, rift-related morphology changes from the narrow fault bound graben as in central Afar to the topographically prominent axial volcanic range in the Danakil depression where the relatively large volumes of erupted material show that a marked increase in rate of basaltic volcanism characterises the final stages of continental break-up. Here, the accumulation of thick sequences of basalt flows and evaporites represents a modern analogue for the processes responsible for the development of volcanic rifted margins worldwide.

References

Acocella V (2006) Regional and local tectonics at Erta Ale caldera, Afar (Ethiopia). J Struct Geol 28:1808–1820

Acocella V (2010) Coupling volcanism and tectonics along divergent plate boundaries: collapsed rifts from central Afar, Ethiopia. Geol Soc Am Bull 122:1717–1728

Acocella V, Abebe B, Korme T, Barberi F (2008) Structure of Tendaho Graben and Manda Hararo Rift: implications for the evolution of the southern Red Sea propagator in Central Afar. Tectonics 27: TC4016. http://dx.doi.org/10.1029/2007TC002236

Acocella V, Bekele A, Coltorti M (eds) (2011) Pre-conference excursion guide: Tectonic Landforms and Volcanism in the Southern Afar. Iag regional conference 2011 geomorphology for human adaptation to changing tropical environments Addis Ababa, Ethiopia, 18–22 Feb 2011

Audin J, Vellutini PJ, Coulon C, Piguet P, Vincent J (1990) The 1928–1929 eruption of Kammourta volcano. Evidence of tectono-magmatic activity in Manda-Inakir rift and comparison with the Asal rift, Afar Depression, Republic of Djibouti. Bull Volcanol 52:551–561

Ayele A, Keir D, Ebinger C, Wright TJ, Stuart GW, Buck RW, Jacques E, Ogubazghi G, Sholan J (2009) September 2005 mega-dike emplacement in the Manda–Harraro nascent oceanic rift (Afar Depression). Geophys Res Lett 36: L20306. http://dx.doi.org/10.1029/2009GL039605

Ayalew D, Marty B, Barbey P, Yirgu G, Ketefo E (2006) Temporal compositional variation of syn-rift rhyolites along the western margin of the southern Red Sea and northern Main Ethiopian Rift. In: Yirgu G, Ebinger CJ, Maguire PKH (eds) The Afar Volcanic Province within the East African Rift System. Geol Soc Spec Pub 259:121–130

Ayele A, Stuart G, Bastow I, Keir D (2007) The August 2002 earthquake sequence in north Afar: insights into the neotectonics of the Danakil microplate. J Afr Earth Sci 40:70–79

Barberi F, Varet J (1970) The Erta Ale volcanic range (Danakil depression, northern Afar, Ethiopia). Bull Volcanol 34:848–917
Barberi F, Varet J (1977) Volcanism of Afar: small-scale plate tectonics implications. Geol Soc Am Bull 88:1251–1266
Barberi F, Ferrara G, Santacroce R, Varet J (1975) Structural evolution of the Afar triple junction. In: Pilger A, Rosler A (eds) Afar Depression of Ethiopia. Schweizerbart, Stuttgart, pp 39–54
Bastow I, Keir D (2011) The protracted development of the continent-ocean transition in Afar. Nat Geosci 4:248–250
Bastow ID, Nyblade AA, Stuart GW, Rooney TO, Benoit MH (2008) Upper mantle seismic structure beneath the Ethiopian hot spot: rifting at the edge of the African low-velocity anomaly. Geochem Geophys Geosyst 9:Q12022. doi:10.1029/2008GC002107
Bastow ID, Keir D, Daly E (2011) The Ethiopia Afar Geoscientific Lithospheric Experiment (EAGLE): probing the transition from continental rifting to incipient seafloor spreading. Geol Soc Am Spec Pap 478:51–76
Benoit M, Nyblade A, VanDecar J (2006) Upper mantle P wavespeed variations beneath Ethiopia and the origin of the Afar hotspot. Geology 34:329–332
Beyene A, Abdelsalam MG (2005) Tectonics of the Afar Depression: a review and synthesis. J Afr Earth Sci 41:41–59
Bilham R, Bendick R, Larson K, Mohr P, Braun J, Tesfaye S, Asfaw L (1999) Secular and tidal strain across the Main Ethiopian Rift. Geophys Res Lett 26:2789–2792
Boccaletti M, Bonini M, Mazzuoli R, Abebe B, Piccardi L, Tortorici L (1998) Quaternary oblique extensional tectonics in the Ethiopian Rift (Horn of Africa). Tectonophysics 287:97–116
Bonatti E, Emiliani C, Ostlung G, Rydell H (1971) Final desiccation of the Afar rift, Ethiopia. Science 172:468–469
Cattin R, Doubre C, de Chabalier J-B, King G, Vigny C, Avouac J-P, Ruegg J-C (2005) Numerical modelling of quaternary deformation and post-rifting displacement in the Asal-Ghoubbet (Djibouti, Afar). Earth Planet Sci Lett 239:352–367
Chorowicz J, Collet B, Bonavia F (1999) Left-lateral strike-slip tectonics and gravity induced individualisation of wide continental blocks in the western Afar margin. Eclogae Geol Helv 92:149–158
Corti G (2012) Evolution and characteristics of continental rifting: analogue modeling-inspired view and comparison with examples from the East African Rift System. Tectonophysics 522–523:1–33
Doubre C, Manighetti I, Dorbath L, Dorbath C, Bertil D, Delmond JC (2007) Crustal structure and magmato-tectonic processes in an active rift (Asal-Ghoubbet, Afar, East Africa): 2. Insights from the 23-year recording of seismicity since the last rifting event. J Geophys Res 112:B05406. doi:10.1029/2006JB004333
Ebinger C, Sleep NH (1998) Cenozoic magmatism in central and east Africa resulting from impact of one large plume. Nature 395: 788–791
Ebinger C, Casey M (2001) Continental breakup in magmatic provinces: an Ethiopian example. Geology 29:527–530
Ebinger C, Ayele A, Keir D, Rowland J, Yirgu G, Wright T, Belachew M, Hamling I (2010) Length and timescales of rift faulting and magma intrusion: the Afar rifting cycle from 2005 to present. Ann Rev Earth Planet Sci 38:439–466
Ferguson DJ, Barnie TD, Pyle DM, Oppenheimer C, Yirgu G, Lewi E, Kidane T, Carn S, Hamling I (2010) Recent rift-related volcanism in Afar, Ethiopia. Earth Planet Sci Lett 292:409–418
Field L, Barnie T, Blundy J, Brooker RA, Keir D, Lewi E, Saunders K (2012) Integrated field, satellite and petrological observations of the November 2010 eruption of Erta Ale. Bull Volcanol 74:2251–2271
Fubelli G, Dramis F (2011) Coseismic surface faulting in the Kara Kore area (Wollo) caused by the May 29, 1961 Earthquake. In: Acocella V, Bekele A, Coltorti M (eds) Pre-conference excursion guide: tectonic landforms and volcanism in the Southern Afar. Iag regional conference 2011 geomorphology for human adaptation to changing tropical environments Addis Ababa, Ethiopia, 18–22 Feb 2011, pp 38–42
Furman T, Bryce J, Hanan B, Yirgu G, Ayalew D (2006) Heads and tails: 30 years of the Afar plume. In: Yirgu G, Ebinger CJ, Maguire PKH (eds) The Afar Volcanic Province within the East African Rift System. Geol Soc Lond Spec Publ 259:95–119
Gasse F (1991) Tectonic and climatic controls on lake distribution and environments in Afar from Miocene to Present. In: Katz B (ed) Lacustrine basin exploration: case studies and modern analogs. AAPG Mem 50:19–41
Gouin P (1979) Earthquake history of Ethiopia and the Horn of Africa. Int Develop Res Centre, Ottawa, Publ 118
Grandin R, Socquet A, Binet R, Klinger Y, Jacques E, de Chabalier J-B, King GCP, Lasserre C, Tait S, Tapponnier P, Delorme A, Pinzuti P (2009) September 2005 Manda Hararo-Dabbahu rifting event, Afar (Ethiopia): constraints provided by geodetic data. J Geophys Res 114:B08404. doi:10.1029/2008JB005843
Grandin R, Socquet A, Jacques E, Mazzoni N, de Chabalier J-B, King GCP (2010) Sequence of rifting in Afar (Manda-Hararo rift, Ethiopia, 2005–2009): timespace evolution and interactions between dikes from InSAR and static stress change modeling. J Geophys Res 115:B10413. doi:10.1029/2009JB000815
Hansen S, Nyblade A, Benoit M (2012) Mantle structure beneath Africa and Arabia from adaptively parameterized P-wave tomography: implications for the origin of Cenozoic Afro-Arabian tectonism. Earth Planet Sci Lett 319:23–34
Hayward N, Ebinger C (1996) Variations in the along-axis segmentation of the Afar rift system. Tectonics 15:244–257
Hofmann C, Courtillot V, Féraud G, Rochette P, Yirgu G, Ketefo E, Pik R (1997) Timing of the Ethiopian flood basalt event and implications for plume birth and global change. Nature 389:838–841
Hutchinson RW, Engels GG (1972) Tectonic evolution of the southern Red Sea and its possible significance to older rifted continental margins. Geol Soc Am Bull 83:2989–3002
Hussein B, Chandrasekharam D, Chandrasekhar V, Jalludin M (2013) Geochemistry of thermal springs around Lake Abhe, Western Djibouti. Int J Sustain Energ. doi:10.1080/14786451.2013.813027
Jacques E, Kidane T, Tapponnier P, Manighetti I, Gaudemer Y, Meyer B, Ruegg JC, Audin L, Armijo R (2011) Normal faulting during the August 1989 Earthquakes in Central Afar: sequential triggering and propagation of rupture along the Dôbi Graben. Bull Seis Soc Am 101:994–1023
Kazmin V (compiler) (1972) Geological map of Ethiopia (1:2,000,000). Ethiopian geological survey, Ministry of Mines and Energy Resources, Addis Ababa
Kazmin V, Seife MB, Nicoletti M, Petrucciani C (1980) Evolution of the northern part of the Ethiopian Rift. In: Geodynamic evolution of the Afro-Arabian Rift System, Accademia Nazionale Dei Lincei, Atti dei Convegni Lincei 47:275–292
Keir D, Stuart G, Jackson A, Ayele A (2006a) Local earthquake magnitude scale and seismicity rate for the Ethiopian Rift. Bull Seis Soc Am 96:2221–2230
Keir D, Ebinger C, Stuart G, Daly E, Ayele A (2006b) Strain accommodation by magmatism and faulting as rifting proceeds to breakup: seismicity of the northern Ethiopian rift. J Geophys Res 111(B5):B05314. doi:10.1029/2005JB003748
Keir D, Hamling IJ, Ayele A, Calais E, Ebinger C, Wright TJ, Jacques E, Mohamed K, Hammond JOS, Belachew M, Baker E, Rowland JV, Lewi E, Bennati L (2009) Evidence for focused magmatic accretion at segment centers from lateral dike injections captured beneath the Red Sea rift in Afar. Geology 37:59–62
Keir D, Pagli C, Bastow I, Ayele A, Ababa E (2011) The magma-assisted removal of Arabia in Afar: evidence from dike injection in the Ethiopian rift captured using InSAR and seismicity. Tectonics 30 (2):TC2008. doi:10.1029/2010TC002785

Keir D, Bastow I, Pagli C, Chambers E (2013) The development of extension and magmatism in the Red Sea rift of Afar. Tectonophysics 607:98–114
Kurz T, Gloaguen R, Ebinger C, Casey M, Abebe B (2007) Deformation distribution and type in the Main Ethiopian Rift (MER): a remote sensing study. J Afr Earth Sci 48:100–114
Lahitte P, Gillot PY, Courtillot V (2003) Silicic central volcanoes as precursors to rift propagation: the Afar case. Earth Planet Sci Lett 207:103–116
Le Gall B, Ahmed Daoud M, Rolet J, Moussa Egueh N (2011) Large-scale flexuring and antithetic extensional faulting along a nascent plate boundary in the SE Afar rift. Terra Nova 23:416–420
Leroy S, Razin F, Autin J, Bache F, d'Acremont E, Watremez L, Robinet J, Baurion C, Denèle Y, Bellahsen N, Lucazeau F, Rolandone F, Rouzo S, Serra Kiel J, Robin C, Guillocheau F, Tiberi C, Basuyau C, Beslier M-O, Ebinger C, Stuart G, Ahmed A, Khanbari K, Al Ganad I, de Clarens P, Unternehr P, Al Toubi A, Al Lazki A (2012) From rifting to oceanic spreading in the Gulf of Aden: a synthesis. Arab J Geosci. http://dx.doi.org/10.1007/s12517-011-0475-4
Makris J, Ginzburg A (1987) The Afar Depression: transition between continental rifting and sea floor spreading. Tectonophysics 141: 199–214
Manighetti I, Tapponnier P, Gillot PY, Jacques E, Courtillot V, Armijo R, Ruegg JC, King GCP (1998) Propagation of rifting along the Arabia-Somalia boundary: into Afar. J Geophys Res 103(B3):4947–4974
Manighetti I, Tapponnier P, Courtillot V, Gallet Y, Jacques E, Gillot P-Y (2001) Strain transfer between disconnected, propagating rifts in Afar. J Geophys Res 106:13613–13665
McClusky S, Reilinger R, Ogubazghi G, Amleson A, Healeb B, Vernant P, Sholan J, Fisseha S, Asfaw L, Bendick R, Kogan L (2010) Kinematics of the southern Red Sea-Afar Triple junction and implications for plate dynamics. Geophys Res Lett 37:L05301. doi:10.1029/2009GL041127
Medynski S, Pik R, Burnard P, Williams A, Vye-Brown C, Ferguson D, Blard P-H, France L, Yirgu G, Seid JI, Ayalew D, Calvert A (2013) Magmatic cycles and development of rift topography of the Manda-Hararo segment (Afar): Insights from cosmogenic 3He. Earth Planet Sci Lett 367:133–145
Mohr P (1978) Afar. Ann Rev Earth Planet Sci 6:145–172
Mohr P (1987) Patterns of faulting in the Ethiopian Rift Valley. Tectonophysics 143:169–179
Morton WH, Black R (1975) Crustal attenuation in Afar. In: Pilger A, Rosler A (eds) Afar Depression of Ethiopia. Schweizerbart, Stuttgart, pp 55–61
Nobile A, Pagli C, Keir D, Wright TJ, Ayele A, Ruch J, Acocella V (2012) Dike-fault interaction during the 2004 Dallol intrusion at the northern edge of the Erta Ale Ridge (Afar, Ethiopia). Geophys Res Lett. http://dx.doi.org/10.1029/2012GL053151
Nyblade AA, Robinson SW (1994) The African superswell. Geophys Res Lett 21:765–768
Pagli C, Wright TJ, Ebinger CJ, Yun S-H, Cann JR, Barnie T, Ayele A (2012) Shallow axial magma chamber at the slow-spreading Erta Ale Ridge. Nat Geosci 5:284–288
Pagli C, Hua W, Wright TJ, Calais E, Lewi E (2014) Current plate boundary deformation of the Afar rift from a 3D velocity field inversion of InSAR and GPS. J Geophys Res (in press)
Pik R, Marty B, Carignan J, Lavé J (2003) Stability of the Upper Nile drainage network (Ethiopia) deduced from (U-Th)/He thermochronometry: implications for uplift and erosion of the Afar plume dome. Earth Planet Sci Lett 215:73–88
Pizzi A, Pontarelli L (2007) Geometria del rift nel margine meridionale dell'Afar: Dire Dawa (Etiopia). Rend Soc Geol It 4:288–289
Ritsema J, Deuss A, van Heijst H, Woodhouse J (2011) S40RTS: a degree-40 shear-velocity model for the mantle from new Rayleigh wave dispersion, teleseismic traveltime and normal-mode splitting function measurements. Geophys J Int 184:1223–1236
Rogers N, Macdonald R, Fitton J, George R, Smith R, Barreiro B (2000) Two mantle plumes beneath the East African rift system: Sr, Nd and Pb isotope evidence from Kenya Rift basalts. Earth Planet Sci Lett 176:387–400
Rooney T, Herzberg C, Bastow I (2012a) Elevated mantle temperature beneath East Africa. Geology 40:27–40
Rooney T, Hanan BB, Graham DW, Furman T, Blichert-Toft J, Schilling J-G (2012b) Upper mantle pollution during Afar Plume-Continental rift interaction. J Petrol 53:365–389
Rooney T, Mohr P, Dosso P, Hall (2013) Upper mantle pollution during Afar Plume-Continental rift interaction. Geochim Cosmochim Acta 102(1):6588. doi:10.1016/j.gca.2012.08.019
Rowland JR, Baker E, Ebinger C, Keir D, Kidane T, Biggs J, Hayward N, Wright T (2007) Fault growth at a nascent slow spreading ridge: 2005 Dabbahu rifting episode, Afar. Geophys J Int 171:1226–1246
Ruegg JC, Lépine JC, Tarantola A (1979) Geodetic measurements of rifting associated with a seismo-volcanic crisis in Afar. Geophys Res Lett 6:817–820
Siebert L, Simkin T, Kimberly P (2010) Volcanoes of the world, 3rd edn. University of California Press, Berkeley
Soliva R, Schultz RA (2008) Distributed and localized faulting in extensional settings: insight from the North Ethiopian Rift–Afar transition area. Tectonics 27:TC2003. doi:10.1029/2007TC002148
Souriot T, Brun J-P (1992) Faulting and block rotation in the Afar triangle, East Africa: the Danakil "crank-arm" model. Geology 20:911–914
Talbot CJ (2008) Hydrothermal salt—but how much? Mar Petrol Geol 25:191–202
Tapponnier P, Armijo R, Manighetti I, Courtillot V (1990) Bookshelf faulting and horizontal block rotations between overlapping rifts in southern Afar. Geophys Res Lett 17:1–4
Tarantola A, Ruegg JC, Lépine JC (1979) Geodetic evidence for rifting in Afar: a brittle-elastic model of the behaviour of the lithosphere. Earth Planet Sci Lett 45:435–444
Tesfaye S, Harding D, Kusky T (2003) Early continental breakup boundary and migration of the Afar triple junction, Ethiopia. Geol Soc America Bull 115:1053–1067
Thurmond AK, Abdelsalam MG, Thurmond JB (2006) Optical-radar-DEM remote sensing data integrated for geological mapping in the Afar Depression, Ethiopia. J Afr Earth Sci 44:119–134
Waltham T (2010) Afar triangle: rift valleys and volcanoes over plate divergence. In: Migoń P (ed) Geomorphological Landscapes of the World. Springer Science+Business Media, pp 183–190
Williams FM, Williams MAJ, Aumento F (2004) Tensional fissures and crustal extension rates in the northern part of the Main Ethiopian Rift. J Afr Earth Sci 38:183–197
Wolfenden E, Ebinger C, Yirgu G, Deino A, Ayalew D (2004) Evolution of the northern main Ethiopian rift: birth of a triple junction. Earth Planet Sci Lett 224:213–228
Wolfenden E, Ebinger C, Yirgu G, Renne P, Kelley S (2005) Evolution of a volcanic rifted margin: Southern Red Sea, Ethiopia. Geol Soc Am Bull 117:846–864
Wright T, Ebinger C, Biggs J, Ayele A, Yirgu G, Keir D, Stork A (2006) Magma-maintained rift segmentation at continental rupture in the 2005 Afar dyking episode. Nature 442:291–294
Wright TJ, Sigmundsson F, Pagli C, Belachew M, Hamling IJ, Brandsdóttir B, Keir D, Pedersen R, Ayele A, Ebinger C, Einarsson P, Lewi E, Calais E (2012) Geophysical constraints on the dynamics of spreading centres from rifting episodes on land. Nat Geosci 5:242–250
Zanettin B, Justin-Visentin E (1975) Tectonical and volcanological evolution of the western Afar margin (Ethiopia). In: Pilger A, Rosler A (eds) Afar Depression of Ethiopia. Schweizerbart, Stuttgart, pp 300–309

Morphometric Characteristics and Hydrology of Selected Ethiopian Rift Lakes

16

Tenalem Ayenew and Merhawi GebreEgziabher

Abstract

In the Main Ethiopian Rift (MER), intra-rift faulting and associated volcanic activities resulted in the formation of volcano-tectonic structural depressions, responsible for the formation of many lakes on the rift floor. These lakes are bordered to the east and west by elevated highlands where the main tributary feeder rivers originate. The MER is occupied by seven major lakes, namely, from south to north, Chamo, Abaya, Awasa, Shala, Abijata, Langano, and Ziway. Most of these lakes are closed basins and interconnected by groundwater, through a NE–SW-aligned regional fault system, whereas Abijata Lake receives surface water from Ziway and Lagano lakes by two small streams. The geomorphic characteristics and size of these lakes changed on time scales of decades to millennia in response to different factors such as volcanic activity, tectonics, climate change and, very recently, human impact (mainly water withdrawal, deforestation, and irrigation). The most prominent example is the dramatic change of Abijata, the size of which decreased dramatically in the last decades.

Keywords

Rift lakes • Groundwater • Hydrological budget • Tectonics • Ethiopia

16.1 Introduction

The distribution of Ethiopian lakes is intimately linked to the geomorphological setup of the country. Ethiopia has three major physiographic regions: the highlands (plateaus), the rift valley and the peripheral lowlands which are often wide plains devoid of lakes. The majority of the Ethiopian lakes are confined within the rift valley (Fig. 16.1).

The geomorphological setting is highly controlled by different episodes of volcano-tectonic activity that shaped the country in the Cenozoic, with the influence continuing till the present day. The volcano-tectonic activities resulted in the formation of a spectacular rift system and one of the largest volcanic plateau of the world. There is a wide variation in altitude that ranges from the highland mountain ranges, with precipitous edges dissected by numerous rivers, to the rift valley which is occupied by a series of lakes. The altitude ranges from 4,533 m a.s.l. at Ras Dejen (Simen Mountains) in the northern highland mountains to about 120 m below sea level in the Dallol depression which is a part of the northern end of the Ethiopian Rift valley.

The Ethiopian Rift can be divided into five hydrological systems: Turkana, Lake Chew Bahir, Lake/swamps Main Ethiopian Rift (MER), Awash River basin and the Danakil basin. In these basins, there are over 15 natural lakes. The MER transects the uplifted Ethiopian highlands for a distance of around 1,000 km, extending from southern Afar across the broad basins and volcanic ranges to the southern watershed of Chamo Lake in the center of the country. The MER floor is occupied by a series of lakes that display wide variations in chemistry and morphometric characteristics (Ayenew 2009a, b).

In this study, emphasis is given to seven major lakes occupying the floor of the MER. These include Chamo, Abaya, Awasa, Shala, Abijata, Langano, and Ziway. These

T. Ayenew (✉) · M. GebreEgziabher
School of Earth Science, Addis Ababa University, P.O. Box 1176, Addis Ababa, Ethiopia
e-mail: Tenalema@yahoo.com

P. Billi (ed.), *Landscapes and Landforms of Ethiopia*, World Geomorphological Landscapes,
DOI 10.1007/978-94-017-8026-1_16,

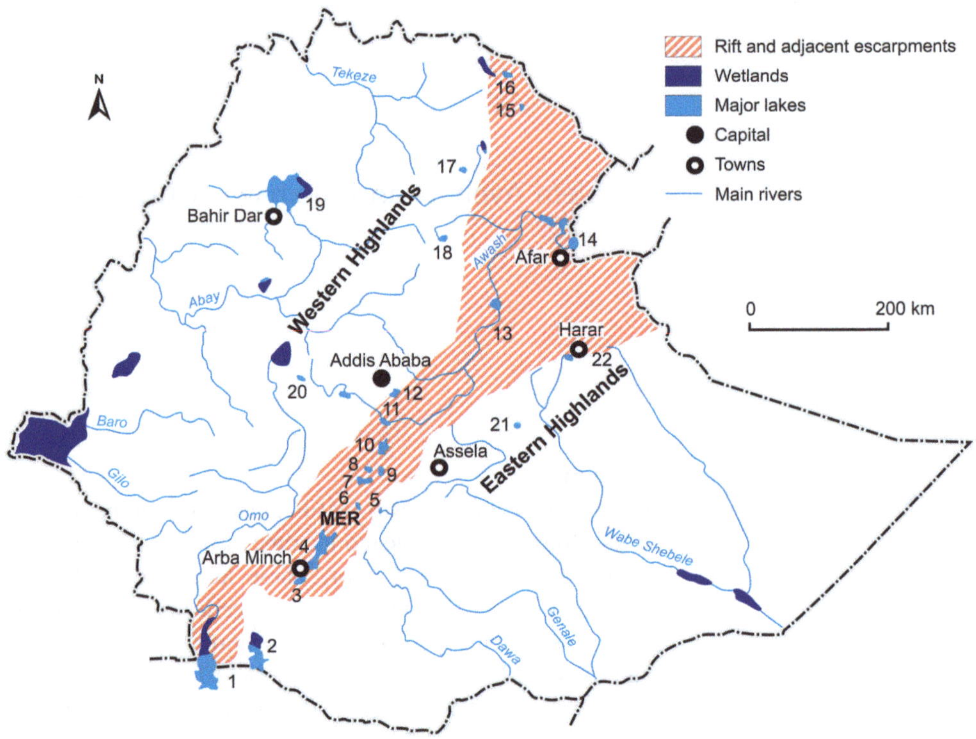

Fig. 16.1 Location of Ethiopian lakes and wetlands associated with the Great East African rift and adjacent highlands (key to wetlands: *1* Turkana, *2* Chew Bahir, *3* Chamo, *4* Abaya, *5* Awasa–Shallo, *6* Chitu, *7* Shala, *8* Abijata, *9* Langano, *10* Ziway, *11* Bishoftu crater lakes, *12* Beseka, *13* Afambo–Abe group, *14* Afrera, *15* Asele, *16* Ashenge, *17* Hayk-Ardibo, *18* Tana, *19* Wonchi–Dendi, *20* Lakes of the Bale Mountains, *21* Ciarciar, *22* Haromaya–Adele–Finkle)

lakes occupy the rift floor in three separate hydrological basins: Abaya–Chamo, Awasa and Ziway–Shala (ZSB). Particular emphasis is given to their morphometric characteristics and hydrology.

16.2 Distribution of Lakes and Major Wetlands

The Ethiopian rift lakes are nested in en echelon fashion, extending in the south from Chew Bahir Lake on the Ethiopia and Kenya border to the extreme north of the Danakil Depression. Figure 16.1 shows the distribution of major Ethiopian lakes and wetlands. The major lakes are confined in the rift floor and are situated within different hydrological basins; from north to south they are: (1) the lakes of the Danakil basin (Asele and Afrera); (2) Awash river basin (Abe, Gamari, Afambo, Beseka); (3) ZSB (Ziway, Langano, Abijata and Shala); (4) Awasa Lake basin; (5) Abaya–Chamo; (6) Chew Bahir and Omo basin (Turkana Lake). Apart from these, there are very small crater lakes scattered on the rift floor and escarpment areas. The crater lakes have small catchment areas and steep crater rim slopes. Crater lakes are not included in this study.

In terms of surface water, most of these lakes are not interconnected owing to topographic and geological factors, whereas their groundwaters are interconnected by a NE–SW-aligned regional fault system associated with tectonically active region (Ayenew 1998, 2009a, b; Alemayehu et al. 2006) and form a unique system within the rift. Similar hydrogeological setup is evident in the Kenyan rift valley where major faults act as subsurface conduits connecting the rift lakes (Ayenew and Becht 2008).

16.3 Overview of the Evolution, Geomorphology, and Climate of the MER

16.3.1 Evolution and Geomorphology

The Ethiopian Rift is part of the Great East African Rift which is one of the most extensive rift systems on earth. It is an impressive geological feature extending from the Jordan River in Middle East southward to Mozambique in southeast Africa, approximately 6,400 km long and 48–64 km wide. The Ethiopian Rift provides a unique opportunity to study the transition between continental rifting to the south in Kenya and new seafloor spreading to the north in Afar in the onshore extension of the Red Sea and Gulf of Aden (Mackenzie et al. 2005).

Intra-rift faulting and associated volcanic activities resulted in the formation of volcano-tectonic structural depressions, responsible for the formation of many lakes in the rift floor (Ayenew 2009a, b). The rift lakes are bordered to the east and west by elevated highlands where the main tributary feeder rivers originate. Magnificent volcanic centers dot the floor of the rift valley (Fig. 16.2), separating the

Fig. 16.2 Ziqwala volcano, 2,989 m a.s.l., dominating the landscape around the Koka reservoir

different lakes. The volcanic centers are associated with hydrothermal activities. In some cases, hot thermal waters feed the lakes at the foot of volcanic centers and caldera rims (Ayenew and Becht 2008).

The MER contains lakes of different hydrological and morphometric characteristics. Most lakes are aligned along the NE trending tectonic depressions characterized by active fault systems and volcanic centers originated during the Cenozoic (since 65 million years ago). Figure 16.3 shows the major lakes of the MER and selected meteorological stations.

The general regional geological setting of the MER involves volcanism, rifting, Quaternary lake level fluctuation and deposition of fluvial and volcano–lacustrine sediments. At the beginning of the Cenozoic and mainly during the Oligocene (Chap. 2 this volume), the region was deluged under molten lava and thick beds of explosive rocks with intense tectonic activity, which initiated the formation of the rift valley. The wide areas buried under volcanic materials, the thickness and variety of the rock sequences and the prolonged duration of the eruptions make the Ethiopian plateaus and the rift one of the world's most notable volcanic regions. Although these episodes of volcanic activity predate the formation of the rift itself, they still played an important role in sculpting and enriching the magnificent scenery, renowned for imposing escarpments, some of them relatively old and weathered while other still fresh, sharp, and steep.

The size of the lakes changed on time scales of decades to millennia. Large rift lakes covered wide depressions in closed basins in the Pleistocene wet intervals (Grove et al. 1975; Chap. 17, this volume). The present-day rift lakes are remnants of large lakes evolved into separate systems as they receded through time (Street 1979; Chap. 17, this volume). Little is known about the evolution and the mode of origin of the majority of Ethiopian lakes, whereas the late Pleistocene and Holocene hydrological and geomorphological changes of the ZSB are well documented (Benvenuti et al. 2002, 2005; Chap. 17, this volume). From lacustrine deposits, it appears that Chamo and Abaya were one single lake in recent geological past. The eruption of central volcano and different episodes of tectonism separated these lakes in the Pliocene. Most of the MER lakes are occupying volcano-tectonic depressions. Some of them are the center of tectonically affected caldera. Typical examples are Awasa and Shala. The volcano-tectonic setting controls the considerable variations in volume and size of these lakes. Table 16.1 displays clearly this variation.

Most of the rift lakes are localized within a closed basin fed by both perennial rivers and seasonal streams. Large highland rivers are the source of sustained supply to the major rift lakes. The amount and distribution of highland rainfall strongly control the level and the size of these lakes (Street 1979; Ayenew 1998).

16.3.2 Climate

Generally, the climate of the MER is semiarid to sub-humid, with distinct wet and dry seasons (Chap. 3, this volume). The adjacent highlands experience high rainfall and lower evaporation. During the wet season, northwesterly monsoon winds bring high rainfall. In general, the high rainfall lasts from June to September (Chap. 3, this volume). The dry

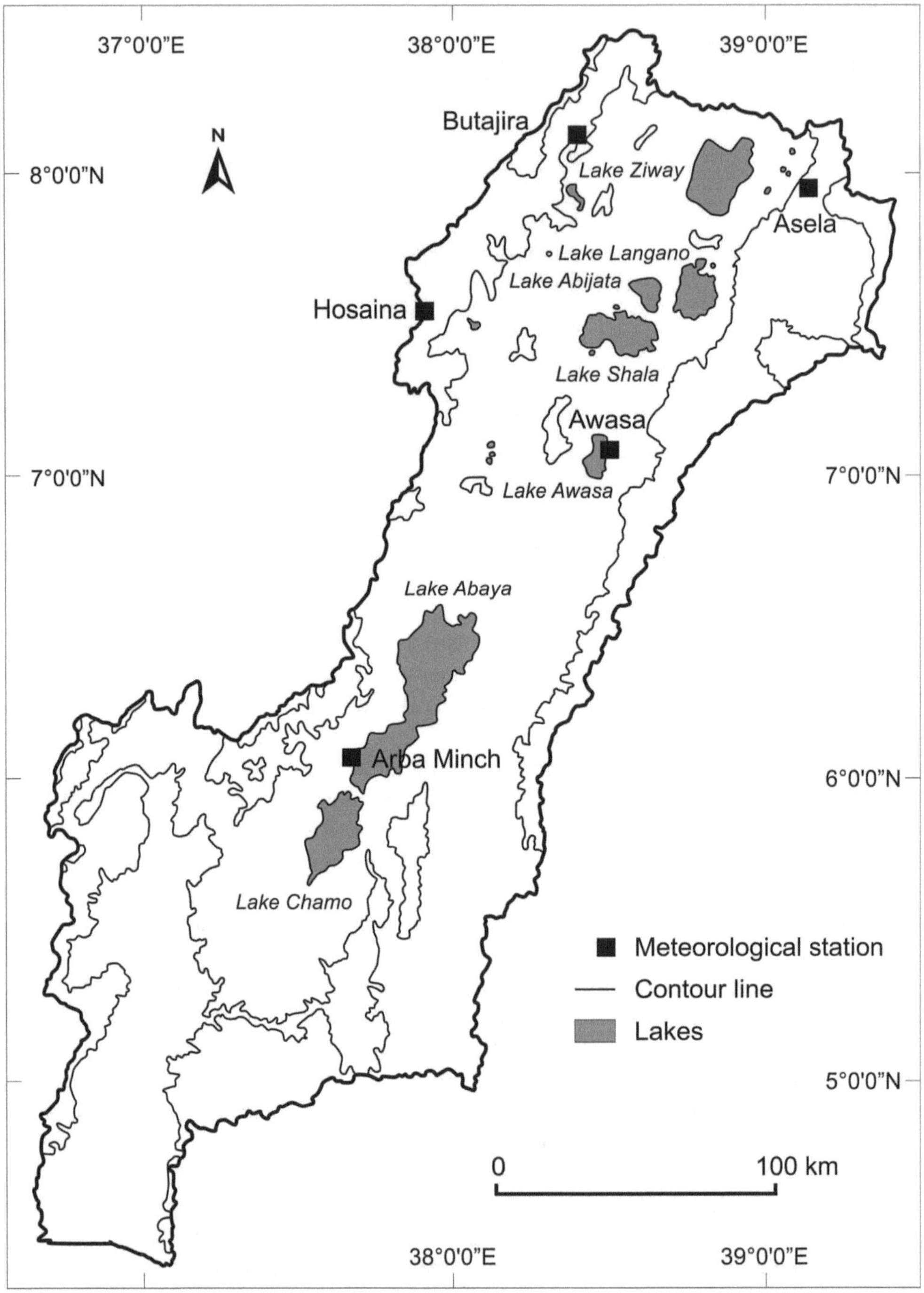

Fig. 16.3 Location of selected meteorological stations and the main MER lakes (key to major lakes: *1* Ziway, *2* Langano, *3* Abijata, *4* Shala, *5* Awasa–Shallo, *6* Abaya, *7* Chamo)

season is from December to April. Table 16.2 shows the long-term average rainfall and temperature in selected stations on the rift floor and adjacent highlands.

Average annual temperature varies from over 20 °C in the central MER to less than 18 °C in the adjacent highlands. Annual rainfall varies from around 630 mm in the central MER to more than 1,500 mm in the southwestern and western highlands. The national average annual precipitation is 744 mm (NMSA 2005). The highland rainfall is the ultimate source of replenishment of the rift lakes in the form of groundwater, river inflows, and surface runoff. Generally, different climatic conditions characterize the highlands, the escarpment, and the rift valley. The lake level is strongly controlled by highland rainfall. Inter-annual variation in the lake levels is the direct reflection of the changes in the highland rainfall, with the exception of a few lakes which are being used for irrigation and those influenced by irrigation return flows and soda ash abstraction (Ayenew 2009a, b).

Table 16.1 Basic morphometric data of Ethiopian rift lakes

No.	Lakes	Altitude (m a.s.l.)	Area (km^2)	Maximum depth (m)	Average depth (m)	Catchment area (km^2)	Volume (km^3)
1	Ziway	1,636	440	8.9	2.5	7,414	1.5
2	Langano	1,585	230	47.9	17	2,000	38
3	Abijata	1,578	135	14.2	7.6	10,740[a]	0.954
4	Shala	155	370	266	86	2,300	37
5	Awasa	1,680	80	22	11	136	1.3
6	Abaya	1,285	1,140	13.1	8.6	16,342	8.2
7	Chamo	1,235	317	13	10.3	18,575[a]	–

[a] The catchment areas of Chamo and Abijata include Abaya and Ziway catchment, respectively. (Modified from Ayenew 2009a, b)

Table 16.2 Summary of long-term average monthly precipitation and temperature recorded at selected stations in the three main physiographic regions (rift, escarpments, and highlands)

Station	Location	Year	Jan	Feb	Mar	Apr	May	Jun	July	Aug	Sept	Oct	Nov	Dec	Annual
A. Precipitation (mm)															
Asela	Central MER	1982–2002	27	50	92	127	126	98	116	114	133	86	33	25	1,027
Awasa	Eastern highland	1970–2004	61	0.1	38	113	44	104	147	172	154	158	21	50	1,061
Butajira	Western highland	1970–2004	38	67	138	127	114	121	172	174	120	46	11	14	1,140
Ziway	Central MER	1970–2004	17	31	56	75	72	83	145	122	86	39	2.4	4	732
B. Temperature (°C)															
Arba Minch	Southern MER	1960–2002	23	24	24	24	22	22	22	22	23	23	22	22	22.7
Asela	Central MER	1988–2003	20	20	21	21	20	20	19	19	19	19	18	19	19.5
Awasa	Eastern highland	1960–2004	14	16	16	16	17	16	15	15	16	15	14	14	15.3
Butajira	Western highland	1972–2004	18	19	19	19	19	19	18	18	18	18	18	18	18.4
Ziway	Central MER	1970–2004	19	20	21	22	22	21	20	20	20	20	19	19	20.2

16.4 Morphometric Characteristics and Hydrology of the MER Lakes

16.4.1 Morphometric Characteristics

Table 16.1 summarizes the basic morphometric characteristics of the seven major lakes. Due to differences in geomorphological setting resulting from volcano-tectonic processes in the Cenozoic and Quaternary, the rift lakes display large variations in size and volume. They have also large differences in their hydrochemistry (Chap. 18, this volume). Most of the closed lakes are highly saline, while the open lakes are fresh. The deepest lake in the Ethiopian rift is Shala (266 m), while the shallowest is Ziway (less than 10 m) if we do not consider Chew Bahir, which used to be a vast lake when it was discovered in 1888 by Samuel Teleki, and now is partly a swamp and for the most part a muddy salt plain.

16.4.2 Hydrology

Over the last few decades, a number of studies have been carried out to study the hydrology of the major Ethiopian lakes (Ayenew 1998, 2009a, b). These studies provided the general picture of the different fluxes and water balance of the dynamic rift valley lakes. The methodologies used to quantify the water balance of each lake are addressed in detail in the respective referred works. Integrated hydrogeological investigations involving hydrogeological mapping, remote sensing techniques, isotope, and hydrochemical data helped to recognize the interaction of the lakes with groundwater and flow paths. The source of the fluxes and the relationships between the lakes with the groundwater and the hydrological conditions of the three main lakes basins (Ziway–Shala, Awasa, Abaya–Chamo) are described below.

Table 16.3 summarizes the water balance of some of the rift lakes. Due to the availability of pertinent data, the central MER lakes (Ziway, Abijata, Langano, Shala, and Awasa)

Table 16.3 Long-term average annual water balance of selected rift lakes ($10^6 m^3$)

Lake	Area (km^2)	Inflow				Outflow					Source
		P_l	R_i	G_i	S_r	E_l	R_o	Go	A	V	
Abaya	1,162	556	Very high	Very high	Very high	1,900	Rare outflow	Medium	–	–	3, 5
Chamo	551	406	High	High	High	900.9	Rare outflow	High	Ng	–	3, 5
Awasa	100	106	83.1	High	83.7	132	Closed	58*	Ng	82.8	6,7
Abijata	180	113	230	26.8	15	372	Closed	1.2	13	−1.4	2,4
Langano	230	186	212	135.4	Very high	463	46	18.9	–	5.5	2,4
Shala	370	232	245	237.6	40	781	Closed	None	Ng	−26.4	2,4
Ziway	440	323	656.5	80.5	48	890	184	14.6	28	−8.6	2,4

P_l precipitation on the lake; R_i inflow from rivers; G_i groundwater inflow; S_r inflow from surface runoff; E_l lake evaporation; R_o outflow in river outlets; *A* abstraction; *V* net flux; *Ng* negligible [*Sources* (1) Tessema (1998); (2) Ayenew (1998); (3) Bekele (2000); (4) Ayenew (2001); (5) Woldemariam (2004); (6) Gebreegziabher (2005); (7) Ayenew and Gebreegziabher (2006)]

are relatively better studied. The water balance assessment was made on the basis of extensive long-term hydrometeorological data (National Meteorological Database from 1965 to 2003) and numerical modeling (Ayenew 2001; Ayenew and Gebreegziabher 2006). In addition, hydrochemical and isotope studies provided information on the subsurface hydraulic links of the lakes (Darling et al. 1996; Tessema 1998; Ayenew 2003; Fikre 2006). Most of the data is extracted from Ayenew 2009a, b.

In recent years, the size of the lakes has changed significantly (Fig. 16.4). The most prominent example is the dramatic change of Abijata Lake (Fig. 16.5). Ziway, Chamo, and Abaya lakes have also shown slight changes. On the contrary, irrigation return flows resulted in expansion of other lakes. The typical example is Beseka Lake which is located in the Awash basin, north of Ziway Lake (Ayenew 2011), whereas the expansion of Awasa Lake in the recent years has not yet been completely understood.

Some Ethiopian Rift lakes, particularly those located in a terminal position, have undergone significant changes (Ayenew 2011). The major rift valley lakes show contrasting lake level trends, the most extreme of which is Lake Abijata. The lakes showing the most dramatic changes are those where the water is used for agricultural and industrial purposes. In a few cases, land degradation and neotectonic movements have caused changes (Ayenew and Legesse 2007). The reduction of the level of Lake Abijata and the complete disappearance of Lake Haromaya (located in the Eastern Ethiopian highlands Fig. 16.1) are examples of the severe consequences of excessive pumping of water.

Abijata's shallow depth and terminal position make it particularly susceptible to changes in climate and pumping.

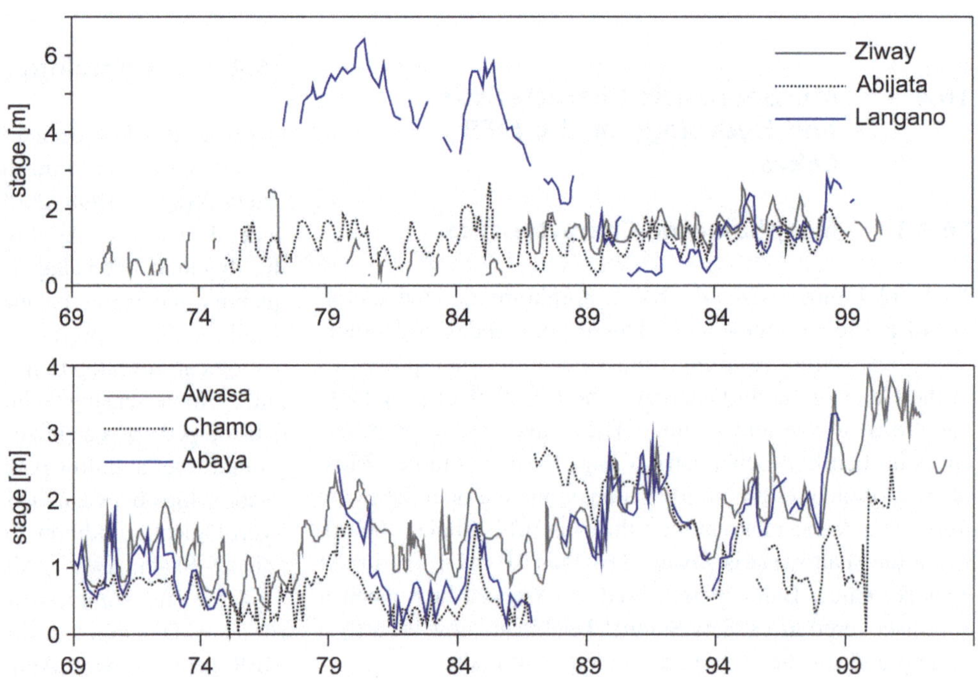

Fig. 16.4 Lake level variation through time. All lakes show a moderate increasing trend with Abijata recovering from a few years of marked decrease (Ayenew 2004)

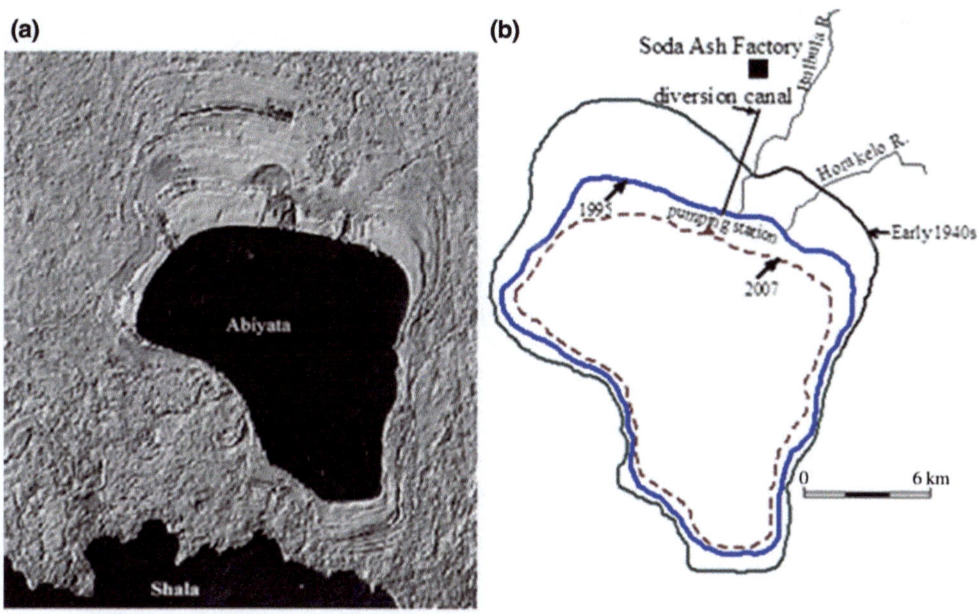

Fig. 16.5 Recession of Abijata Lake: **a** enhanced satellite image showing different strandlines representing shorelines at different times; **b** reconstructed shoreline positions at different years

It receives most of its water from the Ziway Lake catchment through the Bulbula River and a smaller amount from Langano Lake through the Horakelo River (Figs. 16.5 and 16.6). Being a closed lake, the only significant natural water loss is through evaporation. In the absence of human interference, the lake maintained the natural balance between inflow and evaporation until the mid-1970s; that is before irrigation activities started in the Ziway area and the establishment of the Abijata Soda Ash Factory in the mid-1980s. The reduction of the level of Abijata is clearly visible from old shorelines seen in satellite images that make it possible to reconstruct the size of the lake in different years (Fig. 16.5, but see also Fig. 17.2 in Chap. 17, this volume). The maximum reduction in the area, and hence the drop of the level of Abijata Lake, coincided with the large-scale water abstraction for soda production from the lake itself and for irrigation from Ziway Lake and its basin after the 1980s (OEPO 2005). Data from a few meteo-stations in the MER floor and margins show also that during the last three-four decades the air temperature has remarkably increased (Billi and Dramis 2000, Chap. 3, this volume).

The ZSB is a closed basin with a catchment area of about 13,000 km^2, out of which 1,443 km^2 is covered with permanent open water bodies (Ayenew 1998). Ziway, Langano, Abijata and Shala lakes occupy volcano-tectonic depressions at the center of the basin. There are also small lakes and swampy areas in the rift and within its bordering escarpments (Fig. 16.6). Lakes Ziway and Langano are open, while Shala and Abijata are closed. Abijata is fed by outflows from Ziway and Langano lakes through the Bulbula and Horakelo rivers, respectively. Ziway Lake is the freshest of all the four lakes, while Shala and Abijata are highly alkaline and saline due to evaporative enrichment and the supply of saline water from hydrothermal springs. Larger rivers feed Ziway and Shala lakes, whereas smaller streams feed Langano Lake. Shala, Langano and Abijata lakes have also a centripetal network of ephemeral streams and gullies draining the inner side of the caldera.

16.4.2.1 Ziway Lake

The main input to Ziway Lake comes from the Ketar and Meki rivers and direct precipitation. The main outputs are evaporation, discharge via the Bulbula River and groundwater leakage through faults and paleo-channels. Hydrochemical and isotopic evidence reveals southward groundwater migration from Ziway to Abijata (Ayenew 2003).

The annual direct precipitation over the lake is 323 mcm (million cubic meter). Long-term average annual inflow from Ketar and Meki rivers is 392 and 265 mcm, respectively. The remaining ungauged catchments are flat and located in the rift, with low precipitation. However, during high precipitation events overland flow reaches the lake from the northern and western part of its catchment. The contribution from the ungauged catchments is estimated at 48 mcm annually.

Annual lake evaporation and the Bulbula River outflow are 890 and 184 mcm, respectively. The annual abstraction of water for irrigation is around 28 mcm (Ayenew 1998). Deribessa (2006) indicated that the groundwater outflow and pumping rate for irrigation are higher. The current abstraction from the lake and feeder rivers is expected to be over 50 mcm. In order to attain equilibrium, leakage into the subsurface is considered to play a relevant role, which is consistent with the isotope and hydrochemical studies (Ayenew 2003; Deribessa 2006).

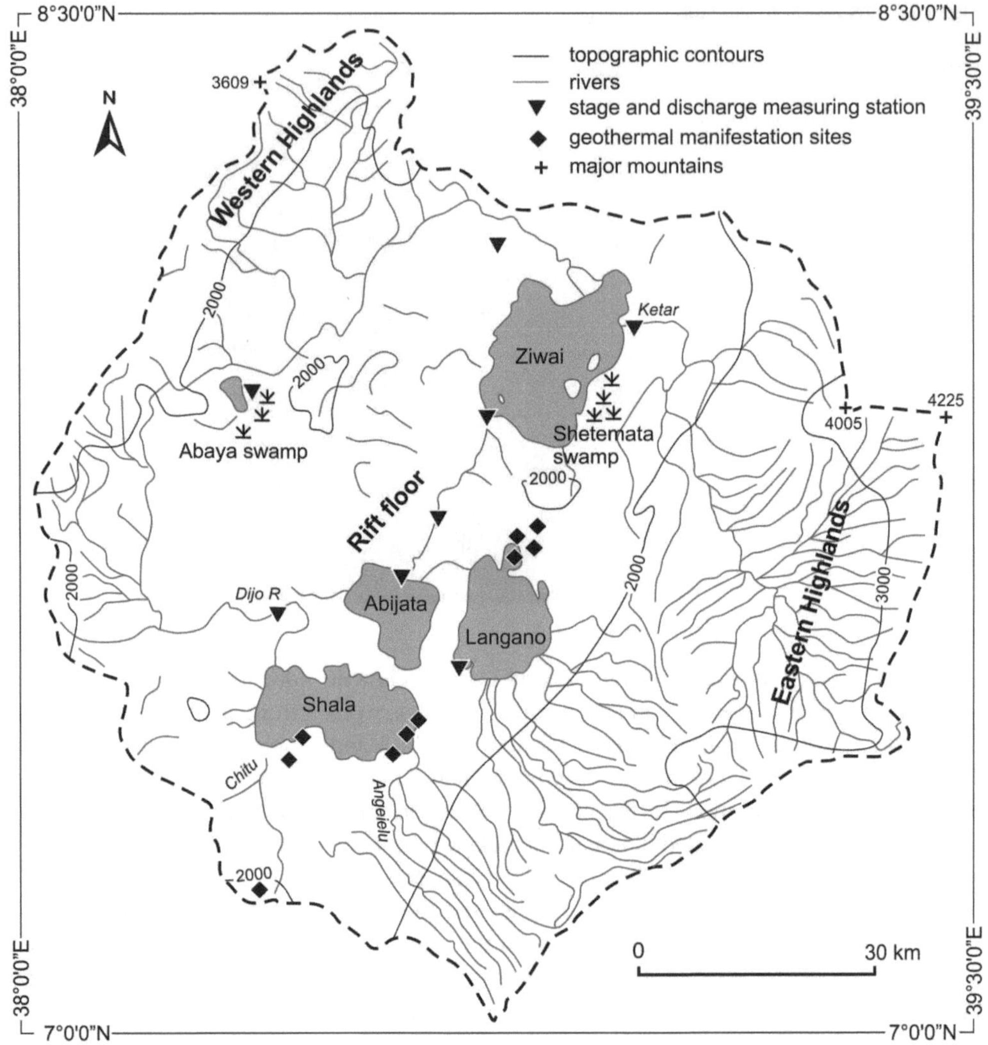

Fig. 16.6 Lakes of Ziway–Shala basin (Ayenew 2009a, b)

16.4.2.2 Abijata Lake

The main input to the lake comes from the Horakelo and Bulbula rivers and direct precipitation. The annual precipitation on the lake accounts to 112 mcm. The long-term average annual inflows from the Bulbula and Horakelo rivers are 184 and 46 mcm, respectively. Surface runoff from the remaining ungauged catchments is 15 mcm. The main loss is through evaporation (372 mcm annually). The lake water is being pumped for soda ash extraction. The recent pumping rate is unknown. In 1998, the total annual pumping rate was estimated at 13 mcm. Using groundwater flow modeling, one may infer insignificant subsurface outflow from the lake (Ayenew 2001). The balance between total inflow and outflow is positive, indicating that there is additional inflow from groundwater, which is likely to be from the northern and western sides.

16.4.2.3 Langano Lake

The annual contribution from precipitation is 146 mcm. There is no discharge data for the few tributary rivers coming from the eastern Arsi Highlands. At least four major perennial rivers (Huluka, Lepis, Gedemso and Kersa) feed the lake. There are also many ungauged seasonal streams and springs, all joining the lake from the eastern side. By extrapolating the discharge record of the gauged rivers, the total annual inflow from the ungauged catchments is estimated at 315 mcm. High-discharge hot springs enter into the lake in the northern shore. These include Oitu, Bole, Tuffa, and the springs east of Langano peninsula. From limited field spring discharge measurements, the annual input from the springs is estimated at 63 mcm.

Annual lake water evaporation and outflow through the Horakelo River are 463 and 46 mcm, respectively. Groundwater flow models revealed the existence of large leakage in the southern shore, which ultimately flows into Lake Shala (Ayenew 2001).

16.4.2.4 Shala Lake

Shala Lake is a groundwater-controlled system, but rivers and precipitation play important roles as well. Annual input from precipitation is 232 mcm. The main tributary rivers are Dijo and Awade, which originate from the western and eastern highlands, respectively. The latter is not gauged. The annual inflow from Dijo is 107 mcm. The recent inflow is quite low due to upstream diversions for local irrigation. The southern and southeastern catchment is similar to that of Ketar (Ziway Lake tributary, see above) in terms of total precipitation, land use, and geology though the topography is flatter. By extrapolating the discharge data from Ketar, the annual contribution from the Awade River could be around 138 mcm. The remaining ungauged catchment generates little runoff since it is located in the flat lowland, with low precipitation. If 5 % of the weighted annual precipitation (785 mm) of the ungauged catchment is assumed to reach the lake as surface runoff, the annual input will be 40 mcm. Many hot springs enter into the lake in the eastern and southern shores. From intermittent discharge measurements, annual inflow from springs is estimated at 18 mcm (Ayenew 1998).

The only loss from the lake is through evaporation, accounting for 781 mcm annually. This brings the difference between the total inflow and outflow to 247 mcm. The high positive residual can be explained in terms of groundwater input. Lakes Shala and Langano are certainly controlled by groundwater more than any other lake in the ZSB. For Shala, the major groundwater input comes from Langano through faults in the northeastern shore and from the elevated areas in the southeast, south, and probably northwest. There is strong evidence of groundwater inflow from the adjacent Hawassa catchment (Darling et al. 1996; Ayenew and Gebreegziabher 2006). The existence of large regional faults and the relative lower topographic position (1,550 m a.s.l.) favors large volumes of groundwater to converge toward Shala Lake from different directions.

16.4.2.5 Awasa Lake Basin

The Awasa basin represents a large collapsed caldera bordered by highlands to the north and east. The center of the caldera is occupied by Awassa and Shallo (Cheleleka) lakes. The elevation of Awasa Lake is 1,680 m a.s.l., representing the culmination of the Ethiopian Rift floor level occupied by lakes. The floor of the caldera is faulted and dotted by volcanic hills. It is believed that this area forms the regional groundwater divide of the entire Ethiopian rift system. Lakes south of Awasa basin leak in the form of groundwater to the south, while Awasa Lake feeds Shala Lake to the north through regional axial faults (Ayenew 1998).

Awasa is the smallest of all the major lakes south of the Awash River Basin. The catchment has no surface water outlet. The lake is about 15 km long and 5.5 km wide with a maximum depth of 22 m and covers 80 km^2 (NUPI 1994). The surface area of Shallo Lake is 12 km^2. The elevation difference between the two lakes is around 5 m (Ayenew and Gebreegziabher 2006). Shallo Lake is a remnant of a much larger lake that included the surrounding swampy plain covering the floor of the caldera northeast of Awasa town (Telford and Lamb 1999). Nowadays, Shallo Lake is on the verge of extinction due to sedimentation and land use changes in the catchment area.

The caldera floor is distinctly separated from the surrounding highlands by scarps and steep mountain slopes. The elevation difference between the floor of the caldera and the bordering scarps and volcanic complexes ranges from 200 to 900 m. Rivers drain into the lakes from the mountainous areas. The rivers from the eastern and southeastern highlands feed Shallo Lake throughout the year. Overflow from Shallo Lake drains into Awasa Lake via the Tikur Wuha (Black Water) river. In the eastern, western, northwestern, and southern sides of the catchment, no perennial rivers reach the lake. Seasonal streams may also terminate in wide-open fractures before reaching the lake (Fig. 16.7).

As Awasa Lake occupies a closed basin, there is no surface water outflow. The input comes from rivers, direct precipitation, and groundwater. The only gauged river is Tikur Wuha connecting Shallo with Awasa Lake. The conventional water balance estimation and hydrological models provided a good picture of the relative importance of the different fluxes (WWSDE 2001; Ayenew and Gebreegizabher 2006). The annual input from surface runoff and precipitation account for 145 and 88 mcm, respectively. Chernet (1982) reported the flow rate of some thermal springs in the Awasa catchment. The contribution of the springs to the hydrological balance of Awasa Lake is likely less than 5 %.

The annual evaporation is 151 mcm (Geremew 2000). On the basis of hydrological models, the net annual groundwater outflow from the lake was estimated to be 58 mcm (Ayenew and Gebreegziabher 2006).

16.4.2.6 The Abaya–Chamo Basin

Abaya (Fig. 16.8) and Chamo (Fig. 16.9) are marvelous lakes ringed by savannah plains bordered by large mountain ranges to the east and west. Abaya and Chamo lakes occupy elongated tectonic grabens and are dotted with many islands. According to the Statistical Abstract of Ethiopia for 1967/68, Abaya Lake is 60 km long and 20 km wide with a surface area of 1,160 km^2. It has a maximum depth of 13 m with an elevation of 1,268 m a.s.l.

Together with the lakes, the total area of the basin is 18,600 km^2, with a freshwater water volume of over 12 km^3. A ridge, called Angels Bridge, separates Abaya from Chamo Lake. Abaya is the largest lake in the Ethiopian Rift. It is relatively less saline and is used for fishing and recreation.

The lake is fed by the Bilate River which flows from the north and other rivers from the eastern and western highlands. The main perennial rivers include Uraye, Shope, Hare,

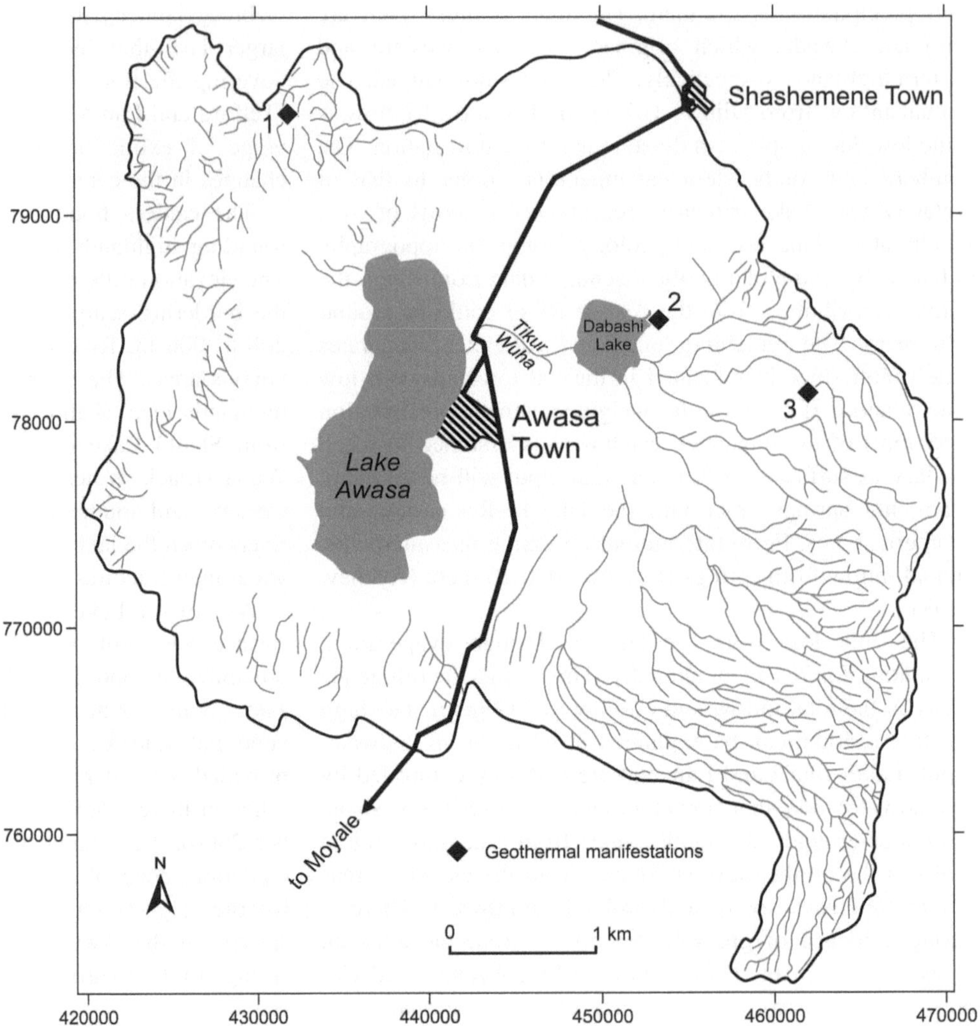

Fig. 16.7 The Awasa Lake catchment (*1* Corbetti volcano; *2* Cheleleka hot springs; *3* Wondo Genet Resort and hot springs)

Kola, Gidabo, and Gelana. Abaya is separated from Chamo Lake by a land isthmus with a vertical offset of 65 m. Occasionally, during high water levels of the wet seasons, the water of Abaya flows into Chamo Lake through a small channel. In the northern part of the basin, there is a large rhyolitic massif, Shire Volcano, west of Bilate River. Moderate temperature hot springs and weak fumaroles exist at its eastern and southern footslopes, respectively. Bodicho hot springs are located on a fault in ignimbrite near the southeast base of Data Volcano. There are also Nech Sar springs near the town of Arba Minch. Dominantly, the hot springs emerge along the western marginal faults near the center of the graben.

Chamo Lake is located south of Abaya Lake, at an elevation of 1,235 m a.s.l. The lake is 26 km long and 22 km wide, with a surface area of 316 km^2 and a maximum depth of 13 m (Woldemariam 2004). The lake water is more saline than that of Abaya Lake; however, the salinity is low as compared to the closed lakes of the ZSB. This indicates groundwater outflow to neighboring basins as in the case of Awasa Lake.

Chamo Lake is fed by Sille and Kulfo Rivers. During extreme wet seasons, it overflows to Chew Bahir Lake through Metenfesha stream. The source of the Segan River, which drains to Chew Bahir, lies east of the southern end of Chamo Lake. There is a water channel connecting Chamo Lake with the Segan River. The overflow occurs when the water level rises by about a few tens of centimeters. The overflow takes place when the precipitation in the region is above long-term average. With increasing diversion of water from tributary rivers for irrigation, the overflow to Chew Bahir and the surface water connection between Abaya and Chamo will likely cease to exist.

There are also three crater lakes in the northern part of the basin, close to the Bilate River. These are Tilo, Mecheferia, and Budemeda lakes (Fig. 16.10). A fourth, dry crater is located 5 km southwest of Mecheferia. These lakes are aligned along a NE–SW trending tectonic line, parallel to the direction of the MER axis. All of them are maars or collapse structures above diatremes, extending to an igneous dyke at greater depth (Lorenz 1986).

Fig. 16.8 Abaya Lake with the volcanic ridge separating it from Chamo Lake

Fig. 16.9 Chamo Lake within the Netch Sar National Park

Comprehensive hydrological studies provided information on the water balance of Abaya and Chamo lakes (Bekele 2000, 2006; Bekele and Horlacher 2001; Woldemariam 2004). The main surface water input to Abaya Lake comes from direct precipitation, rivers (Bilate, Uraye, Shope, Hare Kola, Gidabo, and Gelana) and springs. During wet seasons, Abaya Lake loses water to Chamo Lake. Bekele (2000) estimated the water balance on the basis of a simple hydrological model. Precipitation and runoff contribute 27.2 and 72.8 % of the total input, respectively. Evaporation, pumping for water supply and other usage, such as livestock watering, account for 97.6, 0.2, and 2.2 % of the total annual outflow.

Woldemariam (2004) provided a different picture for the water balance of Abaya Lake. According to his study, the annual lake evaporation is 1,635 mm, which is 1,900 mcm over the 1,162 km^2 total lake area. Precipitation, surface runoff and groundwater contribute 5, 14, and 81 % of the annual total inflow, respectively. The groundwater and lake evaporation account for 86 and 14 % of the annual outflow, respectively. The long-term average annual precipitation of the area is 737 mm.

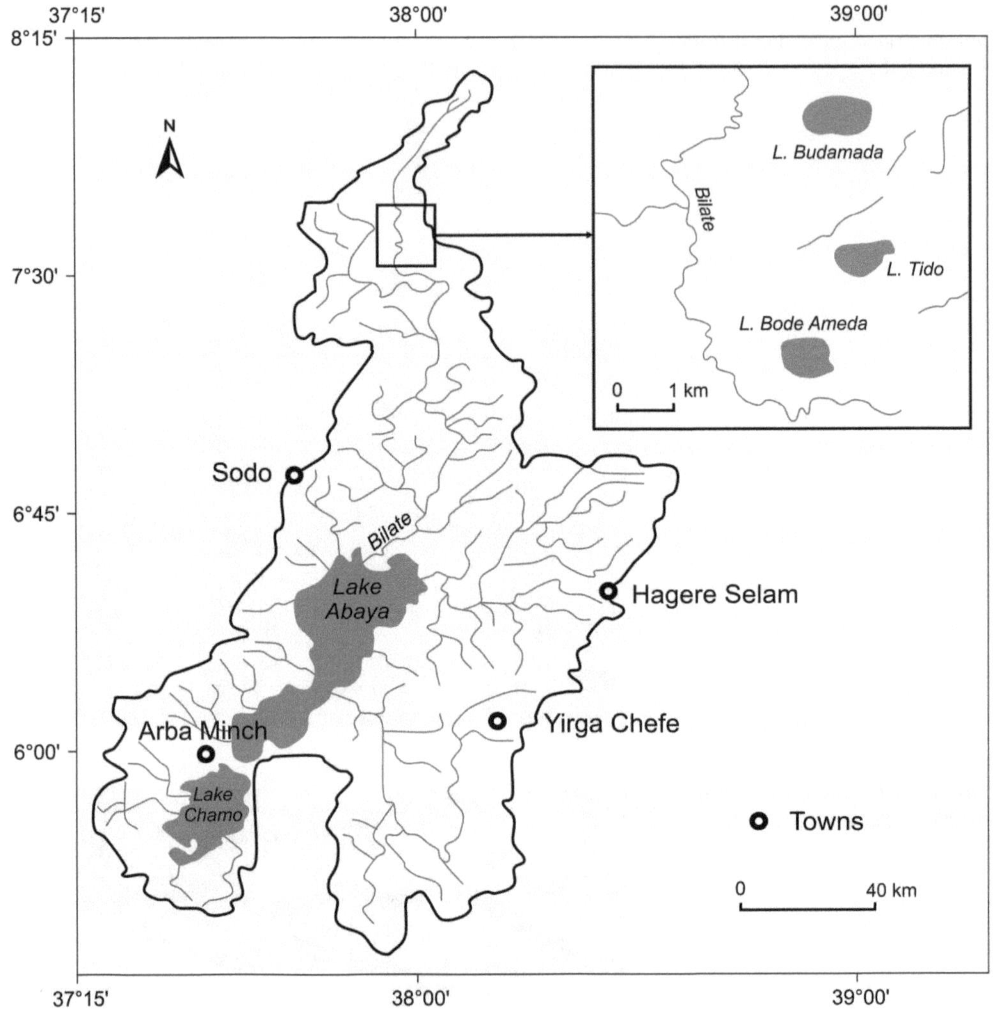

Fig. 16.10 Lakes of the Abaya–Chamo basin

The water balance of Chamo Lake is not well understood. This is especially due to limited knowledge of the seasonal surface inflow and outflow and the groundwater component. The main inputs are precipitation and inflows from Sille and Kulfo Rivers. The amount of groundwater that leaks to the Chew Bahir basin is unknown. Surface water outflow may probably occur during the rainy season (Grove and Goudie 1971).

16.5 Conclusions

Ethiopia is endowed with many natural lakes. Most of these lakes are confined within the Ethiopian Rift system. The volcano-tectonic process in the Cenozoic favored the formation of a chain of lakes aligned along the NE–SW trending rift axis. These lakes have different hydrological and geomorphological settings.

The hydrology of the rift valley lakes is strongly controlled by highland rainfall that feeds the lakes in the form of river flow, surface runoff, and groundwater inflow. The water balance study shows that most lakes are interconnected in the subsurface. The closed lakes are groundwater-dominated systems.

Groundwater flow in the Rift is controlled by geological structures, either directly via flows in the tensional faults or through fluvial and lacustrine deposits whose pattern of occurrence is influenced by tectonics. The major groundwater conduits are faults trending parallel and sub-parallel to the Rift axis. Based on piezometric surveying, hydrogeological field investigations, and ancillary isotope and hydrochemical data, it is concluded that most of the lakes are hydraulically connected through rift fault systems.

The most important result of this study is the elaboration of the intricate nature of the subsurface hydrology of the rift system, as well as the interconnections of the lakes, whose water balances are predominantly controlled by groundwater. These results highlight the importance of studying the details of both the inflow and the outflow groundwater components, including the installation of more piezometers and seepage meters. Accordingly, the future sustainable management of the Rift Valley lakes basins must consider the complex hydrological characteristics of the lakes, and their interrelations through the complex rift fault systems. Indeed, attempting to develop and implement a water management plan for a given lake, without appropriately considering the groundwater fluxes, will almost certainly lead to improper or erroneous water use practices.

Acknowledgments The authors are indebted to Marek Kasprzak for significantly improving many of this chapter figures.

References

Alemayehu T, Ayenew T, Kebede S (2006) Hydrochemical and lake level changes in the Ethiopian rift. J Hydrol 316(1–4):290–300

Ayenew T (1998) The hydrogeological system of the lake district basin, central main Ethiopian rift. Published Ph.D thesis, Free University of Amsterdam, The Netherlands. 259 pp

Ayenew T (2001) Numerical groundwater flow modelling of the central main Ethiopian rift basin. SINET: Ethiopian J Sci 24(2):167–184

Ayenew T (2003) Environmental isotope-based integrated hydrogeological study of some Ethiopian rift lakes. J Radioanal Nucl Chem 257(1):11–16

Ayenew A (2004) Environmental implications of changes in the levels of lakes in the Ethiopian rift since 1970. Reg Environ Change 4:192–204

Ayenew A (2009a) Natural lakes of Ethiopia. Addis Ababa University Press, Addis Ababa, 256 pp

Ayenew T (2009b) Natural lakes of Ethiopia. Addis Ababa University Press, Addis Ababa

Ayenew A (2011) Wetlands of Ethiopia: overview of environmental changes, vol I. Forum for Environment, Adis Ababa

Ayenew T, Becht R (2008) Comparative assessment of the hydrology of selected Ethio-Kenyan rift lakes. J Lakes Reservoirs Res Manag 13:191–196

Ayenew T, Gebreegziabher Y (2006) Application of a spreadsheet hydrological model for computing the long-term water balance of lake Awassa, Ethiopia. Hydrol Sci J 51(3):418–431

Ayenew T, Becht R, Lieshout AV, Gebreegziabher Y, Legesse D, Onyando J (2007) Model-based study of the hydrodynamics of topographically closed lakes in the Ethio-Kenyan rift: the case of lakes Hawassa and Naivasha. J Spat Hydrol 7(1):81–100

Bekele S (2000) Water resources investigation and design guideline for potential exploitation in limited data situation: the case of Abaya-Chamo basin. Ph.D Dissertation, TU Dresden

Bekele S (2006) Investigation of physical and bathymetric characteristics of lakes Abaya and Chamo, Ethiopia, and their management implications. Lakes Reserv Res Manag 11:133–140

Bekele S, Horlacher HB (2001) Modelling the water balance of natural lake under limited data situation for impact assessment: the case of Abaya-Chamo lakes, Ethiopia. IAHR conference, Beijing, China

Benvenuti M, Carnicelli S, Belluomini G, Dainelli N, Di Grazia S, Ferrari GA, Iasio C, Sagri M, Ventra D, Atnafu Balemwald, Kebede Seifu (2002) The Ziway-Shala lake basin (Main Ethiopian Rift, Ethiopia): a revision of basin evolution with special reference to the late quaternary. J Afr Earth Sci 35:247–269

Benvenuti M, Carnicelli S, Ferrari G, Sagri M (2005) Depositional processes in latest Pleistocene and Holocene ephemeral streams of the main Ethiopian rift (Ethiopia). In: Blum MD, Marriott SB, Leclair SF (eds) Fluvial sedimentology VII, Special Publication International Association of Sedimentologists vol 35, p 277–294

Billi P, Dramis F (2000) Recent climatic trends and soil erosion in some selected areas of Ethiopia. In: Feoli E, Pottier D, Zerihun Woldu (eds) Sustainable development in drylands of East Africa, European commission, DG XII, Science Research & Development, pp 145–162

Chernet T (1982) Hydrogeologica map of the lake region (with memo). Ethiopian Institute of Geological Surveys, Addis Ababa

Darling WG, Gizaw B, Arusei MK (1996) Lake-groundwater relationships and fluid-rock interaction in the African rift valley: isotopic evidence. J Afr Earth Sci 22:423–431

Deribessa A (2006) Groundwater and surface water interaction and geoenvironmental changes in Ziway catchment. Unpublished M.Sc. thesis, Addis Ababa University, Addis Ababa

Fikre S (2006) Hydrogeological system analysis in Ziway–Shala lakes area using hydrochemistry and isotope techniques, central Ethiopia. Unpublished M.Sc. thesis, Addis Ababa University

Gebreegziabher Y (2005) Assessment of the water balance of lake Awassa catchmen, Ethiopia. M.SC thesis, International Institute for Geo-Information Science and Earth Observation (ITC), Enschede

Geremew G (2000) Engineering geological study of Hawassa town and its surroundings. M.Sc. thesis, Addis Ababa University, Addis Ababa, 185 pp

Grove AT, Goudie AS (1971) Secrets of Lake Stephanie's past. Geograph Mag 43:542–547

Grove AT, Street FA, Goudie AS (1975) Former lake levels and climate change in the rift valley of southern Ethiopia. Geogr J 141:177–194

Lorenz V (1986) On the growth of maars and diatremes and its relevance to the formation of tuff-rings. Bull Volcanol 48:265–274

Mackenzi GD, Thybo H, Maguire PKH (2005) Crustal velocity structure across the main Ethiopian rift: results from 2-dimensional wide angle seismic modeling. Geophys J Int 162:994–1006

NMSA (2005) National meteorological database. National meteorological Service Agency, Addis Ababa

NUPI (1994) Awassa master plan final report. National Urban Planning Institute, Addis Ababa

OEPO (2005) A review of the current status and an outline of a future management plan for Lakes Abijiata and Ziway. Oromia Environmental Protection Office (OEPO). Unpublished technical report. Addis Ababa, Ethiopia, 80 pp

Street FA (1979) Late quaternary lakes in the Ziway–Shala basin. Ph.D. thesis, University of Cambridge, Southern Ethiopia, 457 pp

Telford RJ, Lamb HF (1999) Groundwater-mediated response to Holocene climate change recorded by the diatom stratigraphy of an Ethiopian Crater Lake. Quatern Res 52:63–75

Tessema Z (1998) Hydrochemical and water balance approach in the study of high water level rise of Lake Beseka. M.Sc. thesis, The University of Birmingham, 90 pp

Woldemariam F (2004) Hydrogeology of Lake Abaya and Chamo basin: with emphasis on the interaction between Lake Abaya and the surrounding aquifers. M.Sc thesis, Addis Ababa University

WWDSE (2001) The study of Lake Hawassa level rise. Southern Nations Nationalities and People Regional State. Water, Mines and Energy Resources Development Bureau. Unpublished report of the Water Works Design and Supervision Enterprise (WWSDE). Main report, vol II. Addis Ababa, Ethiopia, 291 pp

The Geomorphology of the Lake Region (Main Ethiopian Rift): The Record of Paleohydrological and Paleoclimatic Events in an Active Volcano-Tectonic Setting

17

M. Benvenuti and S. Carnicelli

Abstract

This paper illustrates the main geomorphological features of the Lakes Region (Main Ethiopian Rift) which resulted from the interplay of Late Quaternary climatic and hydrological changes with volcanism and tectonics typical of an active continental rift. Studies carried out over several decades demonstrated that the evolution of Late Pleistocene–Holocene fluvio-lacustrine systems, recorded by a plethora of geomorphic and stratigraphic features, was forced by abrupt hydro-climatic events of regional to global extent which occurred at 10^4–10^2 years scales. Besides the widely acknowledged hydro-climatic forcing, the active rift setting concurred to regulate, through volcanism and fault activity, erosion/sedimentation rates, geometry of the lakes basin and of the hydrographic network, and water supply to the lakes. A volcano-tectonic imprint was left particularly during the transition from the Late Pleistocene to the Holocene when dramatic hydrological modifications affected the region.

Keywords

Lake evolution • Paleogeography • Paleohydrology • Climate change • Rift valley tectonics

17.1 Introduction

The Lakes Region in the Main Ethiopian Rift (MER) is a tectonically controlled endorheic basin (Street 1979; Chernet 1982) with four major lakes: Ziway, Langano, Abijata, and Shala (Fig. 17.1), whose Late Quaternary morpho-stratigraphic record (Fig. 17.2; Bacci 1940; Nilsson 1940; Grove and Goudie 1971; Grove et al. 1975; Gèze 1975; Laury and Albritton 1975; Gasse and Street 1978; Street 1979; Street-Perrot 1982; Gillespie et al. 1983; Le Turdu et al. 1999; Benvenuti et al. 2002) is a suitable proxy for paleoclimate reconstructions (Coetzee and van Zinderen Bakker 1989; Street and Street-Perrot 1990; Gasse and Van Campo 1994; Thomas and Thorp 2003). Moreover, the local geomorphic and sedimentary pattern must have been influenced by the Cenozoic–Quaternary magmatic and tectonic activity of the area (Corti 2009). During the last 100 ky, explosive silicic volcanism (Abebe et al. 2007) produced surface uplift and collapse and supplied huge amounts of volcaniclastic sediments to the fluvio-lacustrine systems. Faulting (Abebe et al. 2007; Agostini et al. 2011) modified the basin geometry and the hydrographic network (Benvenuti et al. 2002; Sagri et al. 2008), affecting rates of erosion/sedimentation. All these processes interplayed with climate fluctuations, resulting in complex hydromorphic dynamics that the strictly climatic hypothesis alone cannot account for. Studies carried out in the last 20 years on the geomorphology and stratigraphy of the Lakes Region (Alessio et al. 1996; Sagri 1998; Benvenuti et al. 2002, 2005, 2013; Sagri et al. 2008; Carnicelli et al. 2009) have led to a detailed reconstruction of the Late Pleistocene–Holocene evolution of this lacustrine system and to the discrimination of the roles of climate and volcano-tectonics. This paper reviews the main results of these studies, outlining the geomorphological evolution of the area.

M. Benvenuti (✉) · S. Carnicelli
Dipartimento di Scienze della Terra, Università di Firenze, Florence, Italy
e-mail: oredep@unifi.it

P. Billi (ed.), *Landscapes and Landforms of Ethiopia*, World Geomorphological Landscapes,
DOI 10.1007/978-94-017-8026-1_17, © Springer Science+Business Media Dordrecht 2015

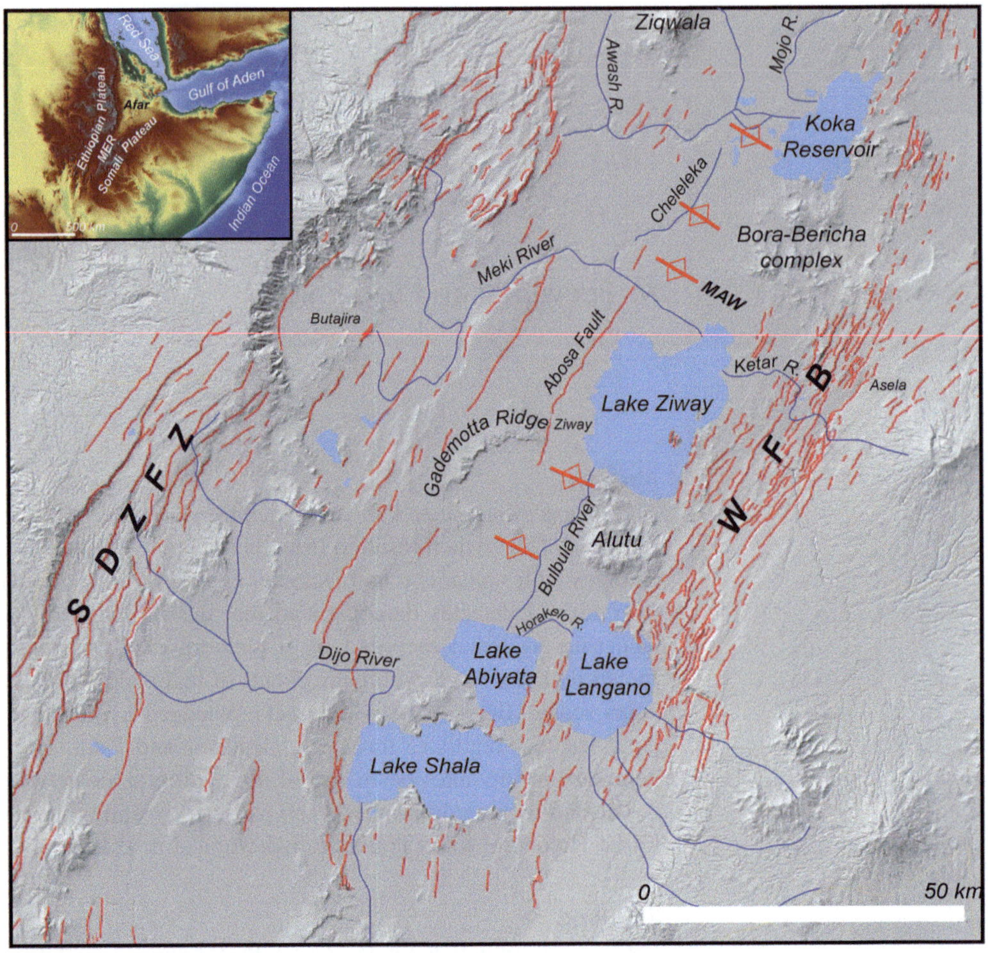

Fig. 17.1 Location and physiography of the Lakes Region in the Main Ethiopian Rift (inset). *Red lines* are the major faults in the region extracted from http://ethiopianrift.igg.cnr.it/data/Main%20Ethiopian%20Rift%20Faults%202.3.1.kmz. *WFB* Wonji Fault Belt; *SDZFZ* Silti-Debre Zeyt Fault Zone. Prominent undulations transversal to the rift trend are shown. *MAW* Meki-Awash watershed

17.2 General Setting

The Lakes Region is located in the central portion of the MER (Fig. 17.1), a 80 km wide and 700 km long NNE–SSW-trending depression bounded by steep escarpments and marked by a stepped morphology, resulting from a dense swarm of NNE–SSW normal faults (Fig. 17.1). Faults cluster within two major belts of intense deformation which are the Wonji Fault Belt and the Silti-Debre Zeyt Fault Zone (Fig. 17.1). The rift floor ranges in elevation from about 1,800 m in the north to about 1,580 m at the Abijata and Shala lake shorelines, and it is not uniformly flat. Transverse, low-lying undulations are common, a prominent one making up the divide between the Lakes and Awash river watersheds (Fig. 17.1). Magmatic activity occurred at different stages during the Quaternary (Mohr 1962; Di Paola 1972; Corti 2009), as testified by inactive or hydrothermally active large volcanoes (Alutu, Bora–Bericha, Ziqwala; Fig. 17.1) which rise above the rift floor by as much as 1,500 m. Prominent in the hydrography of the Lakes Region are the Awash and Mojo rivers, originating from the Ethiopian Plateau and feeding the Koka Reservoir while, to the south, Ziway, Shala, Langano, and Abijata lakes (Fig. 17.1) represent the remnants of a larger Late Quaternary endorheic lake basin (Fig. 17.3). Ziway Lake is fed by the Meki and Ketar rivers, draining the Ethiopian and Somali Plateau, respectively, and forming deltas (Fig. 17.3). Ziway and Langano lakes are connected to Abijata Lake through the Bulbula and Horakelo rivers, respectively (Fig. 17.2), whereas the southernmost Lake Shala is a separate basin, fed partly by the Dijo River (Fig. 17.1) and partly by groundwater seepage from the other lakes (Street 1979; Chernet 1982). The study area is characterized by a marked gradient of annual rainfall from about 1,200 mm along the rift margins to about 700 mm around the Ziway–Shala lakes (Sagri 1998).

17.3 The Late Quaternary Lakes Region: Geomorphology and Stratigraphy of Lake-level Fluctuations

Following the seminal work of Street (1979), who reconstructed the main Late Quaternary fluctuations of the Ziway–Shala lakes from morpho-stratigraphic evidence, further data have been collected in the last decades, to constrain the main

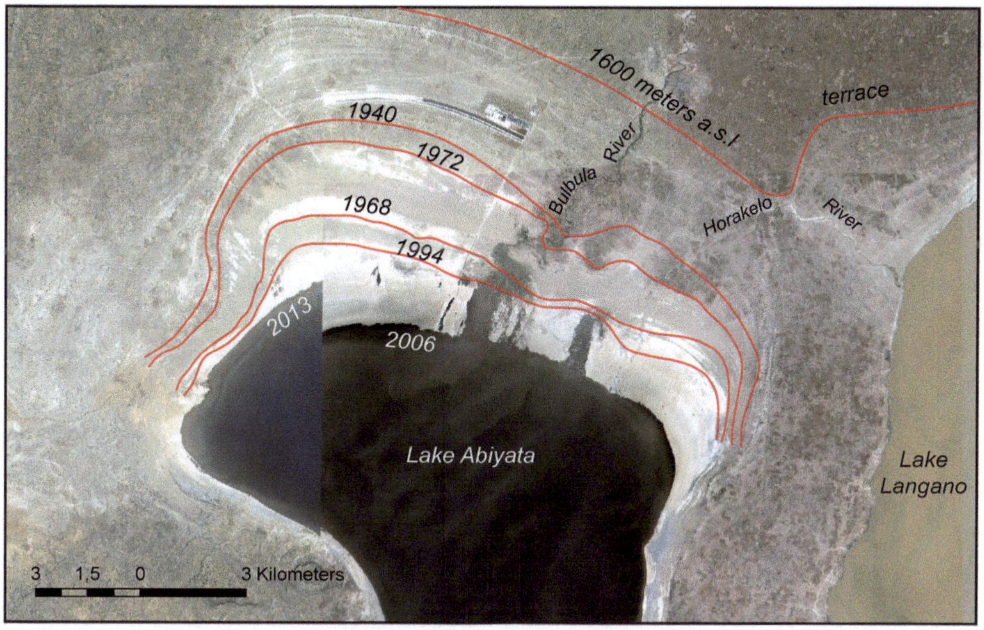

Fig. 17.2 The shore changes of the northern Lake Abijata from annual to millennial scales (Fig. 17.1 for location). The overlapping satellite images of 2013 and 2006 (from Google Maps) outline the dramatic lake-level variations which occurred in recent years. Shorelines from 1940 to mid-1990, compiled from Sagri (1998), attest to lake oscillations at decadal scale. The 1,600 m a.s.l. marks the stillstand at about 2,500 years ago when a single lake, including Abijata, Langano, and Shala lakes, occupied the southern portion of the basin

phases of the Lakes Region evolution (Sagri 1998; Benvenuti et al. 2002). These phases, primarily defined by the maximum extension attained by the lacustrine systems in specific time intervals, are: (1) the Late Pleistocene Megalake, (2) the Reduced Lakes, (3) the Early–Mid Holocene Macrolake, (4) the Late Holocene Separated lakes. Data and evidence of such evolution are shortly reported and discussed herein.

17.3.1 Geomorphology

Evidence of ancient lake levels is widely represented by terraces and stranded paleoshorelines, recognized since long (Nillson 1940), though not homogeneously distributed in the area (Fig. 17.3). Relict stranded shorelines are mainly developed in the central-southern portion of the basin. Particularly, north of Lake Abijata, degraded shorelines, distributed between 1,600 and 1,650 m a.s.l., are separated by a relatively flat surface, standing at 1,590–1,600 m a.s.l. (Fig. 17.3), from fresh shorelines (Figs. 17.2 and 17.4c), grading toward the present lake shores.

Five main lacustrine terraces (terrace V to I, from higher to lower surfaces, after Laury and Albritton 1975; Fig. 17.3), which are smooth undulated surfaces separated by major escarpments, occur west and north of Ziway Lake. Terrace IV is separated from terrace V, resting above 1,670 m a.s.l. (Laury and Albritton 1975), by an escarpment stretching NE–SW (Fig. 17.4a–b). North of the Gademotta Ridge (Fig. 17.1), it coincides with the rift-parallel Abosa fault (Fig. 17.1; Laury and Albritton 1975), a scissor-like fault, the scarp of which ranges in height from more than 20 m to the north to 5–6 m to the south. It is worth mentioning the occurrence of the poorly preserved terrace VI (Laury and Albritton 1975) attesting to the highest level (above 1,700 m a.s.l.) reached by the lakes during the Late Quaternary (see discussion section). Terrace IV and the shoreline-free surface at 1,600 m a.s.l. north of Lake Abijata represent the relict of two significant Holocene lake stillstands (Grove and Goudie 1971; Street and Gasse 1978; Street 1979). Finally, bathymetric surveys of Lake Ziway (AERMAP 1969) detected a 2 km wide and 15 km long depression, crossing longitudinally the lake bottom (Fig. 17.3), interpreted as a paleoriver incision (Sagri et al. 2008).

17.3.2 Late Quaternary Stratigraphy

Integrated analysis of erosional and depositional landforms allowed to establish four unconformity-bounded stratigraphic units (ISSC 1994; synthems 1–4; Fig. 17.5), ascribed to the Upper Pleistocene–Holocene. A wide collection of radiocarbon dates (Haynes and Haas 1974; Geze 1975;

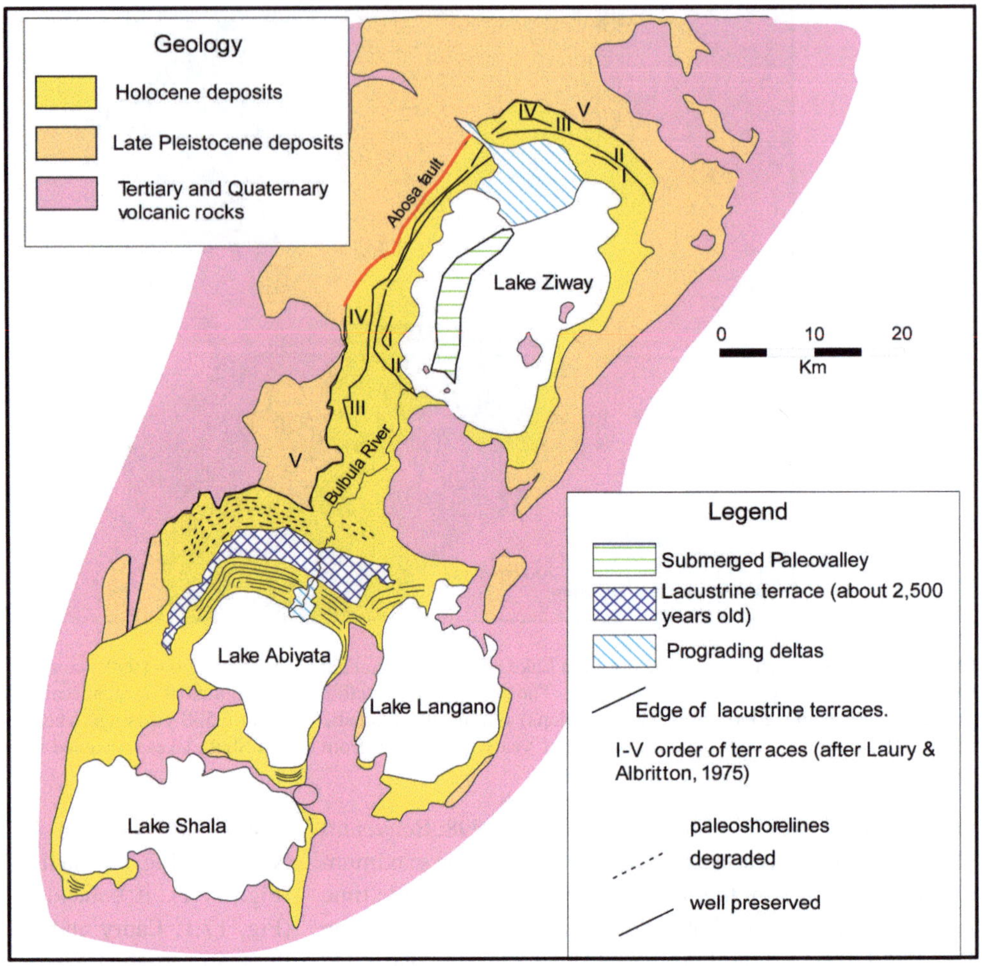

Fig. 17.3 Schematic geomorphology and geology of the Lakes Region: The map specifically shows the main evidence of the Late Quaternary lake-level fluctuations

Street 1979; Gillespie et al. 1983; Le Turdu et al. 1999; Benvenuti et al. 2002; Carnicelli et al. 2009) provided a fairly accurate chronologic calibration of these units, which record the major stages in the lake basin evolution (Fig. 17.5; Benvenuti et al. 2002). In general, these units rest unconformably over Pliocene–Early Pleistocene volcanic and volcaniclastic deposits (Abebe et al. 2005). Locally, as in the area north of Langano Lake and on the slopes of the Gademotta Ridge, the Late Pleistocene–Holocene lacustrine succession overlies reddish-brownish alluvial–colluvial conglomerates, sandstones, and mudstones pointing to a latest Middle Pleistocene age (Laury and Albritton 1975; Coltorti et al. 2002).

The main characteristics of these synthems are the following:

- Synthem 1 consists of colluvial, fluvio-deltaic and lacustrine gravels, sands, and muds, as well as lacustrine diatomites and volcaniclastic materials (Fig. 17.6a, b), deposited during the last glacial period, ca. 100,000–22,000 y BP (Fig. 17.5). This interval corresponds to the Megalake phase, when lake waters inundated wide areas north of Lake Ziway rising over 1,800 m a.s.l.
- Synthem 2 is dominated by alluvial–colluvial and volcaniclastic deposits accumulated in the basin during the last full glacial and late glacial, ca. 22,000–11,000 cal y BP (Fig. 17.7b). Despite the predominance of subaerial volcanic deposits (Fig. 17.6c, e), this unit is topped by 2–3 m thick cross-bedded sands and silts (Fig. 17.6d), referred to a lacustrine nearshore environment dominated by wave reworking of volcaniclastic deposits and standing at about 1,720–1,730 m a.s.l. (Laury and Albritton 1975; Benvenuti et al. 2002, 2005). Synthem 2 records the Reduced Lakes phase, when the lakes underwent a dramatic shrinking (Gasse and Street 1978; Benvenuti et al. 2002). The shoreface deposits on top of the synthem suggest that this phase ended with a short-lived but massive re-flooding of the former lacustrine basin.
- Synthem 3 consists of an articulated alternation of colluvial; fluvio-deltaic; and lacustrine gravels, sands, muds, lacustrine diatomites, and volcaniclastic materials (Fig. 17.6a, f) that were deposited during the Early to Middle Holocene, ca. 11,000–5,000 cal y BP (Benvenuti et al. 2002; Fig. 17.5). It records the Macrolake phase, during which the lake was smaller than recorded by Synthem 1 and characterized by

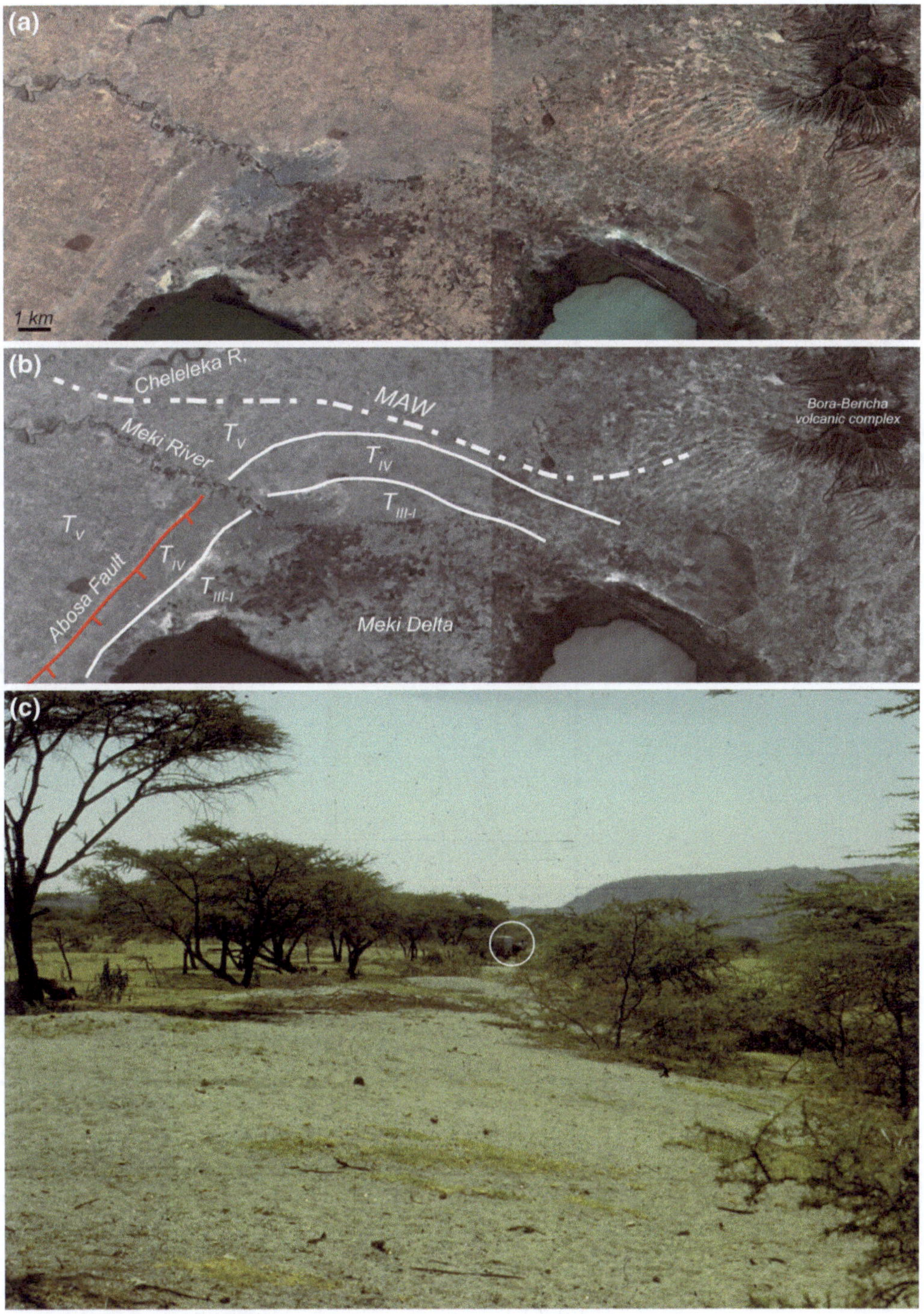

Fig. 17.4 Morphological features related to former lake levels. **a** Oblique aerial view of the northern coast of Lake Ziway (from Bing Maps). **b** Annotated picture in **a** showing some of the lacustrine terraces and other features described in the text. T_{V-I}: terraces after Laury and Albritton (1975); *MAW* Meki-Awash watershed. **c** Ground view looking to the west of a well-preserved paleoshoreline (younger than 2,500 year BP) on the southern coast of Lake Abijata. Car in the circle for scale

frequent fluctuations (Benvenuti et al. 2002, 2005). The maximum elevation of the lake surface during this phase was about 1,670 m a.s.l. (Street 1979; Laury and Albritton 1975; Benvenuti et al. 2002) (Fig. 17.3).

- Synthem 4 is made of colluvial, fluvial, deltaic, and lacustrine sediments (Fig. 17.6g) accumulated over the last 5,000 years. It records the period when the previous single large lake separated into Ziway Lake and a southern lake, itself progressively splitting into the present-day Abyiata, Langano, and Shala lakes (Benvenuti et al. 2002; Fig. 17.5).

Fig. 17.5 Simplified geological map of the Lakes Region (after Benvenuti et al. 2002)

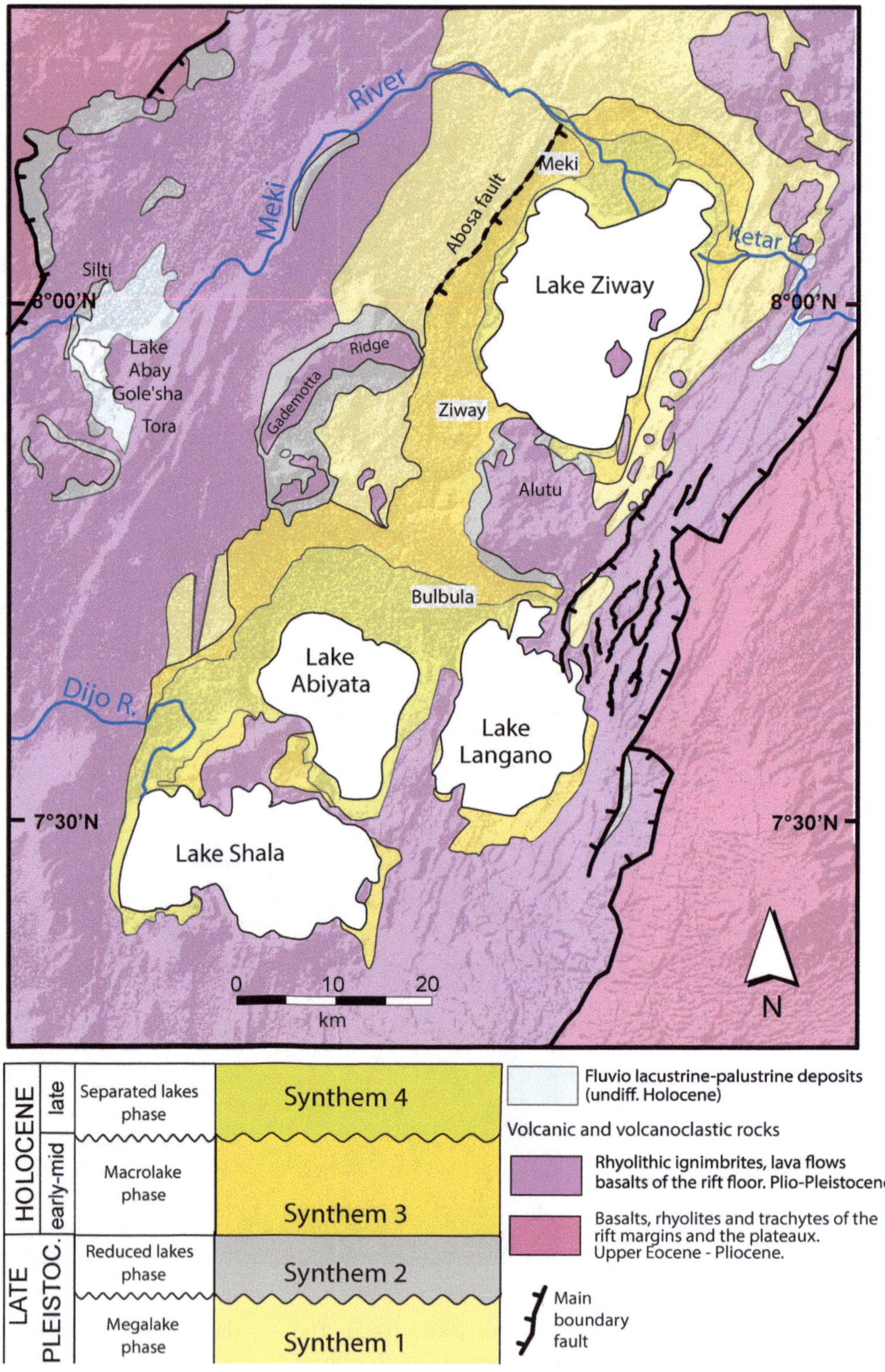

17.4 The Geomorphic and Stratigraphic Record as a Proxy for the Late Quaternary Hydro-climatic Variability

17.4.1 The Late Pleistocene–Holocene Ziway–Shala Lakes Fluctuations

Seven major highstands of the latest Pleistocene–Holocene Ziway–Shala lakes have been identified in previous studies (Street 1979; Gillespie et al. 1983; Fig. 17.7a). Ziway–Shala (Zw-Sh) I was referred at before 50,000 y BP, with a maximum age likely close to 100,000 y BP (Laury and Albritton 1975; Street 1979; Bigazzi et al. 1993; Le Turdu et al. 1999). Zw-Sh II and Zw-Sh III record highstands predating the Last Glacial Maximum (LGM; Fig. 17.7b), whereas stages Zw-Sh IV–VII mark the major Holocene highstands. The Early–Middle Holocene Zw-Sh IV–VI highstands are considered to have reached an elevation of the water surface close to the inferred basin threshold at 1,670 m a.s.l., with possible overflow to the north (Street 1979; Gillespie et al. 1983). The late Holocene Zw-Sh VII stands at about 1,600 m a.s.l., as indicated by the morphological lines; after 5,000 y BP, the lakes never reached the Early–Middle Holocene elevation and progressively reduced their surfaces (Separated Lakes phase). Successive studies (Alessio et al. 1996; Benvenuti et al. 2002; Carnicelli et al. 2009) confirmed the classic Street's pattern of Holocene lake fluctuations (Fig. 17.8a).

17.4.2 Soil Development Between the Latest Pleistocene and the Middle Holocene

The hydro-climatic transitions left clear traces on the geomorphic systems above the lakes and hence less influenced by the lacustrine base level change. An outstanding evidence of that is found in a composite paleosol (Morrison 1977), developed over different volcanic and sedimentary deposits, either included in synthem 1 and 2 or older (Carnicelli et al. 2009), whereas it was never observed over synthems 3–4 sediments. This soil (Fig. 17.9a) covers a wide range of landforms and landscapes, always above 1,670 m a.s.l., i.e., the maximum lake level attained during the Macrolake phase. It is made of up to three soils; the lower soil profile is characterized by the presence of a thick Petrocalcic horizon (Fig. 17.9b) overlain by clay accumulation horizons. One or two pedogenized tephra resting above these horizons (Fig. 17.9a) typically represent the upper soil profile. In places, the paleosol is buried (Fig. 17.9c, d) by younger sediments; radiocarbon dating (Carnicelli et al. 2009) allowed the formal definition of the T'ora Geosol, a pedostratigraphic unit (Morrison 1977; North American Commission On Stratigraphic Nomenclature 1983) of regional significance. The Geosol components, from oldest to youngest referred to as T'ora I to T'ora III, derive from a progressive conversion of volcanic glass into smectite clay along an evolutionary pathway from soils dominated by volcanic glass (Vitrands), through a Luvisol stage to nearly Vertisol characters. A second relevant soil (Wer'ja paleosol; Carnicelli et al. 2009) is found on Synthem 3 slope deposits above 1,670 m a.s.l. (Fig. 17.9e). This soil, locally buried by a Vitrand on more recent, reworked tephra, is very similar to T'ora II with morphological differences mostly due to different drainage conditions (Carnicelli et al. 2009). The genesis of these paleosols indicates a marked seasonality, with alternating, pronounced dry and humid seasons. Such a climate regime was likely generally moister than present. Radiocarbon dating (Carnicelli et al. 2009; Fig. 17.8a) and stratigraphic relations constrained the development of T'ora I in a maximum time span between 19,000 and 8,300 cal y BP. T'ora II and Wer'ja soils turned out to be essentially coeval, developing between 7,000 and 5,500 and between 7,300 and 6,100 cal y BP, respectively. Finally, T'ora III existed as a surface soil since 5,400 cal y BP.

17.4.3 Late Holocene Cut and Fill Cycles in Ephemeral Fluvial Systems Disconnected from Lake Levels

The slopes between 1,720 and 1,670 m a.s.l., such as on the Gademotta Ridge southwest of Lake Ziway (Figs. 17.1 and 17.5), consist of pre-Late Quaternary volcanic rocks and are mantled by synthem 2–4 deposits, also resting unconformably on the lacustrine sediments of synthem 1 (Benvenuti et al. 2002, 2005). Slope and alluvial deposits within synthems 2–4 include a succession of filled box-valleys (Fig. 17.9e) characterized by morphologic and depositional features similar to those of modern ephemeral streams/gullies, disconnected from the lakes (Benvenuti et al. 2005). Independently from their stratigraphic position, the valley fills are made up of stream flow, sheet flow, and mass flow deposits (Fig. 17.9e), stacked into a recurring vertical pattern. Stream flow sands and gravels generally occur at the base of the valley fill, sharply overlain by sheet flow silty sands, including poorly-developed paleosols (Carnicelli et al. 2009; Fig. 17.9e), which form the bulk of valley fills. Mass flow deposits, related to different bank failure mechanisms, are interbedded with alluvial deposits, showing recurrent associations, respectively, between stream flow and debris fall and sheet flow and debris flow deposits (Benvenuti et al. 2005).

Fig. 17.6 The Upper Quaternary deposits of the Lakes Region. **a** Unconformable contact between lacustrine diatomite and volcanics of synthems 1 and 3 visible in the Bulbula River valley, person for scale. **b** Detail of wave-rippled deltaic sandstone of synthem 1. **c** Pumice fall deposits of synthem 2 in a quarry north of Lake Ziway. **d** The uppermost deposits of synthem 2 consist of cross-laminated pebbly sandstone ascribed to delta shoreface environments pointing to a short-lived very high lake level, a quarry north of Lake Ziway. **e** Pumice fall deposits of synthem 2 mantling the slopes of the Bora–Bericha volcanic complex (Fig. 17.1). **f** Typical shell bed in the lacustrine deposits of synthem 3 on the east coast of Lake Shala. **g** Detail of synthem 4 on the east coast of Lake Shala: Foreset beds are sharply overlain by the bottomset of a small backstepping delta during a rise of the Separated Lakes

17.5 Volcano-tectonic Features Related to the Late Quaternary Development of the Lakes

17.5.1 Tectonic Structures Controlling the Macrolake Geometry

In addition to the large evidence of erosional and depositional landforms directly related to lacustrine, alluvial, and slope processes, some specific morphologic features associated with tectonic structures are considered for their implication on the evolution of the lakes. In particular, landforms determining the western and northern limits of the Holocene lakes are described. A significant portion of the escarpment bounding the Holocene lacustrine deposits and landforms (i.e., Terraces V–IV escarpment; Fig. 17.4a–b), west of Lake Ziway, is associated to the Abosa fault (West Ziway Fault in Le Turdu et al. (1999); Figs. 17.1, 17.3 and 17.5). The northern boundary of the Holocene lakes is represented by a WNW–ESE-trending rift floor undulation (Fig. 17.4a–b). This and other similar undulations (Fig. 17.1)

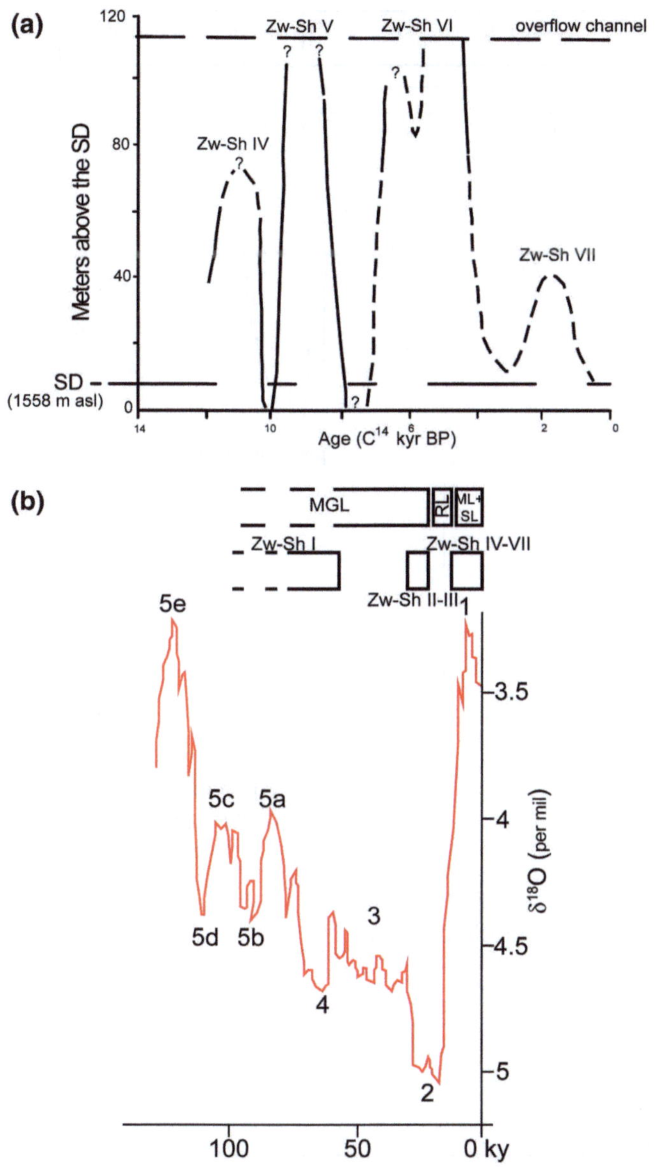

Fig. 17.7 Synopsis of the chronology of the Ziway–Shala lake fluctuations. **a** Classic Street's diagram of Ziway–Shala lacustrine phases: All the radiocarbon dated geomorphic and stratigraphic lines of evidence of lake level are plotted against the elevation. *SD* Shala Datum, Lake Shala elevation in the early 1970 (after Gillespie et al. 1983). **b** The Zw-Sh I–VII lacustrine phases are compared with the phases established in Benvenuti et al. (2002) and plotted on the Late Quaternary MIS chronology (after Porter 1989)

reflect buried tectonic structures, transversal to the main rift fault systems, which outline an orthogonal system of transverse horsts and high-amplitude folds consistent with the extensional model proposed for MER evolution by a few authors (e.g., Rosendhal et al. 1986; Boccaletti et al. 1992; Bonini et al. 1997; Corti 2009).

17.5.2 Explosive Silicic Volcanism During the Reduced Lakes Phase

Synthem 2 consists mostly of volcaniclastic fall and flow deposits that cover a large portion of the Lakes Region (Fig. 17.6c–e). Their reworking by surface processes and the occurrence of weakly developed paleosols (Street 1979; Benvenuti et al. 2002, 2005) suggest subaerial deposition during the Reduced Lakes phase. Marked grain size and thickness gradients hint to sourcing from the main silicic volcanoes and calderas, such as the Bora–Bericcio complex and the Alutu volcano (Fig. 17.1), which produced large volumes of rhyolitic and obsidian lava flows, unwelded pumice flows, pumice falls, and ashes starting from Late Middle Pleistocene (about 0.25 Ma, Di Paola 1972; Mohr et al. 1980; Woldegabriel et al. 1990; Le Turdu et al. 1999; Abebe et al. 2007). During the latest Pleistocene, huge volcanic eruptions delivered large amounts of sediment that mantled the desiccated lakes bottom and congested the drainage network.

17.6 Evidence of the Evolution of the Lakes: A Discussion

17.6.1 Geomorphological Evidence

The chronologic calibration of the lacustrine deposits associated with the various surfaces described above coarsely constrains the development of terraces V and VI to the Late Pleistocene Zw-Sh I–III, corresponding to the Megalake phase. Though not dated, the flooded palaeovalley on Lake Ziway bottom hints to fluvial incision of an almost completely desiccated lake, an event that may be reasonably related to the latest Pleistocene Reduced Lakes phase. A more accurate date calibration of the lacustrine surfaces is provided for the Holocene Macrolake and Separated lakes phase (Fig. 17.7a). The maximum Holocene lake levels, marked by the 1,670 m a.s.l. escarpment that roughly corresponds to the lake overflow threshold (Laury and Albritton 1975; Street 1979), were reached between 9,600 and 9,400 ^{14}C y BP (Zw-Sh IV) and about 5,400–5,100 ^{14}C y BP (Zw-Sh V) (Haynes and Haas 1974; Laury and Albritton 1975). The development of terrace IV occurred during these lake highstands (Fig. 17.10a). The differently preserved, stranded palaeoshorelines, visible north of Abijata Lake, record episodic lake-level stillstands, modulated in a regressive trend of lowering of a southern palaeolake (Fig. 17.10c). Since the escarpment separating terrace IV from terrace III tends to become convex south of Ziway Lake and the minimum age

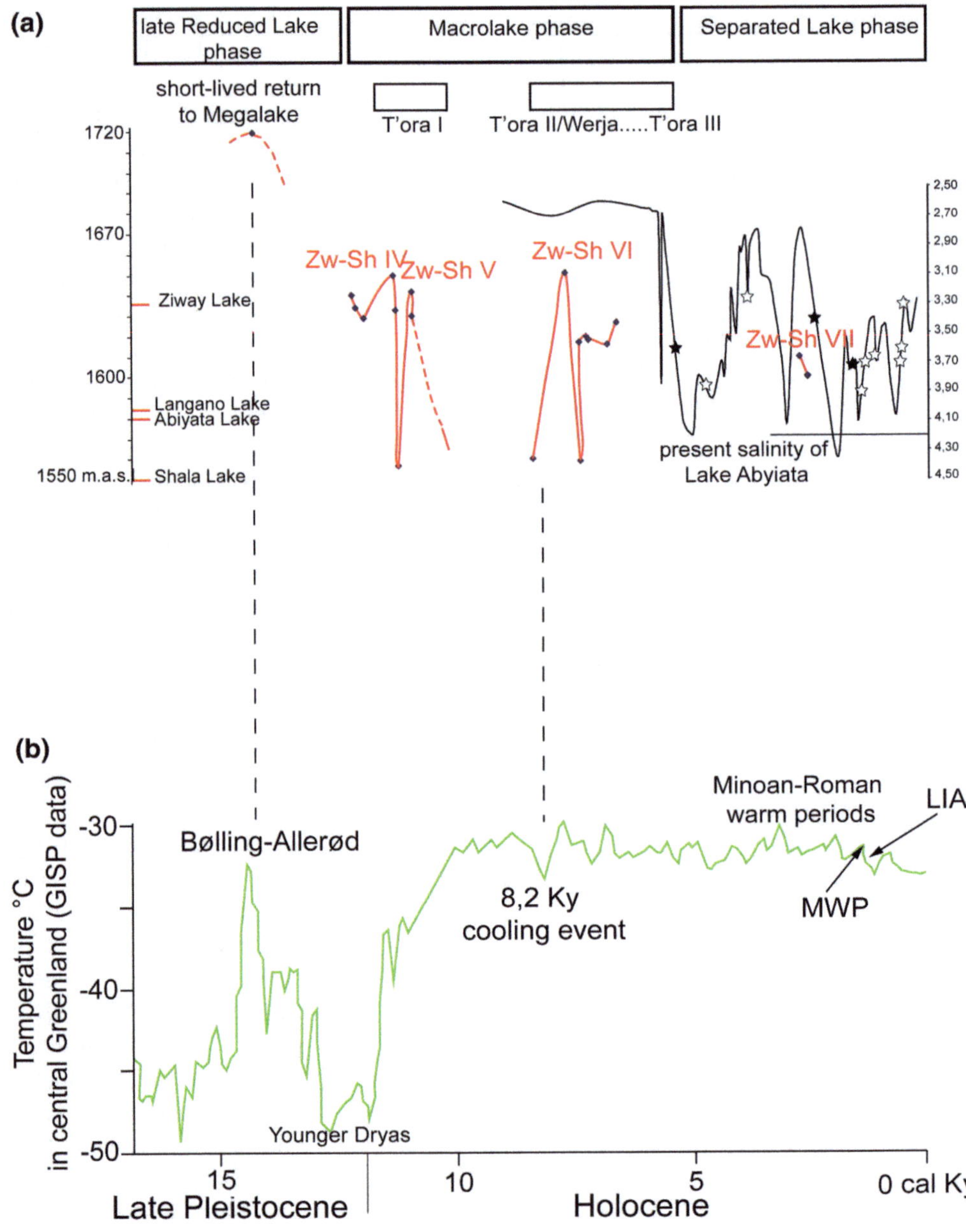

Fig. 17.8 **a** The calibrated chronology of the Lake Shala fluctuations (*right axis, red curve*) and the salinity variation recorded by diatom assemblages of Lake Abijata sediments (*left axis, black curve*, after Chalie and Gasse 2002) as proxies for major hydrologic/climatic events during the latest Pleistocene–Holocene (after Carnicelli et al. 2009). White rectangles indicate the T'ora and Werja soil development during the Megalake phase. *Black* and *white* stars on the salinity curve of Lake Abijata refer to calibrated ages, respectively, of streamflow and sheetflow facies in the box-valley fills formed during the Separated Lakes phase on the lower slopes of the Gademotta Ridge (see Carnicelli et al. 2009 for details). **b** Curve of temperature of Greenland derived from GISP data (after Alley 2000) as a reference for the Lakes Region events. *MWP* Medieval Warm Period; *LIA* Little Ice Age

of terrace III ranges from 5,600 to 5,300 ^{14}C y BP (Haynes and Haas 1974; Laury and Albritton 1975), a palaeo-Ziway lake may have started to separate from a larger southern lake around the Middle Holocene. The palaeoshorelines facing Abijata Lake therefore resulted from progressive shrinking of the southern lake during the Late Holocene. The shoreline-free surface at 1,600 m a.s.l. north of Lake Abijata is associated with the coastline of a southern lake, encompassing the modern Langano, Abijata, and Shala lakes, that still existed around 2,000 ^{14}C y BP (Gillespie et al. 1983; Alessio et al. 1996) during the later Separated Lakes phase (Fig. 17.10c).

17.6.2 Soils and paleosols

The chronologic calibration of paleosols allowed to distinguish distinct stages of soil formation in the stratigraphic and palaeohydrologic framework. T'ora I paleosol apparently encompasses the Reduced Lakes and Early Macrolake phases. Since its development requires a relatively moist seasonal climate, it is unlike that T'ora I started forming during the Reduced Lakes phase, when arid conditions dominated in the region. Also, the short-lived moist episode leading to the huge lake stillstand at about 1,720 m a.s.l. during the later Reduced

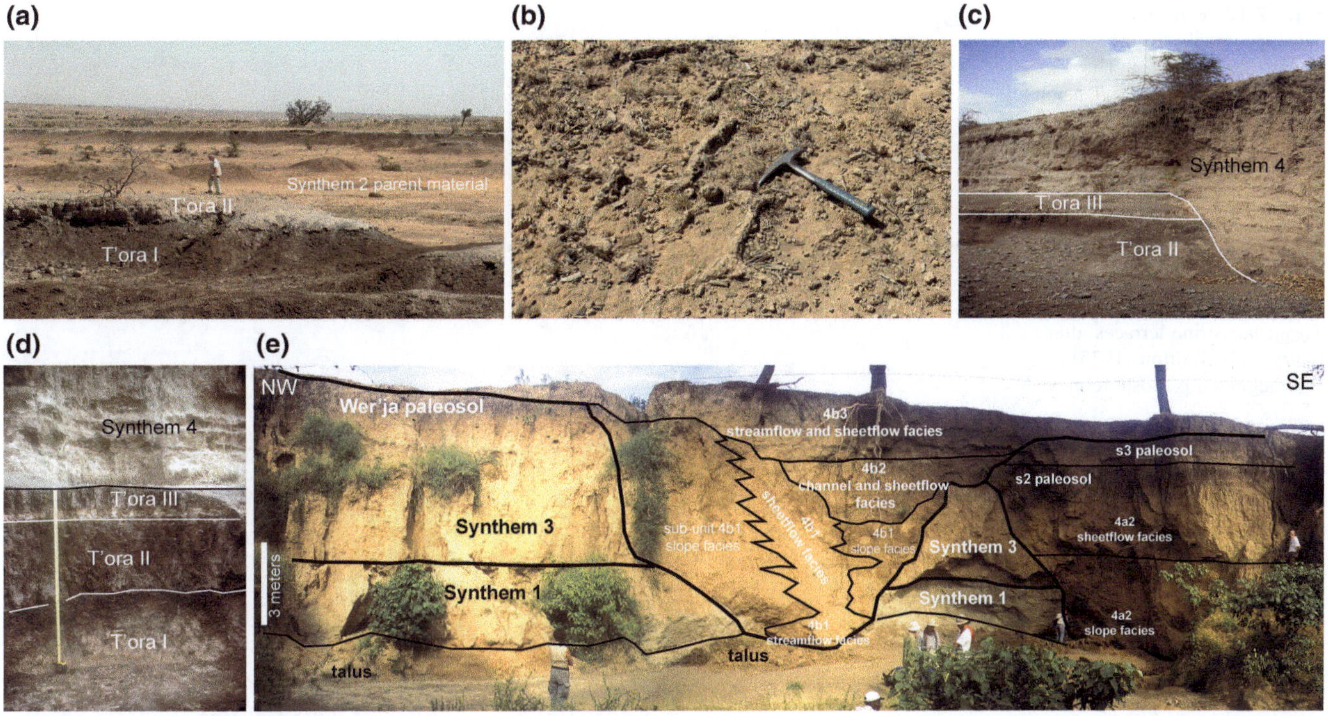

Fig. 17.9 Paleosols and non-lacustrine deposits of the Lakes Region. **a** Panoramic view on the badlands developing on the T'ora Geosol north of Lake Ziway: in the foreground the T'ora I–II. Person for scale. **b** Deeply eroded T'ora Geosol exposing the petrocalcic horizon at the base of T'ora I. **c** The T'ora Geosol unconformably overlain by synthem 4 alluvial deposits at the NE footslopes of the Gademotta Ridge (after Carnicelli et al. 2009). **d** The almost complete T'ora Geosol buried by synthem 4 in the same area of **c**; meter tape for scale. **e** A section in sediments of synthems 1, 3, and 4 exposed on the ENE slopes of the Gademotta Ridge. Two box-valley fills characterize synthem 4, subdivided in different sub-units recording successive stages of valley fill. The valley fills in each sub-unit consist of streamflow facies overlain by sheetflow facies, both flanked by slope facies, on the whole outlining fining-upward trends. Weakly developed paleosols (S2–S3) mark periods of stability of the valley bottom suitable for soil development. Persons for scale (after Benvenuti et al. 2005; Carnicelli et al.2009)

Lakes phase preceded the T'ora Geosol development, as the surfaces where this soil is found today were then flooded. Most likely, the interval of T'ora I genesis was then the earliest Holocene moist stages which determined the Zw-Sh IV–V highstands at the onset of the Macrolake phase. T'ora II and Wer'ja paleosols fully developed during the Macrolake phase, synchronously with the Zw-Sh VI highstand at around 1,670 meters a.s.l. (Gasse and Street 1978; Street 1979). Finally, T'ora III started developing at the end of the Macrolake phase, when a seasonal climate, moister than present, may have still existed in the region.

17.6.3 Late Holocene Cut and Fill Cycles

The cyclic stacking pattern of the alluvial and slope deposits mantling the Gademotta Ridge suggests that the early valley fill phases were dominated by coarse-grained material, transported as bedload by competent and likely permanent water flows. Undermining of the nearly vertical banks commonly determined toppling and debris fall accumulation on valley bottoms. In later stages, as water discharge and transport capacity decreased, the bank failure deposits were redistributed by sheet and debris flows, leading to aggradation, interbedded with paleosols, and eventual overfilling of the valleys (Benvenuti et al. 2005). These box-valley fills are similar to those of discontinuous ephemeral streams of the sub-arid SW USA (Graf 1988; Bull 1997). The Late Quaternary dynamics of these ephemeral streams/gullies has been accounted for by an autocyclic mechanism (Schumm 1973) of valley backfilling and aggradation, resulting in a typical fining-upward sedimentary succession (Packard 1974; Graf 1988; Bull 1997). An alternative interpretation of the incision/aggradation disequilibrium of this kind of streams invokes external controls, such as climatically-driven hydrologic changes (Bull 1997; Waters and Haynes 2001). This latter hypothesis was positively tested by detailed radiocarbon dating of valley fill sediments and buried soils included in synthem 4 (Carnicelli et al. 2009; Fig. 17.8a). A proxy palaeohydrological record for the Separated Lakes phase is provided by the Late Holocene variation of salinity of Lake Abijata, modeled from diatom assemblages found in the lacustrine sediment core (Chalié and Gasse 2002). High–low salinity values approximate

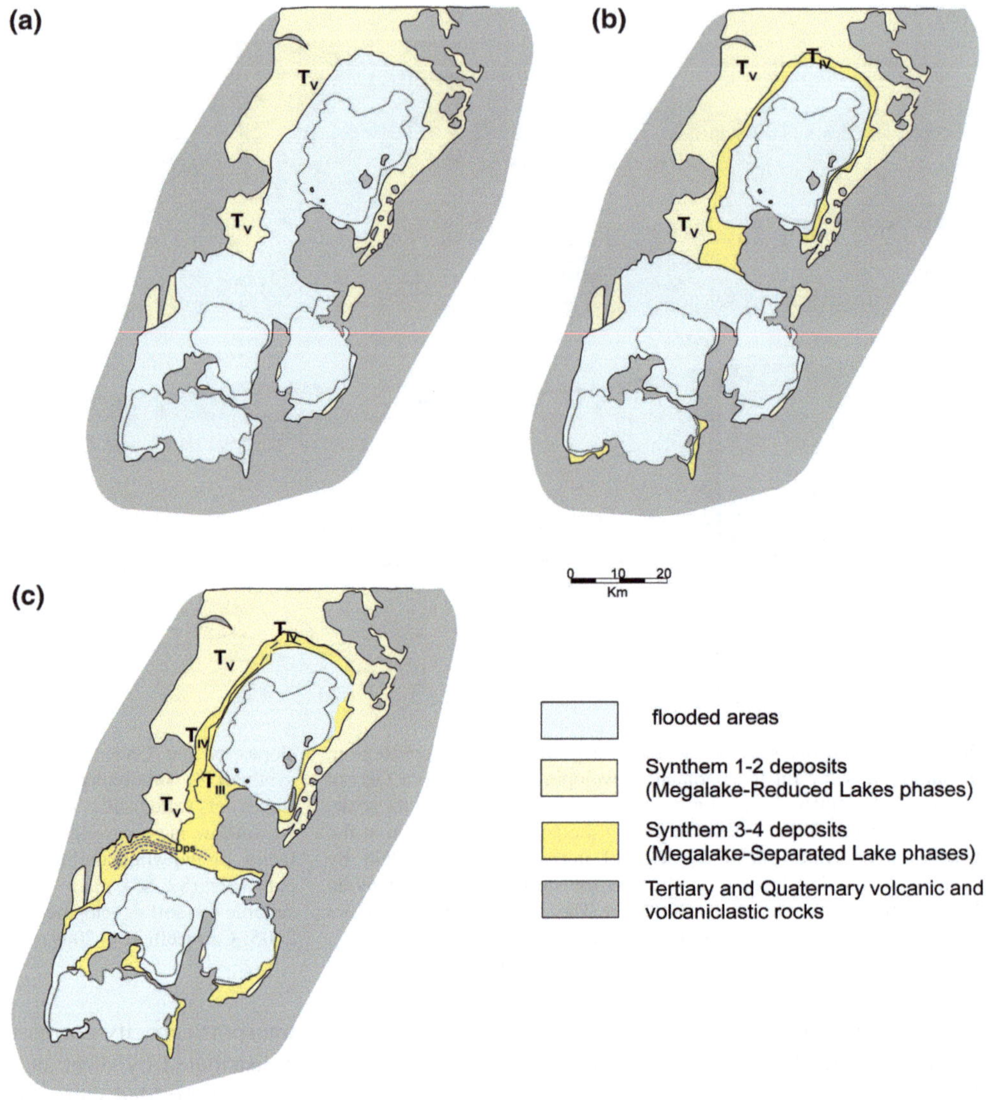

Fig. 17.10 Holocene palaeogeography of the Ziway–Shala basin. **a** The maximum Macrolake extension reached during Zw-Sh IV–VI highstands. **b** Early separation of Lake Ziway from the southern lake at the beginning of the Separated Lakes phase. **c** The lakes during the 2,500 years BP highstand when a southern lake, though reduced, still existed. T_{V-III}: lacustrine terraces after Laury and Albritton (1975); *Dps* degraded paleoshorelines

low–high lake levels in response to decreasing or increasing rainfall. The chronology of synthem 4 valley fills shows that valley aggradation coincides with intervals of high lake salinity (Fig. 17.8a). By contrast, stages of box-valley incision, though not directly dated, are constrained to occur during phases of low lake salinity and higher river water supply and, hence, of larger precipitation (Carnicelli et al. 2009; Fig. 17.8a). Similarly to the modern gullies, their Late Holocene counterparts graded at a base level disconnected from the lakes, which explains the out of phase relation between incision/aggradation and high/low lake status, compared with streams which graded at the lake levels (Fig. 17.11; Benvenuti et al. 2006). During moister periods, the rivers grading at rising lake levels may have experienced aggradation (Fig. 17.12a), while those disconnected from the lakes may have seen valley incision and dominant flushing of transported sediments under high discharges (Fig. 17.12b). During dryer periods, the rivers connected to shrinking lakes may have undergone downcutting, while the others were experiencing valley backfill and aggradation related to reduced water discharge.

17.6.4 Climatic Forcing of the Late Quaternary Lake-level Fluctuations

The Ziway–Shala lake oscillations have been related to the Late Quaternary weakening and strengthening of the monsoonal circulation (Kutzbach and Street Perrott 1985; Coetzee and van Zinderen Bakker 1989; Street Perrott and Perrott 1990; Gasse and Van Campo 1994) that caused alternating drier (off-mode of monsoons) and moister (on-mode of monsoons) conditions over the Lakes Region. In this perspective, the Milankovitch orbital forcing was considered as the primary control on the long-term hydro-climatic dynamics of the Lakes Region and other regions of northern

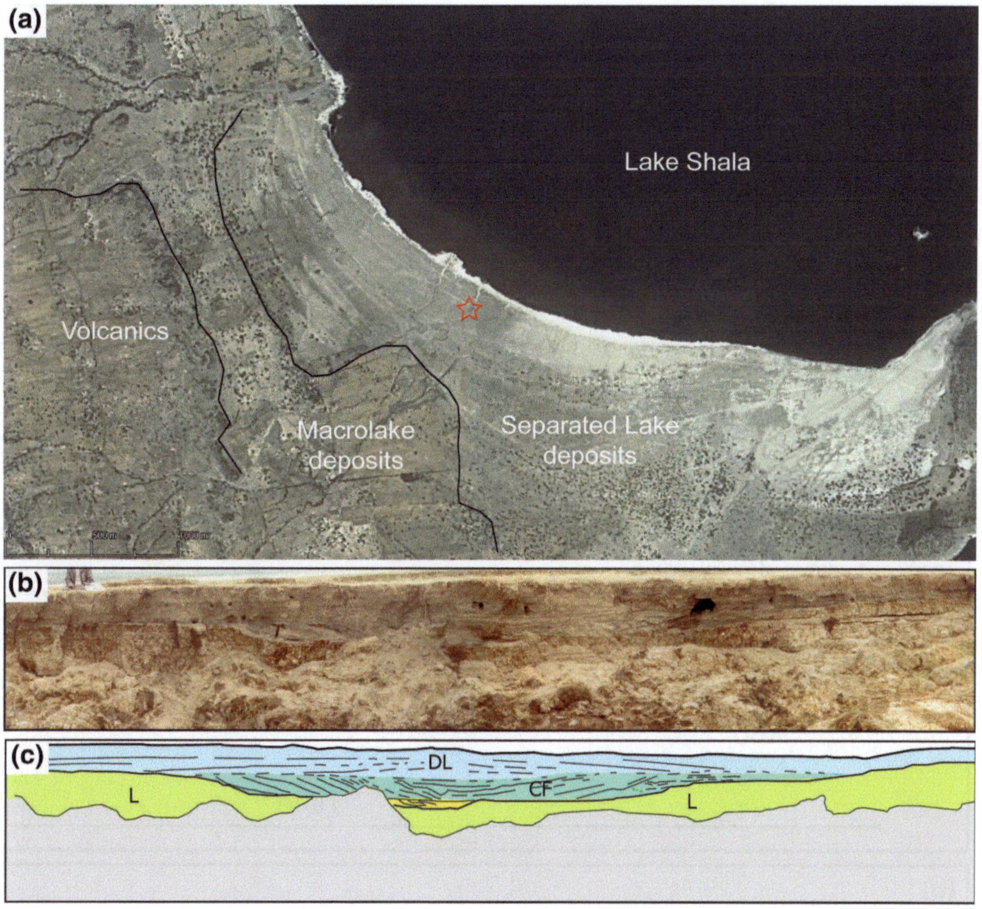

Fig. 17.11 Example of fluvio-deltaic response to high-frequency fluctuations of Lake Shala. **a** The southern shore of Lake Shala with the stranded shorelines formed during the Separated Lakes Phase. **b** A section on lacustrine and fluvio-deltaic deposits exposed on a meander bend of a short stream (red star in A for location), person for scale, the lake is in the background. **c** Interpreted section in B): *L* lacustrine mudstone; *CF* fluvial channel fill sandstone: early (*yellow*), late (*blue*); *DF* delta front sandstone; *gray*: talus scree (see text)

and eastern Africa (Fontes and Gasse 1989, 1991; Hillaire-Marcel et al. 1986; Gasse 1989; Taieb et al. 1991). The Late Pleistocene Zw-Sh I–III stages (Street 1979; Megalake phase) occurred during warm phases (MIS5c–a, MIS3; Fig. 17.7b) punctuating the last glacial period. The missing geomorphic, stratigraphic, and chronologic evidence of lake development between Zw-Sh III and Zw-Sh IV (Reduced Lakes phase) points to a dramatic lacustrine shrinkage centered around 20,000–18,000 y BP and coinciding with the LGM (MIS2; Fig. 17.6b), when aridity dominated in the northern tropics (Coetzee and van Zinderen Bakker 1989). The stratigraphic evidence of lake level standing at about 1,720 m a.s.l. at the end of the Reduced Lakes phase (Benvenuti et al. 2002; Sagri et al. 2008) attests to a huge reflooding of the lakes basin. Despite the lack of direct dating, this high lake level may be reasonably constrained to the post-LGM latest Pleistocene (Benvenuti et al. 2002) when warm–cold fluctuations globally characterized the transition to the Holocene. A warm stage between 14,700 and 12,700 cal y BP (Bølling-Allerød event; Fig. 17.8b) had significant hydrological consequences in the African tropics (Weijers et al. 2007), such as the massive refilling of an almost desiccated Lake Magadi in Kenya (Taieb et al. 1991). Following this short-lived lacustrine episode, the Lakes Region was again affected by arid conditions and lake reduction as indicated by the topmost slope and alluvial deposits of synthem 2 (Benvenuti et al. 2005; Sagri et al. 2008). This stage, predating the Macrolake phase, is considered to record the Younger Dryas cold event (Fig. 17.8b) that determined again arid condition all over the African tropics (Roberts et al. 1993). The return to lacustrine conditions during the Early–Middle Holocene Zw-Sh IV–VI (Macrolake phase; Figs. 17.7a and 17.8a) corresponds to the post-glacial climatic *optimum* (MIS 1). Besides the different interpretation of the forcing mechanisms, it is now widely acknowledged that most of the latest Pleistocene–Holocene hydro-climatic events recorded in the Lakes Region have a regional to global significance (Mayewski et al. 2004; Weijers et al. 2007). Marked lake lowering during the Macrolake and Separated Lakes phase (Figs. 17.7a and 17.8a) occurred around global cooling events, such as the 8,200 cal y BP event (Fig. 17.8b), resulting in arid conditions all over the tropics (Mayewski et al. 2004). The relationship between high lake levels and global warm–moist periods seems not to be satisfied in the last 2,000 years as palaeolimnological data from the Sahel and East Africa (Maley 1976; Verschuren

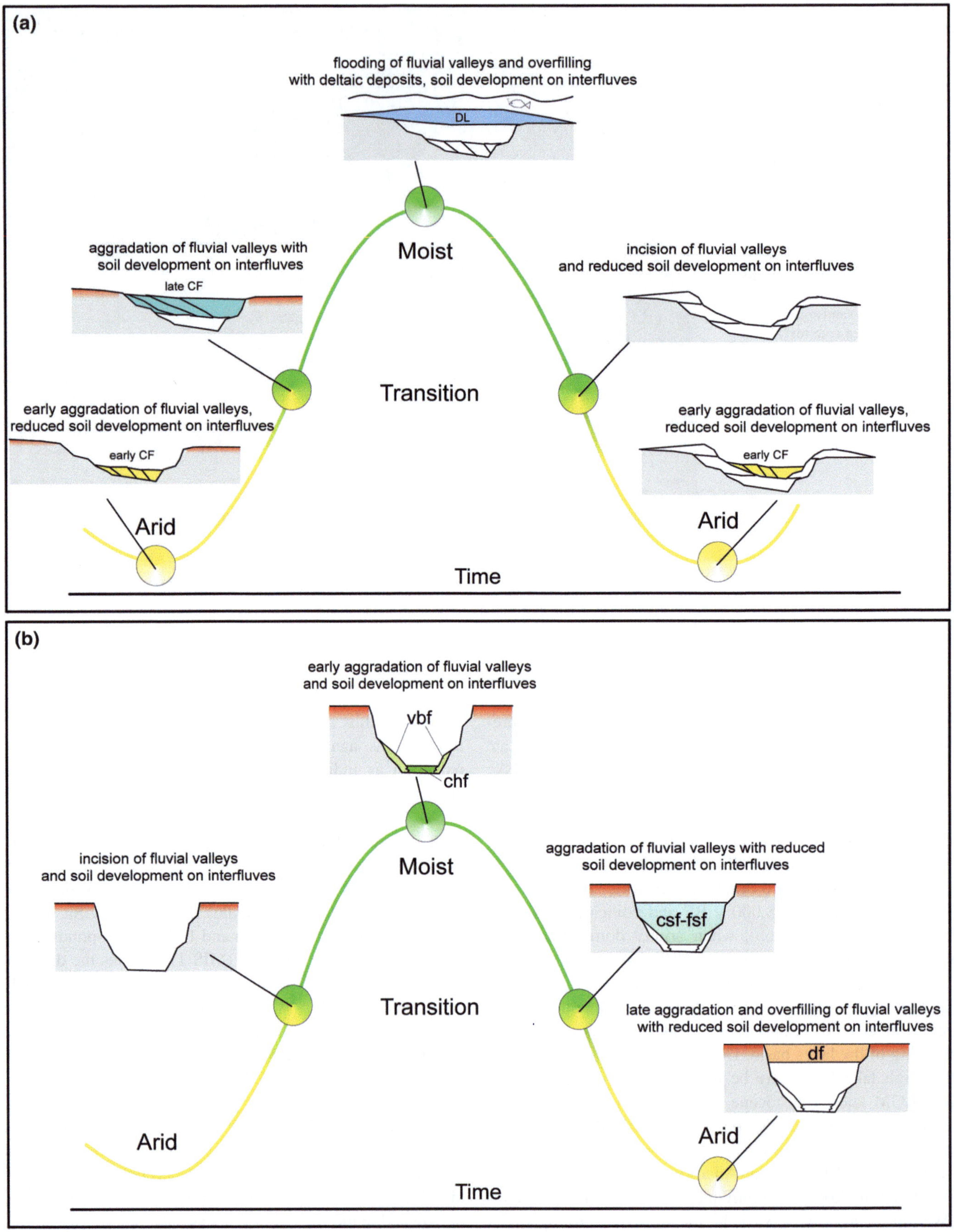

Fig. 17.12 Models of fluvial incision/aggradation in response to moisture fluctuation regulating river discharge and lake levels (after Benvenuti et al. 2006). **a** River connected to lake level (inspired by Fig. 17.11b–c). **b** River disconnected from lake level (inspired by Late Holocene discontinuous ephemeral streams draining the lower slopes of the Gademotta Ridge; after Benvenuti et al. 2005). Codes: *CF* channel fill; *DF* delta front; *vbf* valley bank mass wasting; *csf-fsf* coarse–fine-grained sheetflow deposits; *df* debris flow deposits

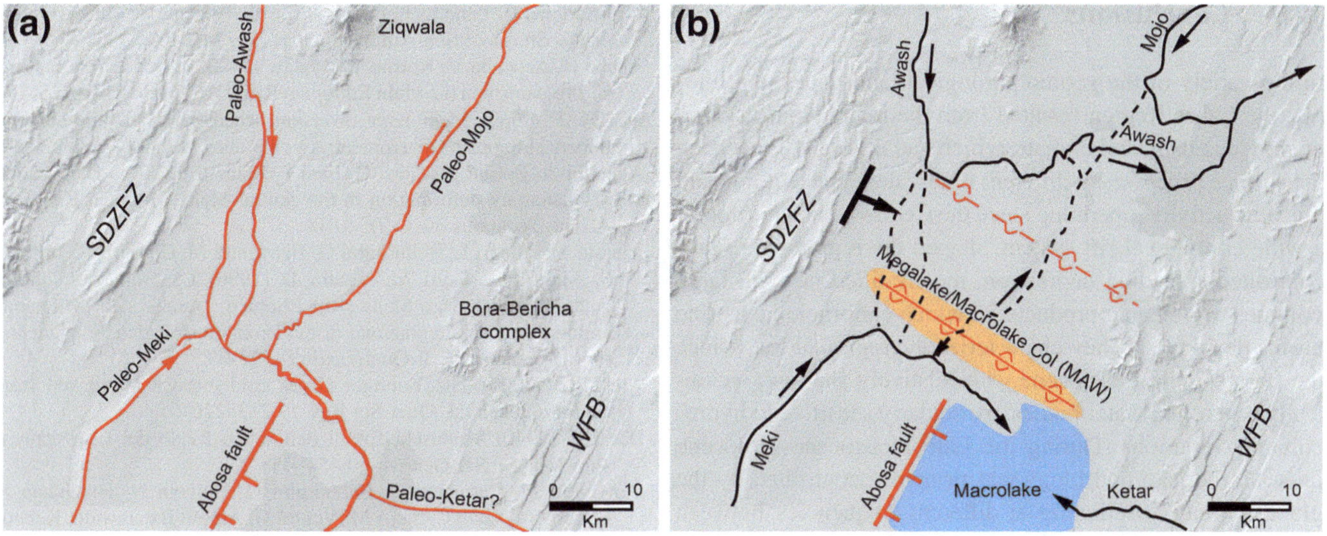

Fig. 17.13 Palaeohydrography north of Lake Ziway. **a** The regional drainage during the Late Reduced Lakes phase (latest Pleistocene) toward the desiccated lake basin, a condition that dominated also during the Megalake phase and ensured high water supply to the lakes. The Abosa fault may have been active during these stages making the central portion of the basin actively subsiding. *WFB* Wonji Fault Belt; *SDZFZ* Silti-Debre Zeyt Fault Zone. Present Lake Ziway and Koka Reservoir for reference. **b** Around the Pleistocene–Holocene transition, the possibly concurrent activity of the Abosa fault and of the transverse high just north the Lake Ziway (presently the Meki-Awash watershed, *MAW*) determined a smaller Macrolake compared to the Megalake maximum extension. These structures, joined to further transverse lines northward (*dotted red line*) and SE-ward tilting possibly due to the activity of the WFB (*thick black arrow*), determined the deviation of the Awash and Mojo rivers toward ENE, subtracting a significant volume of water from the hydrological budget of the Early Holocene Lakes Region (modified from Sagri et al. 2008)

et al. 2000; Chalié and Gasse 2002; Legesse et al. 2002) equate colder to moister periods, such as the Little Ice Age, and warmer to drier periods such as the Medieval Warm Period (Fig. 17.8b).

17.6.5 Tectonic Control on the Basin Geometry and Regional Hydrography

Though the global-scale Late Quaternary climate changes provide a reasonably convincing explanation for the palaeohydrologic evolution of the Ziway–Shala Basin, such a primary control may be, at least partially, questioned when the Megalake and Macrolake extensions are compared. The global increase in temperature and moisture during the LGM–Holocene transition was greater in magnitude than that experienced during any pre-LGM stadial–interstadial transitions (Fig. 17.7b; Dansgaard et al. 1993). In the tropics, such conditions should have resulted in water budgets larger at the onset of Holocene than at the start of any previous interstadial (Thomas and Thorp 1995). In spite of that, the maximum extension of the Holocene Macrolake, regulated by the overflow threshold at 1,670 m a.s.l., was markedly smaller than that of the Late Pleistocene Megalake. Such an apparent paradox may be accounted for by considering the activity of an orthogonal system of tectonic structures during the latest Pleistocene, a period characterized also by intense explosive volcanism. During the Megalake and Reduced Lakes phase, the Abosa fault (Fig. 17.13) may have determined subsidence on its hanging wall, creating future accommodation space for the Macrolake. The geometry of the Early Holocene lake basin was also conditioned by the concurrent activity of the northern transverse high. Nevertheless, the concurrent action of rift-parallel and transverse structures not only created a depocentre for the Early Holocene lakes but also forced cutting off of such large fluvial systems as the Awash and Mojo rivers, which fed the Late Pleistocene lakes from the north (Fig. 17.13a; Sagri et al. 2008). Yet, during the Reduced Lakes phase, the fault-controlled northeastward tilting of the rift floor progressively deviated the Awash River which also captured the Mojo River (Fig. 17.13b; Sagri et al. 2008). At the onset of the Holocene climatic optimum, the lakes were fed exclusively by the Meki and Ketar rivers (Fig. 17.13b) with a significant reduction of water supply due to the latest Pleistocene tectonically controlled reorganization of the river network. The intense fault activity of the Reduced Lakes arid phase was also coupled with intense volcanic activity, as observed in other rifted basins of eastern Africa (Abebe et al. 2007).

17.7 Conclusions

In the variety of the geomorphological landscapes of Ethiopia, the Lakes Region resulted from the interaction of tectonism, volcanism, and climate, which had different impacts on the surface processes at different timescales. The tectonic and volcanic activity, operating since the Cenozoic in a portion of a wide continental rift system, shaped the region as a fault-controlled subsiding depression, mostly filled up with large volumes of volcanic products, in which endorheic lacustrine basins developed. They characterize the region at least since the Late Middle Pleistocene with relatively shallow, tectonically controlled water bodies, particularly sensitive to hydro-climatic variations. During the Late Pleistocene–Holocene interval, the lake hydrology was primarily controlled by the global climate, oscillating at different frequencies between warmer and colder conditions which, in the tropical belt, resulted in alternating moist and arid phases. For this reason, the Lakes Region represents a classic site for exploring the geomorphic and stratigraphic record of these palaeohydrologic and palaeoclimatic fluctuations. The emphasis on a predominant climatic control over the Late Quaternary development could minimize the role of the active regional volcano-tectonic setting that, on the contrary, was particularly significant during the Pleistocene–Holocene transition. In the 18,000–11,000 y BP interval, concurrent fault activity and explosive volcanism modified the basin geometry, supplied high amounts of volcaniclastic sediments, and forced reorganization of hydrography. This critical period, therefore, was characterized not only by arid conditions established in the LGM, which dramatically reduced the lake extension, but also by tectono-magmatic activity. The typical processes of an evolving continental rift were evidently able to leave a clear imprint over the geomorphic evolution, keeping pace with high-frequency climate transitions.

Acknowledgments The authors are indebted with Prof. Mario Sagri and Giovanni Ferrari for their effort and impulse in the field researches in Ethiopia and the constructive contribution in the discussion of the data, for understanding the geomorphological evolution of the Lakes Region during the Late Quaternary. The researches summarized in this paper involved between 1994 and 2005 also a large number of Italian undergraduate and graduate students and Ethiopian colleagues which benefitted from funds from the European Commission (EC-STD3 "Land resource inventory, environmental changes analysis, and their application to agriculture in the Lakes Region, Ethiopia") and the Italian Ministry of University and Research (PRIN 1999 and 2003). Prof. Paolo Billi and Mario Sagri are thanked for their critical reading and review of an early version of this paper.

References

Abebe T, Manetti P, Bonini M, Corti G, Innocenti F, Mazzarini F, (2005) Geological map (scale 1: 200,000) of the northern Main Ethiopian Rift and its implication for the volcano-tectonic evolution of the rift. Geol Soc Am Map Chart Ser, MCH094

Abebe B, Acocella V, Korme T, Ayalew D (2007) Quaternary faulting and volcanism in the Main Ethiopian Rift. J Afr Earth Sci 48:115–124

AERMAP (1969) Meki river diversion scheme: topographic surveys, Imperial Ethiopian Government, Awash Valley Authority, Addis Ababa

Agostini A, Bonini M, Corti G, Sani F, Manetti P (2011) Distribution of quaternary deformation in the central Main Ethiopian Rift, East Africa. Tectonics 30(4) TC4010

Alessio M, Allegri L, Belluomini G, Benvenuti M, Cerasoli M, Improta S, Manfra L, Sagri M, Ventra D (1996) Le oscillazioni tardo-quaternarie del Lago Shala (Rift Etiopico): Analisi dell'evoluzione ambientale dall'integrazione di evidenze morfologiche, sedimentarie e cronologiche. Il Quaternario 9:387–392

Alley RB (2000) The Younger Dryas cold interval as viewed from central Greenland. Quat Sci Rev 19:213–226

Bacci G (1940) Molluschi fossili dell'antico fondo del Lago Ziway. Ann Mus St Nat Genova 60:454–458

Benvenuti M, Carnicelli S, Belluomini G, Dainelli N, Di Grazia S, Ferrari GA, Iasio C, Sagri M, Ventra D, Atnafu Balemwald, Kebede Seifu (2002) The Ziway-Shala lake basin (Main Ethiopian Rift, Ethiopia): a revision of basin evolution with special reference to the Late Quaternary. J Afr Earth Sci 35:247–269

Benvenuti M, Carnicelli S, Ferrari G, Sagri M (2005) Depositional processes in latest Pleistocene and Holocene ephemeral streams of the Main Ethiopian Rift (Ethiopia). In: Blum MD, Marriott SB, Leclair SF (eds) Fluvial Sedimentology VII, Spec Publ int Ass Sediment. vol 35, pp 277–294

Benvenuti M, Carnicelli S, Ferrari G, Sagri M (2006) Coupled evidence of Late Holocene rapid moisture fluctuations in lake and river system proxies, Main Ethiopian Rift. Holivar 2006 Open Science Meeting, Poster Presentation, University College London

Benvenuti M, Bonini M, Tassi F, Corti G, Sani F, Agostini A, Manetti P, Vaselli O (2013) Holocene lacustrine fluctuations and deep CO2 degassing in the northeastern Lake Langano basin (Main Ethiopian Rift). J Afr Earth Sci 77:1–10

Bigazzi B, Bonadonna FP, Di Paola GM, Giuliani A (1993) K-Ar and fission tracks ages of the last volcano-tectonic phase in the Ethiopian Rift Valley (Tulu Moye area). In: Geology and Mineral Resources of Somalia and Surrounding Regions. Ist Agronom Oltremare, Firenze, Relaz Monograf. vol 113, pp 311–322

Boccaletti M, Getaneh A, Tortorici L (1992) The Main Ethiopian Rift: an example of oblique rifting. Annales Tectonicae 6:20–25

Bonini M, Souriot T, Boccaletti M, Brun JP (1997) Successive orthogonal and oblique extension episodes in a rift zone: laboratory experiments with application to the Ethiopian Rift. Tectonics 16:347–362

Bull WB (1997) Discontinuous ephemeral streams. Geomorphology 19:227–276

Carnicelli S, Benvenuti M, Ferrari G, Sagri M (2009) Dynamics and driving factors of late Holocene gullying in the Main Ethiopian Rift (MER). Geomorphology 103:541–554

Chalié F, Gasse F (2002) Late Glacial-Holocene diatom record of water chemistry and lake level change from the tropical East African Rift Lake Abijata (Ethiopia). Palaeogeogr Palaeoclimatol Palaeoecol 187:259–283

Chernet T (1982) Hydrogeology of the Lakes Region, Ethiopia. Ministry of Mines and Energy, Addis Ababa

Coetzee JA, van Zinderen Bakker EM (1989) Palaeoclimatology of East Africa during the last glacial maximum: a review of changing theories. In: Quaternary and environmental research on East African mountains. Balkema, pp 189–198

Coltorti M, Corbo L, Sacchi G (2002) New evidence for the Late Pleistocene and Holocene climatic changes in the Lake Region. In: Dramis F (ed) Proceedings of the symposium international Geomorph Ass, Addis Ababa

Corti G (2009) Continental rift evolution: from rift initiation to incipient break-up in the Main Ethiopian Rift, East Africa. Earth-Sci Rev 96:1–53

Dansgaard W, Johnsen SJ, Clausen HB, Dahl-Jensen D, Gundestrup NS, Hammer CU, Hvidberg CS, Steffensen JP, Sveinbjörnsdottir AE, Jouzel J, Bond G (1993) Evidence for general instability of past climate from a 250-kyr ice-core record. Nature 364:218–220

Di Paola GM (1972) The Ethiopian Rift Valley (between 7°00' and 8°40' lat. North). Bull Volcan 36:517–560

Fontes JC, Gasse F (1989) On the ages of humid Holocene and Late Pleistocene phases in North Africa—remarks on "Late Quaternary climatic reconstruction for the Maghreb (North Africa)" by P. Rognon. Palaeogeogr Palaeoclimatol Palaeoecol 70:393–398

Fontes, J-C, Gasse F (1991) PAHLYDAF (Palaeohydrology in Africa: objectives, methods and major results). Palaeogeogr Palaeoclimatol Palaeoecol 84:191–215

Gasse F (1989) Tectonic and climatic controls on lake distribution and environments in Afar from Miocene to present. AAPG Mem 50:19–41

Gasse F, Street FA (1978) Late Quaternary lake-level fluctuations and environments of the northern Rift Valley and Afar region (Ethiopia and Djibuti). Palaeogeogr Palaeoclimatol Palaeoecol 24:279–325

Gasse F, Van Campo E (1994) Abrupt post-glacial climate events in West Asia and North Africa monsoon domains. Earth Planet Sci Lett 126:435–456

Geze F (1975) New dates on ancient Galla Lake levels. Bull Geophys Obs Addis Ababa 15:119–124

Gillespie R, Street-Perrot FA, Switsur R (1983) Post-glacial arid episodes in Ethiopia have implications for climate prediction. Nature 306:680–683

Graf WL (1988) Fluvial Processes in dryland rivers. Springer, Berlin

Grove AT, Goudie AS (1971) Late quaternary Lake levels in the Rift Valley of Southern Ethiopia and elsewhere in tropical Africa. Nature 234:403–405

Grove AT, Street FA, Goudie AS (1975) Former lake levels and climatic changes in the rift valley of Southern Ethiopia. Geogr J 141:177–202

Haines V, Haas H (1974) Southern Methodist University radiocarbon date list. Radiocarbon 16:368–380

Hillaire-Marcel C, Carro O, Casanova J (1986) 14C and Th/U dating of Pleistocene and Holocene stromatolites from east african paleolakes. Quat Res 25:312–329

ISSC—International Subcommission On Stratigraphic Classification (1994) A guide to stratigraphic classification, terminology, and procedure, 2nd edn. IUGS, Boulder

Kutzbach JE, Street-Perrott FA (1985) Milankovitch forcing of fluctuations in the level of tropical lakes from 18 to 0 kyr BP. Nature 317:130–134

Laury RL, Albritton CC (1975) Geology of the middle stone age archeological sites in the Main Ethiopian Rift valley. Geol Soc Am Bull 86:999–1011

Le Turdu C, Tiercelin JJ, Gibert E, Travi Y, Lezzar K, Richert J, Massault M, Gasse F, Bonnefille R, Decobert M, Gensous B, Jeudy V, Tamrat E, Mohamed MU, Martens K, Balemwal A, Chernet T, Williamson D, Taieb M (1999) The Ziway-Shala lake basin system, Main Ethiopian Rift: influence of volcanism, tectonics, and climatic forcing on basin formation and sedimentation. Palaeogeogr Palaeoclimatol Palaeoecol 150:135–177

Legesse Dagnachew, Gasse F, Radakovitch O, Vallet-Coulomb C, Bonnefille R, Verschuren D, Gibert E, Barker P (2002) Environmental changes in a tropical lake (Lake Abijata, Ethiopia) during recent centuries. Palaeogeogr Palaeoclimatol Palaeoecol 187:233–258

Maley J (1976) Les variations du lac Tchad depuis un millénaire: conséquences paleoclimatiques. Palaeoecol Afr 9:44–47

Mayewski PA, Rohling EE, Stager JC, Karlen W, Maasch KA, Meeker LD, Meyerson EA, Gasse F, van Kreveld S, Holmgren K (2004) Holocene climate variability. Quat Res 62:243–255

Mohr PA (1962) The Ethiopian Rift System. Bull Geophys Obs Addis Ababa 5:33–62

Mohr PA, Mitchell JC, Raynolds RGH (1980) Quaternary volcanism and faulting at O'A Caldera, central Ethiopian Rift. Bull Volcan 43:173–189

Morrison RB (1977) Quaternary soil stratigraphy, concepts, methods and problems. In: Mahaney WC (ed) Quaternary soils. Geo-Abstracts, Norwich, pp 77–108

Nilsson E (1940) Ancient changes of climate in British East Africa and Abyssinia; a study on ancient lakes and glaciers. Stockholm Geogr Ann 22:1–79

North American Commission on Stratigraphic Nomenclature (1983) North American stratigraphic code (NASC). AAPG Bull 67:841–875

Packard FA (1974) The hydraulic geometry of a discontinuous ephemeral stream on a Bajada near Tucson, Arizona. Unpublished Ph.D. Thesis, Univ Arizona

Porter SC (1989) Some geological implications of average quaternary glacial conditions. Quat Res 32:245–261

Roberts N, Taieb M, Barker P, Damnati B, Icole M, Williamson D (1993) Timing of the Younger Dryas event in East Africa from lake-level changes. Nature 366:146–148

Rosendahl BR, Reynolds DJ, Lober PM, Burgess CF, McGill J, Scott D, Lambiase JJ, Derksen SJ (1986) Structural expression of rifting: lesson from Lake Tanganika, Africa. Geol Soc Spec Publ 25:29–43 Sedimentation in the African Rifts

Sagri M (ed) and the staff of EU Project (1998) Land resources inventory, environmental change analysis and their application to agriculture in the Lakes Region (Ethiopia) Final report. European Commission, DG XII, Bruxelles

Sagri M, Bartolini C, Billi P, Ferrari G, Benvenuti M, Carnicelli S, Barbano F (2008) Latest Pleistocene and Holocene river network evolution in the Ethiopian Lakes region. Geomorphology 94:79–97

Schumm SA (1973) Geomorphic thresholds and complex response of drainage systems. In: Morisawa M (ed) Fluvial Geomorphology. Geomorphology, Binghamton State University, Binghamton, pp 299–310

Street FA (1979) Late Quaternary lakes in the Ziway-Shala Basin, Southern Ethiopia. Unpublished Ph.D. Thesis, University of Cambridge

Street-Perrott FA (1982) Twentieth century fluctuations in lake level in the Ziway-Shala basin, Ethiopia. Palaeoecol Afr 14:99–110

Street-Perrott FA, Perrott RA (1990) Abrupt climate fluctuations in the tropics: the influence of Atlantic Ocean circulation. Nature 343:607–612

Taieb M, Barker P, Bonnefille R, Damnati B, Gasse F, Goetz C, Hillaire Marcel C, Icole M, Massault M, Roberts N, Vincens A, Williamson D (1991) Histoire paléohydrologique du lac Magadi (Kenia) au Pléistocène supérieur. C R Acad Sci Paris 313:339–346

Thomas MF, Thorp MB (1995) Geomorphic response to rapid climatic and hydrologic change during the Late Pleistocene and Early Holocene in the humid and sub-humid tropics. Quat Sci Rev 14:193–207

Thomas MF, Thorp MB (2003) Palaeohydrological reconstructions for tropical Africa- evidence and problems. In: Benito G, Gregory KJ (eds) Palaeohydrology: understanding global change. Wiley, Chichester, pp 167–192

Verschuren D, Laird KR, Cumming BF (2000) Rainfall and drought in equatorial East Africa during the past 1,100 years. Nature 403:410–414

Waters MR, Haynes CV (2001) Late quaternary arroyo formation and climate change in the American Southwest. Geology 29:399–402

Weijers JWH, Schefuß E, Schouten S, Sinninghe Damsté JS (2007) Coupled thermal and hydrological evolution of tropical Africa over the last deglaciation. Science 315:1701–1704

Woldegabriel G, Aronson J, Walter RC (1990) Geology, geochronology and rift basin development in the central sector of the Main Ethiopian Rift. Geol Soc Am Bull 102:439–458

18 Water–Rock Interaction and Lake Hydrochemistry in the Main Ethiopian Rift

Azeb Belete, Luigi Beccaluva, Gianluca Bianchini, Nicolò Colombani, Massimiliano Fazzini, Chiara Marchina, Claudio Natali, and Tewodros Rango

Abstract

This study investigates the halogen budget of the Main Ethiopian Rift (MER) lithologies and water, and on the basis of new laboratory experiments gives insights on the water–rock interaction processes which ultimately mobilize fluoride in the environment. The halogen composition, and in particular, the chloride content of MER lakes is also taken into consideration to evaluate compositional variation occurred during the last 80 years that have to be mainly related evaporative effects. The evaporation trends are also investigated on the basis of new analyses of stable isotopes of oxygen and hydrogen that are compared with those available in the literature since the 1970s. In such complex scenario, although the average annual temperature increased ~ 1 °C in 30 years, we did not observe systematic trends valid for all the investigated lakes. The record defined for the last 30 years by $\delta^{18}O$–δD denote fluctuations of the climatic parameters with extreme evaporation preceding the year 2005, then declining to more "normal" conditions. The relation between the observed climatic parameters and the water isotopic composition suggests that the study lakes quickly respond to the environmental changes, possibly within one (or two) year(s). We therefore suggest to continue the data acquisition of climatic and hydrochemical parameters in order to implement the existing hydro-archive that could be useful to point out possible environmental changes.

Keywords

Ethiopian lakes • Hydro-archive • Geochemistry • Stable isotopes

18.1 Introduction

Ethiopia is characterized by a great variety of aquatic ecosystems, including several lakes that are of great scientific interest and economic importance. The majority of these lakes are confined within the Main Ethiopian Rift (MER) valley, which is part of a tectonic depression extending from the Afar region in the north to the Kenyan border in the south (see Chaps. 16 and 17, this volume, for background information about the origin, hydrology, and evolution of the lakes). The water geochemistry within this rifted belt is extremely variable because meteoric water recharge from the surrounding highlands is affected by different degrees of water–rock interaction and evaporation processes. The widespread uprising of thermal water adds to the complexity of the rift hydrological system. Moreover, anthropogenic

A. Belete · L. Beccaluva · G. Bianchini (✉) · M. Fazzini · C. Marchina · C. Natali
Department of Physics and Earth Sciences, University of Ferrara, Ferrara, Italy
e-mail: gianluca.bianchini@unife.it

A. Belete
Department of Earth Sciences, Addis Ababa University, Addis Ababa, Ethiopia

G. Bianchini · C. Natali
Institute of Geosciences and Earth Resources–National Research Council, Pisa, Italy

N. Colombani
Department of Earth Sciences, "Sapienza" University of Rome, Rome, Italy

T. Rango
Nicholas School of the Environment, Division of Earth and Ocean Sciences, Duke University, Durham, USA

P. Billi (ed.), *Landscapes and Landforms of Ethiopia*, World Geomorphological Landscapes, DOI 10.1007/978-94-017-8026-1_18,

activities that altered the water balance during the last decades have also to be taken into account (Deribe et al. 2011; Yohannes et al. 2013; Chap. 16, this volume).

The hydrological cycle is regulated by tectonic activity that could influence the drainage systems at various temporal/spatial scales modifying the basin input/output, and also by volcanism and associated geothermal processes that provide significant geochemical fluxes (Goerner et al. 2009). Hydrology is also influenced by changing climatic conditions that variously affect the lacustrine systems, some of which are endhoreic. Accordingly, geochemical variations in the different water components can be associated with global (or local) climatic variations, implying changes of temperature and rainfall (Wagesho et al. 2013). These variations can be monitored through the isotopic composition of oxygen and hydrogen in the meteoric water (generally expressed as $\delta^{18}O$ and δD) which is site-specific and related to temperature-dependent fractionation processes occurring during the hydrological cycle (Clark and Fritz 1997). In the lakes, the isotopic fingerprinting is remarkably affected by evaporation processes (Gat 1995; Rozanski et al. 2000; Froehlich et al. 2005; Kebede et al. 2009), because evaporation preferentially taps the lightest (^{1}H and ^{16}O) with respect to the heavier (^{2}H, ^{18}O) isotopes, leading to progressively higher $^{2}H/^{1}H$ and $^{18}O/^{16}O$ proportions in the residual lake water. Within this framework, we present and discuss new original $\delta^{18}O$–δD analyses of MER water (Fig. 18.1). Further detail on the limnological processes that affect the chemical composition of MER lakes is based on observed hydrogeochemical changes in the lakes of Ziway, Langano, Shala, Abijata, Awasa, Abaya, and Chamo which occur in response to ongoing environmental variations in the area (e.g., Sagri 1998; Legesse et al. 2004; Dessie and Kleman 2007). The dataset considered includes both the present authors' original data and literature analyses (Omer-Cooper 1930; Loffredo and Maldura 1941; Chernet 1982; Von Damm and Edmond 1984; Wood and Talling 1988; Gizaw 1996; Zinabu et al. 2002; Kebede et al. 2009; Kebede and Travi 2012), focusing on the concentration of halogens such as the chloride and fluoride in the water system (Fig. 18.1). It must be stressed that chloride is a highly conservative tracer (Apelo and Potma 2005), which is commonly used to monitor evaporation processes in surface water (Manno et al. 2007; Kirchner et al. 2010), groundwater (Gianbastiani et al. 2013) and in the vadose zone (Scanlon et al. 2006), whereas fluoride is an anion marking the water–rock interaction processes (Saxena and Ahmed 2001; Tirumalesh et al. 2007). This study also addresses the geochemical source(s) and water–rock interaction(s) on the basis of new laboratory experiments that are focused to understand the anomalous concentration of fluoride which is typically high in the natural water of the rift valley, causing serious health problems in the local populations (Rango et al. 2009, 2010a, b, c, 2012, 2013).

18.2 Physiographic and Hydrological Framework

Remarkable rifting processes, originated by active faulting, formed various volcano-tectonic depressions in the rift floor, in turn favoring the development of closed lacustrine basins fed by surface- and groundwater. Accordingly, the major rivers draining the surrounding plateaus on both sides of the rift convey water to lakes and swamps located at different elevations. The hydrological regime is obviously influenced by climatic parameters such as the monthly distribution of rainfall and temperature (Ayenew 2003; Bewketu 2010). Within the rift valley, climatic conditions are semiarid/semi-humid, with rainfall ranging between 800 and 1,100 mm/yr (uncommonly as much as 1,500 mm/year). The precipitation regime generally shows a first maximum in the summer months. Only in the area of Arba Minch, the regime becomes definitely bimodal, with an absolute maximum in spring and a second peak in late summer. The winter is very dry everywhere. The average annual temperature is about 20 °C, and the maximum is above 25 °C, with a thermal trend of the last 30 years showing an increase in average temperatures of about 1 °C. The potential evaporation is around 1,000 mm/year (Billi 1998; Chap. 4, this volume). Therefore, rivers and lakes recharge is mainly due to water deriving from the highlands, where the rainfall is up to 1,800 mm/year and the mean annual temperature between 14 and 16 °C (Billi et al. 2003; Kebede et al. 2009), while potential evaporation is less than 700 mm. The series of measurements of rainfall and temperature in the investigated sectors of the rift reported in Fig. 18.2 will be discussed in the final section in relationships with the measured geochemical data.

For the sake of simplicity, the MER lakes are subdivided in three groups: those located in the Awash basin (lakes Koka, Beseka, Gemeri and Abe), those in the central part of MER (lakes Ziway, Langano, Shala, Abijata, Awasa, Abaya, and Chamo), and those in the southern MER, close to the Kenyan border (e.g., Lake Chew Bahir that stretches into northern Kenya). As the first group of lakes is concerned, Koka Lake is a hydropower reservoir, resulting from the impoundment of the Awash River, whereas Beseka Lake is mainly fed by hot springs outpouring in an area crisscrossed by active faults (Goerner et al. 2009); Abe and Gemeri lakes correspond to the terminal base level of the Awash River in the Afar region. The second group of lakes, located in the central part of MER, were sampled and investigated by the authors and are the specific object of this paper:

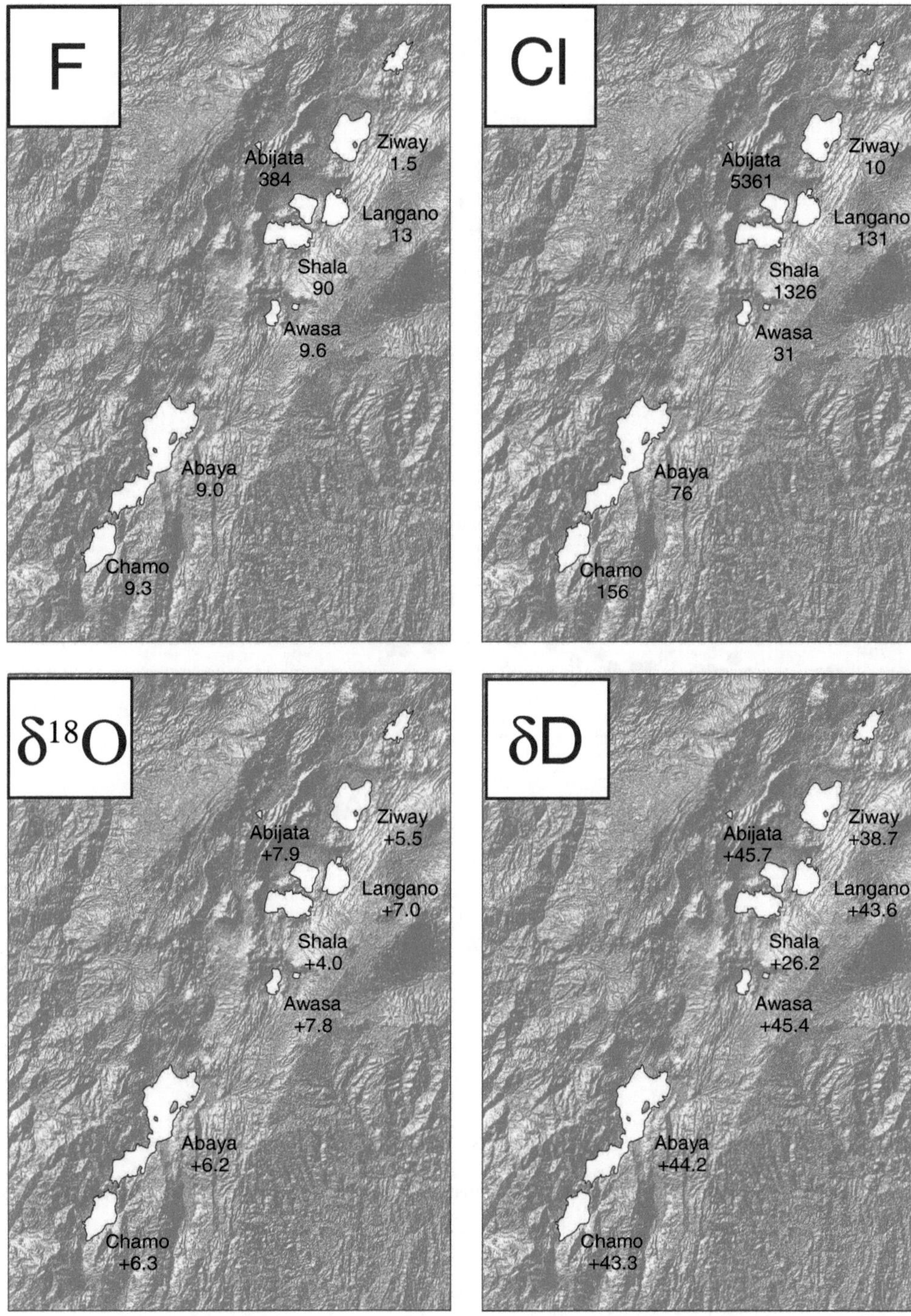

Fig. 18.1 Maps of the studied lakes located in the Main Ethiopian Rift (MER), reporting recent geochemical data on their waters. The data (Table 18.1) include samples collected in January 2006 for lakes Ziway, Langano, Shala, and Abijata (Rango et al. 2009) and new analyses carried out on samples collected in January 2012 for lakes Awasa, Abaya, and Chamo. Emphasis is given to the following tracers: chloride (**a**) which is extremely conservative and describes enrichment by evaporative processes, fluoride (**b**) which is a marker of water–rock interaction processes and limits the use of MER water for drinking purposes, and $\delta^{18}O$–δD isotopic values (**c**, **d**) that can help to understand the origin of waters and the ongoing hydrological processes

- Lakes Abijata, Langano, Shala, and Ziway form an interconnected lakes system, within a closed basin covering an area of $\sim 13{,}000$ km^2. Ziway is characterized by freshest water (conductivity ~ 400 μS cm^{-1}) and receives water supply from the Meki and the Ketar rivers, which flow from the west and the east, respectively. Ziway Lake outflows through the Bulbula River conveying water into Lake Abijata, which is a shallow, small, closed vlake, characterized by very saline water (conductivity >20,000 μS cm^{-1}). Langano Lake, fed from the southeast by five small streams and by several hot springs, is characterized by water with intermediate salinity (conductivity $\sim 1{,}700$ μS cm^{-1}), which is partially conveyed to Abijata Lake by the Horakelo River.

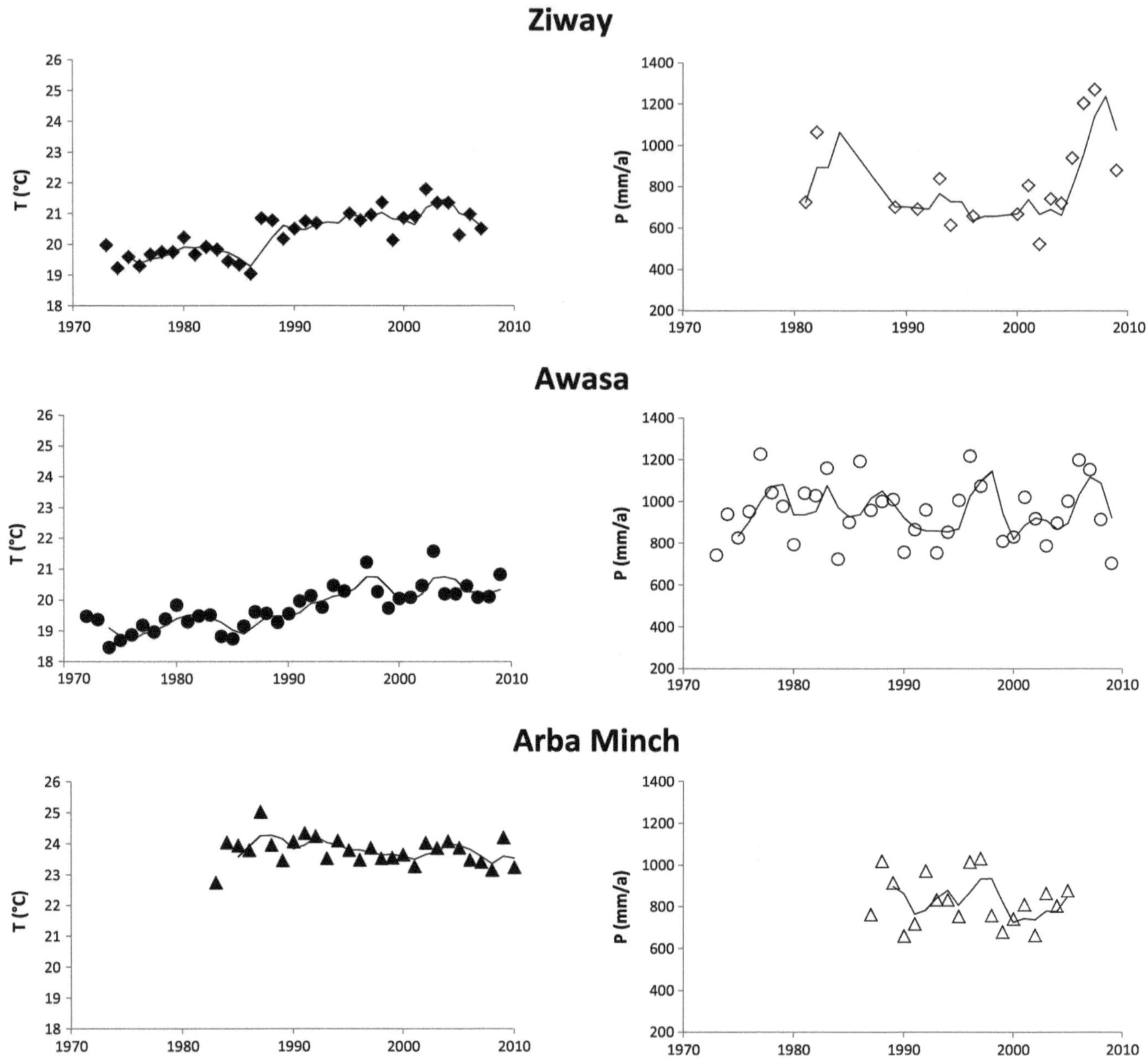

Fig. 18.2 Average annual temperature (°C) and precipitation (mm/yr) in areas close to the studied lakes. *Lines* represent mobile averages in respect to a 3-year period

Lake Shala (Fig. 18.3, characterized by extremely saline water (conductivity >20,000 μS cm^{-1}), is the deepest lake of Ethiopia (up to 266 m) and occupies a down-faulted caldera as do Langano and Abijata lakes; it is fed by the Dijo River that drains the western escarpment and by several hot springs.

- Lake Awasa (Fig. 18.4) is located within a collapsed caldera and receives surface contributions from the intermittent Tikur Wuha stream, originating from the Cheleleka swamp, which only a century ago was included in a larger paleo-Awasa Lake; some other small ephemeral streams contribute to the lake, which is characterized by freshwater (conductivity ~800 μS cm^{-1}) although no visible outlet is observed. Thus, most probably subsurface inflows and outflows occur in this lake.
- Abaya (Fig. 18.5) and Chamo lakes are closely associated within the same drainage basin. In the past, these lakes were interconnected, with Abaya Lake water spilling into Chamo Lake during high-stand intervals (Von Damm and Edmond 1984). Presently, this is no longer so common, though during the rainy season water from Abaya Lake spills through a small river into Chamo Lake. Lake Abaya is fed by the Bilate and Hamesa rivers and by several hot springs (Chernet 2011) and is characterized by relatively

Fig. 18.3 Shala Lake is the deepest lake of Ethiopia (266 m) and formed within a large caldera

Fig. 18.4 Awasa Lake is a shallow (maximum depth 22 m) body of freshwater in the middle of a very large caldera

low salinity (conductivity ~800 μS cm^{-1}). The neighboring Lake Chamo, fed by the Kulfo River and by other minor streams, has a more elevated salinity (conductivity ~1,600 μS cm^{-1}). Generally, the Chamo Lake is closed, though occasionally overflowing into the Sagan River and, in turn, in the ephemeral terminal lake of Chew Bahir close to the border with Kenya. This overflow takes place only when precipitation in the region is above long-term average (Von Dam and Edmond 1984; Kebede et al. 1994). With increasing diversion of water from tributary rivers for irrigation, the overflow toward Chew Bahir and the surface water connection between Abaya and Chamo lakes are expected to definitely disappear.

The study lakes are roughly aligned N–NE/S–SW, at altitudes varying from 1,680 to 1,170 m above sea level. According to the literature data (Table 18.1), their surface areas are ~90 km^2 for Awasa, ~160 km^2 for Abijata, ~230km^2 for Langano, ~320 Km^2 for Chamo, ~330 for Shala, ~440 km^2 for Ziway, and 1,130 km^2 for Abaya. The lake surface extension of Ziway, Langano, Shala, and Abaya remained almost constant in the last 30 years, whereas Awasa expanded and Abijata and Chamo shrunk (Ayenew 2004). The average water depth is less than 20 m for all lakes, with the exception of Lake Shala which is a crater lake having an average water depth of ~90 m (maximum depth about 266 m; Brunelli et al. 1941; Chernet 1982; Le Tardu et al. 1999; Alemayehu et al. 2006).

18.3 Geological and Geochemical Processes Influencing the Water Chemistry

The study area is characterized by volcano-tectonic activity. The paroxysmal phase occurred in the Oligocene, with lithospheric bulging and uplift of the whole region and a

Fig. 18.5 Abaya Lake is the second largest lake of Ethiopia (1,140 km^2). Its water is fresh and very shallow (maximum depth 13 m)

Table 18.1 Physical details of the studied lakes

	Altitude (m)	Mean Depth (m)		Max Depth (m)				Surface area (km^2)		
Lake name	Wood and Tailing	Wood and Tailing	Kebede et al.	Wood and Tailing	MCE	Kebede et al.	FAO	Wood and Tailing	MCE	Kebede et al.
	1988	1988	2009	1988	2001	2009	1982	1988	2001	2009
Ziway	1,639	2.5	4	8.95	8.9	9	434	442	440	442
Langano	1,582	17	20	47.9	47.9	47	230	241	230	241
Abijata	1,578	7.6	5	14.2	14.2	10	205	176	180	159
Shala	1,558	87	103	266	266	266	409	329	370	329
Awasa	1,680	10.7	10	21.6	22	23	130	129	92	92
Abaya	1,285	7.1	–	13.1	24.2	–	1,161	1,162	1,140	1,127
Chamo	1,233	–	–	13	14.2	–	551	551	317	347

Data sources are as follows: Wood and Talling 1988; Kebede et al. (2009 and references therein), Awulachew et al. (2007 and references therein). The label FAO and MCE stand for "Food and Agriculture Organization" and "Metaferia Consulting Engineers"

huge emission of continental flood basalts (Beccaluva et al. 2009). Then, differentiated rhyolite magmas appeared late in the sequence marking the onset of rifting and block tilting (Natali et al. 2011), which progressively continued leading, ultimately, to the origin of the East African Rift (Corti 2009).

Within this framework, the rift floor is mainly paved by volcanic products (felsic tuffs, ignimbrites, and subordinate lavas). Locally, these volcanic rocks have been weathered and the resulting particles have been displaced by surface runoff, leading to fluvial-lacustrine sedimentary formations. The study of these deposits has been useful to understand the evolution of the rivers/lakes network (Gasse and Street 1978; Le Tardu et al. 1999; Sagri et al. 2008; Benvenuti et al. 2013), giving insights into the climatic fluctuation during the Quaternary (Ghinassi et al. 2012). Benevenuti et al. (2002 and Chap. 14, this volume) have shown that these lacustrine basins underwent four main lake stand phases, with the oldest one in the early Pleistocene when the existing lakes were united in singular, large paleo-lake (Geze 1975).

The above mentioned weathering processes contribute to release dissolved ions in the surface- and groundwater interaction, as well as sediment load to rivers and lakes. The amount of the TDS (total dissolved salts) can be easily assessed in the field by measuring the electrical conductivity and calculating the salinity by the use of algorithms, which were properly defined for MER water by Wood and Talling (1988) and Zinabu et al. (2002).

Surface and groundwater supplied from the highlands is typically characterized by low conductivity, low TDS, and a calcium bicarbonate facies. Pristine water progressively migrating downhill reacts with the interacting lithologies increasing the TDS. This evolution ultimately leads to more "mature" water types characterized by a sodium bicarbonate hydrochemical facies, which is typical of the rift valley

(Rango et al. 2009). This hydrochemical evolution, resulting from progressive water–rock/sediment interaction, is accompanied by an increase of pH and the appearance of dissolved fluoride (and arsenic), which are affecting water quality of springs and wells located in the rift valley, causing severe health problems to the local populations (Rango et al. 2009, 2010a, 2012, 2013). The chloride/fluoride ratio indicates the water "maturity," as the halogen content of meteoric components is mainly influenced by marine aerosol where chloride is distinctly predominant, while fluoride is mainly leached from the local rocks that are particularly enriched in this element. Coherently, the chloride/fluoride ratio of MER rainwater (~ 12 and 0.05 mg/l of Cl and F, respectively, have been measured in a rainwater collected in Arba Minch during January 2012) is extremely high with Cl/F up to 200, whereas Cl/F progressively diminishes to values of 1 in some hot springs, as result of pervasive water–rock/sediment interactions.

Better understanding of these processes necessarily requires the geochemical study of the lithologies outcropping in the rift valley, which are mainly constituted by volcanic glass (amorphous silicate phases) or their weathering products developed during supergene processes. In Fig. 18.6, we report a SEM picture showing the particles of a volcanoclastic deposit sampled close to the Ziway Lake and the significant presence of a crystal of fluoro-apatite among the vesciculated glass shards. The high fluorine activity of the original magmas is supported by the SEM chemical analyses of both glass and fluoro-apatite (F up to 0.38 and 8.2 wt%, respectively) (Table 18.2). To constrain the reactivity of MER lithologies to the atmospheric agents, we carried out water extraction tests in which 10 g of powdered rock was leached by 50 ml of water, investigating both volcanic rocks and the fluvio-lacustrine sediments. The leaching experiments carried out for duration of 15, 30, and 90 days with both deionized water and a calcium bicarbonate water having TDS ~ 130 mg/l that simulate the recharge water conveyed from the highlands, demonstrated that fluoride is efficiently removed from MER lithologies (Table 18.3). In particular, experiments show that among the distinct MER lithologies, the fluvio-lacustrine sediments contain (and release) higher amount of fluoride which is plausibly trapped and concentrated by clay minerals (i.e., phyllosilicates in which fluoride can vicariate the oxydril radicals). This was further constrained by column displacement tests on volcanic ashes having different grain size, in which the fluoride concentration increased with decreasing the grain size (Rango et al. 2010c). In addition, an elevated mobility of fluoride was observed in saturated conditions, with much of the available budget leached in a few column pore volumes, similar to chloride.

The real MER waters, and especially those of the study lakes, are also subject to effective evaporation processes that further concentrate the halogens.

18.4 Lake Water Composition

The open or closed nature of each drainage system and the effective in/out seepage obviously influence the salinity of the study lakes that is generally expressed as milligrams per liter of TDS. Accordingly, in the study lakes, TDS is extremely heterogeneous, with values of 0.3–0.4 g/l in Ziway, ~ 1.4 g/l in Langano, ~ 11.5 g/l in Shala, ~ 52.7 g/l in Abijata, ~ 0.7 g/l in Awasa, ~ 0.8–0.9 mg/l in Abaya, and ~ 1.0–1.2 mg/l in Chamo; pH is generally higher than 8,

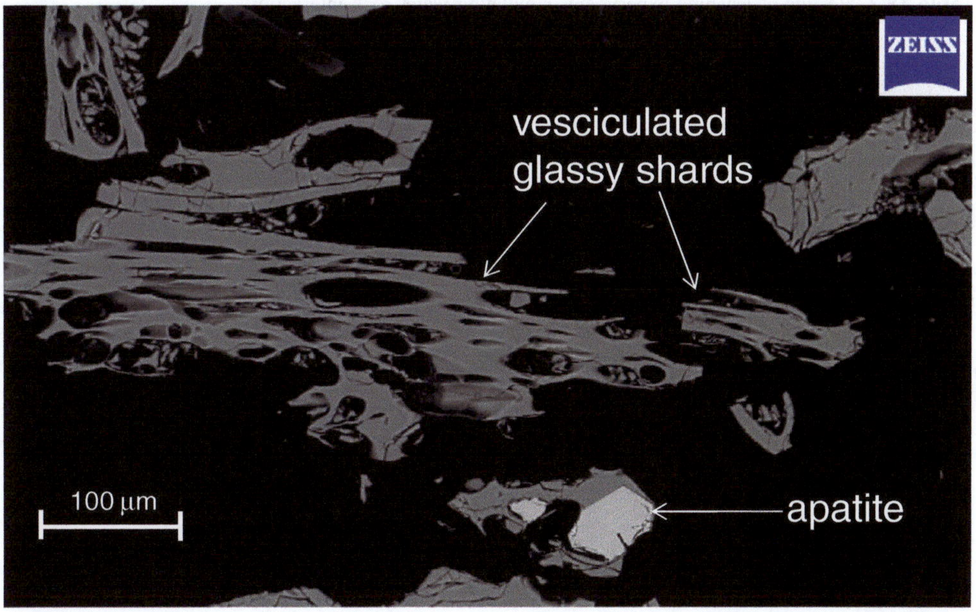

Fig. 18.6 Backscattered SEM image of a MER volcanoclastic rock. Vesciculated particles characterized by *light gray color* consist of glass, whereas the euhedral white crystal is fluorine-rich apatite

Table 18.2 SEM analyses of glassy particles and apatite crystals recorded within the volcanoclastic deposit shown in Fig. 18.3

	EDS analyses of glassy particles in TW39. Analytical conditions were adjusted for the analyses of halogens, and F was detected by WDS														
SiO_2	71.65	71.88	72.30	72.24	70.15	69.90	69.07	68.96	71.42	70.95	69.51	69.36	69.69	69.42	72.85
TiO_2	0.32	0.29	0.32	0.30	0.28	0.27	0.28	0.31	0.33	0.32	0.30	0.31	0.28	0.31	0.31
Al_2O_3	8.12	8.07	8.12	8.13	7.93	7.83	7.77	7.74	8.07	7.96	7.69	7.87	7.80	7.76	8.20
FeO	6.03	6.11	6.09	6.21	5.93	5.92	5.79	5.78	6.04	5.96	5.94	5.87	5.92	5.82	6.15
MnO	0.27	0.20	0.25	0.24	0.22	0.25	0.23	0.25	0.24	0.24	0.24	0.23	0.26	0.25	0.27
MgO	n.d.	n.d.	n.d.	n.d.	n.d.	n.d.	n.d.	n.d.	n.d.	n.d.	n.d.	n.d.	n.d.	n.d.	n.d.
CaO	0.00	0.25	0.28	0.24	0.27	n.d.	n.d.	n.d.	n.d.	n.d.	n.d.	0.27	0.26	n.d.	0.24
Na_2O	0.79	0.70	0.61	0.36	3.93	3.14	4.39	3.43	1.34	1.18	3.81	3.09	2.19	3.28	0.43
K_2O	3.00	2.88	2.77	2.21	4.14	4.12	4.07	4.17	3.76	3.44	4.10	4.18	4.14	4.09	1.99
P_2O_5	n.d.	n.d.	n.d.	n.d.	n.d.	n.d.	n.d.	n.d.	n.d.	n.d.	n.d.	n.d.	n.d.	n.d.	n.d.
F	0.37	0.43	0.30	0.29	0.30	0.31	0.38	0.36	0.34	0.30	0.35	0.31	0.26	0.30	0.30
Cl	0.19	0.19	0.22	0.21	0.20	0.20	0.19	0.17	0.20	0.19	0.19	0.22	0.18	0.20	0.20
	EDS analyses of apatite crystals in TW39														
SiO_2	1.43	1.34	1.36	1.26	1.13	1.20	1.44	1.42	1.66						
TiO_2	n.d.	n.d.	n.d.	n.d.	n.d.	n.d.	n.d.	n.d.	n.d.						
Al_2O_3	n.d.	n.d.	n.d.	n.d.	n.d.	n.d.	n.d.	n.d.	n.d.						
FeO	n.d.	n.d.	n.d.	n.d.	n.d.	n.d.	n.d.	n.d.	n.d.						
MnO	n.d.	n.d.	n.d.	n.d.	n.d.	n.d.	n.d.	n.d.	n.d.						
MgO	n.d.	n.d.	n.d.	n.d.	n.d.	n.d.	n.d.	n.d.	n.d.						
CaO	52.48	53.53	53.71	53.58	54.02	54.13	53.60	53.71	53.61						
Na_2O	n.d.	n.d.	n.d.	n.d.	0.37	0.48	0.34	0.37	n.d.						
K_2O	n.d.	n.d.	n.d.	n.d.	n.d.	n.d.	n.d.	n.d.	n.d.						
P_2O_5	36.52	37.37	37.69	37.39	37.89	37.71	37.38	38.15	38.43						
F	7.64	6.91	7.27	7.76	7.57	8.15	7.47	6.76	6.63						
Cl	n.d.	n.d.	n.d.	n.d.	n.d.	n.d.	n.d.	n.d.	n.d.						

Glass shows dacite and trachydacite compositions with F/Cl ratio always greater than one. Apatite is extremely enriched in F with negligible amount of Cl; n.d. = not detected

Table 18.3 Results of leaching experiments simulating water-rock interaction

			Deionized water leachates (mg/l)			Calcium bicarbonate water leachates (mg/l)		
	Bulk analysis	(mg/kg)	15 days	30 days	90 days	15 days	30 days	90 days
Ted 4	Pumice deposit	1,690	7.73	7.48	8.93	6.42	6.79	7.79
TW31	Ignimbrite deposit	985	3.16	3.66	3.66	2.94	3.22	3.53
TW39	Pyroclastic deposit	1,610	1.62	1.82	2.35	1.77	1.98	2.41
TW22	Sediment	2,510	8.00	8.01	8.52	7.80	7.76	8.24
Ted 38	Sediment	1,590	8.73	7.64	n.a.	5.97	6.68	8.26

The first column reports the fluorine content of MER lithologies (mg/kg), including volcanic rocks and sediments which results from the weathering, reworking, and deposition of the volcanic rocks. The analysis has been carried out by the ion-selective electrode (ISE) NexSens WQ-FL after alkaline fusion (0.4 g of sample + 0.3 of KNO_3 + 2 g of Na_2CO_3 heated at 1,000 °C for 40 min) and dissolution in deionized water. Other columns report the fluoride content of solutions (mg/l) obtained by the interaction of the above mentioned MER lithologies (10 g of powder) with water (50 ml). The leaching experiments were performed with deionized water and with a Ca-HCO_3 water simulating the water conveyed in the rift from the highlands, analyzing the solutions by IC (Ion chromatography; Dionex ICS 1000) after 15, 30, and 90 days; n.a. = not analyzed

Table 18.4 Physico-chemical parameters of MER water (pH, TC, and EC) measured in situ, laboratory analyses of halogens (Cl, F Ion chromatography; Dionex ICS 1000) and isotopic ratios of oxygen and hydrogen

	Code	Year	Coordinates		T (°C)	pH	EC μS/cm	Cl (mg/l)	F (mg/l)	$\delta^{18}O$ ‰	δD ‰	
Meki River	RI30	2006	38.822	8.153	23.3	8.66	530	16	1.5	1.7	6.1	Rango et al. (2010a, b)
Ketar River	RI31	2006	39.003	8.046	21.5	7.83	180	4	1.6	−2.8	−11.3	Rango et al. (2010a, b)
Groundwater flow	WL14	2006	38.966	8.120	24.6	7.48	350	145	3.1	−3.4	−13.7	
toward Lake Ziway	WL18	2006	38.822	8.159	25.2	7.82	490	4	0.9	−2.7	−11.0	Rango et al. (2010a, b)
	WL20	2006	38.532	8.006	36.4	7.89	730	7	3.1	−3.5	−19.7	
	WL23	2006	38.812	8.145	25.1	7.81	2750	176	20	−2.1	−5.2	Rango et al. (2010a, b)
	13	2011/12								−4.2	−22.8	Rango et al. (2013)
	14	2011/12								−3.7	−18.3	Rango et al. (2013)
	15	2011/12								−2.0	0.0	Rango et al. (2013)
	16	2011/12								−3.1	−8.5	Rango et al. (2013)
	17	2011/12								−2.7	−3.9	Rango et al. (2013)
	18	2011/12								−4.2	−24.2	Rango et al. (2013)
	19	2011/12								−4.7	−24.7	Rango et al. (2013)
	20	2011/12								0.2	7.3	Rango et al. (2013)
	21	2011/12								−5.1	−28.1	Rango et al. (2013)
	22	2011/12								−3.4	−11.8	Rango et al. (2013)
	23	2011/12								−3.2	−13.8	Rango et al. (2013)
Ziway Island spring	HS2	2006	38.853	7.934	78.0	6.79	2,040	13	128	−3.0	−15.7	Rango et al. (2010a, b)
Lake Ziway	LW29	2006	38.736	7.919	25.2	8.60	420	10	1.5	4.9	33.8	Rango et al. (2010a, b)
		2011/12								6.1	43.6	Rango et al. (2013)
Spring at the	HS1	2006	38.773	7.690	96.0	8.36	4,530	429	24	−0.4	−1.0	Rango et al. (2010a, b)
Langano shore	HS9	2006	38.811	7.664	38.8	7.33	630	27	1.9	−3.4	−15.9	Rango et al. (2010a, b)
	HS10	2006	38.773	7.712	62.0	6.96	3,900	435	23	−1.2	−3.4	Rango et al. (2010a, b)
Lake Langano	LW25	2006	38.684	7.538	24.2	9.03	1,730	131	13	7.0	43.6	Rango et al. (2010a, b)
Spring at the	HS3	2006	38.634	7.477	93.4	8.16	14,440	1,462	97	2.5	20.7	Rango et al. (2010a, b)
Shala Shore	HS4	2006	38.634	7.477	57.3	7.81	10,570	1,324	64	3.7	24.2	Rango et al. (2010a, b)
	HS5	2006	38.634	7.477	91.4	8.00	13190	1,148	55	2.3	20.1	Rango et al. (2010a, b)
	HS6	2006	38.637	7.477	52.6	8.15	1,780	17	4.5	−0.4	2.8	Rango et al. (2010a, b)
	HS7	2006	38.723	7.402	59.3	7.30	4,630	321	18	1.4	12.3	Rango et al. (2010a, b)
Lake Shala	LW27	2006	38.435	7.423	25.8	9.60	>20,000	1,326	90	4.0	26.2	Rango et al. (2010a, b)
Lake Abijata	LW28	2006	38.595	7.671	27.6	9.67	>20,000	5,361	384	7.9	45.7	Rango et al. (2010a, b)
Springs in the Lake												
Awasa catchment	AZ1	2012	38.522	6.993	60.9	6.2	1,200	36	n.d.	−1.0	0.3	Original data
	AZ2	2012	38.523	6.991	>80	6.9	2,330	73	14.6	−0.4	3.4	Original data
	AZ17	2012	38.638	7.079	65.6	7.4	1,160	25	1.6	−3.0	−7.4	Original data
	AZ18	2012	38.644	7.071	59	7.9	1,120	27	1.8	−2.9	−5.8	Original data
	AZ19	2012	38.643	7.076	59	7.7	1,020	25	1.5	−3.0	−6.6	Original data
Lake Awasa	AZ3	2012	38.458	7.056	24.6	9.9	810	31	9.6	7.8	45.4	Original data
Springs in the Lake	AZ4	2012	38.129	6.935	36.6	7.2	1,030	20	16.0	−2.9	−14.7	Original data
Abaya catchment	AZ5	2012	38.079	6.935	39	6.9	950	17	17.0	−2.5	−14.0	Original data
	AZ6	2012	38.121	6.938	39.2	6.8	960	18	17.0	−2.6	−12.8	Original data
	AZ11	2012	38.134	6.928	56.5	8.3	1,150	37	26.5	−2.5	−10.2	Original data
	AZ12	2012	38.088	6.810	45.3	7.8	1,340	51	18.4	−1.9	−3.4	Original data
	AZ13	2012	38.087	6.811	56	7.5	1,480	52	21.0	−1.6	−3.1	Original data
	AZ14	2012	38.046	6.755	66.5	7.6	5,530	147	45.2	1.7	2.5	Original data
	AZ15	2012	38.046	6.756	63	7.8	5,500	162	43.5	0.7	4.4	Original data
	AZ16	2012	38.045	6.723	> 80	7.9	6,200	144	44.9	−2.3	0.4	Original data
	AZ22	2012	37.909	6.636	65	6.6	2,700	65	18.1	−3.1	−11.1	Original data
	AZ23	2012	37.910	6.636	60.8	7.3	2,560	70	18.3	−3.0	-9.7	Original data
	AZ31	2012	37.577	6.243	16.8	6	60	9	0.6	−2.8	−6.0	Original data
	AZ32	2012	37.577	6.241	18.3	5.3	10	1	0.1	−2.4	−3.9	Original data
Hamesa River	AZ24	2012	37.825	6.568	35.5	6.4	420	4	2.3	−2.7	−7.5	Original data

(continued)

Table 18.4 (continued)

	Code	Year	Coordinates		T (°C)	pH	EC μS/cm	Cl (mg/l)	F (mg/l)	$\delta^{18}O$ ‰	δD ‰	
Lake Abaya	AZ33	2012	37.629	6.119	25.5	9.1	970	76	9.0	6.2	44.2	Original data
Springs in the Lake	AZ25	2012	37.691	5.887	40.1	7.9	2,270	116	11.0	−2.0	−1.7	Original data
Chamo catchment	AZ26	2012	37.691	5.884	45.5	8.0	2,400	115	11.1	−1.8	−1.5	Original data
	AZ27	2012	37.692	5.886	45.0	7.5	2,450	118	11.4	−1.8	−2.1	Original data
	AZ28	2012	37.692	5.886	55.6	7.7	2,600	115	10.9	−2.0	−1.5	Original data
Lake Chamo	AZ34	2012	37.535	5.929	28.8	9.3	1,640	156	9.3	6.3	43.3	Original data

In particular, the new isotopic analyses ($^{18}O/^{16}O$ and $^{2}H/^{1}H$) were determined by laser absorption spectrometry using the CRDS LOS GATOS LWIA 24d isotopic analyzer, reporting results as $\delta^{18}O$ and δD ‰ relative to the Standard Mean Ocean Water (SMOW)

increasing up to 10 in the terminal lakes such as Shala and Abijata. These characteristics delineate a notable salinity–alkalinity series, in which the waters are invariably characterized by a sodium bicarbonate hydrochemical facies. The water of Abijata Lake, which is subjected to high rates of evaporation (Ayenew 2002; Legesse et al. 2004), represents an exception as the hydrochemical facies is displaying a peculiar sodium chloride pattern.

Some of these lakes do not appear chemically stratified as demonstrated by chemical analysis of Awasa and Shala water before and after detonation causing effective mixing (Klemperer and Cash 2007). Others, like Lake Chamo, present variation of the salinity along the water column (Deriemaecker 2013).

18.5 Results and Discussion

In this paper, particular attention is given to the changes of lake water composition over time, presenting new data (Table 18.4) which are discussed together with those available in the literature. This can be done taking into account the earliest hydrochemical data that are available for most of the considered lakes since the year 1937 (Loffredo and Maldura 1941), and for Lake Abijata since 1926 (Omer-Cooper 1930). In this comparison, an emphasis is focused on chloride that is a conservative element, the variation of which has to be related mainly to evaporation/dilution processes. The comparison of the chloride content of the study lakes in different periods is synthesized in Fig. 18.7.

As the Ziway–Langano–Abijata–Shala system is concerned, we observe that in the last 80 years, the chloride content is constant in Ziway Lake, slightly decreasing in Langano and Shala lakes, and increasing only in Abijata Lake. It is worth noting that the latter has been impacted by anthropogenic activities (Ayenew 2002) and exploited to obtain trona ($Na_2CO_3 \cdot NaHCO_3 \cdot 2H_2O$) and finally soda ($Na_2CO_3$). The chloride content of Awasa and Abaya lakes has remained quite constant, whereas in Chamo Lake, TDS increased during the last 20 years, plausibly reflecting progressive isolation of its basin.

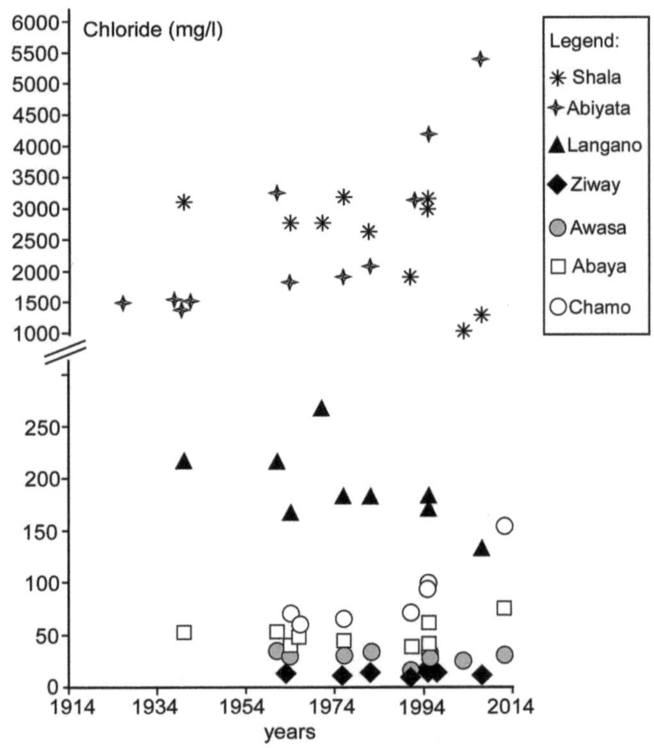

Fig. 18.7 Temporal evolution of the chloride content in the studied lakes. The authors' data (Table 18.4) include samples collected in January 2006 for lakes Ziway, Langano, Shala, and Abijata (Rango et al. 2009) and new analyses carried out on samples collected in January 2012 for lakes Awasa, Abaya, and Chamo, and are compared with older literature data (Omer-Cooper 1930; Loffredo and Maldura 1941; Chernet 1982; Von Damm and Edmond 1984; Wood and Talling 1988; Gizaw 1996; Zinabu et al. 2002) to emphasize possible evolutionary trends

This demonstrates that there is not a unique—ubiquitous—trend valid for the whole region and indicates a more complex framework. The variation of the chloride content through the investigated interval suggests that evaporative rate remained relatively constant in the Ziway–Langano–Abijata–Shala and Awasa systems, whereas evaporation slightly increased in the Abaya–Chamo sector. This evidence implies that microclimatic conditions diversify the water balance in distinct MER sectors. However, other factors have to be considered since

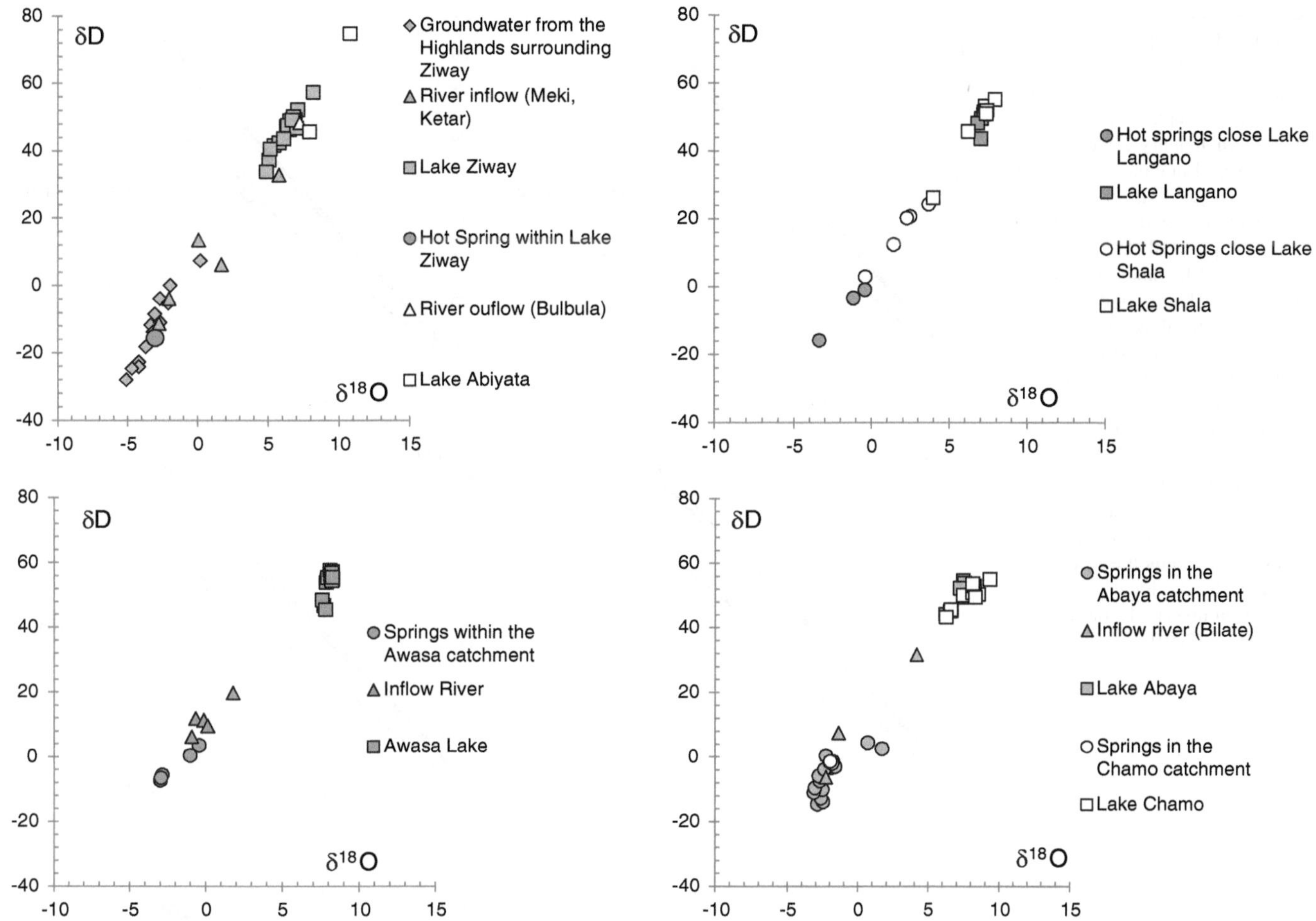

Fig. 18.8 $\delta^{18}O$–δD isotopic compositions of water of the studied lakes and related inflow rivers/springs. The authors' data (Table 18.1) include samples collected in January 2006 for lakes Ziway, Langano, Shala, and Abijata (Rango et al. 2010b; 2013) and new analyses carried out on samples collected in January 2012 for lakes Awasa, Abaya, and Chamo, and are supplemented with data provided by Kebede et al. (2009)

tectonic process and human impact (mainly land use change) could have contributed to overprint the climatic effects on the hydrological budget (Vallet-Coulomb et al. 2001; Ayenew 2002; Gebreegziabher 2004; Legesse 2004).

To provide further indication on the possible relationships between climatic changes and water composition, we focused on the water isotopic ratios of hydrogen and oxygen (Table 18.4). As mentioned in the introductive section, the $\delta^{18}O$–δD of meteoric water is related to temperature-dependent fractionation processes occurring during the hydrological cycle, and each geographic site displays a specific precipitation composition, which in turn is transferred to the local surface waters (Clark and Fritz 1997). In the study area, the inflow rivers and the springs that significantly contribute to the lake water budget have light isotopic composition often characterized by negative $\delta^{18}O$ and δD, whereas the lake water invariably have heavy isotopic composition characterized by positive $\delta^{18}O$ and δD (Fig. 18.8). The most extreme isotopic composition has been measured in the ultra-evaporated water of lake Abijata, showing $\delta^{18}O$ 10.8 and δD 74.8.

The effects of evaporation processes are also emphasized in the $\delta^{18}O$–δD diagrams (Fig. 18.9), in which the lake compositions are compared with the local meteoric lines (i.e., the isotopic composition of the local rains represented by the black lines), specifically proposed for the study area by Rango et al. (2010b) and Kebede and Travi (2012). The isotopic compositions of the lake waters systematically deviate from these meteoric lines, showing a $\delta^{18}O$ displacement toward the right quadrant of the diagram. This deviation led Rango et al. (2010b) to define a peculiar local evaporation line (LEL; dashed line), specific for the MER region.

It is worth to note the comparison between our isotopic data and those reported in the literature for the last 35 years. In particular, our sampling campaigns were carried out in January 2006 for Ziway, Langano, Shala, and Abijata lakes (and related rivers and springs) and in January 2012 for

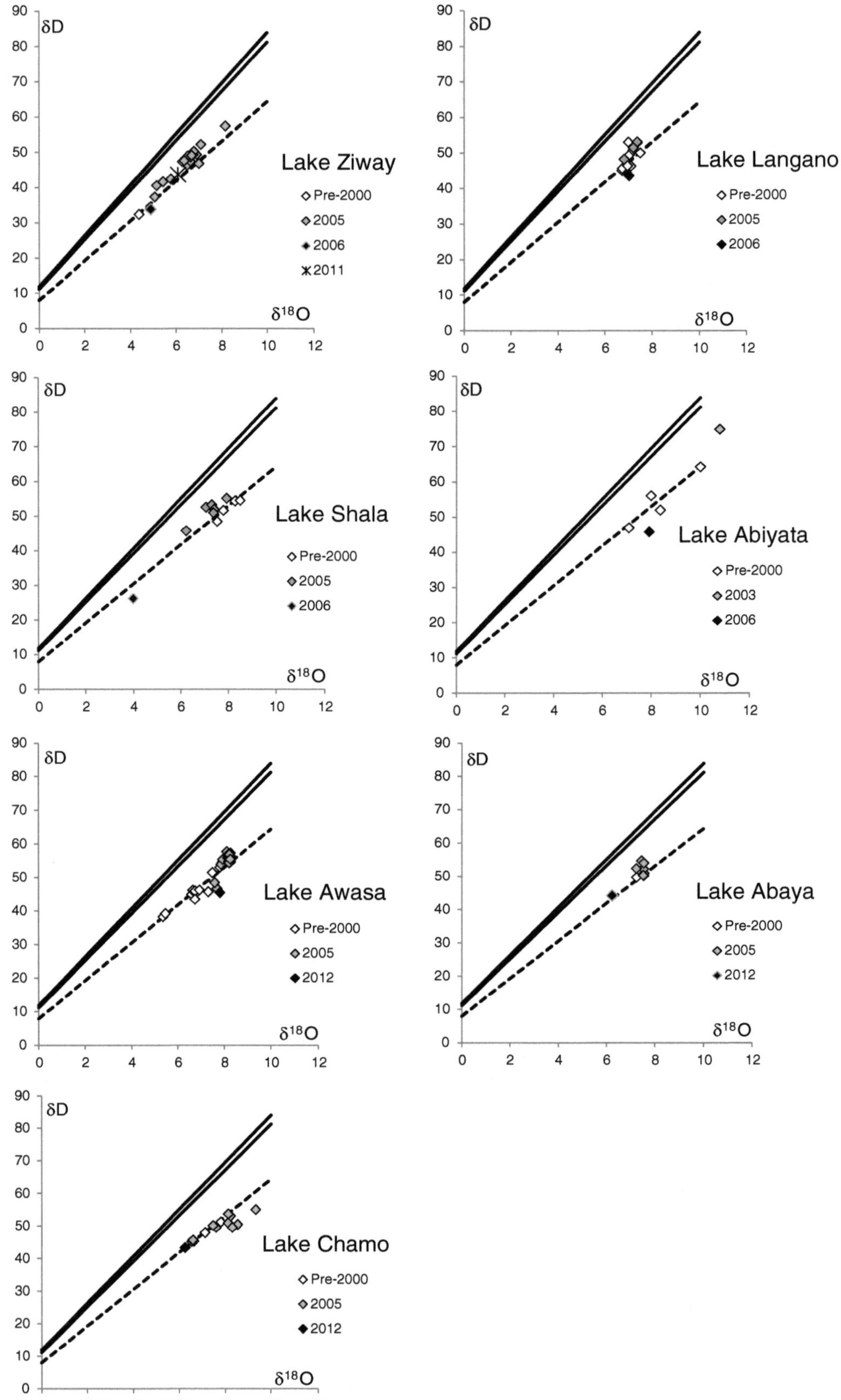

Fig. 18.9 Temporal evolution of the $\delta^{18}O$–δD isotopic compositions of water of the studied lakes. The authors' data (Table 18.1) include samples collected in January 2006 for lakes Ziway, Langano, Shala, and Abijata (Rango et al. 2010b; 2013) and new analyses carried out on samples collected in January 2012 for lakes Awasa, Abaya, and Chamo, and are compared with older literature data provided by Kebede et al. (2009 and references therein). Composition of local rains is delineated by the local meteoric water lines (LMWL; *black lines*) defined for the MER region by Rango et al. (2010b) and Kebede and Travi (2012). Evaporative evolutionary trends are also delineated by the local evaporation line (LEL) defined by Rango et al. (2010b; *dashed line*)

Awasa, Abaya, Chamo lakes (and related rivers and springs). The isotopic data available in the literature for the same lakes are referred to periods before the year 2000 (Craig et al. 1977; Gizaw 1989; Ayenew 1998) and to the year 2005 (Kebede et al. 2009).

An univocal temporal progression of evaporation trends (i.e., an increase of the $\delta^{18}O$–δD values along the LEL line) is not observed. Considering literature data, we observe an increase of the evaporative effects in the year 2005 with respect to the last decade of the twentieth century. By contrast, after the year 2005, our 2006 and 2012 data suggest a decline of the evaporative effects. These trends seem to be supported by the climatological parameters (Fig. 18.2), which emphasize fluctuations and a significant climatic forcing around the years 2004/2005. This, in turn, implies that the study lakes quickly respond to the environmental changes, possibly within one (or two) year(s). For this reason, they represent important hydro-archives (Zuppi and Sacchi 2004) and their continuous isotopic monitoring will be useful to evaluate ongoing climatic changes.

18.6 Conclusions

Evidence of ongoing changes in lake physiography and water chemistry is potentially related to concomitant changes in the climatic conditions. Unfortunately, the climatic influence is disturbed by multiple natural and anthropogenic factors. For example, active tectonic processes could favor the formation of new hydrothermal springs that change the water balance in terms of volumes and/or chemistry and can alter the water inflow/outflow. Human activity (such as deforestation) could increase surface runoff facilitating and speeding water flow toward the lakes, but at the same time, withdrawal of water from the lakes, mainly for agricultural purposes, may occur.

In such a complex scenario, we do not observe systematic trends valid for all the investigated lakes. The only two lakes that show a coherent temporal decrease in the surface extension coupled with an increase of salinization are Lake Abijata and Lake Chamo. The behavior of Lake Awasa is difficult to be deciphered, because Table 18.1 suggests that the surface is shrinking, but other observations reveal that the level has slightly risen (Ayenew 2004) and the water chemistry (Fig. 18.7) does not show substantial variations.

Other lakes show nearly constant areal extension and steady hydrochemistry, with second-order fluctuations that are difficult to be decoded in terms of climatic changes. This incongruence probably suggests that the physiographic data reported in the literature are scarcely constrained and have to be taken with caution. Moreover, the record defined for the last 30 years by stable isotopes also denotes fluctuations of the climatic parameters with a more extreme evaporation just preceding the year 2005 and then declining to more "normal" conditions.

We therefore propose to continue the monitoring of the same climatic and hydrochemical parameters in the next years in order to improve the existing dataset, which could be useful to unravel possible environmental changes.

References

Alemayehu T, Ayenew T, Kebede S (2006) Hydrogeochemical and lake level changes in the Ethiopian Rift. J Hydrol 316:290–300

Appelo CAJ, Postma D (2005) Geochemistry, groundwater and pollution, 2nd edn. Balkema, Rotterdam

Awulachew SB, Yilma AD, Loulseged M, Loiskandl W, Ayana M, Alamirew T (2007) Water resources and irrigation development in Ethiopia. International Water Management Institute, Colombo, 78 p. (Working Paper 123)

Ayenew T (1998) The hydrogeological system of the Lake District Basin, Central Main Ethiopian Rift. PhD Thesis, Free University of Amsterdam, The Netherlands

Ayenew T (2002) Recent changes in the level of Lake Abiyata, central main Ethiopian Rift. Hydrolog Sci J 47:493–503

Ayenew T (2003) Environmental isotope-based integrated hydrogeological study of some Ethiopian rift lakes. J Radioanal Nucl Ch 257:11–16

Ayenew T (2004) Environmental implications of changes in the levels of lakes in the Ethiopian Rift since 1970. Reg Environ Change 4:192–204

Beccaluva L, Bianchini G, Natali C, Siena F (2009) Continental flood basalts and mantle plumes: a case study of the northern Ethiopian plateau. J Petrolol 50:1377–1403

Benvenuti M, Carnicelli S, Belluomini G, Dainelli N, Di Grazia S, Ferrari GA, Iasio C, Sagri M, Ventra ., Atnafu B, Kebede S (2002) The Ziway–Shala lake basin (main Ethiopian rift, Ethiopia): a revision of basin evolution with special reference to the Late Quaternary. J Afr Earth Sci 35: 247–269

Benvenuti M, Bonini M, Tassi F, Corti G, Sani F, Agostini A, Manetti P, Vaselli O (2013) Holocene lacustrine fluctuations and deep CO_2 degassing in the northeastern Lake Langano Basin (Main Ethiopian Rift). J Afr Earth Sci 77:1–10

Bewketu K (2010) Hydrodynamics of selected ethiopian Rift Lakes. MSc thesis, Addis Ababa University

Billi P, Fazzini M, Fratianni S, Dramis F, Biancotti A, Bisc C (2003) Etude de la distribution spatio-temporelle des parametres climatiques dans le territoire de l'Ethiopie: Premiers resultats [Study of the spatiotemporal distribution of climatic parameters in the territory of Ethiopia: initial results]. Dokumentacja Geograficzna 29:49–52

Billi P (1998) Climatic change. In: Sagri M (ed.) Land resources inventory, environmental changes analysis and their application to agriculture in the Lake Region (Ethiopia), Final Report, STD3 Project n. CT92-0076, European Commission, 16–18

Brunelli G, Canicci G, Loffredo S, Maldura CM, Morandini G, Parenzan P, Vatova A, Zolezzi C (1941) Esplorazione dei laghi della Fossa Galla. Collezione Scientifica e Documentaria dell'Africa Italiana III, Vol I, Ministero dell'Africa Italiana, Roma

Chernet T (1982) Hydrogeology of the Lakes Region, Ethiopia (Lakes Ziway, Langano, Abiyata, Shalla and Awasa). Report of the Ethiopian Institute Geological Survey, 97 p

Chernet T (2011) Geology and hydrothermal resources in the northern Lake Abaya area (Ethiopia). J Afr Earth Sci 61:129–141

Clark ID, Fritz P (1997) Environmental isotopes in hydrogeology. CRC Press, Boca Raton

Corti G (2009) Continental rift evolution: from rift initiation to incipient break-up in the Main Ethiopian Rift, East Africa. Earth Sci Rev 96:1–53

Craig H, Lupton J E, Horowiff RM (1977) Isotope geochemistry and hydrology of geothermal waters in the Ethiopian Rift Valley. Scripps Inst. of Oceanography, Report 77–14, 160 p

Deribe E, Rosseland BO, Borgstrøm R, Salbu B, Gebremariam Z, Dadebo E, Norli HR, Eklo OM (2011) Bioaccumulation of persistent organic pollutants (POPs) in fish species from Lake Koka, Ethiopia: the influence of lipid content and trophic position. Sci Tot Env 410–411:136–145

Deriemaecker A (2013) A comparison regarding the physico-chemical variables and zooplankton community characteristics of two Ethiopian Rift Valley Lakes: Lake Chamo and Lake Abaya. Unpublished Master thesis at the KU Leuven autonomous university, 81 p

Dessie G, Kleman J (2007) Pattern and magnitude of deforestation in the South Central Rift Valley Region of Ethiopia. Mt Res Dev 27:162–168

Froehlich K, Gonfiantini R, Rozanski K (2005) Isotopes in lakes studies: a historical perspective. In: Aggarwal PK, Gat JR, Froehlich K (Eds) Isotopes in the water cycle: present and future of a developing science, 139–150

Gasse F, Street FA (1978) Late quaternary lake-level fluctuations and environments of the northern Rift Valley and Afar region (Ethiopia and Djibouti). Palaeogeogr Palaeocl 24:279–325

Gat JR (1995) Stable isotopes of fresh and saline lakes. In: Lerman A, Imboden D, Gat J (eds) Physics and chemistry of lakes. Springer, Berlin, pp 139–162

Gebreegziabher I (2004) Assessment of the water balance of Lake Awassa catchment, Ethiopia. MSc thesis, International Institute for Geo-Information Science and Earth Observation, Enschede

Geze F (1975) New dates on ancient Galla lakes levels. Bulletin Geophysics Observatory University Addis Ababa 15, 119-124

Ghinassi M, D'Oriano F, Benvenuti M, Awramik S, Bartolini C, Fedi M, Ferrari G, Papini M, Sagri M, Talbot M (2012) Shoreline fluctuations of Lake Hayk (northern Ethiopia) during the last 3500 years: geomorphological, sedimentary, and isotope records. Palaeogeogr Palaeocl 365–366:209–226

Giambastiani BMS, Colombani N, Mastrocicco M (2013) Characterization of the lowland coastal aquifer of Comacchio (Ferrara, Italy): hydrology, hydrogeochemistry and evolution of the system. J Hydrol 501:35–44

Gizaw B (1989) Geochemical investigation of the Aluto-Langano Geothermal field, Ethiopian Rift Valley. M. Phil. Thesis, University of Leeds, UK

Gizaw B (1996) The origin of high bicarbonate and fluoride concentrations in waters of the Main Ethiopian Rift Valley, East African Rift system. J Afr Earth Sci 22:391–402

Goerner A, Jolie E, Gloaguen R (2009) Non-climatic growth of the saline Lake Beseka, Main Ethiopian Rift. J Arid Environ 73:287–295

Kebede E, Zinabu GM, Ahlgren I (1994) The Ethiopian rift valley lakes: chemical characteristics of a salinity-alkalinity series. Hydrobiologia 288:1–12

Kebede S, Travi Y, Rozanski K (2009) The $\delta^{18}O$ and $\delta^{2}H$ enrichment of Ethiopian lakes. J Hydrology 365:173–182

Kebede S, Travi Y (2012) Origin of the $\delta^{18}O$ and $\delta^{2}H$ composition of meteoric waters in Ethiopia. Quatern Intern 257:4–12

Kirchner JW, Tetzlaff D, Soulsby C (2010) Comparing chloride and water isotopes as hydrological tracers in two Scottish catchments. Hydrol Process 24:1631–1645

Klemper SL, Cash MD (2007) Temporal geochemical variation in Ethiopian Lakes Shala, Arenguade, Awasa, and Beseka: possible environmental impacts from underwater and borehole detonations. J Afr Earth Sci 48:174–198

Legesse D, Vallet-Coulomb C, Gasse F (2004) Analysis of the hydrological response of a tropical terminal lake, Lake Abiyata (Main Ethiopian Rift Valley) to changes in climate and human activities. Hydrol Processes 18(3):487–504

Le Turdu C, Tiercelin JJ, Gibert E, Travi Y, Lezzar KE, Richert JP, Massault M, Gasse F, Bonnefille R, Decobert M, Gensous B, Jeudy V, Tamrat E, Umer M, Martens K, Atnafu B, Chernet T, Williamson D, Taieb M (1999) The Ziway-Shala lake basin system, Main Ethiopian Rift: Influence of volcanism, tectonics, and climatic forcing on basin formation and sedimentation. Palaeogeogr Palaeocl 150:135–177

Loffredo S, Maldura CM (1941) Risultati generali delle ricerche di chimica limnologica sulle acque dei laghi dell'Africa orientale italiana esplorati dalla Missione Ittiologica. Esplorazione dei laghi della Fossa Galla. Collezione Scientifica e Documentari dell'Africa Italiana III, Vol. I, Ministero dell'Africa Italiana, Roma: 181–200

Manno E, Vassallo M, Varrica D, Dongarrà G, Hauser S (2007) Hydrogeochemistry and water balance in the coastal wetland area of Biviere di Gela, Sicily, Italy. Water Air Soil Poll 178:179–193

Natali C, Beccaluva L, Bianchini G, Siena F (2011) Rhyolites associated to Ethiopian CFB: clues for initial rifting at the Afar plume axis. Earth Planet Sci Lett 312:59–68

Omer-Cooper J (1930) Dr. Hugh Scott's expedition to Abyssinia. A preliminary investigation of the freshwater fauna of Abyssinia. Proc Zool Soc Lond 195–207

Rango T, Bianchini G, Beccaluva L, Ayenew T, Colombani N (2009) Hydrogeochemical study in the main Ethiopian Rift: new insights to source and enrichment mechanism of fluoride. Environ Geol 58:109–118

Rango T, Bianchini G, Beccaluva L, Tassinari R (2010a) Geochemistry and water quality assessment of central main Ethiopian Rift natural waters with emphasis on source and occurrence of fluoride and arsenic. J Afr Earth Sci 57:479–491

Rango T, Petrini R, Stenni B, Bianchini G, Slejko F, Beccaluva L, Ayenew T (2010b) The dynamics of central main Ethiopian Rift waters: evidence from δD, δ18O and $^{87}Sr/^{86}Sr$ ratios. Appl Geochem 25:1860–1871

Rango T, Colombani N, Mastrocicco M, Bianchini G, Beccaluva L (2010c) Column elution experiments on volcanic ash: geochemical implications for the Main Ethiopian Rift waters. Water Air Soil Poll 208:221–233

Rango T, Kravchenko J, Atlaw B, Peter GM, Jeuland M, Merola B, Vengosh A (2012) Groundwater quality and its health impact: an assessment of dental fluorosis in rural inhabitants of the main Ethiopian Rift. Environ Int 43:37–47

Rango T, Vengosh A, Dwyer G, Bianchini G (2013) Mobilization of arsenic and other naturally occurring contaminants in groundwater of the Main Ethiopian Rift aquifers. Water Res 47:5801–5818

Rozanski K, Froehlich K, Mook WG (2000) Surface water. In: Mook WG (Ed) Environmental isotopes in the hydrological cycle: principles and applications. Technical documents in hydrology (IAEA)

Sagri M (ed) (1998) Land resources inventory, environmental changes analysis and their application to agriculture in the Lake Region (Ethiopia), Final Report, STD3 Project n. CT92-0076, European Commission, Bruxelles

Sagri M, Bartolini C, Billi P, Ferrari G, Benvenuti M, Carnicelli S, Barbano F (2008) Latest pleistocene and holocene river network evolution in the Ethiopian Lakes Region. Geomorphology 94:79–97

Saxena VK, Ahmed S (2001) Dissolution of fluoride in groundwaters: a water–rock interaction study. Environ Geol 40:1084–1087

Scanlon BR, Keese KE, Flint AL, Flint LE, Gaye CB, Edmunds WM, Simmers I (2006) Global synthesis of groundwater recharge in semiarid and arid regions. Hydrol Process 20:3335–3370

Tirumalesh K, Shivanna K, Jalihal AA (2007) Isotope hydrochemical approach to understand fluoride release into groundwaters of Ilkal area, Bagalkot District, Karnataka, India. Hydrogeol J 15:589–598

Vallet-Coulomb C, Legesse D, Gasse F, Travi Y, Chernet T (2001) Lake evaporation estimates in tropical Africa (Lake Ziway, Ethiopia). J Hydrol 245:1–18

Von Damm KL, Edmond JM (1984) Reverse weathering in the closed-basin lakes of the Ethiopian Rift. Am J Sci 284:835–862

Wagesho N, Jain M, Goel N (2013) Effect of climate change on runoff generation: application to rift valley lakes basin of Ethiopia. J Hydrol Eng 18:1048–1063

Wood RB, Talling JF (1988) Chemical and algal relationships in a salinity series of Ethiopian inland waters. Hydrobiologia 158:29–67

Yohannes YB, Ikenaka Y, Nakayama SMM, Saengtienchai A, Watanabe K, Ishizuka M (2013) Organochlorine pesticides and heavy metals in fish from Lake Awassa, Ethiopia: Insights from stable isotope analysis. Chemosphere 91:857–863

Zinabu GM, Chapman LJ, Chapman CA (2002) Conductivity as predictor of a total cations and salinity in Ethiopian lakes and rivers: revisiting earlier models. Limnologica 32:21–26

Zuppi GM, Sacchi E (2004) Hydrogeology as a climate recorder: Sahara-Sahel (North Africa) and the Po Plain (Northern Italy). Global Planet Change 40:79–91

The Landscape and Landforms of the Ogaden, Southeast Ethiopia

19

Daniel Mège, Peter Purcell, Stéphane Pochat, and Thomas Guidat

Dedication This chapter is dedicated to the memory of hydrogeologist Dr. Costantino Faillace, who worked extensively on the basalt aquifers of Somalia and India, among many other countries, in a life dedicated to providing clean water supplies to remote communities. He died in 2012.

Abstract

The Ogaden region is located on the Somali Plateau, in southeast Ethiopia. Originally a clan-based term, the Ogaden is now commonly used for the entire region below about 1,500 m a.s.l., an area of some 300,000 km^2 that encompasses most of the Somali Regional State and includes the southwest portion of Oromia. The climate is hot, arid to semiarid, corresponding to the Ethiopian *bereha* and *kolla* climatic zones. Three basic physiographic provinces are recognized: the Genale and Shebele drainage basins and the Eastern Slope and Plains. The two drainage basins include spectacular upstream canyons that witness the vertical movements that have accompanied the succession of rifting events in the Ethiopian Rift, Afar, and the Gulf of Aden. In strong contrast, the Eastern Slopes and Plains is dipping less than 0.4° on average over hundreds of kilometers to the southeast and is mantled by red sands. Several remarkable Ogaden landforms are described and analyzed, including volcanic, fluvial, and gravitational features, some having few equivalents in other areas on Earth. A variety of volcanic landforms are present across the region, reflecting the complex Cenozoic history of the Ogaden's margins. For instance, meandering basalt hills provide a textbook example of inverted topography by fossilizing paleodrainage networks of various ages. The modern drainage network provides information on the genesis of the mega-geomorphology of the Ogaden and documents its uplift history. In western Ogaden, the deep incision has exposed the Cretaceous evaporites and triggered the development of one of the largest gravitational spreading domains on Earth, the Audo Range.

D. Mège (✉)
WROONA Research Group, Institute of Geological Sciences, Research Centre in Wrocław, Polish Academy of Sciences, ul. Podwale 75, 50-449 Wrocław, Poland
e-mail: daniel.mege@univ-nantes.fr

D. Mège · S. Pochat · T. Guidat
Université de Nantes, CNRS UMR6112, , Laboratoire dePlanétologie et Géodynamique de Nantes, Nantes, France

P. Purcell
P&R Geological Consultants, 141 Hastings Street, Scarborough, WA 6019, Australia

D. Mège · S. Pochat
Observatoire des Sciences de l'Univers Nantes Atlantique (OSUNA, CNRS UMS 3281), Nantes, France

P. Billi (ed.), *Landscapes and Landforms of Ethiopia*, World Geomorphological Landscapes,
DOI 10.1007/978-94-017-8026-1_19, © Springer Science+Business Media Dordrecht 2015

Keywords
Ogaden • Wabe Shebele • Audo Range • Marda Range • Basalt • Topographic inversion • Rift flank uplift • Incision rate

19.1 Introduction

The landscape and landforms of Ethiopia, and the peoples and cultures that evolved there, have captured the imagination since biblical times. This dramatic topography is the creation of earth forces uplifting and splitting apart the Afro-Arabian continent for more than 30 Myr. A vast rift complex now cleaves south through the uplifted dome of Ethiopia, separating western (Ethiopian) and southeastern (Somali) plateaus, and creating a landscape that has profoundly influenced the history and culture of the region. The high and dissected western plateau, long known as Abyssinia, sheltered the Coptic Christian kingdom from Islamic and European expansionism. Conversely, the southeastern plateau's easy access to the Islamic coast saw early conversion of Sidama, Oromo, and Somali clans and the establishment of Harar in the thirteenth century as the religious center of the Horn of Africa.

In this chapter, we describe the landscape and landforms of the lower regions of the southeastern plateau, the area known informally as the Ogaden (Fig. 19.1). This poorly accessible and seldom visited region is not as well known or studied as the Ethiopian Plateau and is often dismissed as 'desert.' Certainly, the region has vast red soil plains, but it also has the grand canyons of the Wabe Shebele and its tributary rivers, the spectacular Audo Range gravitational spreading complex, and a variety of landforms, soils, and vegetation that reflect the changing elevation of the plateau. We summarize the geographical and geological information and then focus on three major features of the Ogaden landscape: the volcanic landforms, the drainage system and its evolution, and the very large-scale gravitational spreading structures of the Audo Range. These landforms are described and also discussed from a mega-geomorphology perspective.

There is no formal definition of 'the Ogaden' but, in popular usage, it corresponds approximately to the area of the Somali Plateau from about 1,500 m elevation to the Somalia and Kenya borders. The Ogaden takes its name from the Ogaden clan (Somali, Ogaadeen), a subclan of the Darod, who live in the region. The term originally referred only to the Ogadeni lands but is now commonly used for the entire Somali region, even, albeit rather imprecisely, as an alternative for the Somali Administrative Region (Hagmann and Khalif 2006; Temin 2006). Climatically, it corresponds to the hot arid region of the Ethiopian *bereha* and *kolla* climatic zones. In geological terms, it is the area where sedimentary rocks occur on or near the surface, underlain by a deep sedimentary basin.

The modern use of the Ogaden name is opposed by some people in the region because they feel it implies Ogadeni political leadership, both in general and specifically over non-Ogadeni lands (Hagman 2007). However, there is no useful alternative term and it is now widely used and accepted in many fields of study.

19.2 Landscapes

19.2.1 Human Geography

The Ogaden occupies the southern portion of the Somali Regional State of Ethiopia, and the southeastern portion of the Oromia State. Jijiga is now the regional capital of the Somali State and the administrative center for the region. Adama (or Nazret) is the capital of Oromia. Figure 19.1 shows the Ogaden region, as described in this paper, with the main towns, rivers, and roads. Important towns include Aware (Teferi Ber), Daghabur, Gode, Kebri Dehar, Fik, Filtu, and Werder (Harar, the historical capital of the region, is now designated as a separate state, as is Dire Dawa). The Jijiga/Ferfer road coincides with a major topographic and geological divide between eastern and western Ogaden regions.

The Ogaden region encompasses about 300,000 km^2 and has a population of the order of 5 million, predominantly Somali and Oromo clans, with each constituting near 90 % of the population in their state (Ethiopian Government Portal 2014). Historically, the people of the Ogaden have been primarily nomadic pastoralists, but the population is increasingly gathered into the large and small towns across the region. Agriculture predominates at higher elevations and along the lower Wabe Shebele.

Major roads provide access to the Ogaden from the national capital Addis Ababa (Fig. 19.1). In the north, where numerous major river canyons occur, secondary roads extend south from the main road to Jijiga. Access to southwestern areas is limited from the north by the Wabe Shebele canyon and is primarily via roads linking to the Imi-Gode-Mustahil road. A main road from Jijiga southeast through Kebri Dehar to Ferfer on the Somalia border links with secondary roads in the central and eastern region and provides the road link to Somalia capital, Mogadishu. Secondary roads in the Ogaden, as well as some of the main

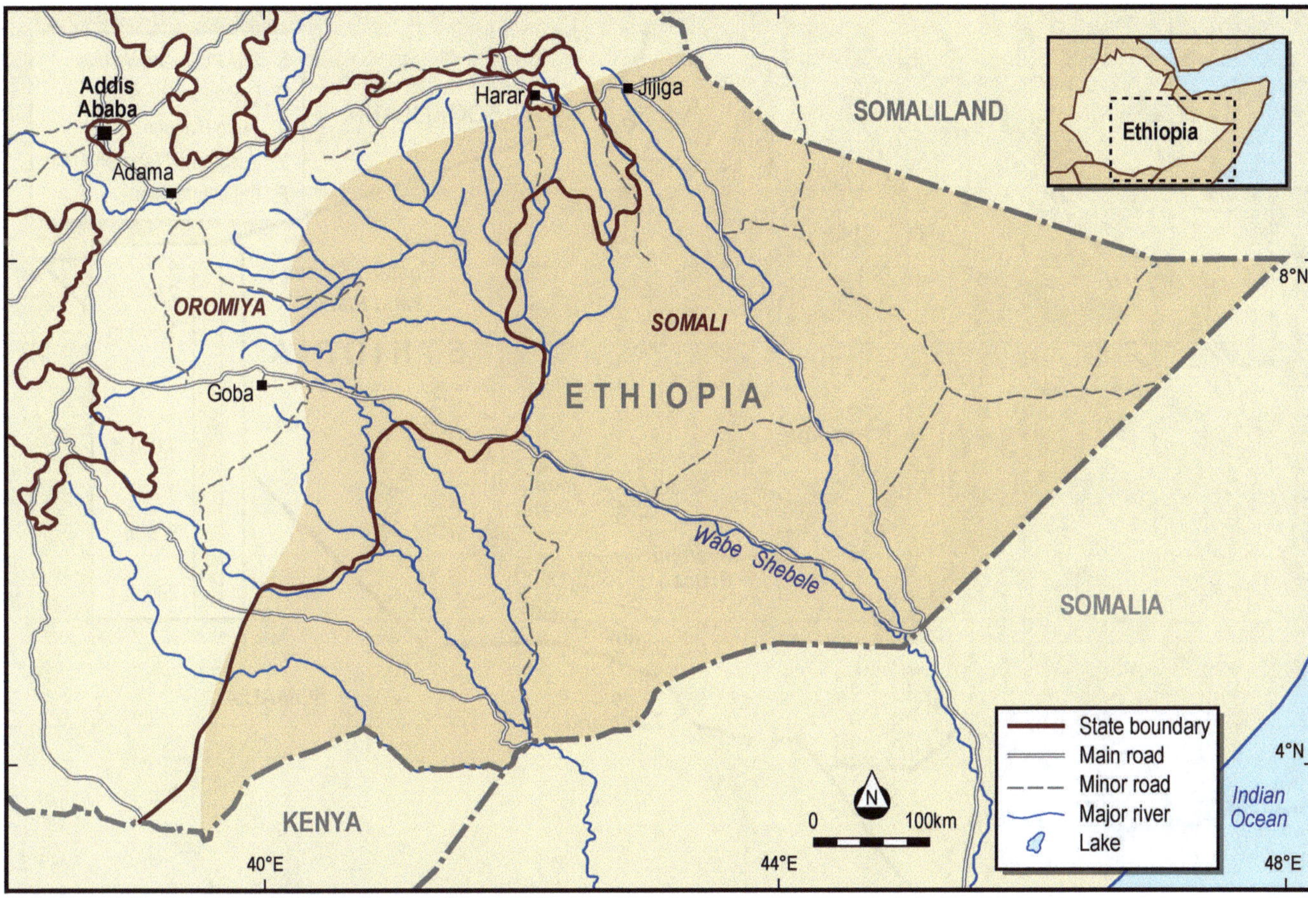

Fig. 19.1 Location map, Ogaden region, Ethiopia

roads, are unsealed and can be in poor condition, with river crossing very difficult, especially in the wet seasons.

19.2.2 Physical Geography

The digital elevation model of southeastern Ethiopia of Fig. 19.2 shows the main features of the Ogaden landscape and the clear subdivision into highly dissected western and subdued eastern regions. The boundary is relatively abrupt, along the eastern side of the uplifted Marda Range, a prominent NW/SE-trending structure thought to have formed by Phanerozoic multiphase reactivation of a Precambrian shear zone (Purcell 1976). In the western Ogaden, the high plateau rises to the rift margin rim and is dissected by the steep canyons of south- and east-draining rivers, creating a rugged topography. The steep river canyons and profiles are evidence of the recent relative uplift along the plateau rim. These rivers join the Wabe Shebele, which flows southeast across the central Ogaden into Somalia, its valley slowly diminishing from a steep-walled high canyon to a broad flood plain. Pronounced NW-trending features, such as the Marda Range complex, are clearly seen.

By contrast, the eastern Ogaden is a very gentle slope, on average dipping less than 0.4° to the southeast, with no major rivers or high topographic features. Comprehensive geomorphological studies remain to be conducted; nevertheless, an analysis of the most prominent landforms, the scattered low basaltic hills, has been undertaken and is discussed in Sect. 19.3.2. Rock outcrops are relatively rare and the plain is covered with alluvial and eolian red sand, commonly a few meters thick but locally as deep as 13 m, based on the results of oil and water bores (Walsh 1976; Faillace 1993).

The main physiographic provinces and features of the Ogaden are shown in Fig. 19.2. The basic subdivision recognizes the main provinces of the Wabe Shebele and Genale watersheds and the Eastern Slopes and Plains. The further subdivision into regions is clearly defined in some areas, such as the Audo Range, but in other cases, the boundaries between regions are diffuse. Figure 19.3 illustrates typical landscapes from several of the physiographic regions,

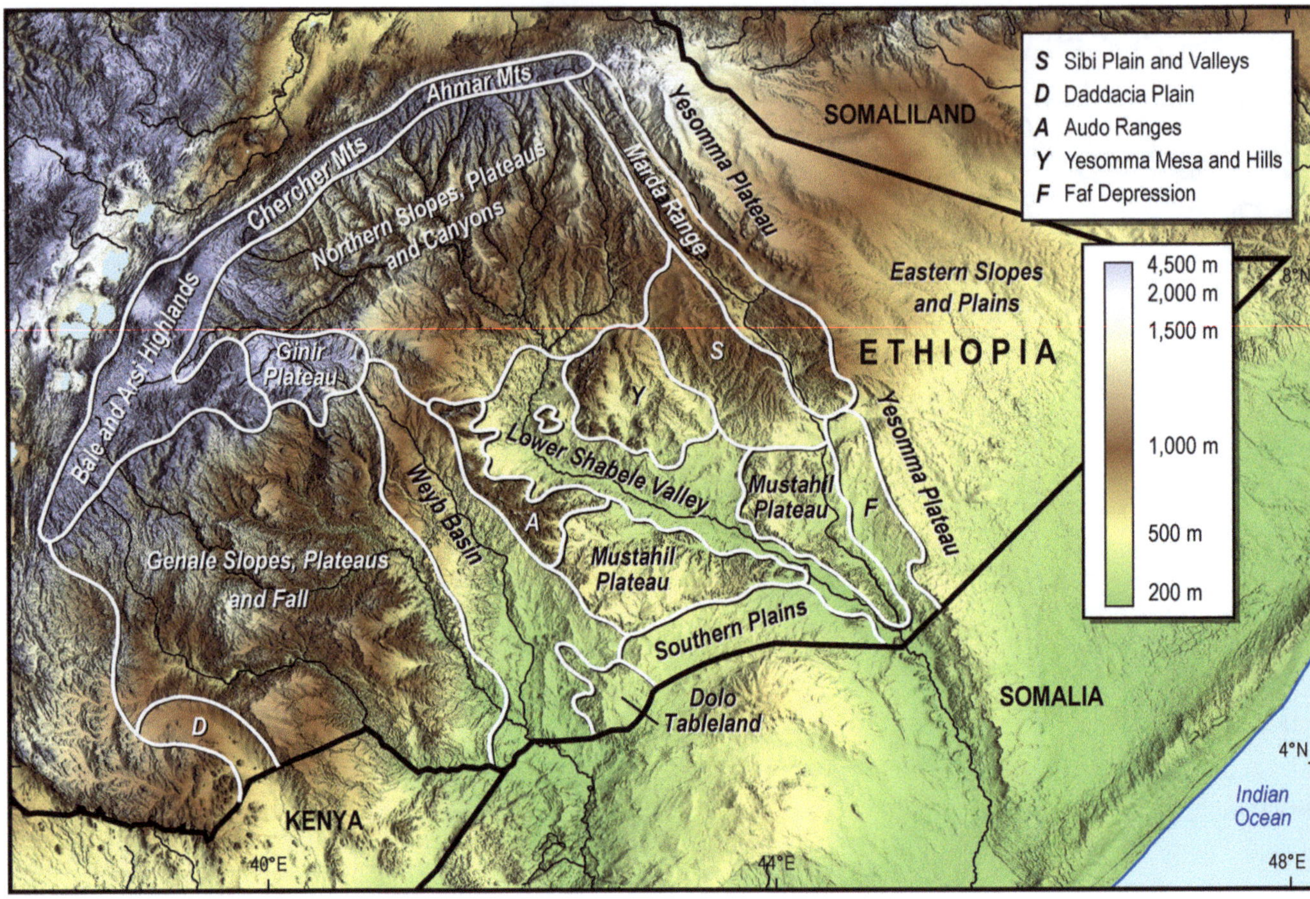

Fig. 19.2 Digital elevation model, southeast Ethiopia, showing informal physiographic subdivisions

including the Mustahil Plateau, the Eastern Slopes and Plains, and the Wabe Shebele valley.

19.2.2.1 Elevation

The Somali Plateau has its highest elevation along the uplifted faulted margins of the Afar Depression and the Main Ethiopian Rift (MER), reaching over 4,000 m a.s.l. in the Arsi and Bale mountains in the west and over 3,000 m a.s.l. in the Chercher and Ahmar mountains in the north/northwest. Except in the west, the areas above 3,000 m in elevation occur only relatively near the plateau edge and, in all areas, are associated with volcanoes or thick Tertiary basalt flows. The elevation declines to the south and east and is about 300 m along the Somalia border (Fig. 19.4).

The present elevation of the Somali Plateau is the result of geodynamic events that produced vertical movements of the crust over the Cenozoic. The Ogaden is located at the edge of a dynamic mantle plume, upwelling for more than 30 million years in eastern Africa and western Arabia (Moucha and Forte 2011). Flood lava eruption in the Ethiopian-Yemenite province, which peaked at 30–29 Ma (Hofmann et al. 1997), was accompanied by vertical displacements in the region, though whether and where it resulted in uplift or subsidence depends on an interplay of parameters (Olson 1994), details of which are limited by the scarcity of adequate geological observations (see, however, Juch 1975). On the northern side of the Ogaden, rifting of the Gulf of Aden (Leroy et al. 2012) started as early as Late Eocene, while uplift peaked at 20–18 Ma and stopped around 16 Ma, when oceanization started (Watchorn et al. 1998; Fournier et al. 2010). Northward propagation of the Main Ethiopian Rift and initiation of the southern Afar started at 11–10 Ma (Wolfenden et al. 2004; Bonini et al. 2005), and rifting has proceeded until the present. Rift-flank uplift at the western edge of the Ogaden is thought to have occurred during this interval. The overall southeastward slope of the Ogaden seems primarily a consequence of the topographic evolution of this tectonism along the northwestern edge of the Somalia Plate, although it does appear that the region had a preexisting southward slope dating to

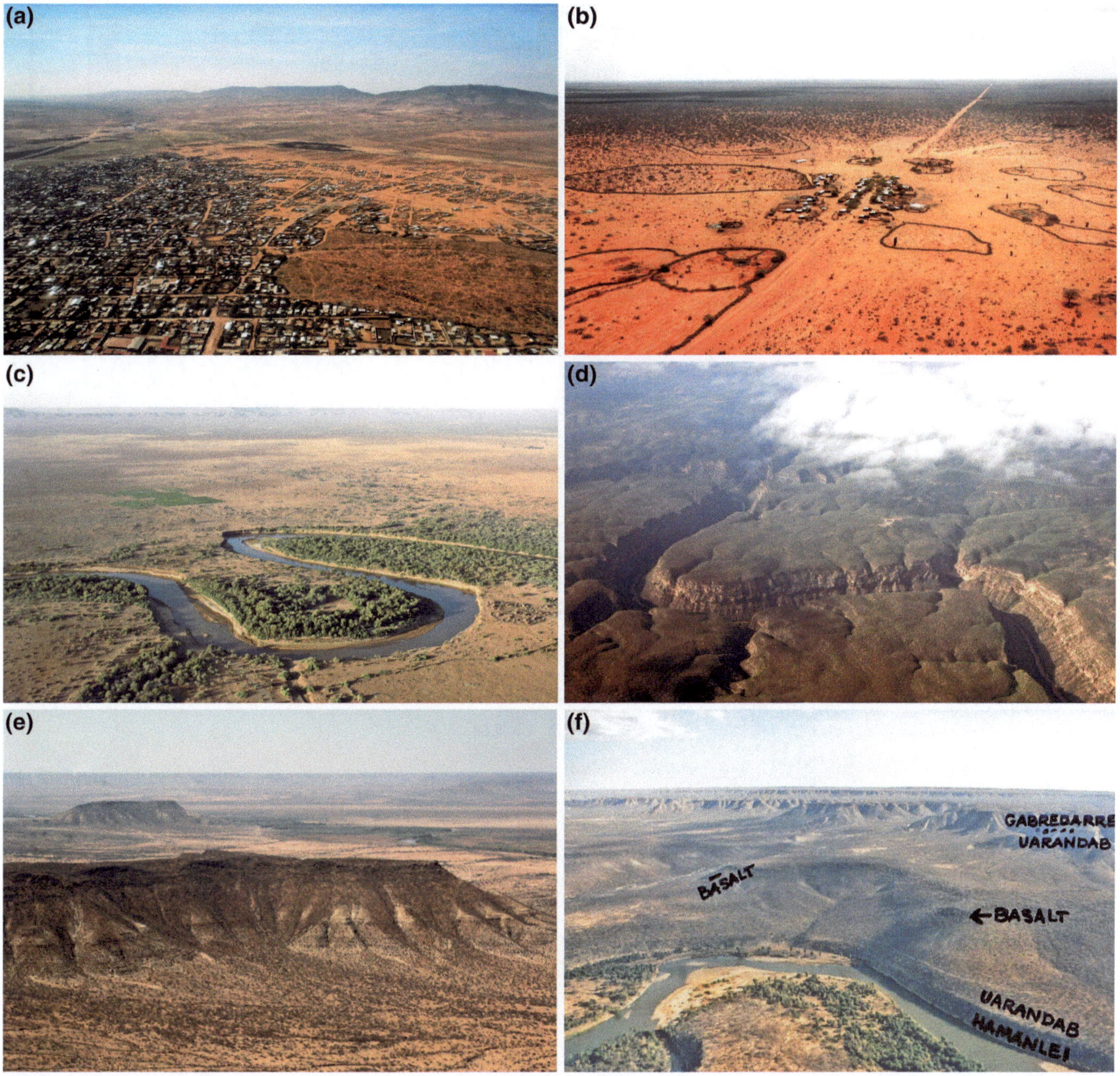

Fig. 19.3 Selected Ogaden landscapes. **a** Somali state capital Jijiga, looking southwest toward Marda Range; **b** Red sand plains in eastern Ogaden, with village in foreground; **c** Wabe Shebele near Gode; **d** Canyons of the Wabe Shebele incising Uarandab Shale plateau, south of Wabe/Ramis junction; **e** Mustahil Plateau, south-central Ogaden, looking across Wabe Shebele floodplain; and **f** Genale Plateau on Gabredarre limestone above Genale River (*Photos* by P. Purcell except **e** which is courtesy of Hunt Oil Company)

the Late Jurassic, as revealed by the thinning and transition to less marine facies of the Upper Jurassic sediments outcropping in the northern and western areas (Purcell 1981).

19.2.2.2 Climate

The climatic zones in the Ogaden and adjacent region are shown using traditional Amharic terminology on Fig. 19.5 (Ethiopian Mapping Authority—EMA 1988; Lemma 1996). The *bereha* zone (hot arid) covers the region below 500 m elevation, where annual rainfall is less than 400 mm, resulting in sparse vegetation with extensive bare ground. The *kolla* zone (warm to hot semiarid) covers the region between 500 and 1,500 m in elevation, where the average annual rainfall is generally around 600–800 mm. The higher regions of the Somali Plateau are classified as *weyna dega* (1,500–2,500 m a.s.l.; warm to cool semihumid) and *dega* (>2,500 m a.s.l.; cool to cool humid). Under the Köppen classification, the Ogaden is classified as hot arid (Bwh) and hot semiarid

Fig. 19.4 Simplified elevation map, southeast Ethiopia

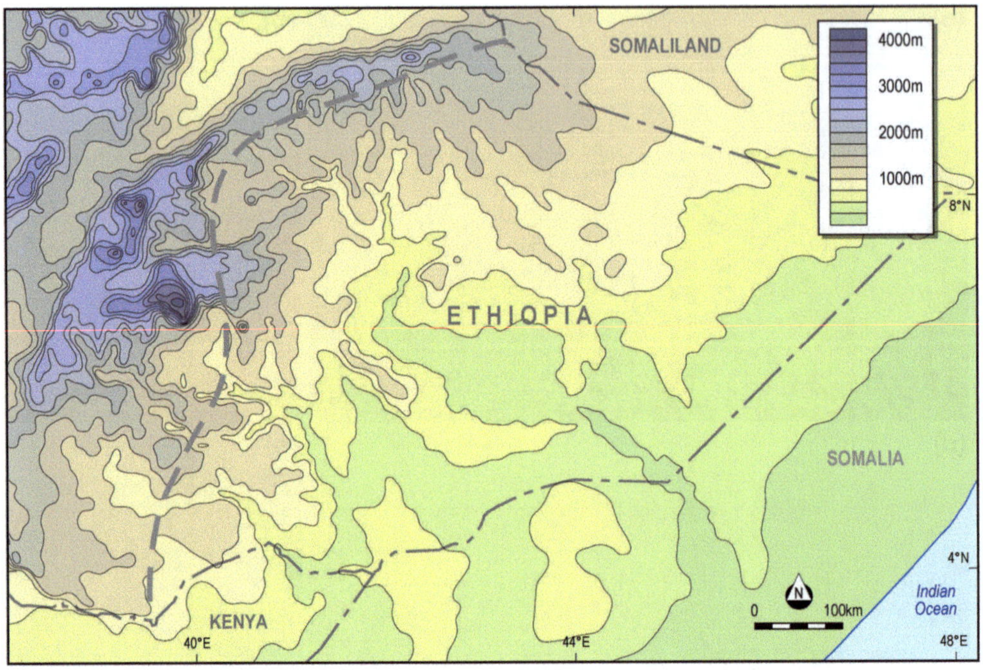

Fig. 19.5 Climate zones, southeast Ethiopia (after Lema 1996; EMA 1988)

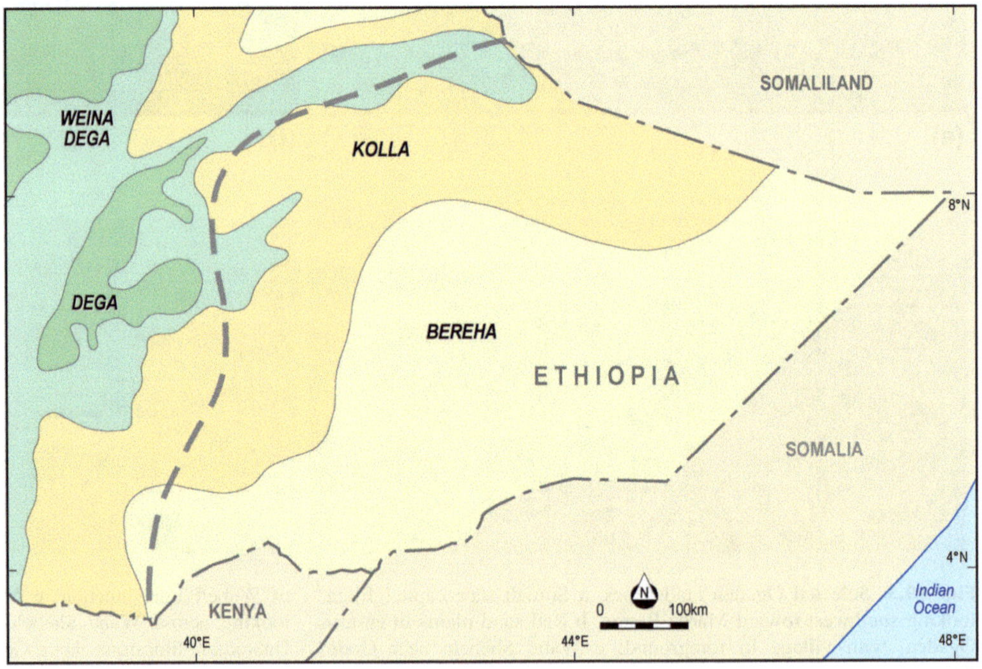

(Bsh). In the high country above 2,000 m (beyond the Ogaden sensu stricto), the climate is warm temperate (Cwb). For all intents and purposes, Koppen zones Bwh and Bsh correspond to the *bereha* and *kolla* zones.

Ogaden aridification dates back to Late Miocene, when, after the Eocene optimum, the Sahara desert formed (Micheels et al. 2009) and the Messinian salinity crisis in the Mediterranean region occurred (Feakins 2013). Aridity has been globally maintained until modern times. The main causes in the Ogaden are, firstly, that the NE and the SE monsoons that blow across the region are relatively dry and carry little rain and, secondly, that the Ogaden is in the rain shadow of the Ethiopian highlands as regards Atlantic moisture carried east by the unstable Congo airstream (Nicholson 1996; Sepulchre et al. 2006). The rainfall over most of the Ogaden falls mainly during the transition period between the monsoons, giving a bimodal rainfall distribution (EMA 1988), with rainy seasons in March–June (main) and September–November (minor), known in Somali as Dehr and Gu, respectively.

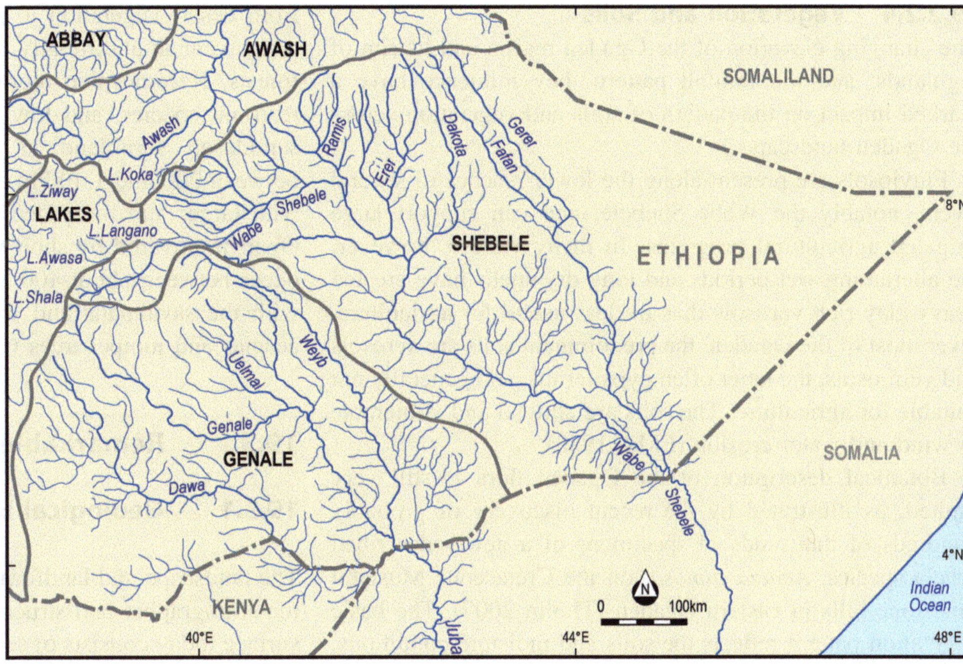

Fig. 19.6 Drainage basins, southeast Ethiopia (after EMA 1988)

Annual rainfall is between 400 and 800 mm over much of the Ogaden (refer to *bereha* and *kolla* zones on Fig. 19.5) but is very irregular and variable in the extreme east and southern areas, often 200 mm or less. Annual variability can be considerable within and between local areas, mainly linked to variability in Gu rains, but recent analyses show no overall decline in annual rainfall in the period 1965–2002 (Cheung and Senay 2008).

19.2.2.3 Hydrology

The Ogaden region contains two vast drainage systems, the Wabe Shebele basin in the northeast and the Genale basin in the southwest, as shown in Fig. 19.6 (EMA 1988). Both systems drain into the Indian Ocean.

The Wabe Shebele (Somali, River of Leopards), known as the 'Second Nile' to early Arabian geographers, is over 1,300 km long in Ethiopia (about 2,000 km overall) and has a catchment area of over 205,000 km^2. The river rises in the Arsi highlands in the west and flows initially northeast in spectacular deep canyons, reaching over 900 m deep near the ancient town of Sheik Hussein, before abruptly swinging southeast and meandering across the Ogaden into Somalia. Near Mogadishu, it is deflected southwestward by coastal dunes and, in the wet season at least, joins the Juba River and enters the ocean near Kisamayo. The main tributaries of the Wabe Shebele are the Galeti, Ramis, Erer, and Dakota rivers, rising in the north in the Ahmar Mountains and cutting deep narrow gorges in the northern slopes of the plateau, as discussed further in Sect. 19.3.3. The easternmost major river in the region is the Fafan, which, with its main tributary, the Gerer (or Yerer), flows southeast along the Marda Range and then southward, drying into the desert, except in heavy wet seasons when it flows into the Wabe Shebele.

The Genale catchment area covers about 168,000 km^2 and contains three main rivers, the Genale, Weyb (or Webi Gestro), and Dawa, all of which meet near Dolo on the Somalia border and continue south as the Juba River. The topographic divide between the Wabe Shebele and Genale watersheds runs through the Audo Range, an alignment of gravitationally unstable mesas, further discussed in Sect. 19.3.4.

In terms of landscape evolution, the rainfall regime is responsible for the development of temporary streams and small river channels throughout the Ogaden lowlands. They are dry most of the time, but are reactivated during the short and intense rainfalls, washing out the surface of the hills and accumulating the erosional products downstream in pan-type depressions. Such depositional areas are easily identified downstream from basaltic hills, because the dark clay accumulations contrast with the surrounding red sands. In the vast Wabe Shebele basin, intense rainfall in the upper catchment produces flash floods that periodically cause loss of life and destruction of buildings, crops, and livestock, by the strength of the flow of the crocodile-infested waters. At the same time, these waters and the soil they carry play a considerable role in the local agro-pastoral economy by making possible flood-recession agriculture in the riverbed in the Gu and Dehr seasons (UNDP 1999). Other major rivers, such as the Fafan and Gerer, are subject to intense floods lasting hours or days (Bauduin et al. 1973) but of much lower dramatic consequences than the Shebele floods.

19.2.2.4 Vegetation and Soils

The changing elevation of the Ogaden region and its rim of highlands, and the rainfall pattern they influence have a marked impact on the pattern of soils and vegetation across the Ogaden landscape.

Fluviosols are present along the lower reaches of several rivers, notably the Wabe Shebele, and can support large irrigated agricultural programs. In many valleys, however, the alternating wet periods and long dry spells have created heavy clay-rich vertisols that are less useful for agriculture. Over most of the Ogaden, the predominant soils are xerosols and yermosols, the latter often gypsiferous, and generally not suitable for agriculture. The soils are alluvial and vulnerable to wind and water erosion (EMA 1988).

Botanical description of the Ogaden flora is still very limited, as illustrated by the recent discovery of probably hundreds of thousands of specimens of a newly described acacia species, *Acacia fumosa*, on the Cretaceous Mustahil limestone hills in eastern Ogaden (Thulin 2007). The basic vegetation pattern reflects the soils and moisture conditions. While a complex interfingering of zones is determined by local conditions, there is an overall change from SE to NW, from desert vegetation to steppe to grassland to woodland and savannah and, finally, in the high country to coniferous forests. Xeromorphic thorny plants and grasses, often salt resistant species, and low acacias dominate the desert and semidesert scrubland in the southeast. In the steppes, between 200 and 1,400 m a.s.l., the vegetation is similar but with larger and more dense growth. The grasslands in the Ogaden, covered by short to medium, fire-resistant grasses, occur between about 400 and 2,000 m a.s.l. and interfinger with the savannahs and woodlands that are dominated by acacias and juniper trees (EMA 1988).

19.3 Remarkable Landforms

19.3.1 Geological Background

The landscape and landforms of the Ogaden are a product of the stratigraphic and structural evolution of the region. The surface today consists of sediments, Mesozoic to Neogene in age, as well as Precambrian basement rocks and Cenozoic volcanics (Fig. 19.7). Jurassic and Cretaceous sediments are

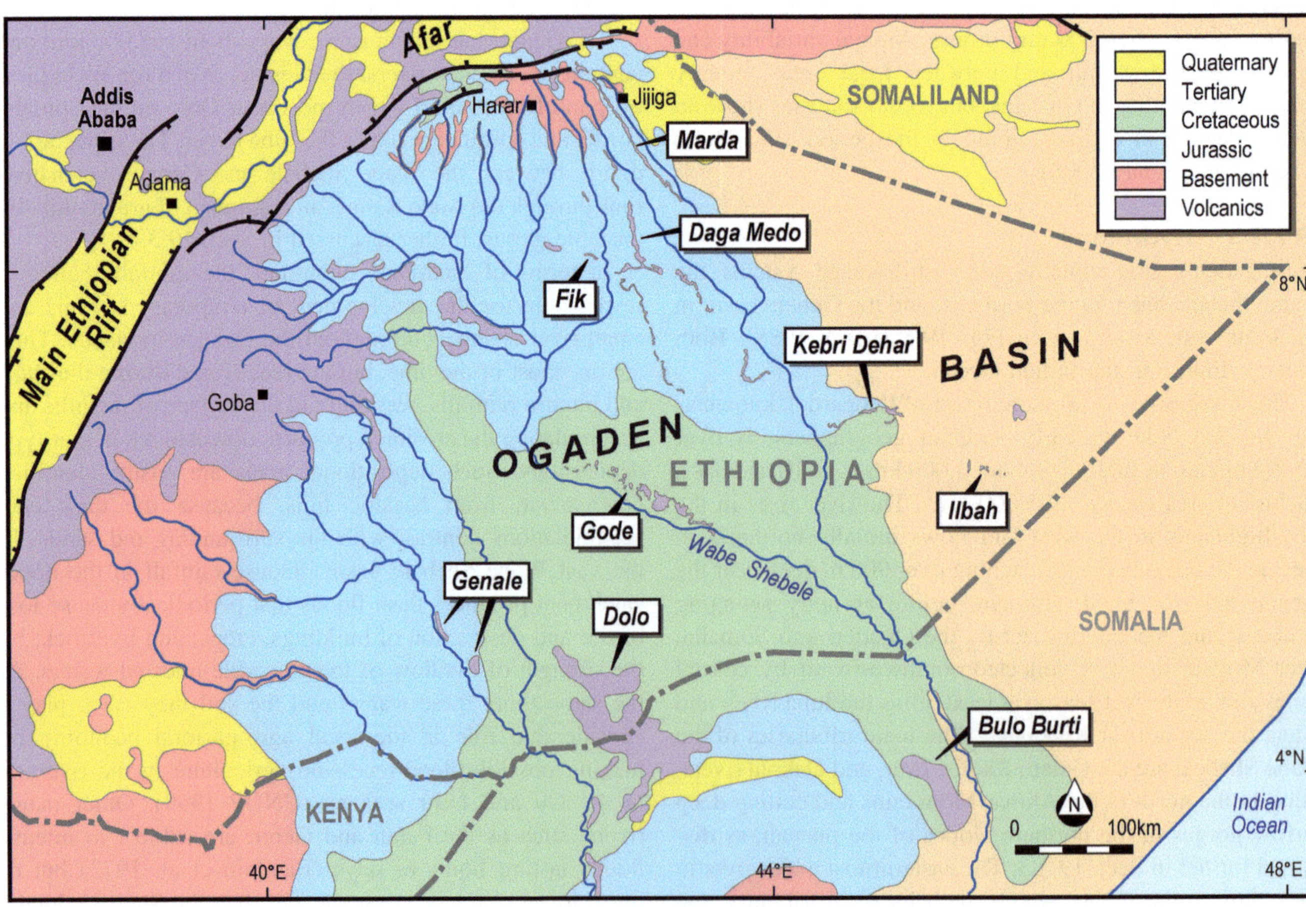

Fig. 19.7 Geology map, southeast Ethiopia, showing location of volcanic features discussed in text

the predominant outcrop in the western Ogaden, while Cenozoic sediments occur in the east; the contact between the two provinces being roughly coincident with the Marda Fault Zone. The underlying Ogaden Basin contains about 8 km of sediments deposited from Permian to Eocene time, approximately 260–35 million years ago.

The Ogaden Basin formed initially as part of the Karoo rifting in the Permian (Purcell 1981). In the Late Triassic to Early Jurassic, it developed into a sag basin and marine waters from the newly opening Indian Ocean flooded across the region, extending to northern Ethiopia, and depositing some 1,500 m or more of predominantly carbonates and evaporates in a vast shallow sea. In the Oxfordian, the seas reached their maximum flooding level and the Uarandab Shale developed. It was overlain, as the seas shallowed, by the carbonate Gabredarre Formation. The alternating cliff-forming limestones and softer shale and evaporite units result in a cliff-and-terrace topography in the eroded valleys in the northern and western Ogaden.

In the Early Cretaceous, silling of the basin, probably by a barrier reef trend near the present coastline, established a broad sabhka environment over the Ogaden, where the thick evaporites (1,600 m) of the Gorrahei Formation (previously, the Main Gypsum) were deposited in the basin center, thinning to the north and west. A marine transgression in the Aptian reestablished open marine conditions and deposited the dense limestones of the Mustahil Formation. Regional uplift in the Santonian exhumed the western Ogaden and reoriented the eastern basin to a NE-trend along the newly formed continental margin (Purcell 1981). During the Maastrichtian–Paleocene, the continental Yesomma Formation was deposited over much of the Ogaden, becoming more marine in the east. The Mustahil and Yesomma formations were to collapse later into the underlying Gorrahei evaporites by block tectonics of gravitational origin, as discussed further in Sects. 19.3.4 and 19.4.

An Eocene transgression deposited the carbonates and evaporities of the Auradu, Karkar, and Taleh formations in the eastern Ogaden and, thereafter, sedimentation was limited to the coastal Somalia area. Further widespread erosion occurred in the Oligocene, driven by uplift of the Somali Plateau and creating a landscape of eroded Tertiary, Cretaceous, and Jurassic sediments, similar in general terms to the present surface.

The early plume volcanism along what is now the Somali Plateau margin is relatively undocumented, but probably commenced during the Oligocene around 30 Ma, as on the Ethiopian Plateau (Hofmann et al. 1997), since the two regions were not then separated by the Afar and the Main Ethiopian Rift (MER). In the eastern Ogaden, the $^{40}Ar/^{39}Ar$ age of volcanic rocks geochemically akin to the plume basalts covers a time span of $\sim$8 million years, between 30 and 22 Ma (Mège et al. 2012a), encompassing the onset of rifting in the Red Sea and Gulf of Aden. Excluding the 2,000 m or more of accumulations at major volcanic centers along the plateau rim, basalt and associated volcanics on the plateau proper are 200–500 m thick in the north and west and more commonly 50 m or less in southern and eastern areas. While not comparable volumetrically with the Ethiopian Plateau volcanics, the Ogaden basalt flows were extensive, reaching over 600 km into coastal Somalia, as revealed by the widely scattered remnants seen on the surface and identified in the subsurface by aeromagnetic surveys and intersections in water bores and oil exploration wells (Purcell et al. 2011). Low viscosity lavas, that flowed for 100–200 km or more, filled river canyons and are preserved today as the inverted topography of basalt hills meandering across the landscape. The volcanic features of the Ogaden are discussed in detail in Sects. 19.3.2 and 19.3.3. A major feature of the Ogaden magmatism was the Late Oligocene Ogaden Dyke Swarm, which can be traced across the Ogaden on outcrop and magnetic data from the Afar margin to the Somalia border (Mège et al. in press).

The tectonic activity at the northern and western edge of the Somalia plate since the Miocene, the volcanic activity since the Oligocene, and contrasts in rheology and strength of the Meso-Cenozoic cover are the main geological influences on the geomorphological evolution of the Ogaden. In particular, they are key to understanding the three prominent landforms of the Ogaden discussed below: the volcanic landforms, the exhumed canyon-filling lava flows and the gravitational spreading structures.

19.3.2 Volcanic Landforms

The geomorphology of the volcanic outcrops in southeast Ethiopia have been discussed by Purcell et al. (2011), based primarily on mapping of the entire region on Landsat Geocover mosaics 1990 and 2000, high-resolution satellite images available on Google Earth™, and helicopter-supported field work in the eastern and southern Ogaden. Volcanics in southeast Ethiopia occur as broad volcanic plateaus, linear outcrops, isolated hill complexes, and meandering ribbons of exhumed paleocanyon basalt fill. The locations of various outcrops discussed in this chapter are shown in Fig. 19.7. Selected examples of Ogaden volcanic landforms are shown in Fig. 19.8. It is interesting to note that the volcanics have locally influenced settlement patterns in the region. The rolling basalt hills at Fik (Fig. 19.8b), for example, provide a natural shelter for the town. At nearby Daga Medo, the town has been built atop a recent flow, protecting residents from flash flooding during occasional heavy rains.

Fig. 19.8 Selected volcanic landforms of the Ogaden. **a** Ilbah Hills, a dyke-related volcanic complex in east central Ogaden; **b** volcanic hills near Fik town; **c** part of a Landsat ETM+ image (Geocover 2000) of basalt hills formed by inversion of canyon-filling lava flows in the ancestral Shebele river. *Arrow* shows direction of view in (**d**); **d** Wabe Shebele and flanking volcanic hills near Gode. North to the upper right; **e** Arid land farming around volcanic cones, east of Jijiga near the Somaliland border; and **f** view southeast down the Marda Range from quarry in hillside above the Marda Pass (*Photographs* by P. Purcell)

19.3.2.1 The Main Ethiopian Rift Shoulder

The most extensive basalt features occur in the west as eroded plateaus flanking the upper Wabe Shebele (Fig. 19.7). These plateaus are considered to be remnants of flows that originally extended continuously from the rift margin far into the Ogaden. The ages of the basalts are not well constrained. The older Trap volcanics overlying Jurassic sediments were dated in the early 1970s using K–Ar as Oligocene to Miocene (Megrue et al. 1972) and need reevaluation. Dating ($^{40}Ar/^{39}Ar$) of a sample from the Gara Mulata area was attempted by the authors; a well-defined but depressed plateau indicates obvious argon loss, which compromises the age determination; nevertheless, an emplacement age for the first basalts >24 Ma can be inferred. The Pliocene (2.2 Ma) age obtained for the overlying Ginir Formation rhyolites (Merla et al. 1973) also needs reevaluation. The columnar-jointed Ginir rhyolite, rising sharply above the eroded slope of the Garbaharre sandstones, forms a prominent north-facing cliff known as the Bilka Ridge (Kibrie and Yirga 2008). The multilayered flows are commonly over 100 m thick and reach a maximum of 365 m and are a prominent landform of the plateau adjacent to the Wabe Shebele canyon.

19.3.2.2 The Genale and Dolo Basaltic Tablelands

Several large, undated basaltic tables occur in southwest Ogaden and adjacent Somalia. Lava flows originating close to the Main Ethiopian Rift shoulder flowed down the Uelmal River, a tributary of the Genale River, and accumulated downstream on the Genale valley floor east of Filtu. They are now manifest as two basaltic tables, both 150 km^2 in surface area (Fig. 19.7). The northern table stands 100–150 m above the Genale River and is dissected into mesas by numerous antecedent rivers. About 20 km downstream, the southern table culminates 300 m above the Genale River. Nearly 100 km to the southeast, the Dolo basaltic table is 100–150 m high and covers some 6,500 km^2. Basaltic outliers, now separated by rivers, testify to an original much broader extent. The Dolo basalt table forms a cliff above the Weyb river on its western side, while on the eastern side its edge is masked by extensive Quaternary deposits and its true extent is not known; aeromagnetic data suggest that it could extend more or less continuously below the Quaternary mantle to the Bulo Burti basalt outcrop in Somalia, and then as far as the Indian Ocean (Bosellini 1989; Purcell et al. 2011). In the south, the geometry of the present outcrops suggests that the Dolo basaltic table could have fed the meandering basalts that filled the Juba paleo river channel, as discussed in Sect. 19.3.3.

19.3.2.3 The Marda and Jijiga Volcanics

The best known volcanic outcrop in the region is the Marda Range, a chain of uplifted and eroded hills of Jurassic limestone, capped by a linear basaltic layer, 150 km long and up to 2 km wide, trending SSE from Afar to the central Ogaden, parallel to the first order drainage pattern (Fig. 19.8f). In tectonic terms, it is referred to as the Marda Fault Zone and described as a Precambrian mylonite zone that has been reactivated several times during the Phanerozoic (Purcell 1976; Boccaleti et al. 1991). In the north, near Jijiga, the Marda basaltic layer is about 200 m thick and consists of a number of thick, columnar basaltic flows (Fig. 19.8f). There is no evidence of a feeder dyke within the limited outcrop area, but recently cut quarries nearby reveal numerous hypovolcanic basaltic intrusions, ranging to over 10 m in width and from which narrow dykes (<1 m) have propagated upward. The rare earth element spectra for the intrusives and the Marda extrusives are similar enough to suggest the same parental magma, in agreement with a similar $^{40}Ar/^{39}Ar$ age of 23–25 Ma (Mège et al. in press).

East of the Marda Range, several complexes of basaltic hills rise prominently above the red sand plains (Fig. 19.8a). Volcanism has been dated using the $^{40}Ar/^{39}Ar$ method, evidencing a 30–24 Ma age range (Mège et al. in press). The hills are generally broad mounds covered with rounded basalt cobbles, and actual outcrops are few and scattered. The topographic linearity of the hills suggest the presence of emergent dykes, but the poor quality of the outcrops makes identification difficult, except at Ilbah Hills where a dyke has been confirmed from its chilled margins (Mège et al., in press). Satellite images and SRTM topography reveal widespread NW/SE-trending lineaments associated with several volcanic hills and confirmed by field studies to be shallow depressions, 100–300 m wide and a few meters deep, marked by contrasting vegetation density and occasional calcrete exposures, but without any surface evidence of associated volcanic activity. Many are coincident with high-frequency linear magnetic anomalies (Purcell et al. 2011, Mège et al. in press) and are considered to be the product of hydrothermal alteration and limestone or gypsum karst development along dykes and dyke-parallel fissures (Mège et al. 2012b).

19.3.3 Inverted Basalt-Filled Paleo River Channels

A prominent feature of Ogaden and Somali volcanism is the presence of long (>100 km) basaltic flows in paleo river channels. In the central western Ogaden, an elegant chain of meandering hills extends south for 120 km from the Daga Medo flow. No recent age dating has been done on this flow; however, Maxus Ethiopia (1993) obtained 27.4 ± 1.4 Ma using the K-Ar method (Table 19.1). Another flow, albeit more eroded, can be seen northwest of Mustahil. It is 90 km long, but with eroded 'gaps' between the basaltic remnants and originally flowed in a gently meandering succession of narrow valleys in the Cretaceous Mustahil limestones.

The Gode lava flow (Fig. 19.8c, d) can be traced for more than 200 km and appears to have flowed into the large meandering channel of the paleo-Wabe Shebele. The Gode and Mustahil flows were sampled during the 2008 helicopter survey. Locally, this flow appears to overflow the main channel to form overbank basaltic pods. The Juba River in Somalia has meandering basaltic outcrops, following a paleo-Juba River channel (Abdirahim et al. 1993; Ali Kassim et al. 1987, 1993), which are geomorphologically very similar to the Gode flow outcrops. Initially thought Oligocene, the geomorphology of both the Juba and Wabe Shebele basalt outcrops argues in favor of a much younger age of emplacement. A 7.5 ± 0.4 Ma ^{40}Ar-^{39}Ar age has been determined for the Wabe Shebele basalts (Table 19.1).

Table 19.1 Location and age of the basalt samples used in this work

Volcanic site	Sample ID	Type	Coordinates (WGS84)		Age	Method	Material	References
			Latitude N	Longitude E	(Ma ± 2σ)			
Marda (north)	MA03	Flow	9°21′52.13″	42°41′57.74″	23.68 ± 0.54	$^{40}Ar/^{39}Ar$	Groundmass	Mège et al. (in press)
Marda (north)	MQR7	Dyke	9°21′48.51″	42°42′7.68″	25.04 ± 0.65	$^{40}Ar/^{39}Ar$	Groundmass	Mège et al. (in press)
Marda (south)	L-2-3	Flow	8°43′50″	43°4′20″	25.5 ± 1.3	K–Ar	Whole rock	Maxus Ethiopia (1993)
Daga Medo	K-2-1	Flow	8°25′40″	42°54′30″	27.4 ± 1.4	K–Ar	Whole rock	Maxus Ethiopia (1993)
Mustahil	20	Flow	5°32′22.12″	44°35′37.80″	28.09 ± 0.81	$^{40}Ar/^{39}Ar$	Whole rock	Mège et al. (in press)
Fik	A-3-1	Flow	8°12′	42°40″	28.4 ± 1.4	K–Ar	Whole rock	Maxus Ethiopia (1993)
Kebri Dehar	K3.2	Flow	6°49′38.00″	44°56′49.45″	27.64 ± 0.40	$^{40}Ar/^{39}Ar$	Whole rock	Unpublished*
Kebri Dehar	K1.6	Flow	6°45′18.34″	44°26′3.39″	27.61 ± 0.59	$^{40}Ar/^{39}Ar$	Whole rock	Unpublished*
Gode	WS	Flow	6° 1′12.79″	43°21′55.28″	7.46 ± 0.47	$^{40}Ar/^{39}Ar$	Groundmass	Mège et al. (in press)

*The ages yet unpublished have been obtained from this chapter's team of contributors. Sample K3.2 has a 78 % plateau age with MSWD (mean square of weighted deviates) 0.61, and sample K1.6 an inverse isochron with intercept 289.3 ± 8.6 and MSWD 1.15

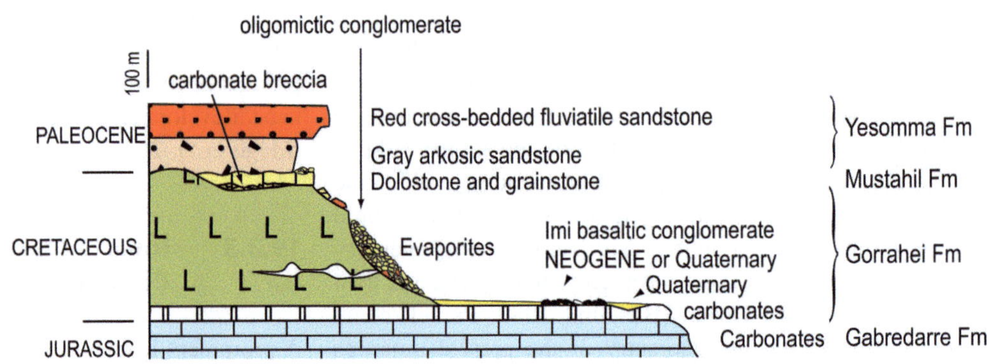

Fig. 19.9 Simplified stratigraphy of the northern Audo Range (Mège et al. 2013)

The current outcrops of the Gode and Juba basalts suggest an original total lava flow length of several hundreds of kilometers, similar to the length of lava flows in other flood basalt provinces, for example, the Columbia River flood basalts (Tolan et al. 1989). Surprisingly, however, and contrary to the other very long basaltic flows, which appear to have formed as part of the main flood basalt emplacement event, the age of the Gode flow very significantly postdates the age of the main Ethiopian flood basalt events. The Juba basalts, which have a similar morphology and similar relationships with the Juba river, might also be of Upper Miocene age, as might the Dolo volcanic table, given the good geomorphological evidence that the Juba basalts flowed from it.

19.3.4 Gravitational Spreading Structures

19.3.4.1 Manifestation and Geographical Extent

The Portlandian–Neocomian Gorrahei Formation outcrops extensively in western Ogaden, where it is dominantly evaporites alternating with shales and over 150 m thick (Fig. 19.9). It overlies the Kimmeridgian–Portlandian Gabredarre Formation inner shelf limestones, which provides a rigid basement on which the spreading has occurred: the Mustahil Formation (limestone) and the Yesomma Sandstone overlying the Gorrahei Formation have spread spectacularly by block faulting and tilting (Figs. 19.10 and 19.11), forming one of the world's largest gravitational spreading domains. Presently covering about 5,000 km^2, the Ogaden complex is 25 times larger than the Canyonlands grabens area in Utah, another spectacular continental-spreading domain.

The spreading zone extends from the Weyb River to the Wabe Shebele (Fig. 19.10a) and is mainly manifest in the Audo Range, an element of the water divide between the Genale and Wabe Shebele basins. The Audo Range is markedly asymmetric, with the most dramatic topographic relief and best-preserved gravitational features along the Weyb River, and more eroded and attenuated landscapes on the Wabe Shebele side. An unnamed ridge north of the Audo Range also displays evidence of normal faulting, producing tilting of post-Gorrahei sedimentary layers up to 25° (Figs. 19.10a and 19.11). East of the Wabe Shebele, the Goja mesa displays tilted blocks bounded by downhill- and uphill-facing normal fault scarps on the

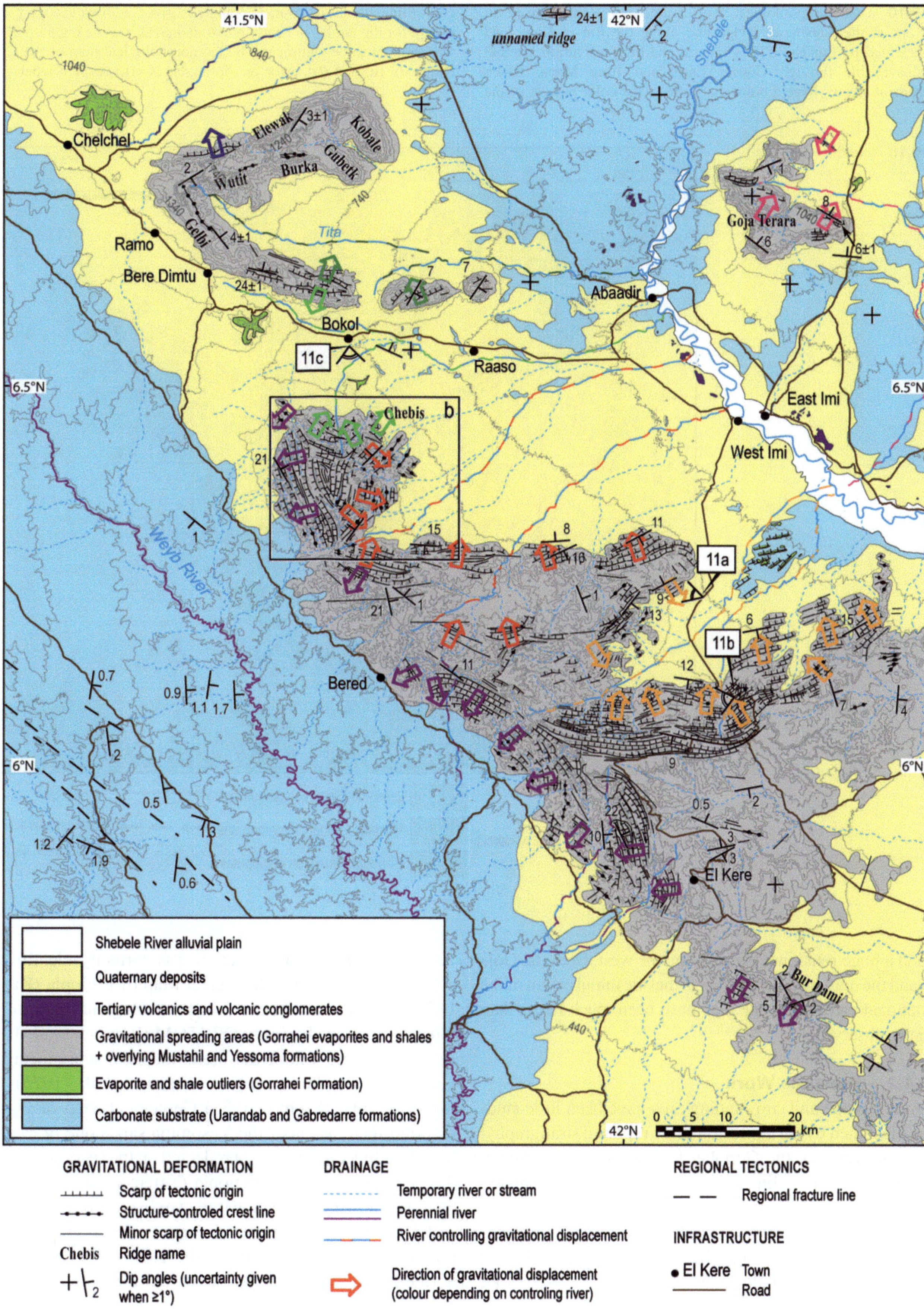
41.5°N
42°N
unnamed ridge
Shebele
Chelchel
Elewak
Kobale
Burka
Guberk
Wutif
Gelbi
Ramo
Bere Dimtu
Tita
Goja Terara
Abaadir
Bokol
11c
Raaso
6.5°N
Chebis
b
East Imi
West Imi
11a
11b
Weyb River
Bered
6°N
El Kere
Bur Dami
Shebele River alluvial plain
Quaternary deposits
Tertiary volcanics and volcanic conglomerates
Gravitational spreading areas (Gorrahei evaporites and shales + overlying Mustahil and Yessoma formations)
Evaporite and shale outliers (Gorrahei Formation)
Carbonate substrate (Uarandab and Gabredarre formations)
0 5 10 20 km
GRAVITATIONAL DEFORMATION
Scarp of tectonic origin
Structure-controled crest line
Minor scarp of tectonic origin
Chebis Ridge name
Dip angles (uncertainty given when ≥1°)
DRAINAGE
Temporary river or stream
Perennial river
River controlling gravitational displacement
Direction of gravitational displacement (colour depending on controling river)
REGIONAL TECTONICS
Regional fracture line
INFRASTRUCTURE
El Kere Town
Road

◀ **Fig. 19.10** **a** Geomorphological map of the Audo Range, based on field work at the Kebenawa Ridge and along the Imi–el Kere road complemented by Landsat ETM+ images and high-resolution satellite images available through Google Earth and Bing Maps. Block displacement arrows have a color that reflects the color pattern of the controlling stream or river. Dip angle is given with an accuracy of 1°, except for dip angles in nearly horizontal areas when a higher precision was thought to be significant. **b** Geomorphological map of the Chebis Ridge. In (**a**) and (**b**), error on dip angles is <0.1, except when given

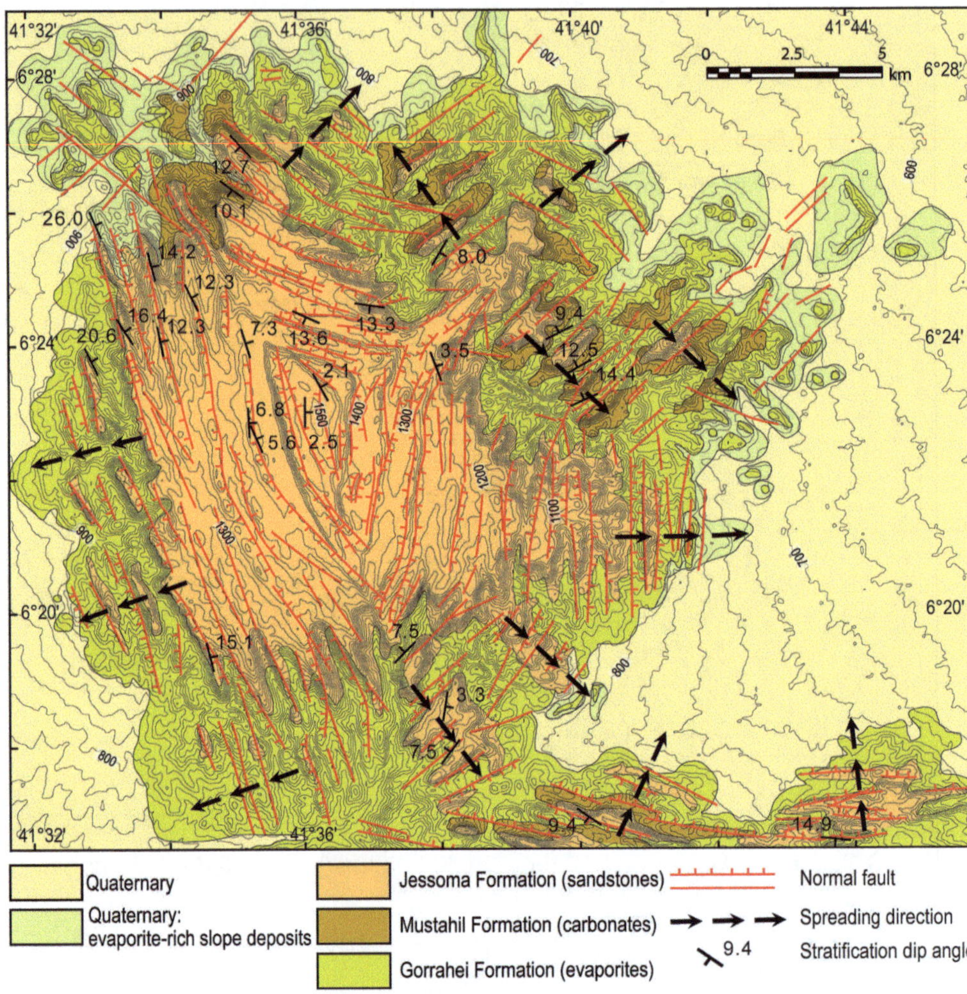

Fig. 19.10 (continued)

east and a huge rotational landslide on the west. Both the Goja mesa and the unnamed ridge indicate that the initial extent of the gravitational spreading zone may have been much larger than presently observed.

19.3.4.2 Previous Work

The Audo Range and surrounding area have been little studied, as illustrated by the confusion in the reported stratigraphy. The rocks overlying the Gorrahei Formation are the Aptian Mustahil Formation limestones and the Maastrichtian–Paleocene Yesomma Formation sandstones (Fig. 19.9), but are shown in many documents as Tertiary basalts. Unrecognized on the earliest geological maps of Ethiopia, where it appears as Jurassic ('Antalo') limestone (Dainelli 1943; Mohr 1963), the Yesomma Formation was correctly identified on the first edition of the Geological Map of Ethiopia (Kazmin 1972) and a contemporary hydrogeological map of the Shebele drainage basin by ORSTOM (Bauduin et al. 1973). However, these sediments were mapped subsequently as volcanics of unknown age by Merla et al. (1973) and, reflecting this, Beicip's (1985) Geological Map of Ogaden reported them as either the Yesomma sandstone or 'volcanics.' This indecision was not reflected on the second edition of the Geological Map of Ethiopia (Tefera et al. 1996) where they are shown as Eocene basalts (Ashangi Formation).

Evidence of halokinetic deformation predating gravitational spreading was found at the Kebenawa Ridge in the northern Audo Range (Mège et al. 2013). Gravitational spreading—in the form of tilted blocks surrounding mesas—was documented at the Chebis Ridge (Lopez-Gonzalez 2006), with faults dipping outward in most cases (Fig. 19.10b). Mège et al. (2013) suggested that gravitational spreading commenced after the Yesomma regression (Bosellini 1989; Purcell 1981), when erosional incision of the uplifted Ogaden surface reached the level of the Gorrahei evaporites. Topographic debuttressing and evaporite rheology then triggered fragmentation and spreading of the overlying formations, similar to the spreading of the Honaker Trail Formation and the Cutler Group over the evaporite Paradox Formation in the Needles district of the Canyonlands Park in Utah (Mège et al. 2013). A significant difference, however, is that the Canyonlands topography was incised by the Colorado River only on one side, leading to a single spreading direction, whereas the Audo Range is wholly surrounded by rivers incising the evaporite formation, leading to multidirectional spreading. Another difference is that spreading has promoted the development of half graben tilted up to 25° in the Audo Range, whereas spreading in the Canyonlands produced nearly symmetric grabens (Moore and Schultz 1999).

19.3.4.3 Landform Development

The direction of displacement of the tilted blocks in a gravitational spreading complex provides important insights into the development of the structure. Because gravitational spreading occurs in response to debuttressing of the evaporite layer, displacement is predictably perpendicular to the orientation of the 'free' boundary, that is, to the side where the river has cut down through the gypsum layer. Comparison between block displacement and the drainage system at the Audo Range suggests that gravitational spreading has occurred in response to incision by tributaries of the Wabe Shebele and by the Weyb River, but not by the Wabe Shebele itself. A correlation between stream or river incision and block displacement direction is proposed on Fig. 19.10a.

The gravitational spreading features are much better preserved on the western side of the Audo Range than on the eastern side, where the tilted blocks are much more dissected by the Wabe Shebele tributaries. This indicates that gravitational spreading is older in the east and that topographic debuttressing of the western side by the Weyb River occurred later. From the fluvial nature of the Cretaceous–Paleocene upper Yesomma Formation sandstones and the paleodeltaic deposits downstream in southern Somalia, the Shebele basin has been dated as old as the Cretaceous–Paleocene in western Ogaden (Bosellini 1989). Hence, there has been considerable time for erosion of the eastern side of the Audo Range. No evidence for such an age has been reported for the Weyb drainage basin, which could be a recent component of the Juba Basin—though the Juba Basin itself is thought to be of similar age to the Shebele Basin, based on the same sedimentary criteria. Fracture lines in the SW corner of the Audo Range map (Fig. 19.10a) which are parallel to the Weyb River and some Genale River tributaries (see Fig. 19.6) may testify to recent moderate tectonic movement related to the opening of the Main Ethiopian Rift (Gani et al. 2009), possibly controlling the incision of the Weyb River and triggering the more recent gravitational spreading on the western side of the Audo Range.

Dip angles were measured at 196 sites in the Audo Range area. The dip angle of the tilted layers is proportional to the quantity of stretching (e.g., Angelier and Colletta 1983) and provides additional information as to the spreading mechanisms. Dip angles were measured using the Orion™ structural analysis software from Pangaea Scientific (Fueten et al. 2005) applied to a co-registered SRTM90 digital elevation model and a Landsat ETM+ image (14.25–28.5 m/pixel). SRTM90 data (87–89 m/pixel) have an error of 5 m in southeast Ethiopia (Farr et al. 2007). A best-fit plane was computed from multilinear regression from manually selected points of known horizontal coordinates and elevation, picked out along a stratigraphic plane on the satellite image. The validity and accuracy of the computed plane is controlled by a series of statistics (see Fueten et al. 2005) and 3D visualization tools in OrionTM, and cross-checked with the high-resolution (0.5–2.5 m/pixel) 2D and 3D views using Google Earth and Bing Maps. Geologically speaking, each reported dip angle corresponds to the mean dip angle of a given strata of constant dip angle, averaged over a measurement site varying in length from a few to several hundred meters (7–30 measurements). Dip errors correspond to a 95 % confidence level and are reported only when >0.1°. Representative dip angles are reported on Fig. 19.10a. Figure 19.10b gives all the dip angles measured on the Chebis Ridge.

Dip angles measured in the tilted layers of the Mustahil and Yesomma formations are in the range 0–26°. The uppermost tilted blocks are connected to remnant Yesomma plateau fragments, which are either horizontal (north of El Kere) or more frequently, tilted by a few degrees (e.g., Elewak Ridge, Chebis Ridge). Dip angles could not be measured in the Gorrahei Formation, due to the difficulty in identifying continuous layers on the satellite imagery. Dip angles measured from the top of the Chebis Ride to its western foot (Fig. 19.10b) increase from the top to the bottom, suggesting that spreading proceeded inward.

Dip angles in the underlying Gabredarre Formation are usually close to zero and were never found to exceed 2–3°. In the southernmost area, dip angles were found to be consistently 0.4–0.7° to the east, accordant with the general 0.8°E dip angle calculated for the underlying Jurassic sequences from seismic reflection data (Line 93-GR-07E) about 70 km south of Fig. 19.10a (Beicip-Franlab 1998). These low dip angles show that gravity gliding along a regional structural slope (in the sense given by Schultz-Ela

Fig. 19.11 Tilted block in the eastern Audo Range: **a** in the Yesomma sandstones; **b** in the Mustahil carbonates; **c** at the Chebis Ridge. Location on Fig. 19.10

Fig. 19.12 Topographic profile along the Audo range water divide

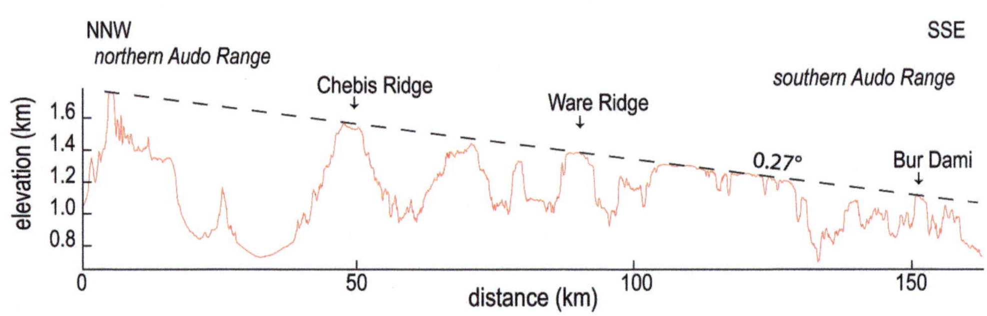

2001) over the rigid pre-Cretaceous substrate did not play a significant role in the development of the observed landscape.

It is significant to note that the various plateaus of sub-horizontal Yesomma sandstone that constitute the summit blocks of the Audo Range Yesomma are fragments of the same dipping surface. A topographic profile along the water divide shows a linear decrease of plateau surface elevation from NNW to SSE, with an inclination of 0.27° (Fig. 19.12). An analysis of thirty-one points on the summit plateau surfaces throughout the Audo Range, plus 3 points on the Goja mesa, shows they fit a single plane of strike N003°E ± 016 and dip 0.4 ± 0.2° (at a 95 % confidence level), with a goodness of fit of 98.2 %. This remarkable fit indicates that the many geodynamic events that occurred in the Cenozoic after Yesomma sandstone deposition, including flood lava emplacement, and the rifting in the Gulf of Aden, Ethiopian Rift, and Afar, had very little influence on deformation in western Ogaden, beyond the uplift of the plateau itself.

19.4 Landform Evolution at Regional Scale

19.4.1 Adaptation of Rivers to Topography

Analysis of drainage systems provides valuable information regarding vertical motions of the surface at a regional scale (Snyder et al. 2000; Schumm et al. 2002; Duvall et al. 2004; Whipple 2004; Whipple et al. 2013) and can help understanding of the evolution of landforms in relation to geodynamics. In this section, the development of the present drainage network and the current topography is analyzed.

The general organization of the Ogaden basin shows large-scale structural and geodynamic control. The regional geometry of the Genale and Wabe Shebele river systems indicates two main directions of flow (Fig. 19.13). A dominant N130E orientation is exemplified by the main orientation of both the Genale and Shebele basins and by tributaries such as the Gerer, Fafan, and Dawa rivers. A second dominant orientation is N050-060E and is especially exemplified by a 350-km-long segment of the upper Wabe Shebele in the Arsi highlands. It is also observed in the lower part of the Dakota river. The N130°E trend is the orientation of the tilt of the Somali Plateau. The N050°–060°E trend is the orientation of the Main Ethiopian Rift

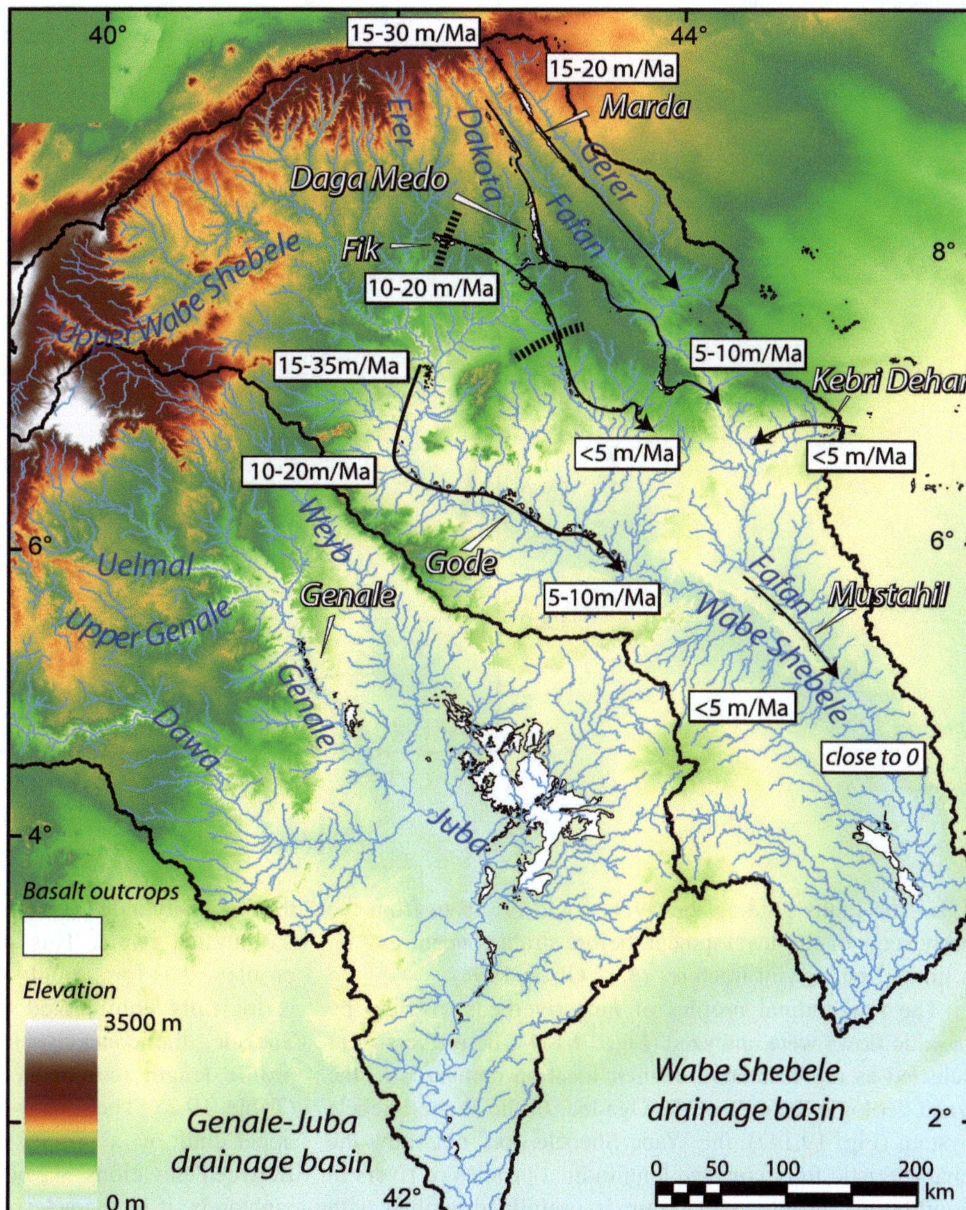

Fig. 19.13 Drainage system map, with the location of the basalt outcrops in the Ogaden (*white* surrounded with *black line*). The rivers are in *blue* and the watersheds in *dark grey*. The basalt flows used in the morphometric analysis are named. The *black arrows* indicate lava flow directions. Local mean incision rates since basalt outpouring are indicated in the boxes

Fig. 19.14 Longitudinal profiles of rivers of the Wabe Shebele and Genale systems discussed in the text, and comparison with profiles of rivers at equilibrium (Darling river tributary, Australia); in equilibrium except in localized segments, where they are slightly disturbed by tectonics or localized abrasion (Rhine and Loire rivers, respectively, Europe); and in disequilibrium owing to active tectonic uplift and resulting strong fall of base level (Amazon tributary, Andes). The profiles were extracted from ASTER GDEM with RiverTools® (Peckham 1998). **a** Wabe Shebele system; **b** Genale system. The profiles are leveled at the elevation value at the end of the Wabe Shebele and Genale

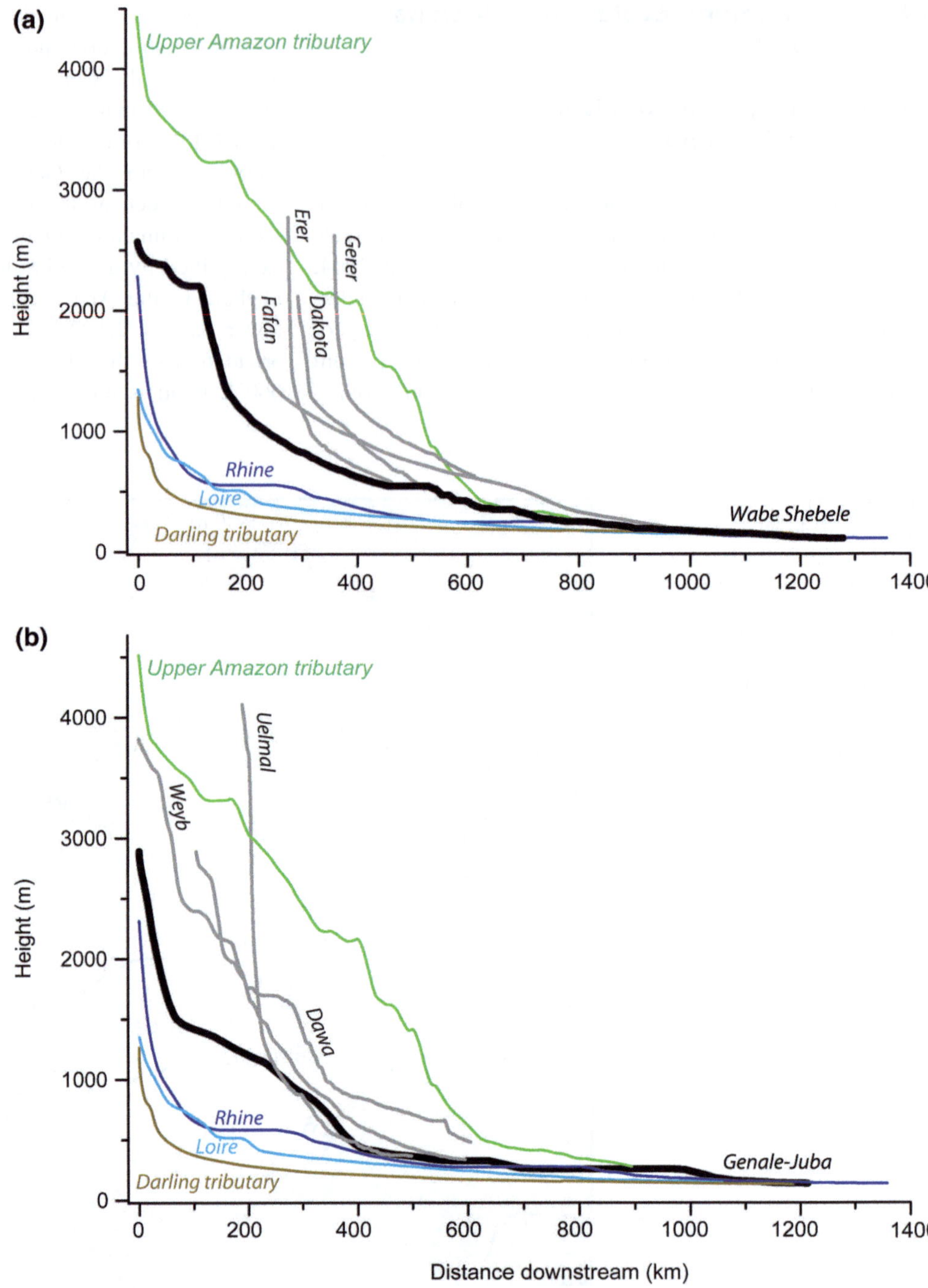

axis. The faults guiding the rivers could originate from the early stage of rift development, before strain concentrates on a limited number of fractures (e.g., Olson 1993).

The longitudinal profiles of nine riverbeds overlain by basaltic flows were analyzed (Fig. 19.14). These rivers were selected as representative of their location compared to the main geologic features of the Ogaden. In the Wabe Shebele system (Fig. 19.14a), the Wabe Shebele does not show the characteristic fully concave longitudinal profiles of rivers in equilibrium. Profile equilibrium is usefully described with the concavity index (Langbein 1964), given in Table 19.2 for the studied rivers. This index is zero for a straight river profile; −1 is for strongly unequilibrated (convex) profile; 1 is for fully equilibrated (concave) profile. For the Wabe Shebele, the concavity index cumulated over the whole profile length (cumulative concavity index, CCI) is 0.83 (Table 19.2). The Wabe Shebele is characterized over its upper half by a succession of vertical-step knick-points. Between elevations ~2,600 and ~2,300 m a.s.l. in the Arsi highlands, it is characterized by an alternation of relatively

Table 19.2 Geomorphometric parameters for the rivers on Figs. 19.14 and 19.15 (graphical definition on Fig. 19.15)

River name	h_{max}	h_{min}	l	Concavity indexes	h			
	(m)	(m)	(km)		25 %	50 %	75 %	100 %
Main Ethiopian rivers								
Genale	2,825	50	1,315	*D*	*2 %*	*5 %*	*23 %*	*100 %*
				CCI	0.04	0.27	0.48	0.87
				ICI	0.04	0.3	**−0.2**	0.54
Wabe Shebele	2,685	165	1,505	*D*	*8 %*	*10 %*	*23 %*	*100 %*
				CCI	**−0.12**	**−0.49**	0.42	0.83
				ICI	**−0.12**	**−0.01**	0.16	0.63
Tributaries								
Uelmal	4,040	300	305	*D*	*6 %*	*13 %*	*10 %*	*100 %*
				CCI	**−0.42**	**−0.82**	0.29	0.87
				ICI	**−0.42**	0.34	0.32	0.68
Dakota	2,735	740	250	*D*	*0.8 %*	*4 %*	*22 %*	*100 %*
				CCI	0.48	0.58	0.81	0.72
				ICI	0.48	0.32	0.60	0.12
Dawa	2,825	415	500	*D*	*8 %*	*24 %*	*46 %*	*100 %*
				CCI	**−0.48**	0.43	0.11	0.60
				ICI7	**−0.48**	0.18	**−0.71**	0.07
Weyb	3,750	220	700	*D*	*10 %*	*25 %*	*35 %*	*100 %*
				CCI	**−0.49**	0.56	0.17	0.69
				ICI	**−0.49**	0.28	0.18	0.74
Erer	2,235	615	250	*D*	*4 %*	*8.5 %*	*44 %*	*100 %*
				CCI	0.24	**−0.05**	0.59	0.57
				ICI	0.24	**−0.13**	0.1	0.2
Fafan	2,230	280	800	*D*	*1 %*	*12 %*	*45 %*	*100 %*
				CCI	0.5	0.65	0.6	0.53
				ICI	0.5	0.47	0.2	0.26
Gerer	1,875	855	300	*D*	*11 %*	*33 %*	*66 %*	*100 %*
				CCI	0.27	0.23	0.32	0.22
				ICI	0.27	0.16	0.04	0.004
Other rivers								
Darling (Australia)	1,245	105	1,225	*D*	*0.4 %*	*2.5 %*	*7 %*	*100 %*
				CCI	0.67	0.52	0.75	0.91
				ICI	0.67	0.25	0.43	0.78
Rhine (Europe)	1,360	0	2,170	*D*	*0.07 %*	*2 %*	*7 %*	*100 %*
				CCI	0.18	0.36	0.7	0.82
				ICI	0.18	0.21	0.25	0.54
Loire (France)	1,200	0	1,055	*D*	*2 %*	*8 %*	*20 %*	*100 %*
				CCI	0.27	0.33	0.53	0.81
				ICI	0.27	0.53	0.53	0.44
Amazon tributary (Andes)	4,490	255	890	*D*	*11 %*	*33 %*	*55 %*	*100 %*
				CCI	0.69	0.16	0.11	0.33
				ICI	0.69	**−0.19**	**−0.51**	0.74

h_{max} = maximum river elevation a.s.l.; h_{min} = minimum river elevation a.s.l.; l = river length; h = river height decrease since its source; D = horizontal distance along the river taken from its source; CCI = Cumulative concavity index; ICI: Incremental cumulative index. Cumulative index (Langbein 1964) is defined as 2A/H, with A the vertical difference between the profile midterm and a straight line joining the two ends (or any point) of the longitudinal profile; H is the elevation difference between the uppermost and lowermost points of the straight line. CCI measures concavity of a river profile from its source to a given D. ICI measures concavity between two given D's, as illustrated in the insert of Fig. X.15. The indexes are 0 for straight river profile; −1 is for strongly unequilibrated (convex) profile; 1 is for fully equilibrated (concave) profile. Indexes are bold when convex. In each category, the rivers are ranked by cumulative concavity index, from higher to lower

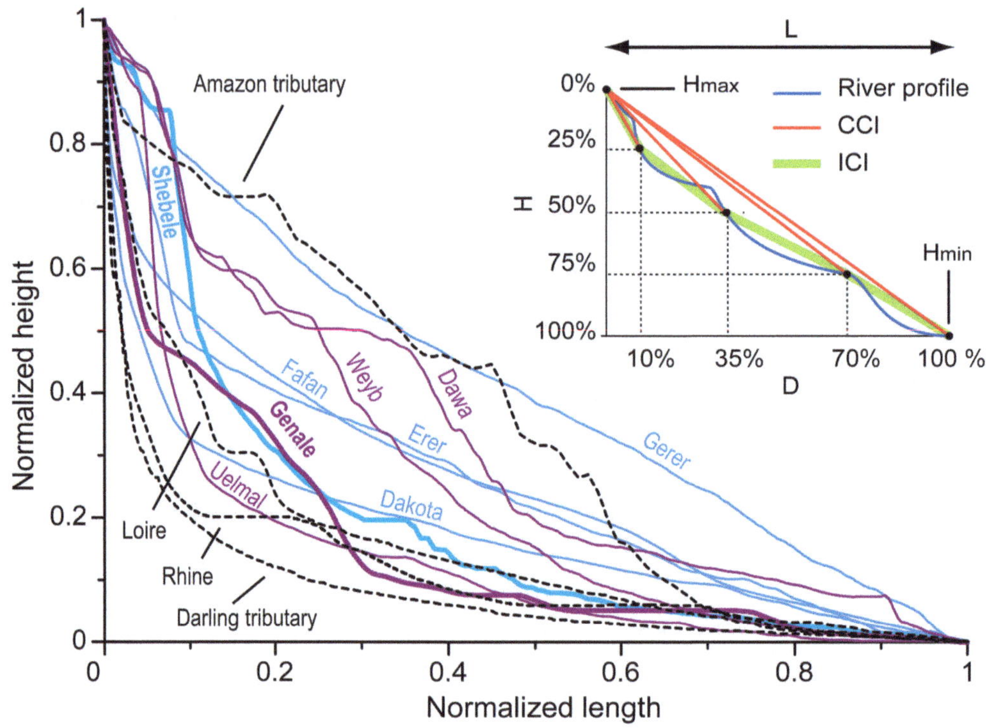

Fig. 19.15 Normalized river profile plot for the rivers displayed on Fig. 19.14. The *inset* shows relationships between geomorphometric parameters used in Table 19.2

flat areas and slopes with a gradient of 5–7 m/km, giving an overall gradient of about 3 m/km. However, at about 110 km, the slope increases abruptly to 20 m/km over several tens of kilometers and only recovers its average slope of 3 m/km after 340–350 km, at 660 m of altitude. This huge vertical-step break in slope creates the spectacular deep canyons for which the river is famous (Fig. 19.3d). In its lower reaches, the Wabe Shebele is again marked by a succession of vertical-step knick-points but of one order of magnitude less prominent. Eventually, the Wabe Shebele ends its Ethiopian course in a wide floodplain (10–20 km) with a slope of only 0.2 m/km. Even if its upper part is highly disrupted by knick-points, its overall profile is close to those of Rhine or Loire rivers, being smooth concave with irregularities: a near equilibrium profile (Fig. 19.15), with a CCI of 0.82 and 0.81, respectively (Table 19.2).

The Fafan River has a much lower CCI (0.53), denoting a significantly different evolution. Except in the uppermost 20 km, where a high slope (25 m/km) is manifest (incremental cumulative index ICI = 0.5), the Fafan has a linear profile for approximately 700 km, with a slope of 1.5 m/km and without any remarkable knick-points (ICI = 0.2–0.26). The same evolution is observed for the Gerer (linear profile at 3 m/km along 300 km), Dakota (2.5 m/km along 200 km) and Erer rivers (3.5 m/km along 225 km). High-gradient slopes of ~35–40 m/km are present in these rivers only in the initial 25 km. None of those valleys show an equilibrium profile. The Fafan and Gerer valleys are spectacularly linear and long compared to the other tributaries of the Wabe Shebele, which is attributed to the structural control of the Marda Range.

In the Genale system (Fig. 19.14b), the Genale River (1,300 km) has a linear profile upstream (20 m/km upstream, CCI = 0.04, Table 19.2) but changes abruptly at km 70 to a spectacular convex profile, with the slope decreasing sharply to 2 m/km and then increasing regularly to 5 m/km between km 350 and 400 (ICI = −0.2). Another abrupt change occurs downstream, with the slope decreasing to around 0.3 m/km and staying constant for the next 900 km.

The Weyb and Dawa rivers are both marked by very irregular profiles, with a succession of major slope breaks in their upper reaches giving convex profile sections (ICI ≤ −0.48) and very high average slopes: 10 m/km for the upper 300 km for the Weyb (CCI = 0.69) and 12 m/km during the upper 250 km for the Dawa (CCI = 0.6). In contrast, the Uelmal River is close to an equilibrium profile (ICI = 0.87). Its slope decreases gradually from the very high value of 100 m/km down to 1 m/km during the 300 km course, albeit marked by knick-points at km 50 and km 100 (CCI ≤ −0.42).

The highland part of both Genale and Wabe Shebele is not in equilibrium with their base level. Comparison with profiles from the Andes, where uplift is active (CI = 0.33 but negative over much of its profile ICI), and Australia, where the selected river is in equilibrium (Fig. 19.15) (CCI = 0.91), reveals that the Wabe Shebele, Genale, and Uelmal are close to an equilibrium profile, but the presence of irregularities indicates that none of them is a perfect equilibrium profile.

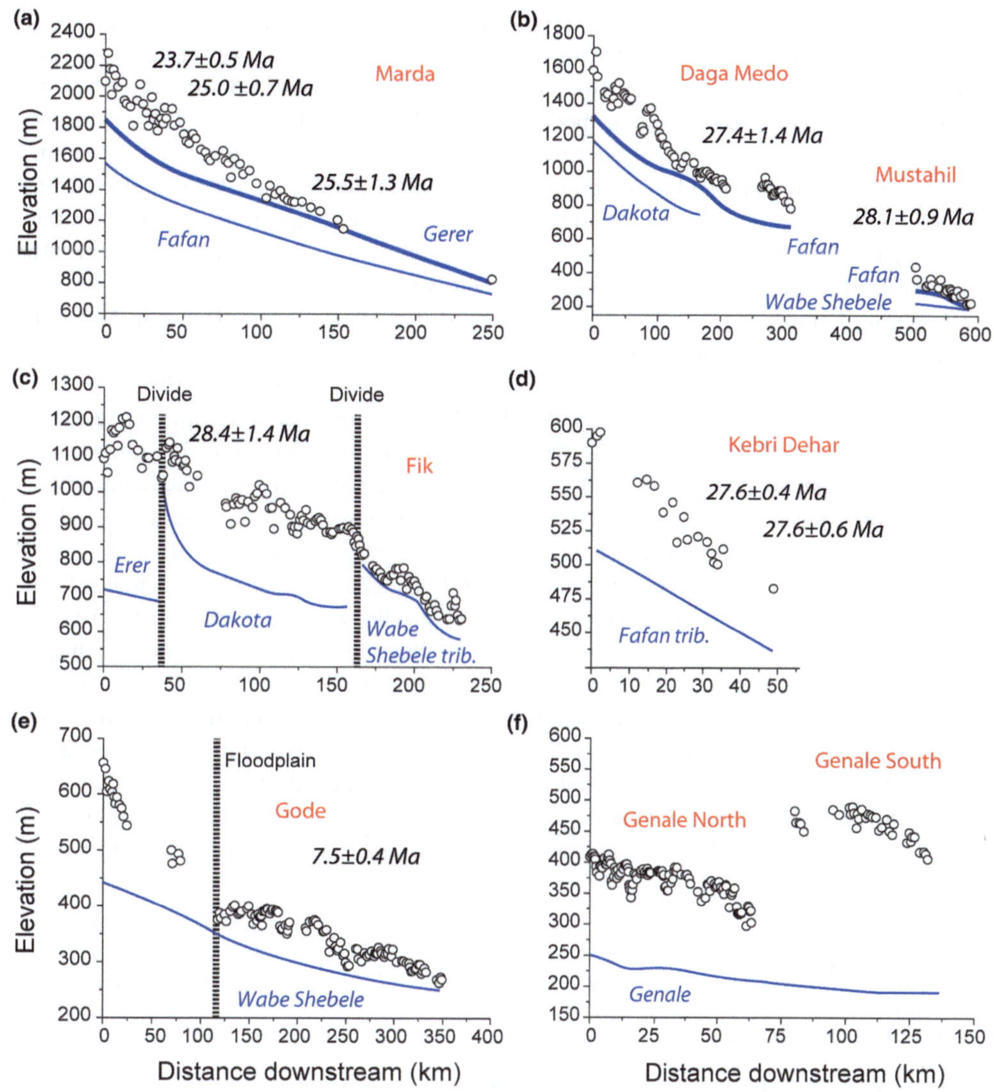

Fig. 19.16 Profiles of modern rivers (*blue lines*) and the base of basalt flows (*black dots*, names in *red*). Age details are on Table 19.1. The water divides in (**c**) are located on Fig. 19.13. The two data clusters in (**f**) are for the two Genale basalt tables located on Fig. 19.7. The difference in elevation reveals the depth of incision

The Dawa and Weyb profiles are closer to the Andean profile, whereas the Erer, Dakota, and Fafan show exotic bimodal linear profiles, and the Gerer has a unimodal linear profile, the latter being very uncommon. This analysis of the profiles of the Genale, and Wabe Shebele and their tributaries suggests a major base-level fall in the Ogaden, most likely due to tectonic uplift of one of its margins. In the next section, we discuss the use of age-dated basalts in paleo river valleys to help constrain uplift amplitude and timing.

19.4.2 Insight into the Evolution of Ogaden Topography Since the Upper Oligocene

The presence of age-dated basaltic flows, which filled paleo river valleys and presently are only a short distance from the modern river valley, allows the determination of average incision rates. For example, the elevation of the base of the Marda volcanics can be compared with the elevation of the beds of the Gerer and Fafan rivers, and the Gode flow, with the Wabe Shebele (Fig. 19.13).

This difference in elevation reveals the depth of incision since the basalt outpouring and that can be converted to a mean incision rate. The reliability of such an analysis requires that the horizontal distance between paleo- and present rivers is minimal and both sites are located in the same geological domain, without intervening deformation subsequent to the lava flow.

The elevation of the longitudinal profiles of six basalt-filled paleo river channels in the Wabe Shebele drainage basin was analyzed and compared with the elevation of the present river profiles. In addition to the Marda and Gode flows mentioned above, the Daga Medo volcanics were compared with the Fafan river, and the Fik volcanics with the Erer and Dakota rivers in the north and a Wabe Shebele tributary in the south. The elevation of the two Genale basaltic tables was also compared with the elevation of the Genale River (Figs. 19.13 and 19.16). In all cases, the area between the volcanic flow and the modern river valley has been geologically stable since the time of eruption. The Marda volcanic flow, being located on the eastern side of the prominent Marda Precambrian deformation zone, warrants some caution in that regard but the excellent and continuous rock exposures through the Marda volcanics and across the neighboring Fafan and Gerer rivers on the Harar-Jijiga road show no significant postflow deformation.

The elevation of the base of every basalt outcrop was measured using elevation data from ASTER DEM, with geographical control from Landsat ETM+ images and high-resolution images available on Google Earth™. Uncertainty in elevation control is estimated to 10–15 m. Some of these flows (Marda, Mustahil, Kebri Dehar, Gode) have been dated recently using argon radiochronology and their age is considered reliable. The Marda, Daga Medo, and Fik flows were dated in the 1990s by the K-Ar method but have the same age span as most of the argon-dated samples, suggesting these ages are plausible. The Genale tables have not been dated. More details are given on Table 19.1.

Figure 19.16 presents the results of these comparisons and shows clearly that incision is intense at the river head and decreases downstream. This is true for the older basalts at Marda, Daga Medo, Fik, and Kebri Dehar (Fig. 19.16a–d), and also for the younger Gode flow. Moreover, the Fafan has incised faster than the Gerer (Fig. 19.16a); the Dakota than the Gerer (19.16b), and the Erer than the Dakota (19.16c), revealing that the mean incision rate has increased from east to west. The incision rate has also increased from south to north, mean incision rate for the middle and upper Dakota being 7–11 m/Myr. (Figure 19.16c) and 13–20 m/Myr, respectively (Fig. 19.16b). For the Fafan it has increased, south-to-north, from 10–13 m/Myr. (Figure 19.16b) to 16–36 m/Myr. (Fig. 19.16c). The maximum mean incision rate of about 30 m/Myr (Fig. 19.16e) was obtained for the Wabe Shebele in its upstream course at the northernmost outcrop of the Gode flow. Incision rates calculated for the different rivers and the age of the associated lava flows are given on Fig. 19.17.

The base of the Fik flow has a linear profile but crosses two water divides (Fig. 19.16c), showing that the drainage network has changed since the basalt outpouring, with the Erer and Dakota rivers achieving their present geometry more recently than the Fafan and Gerer. This is well illustrated on the comparative river profile plot (Fig. 19.15), in which the Erer and Dakota river profiles are close to recently equilibrated profiles, in contrast to the linear Fafan and Gerer profiles, which indicate a relict transient profile (Kirby and Whipple 2012 and references therein). The Erer and Dakota have their convergence point with the Wabe Shebele much

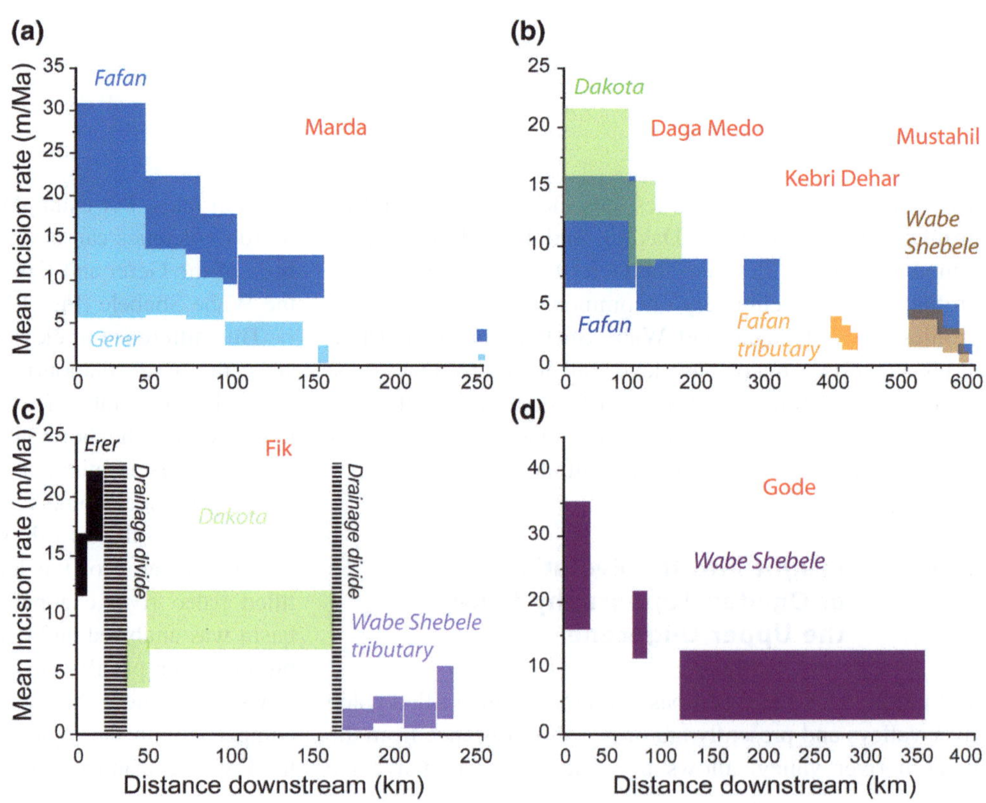

Fig. 19.17 Mean incision rate calculated from the data presented in Fig. 19.16. Incision after the Marda (**a**), Daga Medo, Kebri Dehar, and Mustahil (**b**), and Fik flows (**c**), starts at 28–22 Myr, i.e., as soon as the lavas are erupted. For Gode (**d**) erosion starts at 8–5 Myr

higher upstream than the Fafan and have been therefore more sensitive to the geodynamic events occurring in the upper Shebele region.

Downstream, the Fik flow follows a Wabe Shebele tributary with a constant vertical base-difference of 50 m. This tributary converges on the Wabe Shebele 100 km southward, cutting through the Gode basalt outcrops. This moderate incision reflects the lower uplift rate of this area.

Similarly, in the easternmost Shebele basin, where the Kebri Dehar flow is located, incision in the Fafan tributary that parallels the Kebri Dehar flow is slow (2.5 m/Myr) and nearly constant. This is consistent with the general incision pattern in the basin, with the mean incision rate increasing from east to west and south to north, toward the upper Wabe Shebele (Fig. 19.13). An extreme and remarkable case is the Mustahil flow (Fig. 19.16b) on the southeasternmost border of the Ogaden (that is, as far as Ogaden can be from Afar), which, in spite of its 28 Ma age, ends at the Wabe Shebele–Fafan confluence, at appreciably the same elevation as the present Wabe Shebele, pointing to the vertical stability of this confluence. This specific location along the Wabe Shebele may be close to the point in the river profile where incision stops and deposition starts. Perhaps not coincidentally, with the exception of the northeastern Ogaden (Mège et al. in press), the Cenozoic volcanic rocks in the Ogaden are outcropping and being eroded, whereas further east in Somalia the volcanic rocks are covered by Miocene to Quaternary sediments (e.g., Bosellini 1989; Faillace 1993). More generally, there is a marked absence of Neogene rocks throughout the Ogaden, except its northeastern part, while sediments of this age are widespread in Somalia (Abbate et al. 1994).

The Wabe Shebele floodplain begins near the latitude of Gode town. Its beginning is apparent on Fig. 19.16e, at the convergent point between the upstream profile, with a steep longitudinal slope (2 m/km) and a high mean incision rate (30 m/Myr), and the downstream profile, where the slope is gentle (0.5 m/km) and the mean incision rate is between almost 15 m/Myr and zero, the latter occurring at the convergence point with the Fafan River.

Floodplains usually develop in subsidence areas (Syvitski et al. 2012), in which detrital deposits accumulate, elevating the slope profile. As a result, the Wabe Shebele incision may be underestimated. The sharp transition in the slope profile of the base of the Gode basalts observed on Fig. 19.16e may be due to either a variation in the ancestral Wabe Shebele profile, or a difference in uplift rate, with an increasing mean uplift gradient from south to north.

The analysis of the paleotopography preserved by the basalts indicates that the present-day Wabe Shebele catchment is divided into two main parts that had contrasting responses to uplift. In the west, river profiles of Wabe Shebele-type have nearly completed a return to an equilibrium state, even though the entire profile is not yet smooth and concave. In the east, Fafan-type rivers have preserved a tilted slope in the uppermost part of the profile, but most of the profile has not changed. The Gerer River is an extreme case in which the tilted slope is atrophied and much of the remaining slope is constant, denoting geodynamic stability and probably an increasingly dry climate and subsequent lower erosion rate since the Miocene (e.g., Feakins 2013).

19.4.3 Uplift Rates

Commonly, large-scale river incision is the result of base-level fall, which can be caused by either tectonic uplift or sea-level (or lake-level) fall. In the Wabe Shebele valley, the geomorphological analyses indicate that the Wabe Shebele valley profile has nearly reached its equilibrium profile downstream. All the perturbations in valley profiles are located in the very upstream part of the system. Consequently, the calculated mean incision rates (Fig. 19.17) can be taken to approximate the uplift rates, with only minor, if any, contribution coming from sea-level variations.

Vertical movements of the surface (and underpinning crust) in the Ogaden region are a result of the complex geodynamics in the Horn of African since the Oligocene. The calculation of river incision rates helps in the understanding of the present topography at a regional scale, but does not inform regarding the uplift rate variations. On the Somali Plateau, such variations were probably not negligible, as discussed above in Sect. 19.2.2 with regard to the origin of the plateau elevation, and mean uplift rates need to be interpreted with caution when reconstructing paleotopographic evolution stages. Nevertheless, for young ages, additional information is given by the slope break in the paleo-Wabe Shebele profile (Fig. 19.16e), which shows that at least part of the uplift postdates 7 Ma. Since that time, the mean incision rate has been ~ 30 m/Myr upstream, decreasing to zero downstream near the Somalia border. This incision and the causative uplift (as argued above) are related to the development of the Main Ethiopian Rift. These measurements have been obtained at a minimum distance of 250 km from the rift margin, and the uplift rate of the westernmost Ogaden along the rift shoulder must have been much higher, given the usual rift-flank concave curvature. Furthermore, if rift-flank uplift has been the dominant mechanism of topographic building, then visco-elastic models predict that this ~ 30 m/Myr uplift rate could have been relatively constant since the beginning of rifting (Sachau and Koehn 2010). The mean incision rates obtained for the Wabe Shebele would correspond to a constant uplift rate since that time.

The Fafan River is located on the western side of a major crustal discontinuity, now manifest at the surface in the Marda Range; the Gerer River is located on its eastern side, and both rivers are parallel to it. The Marda Range serves, therefore, as the boundary between the northern uplifted Ogaden, with its rivers responding according to their setting, and the eastern stable Ogaden for which the evolution of the Main African Rift appears, perhaps deceptively, to have been only a remote influence.

19.5 Concluding Remarks

The landscape and landforms of the Ogaden reflect the geological development of the region. This is clearly manifest at a regional scale: the steep, nearly impenetrable canyons in the north, for example, are a consequence of the relative uplift of the Somali Plateau margin. Similarly, the contrast between the eastern and western Ogaden landscapes owes much to the pattern of Jurassic and Cretaceous sedimentation in these areas, and the impact of erosion during Cretaceous, Tertiary and Recent times. This interplay of Mesozoic sediments and Tertiary–Recent erosion is also seen on a local scale in the Audo Range, where erosional debuttressing of the Gorrahei evaporites has triggered gravitation collapse and spreading of blocks of the overlying Mustahil and Yesomma formations.

In spite of its diversity of landforms, somewhat amazingly, the Ogaden forms a continuous and coherent mega-geomorphological unit: a long eastward-dipping slope of dominantly thermal origin that docilely recorded the tectonic jolts from its neighborhood, mainly the many volcanic and rifting events. These resulted in dramatic upwarping on the western and northern side of the region, while the eastern and southern regions underwent negligible deformation. Vertical motions aside, the huge size of the area has allowed landforms to develop over broad surfaces: many identified lava flows are >100 km long and the Audo Range gravitational spreading complex could be the largest gravitational spreading domain on Earth. Other landforms not described in this chapter include the karstic landscape of the Gabredarre limestones south of Ginir, where the splendid Sof Omar caves are located.

A better understanding of the geomorphology of Ogaden will require considerably more exploration and study, perhaps especially the easternmost region. That the landscape of the Ogaden and its flora and fauna have not been better studied to date is, in part at least, a product of the remoteness and limited infrastructure but also of the difficult security conditions that have prevailed over large areas for many years. Current issues are the modern guise of conflicts centuries old and have their origins in that landscape and the rift that developed between the cultures on the Somali and Ethiopian plateaus.

Acknowledgments Part of this work was funded by the CNRS/INSU Marges programme. Field work in 2008 was conducted in cooperation with Pexco Ethiopia Exploration (East Africa) N.V. $^{40}Ar/^{39}Ar$ dating for this project is performed by Dr Fred Jourdan at the Western Australian Argon Isotope Facility, Curtin University, Perth, Western Australia. ASTER GDEM is a product of METI and NASA.

References

Abbate E, Sagri M, Sassi FP (1994) Geological map of Somalia, 1:1,500,000. Somali National University, Mogadishu

Abdirahim MM, Ali Kassim M, Carmignani L, Coltorti M (1993) The geomorphological evolution of the upper Juba valley in southern Somalia. In: Abbate E, Sagri M, Sassi FP (eds) Geology and mineral resources of Somalia and surrounding regions. Florence, Italy, vol 113 (Ist. Agron Oltremare Relaz E Monogr), pp 241–250

Ali Kassim M, Carmignani L, Fazzuoli M (1987) Geology of the Luuq-Mandera basin, geology of Somalia and surrounding regions, excursion A, GEOSOM 87. Somali National University, Department of Geology, Mogadishu, pp 1–43

Ali Kassim M et al (1993) Flood basalts of the Gedo Region (Southern Somalia): geology, petrology and isotope geochemistry. In: Abbate E, Sagri M, Sassi FP (eds) Geology and mineral resources of Somalia and surrounding regions. Florence, Italy, (Ist. Agron. Oltremare, Relaz. E Monogr) vol 113, pp 311–334

Angelier J, Colletta B (1983) Tension fractures and extensional tectonics. Nature 301:49–51

Bauduin D et al (1973) Projet du Wabi Shebele: étude hydrologique. ORSTOM, fdi:06586, Paris

Beicip (1985) Geological map of the Ogaden and surrounding area, 1:1,000,000. Geological Survey of Ethiopia, Addis Ababa

Beicip-Franlab (1998) Petroleum potential of Ethiopia. The Ministry of Mines and Energy of Ethiopia, Addis Ababa

Boccaleti M, Getaneh A et al (1991) The Marda fault: a remnant of an incipient aborted rift in the paleo-African Arabian plate. J Pet Geol 14:79–92

Bonini M et al (2005) Evolution of the main Ethiopian rift in the frame of Afar and Kenya rifts propagation. Tectonics 24, TC1007. doi:10.1029/2004TC001680

Bosellini A (1989) The continental margins of Somalia: their structural evolution and sequence stratigraphy. Mem Sci Geol 41:373–458

Cheung WH, Senay GB (2008) Trends and spatial distribution of annual and seasonal rainfall in Ethiopia. Int J Climatol 28 (13):1723–1734. doi:10.1002/joc

Dainelli G (1943) Geologia dell'Africa orientale. Centro Studi per l'Africa Orientale Italiana, Rome

Duvall A, Kirby E, Burbank D (2004) Tectonic and lithologic controls on bedrock channel profiles and processes in coastal California. J Geophys Res 109(F3). doi:10.1029/2003JF000086

Ethiopia Government Portal (2014) www.ethiopia.gov.et/stateoromia & /statesomali. Accessed Jan 2014

Ethiopian Mapping Authority (1988) National atlas of Ethiopia. Addis Ababa, Ethiopia

Farr T et al (2007) The shuttle radar topography mission. Rev Geophys 45:1–33

Faillace C (1993) Hydrogeological importance of the sub-surface basalts in the Mudug-Galgadud plateau. In: Abbate E et al (eds) Geology and mineral resources of somalia and surrounding regions, vol B, Mineral and water resources. (Ist Agro l'Oltremare, Rel Mon, Firenze, Italy) vol 113, pp 649–664

Feakins SJ (2013) Pollen-corrected leaf wax D/H reconstructions of northeast African hydrological changes during the late Miocene. Palaeogeogr Palaeoclimatol Palaeoecol 374:62–71. doi:10.1016/j.palaeo.2013.01.004

Fournier M et al (2010) Arabia-Somalia plate kinematics, evolution of the Aden-Owen-Carlsberg triple junction, and opening of the Gulf of Aden. J Geophys Res 115, B04102. doi:10.1029/2008JB006257

Fueten F, Stesky RM, MacKinnon P (2005) Structural attitudes of large scale layering in Valles Marineris calculated from mars orbiter laser altimeter data and mars orbiter camera imagery. Icarus 75:68–77

Gani NDS, Abdelsalam MG et al (2009) Stratigraphic and structural evolution of the Blue Nile Basin, Northwestern Ethiopian Plateau. Geol J 44:30–56

Hagmann T (2007) The political roots of the current crisis in region 5. Social Science Research Council, Web forum Crisis in the Horn of Africa, New York

Hagmann T, Khalif MH (2006) State and politics in Ethiopia's Somali Region since 1991. Bildhaan. Int J Somali Stud 6:25–49

Hofmann C, Courtillot V et al (1997) Timing of the Ethiopian flood basalt event and implication for plume birth and global change. Nature 389:838–841

Juch D (1975) Geology of south-eastern escarpment of Ethiopia between 39° and 42° long. East. In: Pilger A, Rosler A (eds) Proceedings of an international symposium on the afar region and related rift problems. Stuttgart, E. Schweizerbart'sche Verlagsbuchhandlung

Kazmin V (1972) Geological map of Ethiopia, 1st edn, 1:2,000,000. Geol Surv Ethiopia, Addis Ababa

Kibrie T, Yirga T (2008) The geology of Bedesa area (NC37-16) Geol Surv Ethiopia Memoir 10, Addis Ababa

Kirby E, Whipple KX (2012) Expression of active tectonics in erosional landscapes. J Struct Geol 44:54–75

Lemma G (1996) Climatic classification of Ethiopia. National Mapping Authority, Addis Ababa

Leroy S et al (2012) From rifting to oceanic spreading in the Gulf of Aden: a synthesis. J Arab Earth Sci 5:859–901. doi:10.1007/s12517-011-0475-4

Langbein WB (1964) Profiles of rivers of uniform discharge. US Geol Surv Prof Pap 501-B:119–122

Lopez-Gonzalez T (2006) L'étalement gravitaire de la ride de Chebis dans le bassin de l'Ogaden, Ethiopie. MSc. Thesis, Laboratoire de Planétologie et Géodynamique, Nantes University

Maxus Ethiopia (1993) Ogaden concession final report. Addis Ababa, Ethiopia

Mège D, Le Deit L, Rango T, Korme T (2013) Gravity tectonics of topographic ridges: Halokinesis and gravitational spreading in the western Ogaden, Ethiopia. Geomorphology 193:1–13

Mège D, Purcell, PG, Jourdan F (2012a) Volcanism in southeast Ethiopia and the Ogaden Dyke Swarm. Magmatic rifting & active volcanism conference, Addis Ababa, Abstract

Mège D, Purcell P, Jourdan F (2012b) Dykes and linear troughs: new observations on the Somali Plate. Lunar and planetary science conference 43, Lunar and Planetary Institute, Houston, Texas, Abstract 1317

Mège D, Purcell P et al A major dyke swarm in the Ogaden region south of afar and the early evolution of the afar triple junction. In: Wright TJ, Ayele A et al (eds) Magmatic Rifting and Active Volcanism. Special Publication Geological Society of London, 420, in press

Megrue GH, Norton E et al (1972) Tectonic history of the Ethiopian rift as deduced by K-Ar ages and palaeomagnetic measurements of basaltic dykes. J Geophys Res 77(29):5744–54

Merla G, Abbate E et al (1973) Geological map of Ethiopia and Somalia, 1:2,000,000. Consiglio Nazionale Della Ricerche, Florence

Micheels A, Eronen J, Mosbrugger V (2009) The late Miocene climate response to a modern Sahara desert. Glob Planet Change 67:193–204. doi:10.1016/j.gloplacha.2009.02.005

Mohr P (1963) Geologic map of African Horn of Africa. Philip and Tacey, Fulham

Moore JM, Schultz RA (1999) Processes of faulting in jointed rocks of Canyonlands National Park, Utah. Geol Soc Am Bull 111:808–822

Moucha R, Forte AM (2011) Changes in Africa topography driven by mantle convection. Nat Geosci 4:707–712. doi:10.1038/NGEO1235

Nicholson SE (1996) A review of climate dynamics and climate variability in East Africa. In: Johnson TC, Odada E (eds) The limnology, climatology and paleoclimatology of the East African lakes. Gordon and Breach, Amsterdam, pp 25–56

Olson JE (1993) Joint pattern development: effects of subcritical crack growth and mechanical crack interaction. J Geophys Res 98 (B7):12,251–12265

Olson P (1994) Mechanics of flood basalt magmatism. In: Ryan MP (ed) Magmatic systems. Academic Press, New York, pp 1–18

Peckham SD (1998) Efficient extraction of river networks and hydrologic measurements from digital elevation data. In: Barndorff-Nielsen OE et al (eds) Stochastic methods in hydrology: rain, landforms and floods. World Scientific, Singapore, pp 173–203

Purcell P (1976) The Marda Fault Zone, Ethiopia. Nature 261 (5561):569–571. doi:10.1038/261569a0

Purcell P (1981) Phanerozoic sedimentary history and petroleum potential. In: Chewaka S, de Wit MJ (eds) Plate tectonics and metallogenesis: some guidelines to Ethiopian mineral deposits. Ethiopian Inst Geol Surv Bull 2:97–114

Purcell P, Mège D, Jourdan F (2011) Volcanic geomorphology of Southeast Ethiopia. In: Asrat A et al (eds) Geomorphology for human adaptation to changing tropical environments. IAG/IAG regional conference, Addis Ababa, 127

Sachau T, Koehn D (2010) Faulting of the lithosphere during extension and related rift-flank uplift: a numerical study. Int J Earth Sci 99:1616–1632. doi:10.1007/s00531-010-0513-6

Schultz-Ela DD (2001) Excursus on gravity gliding and gravity spreading. J Struct Geol 23:725–731

Schumm SA, Dumont JF, Holbrook JM (2002) Active tectonics and alluvial rivers. Cambridge University Press, Cambridge

Sepulchre P, Ramstein G, Fluteau F, Schuster M, Tiercelin J-J, Brunet M (2006) Tectonic uplift and eastern Africa aridification. Science 313:1419–1423

Snyder NP, Whipple KX, Tucker GE, Merritts DJ (2000) Landscape response to tectonic forcing: digital elevation model analysis of stream profiles in the Mendocino triple junction region, northern California. Geol Soc Am Bull 112:1250–1263

Syvitski JPM, Overeem I, Brakenridge GR, Hannon M (2012) Floods, floodplains, delta plains—a satellite imaging approach. Sedim Geol 267–268:1–14

Tefera M, Chernet T, Haro W (1996) Geological map of Ethiopia, 2nd edn, 1:2,000,000. Geol. Survey Ethiopia, Addis Ababa

Temin J (2006) Grassroots conflict assessment of the Somali Region, Ethiopia. CHF International

Thulin M (2007) Acacia fumosa sp. nov. (Fabaceae) from eastern Ethiopia. Nordic J Bot 25:272–274

Tolan TL, Reidel SP et al (1989) Revisions to the estimates of the areal extent and volume of the Columbia River Basalt Group. In: Reidel SP, Hooper PR (eds) Volcanism and tectonism in the Columbia River Flood-Basalt Province, US Geol Surv Sp Pap 239:1–20

UNDP (1999) Drought and floods stress livelihoods and food security in the Ethiopian Somali Region. UNDP Emergencies Unit for Ethiopia report

Walsh J (1976) Oil and gas in Ethiopia. Geol Surv Ethiopia Min Circ 4

Watchorn F, Nichols GJ, Bosence DWJ (1998) Rift-related sedimentation and stratigraphy, southern Yemen (Gulf of Aden). In: Purser

BH, Bosence DWJ (eds) Sedimentation and tectonics of rift basins: red Sea-Gulf of Aden. Chapman and Hall, London, pp 165–189
Whipple KX (2004) Bedrock rivers and the geomorphology of active orogens. Ann Rev Earth Planet Sci 32:151–185
Whipple KX, DiBiase RA, Crosby BT (2013) Bedrock rivers. In: Shroder J Jr, Wohl E (eds) Treatise on geomorphology, vol 9., Fluvial Geomorphology. Academic Press, San Diego, pp 550–573
Wolfenden E, Ebinger C et al (2004) Evolution of the northern Main Ethiopian rift: birth of a triple junction. Earth Planet Sci Lett 224:213–228

Author Biographies

Daniel Mège Daniel Mège obtained a M.Sc. degree in geophysics and internal geodynamics in 1991 in Paris at University of Paris-Sud, Orsay, Ecole Normale Supérieure-Ulm, and Pierre and Marie Curie University. In 1994, he obtained a Ph.D. degree in planetary sciences at University Paris-Sud, Orsay, France. In charge of the NASA Regional Planetary Image Facility at University of Paris-Sud in 1995–96, he was also a postdoc at DLR and at University of Nevada, Reno. He was then appointed as assistant professor at Blaise-Pascal University, Clermont-Ferrand, France, before obtaining an associate professor position at Pierre and Marie Curie University in 1997. In 2001, he obtained his habilitation in comparative planetology and moved to University of Nantes, France, in 2004. Since 2011, he is a Visiting Professor at the Institute of Geological Sciences at the Polish Academy of Sciences, where he is in charge of developing a research group working on planetary sciences. He conducts research on the geology of the Gojjam lowlands in western Ethiopia and the Ogaden in order to contribute to document the evolution of the Ethiopian Large Igneous Province and the Afar triple junction, and of various bodies of the solar system. Daniel is a member of the Geological Society of America, the American Geophysical Union, the Geological Society of Africa, and IEEE.

Peter Purcell Peter Purcell is an Australian geologist with an MSc from the University of Sydney. He is General Manager of P&R Geological Consultants Pty Ltd, with over 40 years experience working in Australia, SE Asia, and East Africa. Peter first worked in Ethiopia in 1973–1977 and continues to be involved with exploration and research there. He has edited several books on Australian geology and is the author of numerous articles on geology and petroleum exploration in Australia and East Africa, as well as environmental and social issues. Peter is a fellow of the Geological Society of London and a member of AAPG, PESA, and ASEG.

Stéphane Pochat Stéphane Pochat received the MSc degree in sedimentology and paleoclimatology in 1999 and a PhD degree in sedimentology and tectonics at University of Rennes 1 (Geosciences Rennes), France. He was appointed as assistant professor, then associate professor in 2007 at University of Nantes (LPG Nantes), France. He shares his time between paleoclimatic and sedimentary basins dynamic during the Paleozoic, the study of marine terraces, fluvial geomorphology in Ethiopia and Somalia, and fluvial and glacial geomorphology on Mars.

Thomas Guidat Thomas Guidat is a geomorphologist who obtained a MSc degree in planetology in 2013 at University of Nantes (LPG Nantes), France. He worked on geomorphometric analysis of basalt-filled paleorivers in Ogaden and Somalia, and glacial evolution of Isidis Planitia on Mars. He is currently involved in the PhD program at Trinity College, Dublin, where he is studying eolian dynamics on Mars. Thomas is a member of the British Society of Geomorphology.

Part III

Applied Aspects

Geo-hazard in Ethiopia

20

Giandomenico Fubelli and Francesco Dramis

Abstract

This paper presents an overview of the potentially dangerous geological–geomorphological processes that characterize the territory of Ethiopia. Due to the active crustal mobility causing the ongoing rifting between the Arabian and African plates, this country is particularly prone to high magnitude earthquakes and volcanic eruptions. Moreover, the climate, characterized by long-lasting and intense precipitation during the rainy seasons, and the recurrent earthquake shocks are the triggering factors of a variety of landslides that are particularly favoured by the rugged topography, marked by deep valleys and high steep slopes. All these processes may induce different levels of risk in relation to the different hazards and vulnerabilities, especially after the recent development and planning of road networks, dams, reservoirs, and the establishment of new urban settlements in previously uninhabited areas. Finally, an overview of the measures taken by the Ethiopian authorities so far to mitigate geological hazards is presented along with some proposals for future mitigation initiatives.

Keywords

Geo-hazards • Earthquakes • Volcanoes • Landslides • Risk • Ethiopia • East Africa

20.1 Introduction

Geo-hazards, including earthquakes, ground fissuring, volcanic eruptions, and landslides, are crucial problems for Ethiopia (Fig. 20.1), causing heavy limitation for urbanization and infrastructural projects and, more in general, for the socioeconomic development of the country.

Two main factors are responsible for the occurrence of geo-hazards in the Horn of Africa: (1) The regional geodynamics causing the ongoing progressive separation of the Arabian and African plates, i.e., the East African Rift System (EARS), and (2) the climate characterized by long-lasting and intense precipitation during the rainy seasons.

Earthquakes and volcanic eruptions are related to crustal diastrophism and, in particular, to different systems of fractures and faults, the modern activity of which is testified by the dislocation of Late Quaternary volcanics and alluvial deposits (Faure 1975; Abebe et al. 2007).

Seismic events with magnitudes greater than 5 are not uncommon along the margins of the Main Ethiopian Rift (MER), the northernmost sector of the EARS which divides the Ethiopian plateau into two uplifted Blocks (the northwestern Plateau and the southeastern Plateau), and in the Afar Triangle, where the MER joins with the Gulf of Aden and Red Sea rifts, thus splitting the African Plate into the Nubian Plate to the north and the Somali Plate to the south (Fig. 20.2).

Widespread evidence of seismically induced landslides, mostly first generated rapid movements, is provided by the historical record of earthquakes and related surface effects (Gouin 1979). However, the main triggering factor of slope failures in Ethiopia is heavy rainfall, mostly concentrated in July and August (Gamachu 1977; Ethiopian Mapping Authority 1988; Chap. 3, this volume).

G. Fubelli (✉) · F. Dramis
Department of Sciences, Roma Tre University, Rome, Italy
e-mail: giandomenico.fubelli@uniroma3.it

F. Dramis
e-mail: dramis@uniroma3.it

P. Billi (ed.), *Landscapes and Landforms of Ethiopia*, World Geomorphological Landscapes,
DOI 10.1007/978-94-017-8026-1_20, © Springer Science+Business Media Dordrecht 2015

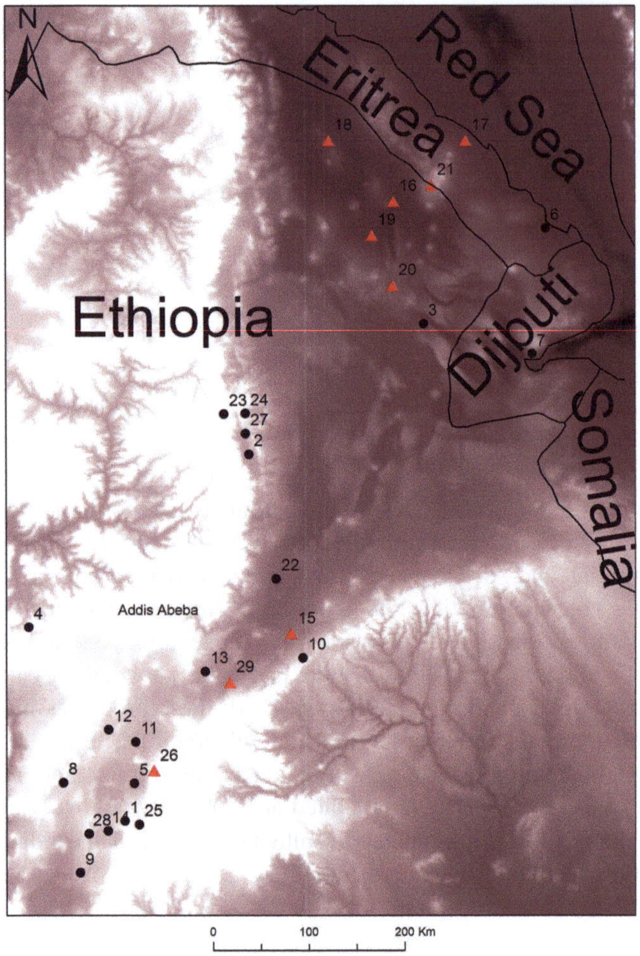

Fig. 20.1 Location map of the cited localities: *1* Awasa; *2* Kara Kore-Mejete; *3* Serdo-Dôbi graben; *4* Shewa; *5* Langano; *6* Massawa; *7* Djibouti; *8* Hosaina; *9* Yirga Alem-Moto; *10* Metehara; *11* Ziway; *12* Butajira; *13* Welenchiti; *14* Melka; *15* Fentale Volcano; *16* Afdera Volcano; *17* Dubbi Volcano; *18* Erta' Ale Volcano; *19* Dabbahu Volcano; *20* Mand Hararo Volcano; *21* Nabro Volcano; *22* Ankober; *23* Dese (Tossa—Azwa Gedel); *24* Kombolcha; *25* Wendo Genet; *26* Mt. Bosetti; *27* Borkena River; *28* Muleti; *29* Kone volcano-Galibardi Pass. For river location, see Fig. 4.2 of Chap. 4 (this volume)

All geodynamic and geomorphic processes may interact with people, buildings, infrastructures, and farmlands, inducing different levels of risk in relation to the different hazards and vulnerability levels (IUGS WGL-CRA 1997).

The recent development and planning of new road networks, dams, reservoirs, and the establishment of new urban settlements in previously uninhabited areas has definitely increased the geological–geomorphological risk in Ethiopia, a developing country where "social vulnerability" to natural disasters is particularly augmented by economic, social, political, and cultural constraints (Alcántara-Ayala 2002).

This chapter presents an overview of the geo-hazards (including earthquakes, ground fissuring, volcanic eruptions, flooding, and landsliding) that affect Ethiopia with particular reference to the case study of Dese, a major Ethiopian town heavily struck by landslides. A short description of the risk mitigation measures undertaken so far by the Ethiopian government is also given.

20.2 Earthquakes

The first studies on the seismicity of the MER and the Afar Triangle are those of Gutenberg and Richter (1954), Sykes and Landisman (1964), Gouin and Mohr (1967), Mohr (1967), Fairhead (1968), Gouin (1970, 1979). More recently, seismic episodes have been described by Asfaw (1988, 1992), Ayele (1995, 1997), Kebede (1996), Ayele et al. (2007). According to Gouin (1979), who published a comprehensive study of earthquakes in Ethiopia and the Horn of Africa up to 1977, at least 15,000 tremors, strong enough to be felt by humans, including a large number of highly destructive events, have been recorded since the earliest known earthquake, dating back to AD 1431, during the kingdom of Emperor Zara Yaqob.

Kebede (1996) recognized a total of 16 earthquakes of magnitude higher than 6.0 during the twentieth century in Ethiopia. Significant among these were the events that affected southern Shewa (1906), Awasa (1960), the Kara Kore-Mejete area (1961), Serdo (1969), Langano (1985), and the Dôbi graben in the Afar (1989). Strong earthquakes also struck Massawa, Eritrea (1921—$M = 5.8$), and French Somalia, Djibouti (1960—$M = 6.0$; 1973—$M = 6.0$; 1989—$M = 6.3$). In this century, moderate to high magnitude earthquakes affected the Afar (2005—$M = 5.6$), Hosaina (2010—$M = 5.3$), Yirga Alem (2011—$M = 5.0$), and the southern Eritrean–Ethiopian border (2011—$M = 5.7$).

In many cases, the earthquake shocks heavily damaged built-up areas and, in some cases, caused fatalities (Gouin 1979). However, until the 1960s, the structural damage to buildings and infrastructure was generally low due to their scarcity in the seismic-prone areas of the country. The recent development of urban settlements close to and within the MER and the Afar Triangle has definitely increased the level of seismic risk in Ethiopia.

In the occasion of major events, coseismic ground effects such as surface faults and fissures, as well as gravitational movements were produced (Gouin 1979; Jacques et al. 2011). This is the case of the strong earthquake that struck the Kara Kore-Mejete area in 1961, with more than 3,500 shocks from the end of May to the end of September as recorded at the Geophysical Observatory in Addis Ababa (Gouin 1979). Two shocks had a magnitude >6.4 and seven were stronger than >5.0. The area affected was estimated to be about 300,000 km^2 and relatively higher intensities were observed in the southeastern sector of the zone. In the

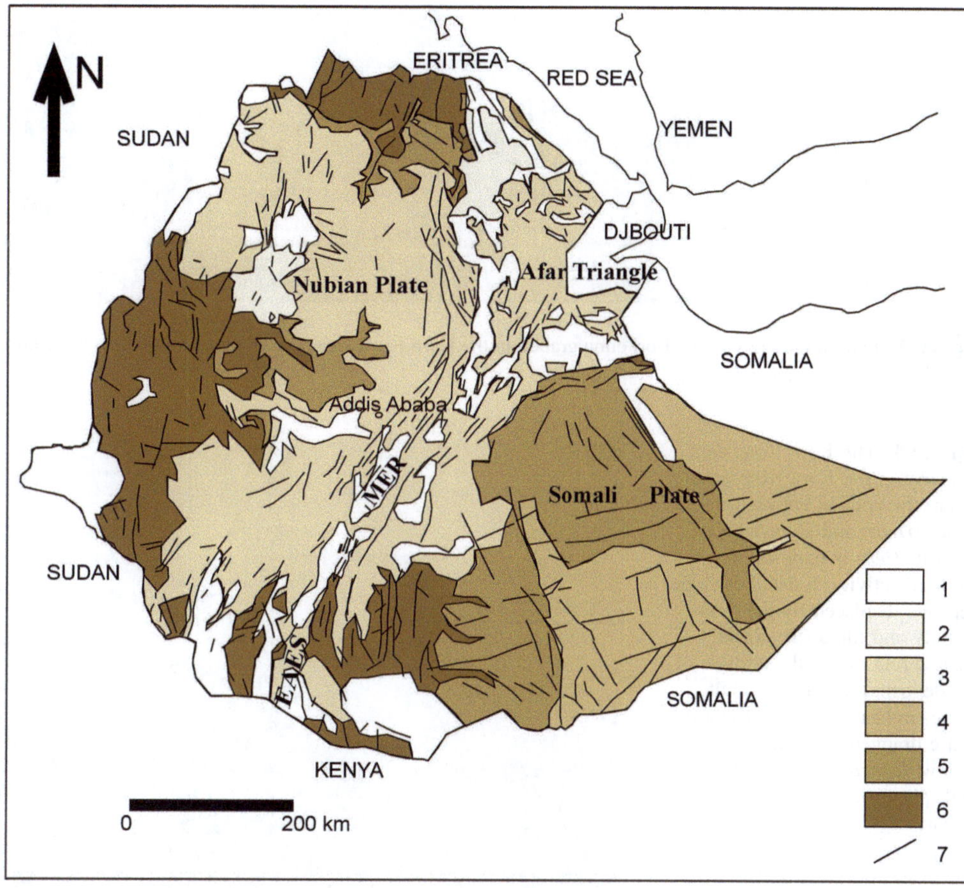

Fig. 20.2 Geological scheme of the Horn of Africa (*1* Quaternary alluvial deposits; *2* Quaternary volcanics; *3* Oligocene–Pliocene volcanics; *4* Cretacic continental and marine successions; *5* Late Jurassic marine and continental successions; *6* Precambrian-Late Proterozoic intrusive rocks and Late Paleozoic-Triassic conglomerates and tillites; *7* Main faults)

epicentral area, the village of Mejete was completely destroyed and the village of Kara Kore was heavily damaged with an estimated intensity of VIII–IX on the Mercalli-modified scale (Gouin 1979). Ground cracks, fissures, rock falls, and subsidence of up to 1 m deep developed on the Addis Ababa—Mekelle road near the town of Kara Kore, heavy damaging culverts and retaining walls. Gravitational movements were observed on steep escarpments and a 15–20 km long fissure, in places 6–7 m deep, formed in unconsolidated soil along the eastern scarp of the Borkena River marginal graben (Fig. 20.3). A more than 2 m high piedmont scarplet formed in unconsolidated materials along the escarpment of the Borkena graben, reactivating a pre-existing north–south trending normal/strike slip fault (Fig. 20.4) over a distance of 12–15 km (Gouin 1979). Considering that this displacement is very large in relation to the earthquake magnitude ($M = 6.6$), it can be explained by the contribution of a strong gravity stress related to the high difference of relief between the Ethiopian Plateau and the Afar (Chorowicz et al. 1999).

Numerous coseismic surface breaks, including scarplets with vertical throws up to 30 cm high and open fissures up to 30 cm wide, were observed in central Afar during the first two days of August 1989 earthquake of Dôbi graben ($M = 6.3$) (Jacques et al. 2011).

20.3 Ground Fissures in the Main Ethiopian Rift

Ground fissures up to some kilometers long (Fig. 20.5) recurrently open in the MER floor consisting of up to more than 100 m thick, unconsolidated lacustrine sediments (Gouin 1971; Asfaw 1982). These features consist of a series of subsidence pits, up to 7 m wide and several tens of meters deep, connected by cracks ranging in width from a few centimeters to a few meters.

Taking the orientation of fissures into account, generally parallel (NE–SW) or slightly oblique (NNE–SSW) to the Rift axis, their occurrence may be related to the present NE–SW extension of the MER. However, only in few cases, the genesis of fissures in the MER floor has been related to seismic events (Gouin and Mohr 1967; Asfaw 1982) and, even in these cases, their opening during earthquakes has never been directly observed (Asfaw 1998) nor confirmed.

A much more frequent and immediate triggering factor of ground fissures is heavy, long-lasting rain (Asfaw 1998) that induces ground collapse, pit subsidence, and enlargement of cracks on top of underground tunnels excavated by groundwater running along tectonic fractures possibly affecting the deep-seated rocky basement. In many cases, the

Fig. 20.3 Panoramic view of the Borkenna graben in the Kara Kore area. Coseismic deformation affected the road on the right of the image due to the 1961 seismic sequences

Fig. 20.4 The Kara Kore fault: **a** quarry in the Kara Kore area where the contact between the basalt (*right*) and slope/alluvial deposits (*left*) is exposed due to the fault activity; **b** detail of slope deposits displaced by the fault activity and tilted (the *white dashed line* marks the altitude of the deposits) toward the fault plane (indicated by *red arrows*) and **c** dragged (as indicated by *yellow dashed lines*) along the fault plane (indicated by *red arrows*)

fissures have been associated with groundwater flow as inferred from the noise of rushing water at depth. On the other hand, fissures in hard rocks, associated to Quaternary faults parallel to the Rift (Fig. 20.6), have been reported from several places of the MER (Gibson 1967; Williams et al. 2004).

Quite different is the explanation provided by Ayalew et al. (2004), according to whom the ground fissures form by the combined action of groundwater flow, aseismically released elastic strain, piping, and hydrocompaction.

The first reports on damage caused by ground fissures in the MER were provided in 1956 by railway engineers who observed them across the Addis Ababa—Djibouti railroad near the town of Metehara (Gouin 1971) and in 1966, when the same railroad was interrupted for a few days near the town of Wolenchiti (Gouin and Mohr 1967). In the following years, other relevant fissure opening affected communication lines (such as the road from Ziway to Butajira) and urban settlements (such as the village of Moto, the Awasa Melka area, the surroundings of Meki, and the town of Muleti in the Awasa basin) causing destruction of houses (Fig. 20.7) and loss of property (Asfaw 1982, 1998).

There is no doubt that the occurrences of fissures in the MER are actually many more than those reported so far since most of these features occurred in desert areas. As far as settlement programs develop and communication lines

Fig. 20.5 Ground fissure at Muleti in the Awasa basin MER

infringe on previously uninhabited areas, the reports will likely become more and more frequent.

20.4 Volcanic Eruptions

Also, volcanic activity is associated with the rifting. This is the case of the traps, an extensive pile of basalt lava flows, covering an area of ca. 600,000 km^2 and up to more than 2,000 m thick, emplaced during the Oligocene on a large part of Ethiopia. Volcanic activity has followed during Miocene, Pliocene, and Quaternary, with the eruption of basalt and subordinate rhyolitic, trachytic, and phonolitic materials, resulting in major topographic changes and the formation of a number of lake basins. Linear basaltic fissure vents dominate the rift axis while point source basaltic vents, such as those located around Debre Zeit, are located as far as 7 km from the rift axis. Off axis central volcanic complexes host the products of young eruptions that are as far as 12 km from the center.

In the last centuries, several volcanic eruptions have been recorded in the MER and the Afar Triangle (Mohr 1962). According to a local tradition, a huge basalt flow originated in 1810 from the southwestern flank of Mt. Fentale, a presently dormant stratovolcano located in the northernmost part of the MER (Fig. 20.8). Other volcanic products, recognized in the MER at Galibardi Pass and Mt. Bosetti, could be related to the same period (Mohr 1962). In June 1907, a relevant eruptive episode occurred in the Afdera Volcano, an isolated stratovolcano in northeastern Ethiopia. The emitted

Fig. 20.6 Fissure in volcanic rocks in the MER near Metehara (*Photograph* by V. Acocella)

Fig. 20.7 House damaged by a ground fissure at Meki (*Photograph* by V. Acocella)

lava flow was about 5 m thick and was accompanied by a sequence of seismic shocks. The most hazardous of these events was the explosive eruption of the Dubbi Volcano which occurred on May 7, 1981, in southern Eritrea, close to the Ethiopian border, killing more than one hundred people.

One volcanic center erupted in the Afar Triangle during the first decade of this century: the Erta' Ale (Fig. 20.9), a basaltic shield volcano with a NNW–SSE elongated caldera hosting a rising lava lake. A major lava overflow occurred on September 25, 2005, killing 250 heads of livestock and forcing thousands of nearby residents to move away. A further lava flow occurred in August 2007, forcing the evacuation of hundreds of people and probably killing two persons. More recent eruptions were reported on November 4, 2008, and November 27–28, 2010. The Dabbahu and Mand Hararo volcanoes started to erupt basalt flows, ash, and smoke plumes along a more than 70 km long north–south fissure. The Nabro Volcano, a stratovolcano thought to be extinct, started erupting on June 12, 2011 in southern Eritrea, not far from the Ethiopia borderline, with the emission of a huge amount of sulfur dioxide and a high reaching ash plume that disrupted the airline traffic in the region. The eruption killed seven persons and caused the displacement of thousands of people. A series of earthquake shocks, including two with magnitude 5.7, struck the region in the hours preceding the eruption.

20.5 Flooding

Floods frequently occur in Ethiopia during the main rainy season (*Kiremt* season—See also Chap. 3, this volume) which extends from mid-June to mid-September. Overflow of rivers from their banks usually happen as a result of heavy rains (Fig. 20.10). This type of flooding usually grows up slowly with magnitude depending on many factors such as watershed area, rainfall intensity and amount, topography, sediment deposition in river beds, presence/absence of vegetation, and human activity.

Where there is little settlement and the main economic activity is pastoralism, the damage is low and floods are even well accepted by the population since they improve grazing and water supply for livestock and people. The only major problems are related to road interruptions, sometimes coupled with the collapse of bridges (Fig. 20.11). On the other hand, flooding may be extremely dangerous in densely inhabited areas as in the case of the Awash River basin whose flood plain includes about 70 % of the country's large-scale irrigated agriculture. It is estimated that almost all of the area delineated for irrigation development may be subject to floods with a submergible surface of ca. 200,000–250,000 ha (Alemayheu 2007). Uncontrolled deforestation and expansion of farmlands have induced in this basin widespread soil erosion (Bishaw 2001; Tadesse 2012) resulting in heavy sediment deposition in the stream beds. For this reason, the Awash River in the Lower Plains has a very unstable course tending to change its channel pattern and to branch out into the adjoining alluvial plains. Siltation has caused a reduction of more than 40 % (Kefyalew 2003) of the capacity of Koka Dam reservoir, which is used to supply water for irrigation, for hydropower generation, and also for flood lamination to protect the downstream plains. Other areas where significant river flooding occurs are the lower reaches of the Wabe Shebele, Genale, and Dawa rivers near the Somali order, the lowlands areas bordering the Baro, Gilo, and Akobo rivers in western Ethiopia, the downstream areas of Omo River, the extensive fields surrounding Lake Tana and the Gumara and Rib rivers.

Flooding in urban settlements, such as Addis Ababa, causes recurrent damages to built-up structures and properties along streams coming down from the adjacent mountains.

Much more dangerous for human lives and properties, because of their suddenness and high speed, are the flash floods that recurrently occur in many parts of Ethiopia as the

Fig. 20.8 The Fentale volcano (*Photograph* by V. Acocella)

Fig. 20.9 The Erta' Ale volcano (*Photograph* by V. Acocella)

result of localized intense rainfall capable to generate high peak discharges in river channels. These events are favored by the rugged topography of the country characterized by steep slopes and river channels, sparse vegetation, and poor infiltration capacity of soil (Abebe 1997).

In the town of Dire-Dawa, the flash flood event of August 5–6, 2006, killed more than 250 people and caused heavy damage to public infrastructures including flood defenses, housing, and livelihoods (FDPPA 2007). Other major floods occurred in the same town in April 1981, April 1994, and May 2005 (Alemayheu 2007, Billi et al. 2015).

20.6 Landslides

Landslides are widespread in Ethiopia (Ethiopian Institute of Geological Surveys 1994, 1995; Asrat et al. 1996; Gezahegn 1998; Ayalew 1999, 2000; Ayalew and Vernier 1999;

Fig. 20.10 Flooding of a small river in the Rift valley, south of Langano Lake during the rainy season (*Photograph* by P. Billi)

Fig. 20.11 Bridge collapsed because of flooding (Golina River)

Temesgen et al. 1999, 2001; Ayalew and Yamagishi 2002, 2004; Nyssen et al. 2002, 2003; Woldearegay et al. 2004; Fubelli et al. 2008a; Moeyersons et al. 2008; Abebe et al. 2010; Broothaerts et al. 2012) where they cause significant damage. From 1993 to 1998, more than 200 houses were destroyed, more than 500 km of roads were interrupted, and about 300 people were killed (Ayalew 1999).

The main predisposing factor of landslides is high relief (ranging on average between 2,000 and 3,000 m a.s.l.) induced by the Pliocene–Quaternary uplift (Almond 1986; Mohr 1986) and the resulting rugged morphology with deep valleys and gorges, where river incision causes further slope steepening and instability. Steep, unstable escarpments, up to several hundred meters high, are also produced by active faults on the rift margins (Chorowicz et al. 1999; Abebe et al. 2007; Fubelli et al. 2008a).

Triggering factors are heavy rainfalls and, to a lesser extent, earthquakes. Intense, long-lasting rains commonly occur in July and August (Gamachu 1977; Ethiopian Mapping Authority 1988; Chap. 3, this volume) resulting in high rates of water infiltration into the ground, widespread slope wash, overflooding, and riverbank erosion. Rain infiltration plays a primary role in triggering landslides by rising the water table, thus increasing the total weight of the slope material and the pore pressure in fine-grained deposits (Brand et al. 1984; Ayalew 1999; Gabet et al. 2004). The infiltration of rain water into fractured bedrock may be enhanced by the presence of deep desiccation cracks in the covering soil, which develop especially at the end of the dry season (Billi and Dramis 2001, 2003).

Landslides are the most common surface effects of earthquakes, even in case of moderate magnitudes (Youd 1978; Harp et al. 1986; Ishihara and Hsu 1986; Cotecchia 1987). The main triggering agents of coseismic landsliding are: oriented accelerations; earthquake induced changes in the groundwater table; cyclic changes of pore pressures and liquefaction; seismo-tectonic surface fracturing and faulting. The latter agents seem also able to induce landsliding even long after the seismic event, by favoring rain water infiltration (Chiodo et al. 1999; Dramis and Blumetti 2005). The triggering of landslides, as well as other seismo-gravitational effects, is also favored by litho-structural contacts and geomorphological settings (e.g., narrow ridges and valleys, and peaks), which may locally amplify seismic shaking (Dramis and Blumetti 2005). Seismically induced landslides in Ethiopia are reported by Gouin (1979), among which were the huge slope failure which destroyed the town of Ankober in 1842 and the fast-moving landslides that killed two people in Dese town during the 1977 earthquake.

The nature of the outcropping bedrock, together with the intensity and duration of rainfall/earthquake shaking, controls the typology and size of failure: (1) on steep slopes in hard bedrock, such as basalt, welded ignimbrite, limestone and sandstone, fast-moving mass movements like rock falls (Fig. 20.12), topplings, and rock slides/avalanches (Varnes 1978; Dikau et al. 1996; Ayalew 2000) are common; (2) on steep slopes in weathered pyroclastics and volcanic agglomerates first generation rapid landslides, such as debris slides (Fig. 20.13) and avalanches, debris flows, earth flows and mudflows frequently occur mobilizing the soil cover (Varnes 1978; Dikau et al. 1996; Temesgen et al. 1999, Fubelli et al. 2008a); (3) rapid failures, such as topples and slumps (Varnes 1978; Dikau et al. 1996), typically affect the alluvial banks of deeply incised rivers and gullies; (4) on clayey materials, such as shaly–marly formations, alluvial–colluvial deposits and thick weathered layers on volcanic rocks, simple and multiple retrogressive rotational slides, sometimes passing to earthflows or mudflows (Varnes 1978; Dikau et al. 1996) are commonly generated; (5) fast-moving translational rock slides moving along predisposed yielding surfaces (Varnes 1978; Dikau et al. 1996) can be expected where soft clayey beds, dipping downslope, are overlain by hard rocky layers; (6) slow-moving translational slides with a step-like evolution, characterized by long quiescence periods and short reactivation phases after heavy rainfalls, may involve thick alluvial–colluvial deposits with intercalated

Fig. 20.12 Rockfall-derived boulder from the rim of the Wendo Genet caldera

Fig. 20.13 Debris slide at Wendo Genet. *Note* people climbing the talus for scale

finer horizons (Ayalew 2000); (7) first generation, fast-moving shallow landslides, such as soil slips or mud flows, commonly mobilize eluvial–colluvial material (Abebe et al. 2010); (8) where thick layers of hard rock overlie clayey formations, large-scale lateral spreading phenomena, moving at an extremely slow rate, are present (Dramis and Sorriso-Valvo 1994; Dikau et al. 1996; Coltorti et al. 2009); (9) clayey/marly formations overlain by fractured rocky aquifers (such as basalt, limestone, or sandstone) that may undergo sudden water table fluctuations during heavy rain are particularly prone to failure (Ayalew 2000); (10) groundwater pressure and seepage may trigger mass movements in colluvial deposits covering the hydrogeological contact between a hard rock cliff and an underlying clayey/marly aquiclude (Ayalew 2000; Nyssen et al. 2003).

Finally, an increasing impact of human activities, such as intensive agriculture, quarrying, road construction, and urbanization, is also inducing slope instability by overloading the slope, modifying the slope profile, and increasing water infiltration into the ground from leaking aqueducts, sewerages, and irrigation ditches on cultivated slopes (Temesgen et al. 1999; Ayalew 2000; Nyssen et al. 2002, 2003; Abebe et al. 2010).

20.6.1 The Case of the Dese Urban Area

One Ethiopian area heavily affected by landslides is the urban territory of Dese (Welo) (Ethiopian Institute of Geological Surveys 1995; Ayalew and Vernier 1999; Terefe 2001; Ayenew and Barbieri 2005; Fubelli et al. 2008a), a medium-sized urban settlement with about 200,000 inhabitants located at ca. 1,800 m a.s.l. in a small tectonic depression "hanging" 800 m over the more eastern depression of Kombolcha (Figs. 20.14 and 20.15). This depression is part of a basin and range landscape produced on the eastern margin of the northern Ethiopian Plateau by the Cenozoic extension of northeast Africa (Boccaletti et al. 1998; Ukstins et al. 2002). The outcropping bedrock consists of a sequence of variably weathered Tertiary ignimbrites and lava flows (Tefera et al. 1996), mantled by colluvial and alluvial deposits mainly derived from the two north–south trending fault escarpments that border the basin (Tossa and Azwa Gedel escarpments). Besides the main border faults, southwest–northeast trending transfer faults are also present on the northern and southern edges of the basin (Fig. 20.16). An up to 100 m thick sedimentary sequence of alluvial-swampy-lacustrine deposits fills the lower part of the basin floor (Ayenew and Barbieri 2005). These deposits are deeply incised by the Borkena River, forming a wide terrace over which a wide sector of Dese town is located. Two alluvial fans, formed by right-side tributaries of the Borkena River, are also occupied by a part of the Dese settlement. Both the terrace and the alluvial fans are connected to the Tossa and Azwa Gedel escarpments by up to 15 m thick talus belt.

A large number of landslide events have repeatedly struck this area since long, with heavy impact on buildings and infrastructure and loss of human lives (Fubelli et al. 2008a). The main triggering factor is long-lasting intense rainfall (Ayalew 1999; Ayenew and Barbieri 2005) and, to a lesser extent, the earthquakes which recurrently strike the area (Gouin 1979). The latter triggers rapid slope failures such as rock falls and debris flows, as those occurred during the seismic event of July 8, 1988, causing the death of 2 persons (Gouin 1979). Man-made activities, such as the construction of houses, roads, bridges, or leakage of water from aqueducts, also induce slope instability by adding load to incipient landslides, modifying slope profiles, or changing groundwater levels and flow.

A detailed GIS-supported geomorphological analysis of the Dese area (Fubelli et al. 2008b) shows that landslide incidence and typology vary with the different landform

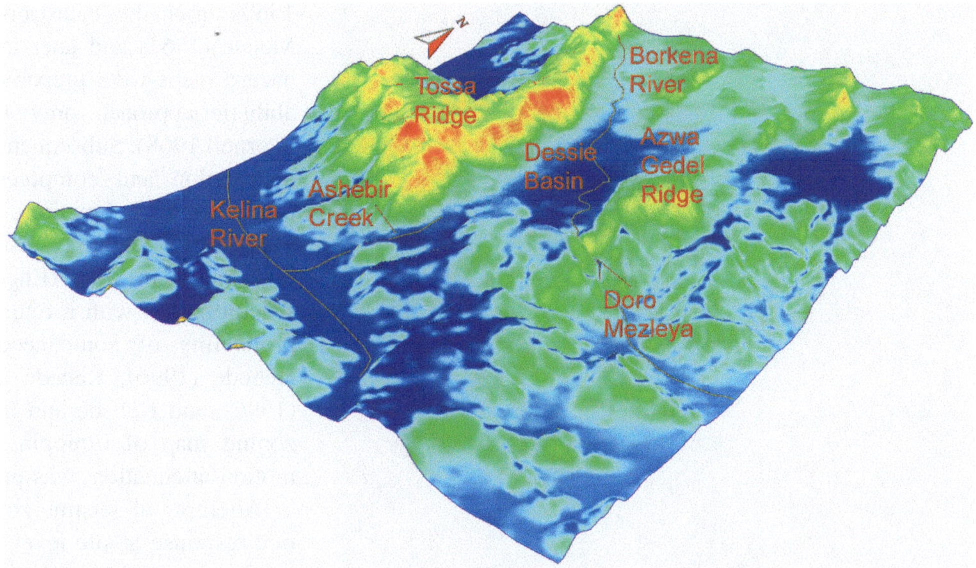

Fig. 20.14 Location map of the Dese basin (after Fubelli et al. 2008a, modified)

Fig. 20.15 An overview of the Dese basin. The escarpment in the back is ca. 400 m high

units making up the basin. Rapid mass movements such as rock falls and topplings, rock slides (Fig. 20.17), debris flows, and debris slides triggered by heavy rain or earthquakes, characterize the major north–south and southwest–northeast fault escarpments, the upper parts of the talus belts placed at the foot of the escarpments, and the upper catchment slopes of the Borkena River tributaries. The escarpment talus belts are characterized by high hazard levels due to deep-reaching translational earth slides (Fig. 20.18). Finally, rotational slides (Fig. 20.19) affect the erosional escarpments of the terraced area and the incised alluvial fans where a large part of the town is located.

20.7 Hazard Assessment and Risk Mitigation Measures

Notwithstanding the growing information on the occurrence of catastrophic earthquakes, volcanic eruptions, and landslides in Ethiopia and the valuable papers published on these topics, the response of public administrators and decision makers is still inadequate though some research projects and new regulations have been launched on the assessment of geo-hazards and related risk at a regional scale (IUGS WGL-CRA 1997).

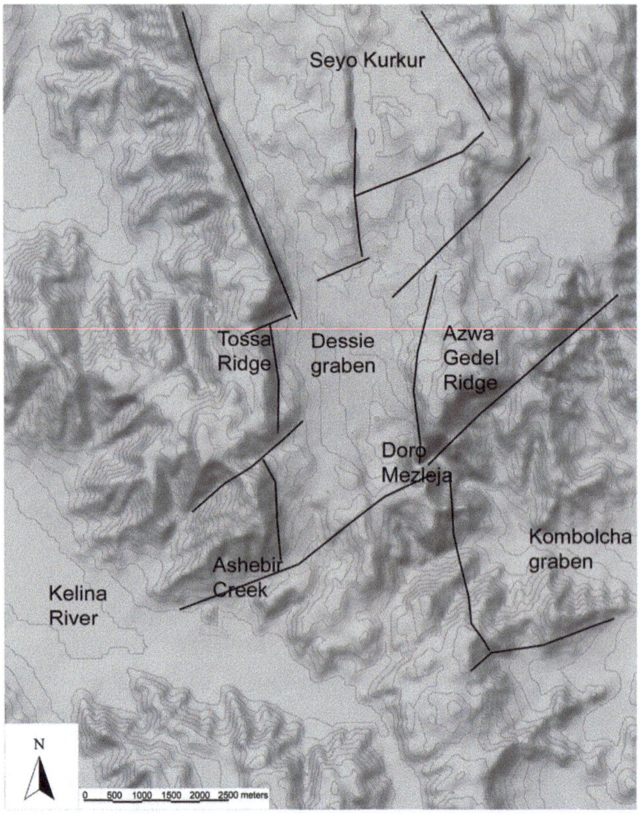

Fig. 20.16 Fault pattern of the Dese basin (after Fubelli et al. 2008a, modified)

20.7.1 Earthquakes

The need for earthquake hazard assessment in Ethiopia was initially recognized by the scientific community in the early 1960s after the catastrophic earthquakes of Kara Kore-Mejete (1961) and later in Serdo (1969). The first seismic hazard maps were prepared by Gouin (1976) using a probabilistic approach previously applied in North America (Cornell 1968). Subsequently, maps of earthquake epicenters distribution and computerized analyses of intensity and acceleration as functions of time were produced (Asfaw 1988, 1990). The first seismic building code was introduced in Ethiopia in 1980 (Engeda et al. 2011). It defines four seismic zones with a return period of 100 years and 90 % probability of non-exceedance. Following the works of Kebede (1996), Kebede and Asfaw (1996), Panza et al. (1996), and Kebede and Eck (1997), a new seismic hazard zoning map of Ethiopia, taking into account the ground motion attenuation, was produced.

Attempts of seismic zoning at a detailed scale and seismic response at site level were performed for some important areas such as the city of Addis Ababa (IDNDR RADIUS Project 1999; Mammo 2005; Yoseph and Ramana 2008). Between 2001 and 2003, the Ethiopia Afar Geoscientific Lithospheric Experiment (EAGLE) international research project was carried out to explore the kinematics and dynamics of the lithosphere using a broadband seismic array (Nyblade and Langston 2002; Bastow et al. 2011). Another noteworthy research project involves the Afar Rift Consortium, funded by the UK Natural Environment Research Council with partners from UK, Ethiopia, France, New Zealand, and the USA. Its aim was to conduct experiments in the Afar area to better understand the seismo-tectonic processes active there. More in general, an accurate delineation of seismogenic belts associated with active faults, fissures, subsidence, and volcanism is an indispensable basis for future hazard assessments.

Fig. 20.17 The crown of a rock slide in weathered basalt near Dese

Fig. 20.18 Trench of a deep-reaching translational debris slide at Dese

Fig. 20.19 Rotational shallow earth slide in the Dese basin

20.7.2 Ground Fissures

The random occurrence of ground fissures after heavy rains and their destructive potential clearly characterizes them as natural hazards. In this context, the identification and mapping of ground cracks and incipient fissures by air photo analysis, detailed field survey, and geophysical methods is an indispensable task to be performed before establishing new settlements and transportation lines (Asfaw 1998).

20.7.3 Volcanic Eruptions

Concerning volcanic hazard and risk mitigation, very little has been done so far in Ethiopia even if more than 9 million people live within 30 km of a potential eruptive center and at least 5–10 of the 65 Holocene volcanoes recognized in the country have been classified as "high risk" by the Global Facility for Disaster Risk Reduction (GFDRR); a World Bank funded pilot study that aims to establish science-based evidence for better integration of volcanic risks (Sparks et al. 2012). At least in some cases, explosive activity with pyroclastic flows and gas emission is expected. In particular, there are, so far, no systematic studies, monitoring, and mitigation plans for active volcanic centers and zones of the MER and the Afar Triangle, notwithstanding the growing development of urban settlements, infrastructure, and economic activities in these areas.

20.7.4 Flooding

Flood management and control in Ethiopia is not yet at an advanced stage. The only major interventions have been carried out in the Awash River Basin with the implementation of irrigation networks protected from floods by dykes and the establishment of a *Project Control Center*, equipped with adequate maintenance facilities. Moreover, the Government established a *River Training Unit* for the Lower Awash Plains, equipped with facilities to remove silt from the river beds. The Unit also constructs and maintains flood protection dykes along the river.

Watershed management projects have been recently started for different parts of Ethiopia. They include the emplacement of check-dams and weirs for water conservation and retaining floods, the construction of terraces along mountain slopes to reduce runoff and favor ground water recharge.

Moreover, in the last years, an extensive project entitled *Flood Preparedness and Early Warning System* has been launched in the Eastern Nile countries (Sudan, Ethiopia, and Egypt). The planned activities, when accomplished, should enable these countries to improve their capacity of mitigating flood risk.

20.7.5 Landslides

Very high-risk levels, due to the possible occurrence of rapid mass movements related to heavy rainfall, are typical for settlement areas located at the foot of steep slopes in weathered volcanic rocks or at the outlet of mountain streams, from which concentrated flood flows and debris can suddenly burst out. Settlements and infrastructures, located at the base or at the edge of high fault escarpments in earthquake prone areas, are also exposed to risk.

In the last years, detailed geomorphological–geotechnical studies on landslide susceptibility/hazard of selected areas in Ethiopia have been published (Temesgen et al. 2001; Ayenew and Barbieri 2005; Fubelli et al. 2008b; Abay and Barbieri 2012; Zvelebil et al. 2010) or performed as Master and PhD thesis (Terefe 2001; Woldearegay 2005; Weldegiorgis 2008; Asfaw 2010; Jbrahim 2011; Suyum 2011; Hagos 2013).

A preliminary assessment of landslide risk in the Dese urban area has been recently published by Fubelli et al. (2013). These works are, however, just a small step in a large populated country such as Ethiopia, where the official geological maps, published by the Geological Survey, cover the whole territory only at the 1:2,000,000 scale (Kazmin 1972) and only a few parts of the country are mapped at the 1:25,000 scale.

20.8 Conclusions

The authors hope that this general overview of geo-hazards in Ethiopia could help to make people more aware of their impact on the environment, built-up areas, infrastructure, economy, and human lives. Further investigation on seismicity, volcanism, and slope stability, with particular reference to hazard zoning and risk mitigation at a local scale, are mandatory. A very weak point in this context is the absence of medium-scale geological maps covering the entire Ethiopian territory, notwithstanding the availability of good 1:50,000 topo-sheets.

Other imperative actions include making use of institutional monitoring facilities with state-of-the-art instruments; training specialized personnel (engineers; geologists, geophysicists; geomorphologists) at the MSc and PhD level to be employed in public institutes and agencies; improving research activity also in cooperation with international research groups; organizing scientific conferences; disseminating proper information by school teaching, village seminars, and TV broadcasts able to reach even the most remote villages.

References

Abay A, Barbieri G (2012) Landslide susceptibility and causative factors evaluation of the landslide area of Debresina, in the southwestern Afar escarpment, Ethiopia. J Earth Sci Eng 2:133–144

Abebe B (1997) Flood vulnerability in Ethiopia and needs for preparedness. Paper for workshop vulnerability in Ethiopia from disaster to development. Addis Abeba, June 23–25, 1997

Abebe B, Acocella V, Korme T, Ayalew D (2007) Quaternary faulting and volcanism in the Main Ethiopian Rift. J Afr Earth Sci 48: 115–124

Abebe B, Dramis F, Fubelli G, Umer M, Asrat A (2010) Landslides in the Ethiopian highlands and the Rift margins. J Afr Earth Sci 56:131–138

Alcántara-Ayala I (2002) Geomorphology, natural hazards, vulnerability and prevention of natural disasters in developing countries. Geomorphology 47:107–124

Alemayheu DD (2007) Assessment of flood risk in Dire Dawa town, eastern Ethiopia, using GIS. Unpublished Master thesis, Addis Ababa University

Almond DC (1986) Geological evolution of the Afro-Arabian dome. Tectonophysics 131:301–332

Asfaw EY (2010) Landslide assessment in Blue Nile Gorge, central Ethiopia. Master thesis, Universiteit Gent—Vrije Universiteit Brussel, Belgium

Asfaw LM (1982) Development of earthquake induced fissures. Nature 286:551–553

Asfaw LM (1988) Catalogue of Ethiopian earthquakes. In: Woldeyes G (ed), SAREC-ESTC conference. Addis Abeba, pp 252–279

Asfaw LM (1990) Seismicity and earthquake risk in the Addis Abeba region. SINET -Eth J Sci 13(1):15–35

Asfaw LM (1998) Environmental hazard from fissures in the Main Ethiopian Rift. J Afr Earth Sci 27:481–490

Asrat A, Eshete G, Taddesse S, Getaneh A, Fekede, K (1996) Land mass movement of November 10, 1994 in Goffa District, Northern Omo Zone, Southern Ethiopia. Abstract 3rd Ethiopian Geoscience and Mineral Engineering Congress, 15–17 November 1996, Addis Ababa

Ayalew L (1999) The effect of seasonal rainfall on landslides in the highlands of Ethiopia. B Eng Geol Environ 58:9–19

Ayalew L (2000) Factors affecting slope stability in the Blue Nile basin. In: Bromhead M, Dixon N, Ibsen M (eds) Landslides in research, theory and practice. Thomas Telford, Cardiff, pp 101–106

Ayalew L, Vernier A (1999) Causes and mechanisms of slope instability in Dessie town, Ethiopia. In: Yagi N, Yamagami T, Jang J-C (eds) Slope stability engineering. A.A. Balkema, Rotterdam/Brookfield, pp 1181–1186

Ayalew L, Yamagishi H (2002) Landsliding and landscape development; the case of northern Ethiopia. In: Proceedings of the international congress intrapraevent 2002, Matsumoto, Japan, pp 595–606

Ayalew L, Yamagishi H (2004) Slope failure in the Blue Nile basin, as seen from landscape evolution perspective. Geomorphology 57:95–116

Ayalew L, Yamagishi H, Reik G (2004) Ground cracks in Ethiopian Rift Valley: facts and uncertainties. Eng Geol 75:309–324

Ayele A (1995) Earthquake catalogue of the Horn of Africa for the period 1960–93. Seismology Department, Uppsala, Report 3–95

Ayele A (1997) Spatial and temporal variations of seismicity in the Horn of Africa from 1960–1963. Geophys J Int 130:805–810

Ayele A, Jacques E, Kassim M, Kidane T, Omar A, Tait S, Nercessian A, Chabalier J-B, King G (2007) The volcano–seismic crisis in Afar, Ethiopia, starting September 2005. Earth Planet Sci Lett 255:177–187

Ayenew T, Barbieri G (2005) Inventory of landslides and susceptibility mapping in the Dessie area, Northern Ethiopia. Eng Geol 77:1–15

Bastow ID, Keir D, Daly E (2011) The Ethiopia Afar Geoscientific Lithospheric Experiment (EAGLE): probing the transition from continental rifting to incipient seafloor spreading. Geol Soc Am Spec Paper 478:51–76

Billi P, Dramis F (2001) Recent climatic changes and the spreading of gully erosion in Ethiopia. In: Andah K (ed) Water resources management in a vulnerable environment for sustainable development. UNESCO-IHP-CNR, Grifo Publ, Perugia, pp 131–141

Billi P, Dramis F (2003) Geomorphological investigation on gully erosion in the Rift Valley and northern highlands of Ethiopia. Catena 50:353–368

Billi P, Alemu Y, Ciampalini R (2015) On the increased frequency of flash floods in Dire Dawa, Ethiopia: climate change or human impact?, Natural Hazards, submitted

Bishaw B. (2001) Deforestation and land degradation on the Ethiopian Highlands: a strategy for physical recovery. In: International Conference on African Development Archives, Paper 2. http://scholarworks.wmich.edu/africancenter_icad_archive/2

Boccaletti M, Bovini M, Mazzuoli R, Abebe B, Piccardi L, Tortorici L (1998) Quaternary oblique extensional tectonics in the Ethiopian Rift (Horn of Africa). Tectonophysics 287:97–116

Brand EW, Premchitt J, Phillipson HB (1984) Relationship between rainfall and landslides in Hong Kong. In: Proceedings of the 4th international symposium on landslides, Toronto, September 17, 1984, vol 1, p 377–384

Broothaerts N, Kissi E, Poesen J, Van Rompaey A, Getahun A, Van Ranst E, Diels J (2012) Spatial patterns, causes and consequences of landslides in the Gilgel Gibe catchment, SW Ethiopia. Catena 97:127–136

Chiodo G, Dramis F, Gervasi A, Guerra I, Sorriso-Valvo M (1999) Frane sismo-indotte e pericolosità di sito: primi risultati dello studio degli effetti di forti terremoti storici in Calabria centro-settentrionale. Atti del XVIII Convegno Nazionale GNGTS, Roma, CD-Rom

Chorowicz J, Collet B, Bonavia F, Korme T (1999) Left-lateral strike-slip tectonics and gravity induced individualisation of wide continental blocks in the western Afar margin. Eclogae Geol Helv 92:149–158

Coltorti M, Pieruccini P, Berakhi O, Dramis F, Asrat A (2009) The geomorphological map of Mt. Amba Aradam southern slope (Tigray, Ethiopia). J Maps 2009:56–65

Cornell CA (1968) Engineering seismic risk analysis. B Seismol Soc Am 18:1583–1606

Cotecchia V (1987) Earthquake-prone environments. In: Anderson MG, Richards KS (eds) Slope stability. Wiley, Chichester, pp 287–330

Dikau R, Brunsden D, Schrott L, Ibsen ML (eds) (1996) Landslide recognition. Identification movement and causes. Wiley, Chichester

Dramis F, Blumetti AM (2005) Some considerations about seismic geomorphology and paleoseismology. Tectonophysics 408:177–191

Dramis F, Sorriso-Valvo M (1994) Deep-seated gravitational slope deformations, related landslides, and tectonics. Eng Geol 38(3–4): 231–243

Engeda S, Kebede A, Tessema E (2011) Notes and proposed guidelines on updated seismic codes in Ethiopia. Implications for large-scale infrastructures. J Eth Eng Arch 28:90–110

Ethiopian Institute of Geological Surveys (1994) A report of engineering geological problems of studies of part of the Blue Nile Gorge (Goha Tsion-Dejen). EIGS, Addis Ababa

Ethiopian Institute of Geological Surveys (1995) A report on landslide problems of Dessie town. EIGS, Addis Ababa

Ethiopian Mapping Authority (1988) National atlas of Ethiopia. Addis Ababa

Fairhead JD (1968) The seismicity of the East African rift system 1955 to 1968. MSc Dissertation, University of Newcastle upon Tyne

Faure H (1975) Neotectonics in the Afar (Ethiopia, T.F.A.I.). In: Suggate RP, Creswell MM (eds) Quaternary studies. Royal Society of New Zealand, Wellington, pp 121–126

FDPPA Federal Disaster Prevention and Preparedness Agency (2007) Regional summary of multi-agency flood impact assessment of 2006. Addis Ababa

Fubelli G, Abebe B, Dramis F, Vinci S (2008a) Geomorphological evolution and present-day processes in the Dessie Graben (Wollo, Ethiopia). Catena 75:28–37

Fubelli G, Vinci S, Abebe B, Dramis F (2008b) Cartografia geomorfologica di dettaglio e analisi statistica in ambiente GIS dei processi di versante nel bacino di Dessie (Wollo, Etiopia). Mem Descrittive Carta Geol It 78:92–99

Fubelli G, Guida D, Cestari A, Dramis F (2013) Landslide hazard and risk in the Dessie Town area (Ethiopia). In: Margottini C, Canuti P, Sassa K (eds) Landslide science and practice, vol 6. Springer, Berlin, pp 357–361

Gabet EJ, Burbank DW, Putkonen JK, Pratt-Sitaula BA, Ojha T (2004) Rainfall thresholds for landsliding in the Himalayas of Nepal. Geomorphology 63:131–143

Gamachu D (1977) Aspects of climate and water budget in Ethiopia. Addis Ababa University Press, Addis Ababa

Gezahegn A (1998) Slope stability assessment in the Blue Nile Gorge, Ethiopia. In: Proceedings of the 8th international IAEG congress, September 21–25, 1998, Vancouver, Canada, pp 1437–1442

Gibson IL (1967) Preliminary account of the volcanic geology of Fentale, Shoa. B Geophys Observ, Addis Ababa 10:59–67

Gouin P (1971) Surface cracks and subsidence in Mojo. Report TR-011. Ministry of the Interior, Addis Ababa, 5 p

Gouin P (1970) Seismic and gravity data from Afar in relation to surrounding areas. Philos T Roy Soc A 267:339–358

Gouin P (1976) Seismic zoning in Ethiopia. B Geophys Observ, Addis Ababa 7:1–46

Gouin P (1979) Earthquake history of Ethiopia and the Horn of Africa. International Development Research Center (IDRC), Ottawa

Gouin P, Mohr PA (1967) Recent effects possibly due to tensional separation in the Ethiopian rift system. B Geophy Observ, Addis Ababa 10:69–78

Gutenberg B, Richter CF (1954) Seismicity of the Earth and associated phenomena. Princeton University Press, Princeton

Hagos AH (2013) Remote sensing and GIS-based mapping on landslide phenomena and landslide susceptibility evaluation of Debresina area (Ethiopia) and Rio San Girolamo basin (Sardinia). Doctoral thesis, University of Cagliari (Italy)

Harp EL, Wilson RC, Keefer DK, Wieczorek GF (1986) Seismically induced landslides: current research by the U.S. Geological Survey. In: Proceedings of the international symposium on engineering geology problems in seismic areas, Bari, Italy, vol 2, pp 159–173

IDNDR RADIUS Project (1999) Addis Ababa case study, final report, prepared by the Addis Ababa RADIUS Group

Ishihara K, Hsu H-L (1986) Landslides in natural slopes during earthquakes. In: Proceedings of the international symposium on engineering geology problems in seismic areas, Bari, Italy, vol 5, p 273–298

IUGSWGL-CRA (Working Group on Landslides—Committee on Risk Assessment) (1997) Quantitative risk assessment for slope and landslides—the state of the art. In: Cruden D, Fell R (eds) Landslide risk assessment, Balkema, Rotterdam/Brookfield, pp 3–12

Jacques E, Kidane T, Tapponnier P, Manighetti I, Gaudemer Y, Meyer B, Ruegg JC, Audin L, Armijo R (2011) Normal faulting during the August 1989 earthquakes in central Afar: sequential triggering and propagation of rupture along the Dôbi Graben. B Seismol Soc Am 101(3):994–1023

Jbrahim J (2011) Landslide assessment and hazard zonation in Mersa and Wurgessa, north Wollo, Ethiopia. MSc thesis, Addis Ababa University, Addis Ababa, Ethiopia

Katzmin V (1972) Geological map of Ethiopia. EIGS, Addis Ababa, Ethiopia

Kebede F (1996) Seismic hazard assessment for the Horn of Africa. In: Proceedings of the Ethiopian Association of Seismology and Earthquake Engineering (EASEE)—workshop on seismic hazard assessment and design of structures for earthquake resistance, Addis Ababa, Ethiopia, February 21, 1996, pp 25–38

Kebede F, Asfaw LM (1996) Seismic hazard assessment for Ethiopia and the neighboring countries. SINET Eth J Sci 19 (1):15–50

Kebede F, van Eck T (1997) Probabilistic seismic hazard assessment for the Horn of Africa based on seismotectonic regionalization. Tectonophysics 270(3–4):137–331

Kefyalew A (2003) Integrated flood management. WMO/GWP Associated Programme on Flood Management, Addis Ababa

Mammo T (2005) Site-specific ground motion simulation and seismic response analysis at the proposed bridge sites within the city of Addis Ababa, Ethiopia. Eng Geol 79(3–4):127–150

Moeyersons J, Van Den Eeckhaut M, Nyssen J, Tesfamichael G, Van de Wauw J, Hofmeister J, Poesen J, Deckers J, Mitiku H (2008) Mass movement mapping for geomorphological understanding and sustainable development: Tigray, Ethiopia. Catena 75:45–54

Mohr PA (1962) The geology of Ethiopia. Addis Ababa University College Press, Addis Ababa

Mohr PA (1967) The Ethiopian rift system. B Geophys Observ Addis Ababa 11:1–65

Mohr P (1986) Sequential aspects of the tectonic evolution of Ethiopia. Mem Soc Geol It 31:447–461

Nyblade AA, Langston CA (2002) Broadband seismic experiments probe the East African Rift. EOS Trans AGU 83(405):408–409

Nyssen J, Moeyersons J, Poesen J, Deckers J, Mitiku H (2002) The environmental significance of the remobilisation of ancient mass movements in the Atbara-Tekeze headwaters, Northern Ethiopia. Geomorphology 49:303–322

Nyssen J, Poesen J, Moeyersons J, Deckers J, Mitiku H, Lange A (2003) Human impact on the environment in the Ethiopian and Eritrean highlands—a state of the art. Earth Sci Rev 64:273–320

Panza GF, Vaccari F, Costa G, Suhadolc P, Fäh D (1996) Seismic input modelling for zoning and microzoning. Earthq Spectra 12:529–566

Sparks S, Aspinall W, Auker M, Crosweller S, Hincks T, Mahony S, Nadim F, Pooley J, Syre E (2012) Mapping and characterizing volcanic risk. In: Abstracts of the magmatic rifting and active volcanism conference, Addis Ababa, January 11–13, 2012

Suyum GM (2011) Landslide mapping assessment using GIS techniques in Dessie area, Northern Ethiopia. Unpublished Master thesis, Universiteit Gent-Vrije Universiteit Brussel, Belgium

Sykes LR, Landisman M (1964) The seismicity of East Africa, the Gulf of Aden and the Arabian and Red Seas. B Seismol Soc Am 54:1927–1940

Tadesse Y (2012) Socioeconomic impacts of flooding in Dire Dawa, Ethiopia: flood triggering factors, social, infrastructural and economic impacts of flooding. LAP—Lambert Academic Publishing, Saarbrücken

Tefera M, Chernet T, Haro W (1996) Explanation of the geological map of Ethiopia, 2nd edn. EIGS, Addis Ababa

Temesgen B, Umer M, Asrat A, Berakhi O, Ayele A, Dramis F (1999) Landslide hazard on the Debicho, Wendo Genet area: the case of June 18, 1996 event. SINET Eth J Earth Sci 22:127–140

Temesgen B, Umer M, Korme T (2001) Natural hazard assessment using GIS and remote sensing methods, with particular reference to the landslides in the Wondogenet Area, Ethiopia. Phys Chem Earth PT C 26(9):665–675

Terefe K (2001) Engineering geological mapping and landslide assessment of Dessie town. MSc thesis, Addis Ababa University, Addis Ababa

Ukstins IA, Renne PR, Wolfenden E, Baker J, Ayalew D, Menzies M (2002) Matching conjugate rifted margins: 40Ar/39Ar chronostratigraphy of pre- and syn-rift bimodal flood volcanism in Ethiopia and Yemen. Earth Planet Sci Lett 198:289–306

Varnes DJ (1978) Slope movements types and processes. In: Schuster RL, Krizek RJ (eds) Landslides, analysis and control. Transportation Research Board, Special Report, vol 176, National Academy of Sciences, Washington DC, pp 11–33

Weldegiorgis H (2008) Landslide hazard zonation mapping in Blue Nile gorges. MSc thesis, Addis Ababa University, Addis Ababa
Williams FM, Williams MAJ, Aumento F (2004) Tensional fissures and crustal extension rates in the northern part of the Main Ethiopian Rift. J Afr Earth Sci 38:183–197
Woldearegay K (2005) Rainfall-triggered landslides in the northern highlands of Ethiopia: characterization, GIS-based prediction and mitigation. PhD thesis, Graz University of Technology
Woldearegay K, Riedmueller G, Schubert W, Mogessie A (2004) Controlling parameters and failure mechanisms of a large-scale landslide in Ethiopia. Felsbau 22(3):46–55
Worku A (2011) Recent developments in the definition of design earthquake ground motion. Calling for a revision of the current Ethiopian seismic code. EBCS 8: 1995. J Eth Eng Arch 28:1–15
Yoseph B, Ramana G (2008) Probabilistic seismic hazard analysis at selected sites in Addis Ababa, Ethiopia. Int J Geotech Eng 2 (2008):129–141
Youd TL (1978) Major cause of earthquake damage is ground failure. Civil Eng 48(4):47–51
Zvelebil J, Šíma J, Vilímek V (2010) Geo-risk management for developing countries—vulnerability to mass wasting in the Jemma River Basin, Ethiopia. Landslides 7(1):99–103

Land Degradation in the Ethiopian Highlands

21

Jan Nyssen, Jean Poesen, Sil Lanckriet, Miro Jacob, Jan Moeyersons, Mitiku Haile, Nigussie Haregeweyn, R. Neil Munro, Katrien Descheemaeker, Enyew Adgo, Amaury Frankl, and Jozef Deckers

Abstract

The high soil erosion rates in the Ethiopian highlands find their causes in the combination of erosive rains, steep slopes due to the rapid tectonic uplift during the Pliocene and Pleistocene, and human impact by deforestation, overgrazing, agricultural systems where the open field dominates, impoverishment of the farmers, and stagnation of agricultural techniques. Travelling in the Ethiopian highlands, one can see many soil and water conservation structures. Indigenous knowledge and farmers' initiatives are integrated with these introduced technologies at various degrees. This chapter addresses the status and drivers of land degradation in northern Ethiopia, including changes over the last century.

Keywords

Desertification • Soil erosion • Slope processes • Soil and water conservation

21.1 Introduction

The rugged landscapes of the Ethiopian highlands have been imprinted and partly degraded by agriculture since 3 millennia at least (Nyssen et al. 2004b). This chapter particularly addresses rainfall, runoff, and soil erosion processes. The high soil erosion rates by water and tillage as well as by landsliding in the Ethiopian highlands find their causes in the combination of erosive rainfall, steep slopes due to the rapid tectonic uplift during the Pliocene and Pleistocene, and human impact by deforestation, overgrazing, agricultural systems where the open field dominates, impoverishment of the farmers, and stagnation of agricultural techniques (Ståhl 1974; Girma and Jacob 1988; Ståhl 1990). In flat areas and on stone-covered slopes (Nyssen et al. 2002b; Van de Wauw et al. 2008), soil profiles have not yet been fully truncated by soil erosion that is concomitant to tilled agriculture. Agricultural practices are well adapted to the environment: the *mahrasha* tillage tool (the traditional 'ard' plough) was developed during the high-tech Axumite period; the cropping systems fit seamlessly to soil catenas (Nyssen et al. 2008a); and the farming systems are well adapted to interannual variation in rainfall conditions (Pietsch and Machado 2014). Whereas, technically, under the traditional circumstances, agricultural adaptation to soil and climate variability is nearly optimal, land management has for long been hampered by unequal access to land and prevalent free grazing. Most reports from the first half of the twentieth century (e.g. Giglioli 1938a, b; Joyce 1943) recognised the soil erosion problem but did not consider that it was a major problem. Frankl et al. (2011) have shown that the gullies

J. Nyssen (✉) · S. Lanckriet · M. Jacob · A. Frankl
Department of Geography, Ghent University, Ghent, Belgium
e-mail: jan.nyssen@ugent.be

J. Poesen · R.N. Munro · J. Deckers
Department of Earth and Environmental Sciences, KU Leuven, Leuven, Belgium

J. Moeyersons
Royal Museum for Central Africa, Tervuren, Belgium

M. Haile · N. Haregeweyn
Department of Land Resources Management and Environmental Protection, Mekelle University, Mekelle, Ethiopia

K. Descheemaeker
Department of Plant Sciences, Wageningen University, Wageningen, The Netherlands

E. Adgo
College of Agriculture and Environmental Science, Bahir Dar University, Bahir Dar, Ethiopia

P. Billi (ed.), *Landscapes and Landforms of Ethiopia*, World Geomorphological Landscapes,
DOI 10.1007/978-94-017-8026-1_21, © Springer Science+Business Media Dordrecht 2015

currently visible in the landscape started to develop in the 1960s. This chapter addresses the status and causes of land degradation in northern Ethiopia over the last century.

21.2 Rainfall and Runoff as Driving Forces for Soil Erosion Processes

The climates of Ethiopia are complex: 'Within short horizontal distances, climates from tropical to sub-humid and sub-tropical to arctic can occur' (Krauer 1988). Precipitation and air temperature vary mainly with elevation, but slope aspect also plays an important role. Furthermore, precipitation decreases and seasonality increases with latitude.

21.2.1 Precipitation Patterns

From the end of June onwards, the Intertropical Convergence Zone (ITCZ) is situated at its most northerly position (16°N–20°N). The southern air masses, limited to the lower layers of the atmosphere, bypass the highlands and reach them from the south-west, giving way to the main rainy season (Goebel and Odenyo 1984). Generally, clouds are formed at the end of the morning, as a result of evaporation and convective cloud formation due to daytime heating of the soil, and cause rains in the afternoon. In Afdeyu station, on the Eritrean highlands, 80 % of daily precipitation takes place between 12 and 16 h (Krauer 1988). (All mentioned localities are indicated on Fig. 21.1). This convective nature of rainfall also explains why individual showers have a very local distribution. At the end of the summer, the ITCZ returns quickly to the south, preventing the arrival of monsoon rain. This is the end of the rainy season in the highlands.

Abebe and Apparao (1989) calculated from 241 stations in Ethiopia a mean annual precipitation of 938(±83) mm year^{-1}. For the highlands, annual precipitation varies between 450 mm year^{-1} in Tigray and more than 2,000 mm year^{-1} in the south-west of the country (Krauer 1988). The interaction of latitude and altitude controls total annual precipitation (Troll 1970). At the regional scale, one should, however, also take into account that during the rainy season winds come essentially from the south-west, as well as orographic effects. Valleys are preferred flow paths for the penetration of humid air masses into the highlands (Nyssen et al. 2005) and rainfall distribution is highly erratic (Jacob et al. 2013).

21.2.2 Rainfall Erosivity in the Ethiopian Highlands

High rainfall erosivity is an important factor of soil erosion in the highlands. Data from two automatic rain gauges installed in central Tigray during one year (1975 and 2001, respectively) indicate that 30–70 % of all rain events had an intensity >25 mm h^{-1} (Hunting Technical Services 1976; Nyssen et al. 2005). Krauer (1988) obtained from the rainfall data of six Soil Conservation Research Programme (SCRP) stations mean annual universal soil loss equation (USLE) rainfall erosivity indices R between 166.6 (Afdeyu, Eritrea) and 543.7 J cm m^{-2} h^{-1} year^{-1} (Anjeni, Gojam). Hurni (1979), in an analysis of rainfall erosivity in the Simien Mountains, insisted on two other particularities of Ethiopian mountains: erosivity due to hail (2.5 times more important than erosivity due to rain) and the influence of hillslope aspect. A soil surface unit exposed to wind receives a greater quantity of water than a surface unit with an opposite exposure.

Given that rainfall characteristics in tropical highlands are different from those of more temperate climates, it is difficult to apply erosivity equations, such as those proposed in (R) USLE (Wischmeier and Smith 1978; Renard et al. 1997), which have been developed for North America, to rainfall on the Ethiopian highlands. Based on drop size measurements, Nyssen et al. (2005) showed that for the same rainfall intensity, rainfall erosivity is significantly higher in the Ethiopian highlands compared to elsewhere in the world because of larger raindrop sizes, also during low intensity rain events. Moreover, in the absence of a network of automatic rain gauges, maximum hourly rainfall intensities could be measured only in a small number of research stations in Ethiopia.

Rainfall erosivity is a function of the depths and intensities of the individual rainstorms, and these are not closely related to annual precipitation. However, the United States data indicate that for a given annual precipitation, the range of likely erosivity values can be somewhat narrowed by knowledge of the general climatic conditions in the particular geographic area (Wischmeier and Smith 1978). In East Africa (i.e. Tanzania, Kenya, and Uganda), the relationship between total precipitation and erosivity index improves if rainfall stations are grouped by geographical area (Moore 1979). For Ethiopia, Hurni (1985) and Krauer (1988) elaborated, from monthly data of 6 SCRP stations, correlations between USLE's R-factor and mean annual rainfall and Krauer (1988) presented an isoerodent (rain erosivity) map of Ethiopia. More recently, several studies have reported rain erosivity data for Ethiopian rain stations, as well as for other stations in Africa (e.g. Vrieling et al. 2010; Diodato et al. 2013).

21.2.3 Runoff and Infiltration

In Ethiopia, surface runoff production has been measured at various temporal and spatial scales (from runoff plot to

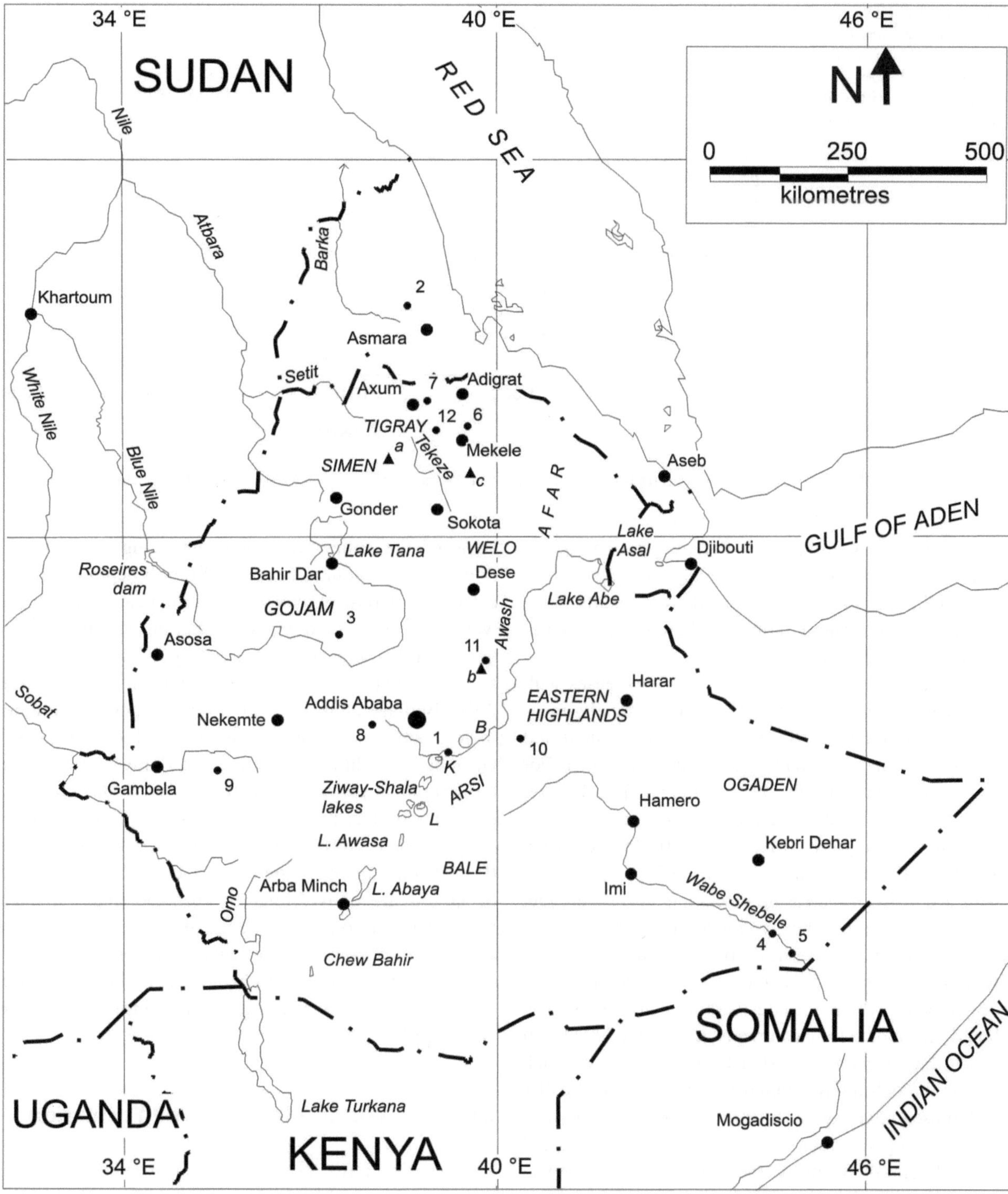

Fig. 21.1 Map of Ethiopia and Eritrea, with indication of localities mentioned in this chapter. Minor localities indicated with numbers: *1* Adama/Nazret + Debre Zeit, *2* Afdeyu, *3* Anjeni, *4* Kelafo, *5* Mustahil, *6* May Makden, *7* Adwa, *8* Ambo, *9* Dizi, *10* Hunde Lafto, *11* Debre Sina, *12* Dogu'a Tembien; summits are represented by *triangles*: *a* Ras Dejen, *b* Ankober, *c* Amba Alage; *open dots* represent lakes: *B* Lake Besaka, *K* Koka reservoir, *L* Lake Langano

catchment). Runoff has been monitored in the SCRP catchments and data series of up to 12 years are available (SCRP 2000). Generally, runoff coefficients (RC) from small (<1,000 m^2) runoff plots are very variable (0–50 %) (Nyssen et al. 2004b), which is attributed to the variable experimental conditions. Besides different slope gradients, local

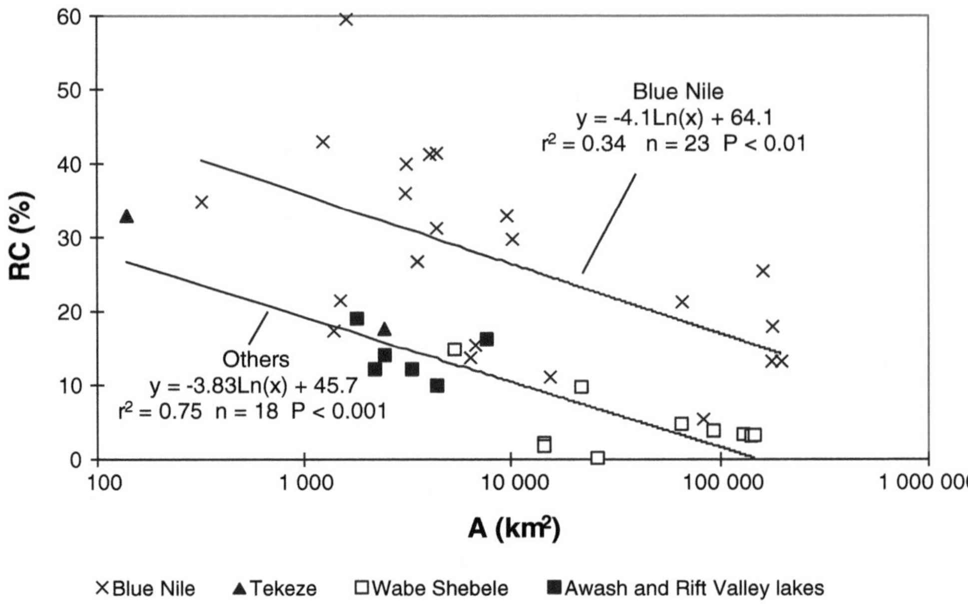

Fig. 21.2 Runoff coefficients (RC) versus drainage area (A) for catchments of the basins of the Blue Nile, Tekeze, Wabe Shebele, Awash, and Rift Valley lakes (after Nyssen et al. 2004b)

differences in soil texture, land use, vegetation cover, organic matter content, or rock fragment cover result in a wide range of infiltration rates obtained from runoff plots. Results on RC from runoff plots are therefore not representative for RC of catchments.

For large catchments ($A \geq 100$ km^2), RC decreases with increasing catchment area (Fig. 21.2). The already mentioned conditions for high RC (presence of open field and high rainfall) are mainly found in the Blue Nile basin. For this reason, two data series can be considered. The Tekeze, Awash, and Wabe Shebele basins are mainly situated in dry sub-humid to arid regions (Mersha 2000). In the Wabe Shebele basin, Bauduin and Dubreuil (1973) explained a decreasing RC with an increasing catchment size by the fact that small catchments are mostly situated in the headwaters where nearly impervious, basalt-derived soils dominate, and also by a smaller mean annual basin precipitation in the larger catchments which include (semi)arid lowlands. The rainfall and runoff data for catchments in the Blue Nile basin suffer, according to its authors (USBR 1964), from the lack of precision in delimiting drainage areas (A) for smaller basins. Representative catchment precipitation data are also difficult to obtain given poor station density and large spatial variability of rainfall (Conway and Hulme 1993). Conway (1997) pointed to short mean observation periods (i.e. 1.5 years) and possible errors in rainfall data. Despite the wide scatter for the Blue Nile basin, it can be observed that RC are larger than those for the other basins but that they still follow a parallel trend (Fig. 21.2). Decreasing RC values with increasing A values in the Blue Nile basin are thought to be a result of (a) runoff transmission losses, due to evaporation and possibly lithological changes, and (b) less rainfall and larger potential evapotranspiration in the western areas of the Blue Nile catchment along the border with Sudan, which reduces the overall runoff depth for the whole catchment (Conway, personal communication 1999). In situ water harvesting and the construction of small reservoirs have both led to strongly decreasing RC at catchment scale, and to increased levels of the water tables (Nyssen et al. 2010; Berhane et al. 2013). However, significant differences in RC between the sub-catchments within the 5,000 km^2 Geba basin could not be demonstrated, most probably due to the overall implementation of soil and water conservation (SWC) activities (Zenebe et al. 2013). Additional research on this topic is currently being conducted in Lake Tana basin (Poppe et al. 2013; Dessie et al. 2014). Particularly, in cases of large-scale conversions of cropland and rangeland to forest, such as on the escarpment upslope from Alamata, effects are very clear, particularly in terms of decreased downstream flooding and changes of hydrogeomorphology (i.e. river channel incision and narrowing) (Gebreyohannes et al. 2014).

21.3 Weathering and Soil Formation

Few studies have been made on weathering of parent material in the highlands. Hövermann (discussion in Bakker 1967) studied the basal Precambrian granites in northern Ethiopia where weathering mantles are up to 120 m deep. No studies exist for Mesozoic sedimentary rocks or for Tertiary volcanics, but the depth of weathering mantle is expected to be much less.

Hurni (1983), through the study of soils developed on periglacial slope deposits, extrapolated soil formation rates for the different agroclimatological zones of Ethiopia.

Zonation in Ethiopia is based on altitude and more specifically on the corresponding local climate. These soil formation rates are mean rates, taking into account rainfall depth and air temperature conditions, but not lithology. They are intended to be compared with soil loss rates, but, to our understanding, cannot be applied to the vast areas where the soil mantle results from sediment deposition rather than from pedogenesis.

21.4 Sheet and Rill Erosion

Most research on soil erosion in Ethiopia focused on sheet and rill erosion (Fig. 21.3). Hurni (1975, 1978, 1979) studied thoroughly the Jinbar valley (3,200–4,000 m a.s.l.) in the Simien Mountains. Andosols occupy the whole valley, which is partially under rangeland and degraded forest and partially under barley. The depth of the A-horizon was measured at some 300 sites in cropland and compared with A-horizon depth in non-cultivated areas for similar slope gradients. Mean total soil profile truncation depth from cropland, occurring between the beginning of permanent human occupation (500–200 years ago) and 1974, was measured as 14.5 ± 2.1 cm, or 950 ± 200 t ha^{-1}, or 2–5 t ha^{-1} $year^{-1}$. Due to elevation and to the proximity of the climatic limit of barley cultivation, deforestation here has started much later than in most other parts of the highlands (Hurni 1982). The variability in soil loss depth is correlated with slope aspect and probably with the age of deforestation (Hurni 1975, 1978). Measurements of sheet and rill erosion rates were conducted in the Ethiopian highlands (Hurni 1985, 1990; Kejela 1992; Herweg and Ludi 1999; SCRP 2000; Nyssen et al. 2009c).

Soil loss occurs mainly at the beginning of the main summer rainy season (*kiremt*). In those regions where spring rains (*belg*) are sufficient for cultivation, these crops have been harvested and the land ploughed again before *kiremt* (Tesfaye 1988). In the northern highlands, spring rain is unreliable and the land is only sown at the beginning of the *kiremt* season, when rains are intensive and their onset more regular. The farmlands have then undergone at least two tillage operations are bare and offer less resistance to splash and runoff erosion (Virgo and Munro 1978). Studies in southern Ethiopia, where deforestation is ongoing, show a tremendous increase in soil loss over the last few decades (Kassa et al. 2013). In the northern highlands, with the advance of the rainy season, soil loss decreases as crop cover increases (Tesfaye 1988). This pattern was also observed and similarly accounted for by Billi (2004) for the suspended sediment concentration in the Meki River, a main tributary to Ziway Lake in the Rift Valley. However, substantial runoff is produced more than one month after the beginning of the *kiremt* rains. In the beginning of the rainy season, most rainfall infiltrates quickly in the dry, tilled farmlands (Gebreegziabher et al. 2009; Zenebe et al. 2013). Furthermore, on Vertisols, which are well represented in Ethiopia (Kanwar and Virmani 1986; Moeyersons et al. 2006), the first rains are well absorbed, because of the deep shrinkage cracks. After some time, with the closing of the cracks, these soils become completely impervious and favour significant runoff production (Bauduin and Dubreuil 1973; Gebreegziabher et al. 2009; Oicha et al. 2010; Araya et al. 2011). Moreda and Bauwens (1998) found the most significant correlation between monthly precipitation and summer flow in the Awash headwaters to occur in August, at the beginning of the second half of the rainy season, when 'there is greater opportunity for flow generation (even for smaller storms) since the catchment is already moist'. Sutcliffe and Parks (1999) estimated that 'early rainfall is required to replenish the soil moisture storage after the dry season'.

Fig. 21.3 Rill erosion at a farmer's field at Wonzima (Blue Nile basin). Rills occur particularly on long and steep slopes without soil conservation structures; here, the depth of the rill is controlled by the tillage pan, on which plough marks of the ard are visible (*Photograph* E. Monsieurs, August 2013)

Table 21.1 The Revised Universal Soil Loss Equation (RUSLE)—adapted for field assessments in Ethiopia (Nyssen et al. 2009c)

Equation: **annual soil loss rate** $A = R * K * S * L * C * P$ (Mg ha^{-1} year^{-1})					
1. ***R*: annual rainfall erosivity** (MJ mm ha^{-1} h^{-1} year^{-1})					
$R = 5.5\ \mathrm{Pr} - 47$					
Pr = annual precipitation (mm)					
2. ***K*: soil erodibility** (Mg h MJ^{-1} mm^{-1}), including effects of rock fragment cover					
$K = [2.1\ M^{1.14}\ (10^{-4})(12 - a) + 3.25\ (b - 2) + 2.5\ (c - 3)] * e^{-0.04\ (d-10)} * 0.001317$					
M = particle size parameter = (% silt and very fine sand) * (100 − % clay)					
a = percentage of organic matter					
b = soil structure code, ranging between 1 (very fine granular) and 4 (blocky, platy, or massive), with default value 2					
c = permeability class, ranging between 1 (rapid) and 6 (very slow), with default value 3					
d = stone (rock fragment) cover (in %)					
3. ***S*: slope steepness factor** (dimensionless)					
$S = -1.5 + 17/(1 + e^{(2.3-6.1\ \sin\theta)})$					
θ = slope angle (°)					
4. ***L*: slope length factor** (dimensionless)					
$L = 0.232\ \lambda^{0.48}$ (5 m ≤ λ ≤ 320 m)					
λ = slope length (horizontal projection, in m)					
5. ***C*: cover-management factor** (dimensionless)					
Dense forest	0.001	Degraded rangeland (<50 % vegetation cover)	0.42	Badlands hard	0.05
Dryland forest; exclosure	0.004			Badlands soft	0.40
Dense grass	0.01	Degraded grass	0.05		
Sorghum, maize	0.10	Tef (in high rainfall areas)	0.25	Fallow hard	0.05
Cereals, pulses	0.15	Tef (in semi-arid areas)	0.07	Fallow ploughed	0.60
6. ***P*: supporting practices** (dimensionless)					
$P = P_C \cdot P_N \cdot P_M$ (on cropland); $P = P_N$ (on other land)					
Ploughing and cropping practices	P_C	Conservation structures	P_N	In situ conservation practices	P_M
Ploughing up and down	1	No conservation structures	1	Stubble grazing; no mulching	1
Ploughing along the contour	0.9	Stone bund (average condition; smaller value for new s.b. and larger for older s.b.)	0.3	Applying mulch	0.6
Strip cropping	0.8	Grass strip (1 m wide; slope ≤ 0.1 mm^{-1})	0.4	Zero grazing	0.8
Intercropping	0.8	Grass strip (1 m wide; slope > 0.2 mm^{-1})	0.8		
Dense intercropping	0.7				

Source Renard et al. (1997). Adaptations: *R* correlation by Hurni (1985); *K* adjustment for rock fragment cover by Poesen et al. (1994); *L* correlation by Hurni (1985); *C* values by Hurni (1985) and Nyssen et al. (2009c); *P* model by Nyssen et al. (2009c); *P* values by Hurni (1985), Nyssen (2001), Gebremichael et al. (2005), Nyssen et al. (2007a, b, 2008b). Limitations as mentioned in Sect. 21.4

From their research, Hurni (1985) and later Nyssen et al. (2009c) adapted the Revised USLE (Renard et al. 1997) to Ethiopian conditions for use by development agents in the field of SWC (Table 21.1). The soil erodibility factor *K* can be assessed from soil textural data, organic matter content, and rock fragment cover (Table 21.1, Sect. 21.2). We recommend including the rock fragment cover, which is a widespread feature in the Ethiopian highlands, as a correction factor for the *K*-value, rather than in the management factor *P* (Nyssen et al. 2002b).

For the *R*-factor (rainfall erosivity), the Ethiopia-specific equation (Table 21.1, Sect. 21.1) may be used, bearing in mind that additional studies, taking into account the above average drop sizes in the Ethiopian highlands, should be carried out (Nyssen et al. 2005). Calculations of the slope steepness factor (*S*) and the slope length (*L*) factor are shown in Table 21.1 (Sects. 21.3 and 21.4). The use of equations for *L* requires caution, since 'slope length is the factor that involves the most judgement, and length determinations made by users vary greatly' (Renard et al. 1997). In Ethiopian highland conditions, this runoff length is generally longer than one single farm plot and shorter than the whole slope, from ridge to foot. Cover-management *C* values (Table 21.1, Sect. 21.5) have been reported by Nyssen et al. (2009c). The *P* factor (dimensionless) relates to supporting practices and indicates reduced soil erosion potential due to farming practices and conservation measures. Sub-factors yield one composite P-value (Foster and Highfill 1983) for a conservation system (Table 21.1, Sect. 21.6):

$$P = P_C * P_N * P_M \tag{21.1}$$

where

P_C = Sub-factor for ploughing and cropping practices;
P_N = Sub-factor for conservation structures;
P_M = Sub-factor for in situ conservation practices.

21.5 Gullying

Gullying (Fig. 21.4) is not restricted to the highlands of Ethiopia but is widespread at sub-continental scale in Africa (Moeyersons 2000). In Tigray, the increase of runoff response on many hillslopes has been attributed to an overall lowering of the infiltration capacity of the soils due to removal of natural vegetation (Virgo and Munro 1978; Machado et al. 1998). Buried soils indicate advanced deforestation, which in the Ethiopian highlands might have started around 5,000 ^{14}C years BP (Machado et al. 1998; Nyssen et al. 2004b; Pietsch and Machado 2014). Since the twentieth century, however, vegetation removal has also affected shrub and small tree cover, as well as grass strips in between the farmlands and on steep slopes. This removal of vegetation has further lowered the infiltration capacity of the soils, favoured the occurrence of flash floods, and is considered to be the major cause of rapid gullying in many areas (Frankl et al. 2011). One should also stress the importance of cropland abandonment for gully initiation, especially if it is converted into grazing land. The overgrazed soil surface has a higher runoff coefficient than regularly ploughed farmlands; SWC structures are no longer maintained, and bank gullying often starts at places where these structures collapse.

Brancaccio et al. (1997) explained the present-day processes of channel incision in northern Ethiopia by an increasing erosional power of concentrated runoff due to a decreasing sediment load (clear water effect), associated with the advanced phase of soil erosion on the hillslopes where bedrock is now outcropping. Since the late nineteenth century, gullies were present and though they had become stabilised by 1935, a strong incision phase started in the 1960s due to the above-mentioned factors (Frankl et al. 2011).

Gullies in Ethiopia can often be considered as discontinuous ephemeral streams (Bull 1997) comprising a hillslope gully, an alluvial–colluvial cone at the foot of the hill, and renewed incision with gully head formation further downslope in the valley Vertisol. Pediments dissected by gullies are a common feature in many areas (Riché and Ségalen 1973; Berakhi and Brancaccio 1993; Berakhi et al. 1997). In the valley bottoms, initial gully heads often coincide with sinking polygonal structures in Vertisols (Nyssen et al. 2000b), where piping erosion is very active (Frankl et al. 2012).

Fig. 21.4 Gullies, like this one in Harena (Dogu'a Tembien), do not only result in soil loss, but also drain out the landscape (lowering of the water table) and are major obstacles to communication

Active gullying induced by road building on pediments was described by Berakhi and Brancaccio (1993). In a case study along the Mekele—Adwa road, built in 1994, Nyssen et al. (2002a), demonstrated how road building, through the enlargement of drainage areas and the concentration of runoff, induced an artificial exceedance of the critical catchment area at which gully heads are formed for a given slope gradient.

21.6 Tillage Erosion

Soil translocation due to tillage by the ox-drawn ard plough (Fig. 21.5) appears to be an important soil erosion process in the Ethiopian highlands. Assessments of tillage erosion rates indicate that this process contributes on average to half of the sediment deposited behind stone bunds (Nyssen et al. 2000c; Gebremichael et al. 2005). Colluviation occurs in the lower part of the farmland and soil profiles are truncated in the upper part (Herweg and Ludi 1999; Nyssen et al. 2000c). Soil sequences on progressive terraces overlying strongly weathered rock were analysed in the central highlands of Ethiopia, in the Ankober area (Bono and Seiler 1986). At the upper part of the terrace, soils are shallow and water and nutrient storage capacity low. However, in Dogu'a Tembien (Tigray), intra-parcel variability of soil fertility parameters is small. A larger content of soil moisture and of soil organic matter was even observed at the foot of the stone bunds, at the very place where the soil profile has been truncated after stone bund building. Possible effects of soil profile truncation on the values of these two parameters are outbalanced by increased infiltration rates, induced by stone bund building (Vancampenhout et al. 2006). The most common soils in the Ethiopian highlands (i.e. Regosols, Vertic Cambisols and Vertisols) have a quite homogenous composition with depth, which explains low soil fertility gradients in terraced lands.

21.7 Wind Erosion

In the Ethiopian highlands, wind erosion has not been measured and was rarely mentioned. Wind erosion mainly occurs as 'dust devils' in areas with important trampling by humans or cattle, such as market places, footpaths, unmetalled roads, around cattle drinking places or on cropland where post-harvest grazing has taken place. On the numerous isolated mountains or '*inselbergs*', important wind erosion, including the formation of dunes, occurs due to local aerodynamic situation (Uhlig and Uhlig 1989). More research on wind erosion in the Ethiopian highlands seems necessary, as it may have been insufficiently studied.

Wind erosion is especially important in low-lying, dry and hot regions, adjacent to the highlands, such as many places in the Rift Valley. Desert pavements, created by wind erosion, exist around Lake Turkana (Hemming and Trapnell 1957). Wind erosion and deposition contribute to the formation of dunes in the alluvial plains of the Wabe Shebele

Fig. 21.5 Soil tillage by *mahrasha* ard plough, here in Dogu'a Tembien, causes a downslope movement of the topsoil (tillage erosion) (*Photograph* A Roelofs, April 2005)

and to overall deposition of aeolian sediments in that region (Riché and Ségalen 1973). The Eritrean coastal plain is in many places covered by stone mantles produced by deflation as well as by loose sand occurring either as a mantle of variable depth or in the form of mobile dunes (Hemming 1961). Aeolian sediments in the coastal plains can be composed of eroded materials from nearby rocks or brought in by dust storms, which are quite common (Horowitz 1967).

21.8 Mass Movements

Due to steep topography, the presence of lithologies with a low shear strength, torrential rainfall, and in some cases the occurrence of earthquakes, the Ethiopian highlands are also affected by various types of mass movements (e.g. rock falls, debris flows, and slumps). Several studies have mapped landslides in Ethiopia and have analysed their controlling factors (e.g. Moeyersons et al. 2008; Van Den Eeckhaut et al. 2009; Broothaerts et al. 2012). Although many mass movements have been initiated by natural factors, human activities (i.e. land use change, undercutting and overloading during road construction, and improper slope drainage systems) have often contributed to the reactivation of landslides. In south Ethiopia, (Broothaerts et al. 2012) observed many recent landslides along river channels which were triggered by river channel incision due to increased peak flow discharges following deforestation in their catchments. Large landslides redistribute large volumes of sediments in the highlands, hence affecting the spatial patterns of soil types (Van de Wauw et al. 2008).

21.9 Sediment Deposition

On the back- and footslopes of cliffs, a 'classic' sorting of deposited sediment generally occurs, the coarse sediments (rock fragments) being deposited on the debris slope, and the finer material on the footslope, as shown by Riché and Ségalen (1973) in the Wabe Shebele basin. Belay Tegene (1998) emphasised the importance of continuous deposition of colluvium on convergent footslopes which prevents the development of mature soil profiles. Hurni (1985) shows, for a 116 ha catchment in Welo, that the rate of sediment accumulation (17 t ha^{-1} $year^{-1}$) is more important than the rate of sediment export through the drainage system (7 t ha^{-1} $year^{-1}$). In a well-vegetated catchment in southwestern Ethiopia, sediment accumulation rates are 30 t ha^{-1} $year^{-1}$ and sediment export rates through the river only 1.1 t ha^{-1} $year^{-1}$. Here, most of sediment deposition occurs in densely vegetated areas along riverbanks. A sediment budget for a 200 ha catchment in Tigray highlands indicates that 59 % of sediment produced by water erosion is deposited within the catchment (Nyssen et al. 2007b). Reuter (1991) and Descheemaeker et al. (2006b) stressed the magnitude of sediment and organic carbon stored in colluvium on footslopes and reforested areas (exclosures sensu Aerts et al. 2009). Sediment deposition in floodplains and natural lakes is important, but the rates have not been studied systematically in Ethiopia.

21.10 Land Degradation and Desertification

Although climatic conditions (0.05 < annual precipitation/potential evapotranspiration < 0.65) in parts of the northern highlands and in the low-lying parts of the country would justify the use of the term 'desertification' (UNEP 1994), the term 'land degradation' will be used to indicate environmental degradation throughout the country. Two major factors inducing land degradation in the Ethiopian highlands are generally considered: drought and land use changes.

21.10.1 Rainfall Variations and Drought

Attention to famines in Ethiopia has created a popular view of a drought-stricken country, with a tendency towards decreasing annual rainfall. The decline of rainfall in the Sahel observed since about 1965 was also seen to a lesser extent in the north-central Ethiopian highlands (Camberlin 1994; Seleshi and Demarée 1995). However, unlike the Sahel, a comparison between two reference periods (1931–1960 and 1961–1990) yields no significant changes in mean precipitation over Ethiopia, but an increased inter-annual variability (Hulme 1992). Analyses of long-term time series of annual precipitation, both for Addis Ababa and the northern highlands, show that although the succession of dry years between the late 1970s and late 1980s produced the driest decade of the last century in the Ethiopian highlands, there is no evidence for a long-term trend or change in the region's annual rainfall regime (Conway 2000a; Conway et al. 2004).

With respect to the inter-annual rainfall variability, Conway (2000b) found a coefficient of variation below 20 % for the wetter areas, but far above that for drier areas to the north and at lower altitudes (see also Chap. 4, this volume). (Hoffmann 1987) also found annual rainfall variability strongly dependent on climatic region: <10 % in the area around Jimma with a tropical rain climate and >45 % in semi-desert areas. Dry years were observed in 1913–1914, 1937, 1941, 1953, 1957, 1965–1966, 1969, 1973–1974, 1976, 1979, 1983–1984, and 1987 (Camberlin 1994). It is evident that, in an already degraded environment, a dry year has a very negative impact, not only on agricultural production, but also on the environment (i.e. overgrazing, cracking of Vertisols, and groundwater depletion). RC in such a year are higher (Casenave and Valentin 1992; Valentin et al. 2005) and result in increased soil erosion.

Besides yearly precipitation, its seasonal distribution must be considered as well. Unlike West Africa, according to Hulme (1992), the seasonality of rainfall over Ethiopia slightly decreased between 1931–1960 and 1961–1990. However, evenly distributed rains mean also that a larger percentage of precipitation falls outside the crop growth season, or that there is a shift from one rainy season to another, particularly decreased summer (*kiremt*) and increased spring (*belg*) rains in the northern Ethiopian highlands (Camberlin 1994; Seleshi and Demarée 1995; Seleshi and Demarée 1998). Differences between the temporal pattern of spring and summer rains are expected to reflect different levels of influence from the Indian and Atlantic oceans (Conway 2000b).

21.10.2 Human Settlement and Changes in Land Use and Land Cover

Human settlement with concomitant agricultural exploitation induces significant changes in land use and land cover, which in turn alter infiltration and runoff conditions, as well as soil erosion processes (Olson 1981; Bunney 1990). Detailed studies show that settlement decisions were made on a clear ecological basis, especially from the beginning of the pre-Axumite era (700 BCE). Preferred locations were at the margin between Vertisol areas and narrow alluvial valley bottoms which could be irrigated (Michels 1988). Human activity expanded from such preferential places to the present-day occupation of steep slopes for agriculture through a number of stages, including forest clearing and removal of remnant trees and shrubs.

In the Ethiopian highlands, livestock grazes on vast deforested areas, commonly called rangeland, as well as various types of climax grasslands: i.e. at high elevations, on Vertisols and on dry places (Klötzli 1977). Much in the same way as in forests and woodlands, vegetation cover decreases in grass- and rangeland. Most of the above-quoted studies of land use changes show, besides decreasing tree and shrub cover, an increase of the area occupied by 'bare land', 'no vegetation', 'open areas', and the like. Overgrazing of rangeland is a particular problem in the cereal zones of the highlands, where current stocking rates are well in excess of estimated optimum rates (Hurni 1993). Livestock plays a key role in the agricultural system of the highlands, providing energy (traction, manure used as fuel), food, fertiliser, insurance, and status (Kassa et al. 2002). Consequences of overgrazing on the environment are decreased surface roughness, compaction of fine textured soils, increased soil bulk density, decreased soil organic matter content, soil structure decay, and decreased hydraulic conductivity. All these factors contribute to decreased infiltration rates and increased runoff volumes. Mwendera et al. (1997) carried out experiments on grazing land with slope gradients <0.08 mm^{-1} in an area between Ambo and Addis Ababa. Comparing ungrazed, moderately, and heavily grazed land, they found significant differences in runoff volumes for slope gradients in the range of 0.05–0.08 mm^{-1}. Steady-state infiltration rates decreased significantly, even under light grazing intensity, and showed the effect of animal trampling on soil compaction (Mwendera and Saleem 1997). On cropland, stubble grazing (a widespread practice) dramatically decreases the infiltration capacity. Field observations also indicate that topsoil degradation by cattle trampling significantly contributes to soil erosion and sediment delivery to water reservoirs.

Repeat photography has also revealed that in the late nineteenth century, the landscapes were at least as barren as they are nowadays (Nyssen et al. 2009b). In recent years, since 1975, the tree cover has improved in 90 % of the analysed landscapes (Munro et al. 2008; Nyssen et al. 2008b). Exclosures (Aerts et al. 2009) have been established in former communal grazing land with the aim of forest restoration and land conservation. The establishment of exclosures was made possible by an important land tenure change in the 1980s, in which large feudal agricultural lands in the valley bottoms and other level areas were shared among the local farmers and this decreased the need of poor farmers to establish marginal farmlands on hillslopes. In these locations, exclosures could then be established after land reform (Rahmato 1994). Although centrally imposed, the implementation of exclosures is rather a bottom-up process. Participation is enhanced by the implementation of remunerated SWC activities and plantation works at the establishment of the exclosure. Location, area, local by-laws related to restrictions and management, instalment and payment of guards are most often decided at the local community level (Muys et al. 2014). The villagers are overall convincingly participating in reforestation and other conservation activities (Kumasi and Asenso-Okyere 2011). However, the encroachment by eucalypts remains a bottleneck for biodiversity. The benefits of planting these trees are largely for individual farmers, whereas the negative effects of this water-demanding tree are borne by the communities (Muys et al. 2014).

21.10.3 Social and Historical Impulses of Land Use and Cover Changes

It appears that rainfall variability, apart from the catastrophic impact of dry years on the degraded environment, cannot be

Fig. 21.6 Plough marks on large rock fragments and pedestal-supported boulders indicate that this 100–200-year-old *Juniperus* forest at Kuskuam near Debre Tabor has grown on previously degraded farmland. Forest re-growth has taken place, as also evidenced by the well-branched older tree in the centre of the photograph that used to grow in an open area (*Photograph* J. Nyssen, July 2011)

invoked to explain the current land degradation. Causes are to be found in changing land use and land cover, which are expressions of human impact (Reid et al. 2000; Feoli et al. 2002). Though deforestation and removal of other vegetation cover over the last 2,000–3,000 years have probably been a cyclic rather than a linear process (Fig. 21.6), studies on land use and land cover change show that, at present, there is a tendency of increasing removal of vegetation cover.

At this stage, it appears necessary to briefly outline the social and historical causes of this human impact. Under feudalism (until 1974), agricultural techniques stagnated for centuries (Crummey 2000). Until the 1940s, the Agricultural Department's only effective activity was collecting the agricultural tax (Joyce 1943). Investment in agriculture started only in 1950s and was in the beginning mostly oriented towards export crops such as coffee (*Coffea arabica L.*), grown in southern Ethiopia. Therefore, there was limited agricultural investment in the highlands, where subsistence production dominated (Ståhl 1990; Mulugetta 1992). Until the late 1970s, sharecropping prevented the farmers from investing in their farmlands. Impoverishment led them to prefer immediate returns, even if it induced environmental degradation (Taddesse 1995). On the other hand, recent land redistributions in order to allocate landless households had a positive impact on land productivity (Benin and Pender 2001). To increase agricultural production, most trees and shrubs between the farmlands and on steep slopes were cleared during the nineteenth and twentieth centuries, thereby increasing runoff and soil erosion (Ståhl 1974; Girma and Jacob 1988; Ståhl 1990). In short, in situations of poverty and social insecurity, short-term survival prevailed over medium- and long-term conservation issues.

21.11 Human Reaction to Land Degradation

21.11.1 Agricultural Intensification and Land Rehabilitation

Faced with a deteriorating environment, society reacts in order to maintain/improve agricultural production, often leading to changes in the production system (Boserup 1981), an innovative process in which modern science needs to be involved (Blaikie and Brookfield 1987; Ståhl 1990). The present-day rise in food production in Ethiopia (Fig. 21.7) can, besides re-established climatic conditions, also be attributed to a variety of human interventions at different levels (Nyssen et al. 2004a). Extension of cropped area and increased grazing pressure is still possible. However, limited space is left for this and productivity decreases. Giglioli (1938a, b) and Joyce (1943) already reported the widespread use of indigenous SWC technology since a long time. Such indigenous technologies can be used as a starting point, but need improvement in order to increase their ecological efficiency (Hurni 1998; Nyssen et al. 2000a).

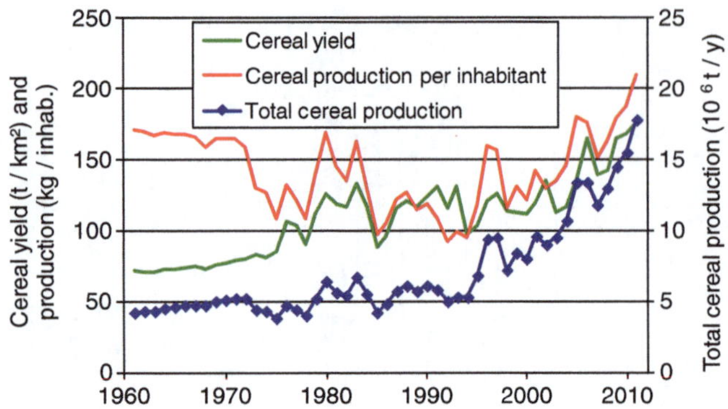

Fig. 21.7 Agricultural intensification in Ethiopia is evidenced by cereal production trends (data retrieved from http://faostat.fao.org)

Nowadays, changes in the agricultural system appear such as haymaking ('cut and carry') (Hurni 1986), partially from exclosures (i.e. land under strict conservation management, often controlled by the community), which are increasingly being organised in the most affected northern highlands (Tekle et al. 1997; Shitarek et al. 2001; Aerts et al. 2004, 2009; Descheemaeker et al. 2006b) and which lead to sediment trapping and enhanced soil fertility status (Elias and Scoones 1999; Descheemaeker et al. 2006a; Mekuria et al. 2007).

Different pathways of agricultural intensification are possible in Ethiopia. Mineral fertilising should not be the overall option, given scarce capital resources. Due to decreased landholdings, a shift in the soil tillage system to gardening and minimum tillage (on self-mulching Vertisols) may be suggested (Astatke et al. 2002; Araya et al. 2012), as well as an extension of the cropping period on Vertisols (Tedla et al. 1999). Asnakew et al. (1994) obtained good maize yields with rock fragment mulching and no-tillage.

Besides these conservation measures, Ethiopia strongly invested in agricultural inputs, particularly fertilisers and improved seeds. As a result, total food production is now higher than ever; also food production per capita in 2005–2010 was 160 % of that in 1985–1990 (Fig. 21.7).

21.11.2 Soil and Water Conservation

The main agricultural intensification observed in Ethiopia is certainly the now widespread catchment management activities (Fig. 21.8). Throughout the Ethiopian highlands, it is apparent that many SWC structures, established during the 1980s, remain in place and are often maintained. Their destruction is not as widespread as stated by Rahmato (1994), often the farmers accept and adopt these structures. Many, probably most of the soil bunds throughout Welo, have evolved into full-grown lynchets. Even in the high rainfall Ankober area, soil bunds have often been 'opened' to allow drainage, but are still in place over most of their length.

Local knowledge and farmers' initiatives are integrated with these introduced technologies at various degrees (Gaspart et al. 1997; Nyssen et al. 2000a, 2004a, 2008b,

Fig. 21.8 Catchment rehabilitation in the sub-humid May Zeg-zeg catchment (Tigray); trenches behind the stone bunds enhance infiltration and decrease catchment runoff response (*Photograph* K. Herweg, May 2005)

2009a; Haile et al. 2006; Gebresamuel et al. 2009). The efficiency of particular techniques cannot be discussed in depth here; the reader is referred to specialised publications (Herweg and Ludi 1999; SCRP 2000; Nyssen et al. 2004c, 2007a, 2009a, 2010; Gebremichael et al. 2005; Haregeweyn et al. 2006; Vancampenhout et al. 2006; Wondumagegnehu et al. 2007; Alemayehu et al. 2009; Reubens et al. 2009; Gebreegziabher et al. 2009; Araya et al. 2011; Lanckriet et al. 2012; Muys et al. 2014; Gebreyohannes et al. 2014).

21.12 Conclusions

Ethiopia is on the map for research on land resources and implementation of sustainable land management (SLM) (Haile et al. 2006). Future research priorities are identified. Cornerstones of SLM include forest development in critical places (Descheemaeker et al. 2006b, 2009), over sufficiently large areas, as demonstrated through the dramatic changes that occurred on the Rift Valley escarpment near Alamata (Gebreyohannes et al. 2014).

Fig. 21.9 Soil and water conservation activities in the Tsinkaniet plain have led to a situation where overland flow strongly decreased in 2006 (**b**—*photograph* J. Nyssen) as compared to 1975 when evidence of flooding is clearly visible at the footslope (**a**—*photograph* R.N. Munro). At the far end, in 2006, a reservoir mainly fed by groundwater is visible

SWC activities also enhance rain infiltration rates during the short but heavy storms and improve the situation with regard to flooding, soil erosion, and groundwater recharge (Nyssen et al. 2009a, 2010). The current land tenure system in which an equality of land holdings is attempted, favours solidarity among the farmers to undertake communal catchment management activities (Kumasi and Asenso-Okyere 2011; Taye et al. 2013). Besides the need for collecting a wide set of original data, conceptually, in all related research, a good comprehension of the hydrological balance is needed. Further, for nutrient, sediment, and water-related processes, it is important to understand the occurrence of sinks and to keep the scale concept in mind. These principles are at the base of the successful implementation of catchment management activities in northern Ethiopia (Fig. 21.9). In this regard, the impacts of the May Zeg-zeg catchment management could be monitored in detail (Nyssen et al. 2009a, 2010; Walraevens et al. 2009) and future development scenarios could be elaborated (Lanckriet et al. 2012).

References

Abebe Y, Apparao G (1989) Annual rain water potential and its variability in drought years over Ethiopia. In: Conference on climate and water, Suomen Akatemian, Helsinki, Finland, pp 219–235

Aerts R, Wagendorp T, November E, Behailu M, Deckers J, Muys B (2004) Ecosystem thermal buffer capacity as an indicator of the restoration status of protected areas in the northern Ethiopian highlands. Rest Ecol 12:586–596

Aerts R, Nyssen J, Haile M (2009) On the difference between "exclosures" and "enclosures" in ecology and the environment. J Arid Environ 73:762–763

Alemayehu F, Taha N, Nyssen J, Girma A, Zenebe A, Behailu M, Deckers J, Poesen J (2009) The impacts of watershed management on land use and land cover dynamics in Eastern Tigray (Ethiopia). Resour Conserv Recycl 53:192–198

Araya T, Cornelis WM, Nyssen J, Govaerts B, Gebregziabher T, Oicha T, Getnet F, Raes D, Haile M, D J, Sayre KD (2011) Effects of conservation agriculture on runoff, soil loss and crop yield under rain fed conditions in Tigray, Northern Ethiopia. Soil Tillage Res 27:404–414

Araya T, Cornelis WM, Nyssen J, Govaerts B, Getnet F, Bauer H, Amare K, Raes D, Haile M, Deckers J (2012) Medium-term effects of conservation agriculture based cropping systems for sustainable soil and water management and crop productivity in the Ethiopian highlands. Field Crops Res 132:53–62

Asnakaw W, Assefa A, Yematawork A, Abera S, Van Straaten P, Groenevelt P, Chesworth W (1994) Report on the results of the Ethiopia-Canada agrogeology project—rock mulch. IDRC-Project 88-1032

Astatke A, Jabbar M, Mohamed Saleem MA, Erkossa T (2002) Technical and economic performance of animal-drawn implements for minimum tillage: experience on Vertisols in Ethiopia. Exp Agric 38:185–196. doi:10.1017/S0014479702000248

Bakker JP (1967) Weathering of granite in different climates, particularly in Europe.In: Symposium international de géomorphologie, 8–16 juin 1966. Les congrès et colloques de l'Université de Liège, Liège-Louvain, pp 51–68

Bauduin D, Dubreuil P (1973) L'inventaire des ressources en eau pour l'aménagement intégré du Bassin du Wabi Shebelle d'Ethiopie. Cah ORSTOM, sér Hydrol 10:307–348

Benin S, Pender J (2001) Impacts of land redistribution on land management and productivity in the Ethiopian highlands. Land Degrad Dev 12:555–568

Berakhi O, Brancaccio L (1993) Some reflections on the origin and land use of pediments on the Ethiopian highlands. Geogr Fis Dinam Quat 16:101–106

Berakhi O, Brancaccio L, Calderoni G, Coltorti M, Dramis F, Tegene B, Umer M (1997) Geomorphological and sedimentary records of Holocene climatic changes and human impact in the highlands of Northern Ethiopia. Geogr Fis e Din Quat Supp 3, T1:77

Berhane G, Martens K, Al Farrah N, Walraevens K (2013) Water leakage investigation of micro-dam reservoirs in Mesozoic sedimentary sequences in Northern Ethiopia. J Afr Earth Sci 79:98–110. http://dx.doi.org/10.1016/j.jafrearsci.2012.10.004

Billi P (2004) Sediment yield of a closed river system: the Meki River, western margin of the Ethiopian Rift Valley. In: Proceedings of the IAHS international symposium on sediment transfer through the fluvial system, Moscow, 2–6 Aug 2004. pp 94–100

Blaikie P, Brookfield H (1987) Land degradation and society. Methuen, New York

Bono R, Seiler W (1986) Soils and soil erosion in the area of Debre Birhan, Eastern High Plateau of Shewa, Ethiopia. Regio Basiliensis 27:39–53

Boserup E (1981) Population and technology. Blackwell, Oxford

Brancaccio L, Calderoni G, Coltorti M, Dramis F (1997) Phases of soil erosion during the Holocene in the highlands of Western Tigray (Northern Ethiopia): a preliminary report. In: Bard K (ed) The environmental history and human ecology of Northern Ethiopia in the Late Holocene. Instituto Universitario Orientale, Napoli, pp 30–48

Broothaerts N, Kissi E, Poesen J, Van Rompaey A, Getahun K, Van Ranst E, Diels J (2012) Spatial patterns, causes and consequences of landslides in the Gilgel Gibe catchment, SW Ethiopia. Catena 97:127–136. http://dx.doi.org/10.1016/j.catena.2012.05.011

Bull WB (1997) Discontinuous ephemeral streams. Geomorphology 19:227–276

Bunney S (1990) Prehistoric farming caused devastating soil erosion. N Sci 125:20

Camberlin P (1994) Les précipitations dans la Corne orientale de l'Afrique: climatologie, variabilité et connexions avec quelques indicateurs océano-atmosphériques. PhD dissertation. Université de Bourgogne

Casenave A, Valentin C (1992) A runoff capability classification system based on surface features criteria in the arid and semi-arid areas of West Africa. J Hydrol 130:231–249

Conway D (1997) A water balance model of the Upper Blue Nile in Ethiopia. Hydrol Sci J 42:265–286

Conway D (2000a) The climate and hydrology of the Upper Blue Nile River. Geogr J 166:49–62

Conway D (2000b) Some aspects of climate variability in the North East Ethiopian highlands—Wollo and Tigray. *Sinet: Ethiop.* J Sci 23:139–161

Conway D, Hulme M (1993) Recent fluctuations in precipitation and runoff over the Nile subbasins and their impact on main Nile discharge. Clim Change 25:127–151

Conway D, Mould C, Bewket W (2004) Over one century of rainfall and temperature observations in Addis Ababa, Ethiopia. Int J Climatol 24:77–91

Crummey D (2000) Land and society in the Christian Kingdom of Ethiopia, from the Thirteenth to the Twentieth Century. University of illinois Press, Addis Ababa

Descheemaeker K, Muys B, Nyssen J, Poesen J, Raes D, Haile M, Deckers J (2006a) Litter production and organic matter accumulation in exclosures of the Tigray highlands, Ethiopia. For Ecol Manag 233:21–35

Descheemaeker K, Nyssen J, Rossi J, Poesen J, Haile M, Moeyersons J, Deckers J (2006b) Sediment deposition and pedogenesis in exclosures in the Tigray highlands, Ethiopia. Geoderma 132:291–314. doi:10.1016/j.geoderma.2005.04.027

Descheemaeker K, Raes D, Nyssen J, Poesen J, Haile M, Deckers J (2009) Changes in water flows and water productivity upon vegetation regeneration on degraded hillslopes in northern Ethiopia: a water balance modelling exercise. Rangel J 31:237–249

Dessie M, Verhoest NEC, Admasu T, Pauwels VRN, Poesen J, Adgo E, Deckers J, Nyssen J (2014) Effects of the floodplain on river discharge into Lake Tana (Ethiopia). J Hydrol 519, part A: 699–710. http://dx.doi.org/10.1016/j.jhydrol.2014.08.007

Diodato N, Knight J, Belolocchi G (2013) Reduced complexity model for assessing patterns of rainfall erosivity in Africa. Global Planet Change 100:183–193

Elias Eyasu, Scoones I (1999) Perspectives on soil fertility change: a case study from southern Ethiopia. Land Degrad Dev 10:195–206

Feoli E, Vuerich L, Woldu Z (2002) Evaluation of environmental degradation in northern Ethiopia using GIS to integrate vegetation, geomorphological, erosion and socio-economic factors. Agric Ecosyst Environ 91:313–325

Foster GR, Highfill RE (1983) Effect of terraces on soil loss—USLE P-factor values for terraces. J Soil Water Conserv 38:48–51

Frankl A, Nyssen J, De Dapper M, Haile M, Billi P, Munro RN, Deckers J, Poesen J (2011) Linking long-term gully and river channel dynamics to environmental change using repeat photography (North Ethiopia). Geomorphology 129:238–251. doi:10.1016/j.geomorph.2011.02.018

Frankl A, Poesen J, Deckers J, Haile M, Nyssen J (2012) Gully head retreat rates in the semi-arid highlands of Northern Ethiopia. Geomorphology 173–174:185–195. http://dx.doi.org/10.1016/j.geomorph.2012.06.011

Gaspart F, Jabbar M, Mélard C, Platteau J-P (1997) Participation in the construction of a local public good: a case study of watershed management in the Ethiopian highlands. Cahiers de la Faculté des Sciences Economiques, Sociales et de Gestion de Namur, Série Recherche 181

Gebreegziabher T, Nyssen J, Govaerts B, Getnet F, Behailu M, Haile M, Deckers J (2009) Contour furrows for in situ soil and water conservation, Tigray, Northern Ethiopia. Soil Tillage Res 103:257–264. doi:10.1016/j.still.2008.05.021

Gebremichael D, Nyssen J, Poesen J, Deckers J, Haile M, Govers G, Moeyersons J (2005) Effectiveness of stone bunds in controlling soil erosion on cropland in the Tigray highlands. North Ethiop Soil Use Manag 21:287–297. doi:10.1111/j.1475-2743.2005.tb00401.x

Gebresamuel G, Haile M, Singh BR (2009) Agronomic and economic performance of reservoir sediment for rehabilitating degraded soils in Northern Ethiopia. Nutr Cycl Agroecosyst 84:23–38

Gebreyohannes T, Frankl A, Haile M, Nyssen J (2014) Catchment rehabilitation and hydro-geomorphic characteristics of mountain streams in the western Rift Valley escarpment of Northern Ethiopia. Land Degrad Dev: Online Early View. doi:10.1002/ldr.2267

Giglioli G (1938a) L'erosione del terreno agrario nei Tropici. Ist Agric Colon Ital Rev 49

Giglioli G (1938b) Le sistemazioni idraulico agrarie dei terreni nell'Africa Italiana. Agric Colon 32:481–496

Girma K, Jacob MJ (1988) Drought, famine and the political economy of environmental degradation in Ethiopia. Geography 318:65–70

Goebel W, Odenyo V (1984) Ethiopia. Agroclimatic resources inventory for land-use planning. Technical report DP/ETH/78/003,

vol I, 208 p; vol II, 95 p. Ministry of Agriculture, Land Use Planning and Regulatory Department, UNDP, FAO, Addis Ababa
Haile M, Herweg K, Stillhardt B (2006) Sustainable land management—a new approach to soil and water conservation in Ethiopia. University of Berne, Berne, Switzerland, Mekelle University, Mekelle, Ethiopia
Haregeweyn N, Poesen J, Nyssen J, De Wit J, Haile M, Govers G, Deckers J (2006) Reservoirs in Tigray: characteristics and sediment deposition problems. Land Degrad Dev 17:211–230
Hemming CF (1961) The ecology of the coastal area of Northern Eritrea. J Ecol 49:55–78
Hemming CF, Trapnell CG (1957) A reconnaissance classification of the soils of the south Turkana desert. J Soil Sci 8:167–183
Herweg K, Ludi E (1999) The performance of selected soil and water conservation measures-case studies from Ethiopia and Eritrea. Catena 36:99–114
Hoffmann R (1987) Die Dürregefährdung äthiopischer Landschaften—ein Beitrag zum Problem der Belastbarkeit tropischer Geoökosysteme. Petermanns Geogr Mitt 131:217–224
Horowitz A (1967) The geology of Museri Island (Dahlak Archipelago, southern Red Sea). Isr J Earth-Sci 16:74–83
Hulme M (1992) Rainfall changes in Africa: 1931–1960 to 1961–1990. Int J Clim 12:685–699
Hunting Technical Services (1976) Tigrai rural development study, Annex 2: water resources, vol 1: hydrology and surface water. Hunting Technical Services Ltd, Hemel Hempstead (G.B.)
Hurni H (1975) Bodenerosion in Semien-Aethiopien (mit Kartenbeilage 1:25 000). Geogr Helvetica 4:157–168
Hurni H (1978) Soil erosion forms in the Simen mountains—Ethiopia (with map 1:25 000). Geogr Bernensia G8:93–100
Hurni H (1979) Semien-Äthiopien: Methoden zur Erfassung der Bodenerosion. Geomethodica 4:151–182
Hurni H (1982) Klima und Dynamik der Höhenstufung von der letzten Kaltzeit bis zur Gegenwart. Hochgebirge von Semien-Äthiopien, Geogr Bernensia G13
Hurni H (1983) Soil erosion and soil formation in agricultural ecosystems Ethiopia and Northern Thailand. Mt Res Dev 3:131–142
Hurni H (1985) Erosion—productivity—conservation systems in Ethiopia. In: Proceedings 4th international conference on soil conservation, Maracay, Venezuela, pp 654–674
Hurni H (1986) Guidelines for Development Agents on soil conservation in Ethiopia. Soil conservation research project. Community Forests and Soil Conservation Development Department, Ministry of Agriculture, Addis Ababa
Hurni H (1990) Degradation and conservation of soil resources in the Ethiopian highlands. Mt Res Dev 8:123–130
Hurni H (1993) Land degradation, famine, and land resource scenarios in Ethiopia. In: Pimentel D (ed) World soil erosion and conservation. Cambridge University Press, Cambridge
Hurni H (1998) Lessons of soil conservation research conducted in Ethiopia. In: Proceedings, soil fertility management workshop. National Fertiliser Industry Agency, Addis Ababa, pp 31–35
Jacob M, Frankl A, Haile M, Zwertvaegher A, Nyssen J (2013) Assessing spatio-temporal rainfall variability in a tropical mountain area (Ethiopia) using NOAAs Rainfall Estimates. Int J Remote Sens 34:8305–8321
Joyce F de V (1943) Notes on agriculture in Ethiopia, Part 3. East Afr Agric J 9:35–38
Kanwar JS, Virmani SM (1986) Management of Vertisols for improved crop production in the semi-arid tropics: a plan for a technology transfer network in Africa. In: Proceedings of the first IBSRAM (International Board on Soil Research and Management) Regional Networkshop in Africa on Improved Management of Vertisols under Semi-Arid Conditions, Nairobi, Kenya, 1–6 Dec 1986. IBSRAM, Bangkok, Thailand, pp 157–172
Kassa H, Gibbon D, Singh B (2002) Livestock improve household food security and sustainability of Ethiopian small farms. J Sustain Agri 21:73–93
Kassa H, Nyssen J, Frankl A, Dondeyne S, Poesen J (2013) The impact of human activities on natural resources in the southwestern highlands of Ethiopia—a state of art (in prep)
Tekle K, Backeus I, Skoglund J, Woldu Z (1997) Vegetation on hillslopes in southern Wello, Ethiopia: degradation and regeneration. Nordic J Bot 17:483–493
Kejela K (1992) The costs of soil erosion in Anjeni, Ethiopia. In: Tato K and H. Hurni (eds) Soil conservation for survival. Soil and Water Conservation Society, Ankeny, pp 219–230
Klötzli F (1977) Wild und Vieh im Gebirgsland Aethiopiens. In: Tuexen R (ed) Symp. der int Ver. für Veg.kunde. Cramer, Vaduz, Rinteln, pp 499–512
Krauer J (1988) Rainfall, erosivity and isoerdent map of Ethiopia. Soil conservation research project, research report 15, University of Berne, Bern
Kumasi T, Asenso-Okyere K (2011) Responding to land degradation in the highlands of Tigray, Northern Ethiopia. IFPRI discussion paper, 01142. International Food Policy Research Institute
Lanckriet S, Araya Tesfay, Cornelis W, Verfaillie E, Poesen J, Govaerts B, Bauer H, Deckers S, Haile M, Nyssen J (2012) Impact of conservation agriculture on catchment runoff and soil loss under changing climate conditions in May Zeg-zeg (Ethiopia). J Hydrol 475:336–349
Machado MJ, Perez-Gonzalez A, Benito G (1998) Paleoenvironmental changes during the last 4000 yr in the Tigray, northern Ethiopia. Quatern Res 49:312–321
Mekuria W, Veldkamp E, Haile M, Nyssen J, Muys B, Gebrehiwot Kindeya (2007) Effectiveness of exclosures to restore degraded soils as a result of overgrazing in Tigray, Ethiopia. J Arid Environ 69:270–284. doi:10.1016/j.jaridenv.2006.10.009
Mersha E (2000) A desertification convention based classification of moisture zones of Ethiopia. Ethiop J Nat Resour 2:1–9
Michels J, 1988. Regional political organization in the Axum-Yeha area during the pre-axumite and axumite eras. In: Paper presented at the Xth international conference on Ethiopian studies, Paris
Moeyersons J (2000) Desertification and Man in Africa. Brussels. Bull Séanc Acad Royal Sci Outre-Mer 46:151–170
Moeyersons J, Nyssen J, Poesen J, Deckers J, Haile M (2006) On the origin of rock fragment mulches on Vertisols: a case study from the Ethiopian highlands. Geomorphology 76:411–429
Moeyersons J, Van Den Eeckhaut M, Nyssen J, Gebreyohannes T, Van de Wauw J, Hofmeister J, Poesen J, Deckers J, Haile M (2008) Mass movement mapping for geomorphological understanding and sustainable development, Tigray, Ethiopia. Catena 75:45–54. http://dx.doi.org/10.1016/j.catena.2008.04.004
Moore TR (1979) Rainfall erosivity in East Africa. Geografiska Annaler Series A 61A:147–156
Moreda F, Bauwens W (1998) Influences of variability of rainfall on flow regimes in central Ethiopia. IAHS Publication 252:297–306
Mulugetta B (1992) Attempts in the transformation of Ethiopia's agriculture. In: Doornbos M, Cliffe L, Abel Ghaffar M and Markakis J (eds) Beyond conflict in the Horn. The Red Sea Press, Trenton, pp 143–153
Munro RN, Deckers J, Grove AT, Haile M, Poesen J, Nyssen J (2008) Soil and erosion features of the Central Plateau region of Tigrai—learning from photo monitoring with 30 years interval. Catena 75:55–64. doi:10.1016/j.catena.2008.04.009
Muys B, Nyssen J, du Toit B, Vidale E, Prokofieva I, Mavsar R, Palahi M (2014) Water-related ecosystem services of forests: learning from regional cases. In: Katila P, Galloway G, de Jong W, Pacheco P, Mery G (eds) Forests under pressure—local responses to global

issues. International Union of Forest Research Organizations, Vienna, pp 423–440

Mwendera EJ, Saleem MAM (1997) Infiltration rates, surface runoff, and soil loss as influenced by grazing pressure in the Ethiopian highlands. Soil Use Manag 13:29–35

Mwendera EJ, Saleem MAM, Dibabe A (1997) The effect of livestock grazing on surface runoff and soil erosion from sloping pasture lands in the Ethiopian highlands. Aust J Exp Agric 37:421–430

Nyssen J (2001) Erosion processes and soil conservation in a tropical mountain catchment under threat of anthropogenic desertification—a case study from Northern Ethiopia. PhD thesis. KU Leuven, Belgium

Nyssen J, Haile M, Moeyersons J, Poesen J, Deckers J (2000a) Soil and water conservation in Tigray (Northern Ethiopia): the traditional daget technique and its integration with introduced techniques. Land Degrad Dev 11:199–208

Nyssen J, Moeyersons J, Deckers J, Haile M, Poesen J (2000b) Vertic movements and the development of stone covers and gullies, Tigray highlands, Ethiopia. Zeitschrift für Geomorphologie 44:145–164

Nyssen J, Poesen J, Haile M, Moeyersons J, Deckers J (2000c) Tillage erosion on slopes with soil conservation structures in the Ethiopian highlands. Soil Tillage Res 57:115–127. doi:10.1016/S0167-1987(00)00138-0

Nyssen J, Poesen J, Moeyersons J, Luyten E, Veyret-Picot M, Deckers J, Haile M, Govers G (2002a) Impact of road building on gully erosion risk: a case study from the Northern Ethiopian highlands. Earth Surf Process Landf 27:1267–1283

Nyssen J, Poesen J, Moeyersons J, Lavrysen E, Haile M, Deckers J (2002b) Spatial distribution of rock fragments in cultivated soils in northern Ethiopia as affected by lateral and vertical displacement processes. Geomorphology 43:1–16

Nyssen J, Haile M, Moeyersons J, Poesen J, Deckers J (2004a) Environmental policy in Ethiopia: a rejoinder to Keeley and Scoones. J Modern Afr St 42:137–147

Nyssen J, Poesen J, Moeyersons J, Deckers J, Haile M, Lang A (2004b) Human impact on the environment in the Ethiopian and Eritrean highlands—a state of the art. Earth Sci Rev 64:273–320. doi:10.1016/S0012-8252(03)00078-3

Nyssen J, Veyret-Picot M, Poesen J, Moeyersons J, Haile M, Deckers J, Govers G (2004c) The effectiveness of loose rock check dams for gully control in Tigray, Northern Ethiopia. Soil Use Manag 20:55–64. doi:10.1111/j.1475-2743.2004.tb00337.x

Nyssen J, Vandenreyken H, Poesen J, Moeyersons J, Deckers J, Haile M, Salles C, Govers G (2005) Rainfall erosivity and variability in the Northern Ethiopian highlands. J Hydrol 311:172–187. doi:10.1016/j.jhydrol.2004.12.016

Nyssen J, Poesen J, Gebremichael D, Vancampenhout K, D'Aes M, Yihdego G, Govers G, Leirs H, Moeyersons J, Naudts J, Haregeweyn N, Haile M, Deckers J (2007a) Interdisciplinary on-site evaluation of stone bunds to control soil erosion on cropland in Northern Ethiopia. Soil Tillage Res 94:151–163

Nyssen J, Poesen J, Moeyersons J, Haile M, Deckers J (2007b) Dynamics of soil erosion rates and controlling factors in the Northern Ethiopian highlands—towards a sediment budget. Earth Surf Proc Land 33:695–711. doi:10.1002/esp.1569

Nyssen J, Naudts J, De Geyndt K, Haile M, Poesen J, Moeyersons J, Deckers J (2008a) Soils and land use in the Tigray highlands (Northern Ethiopia). Land Degrad Dev 19:257–274. doi:10.1002/ldr.840

Nyssen J, Poesen J, Descheemaeker K, Haregeweyn N, Haile M, Moeyersons J, Frankl A, Govers G, Munro RN, Deckers J (2008b) Effects of region-wide soil and water conservation in semi-arid areas: the case of northern Ethiopia. Zeitschrift für Geomorphologie 52:291–315. doi:10.1016/j.catena.2008.04.009

Nyssen J, Clymans W, Poesen J, Vandecasteele I, De Baets S, Haregeweyn N, Naudts J, Hadera A, Moeyersons J, Haile M, Deckers J (2009a) How soil conservation affects the catchment sediment budget—a comprehensive study in the north Ethiopian highlands. Earth Surf Proc Land 34:1216–1233. doi:10.1002/esp.1805

Nyssen J, Haile M, Nauds J, Munro N, Poesen J, Moeyersons J, Frankl A, Deckers J, Pankhurst R (2009b) Desertification? Northern Ethiopia re-photographed after 140 years. Sci Total Environ 407:2749–2755. doi:10.1016/j.scitotenv.2008.12.016

Nyssen J, Poesen J, Haile M, Moeyersons J, Deckers J, Hurni H (2009c) Effects of land use and land cover on sheet and rill erosion rates in the Tigray highlands, Ethiopia. Zeitschrift für Geomorphologie 53:171–197

Nyssen J, Clymans W, Descheemaeker K, Poesen J, Vandecasteele I, Vanmaercke M, Haile M, Haregeweyn N, Moeyersons J, Martens K, Zenebe A, Van Camp M, Gebreyohannes T, Deckers J, Walraevens K (2010) Impact of soil and water conservation on catchment hydrological response—a case in northern Ethiopia. Hydrol Process 24:1880–1895

Oicha T, Cornelis W, Verplancke H, Nyssen J, Deckers J, Behailu M, Haile M (2010) Short-term effects of conservation agriculture on Vertisols under tef (Eragrostis tef (Zucc.) Trotter) in the northern Ethiopian highlands. Soil Tillage Res 106:294–302. doi:10.1016/j.still.2009.12.004

Olson G (1981) Archaeology: lessons on future soil use. J Soil Water Conserv 36:261–264

Pietsch D, Machado MJ (2014) Colluvial deposits—proxies for climate change and cultural chronology. A case study from Tigray, Ethiopia. Zeitschrift für Geomorphologie, Supplementary Issues 58(1):119–136

Poesen JW, Torri D, Bunte K (1994) Effects of rock fragments on soil erosion by water at different spatial scales—a review. Catena 23:141–166

Poppe L, Frankl A, Poesen J, Admasu T, Dessie M, Adgo E, Deckers J, Nyssen J (2013) Geomorphology of the Lake Tana basin, Ethiopia. J Maps 9:431–437. doi:10.1080/17445647.2013.801000

Rahmato D (1994) The unquiet countryside: the collapse of "socialism" and the rural agitation, 1990 and 1991. In: Zegeye A and Pausewang S (eds) Ethiopia in change, peasantry, nationalism and democracy. British Academic Press, London

Reid RS, Kruska RL, Muthui N, Taye A, Wotton S, Wilson CJ, Mulatu W (2000) Land-use and land-cover dynamics in response to changes in climatic, biological and socio-political forces: the case of southwestern Ethiopia. Landscape Ecol 15:339–355

Renard KG, Foster GR, Weesies GA, McCool DK, Yoder DC (1997) Predicting soil erosion by water: a guide to conservation planning with the Revised Universal Soil Loss Equation (RUSLE). Agriculture handbook 703. United States Department of Agriculture, Washington, DC

Reubens B, Poesen J, Nyssen J, Leduc Y, Zenebe A, Tewoldeberhan S, Bauer H, Gebrehiwot K, Deckers J, Muys B (2009) Establishment and management of woody seedlings in gullies in a semi-arid environment (Tigray, Ethiopia). Plant Soil 324:131–156

Reuter G (1991) Bodenkundliche Probleme im Irak und in Äthiopien. Mitteilungen der Deutschen Bodenkundlichen Gesellschaft 66 (2):1181–1184

Riché G, Ségalen P (1973) Les sols et le modelé dans le Nord-Est du bassin du Wabi-Schebelle (Ethiopie). Cahiers ORSTOM, série pédologie 11:237–247

SCRP (2000) Long-term monitoring of the agricultural environment in six research stations in Ethiopia. Soil erosion and conservation database. 7 volumes. Soil conservation research programme, Berne and Addis Ababa

Seleshi Y, Demarée G (1995) Rainfall variability in the Ethiopian and Eritrean highlands and its links with the Southern oscillation index. J Biogeogr 22:945–952

Seleshi Y, Demarée G (1998) Identifying the major cause of the prevailing summer rainfall deficit over the North-Central Ethiopian highlands since the mid-60s. In: Demarée G, Alexandre J and De Dapper M (eds) Proceedings of internatonal conference tropical climate, meteorology, hydrology, Brussels 22–24 May 1996, pp 516–531

Shitarek T, Manaye Sintayehu, Abebe Berihun (2001) Stengthening user-rights over local resources in Wollo. IIED drylands programme, issue paper, Ethiopia 103

Ståhl M (1974) Ethiopia: political contradictions in agricultural development. Stockholm, Rabén and Sjögren

Ståhl M (1990) Environmental degradation and political constraints in Ethiopia. Disasters 14:140–150

Sutcliffe JV, Parks YP (1999) The hydrology of the Nile. IAHS Special Publication 5. International Association of Hydrological Sciences, UK

Taddesse B (1995) Deforestation and environmental degradation in Ethopia: the case of Jam Jam province. Northeast Arican Studies 2 (2):139-156

Taye G, Poesen J, Van Wesemael B, Vanmaercke M, Teka D, Deckers J, Goosse T, Maetens W, Nyssen J, Hallet V, Haregeweyn Nigussie (2013) Effects of land use, slope gradient, and soil and water conservation structures on runoff and soil loss in semi-arid Northern Ethiopia. Phys Geogr 34:236–259. doi:10.1080/02723646.2013.832098

Tedla A, Mamo T, Klaij M, Diedhiou M (1999) Effects of cropping system, seed bed management and fertility interactions on biomass of crops grown on a Vertisol in the central highlands of Ethiopia. Zeitschrift Für Acker Und Pflanzenbau 183:205–211

Tegene B (1998) Pedogenesis and soil-geomorphic relationships on the piedmont slopes of Wurgo Valley, Southern Welo. Ethiop Sinet Ethiop J Sci 21:91–111

Tesfaye M (1988) Soil conservation experiments on cultivated land in the Maybar Area, Wello Region, Ethiopia. Soil conservation research project, research report 16. University of Berne, Berne

Troll C (1970) Die naturräumliche Gliederung Eritreas. Erdkunde 24:249–268

Uhlig S, Uhlig K (1989) On the ecology and vegetation of the plateaus on east Ethiopian isolated mountains. Archiv für Naturschutz und Landschaftsforschung 29:175–179

UNEP (1994) United Nations convention to combat desertification. United Nations Environmental Programme, Nairobi

USBR (1964) Land and water resources of the Blue Nile basin. Appendix III. Hydrology. United States Department of Interior, Bureau of Reclamation

Valentin C, Poesen J, Li Y (2005) Gully erosion: impacts, factors and control. Catena 63:132–153

Van de Wauw J, Baert G, Moeyersons J, Nyssen J, De Geyndt K, Taha N, Zenebe A, Poesen J, Deckers J (2008) Soil-landscape relationships in the basalt-dominated highlands of Tigay, Ethiopia. Catena 75:117–127

Van Den Eeckhaut M, Moeyersons J, Nyssen J, Amanuel Abaha, Poesen J, Haile M, Deckers J (2009) Spatial patterns of old, deep-seated landslides: a case-study in the northern Ethiopian highlands. Geomorphology 105:239–252. http://dx.doi.org/10.1016/j.geomorph.2008.09.027

Vancampenhout K, Nyssen J, Gebremichael D, Deckers J, Poesen J, Haile M, Moeyersons J (2006) Stone bunds for soil conservation in the northern Ethiopian highlands: impacts on soil fertility and crop yield. Soil Tillage Res 90:1–15. doi:10.1016/j.still.2005.08.004

Virgo KJ, Munro RN (1978) Soil and erosion features of the Central Plateau region of Tigrai, Ethiopia. Geoderma 20:131–157

Vrieling A, Sterk G, de Jong SM (2010) Satellite-based estimation of rainfall erosivity for Africa. J Hydrol 395(3):235

Walraevens K, Vandecasteele I, Martens K, Nyssen J, Moeyersons J, Gebreyohannes T, De Smedt F, Poesen J, Deckers J, Van Camp M (2009) Groundwater recharge and flow in a small mountain catchment in Northern Ethiopia. Hydrol Sci J 54:739–753

Wischmeier WH, Smith DD (1978) Predicting rainfall erosion losses: a guide to conservation planning. Agriculture handbook, 537. United States Department of Agriculture, Washington, DC

Wondumagegnehu F, Tsegay A, Ashebir D, Tekie H, Gebre A, Kiros M, Geerts S, Raes D, Nyssen J, Deckers J (2007) Household water harvesting structures in Geba catchment. Tigray Livelihood Papers, vol 5. VLIR—Mekelle University IUC Programme, Mekelle, Ethiopia

Zenebe A, Vanmaercke M, Poesen J, Verstraeten G, Haregeweyn N, Haile M, Amare K, Deckers J, Nyssen J (2013) Spatial and temporal variability of river flows in the degraded semi-arid tropical mountains of northern Ethiopia. Z Geomorph NF 57:143–169

Index

P. Billi (ed.), *Landscapes and Landforms of Ethiopia*, World Geomorphological Landscapes,
DOI 10.1007/978-94-017-8026-1, © Springer Science+Business Media Dordrecht 2015

MIX
Papier aus verantwortungsvollen Quellen
Paper from responsible sources
FSC® C105338

If you have any concerns about our products,
you can contact us on
ProductSafety@springernature.com

In case Publisher is established outside the EU,
the EU authorized representative is:
Springer Nature Customer Service Center GmbH
Europaplatz 3, 69115 Heidelberg, Germany

Printed by Libri Plureos GmbH
in Hamburg, Germany